T0189195

Advances in Intelligent Systems and Computing

Volume 918

Series editor

Janusz Kacprzyk, Systems Research Institute, Polish Academy of Sciences, Warsaw, Poland
e-mail: kacprzyk@ibspan.waw.pl

The series "Advances in Intelligent Systems and Computing" contains publications on theory, applications, and design methods of Intelligent Systems and Intelligent Computing. Virtually all disciplines such as engineering, natural sciences, computer and information science, ICT, economics, business, e-commerce, environment, healthcare, life science are covered. The list of topics spans all the areas of modern intelligent systems and computing such as: computational intelligence, soft computing including neural networks, fuzzy systems, evolutionary computing and the fusion of these paradigms, social intelligence, ambient intelligence, computational neuroscience, artificial life, virtual worlds and society, cognitive science and systems, Perception and Vision, DNA and immune based systems, self-organizing and adaptive systems, e-Learning and teaching, human-centered and human-centric computing, recommender systems, intelligent control, robotics and mechatronics including human-machine teaming, knowledge-based paradigms, learning paradigms, machine ethics, intelligent data analysis, knowledge management, intelligent agents, intelligent decision making and support, intelligent network security, trust management, interactive entertainment, Web intelligence and multimedia.

The publications within "Advances in Intelligent Systems and Computing" are primarily proceedings of important conferences, symposia and congresses. They cover significant recent developments in the field, both of a foundational and applicable character. An important characteristic feature of the series is the short publication time and world-wide distribution. This permits a rapid and broad dissemination of research results.

More information about this series at http://www.springer.com/series/11156

Álvaro Rocha · Carlos Ferrás
Manolo Paredes
Editors

Information Technology and Systems

Proceedings of ICITS 2019

 Springer

Editors
Álvaro Rocha
DEI/FCT
Universidade de Coimbra
Coimbra, Portugal

Carlos Ferrás
Facultad de Geografía
Universidad de Santiago de Compostela
Santiago Compostela, La Coruña, Spain

Manolo Paredes
Departamento de Eléctrica, Electrónica y
Telecomunicaciones
Universidad de las Fuerzas Armadas
"ESPE"
Sangolqui, Ecuador

ISSN 2194-5357   ISSN 2194-5365 (electronic)
Advances in Intelligent Systems and Computing
ISBN 978-3-030-11889-1   ISBN 978-3-030-11890-7 (eBook)
https://doi.org/10.1007/978-3-030-11890-7

Library of Congress Control Number: 2018967946

This Springer imprint is published by the registered company Springer Nature Switzerland AG
The registered company address is: Gewerbestrasse 11, 6330 Cham, Switzerland

Preface

This book is composed by the papers written in English and accepted for presentation and discussion at The 2019 International Conference on Information Technology & Systems (ICITS'19). This conference had the support of the University of Armed Forces (Universidad de las Fuerzas Armadas "ESPE"), IEEE Systems, Man, and Cybernetics Society, and AISTI (Iberian Association for Information Systems and Technologies). It took place at Sangolquí, Quito, Ecuador, February 6–8, 2019.

The 2019 International Conference on Information Technology & Systems (ICITS'19) is an international forum for researchers and practitioners to present and discuss the most recent innovations, trends, results, experiences, and concerns in the several perspectives of information technology and systems.

The Program Committee of ICITS'19 was composed of a multidisciplinary group of 152 experts and those who are intimately concerned with information systems and technologies. They have had the responsibility for evaluating, in a 'double-blind review' process, the papers received for each of the main themes proposed for the conference: (A) Information and Knowledge Management; (B) Organizational Models and Information Systems; (C) Software and Systems Modeling; (D) Software Systems, Architectures, Applications and Tools; (E) Multimedia Systems and Applications; (F) Computer Networks, Mobility, and Pervasive Systems; (G) Intelligent and Decision Support Systems; (H) Big Data Analytics and Applications; (I) Human–Computer Interaction; (J) Ethics, Computers and Security; (K) Health Informatics; (L) Information Technologies in Education; (M) Cybersecurity and Cyber-defense; (N) Electromagnetics, Sensors and Antennas for Security.

ICITS'19 also included several workshop sessions taking place in parallel with the conference ones. They were sessions of the WMETACOM 2019 – 2nd Workshop on Media, Applied Technology and Communication.

ICITS'19 received about 300 contributions from 31 countries around the world. The papers accepted for presentation and discussion at the conference are published by Springer (this book) and by AISTI, and will be submitted for indexing by ISI, EI-Compendex, SCOPUS, DBLP and/or Google Scholar, among others.

We acknowledge all of those that contributed to the staging of ICITS'19 (authors, committees, workshop organizers and sponsors). We deeply appreciate their involvement and support that was crucial for the success of ICITS'19.

February 2019   Álvaro Rocha
                                                                                     Carlos Ferrás
                                                                                  Manolo Paredes

Organization

Conference

Honorary Chair

Álvaro Rocha   University of Coimbra, Portugal

Scientific Committee Chair

Carlos Ferrás Sexto   University of Santiago de Compostela, Spain

Local Organizing Chair

Manolo Paredes   Universidad de las Fuerzas Armadas "ESPE",
                                Ecuador

Local Organizing Committee

Hugo Perez   Universidad de las Fuerzas Armadas "ESPE",
                                Ecuador
Patricio Reyes   Universidad de las Fuerzas Armadas "ESPE",
                                Ecuador
Gonzalo Olmedo   Universidad de las Fuerzas Armadas "ESPE",
                                Ecuador

Scientific Committee

Abdulmotaleb El Saddik	University of Ottawa, Canada
Alexandra González	Universidad Tecnica Particular de Loja, Ecuador
Alexandru Vulpe	University Politehnica of Bucharest, Romania
Amal Al Ali	University of Sharjah, United Arab Emirates
Ana V. Guamán	Universidad de las Fuerzas Armadas "ESPE", Ecuador
André da Silva	IFSP and NIED/UNICAMP, Brazil
André Marcos Silva	University Adventist of São Paulo, Brazil
André Kawamoto	Federal University of Technology, Brazil
Angeles Quezada	Universidad Autonoma de Baja California, Mexico
Ania Cravero	University de La Frontera, Chile
Ankur Bist	KIET Ghaziabad, India
António Augusto Gonçalves	Universidade Estacio de Sá, Brazil
Antonio Raffo	University of Calabria, Italy
Anushia Inthiran	University of Canterbury, New Zealand
Ari Mariano	Universidade de Brasília, Brazil
Benardine Onah	University of Nigeria, Nigeria
Borja Bordel	Universidad Politécnica de Madrid, Spain
Carlos Cares	Universidad de La Frontera, Chile
Carlos Carreto	Polytechnic of Guarda, Portugal
Carlos Grilo	Polytechnic of Leiria, Portugal
Carlos Hernan Fajardo Toro	Universidad EAN, Colombia
Dalila Durães	Technical University of Madrid, Spain
Dália Filipa Liberato	ESHT/IPP, Portugal
Daniela Benalcázar	Universidad Técnica de Ambato, Ecuador
Dante Carrizo	Universidad de Atacama, Chile
Diego Marcillo	Universidad de las Fuerzas Armadas "ESPE", Ecuador
Diego Ordóñez-Camacho	Universidad Tecnológica Equinoccial, Ecuador
Eddie Galarza	Universidad de las Fuerzas Armadas "ESPE", Ecuador
Edgar Serna	Universidad Autónoma Latinoamericana, Colombia
Edison Loza-Aguirre	Escuela Politécnica Nacional, Ecuador
Efraín R. Fonseca C.	Universidad de las Fuerzas Armadas "ESPE", Ecuador
Egils Ginters	Riga Technical University, Latvia
Enrique Carrera	Universidad de las Fuerzas Armadas "ESPE", Ecuador
Ewaryst Tkacz	Silesian University of Technology, Poland
Fabio Gomes Rocha	Tiradentes University, Brazil

Felix Blazquez Lozano	University of A Coruña, Spain
Filipa Ferraz	University of Minho, Portugal
Filipe Sá	Câmara Municipal de Penacova, Portugal
Felipe Machorro-Ramos	Universidad de las Américas Puebla, Mexico
Francesc Gine	University of Lleida, Spain
Francisco Valverde	Universidad Central del Ecuador, Ecuador
Franklim Silva	Universidad de las Fuerzas Armadas "ESPE", Ecuador
Frederico Branco	Universidade de Trás-os-Montes e Alto Douro, Portugal
Gabriel Elías Chanchí Golondrino	Institución Universitaria Colegio Mayor del Cauca, Colombia
Gabriel Pestana	Universidade Europeia, Portugal
George Suciu	BEIA, Romania
Gladys Alicia Tenesaca Luna	Universidad Técnica Particular de Loja, Ecuador
Hector Florez	Universidad Distrital Francisco Jose de Caldas, Colombia
Henrique Lopes Cardoso	University of Porto, Portugal
Ildeberto Rodello	University of São Paulo, Brazil
Isabel Pedrosa	Coimbra Business School - ISCAC, Portugal
Jan Kubicek	Faculty of Electrical Engineering and Computer Science VŠB-TUO, Czech Republic
Javier Criado	University of Almería, Spain
João Paulo Pereira	Polytechnic of Bragança, Portugal
João Vidal de Carvalho	ISCAP/IPP, Portugal
Jorge Buele	Universidad Técnica de Ambato, Ecuador
Jorge Herrera-Tapia	Universidad Laica Eloy Alfaro de Manabí, Ecuador
Jorge Luis Pérez	Universidad de Las Américas, Ecuador
Jose Aguilar	Universidad de Los Andes, Venezuela
José Álvarez-García	University of Extremadura, Spain
José Araújo	SAP, Portugal
José Luís Silva	ISCTE-IUL and Madeira-ITI, Portugal
Juan Jesus Ojeda	University of Almeria, Spain
Juan M. Ferreira	Senate, Paraguay
Júlio Menezes Jr.	Federal University of Pernambuco, Brazil
Jussi Okkonen	University of Tampere, Finland
Justyna Trojanowska	Poznan University of Technology, Poland
Korhan Gunel	Adnan Menderes University, Turkey
Leandro Flórez Aristizábal	Antonio Jose Camacho University Institute, Colombia
Leonardo Botega	UNIVEM, Brazil
Lorena Siguenza-Guzman	Universidad de Cuenca, Ecuador
Mafalda Teles Roxo	INESC TEC, Portugal
Manuel Au-Yong-Oliveira	University of Aveiro, Portugal

Contents

Organizational Models and Information Systems

Intelligent and Decision Support Systems

Big Data Analytics and Applications

Human-Computer Interaction

Ethics, Computers and Security

Health Informatics

Information Technologies in Education

Cybersecurity and Cyber-Defense

Information and Knowledge
Management

Management of Natural Disasters Based on Twitter Analytics. 2017 Mexico Earthquake

Patricia Henríquez-Coronel[✉], Julio García García[✉], and Jorge Herrera-Tapia[✉]

Universidad Laica Eloy Alfaro de Manabí, Manta, Ecuador
{patricia.henriquez, julio.garcia}@uleam.edu.ec,
jorge.herrera@live.uleam.edu.ec

Abstract. Emergency situations generate a high requirement for information, and on the other hand diminish its availability. In the last decade, intellectuals and government authorities have assessed the potential of information circulating through social networks, mainly the one originated from natural disasters. Because of its direct and fast way of communication, and because of the reach of its network, Twitter® is the most used social platform for crisis management. Twitter analytics is a rising area of study. The goal of this research is to analyze the time and content scopes of a significant dataset of tweets in the first 72 h of the 2017 Mexico earthquake around three official profiles. The methodology used is based on text mining techniques; the tweets have been classified into five categories based on the purpose, responses and behavior of both the authorities and the public. The results indicate that the messages about actions, information, and opinion categories predominated over emotions, and technology.

Keywords: Twitter analytics · Crisis management · Social networks · Natural disasters

1 Introduction

The use of social networks, both by governmental and non-governmental institutions in order to address the different phases of emergencies or extreme events, has grown over the past decade [1–5]. Time is critical in emergency care; therefore, Twitter's promptness and reach are the two attributes that make it the most used social network in disaster situations [3]. In times of emergency the number of tweets sent by citizens increases [6, 7] and the agencies involved in disaster relief, whether governmental or not, use Twitter to provide the public with critical information regarding evacuations and other actions aimed at mitigating the effects of the disaster.

Twitter Analytics is an area of study of growing interest; Twitter is being studied as an effective means for the management of natural disasters such as fires [8, 9], earthquakes [1, 10–12], Tsunamis [13–15], or events of public outrage like terror attacks.

Twitter has been used for early warnings [14, 16], for aiding during the impact of the emergency [10] and in the aftermath of the disaster [17, 18]. For Haworth and Bruce [19], and Klonner et al. [20], this social network is mainly used at the time of the

© Springer Nature Switzerland AG 2019
Á. Rocha et al. (Eds.): ICITS 2019, AISC 918, pp. 3–12, 2019.
https://doi.org/10.1007/978-3-030-11890-7_1

emergency. Wang and Ye [21] point out that the preferential use of Twitter during the time of the emergency responds to an increase in the amount of user interaction at that moment and to it being less disaggregated than in the previous and subsequent moments.

Four dimensions of twitter provide sensitive information for the management of natural disasters: Space (GPS coordinates and city name), Time, Network (re-tweets), and Content (text or images that it contains). Recently, Wang and Ye suggested the need to analyze several dimensions simultaneously in order to provide more useful information. This research studies the aspects of; time, network, and content, of the tweets that circulated around the official profiles of two public agencies with competence in emergency management and an NGO during the 2017 earthquake in Mexico.

1.1   Twitter Content Information as a Management Tool

About the participation of citizens, the authors of [15], say that citizens expect governments to appropriately manage emergencies, provide the necessary information, and use social networks. The authors of [22] point to three areas in which social networks are useful for emergencies in local communities: providing information, transmitting information, and responding to emotions. Social networks, and especially Twitter, can play an essential role in all cases.

Texts circulating on Twitter after a natural disaster are potentially a source of data, that mining techniques can turn into critical information of great value within the disaster recovery tasks, undertaken by government agencies and NGOs.

Different studies have analyzed Twitter data in order to improve earthquake management. Morales et al. analyzed twitter content during the Iquique earthquake in Chile to "describe the functions fulfilled by the messages sent by users of the Twitter microblogging service, during the month immediately after the natural phenomenon occurred" (p. 343). The method was a qualitative data analysis around five categories that they established. The authors concluded that users employed Twitter as a platform to express their opinion, to call for social action, and to express their emotions.

In [6] the authors examined the use of the Chinese Twitter "Weibo" during an emergency in 2013, and compared user-generated content with previous findings related to a weather event in North America. A total of 799 tweets were collected. Out of those tweets, 283 were retweets. Two undergraduate coding students were trained in the coding scheme of the content attributes of each tweet, and both the type of content present in the tweet (information about the storm, expressions of affection, spam, humor or insult) and the characteristics of the profile that sent it were identified. The results indicate that, by comparison, the Weibo sample contained proportionally similar degrees of informative and useful content, but that users were less likely to use humor and did not show an increase in emotional response during the crisis.

Qu et al. also studied the popular Weibo, in order to investigate how Chinese citizens used it in response to a major natural disaster: the 2010 Yushu earthquake. They combined multiple methods of analysis in this case study, including analysis of the message content, analysis of trends from different topics, and an analysis on the diffusion process of the messages. This work complements the existing ones with an exploration of a non-Western sociocultural system: use of microblogging services by Chinese users in response to an earthquake.

Sakaki et al. [1] investigated the interaction of Twitter users in real time in events such as earthquakes and proposed an algorithm that classifies tweets based on characteristics such as keywords, word count, and their context. Subsequently, they produced a probabilistic space/time model that can find the center and trajectory of the event location. They consider each Twitter user as a sensor and apply Kalman filtering and particle filtering, which are widely used in computer science to estimate a location. Finally, they built an earthquake report system in Japan that can detect when an earthquake is happening with a high probability (96% of earthquakes from the Japan Meteorological Agency, JMA) only by monitoring tweets. The system then sends an alert email to registered users.

1.2   Twitter Temporal Information

Time is critical in the management of a natural disaster. Citizens and authorities can make safer decisions during emergencies based on real-time information available on social networks.

Regarding the recovery from the disaster, a quick response can lessen the effects caused by it. Regarding the public, timely information can set an adequate perception and a sense of trust from citizens about the fact that the official organisms are adequately addressing the emergency. Less than an hour is an acceptable time for the first pronouncement on the crisis to be made by organizations, according to Zoeteman.

According to Wang et al. [9], the use of the time field in a tweet has been analyzed to find: (a) Evolution of the event in time intervals, usually every hour (b) Cyclical variations, and (c) Causes of variation.

Gurban et al. studied the time variable in a set of 2616 tweets from six official organizations with disaster competence, in the aftermath of the 2010 earthquake in Haiti. The results of the study indicate that the way in which the six organizations used Twitter changed constantly over time. Chi-square analyzes showed how organizations decreased the use of certain strategies to disseminate information through Twitter, such as the use of links; and on the other hand, kept the use of techniques that seek user involvement, such as retweets and calls to action.

Qu et al. Found that Weibo messages reach their peak immediately after the earthquake and then gradually decrease. Mendoza et al. analyzed the variations of Twitter activity during the four days after the 2010 Chile Earthquake. If the temporal analysis was performed around four days, it was observed that the highest volume of tweets happened on the last day, because that's when the communications were restored in most of the country. Later, an hour-by-hour analysis was made for each day, and found for example two peaks on the first day and multiple interruptions due to the failures of the Internet service.

2 Methodology

This research analyzes the use of Twitter for emergency management purposes during the Mexico earthquake, specifically in the initial phase of the seismic activity [23], where the highest concentration of tweets is available. The earthquake had a magnitude of 7.1 Mw, it happened in the State of Puebla in Mexico, on September 19, 2017, at 1:14 p.m. local time (UTC-5)[1].

The analysis of the use of Twitter in natural disasters can be done using information circulating around hashtags [24], centered on official profiles, or with a combination of the two [2]. In the case of the earthquake in Mexico, both criteria were used, the *#Terremoto* hashtag and the data surrounding three official profiles. Two profiles match to organizations with competence in emergency management in Mexico City: @gobMX and @UCS_CDMX, while the third one answers to the NGO Cruz Roja @CruzRoja_CDMX.

Raw data was provided by Twitter®, a total of 153,215 tweets, retrieved between 09/19/2017 at 1:00 p.m. and 9/21/2017 at 12:59 p.m. The automated analysis of the data was processed with data mining tools such as Excel and RapidMiner. The processing of the data had several phases: (a) preparation of the data by deleting special characters except (@ and #), (b) deleting url's, (c) replacing accent-mark vowels, and (d) removing words that do not add meaning.

Content analysis was computed with the classic method of codebook advised by the theory of Glaser and Strauss. Specifically, it started with the list of codes proposed by Morales et al. (2018) for the analysis of the tweets of the Iquique earthquake in Chile. This list classifies the contents into five groups: *Opinion, Information, Emotion, Actions, and Technology*. The list suggested by the authors was adapted to the data set of the Mexico earthquake.

Ten coders (trained students) performed manual coding, and double verification was performed, before the massive analysis of the data, 68 tweets were randomly selected and coded by each person, then the assigned codes were compared, and the criteria of the coders refined. Finally, the manual coding of the 153,215 tweets was undertaken. This manual coding has also served for the training of a model based on deep learning.

The research questions were: What are the uses authorities and citizens make of Twitter during the first three days after the Mexico earthquake in the analyzed data set? And how do citizens' responses change over time?

3 Results

In this section, we analyze the role played by the tweets that circulated in the first three days following the earthquake around the official profiles of the three accounts of selected organizations: @gobMX, @UCS_CDMX, and @CruzRoja_CDMX.

---

[1] Excelsior news paper, 09/19/2017

Day 1

Table 1 shows that Day 1 concentrated the highest amount of tweets and retweets generated during the crisis, with approximately 62%; considering it, we can assure that communication channels were functioning despite the strong earthquake.

Table 1. Tweets and retweets generated.

Day	Date and time	Original tweets	Total retweets	Total tweets
1	19 Sept. 13H00 to 20 Sept. 12H59	6,620	87,647	94,267
2	20 Sept. 13H00 to 21 Sept. 12H59	5,525	30,892	36,417
3	21 Sept. 13H00 to 22 Sept. 12H59	4,375	18,156	22,531
	Total→	16,520	136,695	153,215

An approach to the types of content that was shared during the three days after the earthquake was made based on the five previously established categories: *Opinion, Information, Emotion, Actions, and Technology.*

Figure 1 shows the total of tweets/retweets generated in each category during the first 24 h after the event.

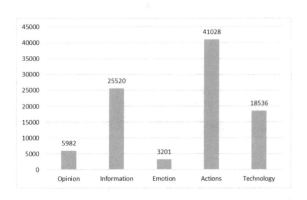

Fig. 1. Total tweets by content type - Day 1

As shown in the above figure, in the initial moments of the emergency, authorities used their official Twitter profiles to suggest citizens with the immediate actions to be taken in order to avoid further damage. Messages such as: *keep calm, do not use a landline to communicate,* and *evict the risk areas* were the most frequent. Regarding citizens, messages such as *the request for resources and help, the search for people, suggestions for measures such as keep calm,* are the most frequent.

Figure 2 shows the details of these actions.

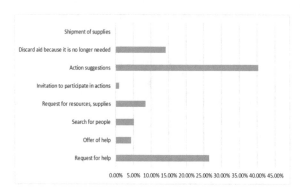

Fig. 2. Actions from a manually coded sample - Day 1

An hourly analysis of the types of content tweeted during the first 24 h after the earthquake, reveals that tweets related to actions began an escalation on the 13th hour and reached their peak at the 21st hour. A similar pattern is shown on tweets related to information, which started their ascent at the 15th hour and reached their peak at the 17th hour, see the Fig. 3.

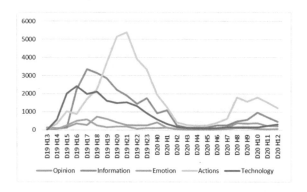

Fig. 3. Total tweets by hour and content type - Day 1

Day 2

On the second day, as shown in Table 1, the amount of related tweets/retweets dropped substantially. The total of tweets/retweets on this day represents 23.8% of the total of tweets analyzed. The type of content of the tweets generated during that day is shown in Fig. 4, organized in the five categories previously named.

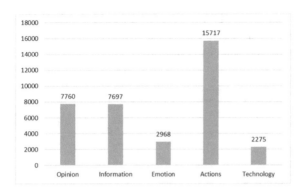

Fig. 4. Total tweets by content type - Day 2

The pattern of Day 1 is repeated, as tweets with content related to the *actions* had a more significant presence in the data set, followed by the *opinion* and *information* categories whose statistical difference is minimal.

The results of the manual coding carried out give us a closer look of the actions that prevailed on that second day after the earthquake, as shown in Fig. 5. This figure shows that on the second day *Action Suggestions* topped the list of the most retweeted content in relation to the *Actions* category; however, and unlike on day one, *Search for people* is in second place, followed by the *Requests for help* that on this day falls to the third place.

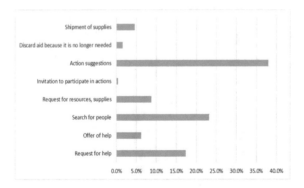

Fig. 5. Actions from a manually coded data set - Day 2

Regarding the type of content of the tweets on Day 2, the *Opinion* category occupied the second position. This category includes general opinions; comments on the actions of the authorities, the people, and the media; as well as acknowledgments and comments about the operation of essential services.

To complete the analysis of Day 2, an hourly view shows how the tweets related to the *Actions* category continue to drop from the previous day, reaching its minimum point at the 12th hour of the 21st day, after showing a peak at the 17th hour of day 20. It is interesting to note that from hour 22 of day 20 to the hour 17 of day 21; all types of content show almost the same trend (Fig. 6).

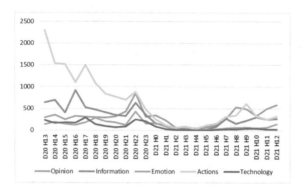

Fig. 6. Total tweets by hour and content type - Day 2

Day 3

Following the trend of the first two days, the third day showed the lowest number of tweets/retweets, specifically 14.7% of the total number of tweets included in this study. The content type categories for this day are shown in Fig. 7.

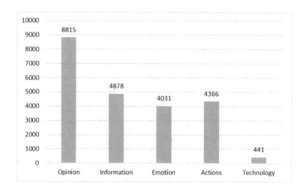

Fig. 7. Total tweets by content type - Day 3

On this day, unlike in the previous two days, tweets with content related to the *Opinion* are the ones taking the first place, followed by tweets linked to the *Information* category.

General opinions hold the first place, however and unlike on day 2, the tweet load related to comments on the actions of the authorities and the acknowledgments is much higher, going from 10% to 25% in the first case and 6% to 19% in the second. In the category of tweets related to *Information*, the subcategories that stand out are information on actions carried out by the authorities and responses from the authorities, with 37% and 39% respectively.

4   Discussion

The time pattern of the tweets analyzed in this data set, show an initial high traffic of tweets around the profiles of the three official agencies that were analyzed on day 1 of the emergency, which then decreases until it reaches its lowest point by the third day. This result confirms those of Qu et al. who found that Weibo messages reached their peak immediately after the China earthquake and then fell gradually. This supports the statements of Planella et al. regarding that time is critical for authorities to address an emergency correctly. From the citizen's point of view, rapid attention in the early moments increases the confidence that the crisis is being well managed and increases resilience.

Regarding the use that has been given to Twitter as a communicative tool, the first days of the earthquake emergency have been fundamental in three categories: calls to action, information transmission, and opinion. On the other hand, the expression of emotions does not seem to have much presence in the analyzed data set. Lin et al. found proportionally similar degrees of informative and emotional content in the tweets. Morales et al. found that in tweets the expression of opinion predominates, to summon social action and express their emotions.

As a future work, new studies will have to be carried out in order to determine whether the fact that the data set analyzed corresponds to official profiles has an influence on the predominance of Information and Action categories.

References

1. Sakaki, T., Toriumi, F., Matsuo, Y.: Tweet trend analysis in an emergency situation. In: Proceedings of the Special Workshop on Internet Disasters - SWID 2011, no. July 2014, pp. 1–8 (2011)
2. Acar, A., Muraki, Y.: Twitter for crisis communication: lessons learned from Japan's tsunami disaster. Int. J. Web Based Commun. 7(3), 392 (2011)
3. Chatfield, A.T., Scholl, H.J.J., Brajawidagda, U.: Tsunami early warnings via Twitter in government: net-savvy citizens' co-production of time-critical public information services. Gov. Inf. Q. 30(4), 377–386 (2013)
4. de Albuquerque, J.P., Herfort, B., Brenning, A., Zipf, A.: A geographic approach for combining social media and authoritative data towards identifying useful information for disaster management. Int. J. Geogr. Inf. Sci. 29(4), 667–689 (2015)
5. Li, H., Caragea, D., Caragea, C., Herndon, N.: Disaster response aided by tweet classification with a domain adaptation approach. J. Contingencies Cris. Manag. 26(1), 16–27 (2018)
6. Qu, Y., Huang, C., Zhang, P., Zhang, J.: Microblogging after a major disaster in China. In: Proceedings of the ACM 2011 Conference on Computer Supported Cooperative Work - CSCW 2011, no. July 2016, p. 25 (2011)

7. Hormazábal, C.: Comunicación e imagen en crisis: análisis de empresas de telecomunicaciones en Chile tras el 27F de 2010. Cent. Estud. en Diseño y Comun. **40**, 65–78 (2012)
8. Starbird, K., Palen, L.: Pass it on?: Retweeting in mass emergency. In: Proceedings of 7th International ISCRAM Conference, no. December 2004, pp. 1–10 (2010)
9. Wang, Z., Ye, X.: Social media analytics for natural disaster management. Int. J. Geogr. Inf. Sci. **32**(1), 49–72 (2018)
10. Wei, Z., Qingpu, Z., Wei, S., Lei, W.: Role of social media in knowledge management during natural disaster management. Int. J. Adv. Inf. Sci. Serv. Sci. **4**(4), 284–292 (2012)
11. Wu, Y., Liu, S., Yan, K., Liu, M., Wu, F.: OpinionFlow: visual analysis of opinion diffusion on social media. IEEE Trans. Vis. Comput. Graph. **2626**(c), 1 (2014)
12. Morales, M.C., Cabezas, N.G., Jaime, K., Rendić, F.: Uso De Twitter En Desastres: El Terremoto De Iquique. Interciencia **43**(May), 343–350 (2018)
13. Lchiguchi, T.: Robust and usable media for communication in a disaster, pp. 44–55 (2011)
14. Online, R., Chatfield, A., Brajawidagda, U.: Twitter tsunami early warning network: a social network analysis of Twitter information flows. In: 23rd Australasian Conference on Information Systems, pp. 1–10 (2012)
15. Jin, Y., Liu, B.F., Austin, L.L.: Examining the role of social media in effective crisis management. Commun. Res. **41**(1), 74–94 (2014)
16. Cheong, F., Cheong, C.: Social media data mining: a social network analysis of tweets during the Australian 2010–2011 Floods. In: 15th Pacific Asia Conference … (2011)
17. Jaen-Cortés, C.I., Rivera-Aragón, S., Reidl-Martínez, L.M., García-Méndez, M.: Violencia de pareja a través de medios electrónicos en adolescentes mexicanos. Acta Investig. Psicológica **7**(1), 2593–2605 (2017)
18. Xu, T., Chen, Y., Jiao, L., Zhao, B.Y., Hui, P., Fu, X.: Scaling microblogging services with divergent traffic demands. Lecture Notes Computer Science (Including Subseries Lecture Notes Artificial Intelligence Lecture Notes Bioinformatics). LNCS, vol. 7049, pp. 20–40 (2011)
19. Haworth, B., Bruce, E.: A Review of Volunteered Geographic Information for Disaster Management. The University of Sydney (2015)
20. Klonner, C., Marx, S., Usón, T., Porto de Albuquerque, J., Höfle, B.: Volunteered geographic information in natural hazard analysis: a systematic literature review of current approaches with a focus on preparedness and mitigation. ISPRS Int. J. Geo-Inf. **5**(7), 103 (2016)
21. Wang, Z., Ye, X., Tsou, M.H.: Spatial, temporal, and content analysis of Twitter for wildfire hazards. Nat. Hazards **83**(1), 523–540 (2016)
22. Hernández-Hernández, M.E., de la Roca Chiapas, J.M., Barragán, L.F.G.: Measurement of the Jungian psychological types in Mexican university students. Acta Investig. Psicológica **7**(10), 2635–2643 (2017)
23. Méndez, M.D., Leiva, M.C., Bustos, C.B., Ramos, N.A., Moyano-Díaz, E.: Mapa exploratorio de intervenciones psicosociales frente al terremoto del 27 de Febrero de 2010 en la Zona Centro-Sur de Chile. Ter. Psicol. **28**(2), 193–202 (2010)
24. Crist, C., Winn, J.: Educational Facilities Disaster and Crisis Management Guide Book 2006–2007. Florida Department of Education (2007)

E-Government and the Quality of Information in Web Portals of the GADM of Ecuador

Patricia Henríquez-Coronel$^{(\boxtimes)}$ ⓘ, Jennifer Bravo-Loor ⓘ,
Enrique Díaz-Barrera ⓘ, and Yosselin Vélez-Romero

Universidad Laica Eloy Alfaro, Manta, Manabí 130802, Ecuador
patricia.henriquez@uleam.edu.ec,
jenimarb@gmail.com, enrique_diazb@hotmail.com,
lupitavelezromero@gmail.com

Abstract. This paper shows the results of a research that evaluated the quality of the information provided by the Ecuadorian local government in their web portals. The research fits a quantitative approach; and it has a descriptive scope. A sample of 121 web portals was analyzed through a rubric designed by researchers and evaluated by experts. The rubric evaluated the quality and periodicity of the information through seventeen indicators. The leading results indicate that the published information complies with the Organic Law of Transparency and Access to Public Information, and the periodicity with which the information is updated is relatively good. Options such as the translation of the page appear in a few municipal portals of the country. The municipal Ecuadorian e-government is in an improved stage, which is characterized by UNESCO as the publication of elementary information to improve communication between the government and citizens.

Keywords: Quality of the information · Local government · E-government · Ecuador

1 Introduction

Electronic Government is related to the modernization of the public administration, it ensures efficiency and effectiveness in processes offered by State institutions to its citizens, providing tools such as public information, online payments, government procurement, and mechanisms to promote citizen participation from anywhere in the world [1]. At the same time another of the purposes of the Electronic Government is to speed up processes that are handled within the institution. Some public entities have already been able to implement tools such as QUIPUX, online training, and the on-line publication of municipal documents. E-Governments that are properly managed promote transparency of processes; moreover, governments take a greater commitment with its citizenship, since important information is made public and accessible. If it were not for E-Government, such information would be hard to broadcast to the public and could cause confusion and lack of trust.

© Springer Nature Switzerland AG 2019
Á. Rocha et al. (Eds.): ICITS 2019, AISC 918, pp. 13–20, 2019.
https://doi.org/10.1007/978-3-030-11890-7_2

To improve organization, production, and transmission of information and services, information management is incorporated to the Electronic Government; that way citizens can properly use information in a search process and transparent access framework. The benefits that are promoted through information management are: informative interaction between government institutions and citizens, reception of documentation such as, requests, suggestions, inquiries, online transactions, and comments or concerns, transparency and effective information, confidence in the citizen with the production and transmission of information, providing timely, accurate, accessible, reliable and sufficient information, all the time [2].

Despite all the benefits that E-Governments could deliver, in many cases of public institutions in Latin America they do not take advantage of all the tools that are provided to them, on the contrary, they are used only to provide basic or incomplete information. In some cases, the heads of public institutions use E-Governments as a means to broadcast political propaganda of their administration [3], and the use of participatory mechanisms or process transactions is not promoted. On the other hand, if those institutions would provide easier access to information, transactions and use of mechanisms for participation, the citizenship would incorporate this system into their daily lives and assume co-government functions.

In Ecuador, since the Electronic Government Plan was implemented in 2011, all public institutions began to subject themselves to this new policy. To this date, all the Autonomous Decentralized Municipal Governments of Ecuador (GADM), have at least created an institutional website, in which public information under the Article 324 of the Organic Code of Territorial Organization, Autonomy and Decentralization (hereinafter COOTAD) [4], and also in Article 7 of Organic Law of transparency and access to Public Information (LOTAIP) should be displayed [5]. However, at the time, there are Autonomous Decentralized Municipal Governments that do not offer this information on their web portals.

The information provided through the web portals must meet a quality criteria, should be useful and sufficient, timely (frequently updated) and must be easily located on the website, so that citizens can locate it quickly. If the information meets that quality criterion, citizens can actively participate in decision-making, comply with the payment of taxes and fees and improve the collection of revenues by the municipal autonomous governments of Ecuador. In addition, information dissemination and public processes could improve the function of social control in the fight against corruption. In short, quality information promotes co-government of citizens and transparent democracy.

This research analyzes the quality of information offered by the web portals of the Autonomous Decentralized Municipal Governments of Ecuador taking into account quality indicators and the Ecuadorian legal framework for e-government.

2   Method

The partial results shown respond to a quantitative approach [6]. Its intention was to analyze a sample of 121 GADM websites out of a universe of 221, to describe e-government services in three big areas according to Cerezo [7]: information, public participation, and transactions. This article shows the results in the information area.

Stratified random sampling answered to the four regions in which Ecuador is divided: Sierra, Costa, Amazon and Galapagos. In the Sierra region a total of 47 GADM were selected, on the Costal region a total of 45 GADM, 29 GADM in the Amazon and a single Galapagos GADM. Information on the total number of municipal governments on each area was obtained from the National Institute of Statistics and Census (INEC) [8], and listings containing the names of municipalities and the URL of their official website were configured. All of the selected GADM websites were downloaded and housed in a local storage in order to ensure that the traits being analyzed were stable over the period of the analysis.

The tool used for the analysis was a rubric made out of the proposed indicators identified in Cerezo [7], PNGE [9], Esteves [10], LOTAIP [5], Law Organic Citizenship [11], Naser [12], COOTAD [4], Líppez-De Castro and Alonzo [13], Bersano [14]. The rubric has a total of 32 indicators, distributed as follows (Table 1):

Table 1. Electronic Government areas assessed in the rubric.

Information	17 indicators
Citizen participation	9 indicators
Online payments	6 indicators

Once developed, each rubric was approved by three experts in E-government. The final version was used by investigators over four weeks to observe each of the aspects evaluated. A total of 160 h were used for the analysis of the aforementioned portals. Regarding the information category that is reported in this article, it took into account classic information services, such as personalized attention, Publication of Government ordinances, Agreements, Contact Information, Page Translation, Accountability, understandable information for ordinary citizens, Recreation and Leisure Guides, Touristic Information, News (Concerning the work of the institution), Institutional Information, functional organizational structure of the organization, complete directory of the institution, results of both internal and governmental audits, and Information on public transport. These indicators are showed in Table 2.

The data collected with the rubric was processed with software for descriptive statistics.

Table 2. Indicators applied assessing information area in GADM web portals

The web portal:
Offers personalized attention
Publishes Local government decrees
Publishes agreements
Give sufficient contact information
Is translated to other idioms
Presents accountability reports
Offers comprehensible information
Includes leisure guide
Includes tourist information
Includes local breaking news
Offers institutional information as vision, mission and values
Publishes the GADM organization chart
Offers the GADM directory staff
Offers local requirements & Planning Application Forms
Publishes information on the annual budget administered by the institution
Publishes the results of the internal and governmental audits
Offers urban transport information

3 Results

3.1 Accountability

Accountability is the process in which the competent authority reports, justifies, and takes responsibility for public spending, showing the public the results. Publication of accountability mechanisms in web portals is mandatory according to the LOTAIP art. 7.

It was found 79% of the portals provide accountability to its citizens; on the contrary, a small 2% does not, and 19% that presents an outdated information Some portals such as San Cristobal of Galapagos GAD present their accountability in video format; others do it in a digital magazine, such as the Gualaquiza GAD from the Morona Santiago province, opposite to the most usual publication format for GAD accountability which is PDF.

There were cases such as the municipal GAD from Quevedo and Vinces, both from the province of Los Rios, who had not published accountability since year 2013 to the day the review was made (September 25, 2017). However, on October 6th it was revised again and the information had already been updated (Graphic 1).

3.2 Page Translation

Translation of a municipal page is important so that people from other countries can understand the information; especially information concerning tourism which is one sector that drives the economy of a country. Article 2 of the Constitution of the Republic of Ecuador also recognizes the Quichua and Shuar as official languages of

Graphic 1. Accountability.

intercultural relations. The other ancestral languages are for official use for indigenous peoples in the areas in where they live and in the terms established by the Law.

The rule states that the State shall respect and encourage its conservation and use. A 13% of the websites analyzed are translated into English, Shuar and/or Quichua, and only 1% to Quichua or Shuar.

Note that the GAD of Lago Agrio in Sucumbios province provides a native translation of its webpage into 20 languages: German, Korean, Haitian Creole, Spanish, French, Welsh, Greek, Dutch, Hungarian, Indonesian, Italian, Japanese, Portuguese, Russian, Turkish, Ukrainian, Urdu, Vietnamese, and Yiddish.

GAD of Saquisili in the province of Cotopaxi is translated into Quechua, important for indigenous peoples where the ancestral language prevails.

GAD of Portovelo in the province of El Oro has the option to translate into English, French, and Italian; however, the option does not work.

3.3 Tourist Information

Tourist information is important for foreign and domestic travelers, this way they can check tourist attractions in the regions, and their weather in order to plan their trips.

It was found that 43% of web portals contain detailed information on tourist attractions and have created tourist routes or specific trip offers. Another 33% only presents the attractions of the area, and 24% of GADM show no tourist information. The latter figure represents a negative impact on tourism since it makes it harder to strengthen and promote tourism at local, provincial, national and international levels.

In the Jipijapa GAD, of the province of Manabí, offering for tourism routes was found; however, when corroborating information, the page was not available. The Portoviejo GAD, in this same province presented a description of routes and videos.

Francisco of Orellana GAD, of the Orellana province offers tips on clothing that tourists should bring to venture into the jungle, also information on the weight of the suitcase in the case of air travel, among other useful information.

In the Sierra area, the GADM of Azogues presented a custom map showing all sorts of attractions such as restaurants, parks, access roads, nightclubs (Graphic 2).

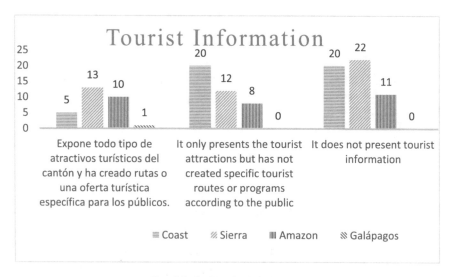

Graphic 2. Tourist information.

3.4 Current Information

This indicator represents the place where the public can be informed about news on what is happening in the municipality and in the city: sports, cultural, and entertainment news should be reported on the website of GADM.

Despite the importance of this indicator, not all GADM fulfill it, only 38% of the municipalities provides monthly information. A worrying case is that of two local governments, in which at the time of the review, the information had not been updated for over a year. The GADM of Chinchipe was last updated in 2014 and the GADM Zamora in 2016. On the other hand, 27% of GADM provide weekly information, and a 35% does so periodically to its citizens.

When the research was conducted, the GADM of Machala updated news weekly, but when corroborating data, it was shown that news were being daily updated.

3.5 Institutional Information

The mission and vision is a fundamental part of any institution, since they reflect the reason why the institution must exist and the ideal of the institution, in this way its citizenship can feel solidarity and contribute to carrying out these ideals. The profile of the Mayor is important for citizens to know who the person leading the municipality is.

A 47% of web portals provide complete information on the institution, including the mission, vision, principles, and the Mayor's Profile. Some 28% contains only the institutional mission and vision, and 25% does not provide any information.

The GADM of San Cristobal in the Galapagos, the GADM of Sucúa in the province of Morona Santiago, and other GADM not only present information on the mayor, but also profiles of the Municipal Council members (Graphic 3).

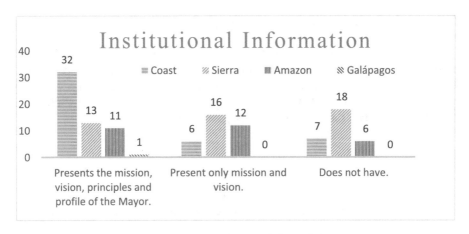

Graphic 3. Institutional information

4 Discussion

The results obtained in this study confirm the findings of Líppez-De Castro and Alonzo [13] in Colombia, regarding that the further progress made on e-government focuses on giving of information to citizens; in contrast, there is a slow progress on tools for citizenship participation and online payments. Which makes it clear that the Municipal web portals in Ecuador are effectively complying with the provisions of LOTAIP [5] and the Article 118 of the Constitution of the Republic of Ecuador, which states that all state institutions that make up the public sector and institutions linked to it, must publish in their information portals, websites, or any other media made available to the public by the same institution, a series of mandatory information, which has been shown throughout this investigation.

While in the field of information laws are met regarding the amount of published information, there is still a lot of improvement to be made, especially in terms of quality and updating. Published information should help improve services and local government processes. The delay in the publication of a certain type of information on GADM webpages detracts value from them. In a digital environment that allows constant updating, it does not make sense that news are updated once a month. Another aspect is that information is static and presented in unattractive formats, also few portals provide opportunities for users to create content. Published information can't be customized by users. This is the case of information regarding cultural, sports, and touristic events.

As Zamora, Arrobo and Cornejo [15] claim, the actions taken by the Ecuadorian Government, far from perfect and Utopian try and initiate a long way to transform and modernize the State; however, there is still a lot of room for improvement.

In relation to information, probably the most important challenges is publishing actualized information daily. Other challenges refer to visual attractive, usability and accessibility of information. In terms of interactivity, it's very important local governments achieve engagement and empathy with users through their web portals.

Change is inevitable, society changes thus reforms are fully relevant to achieve improvements in public service.

According to the characteristics of the information posted on GADM websites, the current state of e-government seems to fit in with what the ONU [16] identifies as Improved Stage. The content of the web portals is rudimentary and promotes one way communication between GAD and the citizen, providing basic information and downloads of certain forms or requests for paperwork.

References

1. Naser, A., Concha, G.: The e-government in public administration. United Nations Publication (Spanish) (2011). https://repositorio.cepal.org/bitstream/handle/11362/7330/1/S1100145_es.pdf
2. Torres, M., Viloria, A., Vasquez, C.: Management and information quality in e-government. Univ. Sci. Technol. **14**(54), 55–64 (2010)
3. Andrade, J., Yedra, Y.: Transparent systems for efficient electronic governments (Spanish). Enl@ce: Venezuelan J. Inf. Technol. Knowl. **4**(2), 81–95 (2007). http://www.redalyc.org/comocitar.oa?id=82340206
4. Act No. 0: Organic Code of Territorial Organization, Autonomy and Decentralization (Spanish). Asamblea Nacional, Quito (2010)
5. Act No. 24: Organic Law of Transparency and access to Public Information (Spanish). Asamblea Nacional, Quito (2004)
6. Hernandez, R., Fernandez, C., Baptista, P.: Investigation Methodology. McGraw-Hill Education, Mexico City (2014)
7. Cerezo, J.M.: The challenges of the administration in Spain (Spanish). Paper presented at the II Jornada Democracia digital, eAdministración y participación ciudadana. Barcelona, España, marzo (2005)
8. National Institute of Statistics and Census: Ecuador Statistical National Institute of Statistics and Census. http://www.ecuadorencifras.gob.ec/estadisticas/?option=com_content&view=article&id=300
9. National Secretariat of the Public Administration: Electoral Plan 2014–2017 Version 1.0. National Secretary of Public Administration (Spanish) (2014). http://www.buenvivir.gob.ec
10. Esteves, J.: Analysis of the development of municipal e-government in Spain. IE Working Paper, Madrid (2005)
11. Act No. 0: Act of Citizen Participation (Spanish). Asamblea Nacional, Quito (2011)
12. Naser, A.: Electronic Government, Indicators. Latin American and Caribbean Institute for Economic and Social Planning (ILPES) (Spanish) (2009). https://www.cepal.org/ilpes/noticias/paginas/0/40660/alejandra_naser_INDICADORES.pdf
13. Líppez-De, C.S., Garcia, A.R.: Citizens and e-government: citizen orientation of municipal web sites in Colombia to promote participation. Univ. Human. **82**(82), 279–304 (2006)
14. Bersano, L.: E-Government best practices in Latin America. What indicators are used to describe them? What is its relevance to measure the level reached on the way to the E-society? University of Buenos Aires, Argentina (2006)
15. Zamora, C., Arrobo, N., Cornejo, G.: The Electronic Government in Ecuador: innovation in public administration (Spanish) (2018). http://www.revistaespacios.com/a18v39n06/a18v39n06p15.pdf
16. United Nations Organization: United Nations Study on Electronic Government, 2012 (Spanish) (2013). https://publicadministration.un.org/egovkb/Portals/egovkb/Documents/un/2012-Survey/Complete-Survey-Spanish-2012.pdf

The Contribution of Knowledge Engineering in Supply Chain: A Literature Review

Fatima Ezzahra Ettahiri[✉] and Mina Elmaallam

Rabat, Morocco
Fatimaezzahraettahiri@gmail.com

Abstract. Using Knowledge Engineering in Supply Chain can assure a high level of performance. Actually, supply chain is a favorable area of knowledge flows, explicit knowledge can be easily identified, capitalized and transferred but the real challenge is how to capture tacit knowledge in the different steps of a supply chain. That's why it's important to lead a research concerning a literature review about the contribution of Knowledge Engineering in supply chain. The aim of this article is to give a synthesis of the different methods of knowledge engineering that can be used to manage knowledge in supply chain.

For this purpose, we will present in the first part the link between cognitive technology and supply chain, while the second part is dedicated to the explanation of knowledge engineering methods. By the end of the article, we will have an overview of organizational methods and knowledge engineering methods that can be used in supply chain.

Keywords: Knowledge engineering · KBE · Supply chain · Knowledge flows · Supply chain management

1 Introduction

There is no doubt that knowledge technologies have an effective power, this power comes from combining several points of view concerning managerial, cultural, socio-organizational and engineering strategies. The aim of this paper is to explain the contribution of knowledge engineering into supply chain management.

Stevens (1989) defined the supply chain as "the connected series of activities which is concerned with planning, coordinating and controlling material, parts and finished goods from suppliers to the customers. It is concerned with two distinct flows through the organization: material and information Stevens (1989)". Further, the notion of Supply Chain Management appears In their article, Supply chain management: the elusive concept and definition, LeMay et al. (2017) defined this concept as "the design and coordination of a network through which organizations and individuals get, use, deliver, and dispose of material goods; acquire and distribute services; and make their offerings available to markets, customers, and clients". It's a set of activities required so as to plan, control and execute a product's flow from raw materials acquisition to production through distributions to the final customer, respecting cost-effectiveness and simplifying processes.

© Springer Nature Switzerland AG 2019
Á. Rocha et al. (Eds.): ICITS 2019, AISC 918, pp. 21–31, 2019.
https://doi.org/10.1007/978-3-030-11890-7_3

In order to simplify innovation and creativity required within an unpredictable environment of business, researchers chose to introduce the logic and techniques of Knowledge Management. In fact, (Spekman et al. 2002) argue that an effective KM is recommended to realize an effective SCM, they advanced the idea that KM, when it's extended through all supply chain levels, is the key success of competitive advantage. That is to say that Knowledge management is becoming fundamental for the firm survival.

For this purpose, our research presents a state of art concerning the contribution of knowledge engineering in supply chain management. The first part concerns research methodology, it contains 4 sections which are; research questions, techniques and procedure of collecting data, background and related work and research value. The second part discusses the role on cognitive technology into supply chain management, it contains 2 sections that aim to analyze the role of ontologies, knowledge sharing and KBE in logistic chain. In the third part we present in a detailed way several methods of knowledge representation dedicated to knowledge engineering and organizational memory.

2   Research Methodology

2.1   Research Questions

In order to give an indication of what problem we will discuss in the present article we proposed two research questions and indicated the purpose of each one of them (Table 1).

Table 1. Research questions and objectives

Research question	Objective
What contribution the cognitive technology has on supply chain?	Realize an overview on articles on Ontology Engineering and Knowledge Based Systems
What are the knowledge engineering methods that can be applied on supply chain?	Present articles concerning organizational methods and knowledge engineering methods and their impact on supply chain

Supply chain is a real atmosphere of sharing explicit and tacit knowledge, this knowledge give place to several flows of information that should be managed using cognitive technology and knowledge engineering methods, that is the purpose of research questions proposed in the previous table.

2.2   Techniques and Procedure of Collecting Data

In order to realize this research, we collected journal articles from Science Direct academic database and using Google Scholar. Results were obtained using the key words "Knowledge Engineering AND Supply chain management", "Knowledge Flows" AND "Logistic Chain", "Artificial intelligence" AND "Supply Chain

Management", "Knowledge based system" AND "supply chain" Using these key words allowed us to realize a restrictive research so as to assemble the literature concerning the contribution of knowledge engineering methods on supply chain. We organized the references assembled using the software "Zotero" which is an open source tool that allows us to collect, organize and cite our bibliography. The literature collected concerns the way cognitive technology impacts supply chain management, using ontology engineering and Knowledge based engineering (KBE). Also, one of the research's objectives is to define each method of knowledge capitalization including methods of organizational memory and methods of knowledge engineering.

2.3   Background and Related Work

This section aims to highlight work done by other authors and ties in with our research. In this respect we will present journal articles that we have based our research on (Table 2).

Table 2. Overview of related researches

Article	Authors	Main idea
Uncovering the knowledge flows in supply chain relationships	Kurtz et al. 2012	Identifying and realizing a mapping of knowledge flows in supply chain between two types of organizations; industries and producers
Implementing knowledge management in supply chain: literature review	Outahar et al. 2013	Summarizing most of the theoretical and methodological characteristics developed to focus on the way in which knowledge management applications can be implemented in the supply chain context
Supply chain knowledge management: a literature review	Marra et al. 2011	This paper contributes in the debate on the role of knowledge management in supply chain using the published literature
The role of knowledge management in supply chain management: a literature review	Rosario et al. 2017	This paper aims to examine the state of research concerning knowledge management in the area of supply chain management from 3 points of view; methodological approach, supply chain management area and knowledge management processes
Knowledge management in startups: systematic literature review and future research agenda	Centobelli et al. 2017	By this article authors conducted a systematic literature review on knowledge management (KM) in the context of startups in order to analyze the state of the art, identify research gaps and define a future research agenda

2.4   Research Value

The articles presented in the previous section concern the impact of knowledge management implementation into a supply chain and how it can assure a high level of performance. In fact the linkage between knowledge flows and supply chain management has appeared since the last decade and researches concerning this theme are numerous. But an imperative distinction must be established between two concepts which are; knowledge management and knowledge engineering. That is to say that knowledge management means the direction the process should take whereas knowledge engineering develops the means to accomplish that direction. The value of this paper is to treat knowledge flows in supply chain from the point of view of knowledge engineering, using ontologies and knowledge based systems, and find out methods that can be applied to establish this linkage.

3   Cognitive Technology and Supply Chain

3.1   The Role of Ontology Engineering and Knowledge Sharing in a Supply Chain

A supply chain comprises the following three functions: the supply of materials to a manufacturer; the manufacturing process; and the distribution of finished goods through a network of distributors and retailers to a final customer. From these functions, a set of flows are supported by the supply chain; financial flows, material flows and information flows. Managing these flows is becoming crucial because they are considered as a source of knowledge that can be used to realize effective supply chain management LeMay et al. (2017).

"Knowledge sharing allows trading partners to orchestrate the operation of supply chain and capture positions of advantage. Yet, lack of knowledge sharing has been consistently found to be the most critical failure factor in supply chain management". Outahar et al. (2013) That is to say that knowledge sharing within a supply chain is a way to integrate their knowledge in order to identify opportunities and realize competitive advantage.

The article (the role of Knowledge sharing in supply chain) aims to measure the effect of knowledge sharing on inter-organizational information systems use (IOISs), on process innovation and firms' out-put performance. The result is that sharing knowledge has always a positive influence on managing all types of flows through a supply chain.

In their article "Ontology Engineering for Knowledge Sharing in Supply Chains", Smirnov and his co-workers tend to use ontologies as a basic element of supply chain configuration. Not to mention that using ontologies is a technique for representing knowledge in an application domain and modeling competences of enterprises. This research is based on four elements which are; capture all relevant characteristics of the overall application domain of SC with domain ontology, represent the competences of SC members with enterprise ontologies, identify candidates for SC networks based on matching between enterprise ontologies and domain ontology and configure the SC network by matching enterprise ontologies of the identified candidates Marra et al. (2011).

3.2   KBE and Supply Chain

To explain the meaning of Knowledge Based Engineering, we suggest the definition proposed by (Guo et al. 2015) which describe the KBE as "a rapidly developing technology with an enormous potential for engineering design applications. It stands at the point of diverse fundamental disciplines, such as artificial intelligence (AI), computer aided design (CAD), and computer programming cognitive technologies such as Artificial Intelligence (AI), machine learning and Natural Language Processing (NLP), are redefining the logic of supply chain management. Knowledge engineering is the discipline that can enhance supply chain functions using these cognitive technologies and relying on several methods, these methods will be discussed in the following sections." Guo et al. (2015), Implementation of knowledge- based engineering methodology in hydraulic generator design. Another definition of KBE is "The use of dedicated software language tools in order to capture and re-use products and process engineering knowledge in a convenient and maintainable fashion."

"The ultimate objective of KBE is to reduce the time and cost of product development by automating repetitive, non-creative design task by supporting multidisciplinary integration in the conceptual phase of the design process and beyond." Cooper et al. (2007). The steps of KBE application are; Knowledge acquisition, Knowledge representation, Knowledge reasoning and establishment of knowledge base Guo et al. (2015).

In the purpose of achieving high performance and low cost, products require the integration of manufacturability and supply chain knowledge in a way which is earlier than usual in the design process. "Knowledge-Based Engineering (KBE) applications can respond to this need through the creation of digital Product Model that informs designers about manufacturability aspects and expected performance" Overend et al. (2017).

4   Supply Chain and Knowledge Engineering Methods

4.1   Knowledge Capitalization Approach

Nowadays the value of knowledge is increasing and is becoming an important asset for companies. Storing, capitalizing and sharing companies experiences and know- how is the way to constitute what (Rosario 1996) called "intellectual capital".

In the point of view of (Simon 1996), the knowledge capitalization cycle means "to reuse, in a relevant way, the knowledge of a given domain previously stored, and modeled in order to perform new tasks" Guerrero and Pino (2001). The principle of capitalization is described in (Grundstein 1992) and it concerns the process that premise us to "locate and make visible the enterprise knowledge, be able to keep it, access it and actualize it, know how to diffuse it and better use it, put it in synergy and valorize it" Malvache and Prieur (1993).

These steps can be applied on the supply chain cycle using the different flows of knowledge that result from it.

4.2   Knowledge Capitalization Methods

In terms of knowledge capitalization, there is two essential classes of methods; methods that concern the corporate memory and methods of knowledge engineering.

4.2.1   Methods of Organizational Memory

Guerrerp and Pino in their article (Understanding Organizational Memory) define the concept of (OM) as "an aggregation of the human memories of all employees of the organization. Clearly this model lacks the information which belongs to the organization itself or the one which employees are not eager to keep in their private memories but it is important to the organization" Guerrero and Pino (2001). Furthermore, (Walsh 1991) said that "Organizational Memory refers to stored information from organization's history that can be brought to bear on present decisions". From Walsh's point of view this concept relies on five components which are; individuals, cultures, transformations, structures and ecology.

The organizational memory starts from the individual level. What is learned in suppliers' relationship, manufacturing or customers' relationship can build an individual memory which can be transformed into an organizational memory by the use of some knowledge engineering methods dedicated to organizational memory, these methods are explained in the following paragraph.

REX method, (Malvache and Prieur 1993) concerns the capitalization of experiences achieved while exerting enterprise's activities. This method is based on three essential steps which are;

- Analyze and identify needs and information resources,
- Build experience elements,
- Develop and exploit the knowledge management system.

Experience elements (documented knowledge elements, experience elements that came from interview with expert and know-how elements from enterprise's activities) are stored in what we call "experience memory" this memory contains a semantic net and a net of terminology used to introduce a vocabulary and viewpoints used in the enterprise.

CYGMA method, (cycle de vie et Gestion des Métiers et des Applications) (Bourne 1997) premise us to constitute a profession memory in manufacturing industry Dieng et al. (1999). In this method, six categories of industrial knowledge for design activity are defined; Singular knowledge, structural knowledge, terminological knowledge, behavioral knowledge, strategic knowledge and operating knowledge

As a result, this method produces four different documents; the profession glossary concerns singular and terminological knowledge, the semantic catalogue describes structural knowledge, the rule notebook for behavioral knowledge and operating manual which contains strategic and operating knowledge. This method is nowadays applied in several enterprises such as Rolls-Royce and Fiat, and can be applied in general in all steps of a supply chain.

4.2.2 Methods of Knowledge Engineering

Knowledge engineering is a domain that allows us to study concepts methods and techniques of knowledge acquisition and modeling. Knowledge management proposes several methods concerning conceptual modeling, cognitive modeling and languages representations. This section aims to present the most known methods of knowledge engineering.

KADS method (Knowledge Acquisition and Design Structuring) proposed in (Weilinga et al. 1992) in the purpose of developing a comprehensive and commercially viable methodology for Knowledge Based System (KBS) construction. In other words, the knowledge extraction process which means traditionally extracting knowledge from human expert and introduce this knowledge into the Knowledge Based System is rarely achieved in the way it should be done, that's why we need a method that allows us to share a common view of problem solving process between the expert, the knowledge engineer and the KBS in order to "make knowledge transfer a viable way to knowledge acquisition". The principle of multiple models means that KADS method distinguishes three steps to construct a KBS, these steps are: defining the problem that KBS should solve in the organization, defining the function that allows the problem solving and define the actual tasks that the KBS is supposed to perform. For each one of these three steps there is a specific model. Thus, KADS is based on an organizational model that takes into account the socio- organizational environment of the KBS system. The second model is the application model that defines the problem that the system should solve and the function of the system able to realize this. The third model is the task model; it defines the way the function of the system is realized through several tasks that system is supposed to perform.

MKMS (Method for Knowledge System Management) was proposed in (Ermine 1992) in order to save and capitalize researchers' knowledge. Later this method has been developed so as to become "MASK" (Method for Analysis and Structuring Knowledge). The approach of MKSM/MASK consists on modeling knowledge according to different points of view from knowledge sources in order to organize it into a "knowledge system". In his article (Methods and tools for corporate knowledge management) (Dieng et al. 2004) describes the MKMS method as "a systemic-based decision-support method". According to this method, an organization's knowledge assets are considered as a complex system. The modeling phase in MKSM method relies on several points of view which are; semantic, syntax and pragmatic and the implementation of this method should respect five phases; knowledge system modeling, activity modeling, concept modeling and task modeling. **MOKA (methodology and tools oriented to knowledge based engineering- KBE applications)** was implemented in the purpose of providing a methodology to reduce the cost of developing and maintaining Knowledge based engineering KBE applications. In the typical KBE life cycle, which is composed of six steps; Identify, justify, capture, formalize, package and activate, MOKA method is dedicated to the capture step, that's why an informal MOKA model (ICARE) was introduced in order to structure the raw engineering knowledge Rasovska et al. (2008). This informal model is divided on five forms which are: illustration, constraints, activities, rules and entities.

One the methods that was applied to supply chain is the **SKOS model**. In fact, W3C defines the Simple Knowledge Organization System (SKOS), "as a common data model for sharing and linking knowledge organization systems via the web". In other words, knowledge organization systems such as thesauri, taxonomies and classification schemes have the same structure; the role of SKOS model is providing the possibility of sharing data and technology across diverse application through capturing lot of similarities concerning knowledge organization systems.

Simple Knowledge Organization System (SKOS) is considered as an area of developing specifications and standards so as to support the use of knowledge organization system.

Cristian Aarón Rodríguez-Enríquez and his co-workers (2014) proposed a linked data-based approach using SKOS, that will be applied in supply chain Knowledge management. Introducing a linked data-based approach using SKOS is a way to facilitate knowledge management between supply chain partners. In other words, this will assure lot of benefits; offering open data sources of knowledge (linked open data), automatizing data organization and procurement for non-expert organizations, improving organizations' process, improving operational and organizational performance, and improving the decision-making process.

Relating to the same area of work, Andreas Blumauer affirms that "SKOS playing a key role in order to improve semantic information management, especially in terms of its following capabilities, taxonomy and thesaurus management, text mining and entity extraction, and finally knowledge engineering and ontology management".

Wu proposed a study that concerns the problem of coordination among multi-agent systems. He presented an issue of coordination problems in supply chains, and focused on how to design multi-agent systems to improve information and knowledge sharing Wu (2001). In the other hand, Becker and Zirpoli realized a research on the theme of knowledge transfer in outsourcing activities. And they were interested in designing an outsourcing strategy to improve knowledge integration Becker and Zirpoli (2003).

That is to say that using linked data-based approach using the SKOS is a promising field of researches concerning the enhance of technologies of knowledge management organization systems.

4.2.3 Synthesis of Reviewed Methods of Organizational Memory and Knowledge Engineering

In order to realize a recapitulation of reviewed methods we chose to classify them in the following paragraph, according to certain criteria. These criteria are, structure, models[24] and implementation steps, and they were chosen in the aim of identifying briefly each method by the important points that characterize it (Table 3).

Through this recapitulation we realized a comparison between the different methods according to three criteria; knowledge structure, used models and implementation steps. Rex method and CYGMA method are dedicated to organizational memory and have a different logic in comparison with knowledge engineering methods. Knowledge sources for REX method are documents, experience and experts know-how. On the other hand knowledge engineering methods have several similarities concerning real world description, functional point of view and problem solving.

Table 3. Synthesis of reviewed methods of organizational memory and knowledge engineering

	Method	Citations	Structure	Models	Implementation steps
Organizational Memory	**REX** (Malvache and Prieur)	449 citations	Documented knowledge elements, experience elements, know how	Experience memory: semantic net, net of terminology	Needs analysis, construction of elementary pieces of experience, computer representation of the knowledge domain, installation of a software package
	CYGMA (Bourne)	449 citations	Singular knowledge, terminological knowledge, structural knowledge, behavioral knowledge, strategic knowledge, operating knowledge	Profession memory	Produce a profession glossary, semantic catalogue, rule notebook, operating manual
Knowledge engineering	**KADS** (Weilinga et al.)	1065 citations	Task, inference, domain	Organizational model, application model, task model	Define the problem KBS should solve, the function that allows the problem solving and the actual tasks that KBS is supposed to perform
	MSMK/MASK (Ermine)	449 citations	Syntax, semantic, pragmatic	Knowledge book	Knowledge system modeling, activity modeling, concept modeling, task modeling
	MOKA (MOKA)	40 citations	Structure, function, behavior	ICARE (illustration, constraints, activities rules and entities	Justify, capture, formalise

5   Conclusion

Through this paper we tried to answer to two research questions. The first one concerns the contribution of cognitive technology on supply chain; the aim of this question is to realize an overview on articles concerning ontology engineering and Knowledge Based Systems.

The second question is about the knowledge engineering methods that can be applied on supply chain, inorder to present articles that explain organizational methods and knowledge engineering methods and their impact on supply chain management.

This paper is an overview of methods used concerning knowledge engineering. Surely our work is an opportunity to realize an application of the reviewed methodologies in order to better assimilate them.

In terms of research perspective this research should be strengthened with a field study in order to evaluate the applicability of each method in reality.

References

Stevens, G.C.: Integrating the supply chain (1989)

LeMay, S., Helms, M.M., Kimball, B., McMahon, D.: Supply chain management: the elusive concept and definition. Int. J. Logist. Manag. **28**(4), 1425–1453 (2017). https://doi.org/10.1108/IJLM-10-2016-0232

Spekman, R.E., Spear, J., Kamauff, J.: Supply chain competency: learning as a component. Supply Chain Manag.: Int. J. **7**, 41–55 (2002)

Kurtz, D.J., Santos, J.L., Varvakis, G.: Uncovering the knowledge flows in supply chain relationships. iBusiness **4**(4), 326–334 (2012). https://doi.org/10.4236/ib.2012.44041

Outahar, I., Nfaoui, E., El Baqqali, O.: Implementing knowledge management in supply chain: literature review (2013)

Marra, M., Ho, W., Edwards, J.S.: Supply chain knowledge management (2011)

Centobelli, P., Cerchione, R., Esposito, E.: Knowledge management in startups (2017)

del Rosario Pérez-Salazar, M., Aguilar Lasserre, A.A., Cedillo-Campos, M.G., Hernández Gonzále, J.C.: The role of knowledge management in supply chain management: a literature review. Received October 2016, Accepted September 2017 (2017)

Hadaya, P., Cassivi, L.: The role of knowledge sharing in a supply chain. Ind. Manag. Data Syst. **107**, 954–978 (2007)

Smirnov, A., Levashova, T., Shilov, N.: Ontology engineering for knowledge sharing in supply chains (2009)

Guo, W., Wen, J., Shao, H., Wang, L.: Implementation of knowledge-based engineering methodology in hydraulic generator design (2015)

Cooper, D., La Rocca, G.: Knowledge-based techniques for developing engineering applications in the 21st Century (2007)

Guo, W., Wen, J., Shao, H., Wang, L. (2015)

Overend, M., Pelken, P.M., Sauchelli, M.: Pages 78–94 | Received 26 Apr 2017, Accepted 02 Aug 2017, Published online: 23 Aug 2017

Rosario, J.G.: Much ado about knowledge capital. Business World, Philippines (1996)

Guerrero, L.A., Pino, J.A.: Understanding organizational memory (2001)

Malvache, P., Prieur, P.: Mastering corporate experience with the Rex method. In: Barthès, J.P. (ed.) Proceedings of ISMICK 1993, Compiègne, October, pp. 33–41 (1993)

Dieng, R., Corby, O., Giboin, A., Ribieere, M.: Methods and tools for corporate knowledge management. Int. J. Hum Comput Stud. **51**, 567–598 (1999)

Weilinga, B.J., Schreiber, A.T., Breuker, J.A.: KADS: a modeling approach to knowledge engineering. Knowl. Acquis. **4**, 5–53 (1992)

Dieng, R., Corby, O., Giboin, A., Ribieere, M.: Methods and tools for corporate knowledge management (2004)

Rasovska, I., Chebel-Morello, B., Zerhouni, N.: A mix method of knowledge capitalization in maintenance. J. Intell. Manuf. **19**(3), 347–359 (2008)

Wu, D.-J.: Software agents for knowledge management: coordination in multi-agent supply chains and auctions. Expert Syst. Appl. **20**(1), 51–64 (2001). https://doi.org/10.1016/S09574174(00)00048-8

Becker, M.C., Zirpoli, F.: Organizing new product development: knowledge hollowing-out and knowledge integration–the FIAT auto case. Int. J. Oper. Prod. Manag. **23**(9), 1033–1061 (2003). https://doi.org/10.1108/0144357031049176

The Destination Choice by Generation Z Influenced by the Technology: Porto Case Study

Pedro Liberato[1(✉)], Cátia Aires[1], Dália Liberato[1], and Álvaro Rocha[2]

[1] School of Hospitality and Tourism,
Polytechnic Institute of Porto, Porto, Portugal
{pedrolib,dalialib}@esht.ipp.pt, aires.ca@hotmail.com
[2] Informatics Engineering Department, Coimbra University, Coimbra, Portugal
amrocha@dei.uc.pt

Abstract. This article aims to contribute to the deepening of scientific knowledge about the specificities of the Generation Z behavior regarding the tourism sector. The objective of this research is to evaluate the destination choice by Generation Z and conclude if the reasons for the choice of destination (city of Porto - Portugal) positively influence the choice of technological resources in the pre-trip phase. To achieve the defined objectives, 400 validated questionnaires were gathered from tourist belonging to generation Z at Porto on the main streets of tourist attraction. The results obtained show, thus, globally, we can conclude that the three main reasons for the choice of destination Porto are: to know the city, previous recommendation, and the cultural offer, followed by wine and gastronomy and history and heritage. The reasons for the choice of destination Porto positively influence the choice of technological resources in the pre-trip phase for the significant relationships encountered.

Keywords: Generation Z · Technology · Satisfaction · Destination choice

1 Introduction

The city of Porto is the second largest city of Portugal in economic and social terms, located in the north coast of the country, with about 1.7 million inhabitants in an area of approximately 2,040 km^2 [10, 11]. Porto is one of the oldest cities in Europe, with a historic center classified by UNESCO as World Cultural Heritage since 1996 [22]. As a tourist destination, it has increased its reputation internationally due to the prize for Best European Destination in 2012, 2014 and for the third time in 2017 [6]. Also, in 2017, the municipality of Porto obtained the first place in the dimension living in the Portugal City Brand Ranking [4]. It is a geographical point of Portugal that has been showing a consolidated growth as a tourist destination, as we can see from the registration of the numbers of overnight stays in hotel establishments in the north of Portugal.

The growth of the guests' number in hotel establishments in the North of Portugal is significant. Over the last five years there has been an increase of guests (one that makes at least one overnight stay in a tourist accommodation) for 2 563 644 guests in the last five years (Fig. 1). The organization that manages and promotes tourism in the

Á. Rocha et al. (Eds.): ICITS 2019, AISC 918, pp. 32–44, 2019.
https://doi.org/10.1007/978-3-030-11890-7_4

city of Porto is the Regional Tourism Authority of Porto and North of Portugal, being one of the ways of publicizing the region. In the case of the city of Porto, the regional DMO identifies its intervention in the interactive tourist shops (example, platform lit - TOMI), located in the urban tourism offices and a store in Francisco Sá Carneiro International Airport [21], which receives around 9.4 million passengers per year, located 15 km from the city center, accessed by public transport (metro and bus) and road access [20].

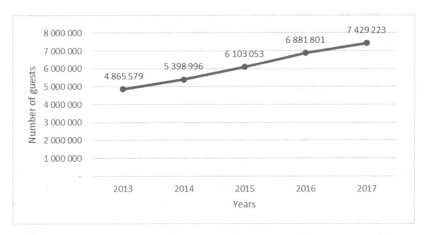

Fig. 1. Overnights in hotel establishments in the north of Portugal Source: INE (2017), adapted

The use of technological resources promotes the sustainability of the city, reduces costs and improves the quality of life of its citizens, becoming an innovative destination with technology, through sustainable development and efficient resources management [9]. Achieving perceived and delineated consumer behavior is the biggest challenge for tourism managers, who often need to adjust their contact with goods and services, with special attention to consumer needs, because in the case of changes of preferences, these must be able to change their destination marketing strategy [13, 16]. Tourism is constantly changing in terms of products, destinations and new communication models. It is a challenge for this sector, open to new opportunities, becoming one of the examples of sectors of the economy where digital transformation is increasingly present. Technology has changed the way that promotion, marketing strategies (mostly online), marketing, media, products, services, and destinations correlate [16]. Technological changes in tourism are increasingly rapid, as tourists have taken a leading role in the use of technology, making the non-digitized resource less appealing without access to the digital world, thus modifying the way tourism is perceived [2, 17]. That's why it's important to analyze the behavior of Generation Z. Their consume is based on the interest in new technologies, the ease of use, the desire to feel safe and, finally, the sensation of escaping to the surrounding reality due to the political, social, technological and economic changes, possess higher expectations, no brand loyalty, given that experience is more important [19]. This change in values has made marketing strategies change, like smartphones, computers, and multi-touch can create more innovative methods with smarter technology to improve the consumer experience [11]. This study presents seeks to evaluate the destination choice by Generation Z and found out if the reasons for the choice of destination (city of Porto) positively influence the choice of technological resources in the pre-trip phase.

This paper is divided in three parts. The first part, the literary review, emphasizes the important role of technology for tourism, and for the generation Z. The second part describes the research methodology, the methodology of data collection and the results of the research are presented and discussed. The third and final part, the conclusion, discusses the innovative perspective introduced in the analysis of the generation Z and the choice of technological resources in the pre-trip phase.

2 Literature Review

According to [5], the use of technology is a constant in tourism, both in terms of its use by destination promotion agencies, services and products, and the use by tourists themselves. [5] argues that the process of online access is focused on: destinations, intermediaries, transport, accommodation, restaurants, other tourist services and that tourist access begins before, during and after the trip, which creates social, followed by technological trends, pursuing innovation [5]. In this sense, technology has significantly changed the way tourism experience is perceived, experience (covering people, space, products, services or cultures) already has a dynamic character, and increasingly the involvement of the consumer, the co-creation and implementation of the technology gain relevance for the promotion of the destination. In sociology, experience is characterized as a subjective and cognitive activity of an individual, where he acquires knowledge and skills in the involvement or exposure to a specific event, linking the emotions, feelings and sensations triggered during the given experience, and can thus be associated with tourist experience [2, 18]. Nevertheless, technologies have played a decisive role not only for the competitiveness of tourism organizations but also for the tourists' experience [16]. Most of the research on travel information and reservations and payments during the preparation phase of a trip is done through the Internet. The same occurs during travel: the Internet, mobile phones and other technologies provide travelers with relevant information sometimes essential to travel, most diverse and useful due to easy accessibility and connection [9, 14]. The technology, in tourism, is an asset to improve the tourist experience, in the scope of the new technologies of information and communication, focused on the research, in order to help the visitors to have access to information on the exhibitions or places that are of interest to them [15]. Also, during the trip-planning phase, tourists can have access to relevant information [15]. For tourists to be able to feel satisfaction regarding the decision of the destination of travel it is necessary to consider four factors in the available information of the destination: quality of the information, the veracity of the sources of information, the interactivity and the accessibility [9].

For tourism purposes, mobile applications have emerged as basic tools for supplementary information to users, but with technological developments new fields have emerged within applications such as reservations, purchases, check-in of airplane tickets, among many other options such as hotel, cars, and restaurants booking, and even information applications as mobile guides to improve the tourist experience before and during their visit at a particular location/destination [3, 8, 16].The availability of mobile applications for tourist use at the destination makes it possible to satisfy the tourist experience with the use of information technology if it is an effective and real-time network that responds to user needs [8].

According to [7], at the European level, Generation Z travelers are gaining more and more relevance to the tourist market in both leisure and business travel, stresses that the traditional method of marketing is not the most effective method for this generation (in all sectors: accommodation, travel, food, activities). Advertised offers must be attractive because it interacts and perceives information distinctly, which must be suitable for several devices and screens (various sizes) because Generation Z use multiple mobile devices [7]. The technology used by the Generation Z influence the way they communicate with other individuals, mostly through the Internet, thus, they tend to be in places where they can access and enjoy the network connection [12]. Regarding the use of the technology to choose the destination, this generation uses technological devices for the preparation of the trip, mainly the mobile phone (smartphone), and portable, to search for tourist destinations and to make reservations [12].

3 Methodology

A quantitative approach was considered appropriate for the research methodology. A survey was used as the data gathering technique for this investigation. For the survey application methodology, the direct or non-probabilistic sampling method was used. The data collection took place in the main tourists' streets of the city of Porto to the tourist belonging to Generation Z. Within the methods of direct or non-probabilistic sampling, the convenience sampling method was used, in which the sample is selected according to the availability of the elements of the target population.

4 Results

The aim of this research is to evaluate the destination choice (city of Porto) by Generation Z. The hypothesis presented in this investigation is: The reasons for the choice of destination Porto positively influence the choice of technological resources in the pre-trip phase. According to [1] (which analyzes data on destination demand, overnight in the municipality and the digital search for it), the reasons most often presented for choosing the city of Porto as a destination were: previous recommendation; to get to know the city; proximity to the place of residence or holiday. The beauty of the city, the cultural offer, the history and heritage, the wine and gastronomy, and the value for money were also indicated [1] (Table 1).

It is important to note that in this issue we could choose more than one option. The results indicate three main reasons: to know the city, recommendation, and cultural offer; followed by wine and gastronomy and history and heritage (Table 1).

Table 1. Frequency and percentage for each reason for choosing Porto

What are the reasons for choosing Porto:	Frequency	Percentage
To get to know the city	268	**67.0**
Have been recommended	208	**52.0**
Proximity to place of residence or holiday	75	18.0
The cultural offer	204	**51**
The beauty of the city	91	22.8
History and heritage	123	**30.8**
The wine and gastronomy	124	**31**
Money value	83	20.8

Source: Compiled by the authors

Table 2. Internal consistency statistics: dimensions under study

Dimension	Cronbach alfa	Number of items
Global importance of online information	**0.824**	9
Use of technological resources for the organization of the trip	**0.828**	11
Evaluation of the use of available applications on the Internet for the trip decision and planning	**0.855**	11
Evaluation of the use of available applications on the Internet for reservations and purchases for the trip	**0.844**	12

Source: Compiled by the authors

In the study of the relationship with the main reason of the trip, as some of the groups in comparison are of small size and the assumption of the normality is not verified, the non-parametric test of Kruskall-Wallis is used (Table 3).

As the information shown in Table 4, this research found statistically significant positive relationships between: the Dimension "Global importance of online information" and the reasons "Previous recommendation" and "other"; The dimension "Technological resources for the organization of the journey" and the reason "other"; The dimension "Applications for the decision and planning of the trip" and the reasons "to know the city", "Previous recommendation" and "the cultural offer"; The dimension "Applications for travel booking" and the reasons "The history and heritage", "The wine and gastronomy" and "quality-price ratio". The results mean that whoever indicates the listed reasons uses more of the related dimensions.

As shown in Table 5, for "Tours", "Accessibility/Maps feature to GPS", "Recommendations" and "Sites of Destination (example: Visitporto.travel)", there are statistically significant differences between the main reasons of the trip. The information on "tourist itineraries" is more important for the reason visit to family and friends and

Table 3. Descriptive statistics and tests of Kruskall-Wallis: relationships between the dimensions and "main reason of the trip"

		N	Average	Standard deviation	KW	p
Global importance of online information	Leisure/holiday	340	3.86	0.56	5.49	0.139
	Visit friends and relatives	34	3.99	0.43		
	Professional reasons	14	**4.02**	0.75		
	Other	12	3.59	0.66		
Technology resources for the organization of the trip	Leisure/holiday	340	3.14	0.72	2.72	0.437
	Visit friends and relatives	34	2.99	0.84		
	Professional reasons	14	**3.40**	0.69		
	Other	12	3.12	0.93		
Applications for the trip decision and planning	Leisure/holiday	340	**3.73**	0.48	3.31	0.347
	Visit friends and relatives	34	3.48	0.71		
	Professional reasons	14	3.63	0.19		
	Other	12	3.71	0.23		
Applications for the bookings and purchases for the trip	Leisure/holiday	338	**3.89**	0.49	1.22	0.748
	Visit friends and relatives	34	**3.88**	0.33		
	Professional reasons	14	3.77	0.54		
	Other	12	3.84	0.29		

Source: Compiled by the authors * $p < 0.05$ ** $p < 0.01$

less for the other reason; the information on "Accessibility/Maps feature to GPS" is more important for the reason leisure/vacation and less for reasons Professionals and other reason; the information on "recommendations" is less important for the other reason; the information on "Sites of the destination (example: Visitporto.travel)" is more important for the reason professional reasons and less for the other reason, with statistically significant differences being observed. In the sample, the other items show the differences illustrated between the main reasons for the trip, which are not statistically significant.

Table 4. Pearson correlation: relations between some dimensions and "what are the reasons for choosing the city of Porto"

		Global importance of online information	Technological resources for the organization of the trip	Applications for travel decision and planning	Applications for travel booking
To know the city	Correlation	0.026	−0.009	**0.171****	0.065
	Proof Value	0.599	0.856	0.001	0.192
	N	400	400	400	398
Previous recommendation	Correlation	**0.145****	−0.031	**0.244****	**0.149****
	Proof Value	0.004	0.537	0.000	0.003
	N	400	400	400	398
Proximity to the place of residence or holiday	Correlation	0.054	0.009	0.053	0.008
	Proof Value	0.280	0.865	0.293	0.869
	N	399	399	399	397
Cultural offer	Correlation	−0.014	0.055	**0.107***	0.069
	Proof Value	0.781	0.271	0.033	0.171
	N	400	400	400	398
Beauty of the city	Correlation	0.058	−0.046	−0.072	−0.088
	Proof Value	0.251	0.358	0.148	0.079
	N	400	400	400	398
History and heritage	Correlation	0.095	0.027	0.083	**0.108***
	Proof Value	0.058	0.592	0.097	0.032
	N	398	398	398	396
Wine and Gastronomy	Correlation	0.070	−0.086	0.087	**0.110***
	Proof Value	0.164	0.085	0.081	0.028
	N	400	400	400	398
Quality-price ratio	Correlation	−0.038	−0.094	−0.003	**0.110***
	Proof Value	0.445	0.062	0.950	0.028
	N	399	399	399	397
Other	Correlation	**0.104***	**0.135****	−0.016	0.038
	Proof Value	0.037	0.007	0.746	0.444
	N	400	400	400	398

Source: Compiled by the authors * $p < 0.05$ ** $p < 0.01$

According to Table 6, this research found statistical significant positive relationships between: the importance of information on "tourist itineraries" and the reasons "have been recommended" and "wine and gastronomy"; The importance of information on "gastronomy and wines" and the reason "wine and gastronomy"; The importance of information on "bus/boat/pedestrian circuits" and the reasons "the beauty of the city", "the Wine and gastronomy" and "other"; The importance of information on "accessibility/maps resource to GPS" and the reasons "to know the city" and "have been recommended"; The importance of information on "main tourist attractions" and

Table 5. Descriptive statistics and tests of Kruskall-Wallis: relations between "what is the importance of this information online for your trip" and "main reason for the trip"

		N	Average	Standard deviation	KW	p
Tourist itineraries	Leisure/holiday	338	**3.83**	0.98	13.94	****0.003**
	Visit friends and relatives	33	**4.33**	0.69		
	Professional reasons	14	3.71	0.99		
	Other	12	3.42	0.79		
Gastronomy and Wine	Leisure/holiday	333	3.77	0.93	5.90	0.117
	Visit friends and relatives	33	**4.06**	0.70		
	Professional reasons	13	**3.85**	0.90		
	Other	10	3.40	0.52		
Buses/boats/pedestrians circuits	Leisure/holiday	312	3.65	0.89	1.80	0.616
	Visit friends and relatives	33	**3.85**	0.91		
	Professional reasons	13	**3.77**	0.83		
	Other	12	3.67	0.98		
Accessibility/maps feature to GPS	Leisure/holiday	331	**4.25**	0.88	11.39	***0.010**
	Visit friends and relatives	33	**3.97**	0.73		
	Professional reasons	13	3.77	1.30		
	Other	12	3.75	0.87		
Main tourist Attractions	Leisure/holiday	335	**3.96**	0.77	4.96	0.175
	Visit friends and relatives	34	3.88	0.64		
	Professional reasons	13	**4.31**	0.85		
	Other	11	3.73	0.90		
Museums	Leisure/holiday	324	3.53	0.86	4.24	0.237
	Visit friends and relatives	33	**3.61**	0.90		
	Professional reasons	13	**3.92**	1.32		
	Other	11	3.36	1.12		

(*continued*)

Table 5. (*continued*)

		N	Average	Standard deviation	KW	p
Recommendations	Leisure/holiday	333	4.17	0.80	10.46	*0.015
	Visit friends and relatives	34	**4.35**	0.77		
	Professional reasons	13	**4.23**	0.93		
	Other	11	3.45	0.82		
Cultural Agenda	Leisure/holiday	316	**3.95**	0.96	4.15	0.246
	Visit friends and relatives	33	**4.15**	0.83		
	Professional reasons	12	3.92	1.00		
	Other	11	3.45	1.13		
Destination websites (visitporto.travel)	Leisure/holiday	297	3.69	0.83	8.94	*0.030
	Visit friends and relatives	32	**3.75**	0.67		
	Professional reasons	12	**4.25**	0.75		
	Other	11	3.36	0.67		

Source: Compiled by the authors * p < 0.05 ** p < 0.01

the Reason "history and heritage"; The importance of information on "museums" and the reasons "proximity to the place of residence or holiday" and "history and heritage"; There is also a statistically significant negative relationship between: the importance of information on "cultural Agenda" and the reason "quality-price ratio", which reveals a high degree of importance attributed to the cultural agenda, also showing a high cultural level of tourism demand in Porto.

Data presented at Table 7 conclude that for "Trivago", "Edreams" and "Booking", there are statistically significant differences between the main reasons for the trip. For the other items, there are no statistically significant differences between the main reasons for the trip. The "trivago" application is more important for the reason leisure/vacation and less for professional reasons, the application "Edreams" is more important for the reason leisure/vacation and less to visit to family and friends, the application "Booking" is more important to the other reason and less to visit to family and friends, with statistically significant differences being observed. In the sample, the other items show the differences illustrated between the main reasons for the trip, which are not statistically significant.

Table 6. Pearson's correlation: relations between "what is the importance of this information online for your trip" and "what are the reasons for choosing the city of Porto"

		Tourist itineraries	Gastronomy and Wine	Buses/boats/pedestrians circuits	Accessibility/maps feature to GPS	Main tourist Attractions	Museums	Recommendations	Cultural Agenda	Destination websites (visitporto.travel)
To know the city	r	0.014	0.046	0.048	0.134**	0.055	−0.046	0.011	−0.052	−0.016
	p	0.780	0.369	0.356	0.008	0.274	0.371	0.831	0.315	0.769
	N	397	389	370	389	393	381	391	372	352
Previous recommendation	r	0.136**	0.063	0.075	0.163**	0.069	0.090	0.072	0.076	0.070
	p	0.007	0.218	0.150	0.001	0.171	0.081	0.155	0.143	0.191
	N	397	389	370	389	393	381	391	372	352
Proximity to the place of residence or holiday	r	0.062	0.011	0.046	0.085	0.034	0.128*	−0.041	0.010	0.049
	p	0.218	0.825	0.380	0.095	0.499	0.013	0.422	0.846	0.360
	N	396	389	370	388	392	380	390	371	352
Cultural offer	r	−0.066	−0.005	−0.005	−0.095	−0.019	0.038	0.015	0.008	0.028
	p	0.187	0.918	0.929	0.060	0.714	0.463	0.769	0.878	0.603
	N	397	389	370	389	393	380	390	372	352
Beauty of the city	r	0.084	0.028	0.103*	0.005	−0.019	0.006	0.022	0.057	0.086
	p	0.095	0.580	0.047	0.927	0.715	0.901	0.671	0.272	0.106
	N	397	389	370	389	393	381	391	372	352
History and heritage	r	0.012	0.038	0.038	0.006	0.141**	0.115*	0.068	0.069	0.017
	p	0.817	0.451	0.468	0.913	0.005	0.026	0.179	0.186	0.752
	N	395	387	368	387	391	379	389	370	350
Wine and Gastronomy	r	0.121*	0.162**	0.115*	−0.011	0.014	0.000	0.007	−0.002	0.064
	p	0.016	0.001	0.027	0.830	0.776	0.997	0.886	0.973	0.229
	N	397	389	370	389	393	381	391	372	352
Quality-price ratio	r	−0.013	0.083	0.044	0.015	−0.014	−0.062	−0.053	−0.111*	−0.095
	p	0.804	0.104	0.400	0.766	0.789	0.227	0.298	0.032	0.075
	N	396	388	369	388	392	380	390	371	351
Other	r	0.035	0.018	0.111*	−0.019	−0.007	0.012	−0.043	0.093	0.063
	p	0.492	0.721	0.034	0.705	0.894	0.811	0.396	0.072	0.241
	N	397	389	370	389	393	381	391	372	352

Source: Compiled by the authors * p < 0.05 ** p < 0.01

Table 7. Descriptive statistics and tests of Kruskall-Wallis: relations between "evaluation of the use of the following applications available on the Internet for the decision and planning of your trip to Porto" and "main reason of the trip"

		N	Average	Standard deviation	KW	p
Instagram	Leisure/holiday	331	4.45	0.76	0.65	0.884
	Visit friends and relatives	33	4.39	0.97		
	Professional reasons	14	4.57	0.51		
	Other	11	4.27	0.90		
Snapchat	Leisure/holiday	319	3.96	0.56	5.79	0.122
	Visit friends and relatives	33	3.58	1.12		
	Professional reasons	14	3.93	0.27		
	Other	10	3.80	0.63		
Storyo	Leisure/holiday	316	4.25	0.69	1.59	0.662
	Visit friends and relatives	33	3.85	1.30		
	Professional reasons	14	4.36	0.50		
	Other	10	4.30	0.48		
Trivago	Leisure/holiday	324	3.56	0.96	14.04	**0.003
	Visit friends and relatives	34	3.12	0.88		
	Professional reasons	14	3.00	0.39		
	Other	11	3.18	0.60		
Tripadvisor	Leisure/holiday	333	3.54	0.99	1.38	0.710
	Visit friends and relatives	34	3.44	0.75		
	Professional reasons	14	3.29	0.73		
	Other	12	3.50	0.90		
Edreams	Leisure/holiday	312	3.94	0.47	9.46	*0.024
	Visit friends and relatives	34	3.56	1.08		
	Professional reasons	14	3.86	0.53		
	Other	12	3.83	0.72		
Skyscanner	Leisure/holiday	320	3.69	0.82	1.03	0.795
	Visit friends and relatives	33	3.52	0.87		
	Professional reasons	14	3.79	0.80		
	Other	12	3.83	0.58		
Momondo	Leisure/holiday	313	3.89	0.55	6.86	0.076
	Visit friends and relatives	34	3.53	1.08		
	Professional reasons	14	3.71	0.73		
	Other	12	3.75	0.97		
Googletrips	Leisure/holiday	317	2.79	1.01	3.23	0.358
	Visit friends and relatives	33	2.64	1.17		
	Professional reasons	14	2.57	0.76		
	Other	11	3.09	0.94		
Booking	Leisure/holiday	327	3.87	0.54	10.00	*0.019
	Visit friends and relatives	34	3.56	1.08		
	Professional reasons	14	3.86	0.36		
	Other	12	4.25	0.45		

(*continued*)

Table 7. (*continued*)

		N	Average	Standard deviation	KW	p
Kayak	Leisure/holiday	325	2.95	0.46	4.05	0.256
	Visit friends and relatives	33	3.03	0.47		
	Professional reasons	14	2.93	0.47		
	Other	11	3.18	0.60		

Source: Compiled by the authors * $p < 0.05$ ** $p < 0.01$

5   Conclusions

The technology is part of generation Z trip, and as we can verify by the results, it is observed the use of the technology in the pre-trip phase. Therefore, overall, we can conclude that the hypothesis "The reasons for choosing the destination Porto positively influence the choice of technological resources in the pre-trip phase" is verified, concerning the meaningful relationships found. Thus, confirming that the technologies are decisive in the search of travel information and reservations and payments during the preparation phase of a trip [9, 14], by Generation Z. In this way it is necessary to appeal to the tourist entities for the constant updating of online information content, such as the implementation and updating of new technologies in the destination. The information can focus on tourist services, such as accommodation, transportation, attractions, museums, tourist packages, among others. In fact, there is a growing trend in the importance of the technological resources of information attributed to the travel preparation, according to the generality of scientific investigations. There are a variety of technologies and channels for more accurate, truthful and personalized information, from early on, through the search and sharing of content anywhere and at any time in the real-time experience, which allow travelers to explore many more alternatives in making trip planning decisions than the traditional information sources [9].

References

1. Associação do Turismo Porto e Norte: Porto European Best Destination 2017. Barómetro do turismo do Porto (2017)
2. Beliatskaya, I.: Understanding enhanced tourist experiences through technology: a brief approach to the Vilnius case. J. Tour. Res./Rev. Investig. Tur. **7**(1) 17–27 (2017). http://revistes.ub.edu/index.php/ara/article/download/20614/22635. e-ISSN 2014-4458
3. Biz, A., Azzolim, R., Neves, A.: Estudo dos aplicativos para dispositivos móveis com foco em atrativos turísticos da cidade de Curitiba (PR). In: Anais do Seminário da ANPTUR (2016). ISSN 2359-6805
4. Blom Consulting City Brand Ranking: Portugal City Branding. https://www.bloom-consulting.com/pdf/rankings/Bloom_Consulting_City_Brand_Ranking_Portugal.pdf
5. Dexeus, R.: Innovación en el sector turístico. Segittur. In: I Congreso Mundial de la OMT de Destinos Turísticos Inteligentes Retrieved (2017). http://www.segittur.es/es/sala-de-prensa/detalle-documento/Innovacin-en-el-sector-turstico-/#.WyGi0qdKjIU

6. European Best Destination. https://www.europeanbestdestinations.com/best-of-europe/european-best-destinations-2017/
7. Expedia. https://info.advertising.expedia.com/european-travel-and-tourism-trends-for-german-british-french-travellers
8. Florido-Benítez, L., Martínez, B., Robles, E.: El beneficio de la gestión de relación entre las empresas y turistas a través de las aplicaciones móviles. J. Tour. Res. **5**(2), 57–69 (2015). e-ISSN 2014-4458
9. Huang, C., Goo, J., Nam, K., Yoo, C.: Smart tourism technologies in travel planning: the role of exploration and exploitation. Inf. Manag. **54**(6), 757–770 (2017). https://doi.org/10.1016/j.im.2016.11.010
10. INE: Anuários Regionais de 2016, Lisboa, Portugal (2017). www.ine.pt
11. INE: Retrato Territorial de Portugal, Lisboa, Portugal (2017). www.ine.pt
12. JIŘÍ, B.: The employees of baby boomers generation, generation X, generation Y and generation Z in selected Czech corporations as conceivers of development and competitiveness in their corporation. J. Compet. **8**(4), 105–123 (2016). https://doi.org/10.7441/joc.2016.04.07. ISSN 1804-171X (Print), ISSN 1804-1728 (On-line)
13. Karimil, M., Pirasteh, H.: Study of consumer buying process in a model tourism destination cotler (case study: Esfahan). Am. J. Mark. Res. **1**(2), 88–92 (2015)
14. Kim, H., Xiang, Z., Fesenmaier, D.: Use of the Internet for trip planning: a generational analysis. J. Travel Tour. Mark. **32**(3), 276–289 (2015). https://doi.org/10.1080/10548408.2014.896765
15. Kuflik, T., Wecker, A., Lanir, J., Stock, O.: An integrative framework for extending the boundaries of the museum visit experience: linking the pre, during and post visit phases. Inf. Technol. Tour. **15**, 17–47 (2014). https://doi.org/10.1007/s40558-014-0018-4
16. Liberato, P., Alén-González, E., Liberato, D.: Digital technology in a smart tourist destination: the case of Porto, J. Urban Technol. **25**(1), 75–97 (2018). https://doi.org/10.1080/10630732.2017.1413228
17. Martin, F.: El reto de las nuevas tecnologías en el sector turístico: Innovación, Tecnología, Valores y Ética. In: Seminario Internacional sobre Nuevas Tecnologías Aplicadas al Turismo, Roatán, Honduras, junio de 2017 (2017). https://www.segittur.es/opencms/export/sites/segitur/.content/galerias/descargas/documentos/Nuevas-tecnologas-aplicadas-al-turismo-discurso-Fernando-de-Pablo-Martn-presidente-de-SEGITTUR-en-seminario-OMT-ok.pdf
18. Neuhofer, B., Buhalis, D., Ladkin, A.: A typology of technology-enhanced tourism experiences. Int. J. Tour. Res. **16**, 340–350 (2014). https://doi.org/10.1002/jtr.1958
19. Priporas, C., Stylos, N., Fotiadis, A.: Generation Z consumers' expectations of interactions in smartretailing: a future agenda. Comput. Hum. Behav. **77**, 374–381 (2017). https://doi.org/10.1016/j.chb.2017.01.058
20. República Portuguesa. http://www.portugalin.gov.pt/
21. Turismo de Portugal. http://www.turismodeportugal.pt/portugu%C3%AAs/turismodeportugal/destaque/pages/novaleidasregioesdeturismo.aspx
22. UNESCO Portugal. https://www.unescoportugal.mne.pt/pt/temas/proteger-o-nosso-patrimonio-e-promover-a-criatividade/patrimonio-mundial-em-portugal/centro-historico-do-porto

A Data Mining Approach for Predicting Academic Success – A Case Study

Maria P. G. Martins[1,3](✉), Vera L. Miguéis[2], D. S. B. Fonseca[3], and Albano Alves[1]

[1] School of Technology and Management, Polytechnic Institute of Bragança, Campus de Santa Apolónia, 5300-253 Bragança, Portugal
prud@ipb.pt
[2] Faculty of Engineering, University of Porto, Rua Dr. Roberto Frias, 4200-465 Porto, Portugal
[3] CISE - Electromechatronic Systems Research Centre, University of Beira Interior, Calçada Fonte do Lameiro, P, 6201-001 Covilhã, Portugal

Abstract. The present study puts forward a regression analytic model based on the random forest algorithm, developed to predict, at an early stage, the global academic performance of the undergraduates of a polytechnic higher education institution. The study targets the universe of an institution composed of 5 schools rather than following the usual procedure of delimiting the prediction to one single specific degree course. Hence, we intend to provide the institution with one single tool capable of including the heterogeneity of the universe of students as well as educational dynamics. A different approach to feature selection is proposed, which enables to completely exclude categories of predictive variables, making the model useful for scenarios in which not all categories of data considered are collected. The introduced model can be used at a central level by the decision-makers who are entitled to design actions to mitigate academic failure.

Keywords: Data mining · Educational data mining · Prediction · Academic success · Random forest · Regression

1   Introduction

The quality of academic training has a paramount role in the growth and development of any country or society. In turn, educational success is closely linked to the efficacy and efficiency of educational institutions. For this reason, the delivery of high-quality training and the definition of strategies which may promote academic success as well as retention recovery have been the subject of deep reflection by the administration board of the Polytechnic Institute of Bragança (IPB – *Instituto Politécnico de Bragança*), a polytechnic higher education institution in inland Portugal. Therefore, the aim to provide a methodology which enables the obtainment of useful knowledge to help and ground decision-making

© Springer Nature Switzerland AG 2019
Á. Rocha et al. (Eds.): ICITS 2019, AISC 918, pp. 45–56, 2019.
https://doi.org/10.1007/978-3-030-11890-7_5

by the IPB management bodies created the need to develop an academic success predictive model, which fits educational data mining field.

This work presents a regression model based on the random forest algorithm [1], developed with the aim to predict at an early stage the global academic performance of IPB undergraduates at the terminus of their academic path.

After this introduction, the present paper is composed of the following sections: Sect. 2 – outline of related studies; Sect. 3 – presentation of the methodology and of the data model developed; Sect. 4 – presentation of results and of the prediction model of performance proposed; Sect. 5 – final discussion of results and respective conclusions.

2   Related Studies

The main goal of educational data mining (EDM) is to generate useful knowledge which may ground and sustain decision-making targeted at improving student communities' learning as well as educational institutions' efficiency. Several systematic literature reviews [2–9] give evidence of the growing importance of EDM throughout time and refer and analyse the main research topics in which EDM has shown a remarkable contribution as a management analysis and support tool. Also, such studies provide evidence of the usefulness, potential and efficacy of the most used data mining methods and algorithms.

Among the typology of tasks where EDM has shown a remarkable contribution, we can find the prediction of academic performance. This research topic is normally approached from three different perspectives: predicting school dropout, predicting retention, or predicting academic success at the end of the degree course. Literature on EDM has shown that within those studies, several authors over time have studied a number of factors which promote academic success or failure. The main goal of the studies [10,11] was to conduct a review on the most used and relevant factors for this kind of predictions. Simultaneously, the authors also intended to determine the main data mining methods and algorithms used in such studies. After analysing a set of 30 studies focusing on the topic, Shahiri et al. [10] conclude that 6 attributes are used most frequently. At the top of the list is the cumulative grade points average (CGPA), almost at the same level as the attributes of internal assessment (marks after entering higher education such as assignments, exams, attendance, etc.). Demographic features and external assessment (pre-university achievement classifications) were the second group of most used attributes. Finally, the third group of attributes that the authors considered to be most used among the set of 6 are the ones related to students' extra-curricular activities and social interaction.

In a similar study to that by Shahiri et al. [10], but with the particularity of focusing only on research related to students attending institutions of the traditional on-site system, Del Río and Insuasti [11] conclude that in order to infer the final mean of the degree course, the authors of a set of 51 studies released between 2011 and August 2016 used, as predictive variables, indicators of academic performance obtained after entering higher education in combination with another type of attribute in 51.8% of the studies. In 37.5% of the papers, they only used information on academic performance within higher education. Among the data mining methods most used in the EDM task, the same authors [11] highlight those regarding classification, reported in 71.4% of the research works mentioned. The methods which followed were those of clustering and association rules, present in 8.9% and 7.1% of the studies.

Regarding the predicting academic success at the end of the degree course, the majority of the works related analyse academic performance restricted to one degree course only, and use relatively small datasets. For example, Natek and Zwilling [12] concluded that the factors most visibly influencing the final mean of the undergraduates of the bachelor degree in Computer Science were connected to information regarding access, demography and extra-curricular activities. The data were processed by the classification algorithms RepTree Model, J48 Model and M5P Model, and concerned 42 students attending the 1st year of the degree course, 32 attending the 2nd year, and 32 attending the 3rd year.

The results of the study conducted by Asif, Merceron, Ali and Haider [13] show that through algorithms Naïve Bayes and random forest Trees, it is possible to predict, with a high level of precision, the global graduation performance of a four-year degree course, by using only pre-university marks and the marks obtained in the course units of the 1st and 2nd years of university.

In the study by Migueis et al. [14], random forest, decision trees, support vector machines, Naïve Bayes, bagged trees and boosted trees were used to predict overall students academic performance based on the information available at the end of the first academic year. Among the algorithms used, random forest was the one that showed the best predictive results, also providing evidence that the most important factors to predict and explain the level of academic success in five-year degree courses in Engineering are the means regarding university access and university access examinations, as well as the mean obtained in the course units of the first academic year. This study also proposes a multiclass segmentation structure, aiming an early classification of students, based on their performance observed at the end of the first academic year and on their propensity for academic success revealed by the predictive model.

3   Methodology and Data Model

For the creation of a predictive model which can predict students' academic success at the terminus of their academic path, we chose to explore data from a universe of students from different educational fields, attending about half a hundred degree courses in an institution made up of 5 schools instead of following the most common procedure of delimiting the prediction to one single specific course. Hence, we intend to provide the institution with one single tool capable of including the heterogeneity of the universe of students as well as educational dynamics. The aim is for this tool to be used at a central level by the decision-makers who are entitled to design actions to mitigate academic failure, thus promoting a better educational experience for their students.

In this study, we chose to base our predictive model on the random forest algorithm proposed by Breiman [1]. It has shown to have surpassed other techniques in similar studies due to its predictive capacity and, more importantly, it allows a good interpretation of its results, in contrast to other techniques, such as Neural Networks and Support Vector Machines, which are considered to be black boxes. In fact, random forest presents the interesting functionality of allowing ordering the importance of the predictors which sustain the model.

Following the procedure commonly adopted, the predictive performance of the models considered was evaluated using a *k-fold* (with $k = 10$) cross-validation. As for the model evaluation metrics, the determination coefficient (R^2) and the root mean squared error (RMSE) were used.

In order to determine students' academic performance, the dependent variable introduced in expression (1) was used as a success indicator,

$$dv = average \times \frac{ects\_aprov}{ects\_aprov + ects\_disaprov}, \tag{1}$$

where **average** is the weighted average of the marks obtained in the completed course units (CUs), **ects_aprov** is the number of ECTS completed successfully and **ects_disaprov** is the number of ECTS in which students enrolled but did not pass. Thus, the metric takes into account not only the students' classification average but also the fraction of matriculations in course units they passed (ratio of successful 'attempts').

It was possible to consider the time period comprised between 2007/2008 and 2015/2016, totalising 9 consecutive academic years. A choice was made to limit the study to the bachelor degree courses since they are the core of the institution's training offer and they encompass a more complete set of data. After a cleaning of data and other pre-processing tasks, the data set that this study focuses on comprised 4530 matriculations in bachelor degree courses concluded in the period between 2007/2008 and 2015/2016 and started in the period between 2007/2008 and 2013/2014.

Regarding the predictive variables of academic success, we considered essentially the same typology of variables used in the related works of reference, namely academic data of sociodemographic nature and of access to higher education. The variables under study can be classified in two important subgroups: variables with cumulative semestral curricular results and 'timeless' variables – variables whose values are unaltered throughout students' academic path. Table 1 presents all the 44 predictive variables considered in this study as well as the dependent variable **dv** which will be used as a success indicator according to Eq. (1).

A vast real data set was used in this study, so care was taken to classify (3rd column of the table) each one of the potential predictive variables of academic success according to their nature into five different categories: *curricular* (C), *matriculation* (M), *demographic* (D), *socioeconomic* (S) and *access* (A). The attributes regarding semestral data (all in category C) are also easily distinguished from the others (timeless data) through the suffix '_s'. Note that the

Table 1. List of variables sustaining the model.

Id	Attribute	Cat	Type	Min..max	Meaning
1	curricular_year_s	C	Discrete	1..4	Student's course year in the a.s. considered
2	academic_year_s	C	Discrete	07..15	Academic year of the a.s. considered
3	scholarship_s	C	Continuous	0..1	Was the student a scholarship holder in the a.s.?
4	union_member_s	C	Continuous	0..1	Was the student a union leader in the a.s.?
5	ects_aprov_s	C	Discrete	0..60	N. of ECTS passed in the a.s.
6	ects_disaprov_s	C	Discrete	0..60	N. of ECTS failed in the a.s. (academic semester)
7	max_s	C	Discrete	0..20	Maximum mark of the CUs passed in the a.s.
8	average_s	C	Continuous	0..20	Average mark of the CUs passed in the a.s.
9	min_s	C	Discrete	0..20	Minimum mark of the CUs passed in the a.s.
10	n_assess_disap_s	C	Discrete	0..18	N. of assessments failed in the a.s.
11	n_courses_aprov_s	C	Discrete	0..10	N. de CUs passed in the academic semester
12	n_courses_disap_s	C	Discrete	0..10	N. of CUs failed in the academic semester
13	dv12_s$^{(a)}$	C	Continuous	−20..20	Diff. in performance from 1st to 2nd semester
14	dv23_s$^{(a)}$	C	Continuous	−20..20	Diff. in performance from 2nd to 3rd semester
15	dv34_s$^{(a)}$	C	Continuous	−20..20	Diff. in performance from 3rd to 4th semester
16	dv45_s$^{(a)}$	C	Continuous	−20..20	Diff. in performance from 4th to 5th semester
17	dv56_s$^{(a)}$	C	Continuous	−20..20	Diff. in performance from 5th to 6th semester
18	enrol_year	M	Discrete	07..13	Year of enrolment
19	cod_degree	M	Nominal	1..51	Code of the degree course
20	cod_school	M	Nominal	1..5	Code of the school
21	cred_ects_tx	M	Discrete	0..100	Fraction of ECTS credited to the student
22	ects_degree	M	Discrete	180..240	Number of ECTS of the degree course
23	enrol_type	M	Nominal	1..9	Type of enrolment in the degree course
24	displaced	D	Binary	0..1	Is the student displaced from usual residence?
25	district	D	Nominal	1..28	Student's district of origin
26	district_n	D	Nominal	1..27	District of birth
27	age	D	Discrete	17..61	Student's age at the time of enrolment
28	nationality	D	Nominal	1..15	Student's nationality
29	gender	D	Nominal	1..2	Gender
30	cod_job_student	S	Nominal	1..12	Student's job
31	cod_job_mother	S	Nominal	1..12	Mother's job
32	cod_job_father	S	Nominal	1..12	Father's job
33	educ_level_mother	S	Ordinal	1..13	Mother's level of education
34	educ_level_father	S	Ordinal	1..13	Father's level of education
35	prof_sit_student	S	Nominal	1..10	Student's employment status
36	prof_sit_mother	S	Nominal	1..10	Mother's employment status
37	prof_sit_father	S	Nominal	1..9	Father's employment status
38	phase	A	Ordinal	1..3	Phase of enrolment
39	access_grade	A	Continuous	0..200	Student's entrance qualification
40	a10_11_grade	A	Continuous	0..200	Mean obtained in the 10th and 11th grades
41	a12_grade	A	Continuous	0..200	Mean obtained in the 12th grade
42	access_option	A	Ordinal	1..6	Order of the option when applying for university
43	access_order	A	Discrete	1..322	Order of entrance among stud. admit. in course
44	access_exams	A	Continuous	0..200	Mean obtained in the entrance examinations
45	dv$^{(b)}$		Continuous	0..20	Dependent var. with student's final performance

$^{(a)}$ dv_{ij}_s $= dv_j − dv_i$, with $i = j − 1$ and dv_n the student's performance in their umpteenth semester, calculated using a metrics similar to that of dv (c.f. Eq. (1)). Each variable dv_{ij}_s is only present in the models with cumulative data of j or more semesters.
$^{(b)}$ Dependent variable used as indicator of success.

curricular data (C) refers to semestral curricular results of academic performance accumulated at the end of each of the student's 6 first semesters.

The selection of the dimensions regarding the student, which explain their academic success, was carried out in two different stages. First, as explained in Sect. 4.1, we selected the student's dimensions which best account for their success, which allowed a first adjustment enabling the exclusion of complete groups of variables. Subsequently, as explained in Sect. 4.2, we fine-tuned the selection of attributes which were not excluded in the first stage. This approach allowed the reduction of the data dimensionality without losing the model's predictive capacity.

Figure 1 depicts a scheme intended to characterise the predictive model designed for this study. It shows the different categories of predictive variables used as input of the random forest algorithm. As we can see, for the group of curricular variables (C), the attributes used are the results accumulated at the end of each one of the student's 6 first semesters. Note that only one of the 6 entries of curricular data (C) is considered in each execution of the algorithm (mutually exclusive entries).

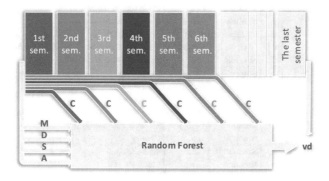

Fig. 1. Scheme depicting the comparative study conducted.

4   Results

In the exploratory analysis of data which follows, we chose to keep the configuration of the random forest algorithm fixed for the study to focus on the set of predictive variables which sustain it.

4.1   Selection of Predictors Categories

As it is known, the assertiveness of a predictive model depends greatly on the set of predictive variables being considered in the analysis. Also, the best model is not always the one which includes all the variables available. Although it is commonly acknowledged that random forests make an internal selection of variables, we still decided to run a test of inclusion, or not, of the categories of variables so as to understand the impact of the several dimensions on the capacity of the model.

Table 2 shows the different categories of attributes chosen in each of those studies. Although the aim is to develop a comprehensive study, it does not seem necessary to include all the possible combinations between the 5 groups of variables, which makes a total of $2^5 - 1 = 31$ possibilities. In fact, since the group of curricular data (C) is clearly the most determinant group of predictors, great difficulties are foreseen in the predictive precision of any model which does not include it. Therefore, only one particular case is considered in which this group of variables is not used: case MDSA (Study 6). With this simplification, the number of studies was reduced to almost a half, or more precisely to $2^4 + 1 = 17$.

Table 2. Categories of predictive variables used in the studies conducted.

Study	Label	Curricular	Matriculation	Demographic	Socioeconomic	Access
1	CMDSA	✓	✓	✓	✓	✓
2	CMDS	✓	✓	✓	✓	
3	CMDA	✓	✓	✓		✓
4	CMSA	✓	✓		✓	✓
5	CDSA	✓		✓	✓	✓
6	MDSA		✓	✓	✓	✓
7	CMD	✓	✓	✓		
8	CMS	✓	✓		✓	
9	CMA	✓	✓			✓
10	CDS	✓		✓	✓	
11	CDA	✓		✓		✓
12	CSA	✓			✓	✓
13	CM	✓	✓			
14	CD	✓		✓		
15	CS	✓			✓	
16	CA	✓				✓
17	C	✓				

Table 3 contains for each of the studies the determination coefficients (R^2) obtained by applying the random forest predictive algorithm to the selected data at the end of the student's 6 first academic semesters. For a better understanding of the results presented, take Study 8 as an example: the input of the random forest algorithm (see Fig. 1) was reduced to the groups of curricular (C), matriculation (M) and socioeconomic (S) variables. In this study, as in all the others (except for Study 6, which does not include the curricular data), the random forest algorithm was run 6 times in order to use the results accumulated at the end of each of the 6 semesters concerning the curricular data. The last column of the table shows the weighted average of the determination coefficients for those

Table 3. Determination coefficient R^2 of the predictive model, for different groups of predictive variables according to students' academic semester.

	Label	1st sem	2nd sem	3rd sem	4th sem	5th sem	6th sem	Average[a]
Study 13	CM	80.4	86.5	92.0	94.3	96.6	97.9	88.4
Study 8	CMS	80.7	86.8	91.7	93.9	96.4	97.7	88.4
Study 4	CMSA	80.7	86.5	91.6	93.9	96.3	97.7	88.3
Study 1	CMDSA	80.3	86.3	91.3	93.7	96.2	97.6	88.1
Study 2	CMDS	80.3	86.4	91.4	93.7	96.3	97.7	88.1
Study 7	CMD	79.8	86.3	91.6	94.0	96.5	97.8	88.1
Study 9	CMA	79.9	86.2	91.6	94.0	96.4	97.7	88.1
Study 3	CMDA	79.7	86.1	91.4	93.8	96.3	97.7	87.9
Study 12	CSA	70.5	78.8	86.6	89.9	94.2	96.6	81.8
Study 15	CS	70.6	78.7	86.5	89.7	94.1	96.5	81.8
Study 5	CDSA	70.5	78.5	86.4	89.8	94.2	96.5	81.7
Study 10	CDS	70.7	78.3	86.2	89.6	94.1	96.4	81.6
Study 17	C	70.3	78.0	86.6	90.0	94.4	96.7	81.6
Study 16	CA	69.6	78.2	86.6	90.1	94.4	96.7	81.5
Study 11	CDA	69.8	78.2	86.3	89.9	94.3	96.6	81.4
Study 14	CD	70.1	77.8	86.3	89.6	94.2	96.6	81.4
Study 6	MDSA	64.4	64.4	64.4	64.4	64.4	64.4	64.4
Average[b]		75.2	82.4	89.0	91.9	95.3	97.2	84.9

[a] Weighted average of semestral values, with weights 6, 5, ..., 2, 1, for the 1st, 2nd, ..., 5th, 6th semesters, respectively.
[b] Average value without considering Study 6.

6 semesters. The choice fell on a weighted average of semestral R^2 in order to value the results of the first semesters at the expense of those obtained in more advanced stages of the student's academic path – note, for example, that for students who complete their training in 3 years, the predictive capacity of the model after the 6th semester is totally irrelevant.

The analysis of the values presented in the table, listed in a descending order of the average value of R^2 (last column), allows the following considerations:

– It was not the model 'feeding' on all the variables (Study 1 – CMDSA) which presented the best predictive capacities. Actually, 6 other models achieved similar or better performances with a lower number of predictive variables.
– There is a visibly big difference in the performance of the 8 best classified models and the remaining ones – note the sudden drop between Study 3 and Study 12. If that sudden drop is clearly due to the loss of the matriculation data (M), while the one witnessed between studies 14 and 6 is due to the loss of the other subgroup of academic data, the curricular data (C).
– Academic data (subgroups CM, Study 13), alone, justify the best study result ($R^2 = 88.4\%$), obtained in *ex-aequo* with Study 8 (CMS).
– As expected, the assertiveness of the model increases consistently in line with the course of students' academic path.
– Although the results of Study 6 (MDSA), the only one not using the attributes of category C, are far behind the others, they clearly confirm that students'

curricular data is the key contributor to the assertive capacity of the model, probably not least because such data is not available at the beginning of the academic path. Still, it is interesting to observe that even at a very early stage of students' academic course, namely in the 1st and 2nd semesters, the average determination coefficient of the model showing the best performance skyrockets from 64.4% to values greater than 80% and 86%, respectively.

In light of the results presented in Table 3 and of the corresponding considerations exposed, it seems pertinent to propose the groups of variables in Study 13 (categories CM) for the predictive model we intend to design. The study achieves the highest determination coefficients using only 2 of the 5 possible categories of variables. Supporting the choice of the group CM (Study 13) as the best study is also the fact that this set of variables also revealed, in additional tests, the lowest Root Mean Square Error (0.966 when the second lowest error was 0.983 and the average 1.105).

The comparative study conducted allowed the exclusion of three irrelevant groups of variables, from a predictive perspective: the demographic, socioeconomic and the access data. In the next stage, we will try to exclude variables presenting a negligible influence in the performance of the predictive model.

4.2 Additional Adjustment of the Model – Selection of Predictors

The performance of any predictive model to be proposed will be all the more valued as the earlier the moment in which it might be applied. In fact, the predictive relevance of a model is based on two crucial aspects: the veracity of its predictions and the degree of anticipation of such predictions. Therefore, it is now important to fine-tune the CM model (which globally showed to be the most assertive) when applied right at the end of the student's 1st semester, as shown in the scheme in Fig. 2. More precisely, we will try to exclude from the set of CM predictive variables (with the subgroup C including only the curricular results of the student's 1st semester) all those which do not contribute positively and significantly to the quality of the model.

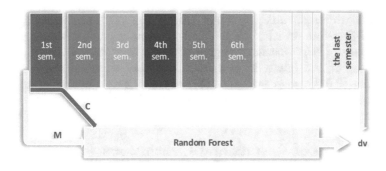

Fig. 2. CM predictive model sustained by data from the 1st academic semester.

Since the CM model did not make use of the socioeconomic and access data, the categories with missing data in a significant number of students, it was possible to use a wider and more comprehensive sample of matriculations containing complete data. Therefore, the size of the data set according to which the CM model will be adjusted increased from 2159 to 4530, the total size of the sample, originating a slight decrease of the model assertiveness in the 1st semester, from $R^2 = 80.4$ to 79.0. We believe that by using a larger sample of matriculations in this second adjustment of the model, it will present greater generalization capacity.

Before moving on to a more systematised process of fine-tuning of the CM model, the random forest algorithm was run for the data set without variables n_courses_aprov_s and n_courses_disap_s, for considering them to have a strong correlation with the attributes ects_aprov_s and ects_disaprov_s, respectively. After confirming the pertinence of excluding these two attributes, the possibility to exclude new variables among those revealing to be less informative, based on random forest ranking of variables in terms of importance, was assessed in successive iterations. The relevant data characterising those several iterations is summarised in Table 4. The data shows that the loss of variables maintained or slightly improved the assertiveness of the model. In a nutshell, we were able to remove 7 out of the 18 attributes from the data set without that affecting negatively the model performance and actually achieving a slight improvement, though of little significance.

Table 4. Removal of variables from the CM data set.

Iter.	Excluded variables		#var	R^2	RMSE
0	n_courses_aprov_s	n_courses_disap_s	16	79.2	1.334
1	scholarship_s	union_member_s	14	79.2	1.335
2	min_s		13	79.2	1.333
3	max_s		12	79.3	1.329
4	n_assess_disap_s		11	79.5	1.326

The 11 variables, which together justify the predictive capacity of the model and therefore reveal to be the most determinant in anticipating the academic success of IPB's bachelor degree students were ordered in a descending order of importance (given in brackets) as follows: ects_disaprov_s (2.387), cod_degree (2.011), average_s (1.785), ects_aprov_s (1.461), cred_ects_tx (1.454), cod_school (1.230), ects_degree (0.359), enrol_type (0.274), academic_year_s (0.241), enrol_year (0.239), curricular_year_s (0.176).

5   Discussion of Results and Conclusions

In this study, the random forest method was used to propose a predictive model of the global academic success of IPB's bachelor degree students at the terminus of their academic path. Instead of following the commonly adopted procedure of delimiting the prediction to one single specific course, the model was developed from a vast real data set involving records of quite heterogeneous undergraduates from over half a hundred degree courses covering a wide variety of educational fields taught in the five schools composing the institution and where each student is characterized by more than four tens explanatory variables. Such specificity allowed studying the influence of a new curricular factor taken into account for the first time in literature: the type of school. The results obtained allowed concluding that students' success also depends on the school they attend. This conclusion indicates that in order to mitigate retention and academic failure, it might be necessary to adopt differentiated strategies of educational promotion according to each school.

The order of importance provided by the random forest algorithm allowed the identification of the factors contributing to students' success or failure. It enabled the observation that the factors regarding the curricular context of students' academic performance are paramount to the intended prediction, which confirms results previously obtained by [15], who stated that such factors can alone account for academic performance. Note that all 11 attributes which revealed to be significant for the prediction belong, without exception, to the curricular or matriculation categories.

The knowledge obtained allows the identification of students at a higher risk of retention and academic failure, which enables the institutional managers to design more assertive educational or tutorial strategies towards educational efficacy and efficiency.

The kind of approach adopted in the identification of students' characteristics which best account for their success seems to differ from that usually used in works related to the same topic. In the present work, the selection of those characteristics was conducted in two different stages. First, the selection of students' dimensions which best explain their success allowed a first adjustment of the model by eliminating complete groups of variables. Later, a fine-tuned adjustment led to the selection of the attributes which were not excluded in the first stage. This approach enabled us, in addition to reduce the 'plague' of the data dimensionality at an early stage, to exclude completely categories of variables, without losing the predictive capacity of the model. This feature is of particular importance since it contributes to the reduction of the multidisciplinarity of the predictors, thereby lowering some of the complexity of the predictive process, and, more importantly, makes it possible to extend the study to other contexts, where not all categories of variables initially considered in this study are available.

Note however, that a significant part of the results obtained in this study cannot be generalized to the whole context of higher education since they were based on a data sample non-representative of that broader context. This study presented a case study focused on IPB, which for being an institution of the polytechnic higher education subsystem and for being located in an inland region with low population density cannot reach the same heterogeneity of students as other institutions located in large coastal urban centres. At best, the results presented here may reflect the reality of higher education institutions with similar conditions to those of the IPB such as other polytechnic institutes located in inland regions of the country far from the big urban centres.

Acknowledgments. This work was supported by the Portuguese Foundation for Science and Technology (FCT) under Project UID/EEA/04131/2013. The authors would also like to thank the Polytechnic Institute of Bragança for making available the data analysed in this study.

References

1. Breiman, L.: Random forests. Mach. Learn. **45**(1), 5–32 (2001)
2. Romero, C., Ventura, S.: Educational data mining: a survey from 1995 to 2005. Expert Syst. Appl. **33**(1), 135–146 (2007)
3. Romero, C., Ventura, S.: Educational data mining: a review of the state of the art. IEEE Trans. Syst. Man Cybern. Part C (Appl. Rev.) **40**(6), 601–618 (2010)
4. Romero, C., Ventura, S.: Data mining in education. Wiley Interdisc. Rev.: Data Min. Knowl. Disc. **3**(1), 12–27 (2013)
5. Baker, R.S.J.D., Yacef, K.: The state of educational data mining in 2009: a review and future visions. JEDM-J. Educ. Data Min. **1**(1), 3–17 (2009)
6. Huebner, R.A.: A survey of educational data-mining research. Res. Higher Educ. J. **19**, 1–13 (2013)
7. Papamitsiou, Z.K., Economides, A.A.: Learning analytics and educational data mining in practice: a systematic literature review of empirical evidence. Educ. Technol. Soc. **17**(4), 49–64 (2014)
8. Peña-Ayala, A.: Educational data mining: a survey and a data mining-based analysis of recent works. Expert Syst. Appl. **41**(4), 1432–1462 (2014)
9. Algarni, A.: Data mining in education. Int. J. Adv. Comput. Sci. Appl. **7**, 456–461 (2016)
10. Shahiri, A.M., Husain, W., Rashid, N.A.: A review on predicting student's performance using data mining techniques. Procedia Comput. Sci. **72**, 414–422 (2015)
11. Del Río, C.A., Insuasti, J.A.P.: Predicting academic performance in traditional environments at higher-education institutions using data mining: a review. Ecos de la Academia. **2016**(7), 185–201 (2016)
12. Natek, S., Zwilling, M.: Student data mining solution-knowledge management system related to higher education institutions. Expert Syst. Appl. **41**(14), 6400–6407 (2014)
13. Asif, R., Merceron, A., Ali, S.A., Haider, N.G.: Analyzing undergraduate students' performance using educational data mining. Comput. Educ. **113**, 177–194 (2017)
14. Miguéis, V.L., Freitas, A., Garcia, P.J.V., Silva, A.: Early segmentation of students according to their academic performance: a predictive modelling approach. Decis. Support Syst. **115**, 36–51 (2018)
15. Manhães, L.M.B.: Predição Do Desempenho Acadêmico De Graduandos Utilizando Mineração De Dados Educacionais. Ph.D. thesis (Tese Doutorado), Universidade Federal do Rio de Janeiro (2015)

Data Analytics on Real-Time Air Pollution Monitoring System Derived from a Wireless Sensor Network

Walter Fuertes[1(✉)], Alyssa Cadena[1], Jenny Torres[3], Diego Benítez[2], Freddy Tapia[1], and Theofilos Toulkeridis[1,2,3]

[1] Departamento de Ciencias de la Computación, Departamento de Seguridad y Defensa, Universidad de las Fuerzas Armadas ESPE, 171-5-231-B Sangolquí, Ecuador
{wmfuertes,akcadena,fmtapia,ttoulkeridis}@espe.edu.ec
[2] Colegio de Ciencias e Ingenierías "El Politécnico", Universidad San Francisco de Quito USFQ, Campus Cumbayá, Casilla Postal 17-1200-841, Quito, Ecuador
dbenitez@usfq.edu.ec
[3] Departamento de Informática y Ciencias de la Computación, Escuela Politécnica Nacional, P.O. Box 17-01-2759, Quito, Ecuador
jenny.torres@epn.edu.ec

Abstract. Air pollution is a problem that causes adverse effects, which tends to interfere with human comfort, health or well-being, and that may cause serious environmental damage. In this regard, this study aims to analyze large data sets generated by real-time wireless sensor networks that determine different air pollutants. Business Intelligence and Data Mining techniques have been applied in order to support subsequent decision-making strategies. For normalization and modeling, we applied the CRISP-DM methodology using the Pentaho Data Integration. Then, the Sap Lumira has been applied in order to acquire models of tables and views. For the data analysis, R-Studio has been used. For validation, Clustering has been applied using the k-means algorithm by the Jambu method, where it has been proceeded to check the consistency of these, being later stored and debugged in PostgreSQL. Results demonstrate that the increase in air pollutants is directly related to the traffic hours, which may cause an increase of asthma or sick related syndrome in the population. This analysis may also serve as a source of information to authorities for improving public policies in such matter.

Keywords: Air pollution · Wireless sensor network · Data analytics · Data Mining · Business Intelligence · Pattern recognition

1 Introduction

According to the World Health Organization (WHO) [1] "Air pollution represents a significant environmental risk to health, because it is one of the causes of strokes, lung cancer and chronic and acute lung diseases, among them asthma". Therefore, it is necessary to find solutions to minimize the causes that lead to this type of diseases in

© Springer Nature Switzerland AG 2019
Á. Rocha et al. (Eds.): ICITS 2019, AISC 918, pp. 57–67, 2019.
https://doi.org/10.1007/978-3-030-11890-7_6

the population. In addition to the obvious, these issues can also serve to better define and improve existing environmental public policies.

The aim of this study has been to conduct the data analytics, Business Intelligence (BI) and Data Mining (DM) of a set of large data flows generated by a wireless sensor network (WSN), which stored information in real time. The data correspond to some polluting agent measurements obtained by the WSN installed in two sectors within the cities of Quito, and Sangolquí, in central Ecuador. In order to address the mentioned concerns, BI and DM techniques have been applied in order to support decision-making. For normalization and modeling, the CRISP-DM methodology has been applied using the Pentaho Data Integration (PDI). Then, Sap Lumira has been applied to acquire models of tables and views. For the data analysis, the R-Studio has been used. For its validation Clustering has been applied using the k-means algorithm using the Jambu method, where it has been proceeded for the verification of the consistency of these, being stored and debugged in PostgreSQL.

The results of our study demonstrate that the hours where the highest amounts of air pollution appeared, correspond to the morning and afternoon traffic hours. This may lead to an increase in asthma-like illnesses in the population. Furthermore, such information may also be used for the definition of public policies by the corresponding agencies.

The rest of the article has been structured as follows: the second section documents the State of the Art in the knowledge of the treated topic, while the third section explains the defined Research Design. In the fourth section, we present the results and the subsequent assessment. Finally, the fifth section finalizes with the conclusions and future work lines.

2 Related Work

Air pollution has been a worldwide concern with a devastating impact on the health of citizens, since it poses an increased risk for respiratory infections and lung cancer among other diseases [2, 3]. Nowadays, there are many factors that can affect the air quality, such as traffic, factory exhaust emissions, weather, and incineration of garbage, besides many others [4]. Measuring air pollution is vital for protecting individual's health [5]. Therefore, air-quality monitoring devices have been deployed for monitoring air-quality parameters which involves the concept of big data, since a big amount of information is stored from monitoring devices. The term big data is used to characterize large data sets that may be complex and difficult to manage by conventional data processing methods. In a wireless sensor network, such huge amounts of data are generated every minute and data need to be collected by the sensor nodes before being transmitted to the base station. The combination of these two technologies brings together an emergent technology that combine the big data challenges with the wireless sensor networks in order to form a new classification based on the requirements of both technologies [6].

In the literature, there are many studies that apply WSN for monitoring air pollution. Concerning indoor environment, authors in [8] have used Fourier transforms for the data analysis, in order to identify patterns on the acquired data. The environmental data have been acquired with a wireless sensor system, the NSensor. The sensing system has been used for indoor environment monitoring, with the capability to store, in a remotely accessed database, air quality parameters such as: temperature, relative humidity, pressure, illuminance, carbon dioxide, and volatile organic components. Data from temperature and relative humidity, in a period of ten months, allowed to identify five main patterns in the temperature. Also, the study on [7] proposed an air monitoring system in household environment which has been connected to a personal health reporting system through a mobile APP. Air quality has been determined using an AQMesh, which is a small sensor based air quality monitoring system constructed with a variety of pollutant sensors for NO, NO_2, O_3, CO and SO_2.

Furthermore, similar to [7], some cloud-based solutions have been proposed [9, 12]. In [12], authors propose a model that uses a web and mobile application interface in order to support users for checking and understanding the air quality at their current location. The mobile application notifies the user about severe toxicity, so people with respiratory problems will be able to receive personalized notifications when poor conditions are given. In [9] the authors proposed a semantic ETL (Extract-Transform-Load) framework on cloud for air quality predictions, using ontologies in order to concretize the relationship of particulate matter (PM) 2.5 from various data sources in order to merge those data into a unified database. The framework included computing nodes, which have been used to execute DM algorithms for the prediction and storage nodes which have been used to store retrieved, preprocessed and analyzed data.

More descriptive studies were stated in [10, 11]. In [10], authors presented a comparison of the Hadoop MapReduce and Spark programing models for air quality simulations. The data set used in the analysis have been air pollution data for a variety of pollutants collected over a fifteen-year period by 179 monitor sites across the state of Texas. The corresponding results revealed that 20–25% performance benefits for the Spark solutions over MapReduce. In [11], the study proposed a big data analytic-based, personalized air quality health advisory model in order to address the issues of sparse air pollution monitoring sites, pollution mixture effects and the lack of personalized air quality health guidance. The authors of that study provided a real example in order to demonstrate the implementation and reasonability of the model based on data collected from Shenzhen city, China. In [13] authors presented a proposal for the integration of Big Data tools for the gathering, storage and analysis of data generated by a WSN that monitors air pollution levels in a city. The study provided a proof of concept that combines Hadoop and Storm for data processing, storage and analysis as well as Arduino-based kits for the construction of sensor prototypes.

Finally, there have been also several studies that combined WSN and Big Data technologies in fields not being necessarily related with air pollution or environmental monitoring systems [14–16]. The authors in [14], for example, discussed the importance of parallel processing using BigTable, Hadoop and MapReduce algorithms, proposing an architecture for storing and processing data obtained from massive WSNs. Hadoop based on Cloud Computing have been proposed in [15]. They used control access and improving the specific features of these services for deploying WSN systems. In [16], authors proposed a ZigBee and big data analysis-based pulse monitoring system, which has been composed of multiple ZigBee based pulse monitoring sensors, customized gateways and back-end system. These authors provided a complete view on a MapReduce framework targeting the implementation of four categories of sensor data analyses: the acquisition, aggregate, range, and space-temporal analyses. With this, they constructed a training model by the standards, such as different age, mood and others.

3   Research Design

This section describes the DM methodology that we have followed in this study. Due to the fact, that the CRISP-DM methodology has become an industry standard for the implementation of DM projects, we have chosen such procedure for our study. Then, the following steps have been followed:

Understanding of Business and Data: In this phase, the relevant information on environmental pollution has been identified and we defined how we may be able to deal with such dilemma through DM.

Data Preparation: Here, we included the initial data collection, exploration, quality verification and the relationships to define the first assumptions. In this project, we used the data obtained by our own WSN, which have been placed in two strategic sectors in the cities of Sangolquí and Quito. The first one represents the baseline, while the second has been situated in a known contaminated site. The data have been provided in the format of excel and csv files. Then Pentaho Data Integration has been used in order to purify, and subsequently upload the data into a source database. Furthermore, the data have been reformatted, cleaned and loaded into another intermediate database. Later we obtained a final database with the data ready to be treated through BI or DM. In this process, the cleaning of null data has been performed, while transformation occurred for dates to days and hours as well as in further, additional validations.

System Architecture: Figure 1 illustrates the architecture of the descriptive model of the data analytics based on measurements conducted to dependent (atmospheric pollutants) and independent (meteorological) variables implemented by BI and validated by DM.

The *Data Source Layer* contains the WSN, consisting of Arduino nodes that, by wireless transmission from the sensor nodes, is an emitter of frames that are processed and stored by the concentrator node (hub), which in turn sends the information to the Gateway, which is being able to be seen by the application. It also includes a Web services solution consisting of an API/REST protocol for sending information between the hub node (Raspberry PI) and the Gateway (server in the cloud). In addition, an application has been implemented for the administration of the WSN receiving the data from the hub node and transferring them to the central database of the model.

However, in the *Storage Layer*, with the previously stored information, we proceeded to apply ETL processes (extraction, transformation and loading), where the data have been purified, cleaned and treated in order to be copied to an intermediate database. Within the ETL processes, the null data have been revised, as well as the transformation of dates to days and hours, in addition to additional validations. Subsequently, we proceeded to select the variables that would be applied in the study,

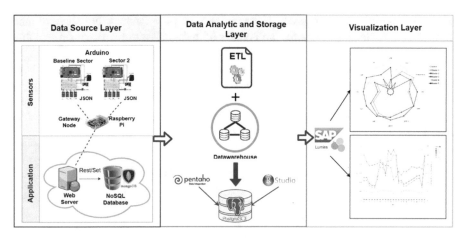

Fig. 1. System architecture. The diagram represents the architecture of the three-layer descriptive model designed as a solution based on data analytics. The data have been provided by a WSN, which has been implemented in order to determine air pollution.

which migrated to a final database, where useful application of BI and DM techniques were followed.

For the *Visualization Layer*, the Sap Lumira tool has been applied for the generation of BI reports, where indicators and graphs corresponding to the analysis of each pollutant gas have been created. In the DM process, the R-studio tool has been applied, with clustering algorithms such as K-means and the relationship between variables such as correlations and component analysis. Once the reports were obtained, the analysis of the polluting gases has been conducted and the respective conclusions and recommendations have been generated.

Modeling of DM: Once the data have been cleaned, we proceeded to define the descriptive models of tables and views in order to explore, classify and analyze the given information. The SAP Business Objects Lumira (i.e. formerly called SAP Visual Intelligence, which is a self-service, data visualization application for business users) tool has been used to prepare the reports.

Evaluation: Once the previous reports and analyzes were obtained, we proceeded to review them to verify their consistency and, if they met the requirements, otherwise they had been returned and the procedure had been repeated.

4   Evaluation of Results

4.1   Data Used and Sample

As stated by [17] air pollutants are any substance in the air that could, in high enough concentration, affect and alter the health of humans and animals alike, vegetation, ecosystems and cause environmental damage. Such pollutants may be present as solid particles, liquid droplets or gases. There are several air pollutants, and some of them have been identified to include compounds of halogen, nitrogen, oxygen, sulphur, besides many others.

In the case of this project, the data captured by the real time air pollution monitoring systems implemented using WSN are described in Table 1. After the cleaning and normalization process of all captured elements, 447477 data records have been acquired for the sensor corresponding to Sector 2, while the baseline Sector received 775184 data records.

Table 1. Air pollutants measured by the WSN used in this study

Pollutant	Unit
Carbon monoxide (CO)	(ppm)
Carbon dioxide (CO_2)	(ppm)
Nitrogen dioxide (NO_2)	(ppm)
Sulfur dioxide (SO_2)	(ppm)
Particulate material of 2.5 microns (PM2.5)	$\mu g/m^3$
Particulate material of 10 microns (PM10)	$\mu g/m^3$

4.2 Data Analysis by Wireless Sensors

According to [18] data analytics is the process of examining data sets in order to derive conclusions about the information they contain, increasingly with the aid of specialized systems and software. These techniques are widely used in commercial industries to enable organizations to make more-informed business decisions, and by scientists and researchers to verify or disprove scientific models, theories and hypotheses. On the basis of this statement, and based on the data collected by the air pollution WSN, a few description are presented below:

Sector 2 (City of Quito): Figure 2 illustrates the data analysis of the air pollutants measured by the WSN located in this sector, which have been previously described in Table 1. Nevertheless, the notable peaks correspond to $PM_{2.5}$ air pollutants. $PM_{2.5}$ are miniature particles in the air that reduce visibility and are able to cause the air to appear hazy when their levels are elevated. Instead, PM_{10} are smaller than $PM_{2.5}$, and their sources include crushing operations and dust stirred up by vehicles. Both vary during the day at different hours, having incremental values during the morning time (6 to 10 am) and at night time (10 to 21 pm). These values do not represent a serious issue for health since they are within the standard tolerable values established by the WHO.

Baseline Sector (City of Sangolquí): Figure 3, on the other hand, illustrates the variations of the obtained data of air pollutants. Obvious peaks of PM_{10} and $PM_{2.5}$ may be observed at the baseline sector under CO_2. These vary during the different hours of the day, having relatively high points during the morning hours (7 to 9 a.m.). In any case, they are less significant than the results obtained in sector 1. Once again, these values do not represent a serious health problem since they are within the standard limit values established by the HWO.

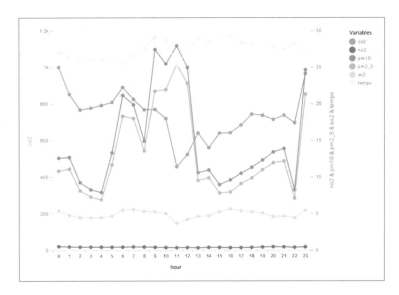

Fig. 2. Air pollutant data analytics (Sector 2)

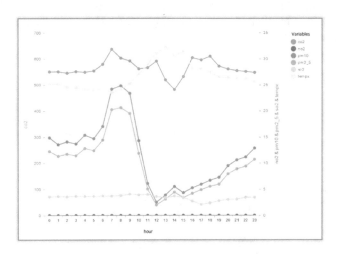

Fig. 3. Air pollutant data analytics (baseline sector)

4.3 Cluster Analysis

In order to corroborate the data analytics, the Cluster Analysis has been chosen. According to [19] it is an exploratory technique of data analysis to organize observed data or cases in two or more groups. Within this context, with the cleaned data, they have been grouped together, according to the closest characteristics. Therefore, the mining data clustering analysis techniques has been used. Specifically the K-means algorithm has been applied using the experimental method named Jambu Code [20].

This has been executed and verified by the methods of Hartigan, Lloyd, Forgy and MacQueen, with a total of 70 iterations, resulting in a K = 5 as reflected in Fig. 4.

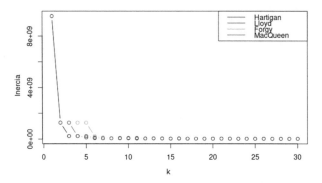

Fig. 4. The K-means Algorithm, which demonstrates the suitable number of five clusters in a dataset using the Elbow method for its validation of consistency

In addition, in Fig. 4, it has been illustrated that clusters 4 and 5 also group the variables of other groups, unlike the rest, since they reflect high values. This information contributes to the understanding about the variables that have high levels of contamination (Fig. 5).

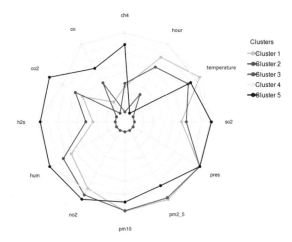

Fig. 5. A Cluster Analysis, which is a radar chart of multivariate data of quantitative variables obtained in the five clusters.

4.4   Discussion

With the design and implementation of the real time air pollution monitoring systems implemented using WSN [21], large raw data flows have been generated. However, it has been necessary to apply diverse techniques in particular with BI and DA, so that such data may serve for decision-making processes of a variety of people and corresponding authorities. This analysis may also serve for competent authorities in order to improve public policies of this context.

The data generated that have been stored and processed through this analysis demonstrated the behavior of polluters in certain sectors. For example, the determinations performed in sector 1 of Quito reveal that there is stronger air pollution than in the baseline of Sangolquí. According to data from the zonal administration of that area, sector 1 is fully urbanized. Some 97% belong to microenterprises, consisting of educational centers, commercial premises, banks, medical centers, clinics, restaurants, fast food sites, trolleybus stations and informal vendors. In the same sector, the roads and sidewalks represent 18% and parks 15%, while lots occupy some 58%.

Various chosen results from our analysis determined that high levels of CO_2 are present in four out of seven days in the week, with values that exceed 0.1% (1000 ppm). Such high values may be one of the factors responsible for asthma or the syndrome of buildings in the adjacent population. Pollutants PM_{10} and $PM_{2.5}$ have already exceeded the maximum local permissible value of 50 μg/m^3. SO_2, which also has exceeded the permitted values of 0.14 ppm per 24 h, which may also cause changes in lung function of asthmatics.

The analysis also requires monitoring and control of PM_{10} and PM_{25} particles, which presented values to be higher than those allowed. According to medical experts, such particles are able to reach the throat and lungs, causing cardiovascular and respiratory diseases up to lung cancer. The analysis also suggests the pertinent placement of pollution sensors at strategic points in the city, which in combination with the application of BI tools, present an ideal information package for decision-making processes. This would facilitate the detection and control of other pollution sources and may avoid harmful effects to the health of the public, animals and also plants.

Finally, this analysis may allow a decisive input for the definition of public policies in such environmental issue and context, or at least the implementation of municipal ordinances and other laws in order to improve the quality of life of the population.

5 Conclusions and Future Work

This study aimed to analyze sets of large data flows generated by a real-time WSN that measures different air pollutants. For normalization and modeling, the CRISP-DM methodology has been applied. We proceeded to apply ETL processes, where the data have been cleaned, filtered and treated. Then, we proceeded to select the variables that would be applied in the study, which have been migrated to a final database, where BI and DM techniques have been applied. For the BI application, we used PDI together with the Sap Lumira, where indicators and graphs corresponded to the analysis of each air pollutant. In DM, the R-studio tool has been used, with clustering algorithms such as k-means and relationship between variables such as correlations and component analysis. Once the reports have been obtained, the analysis of the polluting gases has been performed and the respective conclusions and suggestions have been generated.

For future studies, we plan to perform a real-time analysis of air pollutant behavior, using neural networks and recommendation algorithms, in order to predict the air quality index and alert the population of possible health risks.

Acknowledgments. The authors would like to express their gratitude for the financial support of the Ecuadorian Corporation for the Development of Research and the Academy (RED CEDIA) during the development of this study, under Project Grant CEPRA-XI-2017-13.

References

1. OMS (2018, May 2). Ambient (outdoor) air quality and health. http://www.who.int/en/news-room/fact-sheets/detail/ambient–air-quality-and-health
2. Brauer, M., et al.: Air pollution from traffic and the development of respiratory infections and asthmatic and allergic symptoms in children. Am. J. Respir. Crit. Care Med. **166**(8), 1092–1098 (2002)
3. MacIntyre, E.A., Gehring, U., Mölter, A., Fuertes, E., Klümper, C., Krämer, U., Koppelman, G.H.: Air pollution and respiratory infections during early childhood: an analysis of 10 European birth cohorts within the ESCAPE Project. Environ. Health Perspect. **122**(1), 107–113 (2013)
4. Larson, T.V., Koenig, J.Q.: Wood smoke: emissions and noncancer respiratory effects. Annu. Rev. Public Health **15**(1), 133–156 (1994)
5. Guan, W.J., et al.: Impact of air pollution on the burden of chronic respiratory diseases in China: time for urgent action. Lancet **388**(10054), 1939–1951 (2016)
6. Boubiche, S., Boubiche, D., Bilami, A., Toral-Cruz, H.: Big data challenges and data aggregation strategies in WSN. IEEE Access **6**, 20558–20571 (2018)
7. Ho, K., Hirai, H.W., Kuo, Y., Meng, H.M., Tsoi, K.K.F.: Indoor air monitoring platform and personal health reporting system: big data analytics for public health research. In: 2015 IEEE International Congress on Big Data, New York, NY, pp. 309–312 (2015)
8. Lopes, A.M., Abreu, P., Restivo, M.T.: Analysis and pattern identification on smart sensors data. In: 2017 4th Experiment@ International Conference (exp.at'17), Faro, pp. 97–98 (2017)
9. Chang, Y.S., Lin, K., Tsai, Y., Zeng, Y., Hung, C.: Big data platform for air quality analysis and prediction. In: 2018 27th Wireless and Optical Communication Conference, Hualien, pp. 1–3 (2018)
10. Ayyalasomayajula, H., Gabriel, E., Lindner, P., Price, D.: Air quality simulations using big data programming models. In: 2016 IEEE Second International Conference on Big Data Computing Service and Applications (BigDataService), Oxford, pp. 182–184 (2016)
11. Chen, L., Xu, J., Zhang, L., Xue, Y.: Big data analytic based personalized air quality health advisory model. In: 2017 13th IEEE Conference on Automation Science and Engineering (CASE), Xi'an, pp. 88–93 (2017)
12. Mehta, Y., Pai, M.M.M., Mallissery, S., Singh, S.: Cloud enabled air quality detection, analysis and prediction - a smart city application for smart health. In: 2016 3rd MEC International Conference on Big Data and Smart City (ICBDSC), Muscat, pp. 1–7 (2016)
13. Rios, L.G., Diguez, J.A.I.: Big data infrastructure for analyzing data generated by wireless sensor networks. In: 2014 IEEE International Congress on Big Data, Anchorage, AK, pp. 816–823 (2014)

14. Jardak, C., Riihijärvi, J., Oldewurtel, F., Mähönen, P.: Parallel processing of data from very large-scale wireless sensor networks. In: Proceedings of the 19th ACM International Symposium on High Performance Distributed Computing (HPDC 2010), pp. 787–794. ACM, New York (2010). http://dx.doi.org/10.1145/1851476.1851590

15. Fan, T., Zhang, X., Gao, F.: Cloud storage solution for WSN based on internet innovation union. In: Proceedings of the 2nd International Conference on Cloud-Computing and Super-Computing, vol. 22, pp. 164–169 (2013)

16. Yuan, H., Wang, J., An, Q., Li, S.: Research of WSN and big data analysis based continuous pulse monitoring system for efficient physical training. In: 2016 Future Technologies Conference (FTC), San Francisco, CA, pp. 1137–1145 (2016)

17. Anezakis, V.D., Mallinis, G., Iliadis, L., Demertzis, K.: Soft computing forecasting of cardiovascular and respiratory incidents based on climate change scenarios. In: 2018 IEEE Conference on Evolving and Adaptive Intelligent Systems, Rhodes, Greece, pp. 1–8 (2018)

18. Fotopoulou, E., Zafeiropoulos, A., Papaspyros, D., Hasapis, P., Tsiolis, G., Bouras, T., Mouzakitis, S., Zanetti, N.: Linked data analytics in interdisciplinary studies: the health impact of air pollution in urban areas. IEEE Access 4, 149–164 (2016)

19. Sacha, D., Kraus, M., Bernard, J., Behrisch, M., Schreck, T., Asano, Y., Keim, D.A.: Somflow: guided exploratory cluster analysis with self-organizing maps and analytic provenance. IEEE Trans. Visual. Comput. Graph. 25, 120–130 (2018)

20. Jambu, V., Provine, J., Ranganath, R., Rizvi, A.A.: U.S. Patent Application No. 15/391,697 (2018)

21. Guanochanga, B., Cachipuendo, R., Fuertes, W., Salvador, S., Benítez, D.S., Toulkeridis, T., Torres, J., Villacís, C., Tapia, F., Meneses, F.: Real-time air pollution monitoring systems using wireless sensor networks connected in a cloud-computing, wrapped up web services. In: Arai, K., Bhatia, R., Kapoor, S. (eds.) 2018 Proceedings of the Future Technologies Conference (FTC). FTC 2018. Advances in Intelligent Systems and Computing, vol. 880. Springer, Cham (2018). https://doi.org/10.1007/978-3-030-02686-8_14

Analyzing Scientific Corpora Using Word Embedding

Veronica Segarra-Faggioni[✉] and Audrey Romero-Pelaez

Universidad Técnica Particular de Loja, Loja, Ecuador
{vasegarra, aeromero2}@utpl.edu.ec

Abstract. The bibliographic databases have abstract and citations of scientific articles, the summary being the most consulted section of an article. In order to classify and address the entries in a system of indexing and retrieval of information in the databases of a manuscript, there are keywords, which in many cases this information should not achieve greater dissemination. This paper presents an evaluation of the semantic relatedness between the abstract of scientific papers and their keywords. This analysis will be using word2vec that is a predictive model, and it will find the nearest words. Thus, this study is focused on the metadata quality assessment through the similar semantics between two words that allow the accuracy in relation to metadata of scientific databases.

Keywords: Word embedding · Word2vec · Accuracy ·
Natural language processing

1 Introduction

The bibliographic databases have abstract and citations of scientific articles, the summary being the most consulted section of an article. In order to classify and address the entries in a system of indexing and retrieval of information in the databases of a manuscript, there are keywords, which in many cases this information should not achieve greater dissemination. However, a keyword is a metadata that facilitates the search and discovery resources to reuse it [1].

Many researchers have proposed several word representation techniques. Latent Semantic Analysis is a method for representing words in a low dimensional vector space [1]. Another approach is distributed representations of words that is called word embeddings [2].

This paper presents an evaluation of the semantic relatedness between the abstract of scientific papers and their keywords. This analysis will be using the popular word2vec vector space representation that will find words that have similar semantics allowing the accuracy in relation to metadata of scientific databases.

This paper is organized as follows. Section 2 describes the background of related work about semantic similarity. Section 3 describes the process of mapping between the keywords and each summary written production. Section 4 presents the results of the study.

© Springer Nature Switzerland AG 2019
Á. Rocha et al. (Eds.): ICITS 2019, AISC 918, pp. 68–72, 2019.
https://doi.org/10.1007/978-3-030-11890-7_7

2   Related Work

Nowadays, several works focus on the analysis of the information on scientific publications. Most of them are related in the classification of data. However, the study [3] identify a set of representative words based on word frequency. In fact, the work [4] is focused on word embedding distributed representations are making known deep learning models. Also, distributed word embedding represent the meaning of a word that applying natural language processing algorithms allow extracting a lot of information [4, 5].

There are many methods for evaluating semantic relatedness between words. For example, the similarity between two words or document can be computed using the cosine similarity as the result of applying Latent Semantic Analysis [6]. In our study, we have considered word2vec because it learns relationships between words automatically. Word2Vec was developed by Mikolov [2]. Word2Vec is a predictive model that learn their vectors [7]. In addition, word2Vec is a machine learning algorithm based on neural networks.

As we mentioned before, word2Vec finds words that appeared in similar contexts, that means word2Vec is a distributed word representation technique that learns their vectors to predictive the context in which the corresponding words tend to appear [8, 9]. Taking into account the work of [8], accuracy is considered like quality metadata metrics. It is important for this research because accuracy refers to the degree of closeness between data attributes and values of a specific concept [10].

3   Methodology

The purpose of this study is to analyze semantic relatedness between the abstract of scientific papers and their keywords.

The methodology is described in this section. Figure 1 shows an overview of the methodology.

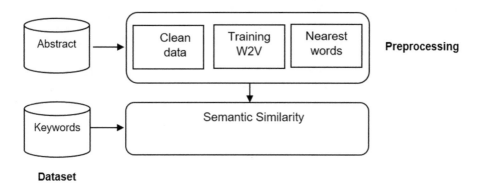

Fig. 1. Methodology overview

3.1   Dataset

Scopus[1] is a bibliographic database that has abstract and citations of scientific articles, the summary being the most consulted section of an article. The data of this research work consisted of scientific production from Scopus database that allows the researcher to search multidisciplinary aspects. We evaluated data about scientific production of computer science area published by researchers from Universidad Técnica Particular de Loja at SCOPUS from 2013 to 2017. This dataset consists of 139 articles, the following information has been considered: title, abstract and keywords. Table 1 shows an example of the articles selected for this study.

Table 1. A sample of selected articles for this study

Article	Title
1	Contribution of big data in E-leaning. A methodology to process academic data from heterogeneous sources
2	The use of ontologies for syllabus representation
3	Integrating OER in the design of educational material: Blended learning and linked-open-educational-resources-data approach
4	Accessibility and usability OCW data: The UTPL OCW
5	Implementation of social technologies for Open Course Ware OCW platforms

3.2   Preprocessing Data

We begin applying NLP techniques to prepare data, to clean data, to remove common words that are called stop-words[2], and to lemmatize the dataset in order to reduce possible redundancies such as concepts with the same meaning. According to [6] lemmatize refers to reduces and represents every word that shares the same root.

The corpus was trained by word2vec model. The experiment has evaluated the embeddings of words in the same semantic space as known embedding vocabulary to the abstract of a scientific paper published in Scopus. For computing coverage, we are going to use similarity measure for identifying the degree to which the two terms are alike. Using word2Vec we can find words that appeared in similar contexts. In addition, word2Vec is a predictive model that learn their vectors in order to improve their predictive ability of Loss [7].

Table 2 shows the relationship between words of the abstract and the nearest neighbors that were identifying using word2vec.

---

[1] www.scopus.com/.

[2] Common words like the, at, which, and others.

Table 2. Nearest words using Word2Vec

Word	Nearest neighbors
Syllabus	Learning, curriculum
Ontology	Taxonomy, schema
Interoperability	Connectivity, collaboration
Accessibility	Usability, operability
Learning	Training, learner

3.3 Word Similarity Evaluation

To determine the semantic relatedness between the keywords of the scientific paper and the nearest words that are identified in the Sect. 3.2 using Word2Vec. We applied cosine distance between the two metadata.

4 Conclusions and Future Works

Finally, this research work is aimed at the evaluation of similarity between the keywords and the abstract of scientific articles published in Scopus, using techniques of natural language processing and text mining.

For future works, a tracking word semantic change could be considered to understand the context of scientific publications. Also, to determine the importance of the application of appropriate keywords in research relationships with common characteristics in the articles as well as to determine trends in the research of teachers of the Universidad Técnica Particular de Loja.

Acknowledgments. The research team would like to thank Universidad Técnica Particular de Loja, especially to Tecnologías Avanzadas de la Web y SBC Group.

References

1. Yih, W., Zweig, G., Platt, J.C.: Polarity inducing latent semantic analysis. In: Conference on Empirical Methods in Natural Language Processing and Computational Natural Language Learning, pp. 1212–1222 (2012)
2. Mikolov, T., Chen, K., Corrado, G., Dean, J.: Efficient estimation of word representations in vector space, CoRR, vol. abs/1301.3 (2013)
3. Yan, E., Zhu, Y.: Tracking word semantic change in biomedical literature. Int. J. Med. Inform. **109**, 76–86 (2018)
4. Ferrone, L., Zanzotto, F.M.: A symbolic, distributed and distributional representations for natural language processing in the era of deep learning: a survey (2017)
5. Goldberg, Y.: A primer on neural network models for natural language processing. JAIR **57**, 345–420 (2016)
6. Deerwest, S.T., Dumais, G.W., Furnas, T.K., Landauer, T.K., Harshman, R.: Indexing by latent semantic analysis. J. Am. Soc. Inf. Sci. **41**, 1212–1222 (1990)

7. Mikolov, T., Le, Q.V., Sutskever, I.: Exploiting Similarities among Languages for Machine Translation, CoRR, vol. abs/1309.4 (2013)
8. Romero Pelaez, A., Segarra-Faggioni, V., Alarcon, P.P.: Exploring the provenance and accuracy as metadata quality metrics in assessment resources of OCW repositories. In: ICETC 2018 (2018)
9. Baroni, M., Dinu, G., Kruszewski, G.: Don't count, predict! A systematic comparison of context-counting vs. context-predicting semantic vectors (2014)
10. ISO 25000 software product quality. http://iso25000.com/index.php/en/iso-25000-standards/iso-25012. Accessed 01 Apr 2018

Planning the Combination of "Big Data Insights" and "Thick Descriptions" to Support the Decision-Making Process

Diana Arce Cuesta$^{(\boxtimes)}$, Marcos Borges, and Jose Orlando Gomes

Graduate Program on Informatics (PPGI), Institute of Mathematics (IM),
Federal University of Rio de Janeiro (UFRJ), Rio de Janeiro - RJ, Brazil
{diana.cuesta,mborges}@ppgi.ufrj.br,
joseorlando@nce.ufrj.br

Abstract. Practitioners and researchers have emphasized the importance of combining "Big Data insights" and "Thick descriptions" to support the decision-making process. However, organizations face challenges in the combination of these approaches. This paper suggests a guideline is necessary to provide clarity to the dynamics between "Big Data insights" and "Thick descriptions". Thus, this paper presents a four steps conceptual Framework to support in the planning stage of "Big Data insights" and "Thick descriptions" combination. This Framework is composed of paradigms which provide ways to ponder the appropriateness of each approach. From the literature review, paradigms detailed in four informative tables were developed. A study case in education management illustrates the Framework application. Results have shown the Framework potential to support the planning of the combination of "Big Data insights" and "Thick descriptions" combination in educational management.

Keywords: Big Data insights · Thick descriptions · Framework · Paradigms

1 Introduction

In recent years, Big Data and the promise of data-driven decision making is being recognized in organizations [1]. Big Data often is described by 3V, volume, velocity, and variety. However, these characteristics focus on Big Data technologies and left the use data effect out, so that; new Big Data definitions have emerged. Thus, Ylijoki and Porras [2], propose the term "Big Data insights" to be used in data usage and related activities. However, critical issues around "Big Data insights" have been raised [3]. In response, the literature [3–9] emphasizes the Ethnography value for compliments Big Data findings. Geerts [10] describe Ethnographic as "Thick descriptions" composed not only of facts but also by comments and interpretations.

The literature shows that the combination of knowledge obtained from "Big Data insights" and "Thick descriptions" allows us to form a complete picture of information [4, 5]. While the combination of these approaches is considered relevant for decision making, today there is not the proper knowledge to apply such combination. That situation is mainly reflected in Latin America organizations where the Big Data knowledge is limited [11]. In the current literature, the combined use of "Big Data

© Springer Nature Switzerland AG 2019
Á. Rocha et al. (Eds.): ICITS 2019, AISC 918, pp. 73–82, 2019.
https://doi.org/10.1007/978-3-030-11890-7_8

insights" and "Thick descriptions" has received little attention. However, in this research were identified two main challenges which organizations face. First, understanding in which situations the combined use of "Big Data insights" and "Thick descriptions" are more appropriate than the single use of these approaches. Second, how organizations can to combine these approaches. From this, we identify the need to provide organizations with information that allows understanding and planning the combined use of "Big Data insights" and "Thick descriptions".

The available literature shows that although the combination of "Big Data insights" and "Thick descriptions" has been applied [12–14], there is not enough guidance, particularly for new practitioners, on how these approaches can be used. Thus, the following problem was raised: how organizations may plan the combined use of "Big Data insights" and "Thick descriptions" to support their decision-making process? Decision makers and analysts should be able both to identify the suitable approach and design the approaches combination when that is necessary. To address the problem above raised, this paper presents a conceptual Framework based on paradigms. For this, the Morgan [15] paradigm definition was adopted: "beliefs and practices that influence how researchers select both the questions they study and the methods that they use to study them". This paper develops and uses paradigms as elements to guide in the planning of the combination of "Big Data insights" and "Thick descriptions". With informative tables, the paradigms provide elements to support the analysis of the organizational situation and the study of the Big Data and Ethnographic approaches. Moreover, these paradigms describe approach combinations forms. By using these paradigms, the challenges discussed earlier were addressed.

The Framework aim is to support organizations in planning the combination of "Big Data insights" and "Thick descriptions" to support the decision-making scenarios and, consequently, improve the quality of decisions. We hypothesized that the information provided in this Framework might help new practitioners in the understanding and planning of the combination of these approaches. This Framework was applied in a study case about the analysis of information in educational management in Ecuador. Therefore, we claim that this Framework can support in a context of educational administration. This paper is organized as follows. Section 2, address the general background of this work and related work. Section 3, detail the Framework. Section 4, describes the Framework experimentation. Section 5, present final considerations and future work.

2 Understanding "Big Data Insights," "Thick Descriptions" and Combined Approaches

Generally, Big Data is defined by three characteristics: volume, velocity, and variety [16]. However, this definition does not consider the effects of Big Data in organizations. Practitioners develop business models on practical Big Data solutions to provide value to their organizations. In this context, new Big Data definitions consider news characteristics known as value and veracity. The veracity refers to the reliability level in specific types of data. Organizations need to regard the veracity as an essential and challenging Big Data requirement [17]. Although value and veracity are considered part of the Big Data definitions, Ylijoki, and Porras [2] suggest that these characteristics are focused on the data use and should be separate from the technical description. Thus, Ylijoki and Porras [2] recommend the use of the term "Big Data insights" to refer to a data use context and related activities. This term was adopted for this research. **Thus, when this paper addresses Big Data, it focuses on the result or interpretation produced**.

In this context, the value and veracity features are essentials for "Big Data insights". Nonetheless, both value and veracity in the results of Big Data analysis have been questioned [18, 19]. According to Boyd and Crawford [3], Big Data change the definition of knowledge due to it leads to rethinking keys issues on the knowledge constitution, processes of search, nature and the categorization of reality. The notion of objectivity is another aspect to consider, since working with large volumes of data is subjective, and having quantifiable data does not necessarily mean to have a demand closer to objective truth. The information may be incomplete since it is not possible to capture cultural and psychological characteristics. Given these critical questions, organizations seek the appropriate form of supporting the decision-making process. Thus, there is a trend to a "Big Data insights" and Ethnography combination [4, 20, 21].

According to Myers [22], Ethnographers immerse themselves in the lives of people who study and seek to place the studied phenomena in their social and cultural context. According to Geerts [10], Ethnography is a "Thick description" which specifies details, conceptual structures, and meanings. Ethnography should produce a "Thick description" composed not only of facts but also of comments, interpretations and, interpretations of the comments. The "Thick description" term was adopted in this research. **Thus, when this paper addresses the Ethnography, it focuses on the result or interpretation produced**.

While Rasmussen and Madsbjerg [19] defend that Ethnography reveal information more valuable than Big Data, Wang [21] argues that "Big Data insights" and "Thick description" combination leads to better results. Ethnography uncovers the meaning behind Big Data visualization and analysis. Big Data reveals ideas with a particular range of data points and, Ethnography reveals the social context and connections between data points. Big Data delivers numbers while Ethnography provides stories, Big Data is based on machine learning and Ethnography in human learning [21]. Therefore, the knowledge combination obtained from Ethnography and Big Data, allows us to lead to a more accurate interpretation of human behavior [4, 5]. However, to combine these approaches is not an easy task. Related research notes that Big Data increase the data set available for analysis, and it allows new approaches and techniques, but it does not entirely replace traditional data studies [23]. The literature examines the connections between qualitative research, Digital Social Science and, Big Data [8, 23], but does not reflect how to combine these approaches. In that sense, Ford [13] describes her work in a collaborative project between Ethnographers and Data Scientists. Data Scientists and Ethnographers have many activities in common, and the key to success is in finding data together instead of compartmentalizing research activities. Laaksonen et al. [14] combine Big Data and Ethnography to examine the interactions between candidates in elections time. Leavitt [24] analyzes mixed methods at scale. Related literature reflects specific ways and experiences of the combined use of Big Data and Ethnography with an emphasis in technical aspects. However, **beyond technical issues**, orientation in the planning of the combination of "Big Data insights" and "Thick descriptions" is scarce. Thus, this paper seeks to support in the planning stage. That is particularly useful for new practitioners.

Due to both "Big Data insights" and "Thick descriptions" are focused on the results, this paper suggests that mixed methods foundations might be useful in the planning of the combination of these approaches. Thus, this Framework was developed based on investigations of quantitative and qualitative methods [25–28] but a "Big Data insights" and "Thick descriptions" context [2, 4, 5, 10, 16, 17, 22].

3 Framework for Planning the Combination of "Big Data Insights" and "Thick Descriptions"

For planning the combination of "Big Data insight" and "Thick descriptions," organizations need to analyze: the problem and main information required; characteristics of every approach; the need for combining the approaches and; how to attend such need [25]. Through four steps, this conceptual Framework provides support mechanisms for such requirements.

3.1 Step1: Analyzing Requirements

For choosing the appropriate approach, it is necessary to analyze the problem and the current situation of the organization, called in this paper as "organizational situation". The organizational situation reflects the resources available and the current information needs. For support such analysis, two paradigms were developed. These paradigms are composed of criteria. Each criterion has assigned one or more questions. Additional information is provided to identify possible answers. These criteria are based on the comparison of quantitative and qualitative approaches by Litchman [27] and Johnson et al. [26]. The paradigm in Table 1, seeks to support the discussion and description of general characteristics of the organization. Venkatesh et al. [28], argue the use of combined approaches must address specific purposes. For this reason, the Specific Analysis Paradigm seeks to delve into the "specific purpose" criterion.

Table 1. General and specific analysis paradigms

GENERAL ANALYSIS PARADIGM			SPECIFIC ANALYSIS PARADIGM		
CRITERIA	QUESTIONS	POSSIBLE ANSWERS	CRITERIA	QUESTIONS	POSSIBLE ANSWERS
Analysis Situation	(1) What is the study domain? (2) What is the problem to be addressed? (3) What are the main characteristics of the situation to be analyzed?	(1) Examples: education, health, marketing. (2) Problem detail. (3) Description of the environment, specific events.	Specific purpose addressed	What is the specific purpose to be addressed?	-----------
			Scientific method	What is the appropriate method (s) of analysis to address the specific purpose?	Inductive method: confirmatory method, test hypotheses and theory with data. Deductive method: explanatory method generates or develops knowledge, hypotheses and reasoned theory of data collected in the field of work.
			Observation Nature	What study environment is appropriate to address the specific purpose?	* Study of human behavior under controlled conditions. * Natural environments.
General Purpose	What is the general purpose of information analysis?	* Understand and interpret social interactions and, or describe and interpret complex societies. * Predict behaviors and trends and, or analyze cultural aspects.	Data Analise	(1) Is numerical information required? (2) Are narratives required? (3) Are real-time analyzes required? (4) Are offline analyzes required? (5) What type of analysis (s) is required?	(1) Quantitative analysis. Example: report of statistical meanings of the results. (2) Qualitative analysis. Example: contextual descriptions and direct citations from the research participants. (3) Analyze how the data are generated. (4) Analysis without high demands in response times. (5) Descriptive, prescriptive, preventive or diagnostic analyzes.
Specific Purpose	What is the specific purpose of information analysis?	* Validate hypotheses or generate hypotheses. * Correct inconsistencies in the data. * Obtain information that is useful and sufficient for the information needs of the organization. * Obtain reliable data.	Results	* Do results need to be generalized?	Generalizable results provide a representation of the population's target point of view.

Through the use of these paradigms, it is expected practitioners to generate information that reflects the current "organizational situation" and guide the selection of subsequent support paradigms.

3.2 Step 2: Selecting an Approach

Once the organization has identified its mains needs of information, it is necessary to select the approach which better support such demands. For this, Creswell [25] suggest analyzing the characteristics of the approaches. Thus, this Framework proposes to compare features of "Big Data insights," "Thick descriptions" and "Combined approach" with the information generated through the Specific Analysis Paradigm in step 1. The comparison aim is to select the approach that better support the information needs. For this, "Big Data insights" and "Thick descriptions" were characterized in paradigms. From that, the combined paradigm was developed. These paradigms are composed of the same criteria than the Specific Analysis Paradigm in Step 1. Based on related literature [2, 4, 10, 16, 17, 22] a detail is provided on every approach. Table 2, show these paradigms. Through the use of these paradigms, it is expected that the practitioners face the next situations. First, "Big Data insights" or "Thick descriptions" are sufficient for their information needs. Therefore, practitioners may proceed to execute the respective approach. Second, "Big Data insights" and "Thick descriptions" combination is necessary. In this case, a more in-depth analysis is needed before to execute the approaches.

Table 2. Big data insights, thick description, and combined approach paradigms

BIG DATA INSIGHTS PARADIGM		THICK DESCRIPTIONS PARADIGM	
CRITERIA	**DETAIL**	**CRITERIA**	**DETAIL**
Purpose Use "Big Data insights" mainly when the organization has the following purposes:	Validate hypothesis. Complete information.	**Purpose** Use "thick descriptions" mainly when the organization has the following purposes:	Generate hypotheses. Verify information.
Scientific method Big Data applies the following scientific method:	Inductive: confirmatory method, prove hypotheses and theories.	**Scientific method** Ethnography applies the following scientific method:	**Deductive:** explicative method, create and development knowledge, hypotheses, and theories substantiate from collected data in the work field.
Observation nature How is the study environment using Big Data?	Study on controlled conditions.	**Observation nature** How is the study environment using Ethnography?	The studies are in natural environments.
Data analysis The data analysis using Big Data has the following characteristics:	* It allows making correlations, comparisons, and report of statistical results. * It allows real-time and online analysis. * It allows descriptive, prescriptive and predictive analyses.	**Data analysis** The data analysis using Ethnography has the following characteristics:	* It finds patterns, and it provides context to data. * It develops narrations with contextual descriptions and direct citations of the research participants. * It allows analyzes in off-line time. * It allows descriptives and diagnosis analysis.
Results	"Big Data insights" may be generalized.	**Results**	Results may not be generalized.

COMBINED APPROACH PARADIGM	
CRITERIA	**DETAIL**
Purpose Use a combined approach mainly when the organization has the following purposes:	* Describing and interpreter complex societies and social interactions considering cultural features. * To predict behaviors and tendencies. * To analyze data inconsistencies. * To get useful and enough information to organization needs. * To obtain reliable data.
Scientific method The combined approach applies the following scientific method:	Inductive and deductive method.
Observation nature How is the study environment using a combined approach?	* It allows analyzed data on natural and controlled environments. * It allows the study of multiple contexts, perspectives, and conditions. * It allows study how several factors operate together.
Data analysis The data analysis using a combined approach has the following characteristics:	* It allows getting causal relationships and correlations. * It allows getting contextual descriptions and statistical meaning of the results. * It allows making descriptions, prescriptions, predictions, and diagnosis analyses. * It allows real-time and offline analyses.
Results	* It allows the integration and presentation of many dimensions and perspectives. * Results may be generalized.

Although a combined approach is valuable, it does not always lead to the discovery, development, or extension of a real theory [25, 28]. For this reason, the use of a combined approach must be analyzed in detail. Following steps seek to support such analysis.

3.3 Step 3: Understanding the Combination of "Big Data Insights" and "Thick Descriptions"

Based on the Venkatesh et al. [28] work, this Framework proposes to use specific purposes like indicators of the need for a second approach. This Framework suggests comparing the "specific purpose" addressed in previous steps with purposes for using mixed methods identified in the literature. The aim of this comparison is identifying similarities between the purposes to confirm the need to combine "Big Data insights" and "Thick descriptions". For guide such comparison, three paradigms were developed. Table 3, show these paradigms. Through the use of these paradigms, it is expected that practitioners identified a paradigm similar to their specific purpose addressed. On the other hand, understanding on which situations using a combined approach, it is relevant, but not sufficient to ensure the combination of these approaches. From that, there is the need for analyzing and define how to combine "Big Data insights" and "Thick descriptions". Step 4, addresses this need.

Table 3. Counteract, completing and verify paradigms

COUNTERACT PARADIGM
Counteract data is necessary for the next cases: *When the organization detects inconsistencies. requires analysing contradictions or requires a different point of view of the same study phenomenon [28]. *When is necessary examining different perspectives to validate information. or to get a better understanding of a particular phenomenon [30].
COMPLETING PARADIGM
Completing data is necessary for the next cases: *To contextualize the results of a particular method and ensure that the complete picture of a phenomenon is obtained [28]. *When the problems are defined by uncertainty and thorough knowledge is required. *To explain data relationships and results. especially when these are unexpected [30]. *To complement the information to the understanding of a research problem or to give clarity to results [31]. *When the organization seeks to generalize the Ethnography results.
VERIFY PARADIGM
Verify data is necessary for the next cases: *When the organization seeks the credibility of inferences returned by a particular approach. *To test a hypothesis or increases the results reliability [28]. *When the organization required exploring relationships between unknown variables. *When the organization needs to correct and test emerging theories [30].

3.4 Step 4: Designing the Combination of "Big Data Insights" and "Thick Descriptions"

The literature does not provide a specific vision on how to combine "Big Data insights" and "Thick descriptions". Therefore, based on the literature on mixed methods quantitative and qualitative [25, 29] three paradigms to support the combination design were developed. The paradigms proposed are based on the integration of three basics designs defined by Creswell [25] and on the principles and practices of mixed methods described by Fetters et al. [29]. In this step, the paradigms are structured into five sections. First, the combination design is described. Second, the design and purpose of combination presented in step 3 are related. Third, according to the study design, relationships between "Big Data insights" and "Thick descriptions" are described. Four, a form of combination is detailed, and finally, an example is presented. Table 4, show these paradigms. Through the use of these paradigms, it is expected that the practitioners select a paradigm to design the approaches combination, to be executing. For that, the Framework proposes to compare the chosen paradigm in Step 3 with the specifics purposes referenced in the second section of the Explanatory, Exploratory and, Convergent Paradigms. Finally, to facilitate the use of the paradigms, it is suggested that the register of the information generated be similar to the format of the paradigms proposed in every step.

Table 4. Explanatory, exploratory and, convergent paradigms

EXPLANATORY PARADIGM
Description: In this design, Big Data is the first approach executing. After, "Big Data insights" are used to generate "Thick descriptions."
Type of combination need: Useful to address the completing or verify needs.
Relationship: How may "thick descriptions" complete or verify "Big Data insights"?
Based on **"Big Data insights"**, structure the Ethnography, so that **"Thick descriptions"** are obtained to explain "Big Data insights"; get answers to why a certain event happened to complement the "Big Data insights"; get cultural and psychological information to contextualize the "Big Data insights"; and to confirm the existence of a problem previously addressed through Big Data.
Combination of approaches: Data analyses are usually connected, and the combination occurs in the interpretation state, specifically in the result discussion
Example: Xerox researchers implemented dynamic parking systems in Los Angeles. Xerox is using Big Data algorithms for answering important questions about parking congestion, and it uses Ethnography to confirm the problem and verify their interpretations [32].

EXPLORATORY PARADIGM
Description: In this design, the problem is explored through of the Ethnography. After, "Thick descriptions" are used to generate "Big Data insights."
Type of combination need: Useful to address the completing or verify needs.
Relationship: How can "Big Data insights" completing or verify "Thick descriptions"?
Based on **"Thick descriptions,"** define Big Data algorithms, so that **"Big Data insights"** are obtained to complement "Thick descriptions"; get a greater reach of the study population, increase the credibility, and generalized the "Thick descriptions"; and to confirm "Thick descriptions." "Big Data insights" provide information about people interaction in a virtual environment which is not possible to get through "Thick descriptions."
Combination of approaches: The combination occurs in the interpretation state, specifically in the result discussion.
Example: Wells Fargo is a financial services company in the United States. Ethnographers visit clients to observe how making businesses with Wells Fargo are. After, these ideas are used to applying Big Data to improve the company site for account corporate [33].

CONVERGENT PARADIGM	
Description: In this design, "Big Data insights" and "Thick descriptions" are generated separately but a similar period. After, results are mixed to make comparisons.	
Type of combination need: Useful to address the completing or counteract needs.	
Relationship: How can "Big Data insights" completing or counteract "Thick descriptions" and vice-versa?	
"Big Data insights" provide a second view of the same study phenomenon, allowing make comparisons and relationships which help to validate and redirect "Thick descriptions."	**"Thick descriptions"** provide a second view of the same study phenomenon, allowing make comparisons and relationships which help to validate and redirect algorithms in a Big Data context.
Combination of approaches: the approaches combination may take place in two contexts: First, initial "Big Data insights" may influence the focus and data type to be considered in the development of "Thick descriptions" or vice-versa. Thus, approaches interact with each other until set up the final results. In this case, it is necessary defining the first approach to be executed. Guidelines to select the approach are provided in the *Convergent Paradigm – Selection.*	
Selection: What is the appropriate approach to the first seeking of information?	
Start using Big Data when: * The problem is evident in the organization. * The initial information available allows doing the information search in a Big Data context.	**Start with Ethnography when:** * There is little information on the problem. It provides clarity to the study situation. * It is necessary to anticipate critical situations. It allows identifying a problem. * It is necessary evaluating the problem exists. * It is necessary evaluating a solution before applying predictive and prescriptive analysis with Big Data.
Second, the results generation of these approaches takes place in parallel, and the combination begins after the generation process has been completed. Thus, a combined plan of generation of "Big Data insights" and "Thick descriptions" is necessary before starting the information search. In this case, similar questions must be raised in both approaches.	
Example: In the paper referred to as "Combining Big Data and Thick Data analyses for understanding youth learning trajectories in a summer coding camp" [12], authors seek to combine reports of automatic programmatic snapshots (Big Data) with social context	

4   Using the Framework in Education Management

To evaluate the Framework, we conducted an experiment in a real environment with two simulated scenarios generally faced by decision makers and analysts of planning in educational management. From that, we illustrate how the Framework supported functionaries in planning the combination of "Big Data insights" and "Thick descriptions" on this application domain. The study scenarios were developed from interviews with two functionaries experienced in educational management from the Ministry of Education of Ecuador (MEE). Also, three functionaries from planning and human talent departments were trained to use this Framework. Next, the experiment was executed in two situations. (a) Functionaries analyzed the first scenario and planed the combination without the Framework only based on their experience. (b) Functionaries examined the second scenario and planned the combination of approaches based on the Framework steps.

Summary of the Simulated Scenario: The MEE promotes the Quality Schools Program (QSP). However, problems of poor school performance continue to persist in areas where this program has been implemented. The respective zone needs to analyze the situation and provide solutions ensuring compliance with the QSP. To identify the suitable approach to attending to their information needs, MEE functionaries studied the scenario and following the steps and paradigms in the Framework. Table 5, shows a part of the information produced by MEE functionaries.

Table 5. Summary of information produced by MEE functionaries

Step 3: Understanding the "Big Data insights" and "Thick descriptions" combination.	Step 4: Designing the combination of "Big Data insights" and "Thick descriptions."
Specific purpose addressed: Correct inconsistencies. Paradigm selected: Counteract Paradigm. Understand the poor school performance is not an easy task. We need a different point of view of the same study phenomena. So, we need analyzing our data register, but also is necessary visit the school to validate if the problem exists and get a better understanding of the situation.	Specific purpose addressed according to the paradigm selected in step 3: Counteract Paradigm. Paradigm Selected: Convergent design. We are going to get Big Data Insight and Thick descriptions in parallel. We are going to raised similar questions in the two approaches. Through the two approaches, we seek identified both where the problem is and where is the problem origin. For that, we plan to make algorithms to analyze the teacher profile; teacher and students interaction on the educative platform; relation number of student vs. infrastructure. On the same time, we plan applied an Ethnography to interact with teachers and students at the school to analyze the problem. To execute the combination plane.

5   Final Considerations and Future Work

This paper suggests a guideline is necessary to provide clarity to the dynamics between "Big Data insights" and "Thick descriptions". Thus, a four steps Framework to supporting practitioners in planning the combination of "Big Data insights" and "Thick descriptions" was developed. The evaluation with users of educational management has shown the Framework potential to support the planning of "Big Data insights" and "Thick descriptions" combination in such domain. From that, it was possible to investigate how the Framework could support practitioners both in the selection of approach and the design of combined use of these approaches. Also, this evaluation indicates the importance of knowledge exchange and negotiation between practitioners with different profiles. Educational managers highlighted that the Framework facilitates the planning of "Big Data insights" and "Thick descriptions" combination, especially to little-experienced practitioners. It provides a suitable detail of information. It is focused on what is of interest to understand the relationships between "Big Data insights" and "Thick descriptions" as well as, its flexibility to be adapted according to the characteristics of the organization. The Framework was considered positive and useful for strategically decisions in educational management.

The evaluation has reflected that Framework information help to plan of "Big Data insights" and "Thick descriptions" combination. That corroborating the hypothesis raised. However, it is important to highlight some limitations still need to be overcome. First, the paradigms developed may be useful in a specific context. Therefore, the Framework may not be generalized. In this case, more evaluations in different domains are necessary to define the usefulness of the Framework in a specific context. Second, the Framework needs to be more practical so that its use does not take long. As future work, a collaborative prototype to assist in the use of Framework and recording of produced information will be developed. Also, we seek to evaluate the Framework in other domains to improve the paradigms and Framework structure.

Acknowledgment. This work was supported by CAPES – Brazilian Federal Agency.

References

1. Labrinidis, A., Jagadish, H.: Challenges and opportunities with big data. In: VLDB Endowment, vol. 5, no. 12. Istanbul Turkey (2012)
2. Ylijoki, O., Porras, J.: Perspectives to definition of big data: a mapping study and discussion. J. Innov. Manag. **4**, 69–91 (2016)
3. Boyd, D., Crawford, K.: Critical questions for big data. Inf. Commun. Soc. **15**, 662–679 (2012)
4. Blok, A., Pedersen, M.A.: Complementary social science? Quali-quantitative experiments in a Big Data world. Big Data Soc. **1** (2014). https://doi.org/10.1177/2053951714543908
5. Chang, R.M., Kauffman, R.J., Kwon, Y.: Understanding the paradigm shift to computational social science in the presence of big data. Decis. Support Syst. **63**, 67–80 (2014)
6. Cowls, J., Schroeder, R.: Causation, correlation, and big data in social science research. Policy Internet **7**, 447–472 (2015)
7. Felt, M.: Social media and the social sciences: how researchers employ Big Data analytics. Big Data Soc. **3** (2016). https://doi.org/10.1177/2053951716645828
8. Smith, R.: Missed miracles and mystical connections: qualitative research, digital social science and big data. In: Big Data? Qualitative Approaches to Digital Research, pp. 181–204 (2014)
9. Swan, M.: Philosophy of big data: expanding the human-data relation with big data science services. In: Big Data Service, pp. 468–477 (2015)
10. Geertz, C.: Thick description: toward an interpretive theory of culture. In: The Interpretation of Cultures: Selected Essays (1973)

11. Gomes, L.: Snapshot of big data trends in Latin America - The Bridge, Revista da National Academy of Engineering (2015)
12. Fields, D.A., Quirke, L., Amely, J., Maughan, J.: Combining big data and thick data analyses for understanding youth learning trajectories in a summer coding camp. In: SIGCSE Proceedings. ACM, New York (2016)
13. Ford, H.: Big data and small: collaborations between ethnographers and data scientists. Big Data Soc. 1 (2014). https://doi.org/10.1177/2053951714544337
14. Laaksonen, S., Nelimarkka, M., Tuokko, M., Marttila, M., Kekkonen, A., Villi, M.: Working the field of Big Data: using Big Data augmented online ethnography to study candidate-candidate interaction at election time. J. Inf. Technol. Politics 14(2), 110–131 (2017)
15. Morgan, D.L.: Paradigms lost and pragmatism regained: methodological implications of combining qualitative and quantitative methods. J. Mixed Methods Res. 1, 48–76 (2007)
16. Laney, D.: 3D data management: controlling data volume, velocity, and variety. META Group Res. NOTE 6, 70 (2001)
17. Schroeck, M., Shockley, R., Smart, J., Romero, D., Tufano, P.: Analytics: the real-world use of big data. Executive report. IBM Global Business Services, IBM, Somers (2012
18. Croft, C.: The limits of big data. SAIS Rev. Int. Aff. 34, 117–120 (2014)
19. Rasmussen, M.B., Madsbjerg, C.: Big data gets the algorithms right but the people wrong (2013)
20. Ackermann, K., Angus, S.D.: A resource-efficient big data analysis method for the social sciences: the case of global IP activity. Procedia Comput. 29, 2360–2369 (2014)
21. Wang, T.: Why Big Data Needs Thick Data. Medium (2016)
22. Myers, M.: Investigating information systems with ethnographic research. Commun. AIS 2 (1999). Article 23
23. Kitchin, R.: Big Data, new epistemologies, and paradigm shifts. Big Data Soc. 1 (2014). https://doi.org/10.1177/2053951714528481
24. Leavitt, A.: Human-centered data science: mixed methods and intersecting evidence, inference, and scalability. In: HCDS Workshop. ACM (2016)
25. Creswell, J.W.: A Concise Introduction To Mixed Methods Research. SAGE Publications (2015)
26. Johnson, R.B., Christensen, L.B.: Educational Research: Quantitative, Qualitative, and Mixed Approaches. SAGE Publications, Thousand Oaks (2017)
27. Lichtman, M.: Qualitative Research in Education: A User's Guide. Sage Publications Inc., Los Angeles (2009)
28. Venkatesh, V., Brown, S.A.: Bridging the qualitative-quantitative divide: guidelines for conducting mixed methods research in information systems. MIS Q. 37, 21–54 (2013)
29. Fetters, M.D., Curry, L.A., Creswell, J.W.: Achieving integration in mixed methods designs —principles and practices. Health Serv. Res. 48, 2134–2156 (2013)
30. Hanson, W., Creswell, J., Clark, V., Petska, K., Creswell, J.: Mixed methods research designs in counseling psychology. J. Couns. Psychol. 52, 224 (2005)
31. Hesse-Biber, N.: Mixed Methods Research: Merging Theory with Practice. Guilford Press, London (2010)
32. Kerschberg, B.: How xerox uses analytics, big data, and ethnography to help government solve big problems (2012)
33. Nash, K.: Wells Fargo Revamps Website after Ethno-Analytics Study. https://blogs.wsj.com/cio/2015/11/10/wells-fargo-revamps-website-after-ethno-analytics-study/

A Proposal for Introducing Digitalization in a City Administration

Pasi Hellsten and Jussi Okkonen[✉]

Tampere University, 33014 Tampere, Finland
`pasi.hellsten@tut.fi`, `jussi.okkonen@uta.fi`

Abstract. A medium sized city decided to uplift its services to the 21st century and take leap to the digital age. An upgrade in the services was decided to be made in ordere to better meet citizens' expectations. The intention was clear; to enable later development towards the ultimate goal of digital channels being the preferred mean for running official errands by the year 2025. To introduce the new ways of operation, a Digiprogram was founded comprising numerous smaller, quicker experiments in various areas of city administration. The program needed an organization and we present pro and con notions of this initiative to act as preliminary report on a way to introduce service innovation to a city administration.

Keywords: Public sector · Digitalization · Service innovation · Trials

1 Introduction

Our children watch foreign TV shows on their smart phones. Our students deliver their presentations from their tablets and use their smart phones to aid them in their studies, to take notes in their thesis supervisory meetings. Our new graduates look for information in their hand held devices constantly. Are they prepared to take time and effort in conducting their official business in the traditional way of visiting the offices involved in establishing oneself in a new situation? Doubtful. Today and today's public sector is about services [1, 2]. This is also something that the case organization, the observed city operation, realized. In order to meet the overall objectives, and to serve their community in best possible and contemporary way by offering modern service on multiple areas a city entails on modern platforms, they decided to venture an entrance into the digital era in this traditional environment. Citizens using digital services in work, studies and leisure have also expectations to be able do so also when facing public sector. In society when in practice all active citizens have access to internet via computer or smartphone it would be a wasted opportunity if public services were still in the dark ages.

The previous examples of the future active citizens bears within the notion that the technologies offer possible solutions to be used in abundance. This in turn requires if not necessarily direct technological know-how and expertise, the right attitude to say the least. Should the amount of possible avenues to pursue seem overwhelming for the user, similarly the choice is to be made by the vendor and developer side too. To supplier this is a matter of great strategic importance. This goes to show that the

© Springer Nature Switzerland AG 2019
Á. Rocha et al. (Eds.): ICITS 2019, AISC 918, pp. 83–92, 2019.
https://doi.org/10.1007/978-3-030-11890-7_9

decisions of technology and even on solution level, however light-hearted they may seem to the end-users, are grave and the experimenting with them is not easy nor cheap.

Also, the resources are under scrutiny everywhere, including the public sector, where there are numerous fields and branches requiring attention, activities and resources to develop but already to maintain the quality of service [3]. Yet the service promise is to be held and operation developed (ibid.). The research problem condenses on the cross section of the areas: public sector services, delivery of the services, i.e. customer service, and ever-expanding digitalization. How, if at all, may these three areas be combined to a functional and well-designed solution? The more precise research question we seek to answer to is "what are the notions in introducing digitalization to a city ICT community by experimentation?" This is about providing user services, as they are ready and willing to acquire them. On the other hand, it is about allocation of scarce resources of providing public service. Both perspectives are about increasing productivity in public sector, yet the effect is somewhat ambiguous and therefore the theme should be inspected thoroughly.

Should an organizations implement such changes they are likely to want their employees to still continue to be active and productive members in their everyday work. It is vital for successful change implementation for the management to deeply understand the dynamics that have an effect to the proceedings [4, 5]. In order to plan for these changes and then to introduce them to the organization, organizations need to win over their employees to adapt willingly to the situation and instead of resistance to show commitment to the changes [6]. Should the members of an organization be curious and actively learning the outcome is bound to be better than in a case where the employees are merely receiving the novelties passively [7], as expected by the organization [8, 9].

The paper is organized as follows: Section two illuminates the theoretical backgrounds for issues addressed in this paper. The following section three presents the research setting and methods. Then the findings are briefly presented in section four. The discussion section five summarizes the results, and the conclusion places the paper in a broader context.

2   Theoretical Background

A city is a complex multifaceted entity with numerous tasks. An administrative entity that has at least four kinds of tasks: democracy, economical, communal, and well-being [10]. This notion serves to point out that the administration of a city needs to take into account various angles from which the development schemes are to, or may, be approached from. Similarly, the development schemes may stem from different functions with different aspirations. Aspirations, which may or may not spiced with political agendas, as the decision makers are in some cases more strongly politically inclined than in others. The Finnish association for municipalities [11] divides the area into sections: legal, economical, education and culture, social and healthcare, community and environment, vitality and employment, democracy and administration. Each area has its own processes, even when comes to their support given by ICT entity as the needed solution differ. Thus to find generally applicable rules is challenging.

Digitalization or digital transformation has a few definitions. It may involve business model renewal, and the change in ways of how employees go on with their routine tasks. It may affect the way the resources are allocated and how the operation is executed [12]. Previous consists various business-related activities and functions but also the processes behind the concrete actions so far that it also starts go alter the organizational culture [13]. Agutter et al. [12] spans the area possibly covered by digitalization from the features of more strategic nature, such as business development, to areas of more traditional doing, such as marketing. How and how widely digitalization is introduced depends on multiple factors. One such factor is organizations and the CIO's eagerness to take part in experimenting and the overall positive attitude towards renewals [14]. These early adopters may receive competitive advantage when they are trailblazing in the area, however it may prove to be an arduous task to make way for the new developments [15].

Digital solutions are used with the expectation of improving the contact to the customers and making the organizational processes more innovative and even fluent [16]. The angle an individual organization takes on the issue may vary, the entering angle may be the cost reduction or equally well customer service improvement or even making new market openings (ibid.). Occasionally the IT community is not entirely up to the task; one or more of the three factors (people, technology, processes) may need improving prior to the endeavor (Hagen et al. 2004). This means that the organization needs to be developed to meet the challenge.

Previously it was stated that the public service has become more customer-oriented service than what it perhaps was before. This applies to ICT community as well. Today is labeled by the convergence of the supplier side and the user side which are narrowing the gap between the two [17]. The former model of planning, building and maintaining is no longer enough. Instead, the users are to be taken along already in and onwards from the defining phase for both products and services (ibid.). As the services are provided with easy access, i.e. services are digitalized; more attention should be paid on user experience. In the case of public services, user experience consists of at least technical usability, access, accessibility, and information ergonomics. It is plausible to assume that the very people using the application and having the background knowledge of the area are also capable of giving insights on how the issues should and could be handled even better. Heads of ICT need to be alert to rethink their operation when need be and also to be open to new ideas regardless where they come from. This is also a matter of organizational culture; how new, emerging ideas are received and accepted. In close contact with the previous is the handling of failures. When operation experiments with new ideas and tries new approaches, one is bound to both succeed and fail. Should the failures be punished or scorned on, the keenness to continue with these is limited to say the least, which in turn may have unwanted effects on the innovativeness and development of the organization. The CIO has a crucial role in both creating and promoting this kind of open culture or the opposite.

The management of this kind of initiative is demanding. The effectiveness of the operation should be measured. According to the proverb in this area or better yet, a rule of thumb: one cannot monitor what cannot be measured and one cannot manage what cannot be monitored [18]. The various areas are to be measured, but is it even possible? So, that the various areas would be comparable with one another.

3 Research Setting

The single case study [19] this paper is based on observes a dedicated group of individuals in a variety of organizational branches and areas of civil service of a city of over 230 000 inhabitants and some 15000 employees in over 2000 sites. The ICT sector, or community, of the city is there to serve all of these. The clients of the ICT community have various needs; the area is vast from the city planning to the employ i.e. the city officials, from the top management to the summer help in the city tourist bureau have different needs but also capabilities when it comes to their use of technologies and tools within. The complexity of the settings presents the management with challenges. It became evident that modern ICT is to be taken on as it might have solutions to the challenges and offer more flexibility to the operation.

Smart Tampere – program has as an ultimate goal creation of digital services to ease the everyday life for the citizens, increase of wellbeing and security and promotion of smoother moving in and of the city. Also creating new business opportunities by enabling cooperation between various stakeholders is listed [20]. This program consists two related sub-programs: Ecosystem program and Digiprogram.

The qualitative data was collected by semi-structured interviews. First the key persons for the Digiprogram were suggested by our contact person, the CIO of the city, who also was interviewed. Further interviewees were invited by these suggestions. In total eighteen interviews were conducted face-to-face at theca se organization premises.

The interviewees of the city were people working on following positions: CIO, Program manager, Productivity controller, ICT manager (AK), Service designer, Enterprise architect, Digimarketing manager, Development manager (city planning), Development manager (customer service management), Development manager (employment services, SO), Project manager (city concept), Project manager (city planning), Project manager (customer service), Project manager (early education, pre-school HK), Project manager (employment services, JT), Project manager (employment services, MV), Project manager (grammar school), Project manager (infrastructure; tram), Project manager (space allocation, JS), Project manager (town planning).

All the interviews were recorded for further elaboration and fact-checking. The interview themes covered issues related to initiation and rationale of the program, resourcing and stakeholders involved, effectiveness and effects, also the process and communications relating to the Digiprogram, and the evaluation of the success.

The data analysis followed interpretive research approach [21]. The researcher went through the material several times to gain an overview of the procurement process, stakeholders involved, and different challenges, and to gather all relevant details. After and already during this, the matters were discussed with another seasoned academic professional but also with the program manager to double check and to ensure that the approach was correct and right. Process diagrams and stakeholder maps were drawn to visually aid the interpretations. These visual maps were further iterated. Due to the size of these visualizations and space limitations, they are omitted from this paper. Finally the findings were compared to the literature.

4   Findings

4.1   General Notions on the Findings

The city decided that the best way to address this issue and to introduce digital transformation to the city was to enable the stakeholders to make experimentations themselves. The experimentation was going to be a joint effort of both the ICT community as well as the employees using the system and applications. To match the 'demand of the times' and to take a stand on the technological developments and to modernize its operation, at least to a degree, the city came up with Smart Tampere program. This program brings together the Ecosystem development scheme aiming at improving and enhancing the overall appeal towards the geographical area both for businesses and people and the Digiprogram looking inwards and aiming at developing the city's operation and processes internally and towards citizens. A program was founded and a manager for it appointed. The objective, or slogan even, for the program is 'Year 2025 the citizens preferred mode of services is digital'.

In the program, there are numerous projects and trials or experiments even, executed in tempo not all that familiar for ICT projects or public sector services, especially in the public sector ICT projects. The pace is quicker than in previous projects. This in turn requires a newer culture to emerge in which such approach is tolerated or even promoted. Quite logically also new expertise is required as there are new technologies and solutions that are introduced to operation. Similarly, the mere way of doing things was somewhat new to people in the city organization thus training took place in various areas, e.g. planning, defining, purchasing of information systems and applications, but also in publishing in social media. Quite a bit attention was given to make the word of the experiments to spread. The project managers, development managers and program manager were encouraged to be active.

However, not all tasks were to be executed by the city officials as the aim was to combine the efforts of third parties and stakeholders also in broader scale, such as vendors and citizens. The vendors invested their time more than usually in development schemes with the ordering party so that they conferred together with the subject specialists about the possible way of solving the issues brought up. This way they together searched for the optimum solution, also in regard of the existing system landscape.

The citizens were also asked whether they would be prepared to take on a digital way of handling their matters, thus giving input in form of justifying even more strongly the implementation of the Digiprogram. The end-users, and in some cases the 'customers', i.e. the citizens, took also part in these experiments for example in form of 'beta-testing'[1] [22] and then giving input on the development work and the functionality of the intended software.

The objective was to alter the operations to better meet the user expectations based on the successful trials and experiments. Already during the experiment, the

---

[1] Beta testing is used to describe the phase of software testing in which a composition of the intended user group tries to use the product under real (life-like) circumstances.

performance was assessed. The actual benefits were evaluated as to give enough material for the decision making regarding whether or not the new way of operating was 'good enough' for making it permanent procedure and thus being scaled up to match the requirements in the particular area in question or even organization-wide use. The assessment proved to be more problematic and challenging than anticipated. The development managers and project managers had their expertise on their corresponding area but not all the features were operationalized to be measured thus making the baseline non-existent and the comparison between the states before and after the development scheme impossible. The person appointed to productivity controller conversed recently actively with the project teams and managers to form a plan for assessing the benefits and/or effects the development had. The need for such a person was acknowledged but the person was recruited and appointed only after the program had started, so that this function has yet to show results.

The assessing of benefits, or actually effects, was performed more on qualitative basis as there were few possibilities to bring on significant quantitative measures. The interviews revealed some reluctance to come up such measures as the argument of saving a certain amount time versus improved quality in meeting a customer for example in the employment services. Seems that trust is an issue also in this context and building trust is advised to be taken seriously.

4.2   Findings on the Experiments

The research project, which this paper is based on observed three areas of the city operation, which are all a part of the Digiprogram: City Planning (CP), Customer Service (CS), and Employment Services (ES). There are other areas in the program, but they are left out of this paper as they were not scrutinized in the study.

In the City Planning the process itself was taken under inspection and later decided to be renewed. Under old normal circumstances to apply for building permit would have taken approximately some 90 days and after the renewal it will have dropped to a third of that amount of days. According to interviews, this entails both technological innovating but more significantly a cultural change where the actual doing was taken under scrutiny and altered to be more customer-oriented one. Similarly, the supervisory function of the building in the city CP aims to deploy new type of material. In practice this means including 3d-modeling and imaging for the city services to aid the operation of this area of city services.

The employment services is a very labour-centered activity where the relationship between two individuals is focal. The official needs to know the applicant, to assess whether there is a possible employment opening or whether the direst need is for benefits (usually monetary) or further training or some entirely different type of service. Consequently, the officials need to get to know the applicants. This is traditionally done by interviewing the applicants, which is quite a time-consuming feature in the process. The city came up with the idea of using a chatbot for this. In effect, the chatbot innovation is said to be able to save up to three quarters of an hour of each customer meeting that are done in thousands. The obvious time saving there is vast. However, not all this time is directly saved as pure plus, as the officials still need to familiarize themselves with the customer records and to plan for actions. Yet such method is practical when conducting large scale interviews. Similarly, the customer needs still time to fill in the questions the chatbot presents him/her with. But still the time slot needed to spent together is cut down.

Another experiment is practical for both Customer Service and Employment Services; an appointment reminder. As simple as it may seem, this saves again tens if not hundreds of hours of time reserve. The customers make a booking for handling their affairs, depending on the busyness of that particular function, the appointment may be days or even weeks later. As the appointment comes closer, the application sends a reminder to the customer that this is the case thus enhancing the possibilities of the customer showing up at the agreed time. The feature seems perhaps a bit mundane, but put into the city scale, the potential saving is substantial.

The program leadership has introduced a 'team day' or a 'program day'; once a week the project managers from various schemes meet to discuss, confer, and learn. This is yet another investment to the program as the project managers are absent from their 'normal' position. During these days the project managers, otherwise in their locations, gather together in the city administration and present their projects current state and hear how the others are doing. This offers peer support in form of listening and sharing experiences. There is also education available in these meetings. The project managers praise the fact that they have enjoyed training in e.g. public sector purchasing; some schemes have a need to invest in new(-er) applications or hardware needed for it. The project managers are not necessarily qualified to take part in these endeavors, as it is not a part of their normal routines. Similarly, the digital marketing or communications is a branch in which there is things and features to be learned in today's ever-changing communication landscape. This applies both to internal and external communications. Thus, the program management offered training also in this area given by the city's professional communicator. This training included some features and publishing policies in channels like Twitter, Facebook to name but a few. The city has a rather liberal attitude towards publishing news on various events within, but some ground rules are there. These were also elaborated. Main focus in this part was still exceedingly on how to address public, what is worthwhile and how one should report on one's doings.

The interviewees that voice their doubts whether the saved amount of time will be invested in making the operation better by their quality offer a discord to this otherwise exceedingly positive appearing reporting. There seems to a slight uncertainty whether the ultimate goal and objective of the Digiprogram is in enhancing the service organization of the city and to improve the quality of the services offered to the citizens. The other optional objective would be merely cutting costs. For the high city officials who need to keep an eye on the both issues this is not an easy task to decide. The resources are increasingly scarce which makes it a plausibly lucrative opportunity to use the developments as a cost cutting method and at its extreme to let people go.

Similarly, a little was said about the individual objectives of the project managers and the measurability of the development schemes. At this point most of the individual objectives are qualitative and as such difficult to be measured exclusively so that they could also be compared with peers. Some voices were of the opinion that a first baseline should have been defined, i.e. what and how the operation was handled prior to the development schemes. Then the operational effectivity after the development work could have been measured to form a picture of the effectiveness of the development scheme and the methods within.

The target group for the development activities, the project managers but also the city officials involved and their readiness to step into the digital era varied. Some were, quite logically, more advanced in their know-how than others. This particular fact alone justifies the claim that the project managers and others are to be treated as individuals and their level of expertise acknowledged when they are read in to the program. The area in which they are to experiment is scarcely directly comparable with their peers. The processes, the systems and applications they use, little is shared or same between the branches. Similarly, the services they require of the organization vary. Some are in need of more training in certain areas than others are. In other areas, the situation may be again different.

The integration of the experiments to the existing ICT infrastructure and enterprise architecture is something that the project managers need not to overly concern themselves with, as there is dedicated personnel in the ICT community to take a stand on this matter. However, should they be experts in their own field may help them to form a picture of the proceedings and support the development even more.

5   Concluding Remarks

When a public sector organization considers such a large development scheme, experimenting with small(-ish) and quick schemes appears to be worth considering as a way to do it. Even if moderate in size and with limited need for resources the experiments quite obviously require careful and capable planning of the actual measures to be taken. It is also notable, that even small success feeds enthusiasm to conduct experiment, yet failure causes critical demotivation. The personnel to be involved in such actions are preferably to be found amongst the existing personnel, as the knowledge of the organization and its ways of working are focal. It could also be argued that should a complete outsider be recruited as a project manager to such an initiative, there is vital tacit knowledge lost, or actually left unused in comparison with the situation where the project manager comes from within and thus the knowledge gap is to be overcome with different means.

An active and strong sponsor, like a CIO, is highly beneficial, if not a necessity, for an endeavor of this magnitude to be successful in this area of operation. There needs to be power and support behind the initiatives. A position of power, like that of a CIO, may also act as a counter measure for the possible resistance for change and to actively promote the initiative. A sponsor's clear vision of the better tomorrow, how operation will be run after certain development procedures help the whole organization to proceed simultaneously defining the course.

The project managers are in a crucial role as they operate in between the change instigators and the organization that the change is implemented in. The project managers may affect significantly on how the initiative is received. Thus these people need to be chosen quite carefully. It is equally important that they are truly won over for the initiative and stand behind it.

It is important to get commitment of various stakeholder groups and even individuals. It is advisable to let them speak up. To ask for their participation and acknowledge it. Quite often, the basic idea behind crowd sourcing may be used to cover unexpected ends and to serve in creating better, more customer-oriented solutions to various organizational challenges.

Communications is a key to success. There are bound to be unclear issues along the way, some angles are not covered, some are neglected for a reason, some by mistake. A clear structure may help to minimize such occurrences and continuous and clear communication delivers the message to the stakeholders.

The finding brought about in the study are initial results of substantial paradigm shift in public administration and services provided to citizens. As presented here major guidelines on user expectations and ways to meet those are easily drawn. As the citizens have devices, skills and willingness to have service through digital platform the case is presented also as a national pathfinder on digitalization of municipal services. To larger extend the findings are transferable to other context too, yet further research is requited to see to sum effect to productivity. It is inevitable that saved resources on supply side and saved time on demand side are motivating shift to 24/7 open digital society with services that are free of spatial or temporal restrictions. Moreover, digitalization of services increases transparency in administration since services are explicitly instructed and there is an event log on actions.

References

1. De Vries, H., Bekkers, V., Tummers, L.: Innovation in the public sector: a systematic review and future research agenda. Public Adm. **94**(1), 146–166 (2016)
2. Higgins, B.: Reinventing Human Services: Community-and Family-Centered Practice. Routledge, Abingdon (2017)
3. Arnaboldi, M., Lapsley, I., Steccolini, I.: Performance management in the public sector: the ultimate challenge. Financ. Account. Manag. **31**(1), 1–22 (2015)
4. Beck, D.E., Cowan, C.: Spiral Dynamics: Mastering Values, Leadership and Change. Wiley, Hoboken (2014)
5. Hekkert, M.P., Negro, S.O.: Functions of innovation systems as a framework to understand sustainable technological change: empirical evidence for earlier claims. Technol. Forecast. Soc. Change **76**(4), 584–594 (2009)
6. Fernandez, S., Rainey, H.G.: Managing successful organizational change in the public sector. Public Adm. Rev. **66**(2), 168–176 (2006)
7. Huitt, W.: A systems model of human behavior. Educ. Psychol. Interact. (2003)
8. Zack, M.H.: Managing codified knowledge. MIT Sloan Manag. Rev. **40**(4), 45 (1999)
9. Zhou, J., George, J.M.: When job dissatisfaction leads to creativity: encouraging the expression of voice. Acad. Manag. J. **44**(4), 682–696 (2001)
10. Jäntti, A.: Kunta, muutos ja kuntamuutos, p. 245 (2016)
11. Kuntaliitto.fi (2018). https://www.localfinland.fi/. Accessed 06 June 2018
12. Agutter, C., England, R., van Hove, S., Steinberg, R.: VeriSM - A Service Management Approach for the Digital Age. van Haren Publishing, Zaltbommel (2017)
13. i-SCOOP: Digital transformation: online guide to digital business transformation (2018)

14. Ding, F., Li, D., George, J.F.: Investigating the effects of IS strategic leadership on organizational benefits from the perspective of CIO strategic roles. Inf. Manag. **51**(7), 865–879 (2014)
15. Rogers, E.M.: Diffusion of Innovations. Simon and Schuster, New York City (2010)
16. Bongiorno, G., Rizzo, D., Vaia, G.: CIOs and the digital transformation: a new leadership role. In: CIOs and the Digital Transformation, pp. 1–9. Springer, Heidelberg (2018)
17. Ahlemann, F.: How digital transformation shapes corporate IT: ten theses about the IT organization of the future. In: 2016 Federated Conference on Computer Science and Information Systems (FedCSIS), pp. 3–4 (2016)
18. Baily, P., Farmer, D., Jessop, D.: Purchasing Principles and Management. Pearson Education, London (2005)
19. Yin, R.K.: Case Study Research: Design and Methods, vol. 5. Sage Publications, Incorporated, Thousand Oaks (2008)
20. Smart Tampere, 19 October 2017. https://www.tampere.fi/smart-tampere.html. Accessed 06 June 2018
21. Walsham, G.: Doing interpretive research. Eur. J. Inf. Syst. **15**(3), 320–330 (2006)
22. What is beta test? Definition and meaning, BusinessDictionary.com. http://www.businessdictionary.com/definition/beta-test.html. Accessed 08 June 2018

Agent-Oriented Engineering for Cyber-Physical Systems

Carlos Cares[1](✉), Samuel Sepúlveda[1], and Claudio Navarro[2]

[1] University of La Frontera, Temuco, Chile
carlos.cares@ceisufro.cl, samuel.sepulveda@ufrontera.cl
[2] TIDE S.A., Temuco, Chile
claudio.navarro@tide.cl

Abstract. Cyber-physical systems are the most relevant mainstream in the contemporary automation of industry, smart cities, and ubiquitous systems. Although most of its components and their behaviors are software-based elements, there are no suitable approaches for specifying, designing, testing and delivering cyber physical systems under a software engineering point of view, i.e. having a systematic, disciplined and measurable approach to its development. This paper describes the suitability of agent oriented software engineering for confronting the development of cyber-physical systems. The contribution is focused on two labels, first, we show a set of requirements for an cyber-physical engineering methodology, that were gathered by means of a literature review, and, second, we present a match between agent-oriented principles and the collected requirements. To illustrate some examples we use Tropos, one of the most disseminated agent-oriented methodologies.

Keywords: Cyber-physical systems · Agent orientation ·
Software engineering · Tropos · AOSE

1 Introduction

The Cyber-physical systems (CPS) are the integration of distributed hardware controlled by distributed software with the aim of controlling physical processes. Therefore they should have both: software-hardware devices for perceiving real world variables and software-hardware devices for acting on the real world [21]. Its impact became a technological revolution which has been namely Industry 4.0 [22]. Their main enablers have been the low cost of IoT devices, mainly sensors and micro-controllers [33] and easy access to LAN and WAN networks. However, beyond industrial applications, the main uses have been on implementing smart-cities product and services [9,14,17], but also a wide set of other applications, also perceived as revolutions, which has been summarized as *smart-everything* [40] and also as *internet-of-everything* [10].

However, the aforementioned revolution, is based on technology, on the easy access to devices and networks, but methodological approaches remain pending.

© Springer Nature Switzerland AG 2019
Á. Rocha et al. (Eds.): ICITS 2019, AISC 918, pp. 93–102, 2019.
https://doi.org/10.1007/978-3-030-11890-7_10

From the initial acknowledgment of the need for a new engineering discipline for cyber-physical systems [21]. If we define Cyber-physical engineering as the application of systematic, disciplined and measurable approaches to the development, operation and maintenance of cyber-physical systems (in similitude to software engineering) then, it is possible to affirm that there are very few approaches in the side of "how" question in cyber-physical engineering.

On the "what" perspective, a wide set of requirements have been formulated as challenges for the "new generation" of cyber-physical systems, such as capability, adaptability, resiliency, safety, security, usability, stability, controllability, accuracy, scalability, transparency, dependability, robustness, autonomy, self-organization, predictability, efficiency, interoperability, trackability among others [30,34,38,42].

However, going back in software engineering history, the topic of real time systems has been an existing topic from the cyber-side and it has effectively considered the monitoring and controlling of physical elements, even including response-time constraints, risk severe consequences, and/or failures [20]. Hence, real-time approaches are software engineering proposals, are part of the scope of its body of knowledge and pretend to cover a part of or the complete system development process [5]. Therefore, the physical-side of cyber-physical systems has effectively been part of software engineering proposals.

A particular approach to the "how" question in software development has been agent-orientation at the level of analysis and design. At an early stage of its development (1998–1999), it was recognized that agents, as autonomous pieces of software provide a natural metaphor for organizing complex distributed systems [43]. Since then, several agent oriented software methodologies have been proposed, e.g. in [2] 75 methodologies have been recognized.

The specific idea of using agent orientation for dealing with cyber-physical systems development has already been proposed by several authors [8,23,24]. In particular, in [7] a map is obtained from a literature review between multi-agent systems as enablers of key features of cyber-physical systems. However, all of them have been formulated under the perspective of agents as technology and not agent-orientation as methodology, it becomes clear that neither of the agent-oriented proposals for cyber-physical systems uses any of the existing agent-oriented methodologies.

In this paper, we present a conceptual analysis on the suitability of using agent-oriented analysis and design principles to develop cyber-physical systems. In order to illustrate this suitability, we use the agent-oriented software methodology namely Tropos [6]. We have chosen this methodology because it has been recognized for having one of the three more active research communities, having the highest quantity of tools and which has been established longer [13].

In order to support this conceptual analysis, in Sect. 2, we present a map of design challenges that we have interpreted as methodology requirements for cyber-physical systems. This was done through the application of a systematic literature review. In Sect. 3, we map the collected set of requirements to agent-oriented principles using some Tropos illustrations. Finally, in Sect. 4 we summarize the reached goals of this article, current gaps between Tropos and cyber-physical modeling challenges which is the basis of our future work.

2   Current Challenges on Cyber-Physical Engineering

Just under a practical perspective, We need to limit the semantic scope of Cyber-physical system science and its corresponding Cyber-physical engineering. Therefore we use as pattern the Software Engineering definition to present the definition of Cyber-physical engineering discipline as a systematic, disciplined, and quantifiable approach to the development, operation, and maintenance of cyber-physical systems.

To get a recent scenario of cyber-physical systems methodologies we conducted a systematic literature review following the steps proposed by Petersen et al. [35]. We selected three primary research sources (Springer, IEEE and Scopus) and, our search string looked for "Cyber-physical" and ("Challenge" OR "Methodology" OR "Roadmap" OR "Modeling" OR "State"). After applying basic exclusion criteria (e.g., duplicated papers) and removing papers that were too specific, we obtained 709 articles (due to the general nature of the topic we did not exclude editorial introductions to workshops or special issues). The full results of each stage can be found in a shared sheet[1].

As a first analysis, we sorted papers by year and we found an ascending curve, starting in 2016 with a few more than 50 papers and achieving more than 150 papers in 2016 and 2017. In 2018, in half a year we got more than 100 papers. The domains of these papers are 40.5% in production systems, 13.5% in smart environments (smart city, smart buildings), 10.8% in automotive systems and also 10.8% in aerospace systems, and, sure by coincidence, a 10.8% in energy systems. Medical systems had 8.1%, and others in 5.4%.

In order to specify engineering stages, Fig. 1 represents the proportion of papers that mention a process stage. The total number of papers that mention general stages are 33% of the total.

Although requirements (and specifications) yielded a low number (17), there are several papers referring to some specific non-functional requirements, some of them being traditional non-functional requirements and, most of them, new non-functional requirements for cyber-physical systems, such as resilience, safety, reliability. We say "new" in the sense that they are not part of traditional quality models such as ISO9126 or ISO25.000. In Fig. 2 we illustrate the proportion of 90 papers mentioning a non-functional requirement in the title. All these results show a general view of research focuses.

In addition, we analyzed the general proposals: those that do not mention either a specific topic or a specific domain. Then, we read the abstract and the result was 79 proposals. From this group we selected only the years 2016 to 2018 (half the year) which resulted in 35 papers. We read these full papers searching for methodological or modeling recommendations, the results were 20 papers.

---

[1] https://goo.gl/2b4hs6.

Table 1. Addressed challenges on recent studies in cyber-physical systems (2016–2018)

	Horváth & Rudas [15]	Abdullah et al. [1]	Kathiravelu & Veiga [18]	McKee et al. [27]	Mueller et al. [32]	Ribeiro & Bjorkman [36]	Ruchkin et al. [37]	Seshia et al. [39]	Aziz & Rashid [3]	Bondavalli et al. [4]	Esterle & Grosu	Giaimo et al. [12]	Khalid et al. [19]	Leitão et al. [23]	Mangharam et al. [25]	Marwedel & Engel [26]	Mezhuyev & Samet [28]	Mosterman & Zander [31]	Thomopoulos [41]	Zeng et al. [45]
Modeling language																				
To enable abstraction layers	✓					✓		✓									✓	✓	✓	✓
Components and system	✓	✓	✓	✓	✓	✓	✓	✓	✓	✓			✓		✓			✓	✓	
Branching and looping	✓										✓	✓								
Time / coordination	✓	✓	✓		✓		✓	✓	✓	✓			✓	✓				✓	✓	
Physical components & behaviors	✓	✓		✓	✓		✓	✓	✓	✓	✓	✓			✓			✓		
Interfaces	✓					✓	✓	✓	✓	✓								✓	✓	
Technology	✓												✓							
Visual notation (expressiveness)	✓															✓	✓			
To represent environment	✓	✓				✓	✓			✓	✓	✓	✓	✓	✓	✓				✓
Unpredictibility	✓	✓																		
Space-time					✓							✓	✓			✓				✓
Human-in-the-loop								✓					✓	✓	✓	✓				✓
Requirements	✓	✓	✓												✓	✓				
Non-functional req.		✓	✓												✓	✓				
Design principles																				
Separation of concerns		✓	✓	✓		✓	✓				✓	✓					✓		✓	✓
Integration of concerns			✓			✓	✓	✓			✓	✓					✓	✓	✓	
To validate by simulation	✓	✓									✓	✓								
Human interaction	✓										✓	✓								
Non-functional issues	✓																			
Discrete/continuous dynamics							✓					✓					✓	✓		✓
Emergence						✓					✓				✓					
Non-functional requirements																				
Human safety			✓				✓				✓	✓			✓	✓		✓		
Evolution			✓	✓	✓						✓				✓					
Adaptability				✓	✓						✓	✓	✓							✓
Security		✓	✓	✓								✓					✓		✓	
Reliability		✓										✓				✓				
Resilience		✓				✓					✓	✓							✓	
Efficiency															✓	✓	✓			
Recommends agent orientation			✓			✓	✓		✓		✓		✓				✓	✓	✓	

These selected general engineering papers on cyber-physical systems present a high commonality on several challenges and desirable focuses for engineering methodologies. We grouped these common aspects into (1) requirements for the modeling language and, (2) requirements for activities to be part of the process,

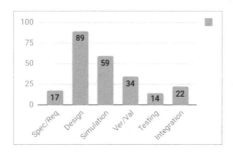
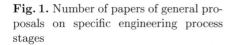
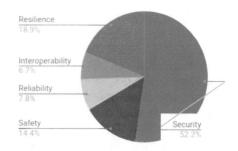

Fig. 1. Number of papers of general proposals on specific engineering process stages

Fig. 2. Number of papers of general proposals mentioning specific non-functional requirements

which is summarized in Table 1. In the last row we illustrate those proposals that have considered agent-orientation as a feasible solution for, at least, some particular aspect of a cyber-physical system.

3   Agent-Oriented Cyber-Physical Systems

Agent-oriented Software Engineering emerges under the hypothesis that it might significantly enhance modeling activities of complex and distributed systems by providing a high abstraction level of autonomous components whose collective behavior produces the desired result [16]. Therefore, multi-agent concepts, like modeling framework, were sustained because: (1) it is a natural metaphor, where interactions, either from humans or from software, could be seen as produced by agents. (2) It allows the distribution of data and/or control, thus each portion could be managed by an agent and the group should be cooperative to control the whole system. (3) Legacy systems, which need to integrate a global system, could be seen as agents with some kind of interaction. (4) Open systems may be integrated as autonomous components [44].

Tropos is a disseminated agent-oriented methodology [6] and it has adopted the i* modeling language, which includes agents, actors, dependencies, resources, tasks, goals, and softgoals [11]. Softgoal is a particular constructor for modeling non-functional requirements and other concepts which require overcoming some involved subjectivity. The aim of i* is to model mainly socio-technical systems. Tropos keeps this focus on its requirements stages and extends the aim of agent modelling to cover only technical components in its architectural stage. In addition, Tropos not only allows to model multi-agent systems but also integrates goal-orientation and other common representations in software engineering (UML interactions diagrams) as part of the Tropos's modeling languages. In the following lines we review how Tropos meets the collected requirements.

Most of the agent-oriented methodologies have structural components such as actors or agents. In the case of Tropos, it has both concepts, and at a high abstraction level, an actor may be a complete software system, an organization, or a human in the sense of a human-role (e.g. operator). Therefore using an actor is possible to model a human in the loop, and, using the concept of *agent*, it is possible to distribute the responsibilities (goals) of the system in different software pieces. The concept of *role* also exists, which is useful to describe common behaviors of different agents. In the sense of procedures (Tropos's plans), Tropos allows goals decomposition and plans decomposition, having the possibility of alternative ways of achieving a goal or implementing a procedure. In addition Tropos integrates UML diagrams for specific purposes, in particular, for expressing procedures (both cyber and physical ones). UML interaction diagrams are suggested for this purpose, thus branching and looping are included as part of Tropos methodology both not only to a high abstraction level, but also to an implementation level.

At a visual level, Tropos fosters most of the istar visual notation. In order to illustrate both the expressiveness of its visual notation and abstraction levels we present the arbiter case study, that is presented in [3]. This case is interesting because the original design is established with an explanation in natural language to a very high level, for example the explanation starts: *"The arbiter system provides a link between 3 master devices and 2 target devices. A link is established between the master and the target devices by the mediator. At a given time, only one master can conduct a read or write transaction and with only one target device.".* However, the resulting design is a very low level specification, for example, the following is a phrase which is part of the explanation of the mechanism: *"It sends an active low pulse on the "req" signal and waits for a "gnt." The "gnt" signal is an active low signal...".* In the middle, a lot of engineering decisions seem missed. The Tropos design for the dependencies between arbiter (as role) and master and target designs are shown in Fig. 3. How the arbiter must be implemented is specified in Fig. 4, thus it represents the internal specification for arbiters, in a different abstraction level, but still to a high level, here we have added a possible justification for a very low level implementation, which is the efficiency, as a non-functional requirement, which also illustrates a sample of requirements traceability in Tropos. In Fig. 5, we show a deployment for the arbiter system, only to this level the design is instantiated to specify that the system will operate with 3 master devices for 2 target devices. Therefore the implementation does not assume the original operation constraints and the resulting system would support other configurations of master and target devices.

On the side of those that we have called design principles, in terms of simulation on Tropos, we may refer to [29] that shows a system which allows modelling and simulation of airports and its airspace to help with decision making related to infrastructure and operations. However, there is a relevant difference between using an agent-oriented model to simulate and to test an agent-oriented design using a simulation of it. We do not know an agent-oriented methodology that includes at the metalevel constructs enabling the simulation of its agents' behaviours. Our conviction is that Tropos modeling language requires an extension for enabling simulation of their designs in a repeatable way.

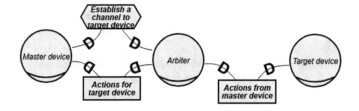

Fig. 3. General design of an arbiter agent

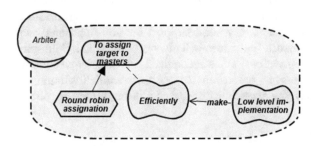

Fig. 4. Rationality of the arbiter design

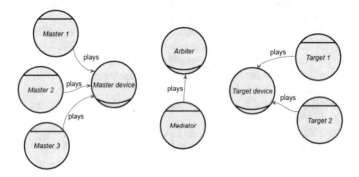

Fig. 5. Deployment with identifiable agents

Also in the case of particular non-functional requirements such as resilience, and human safety, we can affirm that, although Tropos address non-functional requirements at different stages, there are no particular answers to this set of non-functional requirements. A way of addressing this is by design patterns.

4   Conclusions

In this paper, we have shown the relevance of cyber-physical systems in new production and services trends. However, we have also shown that there is no systematic, disciplined and measurable engineering approach to build them. We have conducted a literature review to collect recent challenges for cyber-physical methodologies, which we have interpreted as requirements. As part of this analysis, we affirm that literature has been focused on what it needs more than on how to solve it.

On the solution side, we have shown that software engineering has historically addressed cyber and physical word and that agent orientation constitutes a proper design metaphor for distributed intelligent systems and, moreover, we have shown that Tropos meets most of the required features for developing cyber-physical systems.

On the limitations side, we recognize two relevant gaps, the modeling language does not have particular constructors enabling repeatable simulations and, although Tropos has a very good support for non-functional requirements, it is not enough to ensure that those will always be satisfied. Summarizing, Tropos as is, needs to supply most of the recognized methodology requirements, however, it also needs relevant extensions to both, its modeling language, and detailed phases to verify the accomplishments of non-functional requirements.

References

1. Abdullah, J., Dai, G., Guan, N., Mohaqeqi, M., Yi, W.: Towards a tool: times-pro for modeling, analysis, simulation and implementation of cyber-physical systems. In: Models, Algorithms, Logics and Tools, pp. 623–639. Springer (2017)
2. Akbari, O.Z.: A survey of agent-oriented software engineering paradigm: towards its industrial acceptance. Int. J. Comput. Eng. Res. 1(2), 14–28 (2010)
3. Aziz, M.W., Rashid, M.: Domain specific modeling language for cyber physical systems. In: 2016 International Conference on Information Systems Engineering (ICISE), pp. 29–33. IEEE (2016)
4. Bondavalli, A., Bouchenak, S., Kopetz, H.: Cyber-physical System of Systems. AMADEOS Project (2016)
5. Bourque, P., Fairley, R.E., et al.: Guide to the Software Engineering Body of Knowledge (SWEBOK (R)): Version 3.0. IEEE Computer Society Press (2014)
6. Bresciani, P., Perini, A., Giorgini, P., Giunchiglia, F., Mylopoulos, J.: Tropos: an agent-oriented software development methodology. Auton. Agent. Multi-Agent Syst. 8(3), 203–236 (2004)
7. Calvaresi, D., Marinoni, M., Sturm, A., Schumacher, M., Buttazzo, G.: The challenge of real-time multi-agent systems for enabling IOT and CPS. In: Proceedings of the International Conference on Web Intelligence, pp. 356–364. ACM (2017)
8. Cardenas, A.A., Amin, S., Sastry, S.: Secure control: towards survivable cyber-physical systems. In: 28th International Conference on Distributed Computing Systems Workshops, ICDCS 2008, pp. 495–500. IEEE (2008)
9. Cassandras, C.G.: Smart cities as cyber-physical social systems. Engineering 2(2), 156–158 (2016)
10. Di Martino, B., Li, K.C., Yang, L.T., Esposito, A.: Internet of Everything. Springer, Singapore (2018)
11. Eric, S.Y.: Social modeling and i^*. In: Conceptual Modeling: Foundations and Applications, pp. 99–121. Springer (2009)
12. Giaimo, F., Yin, H., Berger, C., Crnkovic, I.: Continuous experimentation on cyber-physical systems: challenges and opportunities. In: Proceedings of the Scientific Workshop Proceedings of XP 2016, p. 14. ACM (2016)

13. Gomez-Sanz, J.J., Fuentes-Fernández, R.: Understanding agent-oriented software engineering methodologies. Knowl. Eng. Rev. **30**(4), 375–393 (2015)
14. Gurgen, L., Gunalp, O., Benazzouz, Y., Gallissot, M.: Self-aware cyber-physical systems and applications in smart buildings and cities. In: Design, Automation & Test in Europe Conference & Exhibition (DATE), pp. 1149–1154 (2013)
15. Horváth, L., Rudas, I.J.: Engineering modeling for cyber physical systems. In: 2018 IEEE 16th World Symposium on Applied Machine Intelligence and Informatics (SAMI), pp. 000207–000212. IEEE (2018)
16. Jennings, N.R.: On agent-based software engineering. Artif. Intell. **117**(2), 277–296 (2000)
17. Jin, J., Gubbi, J., Marusic, S., Palaniswami, M.: An information framework for creating a smart city through internet of things. IEEE Internet Things J. **1**(2), 112–121 (2014)
18. Kathiravelu, P., Veiga, L.: SD-CPS: taming the challenges of cyber-physical systems with a software-defined approach. arXiv preprint arXiv:1701.01676 (2017)
19. Khalid, A., Kirisci, P., Ghrairi, Z., Thoben, K.D., Pannek, J.: A methodology to develop collaborative robotic cyber physical systems for production environments. Logistics Res. **9**(1), 23 (2016)
20. Laplante, P.A., et al.: Real-Time Systems Design and Analysis. Wiley, New York (2004)
21. Lee, E.A.: Cyber-physical systems - are computing foundations adequate? In: Position Paper for NSF Workshop On Cyber-Physical Systems: Research Motivation, Techniques and Roadmap, vol. 2, pp. 1–9. Citeseer (2006)
22. Lee, J., Bagheri, B., Kao, H.A.: A cyber-physical systems architecture for industry 4.0-based manufacturing systems. Manufact. Lett. **3**, 18–23 (2015)
23. Leitão, P., Colombo, A.W., Karnouskos, S.: Industrial automation based on cyber-physical systems technologies: prototype implementations and challenges. Comput. Ind. **81**, 11–25 (2016)
24. Lin, J., Sedigh, S., Miller, A.: Modeling cyber-physical systems with semantic agents. In: 2010 IEEE 34th Annual Computer Software and Applications Conference Workshops (COMPSACW), pp. 13–18. IEEE (2010)
25. Mangharam, R., Abbas, H., Behl, M., Jang, K., Pajic, M., Jiang, Z.: Three challenges in cyber-physical systems. In: 2016 8th International Conference on Communication Systems and Networks (COMSNETS), pp. 1–8. IEEE (2016)
26. Marwedel, P., Engel, M.: Cyber-physical systems: opportunities, challenges and (some) solutions. In: Management of Cyber Physical Objects in the Future Internet of Things, pp. 1–30. Springer (2016)
27. McKee, D.W., Clement, S., Almutairi, J., Xu, J.: Massive-scale automation in cyber-physical systems: vision & challenges. In: 2017 IEEE 13th International Symposium on Autonomous Decentralized System (ISADS), pp. 5–11. IEEE (2017)
28. Mezhuyev, V., Samet, R.: Metamodeling methodology for modeling cyber-physical systems. Cybern. Syst. **47**(4), 277–289 (2016)
29. Miller, T., Lu, B., Sterling, L., Beydoun, G., Taveter, K.: Requirements elicitation and specification using the agent paradigm: the case study of an aircraft turnaround simulator. IEEE Trans. Softw. Eng. **40**(10), 1007–1024 (2014)
30. Monostori, L.: Cyber-physical production systems: roots, expectations and R&D challenges. Procedia Cirp **17**, 9–13 (2014)
31. Mosterman, P.J., Zander, J.: Industry 4.0 as a cyber-physical system study. Softw. Syst. Model. **15**(1), 17–29 (2016)

32. Mueller, E., Chen, X.L., Riedel, R.: Challenges and requirements for the application of industry 4.0: a special insight with the usage of cyber-physical system. Chin. J. Mech. Eng. **30**(5), 1050 (2017)
33. Norris, D.: The Internet of Things: Do-It-Yourself at Home Projects for Arduino. Raspberry Pi, and BeagleBone Black. McGrow-Hill, San Francisco (2015)
34. Cyber-physical Systems (CPS). Technical report. NSF14-542, National Science Foundation (2014). https://www.nsf.gov/pubs/2014/nsf14542/nsf14542.pdf
35. Petersen, K., Feldt, R., Mujtaba, S., Mattsson, M.: Systematic mapping studies in software engineering. EASE **8**, 68–77 (2008)
36. Ribeiro, L., Björkman, M.: Transitioning from standard automation solutions to cyber-physical production systems: an assessment of critical conceptual and technical challenges. IEEE Syst. J., 1–13 (2017)
37. Ruchkin, I., Samuel, S., Schmerl, B., Rico, A., Garlan, D.: Challenges in physical modeling for adaptation of cyber-physical systems. In: 2016 IEEE 3rd World Forum on Internet of Things (WF-IoT), pp. 210–215. IEEE (2016)
38. Sanislav, T., Miclea, L.: Cyber-physical systems-concept, challenges and research areas. J. Control Eng. Appl. Inf. **14**(2), 28–33 (2012)
39. Seshia, S.A., Hu, S., Li, W., Zhu, Q.: Design automation of cyber-physical systems: challenges, advances, and opportunities. IEEE Trans. Comput.-Aided Des. Integr. Circuits Syst. **36**(9), 1421–1434 (2017)
40. Sutardja, S.: Slowing of Moore's law signals the beginning of smart everything. In: 2014 44th European Solid State Device Research Conference (ESSDERC), pp. 7–8. IEEE (2014)
41. Thomopoulos, S.: Cyber-physical systems challenges with information fusion. In: Signal Processing, Sensor/Information Fusion, and Target Recognition XXVI, vol. 10200, pp. 8–11. International Society for Optics and Photonics (2017)
42. Wolf, W.H.: Cyber-physical systems. IEEE Comput. **42**(3), 88–89 (2009)
43. Wooldridge, M.J., Jennings, N.R.: Software engineering with agents: pitfalls and pratfalls. IEEE Internet Comput. **3**, 20–27 (1999)
44. Wooldridgey, M., Ciancarini, P.: Agent-oriented software engineering: the state of the art. In: International Workshop on Agent-Oriented Software Engineering, pp. 1–28. Springer (2000)
45. Zeng, J., Yang, L.T., Lin, M., Ning, H., Ma, J.: A survey: cyber-physical-social systems and their system-level design methodology. Future Gener. Comput. Syst. (2016)

PSP-CI: A Tool for Collecting Developer's Data with Continuous Integration

Brian Pando[(⊠)] and Tony Ojeda

Universidad Nacional Agraria de la Selva, Carretera Central km. 1.21,
10131 Tingo María, Perú
{brian.pando, tony.ojeda}@unas.edu.pe

Abstract. Personal Software Process (PSP) is a method focused on the production of software from an individual perspective of quality. Its benefits in the improvement of the personal process are demonstrated by different researches. However, demanded documentation has stopped its implementations both in academy and industry, even though there are semi-automatic tools that try to solve the problem. This research work presents a tool based on Continuous Integration (CI), allowing to automate part of the PSP process. A preliminary experiment is carried out to validate the contribution in the software process, obtaining favorable results with almost no effort for the developer.

Keywords: Personal Software Process · Tool · Continuous Integration

1 Introduction

The contribution from PSP [1] both in the academy and the industry, is evidenced by multiple research such as [2, 3]. Demonstrating that there is a strong acceptance for wanting to implement PSP from the professional training processes [4], considering that applying it would double their performance [5]. The method covers a large part of the software process [6], but one of the requirements is the amount of documentation that the professional must collect [7], as well as mistakes can be made when collecting this data manually [8, 9]. Over 20 years, from 1997 [10] to 2018 [11], these tools have been proposed to collect this data manually and semi automatically with good results. Despite these contributions, it has not yet been made a full use of the method in the industry [12].

The cost of developing software falls 70% on the development team [13] and with it on the developers. The role of software engineer is to produce software with quality [14, 15], considering agility and productivity like [16, 17]. Some papers have proposed to measure the individual productivity of a developer from the perspective of agile methods like [18, 19], including a method proposed with Scrum and PSP for companies with only one developer [20].

Agile methods have been a solution strongly accepted by companies and one of the practices that this promotes with great expectation is the continuous integration (CI) [21]. This research paper presents a proposal to collect data in an automated way using CI. It presents in Sect. 2 the literature review and in Sect. 3 a comparison of tools

© Springer Nature Switzerland AG 2019
Á. Rocha et al. (Eds.): ICITS 2019, AISC 918, pp. 103–112, 2019.
https://doi.org/10.1007/978-3-030-11890-7_11

for PSP. Section 4 presents the architecture and description of the proposed tool. In Sect. 5, the results of a preliminary experiment to validate the proposal, and finally its conclusions and future work.

2   Literature Review

An exploratory review was carried out by using the phrase "Personal Software Process" in IEEE Xplore, ACM and Scopus, going through a refinement of repeated articles related to the subject. Resulting a total of 210 works since its publication in the 90s by Humphrey [22], with an average of 8 publications per year, indicating that the research on PSP is constant until today. Most of these investigations evaluate PSP in academics with 31% and some efforts to integrate it into the industry with 8%. 22% includes publications of evaluations, explanations, academic proposals and surveys regarding PSP. 17% of the publications extend the method to integrate it with other practices according to their context. From them, it is highlighted that works have also been found that propose methods for individual development considering agile approaches, like a popular methods XP [23, 24] and Scrum [20] (Fig. 1).

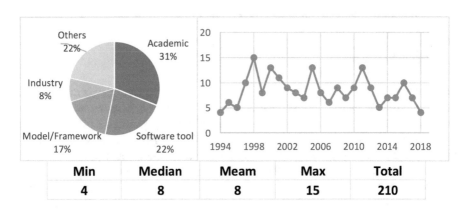

Min	Median	Meam	Max	Total
4	8	8	15	210

Fig. 1. Papers about PSP since the 90s. This shows a constant trend of research about PSP.

A 22%, so it means 47 publications propose or evaluate a software tool to support the PSP process manually and semi automatically, both to collect and analyze data collected. Interest in proposing software tools has been constant to date with an average of 2 publications per year. According to Fig. 2, tools for manual collection have been proposed initially, but since 2003 the semi-automatic collection proposals have started more frequently. In addition, there is a proposed tool for voice collection [25].

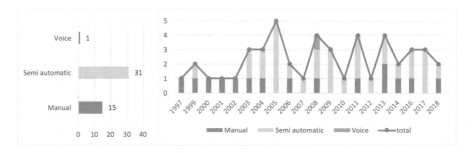

Fig. 2. Papers who propose or use software tools for PSP. This shows the constant effort to discover tools for PSP.

3 Mapping for Existing Tools

Table 1 presents a mapping of highlighted tools from the previous works, in which it indicates that 7 of the 15 tools allow integration with the IDE and some other components. Many of these tools require installation or configuration of a Plugin with the IDE, working with agents that help to collect data automatically, such as [3]. Probably due to incompatibility or extra effort or lack of knowledge, professionals do not use it [7].

Most tools cover the process and the artifacts requested by PSP. 8 tools are prepared for Windows environments, but 5 of them are prepared for web environments, allowing theses last ones, opportunity to work on any platform.

Table 1. Software tools for support PSP.

Software tool	Year	Env.	Coverage (1–5)	Collect	Integration
PSP studio	1997	Win	5	Manual	No
PSP DROPS	1997	Web	3	Manual	No
PSP tool	1998	Lin	3	Manual	No
Process dashboard	2002	Win/Lin	4	Manual	No
PSP palm	2003	Palm	5	Manual	No
PSPA	2005	Win/Lin	3	Semi auto.	IDE
PSP.Net	2005	Win/Lin	4	Semi auto.	No
Jasmine	2007	Win	4	Semi auto.	IDE
PSP student workbook	2009	Win	5	Manual	No
PSP EVA	2009	Web	3	Semi auto.	IDE
Hackystat	2013	Win	3	Semi auto.	IDE
PROM	2013	Web	3	Semi auto.	IDE
WBPS	2014	Web	5	Manual	No
SAPM	2014	Web	5	Manual	No
Alireza plugin	2018	Win	3	Semi auto.	IDE

There is another tool called ProcessPAIR [26], this tool continues to evolve since it appeared in 2012 [27], but its objective is to analyze data already collected by PSP. A tool is also proposed to support only the design phase [28]. Finally, it is mentioned that there are other alternatives to measure the performance of the developer [29].

Therefore, according to the reviewed works, there is a high interest in finding automatic solutions to data collection. Despite the efforts made, the problem has not been solved.

4   Proposed Tool

Most tools use multi-agents for the collection and this makes them dependent on the IDE and consumes resources from the developer's computer. This proposal has two steps for the developer, which consists of registering on the platform and activating a service in its Git repository. Once this is done, the tool does a whole work from inside between the repository, source code and the task manager as shown in Fig. 3.

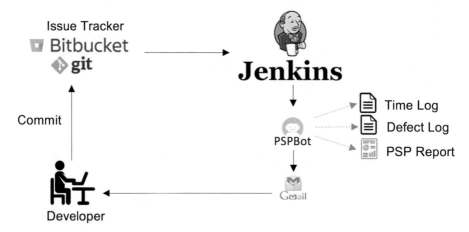

Fig. 3. Deployment of proposal tool. Shows the execution flow when the developer works on the software.

The proposal is based on an Atlassian tool, being one of the ALM leaders with the highest score in the Gartner quadrant [30] and one of the free tools offered is Bitbucket and the Git repository manager. The repository connects with Jenkins CI through Webhooks to notify new commits. Jenkins will connect with an application that synchronizes the Issue Tracker with the Time Log. It runs the unit tests to detect if any part of the code has stopped working and registers the bugs in the defect Log. Then report the new tasks in the Issue Tracker, as well as notify the results with an email to the user. In this way, the developer manages to comply with PSP0 without any effort, in addition to getting the most important PSP artifacts.

Fig. 4. PSP defect log on the platform. This shows an example report of defect log collected.

The information collected on the platform is private and stores the programmer's artifacts as historical information to compare itself over time or process and analyze its performance with other tools such as ProcessPair [26].

	● PSPBot					📋 ⏱ 🖲

PSP Timelog

Hi **tony-ojeda!**. This is the timelog of project process **001-ci-test** :

ID	Name	Description	Phase	Created	Solved	Hours
20	Vista de usuarios	Se codificara toda la vista del usuario, vista registro de material bibliográfico, acceso al material bibliografico, ...	codification	2018-08-28 10:08:48		
19	BasicTest: Failed asserting that false is true. in line 17	BasicTest: Failed asserting that false is true. in line 17	test	2018-08-28 00:04:29	2018-08-28 15:38:51	15 hr 34 min
18	BasicTest: Failed asserting that two strings are equal. in line 29	BasicTest: Failed asserting that two strings are equal. in line 29	test	2018-08-28 00:04:29	2018-08-29 12:38:51	36 hr 34 min
17	Realizar pruebas en vista usuarios	Se realizaran pruebas unitarias para garantizar el buen funcionamiento de las vistas usuarios.	test	2018-08-26 00:16:02		

Fig. 5. PSP time log on the platform. This shows an example report of time log collected.

The platform automatically collects the tasks and bugs that are reported in the issue tracker, as well as update the states if they have been resolved. With the collected data, take control of the phases and times to solve them as shown in Figs. 4 and 5.

Finally, with these data of defects and activities carried out automatically, the PSP report is filled, according to the format it demands, and at the same time be a measurement tool and calculations for the developer and inject a culture of measurement to pursue quality (Fig. 6).

PSP Report

Hi tony-ojedal. This is the report of project process 001-ci-test :

Project Plan Summary

User:	tony-ojeda	Date:	16-09-2018
Software:	001-ci-test	Number:	—
Lenguage:	PHP		

Time in Phase (hr.)	Plan	Current	To date	To date %
Planning		679 hr 23 min	679 hr 23 min	18.75%
Design		2003 hr 14 min	2682 hr 37 min	74.05%
Codification		669 hr 44 min	3352 hr 22 min	92.54%
Compilation		0 hr 0 min	3352 hr 22 min	92.54%
Test		52 hr 8 min	3404 hr 30 min	93.98%
Postmortem		218 hr 15 min	3622 hr 46 min	100%
Total	1 Month 1/2	3622 hr 46 min	—	—

Defects found	Plan	Current	To date	To date %
Planning		0	0	0%
Design		0	0	0%
Codification		0	0	0%
Compilation		0	0	0%
Test		3	3	100%
Postmortem		0	3	100%
Total	30 bug	3	—	—

Defects Removed	Plan	Current	To date	To date %
Planning		0	0	0%
Design		0	0	0%
Codification		0	0	0%
Compilation		0	0	0%
Test		2	2	100%
Postmortem		0	2	100%
Total	30 bug	2	—	—

Fig. 6. PSP report on the platform. This shows an example of the report according to PSP.

The proposal allows that the collection does not depend on the IDE, allowing developers who use Sublime Text, Visual Code or others, to integrate to the quality with PSP, for example, PHP developers.

5 Experiment

To validate the proposed tool, a preliminary experiment was done with six junior professionals with Laravel 5, with at least 1 year of industry experience in software development. Before starting the experiment, an initial talk was given about the benefits of measuring the software process and about unit tests. Then the repositories were created for each one and the Webhook was configured to activate the synchronization with the platform (Fig. 7).

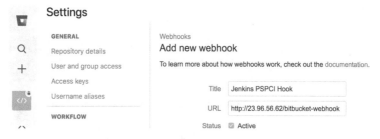

Fig. 7. Webhook on Bitbucket. Shows the configuration for the experiment on each repository.

To execute the experiment, 2 exercises were delivered to solve each professional. The second exercise should generate an impact on the first to ensure the injection of errors. Finally, they were asked to answer a questionnaire adapted from TAM model. The people evaluated responded according to a Likert scale of 1 to 5.

6   Limitations of the Research

The experiment was carried out in a company with 6 members, it is necessary to experiment with a larger number of users.

7   Results and Discussions

47 of 210 reviewed papers propose software tools for PSP, however, no evidence was found that they have generated a strong presence in the software industry. Of them, 31 use automatic processes, mostly based on multi agents, such as [3, 4, 11].

Despite the efforts of [17, 20, 23, 24], to integrate PSP to agile approaches, it is necessary to have tools that automate the data collection work [7], due to the extra effort that requires and errors from manual collection [8, 9]. All the reviewed research supports the importance of PSP in the software process and how it influences quality as presented [14, 15]. However, the use of an automated tool could help fulfill the double production as stated [5]. This work take advantage using the learning of others works presented on Sect. 3, resolving the time for data collection with almost no effort for the developer.

The results show that the perceived utility, the perceived ease of use and attitude toward the use mostly exceed the value of 4, indicating that the tool would add value to their work and that it would be easy to integrate it, in addition to that they all consider that It would be very useful if this tool would be integrated from professional training. Also indicate that users apparently do not perceive any extra effort to use the tool and allows them to increase their productivity in software development (Fig. 8).

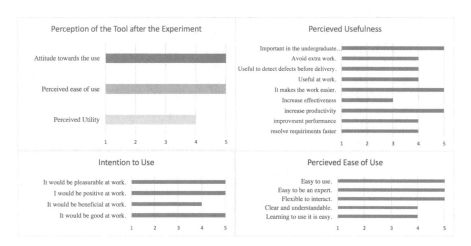

Fig. 8. Valuation by the professionals after the experiment. This shows favorable results about the tool.

In summary, the valuation is positive towards the proposed platform, indicating a rating greater than 4 in the 3 evaluated dimensions, being this result very encouraging to continue complementing actions to the platform to achieve more PSP levels in the most automated and intuitive way possible.

8   Conclusions

For many years, ideas have been promoted to automate the PSP process, but the problem has not been completely solved yet. The works reviewed show that the problem of collecting PSP data persists, being one of the factors for which its use is abandoned.

There are many efforts from the academy to promote the use of PSP [12] but it requires tools that support the process, thus integrating more strongly to the industry. As reviewed in Sect. 3, a feature of most tools is the use of multi agents, and these depend on IDE, however, the current proposal proposes to use the practice of CI to automate the collection process in 2 steps.

The results of the experiment show an encouraging data by indicating that the tool would be very useful for the academy and for the industry and these results should be tested in these scenarios to verify and extend their functionalities until they reach a mature and collaborative tool.

9   Future Works

The next step of this work is to test the tool by an experiment in a software development course to evaluate the impact on student productivity. The proposed tool will continue to evolve to meet more PSP levels in the simplest, most intuitive and automated way possible. Proposals must be generated so that the tool complements personal components such as smartphones to extend functionalities. The partial insertion of PSP should be sought in the curricular meshes of careers related to software engineering to raise awareness among professionals of the benefits of measuring individual performance and increase job opportunities [31] and even entrepreneurship [32]. It is important to continue looking for ways to facilitate the collection of PSP data, whether with applications, methods, frameworks or others, but these proposals must continue to evolve and not remain in their first version.

ProcessPair [26] is an example of which tools must evolve. The proposed tool could be integrated with other applications through micro-services, thus ensuring compatibility with other applications.

References

1. Pomeroy-Huff, M., Cannon, R., Chick, T.A., Mullaney, J.L., Nichols, W.R.: The Personal Software Process (PSP) Body of Knowledge, Version 2.0, Massachuset (2010)
2. Callison, R., MacDonald, M.: A Bibliography of the Personal Software Process (PSP) and the Team Software Process (TSP). Software Engineering Institute (2009). https://resources.sei.cmu.edu/library/asset-view.cfm?assetid=8929. Accessed 11 Aug 2018

3. Marcos, A., Santos, A.B., Valêncio, C.R., Dos, V., Borges, A.: PSP support component integrated into a web project management environment (2014)
4. Nasir, N.M., Fauzi, S.S.M.: Implementation of the Personal Software Process in Academic Settings and Current Support Tools. IGI Global (2014)
5. Paulk, M.C.: Factors Affecting Personal Software Quality (2006)
6. Angarita, L.B., Tovar, J.P.L.: Evaluation of design and code revisions in academic practices of software engineering. In: 2017 6th International Conference on Software Process Improvement (CIMPS), pp. 1–8 (2017)
7. Gasca-Hurtado, G.P., Alvarez, M.C.G., Manrique-Losada, B., Arias, D.M.: Diagnostic on teaching-learning of software desing by using the personal software process framework. In: 2015 10th Iberian Conference on Information Systems and Technologies (CISTI), pp. 1–7 (2015)
8. Disney, A.M., Johnson, P.M.: Investigating data quality problems in the PSP. In: Proceedings of the 6th ACM SIGSOFT International Symposium on Foundations of Software Engineering - SIGSOFT 1998/FSE-6, vol. 23, no. 6, pp. 143–152 (1998)
9. Hairul, M., Nasir, N.M., Yusof, A.M.: Automating a modified personal software process. Malaysian J. Comput. Sci. **18**(2), 11–27 (2005)
10. Syu, I., Salimi, A., Towbidnejad, M., Hilburn, T.: A web-based system for automating a disciplined personal software process (PSP). In: Proceedings Tenth Conference on Software Engineering Education and Training, pp. 86–96 (1997)
11. Joonbakhsh, A., Sami, A.: Mining and extraction of personal software process measures through IDE interaction logs. In: Proceedings of the 15th International Conference on Mining Software Repositories - MSR 2018, pp. 78–81 (2018)
12. Pando, B., Rodríguez, G.: PSP: how well it matches the skill set sought by the software industry in Latin America? A proposal to incorporate PSP into undergraduate SE education. In: Proceedings of Jornadas Iberoamericanas de Ingeniería de Software e Ingeniería del Conocimiento 2018 (JIISIC 2018) (2018)
13. Salinas, E., Cerpa, N., Rojas, P.: A service oriented architecture for the implementation of the personal software process. Rev. Chil. Ing. **19**(1), 40–52 (2011)
14. Mansoor, S., Bhutto, A., Bhatti, N., aamir Patoli, N., Ahmed, M.: Improvement of students abilities for quality of software through personal software process. In: 2017 International Conference on Innovations in Electrical Engineering and Computational Technologies (ICIEECT), pp. 1–4 (2017)
15. Gomez-Alvarez, M.C., Gasca-Hurtado, G.P., Manrique-Losada, B., Arias, D.M.: Method of pedagogic instruments design for software engineering. In: 2016 11th Iberian Conference on Information Systems and Technologies (CISTI), pp. 1–6 (2016)
16. Stark, J.A., Crocker, R.: Trends in software process: the PSP and agile methods. IEEE Softw. **20**(3), 89–91 (2003)
17. Unluturk, M.S., Kurtel, K.: Quantifying productivity of individual software programmers: practical approach. Comput. Inform. **34**(4), 959–972 (2016)
18. Barrera, J.A.H.: Integrating software development frameworks: Scrum, PSP and ISO25000. Vent. Informática **32**, 151–164 (2015)
19. Bernabé, R.B., Navia, I.Á., García-Peñalvo, F.J.: Faat: freelance as a team. In: Proceedings of the 3rd International Conference on Technological Ecosystems for Enhancing Multiculturality - TEEM 2015, pp. 687–694 (2015)
20. Pagotto, T., Fabri, J.A., Lerario, A., Goncalves, J.A.: Scrum solo: software process for individual development. In: 2016 11th Iberian Conference on Information Systems and Technologies (CISTI), pp. 1–6 (2016)
21. Fitzgerald, B., Stol, K.-J.: Continuous software engineering: a roadmap and agenda. J. Syst. Softw. **123**, 176–189 (2017)

22. Humphrey, W.S.: The personal process in software engineering. In: Proceedings of the Third International Conference on the Software Process. Applying the Software Process, pp. 69–77 (1994)

23. Iqbal, N., ul Hassan, M., Rehman Osman, A., Ahmad, M.: A framework for partial implementation of PSP in extreme programming. Int. J. Eng. Res. Appl. **3**(2), 604–607 (2013). http://www.ijera.com/

24. Mihaylov, I., Ivanov, P., Stefanova, E., Eskenazi, A., Ilieva, S.: The expert approach: a case study. In: Proceedings of the 4th International Conference on Computer Systems and Technologies e-Learning - CompSysTech 2003, pp. 101–106 (2003)

25. Ibrahim, A., Choi, H.-J.: A framework for analyzing activity time data. In: 2008 IEEE International Symposium on Service-Oriented System Engineering, pp. 14–18 (2008)

26. Raza, M., Faria, J.P., Salazar, R.: Helping software engineering students analyzing their performance data: tool support in an educational environment. In: 2017 IEEE/ACM 39th International Conference on Software Engineering Companion (ICSE-C), pp. 241–243 (2017)

27. Duarte, C.B., Faria, J.P., Raza, M.: Process PAIR: automated personal software process performance analysis and improvement recommendation. In: 2012 Eighth International Conference on the Quality of Information and Communications Technology, pp. 131–136 (2012)

28. Chaiyo, Y., Ramingwong, S.: The development of a design tool for personal software process (PSP). In: 2013 10th International Conference on Electrical Engineering/Electronics, Computer, Telecommunications and Information Technology, pp. 1–4 (2013)

29. Corley, C.S., Lois, F., Quezada, S.: Web usage patterns of developers. In: 2015 IEEE International Conference on Software Maintenance and Evolution (ICSME), pp. 381–390 (2015)

30. Gartner: Magic Quadrant for Enterprise Agile Planning Tools (2018). https://www.gartner.com/doc/reprints?id=1-4YX5XH0&ct=180509&st=sb. Accessed 08 Aug 2018

31. Swamidurai, R., Umphress, D.: Engaging students through practitioner centered software engineering. In: SoutheastCon 2015, pp. 1–5 (2015)

32. Jirapanthong, W.: Personal software process with automatic requirements traceability to support startups. J. Rev. Glob. Econ. **6**, 367–374 (2017)

Data Management Infrastructure from Initiatives on Photovoltaic Solar Energy

E. Jiménez-Delgado[1(✉)], C. Meza[2], A. Méndez-Porras[1], and J. Alfaro-Velasco[1]

[1] Computer Engineering Department,
Instituto Tecnológico de Costa Rica, Cartago, Costa Rica
{efjimenez,amendez,joalfaro}@itcr.ac.cr
[2] Electronic Engineering Department,
Instituto Tecnológico de Costa Rica, Cartago, Costa Rica
cmeza@itcr.ac.cr

Abstract. Photovoltaic solar energy is a renewable electricity generation resource with very low environmental impact. A photovoltaic installation can be implemented using a variety of brands and equipment models. This becomes a challenge in terms of data collection, storage and analysis for its effective correlation and further generation of information.

This article describes a project which objectives are as follows to contribute to data management in the field of energy resources based on photovoltaic technologies from different manufacturers through a technological infrastructure for Data Management. The approach utilized include three different stages: (1) analysis of project-oriented technologies in the field of energy resources and sustainability, (2) design and development of an infrastructure for data management of photovoltaic panels from different manufacturers, and (3) data collection from the different suppliers that were installed.

Two data collection proposals were designed based on the energy supplying companies (Enphase Energy and SMA Solar Technology) over a one-year term with around two million records. Data is managed through the cloud infrastructure offered by Amazon and a visualization system developed through open source software.

Keywords: Data management · Photovoltaic data ·
Solar energy data · Infrastructure data · Cloud infrastructure ·
Visualization system

1 Introduction

Photovoltaic solar energy is a source for renewable electricity generation, characterized by its low environmental impact. The photovoltaic installations can be modular and scalable, which allows their implementation in different settings. Photovoltaic solar energy has a low energy density, which requires large extensions of space to develop energy. In this way, a photovoltaic solar installation

© Springer Nature Switzerland AG 2019
Á. Rocha et al. (Eds.): ICITS 2019, AISC 918, pp. 113–121, 2019.
https://doi.org/10.1007/978-3-030-11890-7_12

of 10 MW requires thousands of photovoltaic modules and an area of 10 ha approximately. Under these conditions, supervising and identifying failures in the modules becomes an interesting challenge ideally requiring the processing of the electrical and thermal data of the thousands of modules involved in the installations.

The objective of this project is to contribute with data management in the field of energy resources based on photovoltaic technologies through a technological infrastructure for data management. The methodology used is based on an analysis of project-oriented technologies in the field of energy resources and sustainability. This approach was divided into three different stages: (1) analysis of project-oriented technologies in the field of energy resources and sustainability, (2) designing and development of an infrastructure for data management of photovoltaic panels from different manufacturers, and (3) data collection of the different suppliers that were installed. The key contribution of this work is that it proposes the design of a software infrastructure based on data science for the management of information from photovoltaic panels from different manufacturers. In this infrastructure some modules can be added to collect data from panels from different providers and also users can visualize data transparently at the same software platform.

This article is organized as follows: Sect. 2 analyzes related work, Sect. 3 proposes the data science infrastructure, Sect. 4 describes the data collected, Sect. 5 shows the results obtained and the discussion. Finally, Sect. 6 states the conclusions.

2   Related Research Works

Renewable energies play an undeniable role in sustainable practices and carbon neutrality (in the reduction of the carbon footprint, in other contexts), the energy requirements increase actually in every country as their industry and population expand, thus favoring the search for alternatives that supply this growing demand and basically through secure and sustainable sources [1]; projects and solutions that in the past were the most popular initiatives from institutions and government agencies are currently deployed interchangeably in terms of latitudes and economies [2].

Solar, wind, hydro and geothermal energy sources are used for the production of electricity provided to the supply networks, at the same time constituting the advent of diverse solutions in magnitude of the production, physical size of the facilities, technology and location [2]. In the case of photovoltaic systems, these assertions are reflected in plants with hundreds of thousands of panels in large areas in open fields, including designs aimed at urban and residential centers; in the case of technologies, there is also a proliferation of solutions for the capture, conversion, configuration, monitoring and distribution of energy.

Photovoltaic solutions require effective monitoring for their actions [1, 3–6], in this sense, information systems provide opportunities for capturing and analyzing electrical (current, voltage) and meteorological (radiation, ambient temperature, direction-wind speed, humidity, barometric pressure, others) variables.

However, the intervention-interaction of photovoltaic solutions from different manufacturers also involves heterogeneous configurations and technologies [2,4,7].

The courses of action for capturing, monitoring and analyzing data take aspects and multiple variables from the perspective of the mechanisms of capturing information, storaging technology, analysis, visualization as well as implementation costs, setting and maintenance. Rezk et al. [8] propose a data acquisition system (DAQS by its acronym in English) which is effective in terms of costs utilizing[1]. Although the system focuses on obtaining, storing and displaying the variables related to voltage, current and energy in a computer only, it represents an alternative in terms of design and implementation of hardware/software solutions for general application. Mahzan et al. [9] developed a data logger from an Arduino Mega 2560 micro controller[2], allowing the capture and storage of electrical parameters with a general and low cost solution compared to specialized data loggers.

The monitoring systems developed by inverter manufacturers offer several advantages, highlighting an easy configuration, compatibility, reliability and the ability to visualize data through web applications. However, by their nature, these tools commonly known in the field of software development as "closed systems" tend to convert their advantages into disadvantages by being part of photovoltaic solutions that integrate different technologies. The incompatibility with other manufacturers, storage in local medium or non-standardized formats, limited parameterization and inability to consult/export data through external processes are some of the disadvantages.

A proposed system for data acquisition and storage in the cloud for photovoltaic systems uses embedded systems (microcontrollers) and the RS485 standard[3] [10]. This allowed communication with investors, enabling the storage of data in online repositories for effective viewing in web and mobile clients.

Saraiva et al. [11] demonstrate the capacity of platforms based on Arduino microcontrollers and free services for document storage and management in the cloud, for the acquisition and monitoring of photovoltaic systems, as an alternative to dedicated systems (designed exclusively for a technological platform), that are more expensive and limited for the proposed requirements.

Reliability and performance of photovoltaic systems depend on monitoring processes that allow anomalies to be detected due to module failures or suboptimal situations (partial shadows, dirt, hot spots, others). The approaches to the diagnosis of failures are diverse based on their procedures and resources; practical techniques that compare operation variables during regular operation and defective operations [6], use of neural networks [12], inspection by unmanned aerial vehicles [13], use of protocols Wireless communication [14,15], contribute to the improvement of inspection processes.

---

[1] http://www.ni.com/es-cr/shop/labview.html.

[2] https://www.arduino.cc/en/Guide/ArduinoMega2560.

[3] http://www.ti.com/lit/an/slla272c/slla272c.pdf.

In general terms, the conditions derived from the physical size, location and technologies involved, pose challenges of interest in the detection of failures, supervision and integral management of photovoltaic solutions.

3   Data Science Infrastructure

At the Sustainability project at TEC, an inter-departmental and inter-school initiative of the "Tecnológico de Costa Rica", two photovoltaic installations connected to the grid have been designed and implemented at the Central Campus (Cartago province, central region of the country) and at the Regional Headquarters in San Carlos (northern region of the country). These facilities involve the interaction of photovoltaic solutions from different manufacturers, technologies and operating characteristics, continuously generating heterogeneous data volumes. In these photovoltaic installations there are panels manufactured by the companies Enphase Energy[4] and SMA Solar Technology[5].

Figure 1 shows the proposed infrastructure for the management of data from photovoltaic solutions of different technologies. In the center of the figure there is an information repository scheme based on an Amazon Web Service[6] (AWS, for its acronym in English). This service offers computing power, database storage, content delivery and other functionalities that allow the infrastructure to scale and grow. On the right you can see the modules for data collection from panels manufactured by Enphase Energy and SMA Solar Technology. The infrastructure offers the possibility of adding new data collection modules for other types of solar panel technologies. Each of these modules has a custom communication interface (CCI) that allows communication with Amazon Web Services. On the left there is a web platform that allows the visualization of the collected data and the creation of predictions of future behavior of photovoltaic installations.

3.1   Strategy for the Collection and Storage of Information of the Panels Manufactured by Enphase Energy Company

At the main campus of the "Tecnológico de Costa Rica", 32 photovoltaic panels of the company Enphase are installed, each one connected to a micro-inverter. The micro-investors offer the possibility to work on electrical connection on an individual basis or in groups. Therefore, if a micro-inverter fails, it does not produce large losses because the rest of the group's micro-investors continue to work together. Another advantage over traditional investors lies on the ease of identifying problems in the operation of panels. The micro-investors are responsible for storing and sending data to a storage service of the company Enphase Energy.

Through a *script* in the Python language, the data of each of the micro-investors is obtained from the Enphase website and sent to the data deposit scheme to the local server through the customized communication interface.

---

[4] https://enphase.com/en-us.

[5] https://www.sma.de/en/.

[6] https://aws.amazon.com.

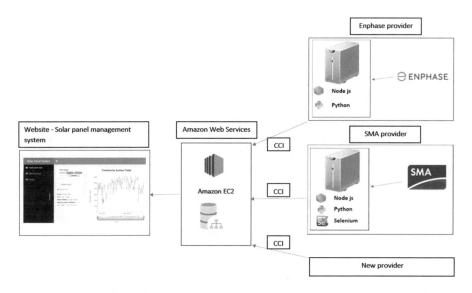

Fig. 1. Diagram of the data infrastructure for information management from photovoltaic solutions of different technologies.

3.2 Strategy for the Collection and Storage of Information of the Panels Manufactured by the Company SMA Solar Technology

At the San Carlos "Tecnológico de Costa Rica" campus, there are 72 photovoltaic panels installed from the manufacturer SMA Solar Technology, connected to 6 inverters. They store and send the production data in kW periodically, in intervals of 5 min. Each investor has a Web Server that responds to requests through the protocol *http* and displays on a Web site the general information of the status and kilo-Watts obtained.

Through a *Script* implemented in Python language and executed automatically through the Selenium tool, data of each of the investors is obtained from the internal website that each of them provides. The data acquired from the devices are stored in a local server located at the university campus, through the customized communication interface (CCI).

4 Data Collection

Data from Enphase Energy and SMA Solar Technology are stored on a PostgreSQL data server. This team is also responsible for storing the proposed CIC strategies and those of new suppliers. The data schema is constituted by the columns: system number, date, inverter number and power generated. These characteristics are those expected for the construction of a personalized communication interface of a new provider.

The local server has a script made in Node.js responsible for sending data every 5 min to the Amazon EC2 server. This has an API-REST service developed in Node.js, which allows universal backup and access to stored data, which otherwise could not be accessed on the local server (via HTTP) due to security restrictions at the university. The Web site performs the data query to the Amazon Server, at the same time when it complements the results with graphics of the relevant information that comes from the CCI. Currently, the infrastructure has more than two million records of the different offices, generated in a period of one year.

Table 1 shows in the first row a total of Megabits of information without database optimization structures with a total of 159 Megabytes with an average of 68.29 bytes per record. The second row shows the total of 198 megabytes used by database indexes to optimize searches with an average of 85.21 bytes per record. In the third row shows the total sum of Megabits in the table with 357 Megabytes and with an average of 153.51 bytes per record.

Table 1. Data storage review from the photovoltaic system.

Description	Total megabytes	Bytes by row (average)
Table with toast size - (compress)	159	68.2989
Indexes size	198	85.2161
Total size table	357	153.5150

The data set contains 2330626 records so far among the 38 inverter that send the information from the ICC during a year. The Amazon cloud storage server has a 30 GB solid state disk dedicated to storage that would allow, if the data production keeps storage constant for at least eleven years. The RAM that this server has is 2 GB, so it would not be able to show reports or predictions where more of this available memory is consumed at this time.

5   Results and Discussion

An initial version of the infrastructure has been implemented, for the storage of data coming from the investors connected to the electricity network at Cartago and San Carlos campuses of the "Tecnológico de Costa Rica".

Despite the initial conditions of operation of the infrastructure, there is evidence of the competence to manage the volume of data from different CCIs, which maintains constant growth. As other CCIs are added to the system, verification of management capabilities will be possible.

An information repository scheme has been designed on the Amazon virtualization cloud, using high availability and backup tools to provide robustness and consolidation of the reports produced by the generation matrix in the photovoltaic panels. This scheme is implemented using the tools PostgreSQL, Node.js, Python and Putty.

The advantage of the design based on Amazon EC2 lies in the possibility of scaling in a controlled basis, without incurring in the purchase of hardware and software resources for data management.

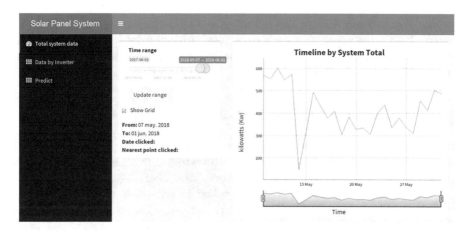

Fig. 2. Screenshot of the website - Timeline by system total

A preliminary visualization process has been developed from the analysis of data present in the storage infrastructure, facilitating the analysis of general variables of the photovoltaic systems, as well as the individual production of the investors.

The visualization system provides administrators with the possibility of showing concise reports of the total and individual production. However, with the possibility of adding new CICs or suppliers, other more complex reports must be designed to help improve decision making process.

The Fig. 2 shows the session of the site that indicate the total production of the energy matrix using a range of dates with the possibility of knowing the highest and lowest values of production during the day or at a specific date and time.

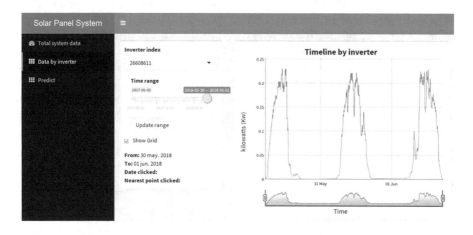

Fig. 3. Screenshot of the website - Timeline by inverter

The Fig. 3 shows the session of the site that indicates the production by inverter using a range of dates with the possibility of knowing the highest and lowest production values during the day or at a specific date and time in addition to selecting the study inverter.

6   Conclusions and Future Work

An initial version of infrastructure for the storage of data has been implemented from the different sources coming from the investors connected to the electricity network at Cartago and San Carlos campus of the "Tecnológico de Costa Rica".

Software technologies are generally applied in renewable energy projects as simple support solutions in the capture and storage of data. Although these types of solutions fulfill their purpose, they are conceptualized from a non-systematic perspective, in order to provide proactive mechanisms for data management from multiple photovoltaic technologies. In this sense, the project aims to contribute to the conceptualization of a generalized solution for the capture, storage, management and analysis of data coming from multiple photovoltaic solution providers.

An information repository scheme has been designed on the Amazon virtualization cloud, using high availability and backup tools to provide robustness and consolidation of the reports produced by the generation matrix in the photovoltaic panels. This scheme is implemented using the tools PostgreSQL, Node.js, Python and Putty. The photovoltaic panels are connected to an inverter that is in charge of storing and sending data. Each investor model has its own method of sending and storing data.

The collection of data generated by photovoltaic panels is complex because each manufacturer has its own communication and access standards. It was necessary to implement a specific data collection technique for each type of panels.

Identifying negative production conditions in correlation to the set of data periodically produced by the investors are variables required for a broader monitoring in order to optimize the energy obtained from the photovoltaic panels.

With the amount of data obtained by the photovoltaic panels, the Big Data application can be applied in the generation of new data analysis models that allow automated monitoring and predictions with more intelligent algorithms.

In the future it is proposed to use artificial vision and automatic learning to determine the physical location of the panels and detect their external failures.

References

1. Madeti, S., Singh, S.: Monitoring system for photovoltaic plants: a review. Renew. Sustain. Energy Rev. **67**, 1180–1207 (2017)
2. Zhou, B., Li, W., Chan, K., Cao, Y., Kuang, Y., Liu, X., Wang, X.: Smart home energy management systems: concept, configurations, and scheduling strategies. Renew. Sustain. Energy Rev. **61**, 30–40 (2016)

3. Dhimish, M., Holmes, V., Mehrdadi, B., Dales, M.: Simultaneous fault detection algorithm for grid-connected photovoltaic plants. IET Renew. Power Gener. **11**(12), 1565–1575 (2017)
4. Rahman, M., Selvaraj, J., Rahim, N., Hasanuzzaman, M.: Global modern monitoring systems for PV based power generation: a review. Renew. Sustain. Energy Rev. **82**, 4142–4158 (2018)
5. Triki-Lahiani, A., Bennani-Ben Abdelghani, A., Slama-Belkhodja, I.: Fault detection and monitoring systems for photovoltaic installations: a review. Renew. Sustain. Energy Rev. **82**, 2680–2692 (2018)
6. Yahyaoui, I., Segatto, M.: A practical technique for on-line monitoring of a photovoltaic plant connected to a single-phase grid. Energy Conv. Manag. **132**, 198–206 (2017)
7. Bottaccioli, L., Patti, E., Grosso, M., Rasconà, G., Marotta, A., Rinaudo, S., Acquaviva, A., Macii, E.: Distributed software infrastructure for evaluating the integration of photovoltaic systems in urban districts. In: 5th International Conference on Smart Cities and Green ICT Systems, SMARTGREENS 2016, pp. 357–362 (2016)
8. Rezk, H., Tyukhov, I., Al-Dhaifallah, M., Tikhonov, A.: Performance of data acquisition system for monitoring PV system parameters. Meas. J. Int. Meas. Confederation **104**, 204–211 (2017)
9. Mahzan, N., Omar, A., Rimon, L., Noor, S., Rosselan, M.: Design and development of an arduino based data logger for photovoltaic monitoring system. Int. J. Simul. Syst. Sci. Technol. **17**(41), 15.1–15.5 (2017)
10. Treter, M., Pietta, L., Xavier, P., Michels, L.: Data acquisition and cloud storage system applied photovoltaic systems. In: IEEE 13th Brazilian Power Electronics Conference and 1st Southern Power Electronics Conference. COBEP/SPEC 2016 (2015)
11. Saraiva, L., Alcaso, A., Vieira, P., Ramos, C., Cardoso, A.: Development of a cloud-based system for remote monitoring of a PVT panel. Open Eng. **6**(1), 291–297 (2016)
12. Chine, W., Mellit, A., Lughi, V., Malek, A., Sulligoi, G., Massi Pavan, A.: A novel fault diagnosis technique for photovoltaic systems based on artificial neural networks. Renewable Energy **90**, 501–512 (2016)
13. Aghaei, M., Grimaccia, F., Gonano, C., Leva, S.: Innovative automated control system for PV fields inspection and remote control. IEEE Trans. Ind. Electron. **62**(11), 7287–7296 (2015)
14. Batista, N., Melício, R., Matias, J., Catalão, J.: Photovoltaic and wind energy systems monitoring and building/home energy management using zigbee devices within a smart grid. Energy **49**(1), 306–315 (2013)
15. Shariff, F., Rahim, N., Hew, W.: Zigbee-based data acquisition system for online monitoring of grid-connected photovoltaic system. Expert Syst. Appl. **42**(3), 1730–1742 (2015)

Fuzzy Knowledge Discovery and Decision-Making Through Clustering and Dynamic Tables: Application in Medicine

Yamid Fabián Hernández-Julio[1]([⊠]), Helmer Muñoz Hernández[1],
Javier Darío Canabal Guzmán[1], Wilson Nieto-Bernal[2],
Romel Ramón González Díaz[1], and Patrícia Ponciano Ferraz[3]

[1] Faculty of Economics, Administrative and Accounting Sciences,
University of the Sinú Elías Bechara Zainúm, Montería, Córdoba, Colombia
yamidhernandezj@unisinu.edu.co
[2] Systems Engineering and Computation,
University of the North, Barranquilla, Colombia
[3] Engineering Department, Universidade Federal de Lavras,
Lavras, Minas Gerais, Brazil

Abstract. The objective of this study was to design, to implement and to validate a framework for the development of decision system support based on fuzzy set theory using clusters and dynamic tables. To validate the proposed framework, a fuzzy inference system was developed with the aim to classify breast cancer and compared with other related works (Literature). The fuzzy Inference System has three input variables. The results show that the Kappa Statistics and accuracy were 0.9683 and 98.6%, respectively for the output variable for the Fuzzy Inference System – FIS, showing a better accuracy than some literature results. The proposed framework may provide an effective means to draw a pattern to the development of fuzzy systems.

Keywords: Clustering · Dynamic tables · Fuzzy sets · Breast cancer

1 Introduction

Cancer is a group of diseases that cause cells in the body to change and spread out of control [1]. In 2018, according to [2], 1,735,350 new cancer cases and 609,640 cancer deaths are projected to occur in the United States. One of the most common and deadly cancer for women is Breast cancer [3]. Approximately 40,920 deaths from invasive female breast cancer are predicted in the United States by the American Cancer Society (ACS) [1]. For those reasons, it is critical to developing some tools (models) that assist in decision-making for early detection, appropriate therapy and treatment [4] with the aim to obtain a real diagnosis. The use of fuzzy logic is on the rise due to the great possibilities it offers in different knowledge areas [5]. Regarding the medicine area, there are very interesting works that aim the management of uncertainties in the field. It has been used for the classification of fibromyalgia syndrome [6], breast cancer classification [7, 8], among others. For the reasons above, the main objective of this study

© Springer Nature Switzerland AG 2019
Á. Rocha et al. (Eds.): ICITS 2019, AISC 918, pp. 122–130, 2019.
https://doi.org/10.1007/978-3-030-11890-7_13

was to design, to implement and to validate a framework for the development of decision system support based on fuzzy set theory using clusters, and dynamic tables for classification problems. To validate the proposed framework, one fuzzy inference system was developed with the aim to classify Breast Cancer and compared with other methods obtained from the literature.

2   Material, Methods, and Results

Framework Development Methodology. With the aim to validate the proposed framework, a case study was developed. Next, each of the phases that are proposed in the framework for the development of decision support systems through clusters and dynamic tables will be explained.

2.1   Datasets

The datasets for this case study were obtained from the UCI Machine Learning repository to evaluate the effectiveness of the proposed classification model using the Wisconsin Breast Cancer Dataset (WBCD) [9, 10]. The data set was collected from the patients of University of Wisconsin-Madison Hospitals. The number of observations of the dataset is 699 data pair. The dataset has missing values. It is comprising of 458 benign cases and 241 malign instances. The descriptive statistics of the dataset can be found in Onan [3]. The input variables are 9 and the output variable is 1 (Breast cancer classification – benign and malign). Samples arrive periodically as Dr. Wolberg reports his clinical cases. The database therefore reflects this chronological grouping of the data. This grouping information appears immediately below, having been removed from the data itself:

> Group 1: 367 instances (January 1989)
> Group 2: 70 instances (October 1989)
> Group 3: 31 instances (February 1990)
> Group 4: 17 instances (April 1990)
> Group 5: 48 instances (August 1990)
> Group 6: 49 instances (Updated January 1991)
> Group 7: 31 instances (June 1991)
> Group 8: 86 instances (November 1991)
> ---
> Total: 699 points (as of the donated database on 15 July 1992).

The attributes of this dataset are:

# Attribute	Domain
1. Sample code number	id number
2. Clump Thickness – CT	1 - 10
3. Uniformity of Cell Size - UCSi	1 - 10
4. Uniformity of Cell Shape - UCSh	1 - 10
5. Marginal Adhesion -MA	1 - 10
6. Single Epithelial Cell Size - SECS	1 - 10
7. Bare Nuclei - BN	1 - 10
8. Bland Chromatin – BC	1 - 10
9. Normal Nucleoli – NN	1 - 10
10. Mitoses - Mi	1 - 10
11. Class:	(2 for benign, 4 for malignant)

2.2   Data Preparation

The first action was to identify all the variables that the researchers worked on, identifying which of them were going to be the fuzzy system inputs and outputs variables. In the case study, all input and output variables were selected. Within this stage, the data had to be processed because the experiment has missing values. When this occurred, the symbol "?" was changed for 0. Another transformation in the dataset was the number of categories for benign and malign instance. Was changed the number 2 for 1 for benign observations and 4 for 2 for malign instances. In this phase, the pre-processed technique used was clustering. This technique will be explained in Sect. 2.6 [11].

2.3   Statistical Software Package Use

After the previous process was completed, the organized and standardized data were copied and pasted into the statistical software IBM SPSS® statistics [12] version 23.0.0.0 64-bit. At this stage, each variable's name was set, and each attribute was registered in the software variable view. Once this step was completed, the next phase proceeded.

2.4   Evaluating the Optimal Number of Clusters

For this case study, the optimal number of clusters was calculated through the use of Criterion proposed by Caliński and Harabasz [13]. According to this criterion, optimal cluster numbers for all the input variables were 10, respectively. These values will be taken as a reference for the next phase.

2.5   Stablishing Number of Clusters (Min and Max)

Considering the above values, the values were 2 as a minimum, and the maximum number of clusters was the optimal value (10) for all input variables. For the output variable, only 2 clusters were used.

The maximum number of clusters was the same clusters' optimal number with the aim to avoid greater fuzzy sets numbers in the output variables or a greater number of rules (input variables interaction). The minimum and a maximum number of clusters represent the number of fuzzy sets that we will use as a knowledge database in the final Fuzzy Inference System – FIS. The main idea is to optimize the performance of the developed fuzzy system with a smaller number of rules.

2.6   Cluster Analysis

In this stage, one of the existing clusters types were performed. For the case study, hierarchical clusters (Ward method) were analyzed with the range of solutions established in the previous stage. The algorithm is implemented using the stored-matrix; the starting point is a matrix of all pairwise proximities between compounds in the data set to be clustered. Each cluster initially corresponds to an individual item (singleton) [14]. As clustering proceeds, each cluster may contain one or more items [15]. For this stage, to classify each variable (input and output) the complete dataset was used. The Ward method shows that the obtained clusters in the previous stage have a high percentage of variation and high values of canonical correlations (close to 1), which means that the clusters have a high discriminant value.

2.7   Dynamic Tables

The idea at this stage is to find (if feasible) a combination that allows promising results without having to reach the maximum or an optimal number of clusters for each variable (input or output). With this process, it is guaranteed that the fuzzy system will be optimized in the rule base number (minimum) and in the number of fuzzy sets for each variable, mainly in the output variables. Another objective with this step is to try reducing the number of input variables (feature extraction). At this stage, the following sub-stages were performed:

2.7.1   Combining Different Datasets Clusters

This step consists in making combinations between input variables and the sets of output variables using dynamic tables. For the case study, Microsoft Excel 2016® software was used. In this case, the results of the clusters and the values of the variables were copied from IBM SPSS® software [12] and pasted into a spreadsheet. This stage becomes in the knowledge base for the fuzzy model rule set elaboration for the case study. Through this, it can be observed that when performing the combination between the input variable Clump Thickness - CT with the other input variables with all clusters is not achieved a positive interaction avoiding overlapping clusters. However, with the combination (9 clusters for CT, 10 for UCSi and 9 for BN), is a combination that represents an important performance accuracy.

2.7.2   Stablishing the Fuzzy Rules for Every Variable

According to the previous stage results, it was possible to define that the optimal number of rules was the combination formed by 9 clusters in the CT input variable, 10 clusters in the UCSi variable, and 9 clusters for BN variable, for a maximum (total) of

810 rules. It should be remembered that the initial idea is to find a combination that, with the least number of rules, explains all the output variables behavior. If it were decided to use the optimum number of clusters for each input variable then, for the case study, it would be $10 \times 10 \times 10 = 1000$ rules. Next, we will give an example of how the set of rules is established using dynamic tables (Table 1).

Table 1. Set of rules using dynamic tables

	BN								
	1	2	3	4	5	6	7	8	9
5 - CT	0.29%	10.87%	1.14%	1.00%	0.57%	0.29%	0.43%	4.01%	18.60%
1 - UCSi	0.14%	8.01%	0.72%	0.00%	0.14%	0.00%	0.00%	0.00%	9.01%
1	0.14%	8.01%	0.72%	0.00%	0.14%	0.00%	0.00%	0.00%	9.01%
2	0.00%	1.00%	0.14%	0.14%	0.00%	0.14%	0.00%	0.14%	1.57%
1	0.00%	1.00%	0.14%	0.14%	0.00%	0.00%	0.00%	0.00%	1.29%
2	0.00%	0.00%	0.00%	0.00%	0.00%	0.14%	0.00%	0.14%	0.29%
3	0.00%	1.14%	0.00%	0.57%	0.14%	0.00%	0.00%	0.57%	2.43%
1	0.00%	1.14%	0.00%	0.00%	0.14%	0.00%	0.00%	0.00%	1.29%
2	0.00%	0.00%	0.00%	0.57%	0.00%	0.00%	0.00%	0.57%	1.14%
4	0.14%	0.29%	0.00%	0.00%	0.00%	0.14%	0.00%	0.57%	1.14%
1	0.14%	**0.14%**	0.00%	0.00%	0.00%	0.00%	0.00%	0.14%	0.43%
2	0.00%	**0.14%**	0.00%	0.00%	0.00%	0.14%	0.00%	0.43%	0.72%
5	0.00%	0.00%	0.00%	0.00%	0.00%	0.00%	0.14%	0.43%	0.57%

Bold values indicate overlapping clusters.

For the example case (Table 1), the columns correspond to the numbers of the groups formed by the BN variable clusters. The rows correspond to the numbers of the groups formed by the CT variable clusters and the numbers of the groups formed for the variable UCSi clusters. The values at the intersections of the columns and rows correspond to the amount of data or cases that this intersection has. The interpretation of the rule base creation would be as follows: If CT is MF5 and UCSi is MF1, then, CLASS is MF1. If CT is MF5 and UCSi is MF2, and BN is MF2, then, CLASS is MF1. If CT is MF5 and UCSi is MF2, and BN is MF3, then, CLASS is MF1. ... If CT is MF5 and UCSi is MF4, and BN is MF1, then, CLASS is MF1. If CT is MF5 and UCSi is MF4, and BN is MF2, then, CLASS is MF1. Etc. Note that, there are some intersections that don't have any cluster value formed for that combination, this means that, within the data, there are no measured or observed values for this interaction between the input variables.

When these cases occur, the inference system calculates these values through the inference engine (existing rule base with the existing database) originated from combining input values with their respective membership degree, by the minimum operator and then by superposition of the rules through the maximum operator [16]. The Defuzzification process uses the Center of Gravity method, which considers all output possibilities, transforming the fuzzy set originated by inference into a numerical value, as proposed by [17, 18].

2.8   Elaborating the Fuzzy System

For the case study, the fuzzy system implementation was carried out in the Matlab® 2017a software [19] using the fuzzy logic toolbox.

2.9   Evaluating the Fuzzy System Performance

For the evaluation of the system performance, can be used some of the following evaluation metrics: The Classification accuracy (ACC), sensitivity, specificity, Function Measure, Area under the curve and Kappa statistics. We proceeded to the validation or fuzzy model comparison with other models that seek the same objective, to predict the output variable through the interaction of the input variables. The selected models for this validation were developed by [3, 8, 20]. These works obtained good performance in the classification task. The results of this process will be presented in the next section.

2.10   Fuzzy Models' Validation

For the framework validation, a fuzzy inference system was developed with the aim to compare it with other models or methods obtained from the literature. To validate the fuzzy inference system were calculated the following evaluation metrics: Classification accuracy, sensitivity, specificity, Function Measure, and Kappa statistics. All these evaluation metrics are explained in detail in [3, 21] with their respective formulae. Table 2 shows the models' performance results.

Table 2. Comparison of performance of the best fuzzy systems or models of the state-of-the-art and this paper using the WBCD dataset.

Main aspects		Authors				
		Pota et al. (2017)	Gayathri and Sumathi (2015)	Onan (2015)	Nguyen et al. (2015)	This work (three variables)
Num of variables		6	4	7	3	3
Num of rules		324	35	-	-	39
Performance	Accuracy (%)	98%	93.60%	99.72%	97.88%	98.57%
	Sensitivity	0.99	-	1.0000	0.9850	0.9793
	Specificity	0.97	-	0.9947	0.9650	0.9891
	F-Measure	-	-	0.9970	-	0.9793
	Area under curve	-	-	1.0000	0.9750	0.9901
	Kappa statistics	-	-	0.9943	-	0.9683

- Not mentioned in the manuscript.

When comparing the results of the implementation of the decision support system based on the fuzzy sets theory using clusters and dynamic tables, by reducing the number of variables to three (with the help of the dynamic tables), the performance of this system was superior to most of the works found in the literature. Only the work done by Onan [3] obtained better performance results, however, it can be seen that these authors worked with 7 of the nine variables.

3   Discussion

For the discussion about the results, four studies found in the literature were chosen with the aim to compare them with those obtained using the Scopus database. The studies were selected because they are recent, and they compared their results with another previous works research showing that they obtained a good performance in the classification task with the same Breast cancer dataset. [20] proposed a procedure based on the application of naïve Bayes hypothesis to fuzzy systems, compared with a full optimization of a rule-based fuzzy system (neural network), since it involves several parameters linear in the number of variables. [8] used one of the feature selection method called Linear Discriminant Analysis (LDA) to select the variables that help to detect benign and malignant cancer with lesser attributes, and to diagnose the patients with less time. They used as input variables CT, UCSi, UCSh and SECs. [22] used different methods to classify the breast cancer dataset, between them, are wavelet transformation, interval type-2 fuzzy logic system. Onan [3] presented a classification model based on the fuzzy-rough nearest neighbor algorithm, consistency-based feature selection and fuzzy-rough instance selection for medical diagnosis.

According to the results (Table 2), it can be observed that the proposed model with three variables exhibited some good performance indices related to the Breast Cancer classifying capacity. The results achieved for specificity, F-measure and Kappa statistics are close to the maximum value. The Kappa statistic value is 0.9683, showing that the fuzzy inference system has an accuracy of 98.57%. This model can be considered as a useful decision-making for helping to diagnose breast cancer.

4   Conclusion

In order to design, implement and validate a framework for the design of decision support systems based on the theory of fuzzy sets using clusters and dynamic tables, a fuzzy inference system - FIS with the objective of classifying breast cancer was developed and compared the performance of the system with other related works found in the literature. Through the obtained results, it can be concluded that:

1. The Fuzzy model with three variables allowed to classify breast cancer with high accuracy, with a Kappa statistic of 0.9863, and an accuracy of 98.57%, allowing a good performance on Breast Cancer classification, showing the best result with a small number of input variables (three of nine for the case study).

2. The use of clusters and dynamic tables in classification problems allows knowledge discovery through the patterns' recognition. Using these two tools, the probability of reducing input variables (feature extraction) is increased, facilitating the decision-making process.
3. The proposed framework may provide an effective means to draw a pattern to the development of fuzzy systems for classification problems.

Acknowledgments. The first author expresses his deep thanks to the Administrative Department of Science, Technology, and Innovation – COLCIENCIAS of Colombia and the *Universidad del Norte* for the Doctoral scholarship. Also expresses their deep thanks to the *Universidad del Sinú Elías Bechara Zainúm* for the scholar and financial support.

References

1. American Cancer Society: Cancer Facts & Figures 2018, p. 76. American Cancer Society Inc., Atlanta (2018)
2. Siegel, R.L., Miller, K.D., Jemal, A.: Cancer statistics, 2018. CA Canc. J. Clinic. **68**(1), 7–30 (2018)
3. Onan, A.: A fuzzy-rough nearest neighbor classifier combined with consistency-based subset evaluation and instance selection for automated diagnosis of breast cancer. Exp. Syst. Appl. **42**(20), 6844–6852 (2015)
4. Hayat, M.A.: Breast cancer: an introduction. In: Hayat, M.A. (ed.) Methods of Cancer Diagnosis, Therapy and Prognosis. Methods of Cancer Diagnosis, Therapy and Prognosis, vol. 1. Springer, Dordrecht (2008)
5. Yazdanbakhsh, O., Dick, S.: A systematic review of complex fuzzy sets and logic. Fuzzy Sets Syst. **338**, 1–22 (2018)
6. Arslan, E., et al.: Rule based fuzzy logic approach for classification of fibromyalgia syndrome. Australas. Phys. Eng. Sci. Med. **39**(2), 501–515 (2016)
7. Nilashi, M., et al.: A knowledge-based system for breast cancer classification using fuzzy logic method. Telemat. Inform. **34**(4), 133–144 (2017)
8. Gayathri, B.M., Sumathi, C.P.: Mamdani fuzzy inference system for breast cancer risk detection. In: IEEE International Conference on Computational Intelligence and Computing Research (ICCIC). IEEE (2015)
9. Bache, K., Lichman, M.: UCI machine learning repository. University of California, School of Information and Computer Science, Irvine (2013)
10. Mangasarian, O.L.: Cancer diagnosis via linear programming. SIAM News **23**(5), 1–18 (1990)
11. Aghabozorgi, S., Teh, Y.W.: Stock market co-movement assessment using a three-phase clustering method. Expert Syst. Appl. **41**(4, Part 1), 1301–1314 (2014)
12. IBM Corp.: Released 2015. IBM SPSS Statistics for Windows. IBM Corp., Armonk (2015)
13. Caliński, T., Harabasz, J.: A dendrite method for cluster analysis. Commun. Stat. Theor. Methods **3**(1), 1–27 (1974)
14. Leach, A.R., Gillet, V.J.: An Introduction to Chemoinformatics. Springer, Dordrecht (2007)
15. Malhat, M.G., El-Sisi, A.B.: Parallel ward clustering for chemical compounds using OpenCL. In: Tenth International Conference on Computer Engineering & Systems (ICCES) (2015)

16. Hernández-Julio, Y.F., et al.: Fuzzy system to predict physiological responses of Holstein cows in southeastern Brazil. Rev. Col. Cienc. Pecu. **28**(1), 42–53 (2015)
17. Tanaka, K.: An Introduction to Fuzzy Logic for Practical Applications, 1st edn. Springer, New York (1996)
18. Sivanandam, S., Sumathi, S., Deepa, S.: Introduction to Fuzzy Logic Using MATLAB, vol. 1. Springer, Heidelberg (2007)
19. The MathWorks Inc.: Design and Simulate Fuzzy Logic Systems. The MathWorks Inc. (2017)
20. Pota, M., Esposito, M., De Pietro, G.: Designing rule-based fuzzy systems for classification in medicine. Knowl. Based Syst. **124**, 105–132 (2017)
21. Ali, F., et al.: Type-2 fuzzy ontology–aided recommendation systems for IoT–based healthcare. Comp. Commun. **119**, 138–155 (2018)
22. Nguyen, T., et al.: Medical data classification using interval type-2 fuzzy logic system and wavelets. Appl. Soft Comput. **30**, 812–822 (2015)

Organizational Models and Information Systems

A Guide for Cascading and Scaling up Green IT Governance Indicators Through Balanced Scorecards: The Case of Datacenter Consolidation

Carlos Juiz[1(✉)], Beatriz Gómez[1], Belén Bermejo[1], Diego Cordero[2], and Andrea Mory[2]

[1] University of the Balearic Islands,
Cra. de Valldemossa, km 7.5, Palma de Mallorca, Spain
{cjuiz,b.gomez,belen.bermejo}@uib.es
[2] Catholic University of Cuenca, Vargas Machuca 6-50, Cuenca, Ecuador
{dcordero,amorya}@ucacue.edu.ec

Abstract. Information Technology (IT) functions and departments around the world are adopting several Green IT practices, but sometimes not connected with the rest of the organization. Thus, a framework is needed to help organizations implement the governance and management of Green IT. In this paper, we propose a framework that allows implementing Balanced Scorecards (BSCs) for Green IT within the context of IT Governance. We illustrate our proposal through the BSC construction with a generalizable example based on the virtualization of a data center. In our example, we focus on the construction of the strategic map to the KPIs (Key Performance Indicators) monitoring needed for Green IT governance. We develop the example from the sustainability objectives that we already include in the strategic map that aims to minimize the negative impact of IT operations on the environment.

Keywords: Governance of IT · Balanced scorecard · Cascading KPI · Green IT · Data center virtualization

1 Introduction

With the advent of the standardization of the Governance of Information Technology, in the last decade, a good number of organizations have been implementing IT governance frameworks. Some of the most useful good practices for implementing IT governance frameworks are about how to direct and to control IT function, particularly, using tools for monitoring the IT strategy performance management. Performance indicators are critical ingredients of performance management, a discipline that aligns performance with strategy. Performance management harnesses information technology to monitor the execution of business strategy and help organizations achieve their goals [1]. On one hand, the activities of any IT governance framework should be the evaluation, direction, and monitoring of IT performance. Therefore, successful IT governance implementations are using different tools as decision support at strategy

© Springer Nature Switzerland AG 2019
Á. Rocha et al. (Eds.): ICITS 2019, AISC 918, pp. 133–142, 2019.
https://doi.org/10.1007/978-3-030-11890-7_14

performance at least on the strategy level. On the other hand, the BSC emerged as one of these tools to monitor business and IT performance.

However, less attention has been paid to Green IT strategy performance through IT governance frameworks. Some of the efforts have been addressed to adapt existing IT governance and IT management processes to catch potential and general Green IT requirements coming from several layers of the organizations [2]. There were also several efforts to use BSC in Green IT problems [3]. However, what has been given less importance is the application of BSC in the monitoring of a Green IT strategy and policy implementation.

This work not only gives an example of BSC applicability for monitoring Green IT with business in IT governance framework implementations, but it also justifies the BSC importance to monitor Green IT actions reporting to the board or senior executive team in a clear way, without the details of the particular implementation framework. In particular, it is proposed one illustrative and generalizable example to show how green IT strategy performance may be implemented with or without green IT implemented processes, different than the ones to monitor the effectiveness of performance indicators. This effectiveness of Green IT direction and control is also provided through bidirectional performance metrics from strategic to tactics and operation levels and backward (cascading and rolling up). This direction and monitoring serve to satisfy the Green IT performance principles of any standardized implementation of Green IT governance framework. The findings are illustrated with the particular problem of consolidation in virtualized data centers to implement one Green IT energy-saving policy. This tactical policy is common in green datacenters deployment [4]. The results show that cascading Green IT strategy performance indicators downwards to provide direction is so as important as monitoring their effectiveness controlling this strategy alignment rolling upwards for Green IT governance framework implementation [1].

Green IT policies need to be implemented through precise objectives and targets expressed by indicators to verify the results of the behavior accomplishment of the IT governance policy. Strategic BSC is an easy way to relate objectives and indicators in several perspectives for Green IT governance [3]. Other possible approximation for Green IT governance may be conducted by modifying internal IT processes within a complete governance and management framework to implement green IT policies [2].

The purpose of this research is to illustrate, through the consolidation of virtualized datacenters example, how the BSC model may be cascaded and rolled up in the implementation of Green IT policies. Moreover, we provide how BSC may be integrated easily as performance monitoring tool into IT governance frameworks.

2 Data Center Virtualization and Green IT

Information technologies have great potential to contribute to the regulation of pollution through automated control in buildings, transportation systems, supply chains, electricity networks [5]. In fact, several authors and researchers have coined a new conceptualization that of Green IT, conceived as the study, the design, the implementation of computer hardware resources efficiently and effectively with minimal impact on the environment [6].

From the point of view of Information Technology, several authors determine areas and fields in which it is possible to contribute to the control of the environment, ranging from the production of hardware technology resources, the operation of this equipment in a productivity environment, and the technological waste when they reach their level of obsolescence, among others. The operational practices of computing resources to promote Green IT, cited in the research model proposed by Molla [7] are summarized in recommendations, where variables are indicated, as follows: Green IT policies, practices for the provision of Green IT, Green IT energy efficiency practices, Green IT monitoring practices, end-of-cycle Green IT practices, physical infrastructure of the Green IT network, technical infrastructure of the Green IT.

For Murugesan [6], Green IT expands the number of possible areas of focus and activities where its application is possible. The hardware and software components of information technologies, within an organization, can be used in different ways and alternatives to contribute to green development [8]. Several investigations on the management of food and electricity consumption are aimed at the management of energy efficiency for data centers, which contributes to the generational Green IT, leading to exponential improvements in performance [9].

For many organizations, today's greatest challenge is to drive more value, efficiency, and utilization from data centers. Virtualization of servers is one way to meet this challenge. Virtualization can be defined as a set of technologies and concepts that aims to provide an abstract environment to tune applications [10]. This set of technologies is a large concept that refers to the installation of a logical version of hardware such as computation, storage and network [11]. Virtualization is also a set of functionalities, such as an abstraction of hardware resources and a simpler access. Also, it provides users to be isolated from one another and supports virtual instances replication, increasing the elasticity of the system [12, 13].

Although the consolidation of virtual machines increases the utilization of the physical resources, and as a consequence, the power is reduced due to the switch-off servers, the energy consumption may be reduced depending on the mean response time of applications [14]. The energy and power consumption of data centers is being a concern for most of the companies. The total consumed electricity by the world's data centers amounts to 0.5%. Also, they drive more carbon emission than both North America and the Netherlands, together [15]. This extremely high energy consumption lies in the inefficient use of these resources, even though the power consumption efficiencies of the hardware.

A green data center is an enterprise-class computing facility that is entirely built, managed and operated on green computing principles. It provides the same features and capabilities of a typical data center but uses less energy and space, and its design and operation are environmentally friendly. It is built to have a minimal effect on the natural environment. The following are primary green data center features [16]: (a) built from the ground up in an environmentally friendly facility; (b) consume minimal power resources for operation and maintenance both for the primary computing infrastructure and supporting electronic resources, such as cooling, backup, and lighting; (c) typically operate with green or renewable energy, such as solar, wind or hydel power; (d) Entire infrastructure is installed with the lowest power and carbon footprint, and (e) minimal e-waste with recyclable or reusable equipment. Particularly, the features (b) and (d) can be achieved through virtual machine consolidation techniques. These techniques improve the energy efficiency of the data center, defined as the ratio of the computation performed and the energy consumed to achieve it.

There are many metrics to measure the green level of a data center. The most common metrics are exposed below: Power Usage Effectiveness (PUE): it is a ratio that describes how efficiently a computer data center uses energy; specifically, how much energy is used by the computing equipment, in contrast to cooling and other overhead [17]. Data Center infrastructure Efficiency (DCiE): it is a performance metric used to calculate the energy efficiency of a data center. DCiE is the percentage value derived, by dividing information technology equipment power by total facility power [18]. Carbon Usage Effectiveness (CUE): it is a metric for measuring the carbon gas a data center emits on a daily basis [19]. Space, Wattage, and Performance (SWaP): it is a metric that assesses the efficiency and effectiveness of rack-optimized server deployments in a data center [20].

Organizations are currently more and more forced to manage resources in a sustainable manner for manifold reasons: legal regulations and societal impact, energy costs, emissions to the atmosphere. Companies invest more efforts and funds in finding new and innovative ways for optimizing energy spending, this implies several tasks, such: reshaping the entire operational model, re-engineering the business processes, removing weak spots in the current business model, optimize IT functionality [21].

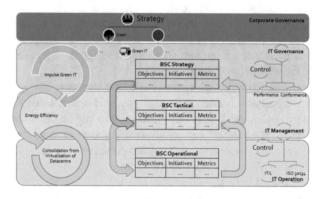

Fig. 1. Cascading strategy green objectives and rolling up green KPIs

Companies develop green policies to cushion the impact of their activities and businesses on the carbon footprint. One of the most recent directions is to reduce electricity consumption and be more environmentally friendly with IT. An example would be Deutsche Bank, where they have developed a green holistic corporate policy for IT [22]. In fact, we can even see the example of virtualization and consolidation in that policy. However, it may be more difficult to establish the cascade of strategic, tactical and operational objectives and, on the way back, to monitor the operational, tactical indices that will make up the value and effectiveness of this cascade of objectives [23]. Figure 1 shows the model of using three-layered balance scorecards (BSCs) to cascade targets and scale green KPIs [24].

When conceiving the BSC, authors in [25] maintained that companies, facing modified conditions of competition in the era of information technology and global markets, need specific financial management and control systems in order to stay competitive. Many firms lack management techniques for intangible assets such as employees, infrastructures, know-how or technologies employed. With the BSC, the potential benefit of investments in intangible assets becomes more visible encouraging all members of the firm to adopt rather a long-term view looking substantially beyond the next annual report [26].

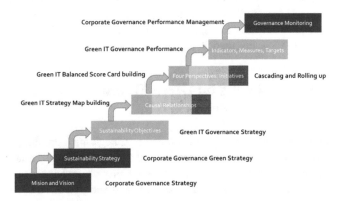

Fig. 2. From corporate governance to Green IT governance using BSC (based on [28]).

The BSC [27] comprising financial as well as non-financial KPIs and being a management tool provides high potential concerning the management of corporate sustainability. Nevertheless, this instrument does not release companies from the definition of sustainability strategies. If a company wants to establish a BSC for corporate sustainability, the setting-up of corporate environmental, social and business strategies must be done beforehand.

The purpose of Andersson and Malmkvist [3] was to investigate if the balanced scorecard model had to be adapted and be used to find objectives that can aid the implementation of Green IT. This was done by using an inductive approach to find Green IT practices through secondary data and with semi-structured interviews with four Swedish companies. However, Andersson and Malmkvist [3] work does not include indicators to measure if targets are accomplished. Moreover, their BSC is only for strategic purposes and does not consider the cascading down into tactics and operation and rolling up. We propose to construct three different BSCs: the strategic, the tactical and the operational.

The steps when defining a BSC for sustainability appears in [28] where the sustainability strategy for the organization should be defined first in order to derive objectives (see Fig. 2). Of course, this strategy clarification should be the result of explicit or implicit vision and mission statements. From this sustainability strategy, the most important step is the establishment of the sustainability objectives. This is especially important when the strategic importance of environmental and social aspects have not been evaluated yet.

After having collected, clarified and classified the objectives, the causal relationships between the objectives can be made transparent. These causal relationships may be used to build a sustainability strategy map [29]. Once the objectives are integrated into the four perspectives (financial, customer, process and learning), the indicators, targets and measures should be defined. Finally, after a consolidation and decision of the top management, the BSC can be put into use for means of monitoring into the governance system [24]. Finally, as strategies being a set of hypotheses of causes and effects, the relevant goals and their KPIs should be linked to each other revealing systematically the architecture of causal relationships. The instrument of the BSC thus ideally provides goal-related KPIs that can be linked in a way of causal relations that emerge from the strategic orientation of the company.

3   Cascading and Rolling up Green IT BSC: The Example of Virtualized Data Centers

Based on the work of [29], we have constructed a strategic map dividing into the same layers of the BSC. In Fig. 3 we can see the example we propose. Starting with the financial layer, we define what produces revenue growth and increase profitability, which in this case has focused on energy reduction and energy efficiency. These drivers lead us to the layer of customers where the discipline of value is defined and how it is achieved.

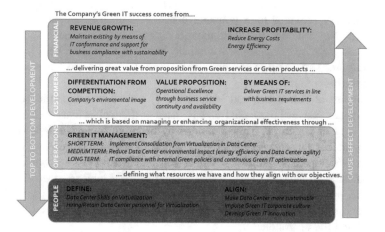

Fig. 3. Green IT strategy map example based on [29]

In our example, it focuses on operational excellence by delivering green services. The next step, in the operation layer, is how energy reduction and energy efficiency are managed internally through green services. The strategic, tactical and operational decision that has been chosen in the example is the virtualization of the data center. Obviously, all these actions depend on the skills and motivation of the employees and the resources in the data center. The strategy map is a reduced version of the complete sustainability strategy but explained from a financial perspective to the bottom. Once the strategy map is built is easier to construct the Strategic BSC taking into account the sustainability objectives from a sustainability strategy (see Fig. 4). In our example, the sustainability objectives are: alignment with sustainability strategy; transparency of Green IT Costs; IT conformance and support for business compliance with sustainability; commitment of top executive management for making Green IT-related decisions and actions; improve Company's environmental image; Deliver Green IT services in line with business requirements; improve business service continuity and availability; IT compliance with internal green policies; Green IT process optimization and agility; make IT assets more sustainable; impulse Green IT corporate culture; develop Green IT innovation processes.

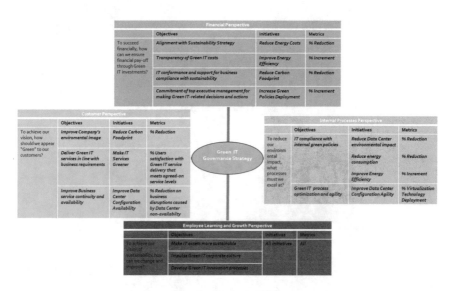

Fig. 4. Strategic BSC

In order to cascade the Green IT governance strategy into tactics, the IT management should convert strategic initiatives into more concrete tactical objectives, tactical initiatives, and tactical metrics. In this case, for space reasons we only develop one tactical initiative that covers all the tactical objectives: implement consolidation from virtualization in the datacenter. The BSC corresponding to the Green IT management is shown in Fig. 5. In this BSC the metrics are used for measuring the usual Green IT indicators, i.e., PUE, CUE, DCiE, and SwaP, among others, in the implementation of consolidation from virtualized data center of the company. The same way Green IT management should be cascaded to Green IT operation. The unique operational objective is the implementation of consolidation from the virtualized data center, but this objective may have different operational initiatives and metrics. In Fig. 6, we show the corresponding BSC for Green IT operation. In this BSC the metrics are related to operational accounting and low-level measurements in the datacenter.

Fig. 5. Tactical BSC

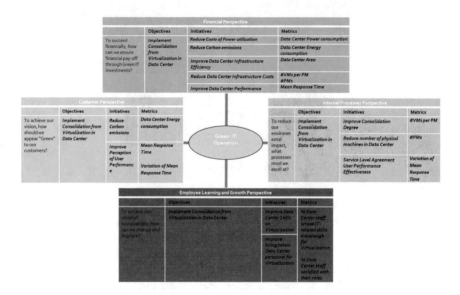

Fig. 6. Operational BSC

Once we have the three cascading BSCs, the company's governance monitoring should roll up the values of indicators from operation to management and from management to governance, i.e., accomplishing the objectives in any layer is contributing to the upper ones. For example, "reducing the number of physical machines (#PM)" (see internal processes perspective) through consolidation at operation in the virtualized data center maybe means improving PUE values in the management of IT and therefore reducing energy costs, which it increases profitability (financial strategy objective).

4 Conclusions

The use of the balanced scorecard for integrating Green IT aspects into the business management of a company could be an appropriate way for understanding and executing a sustainability strategy and can be put from some pieces of policy papers into practice. The gap between top management (governance) and environmental management communication and alignment can be reduced. The green BSC proposal can easily be used as a good IT governance practice because it uses the management, evaluation and control flow that the IT governance standard suggests [1], to help organizations implement the governance and management of Green IT gradually, as well as to improve their maturity level in this area. The effectiveness of Green IT direction and control is also provided through bidirectional performance metrics from strategic to tactics and operation levels and backward (cascading and rolling up). The validation of this proposal by experts and a case study seems to indicate that the proposal can be useful for implementing and improving the Green IT processes in organizations.

References

1. Juiz, C., Toomey, M.: To govern IT, or not to govern IT? Commun. ACM **58**, 58–64 (2015)
2. Patón-Romero, J.D., et al.: A governance and management framework for Green IT. Sustainability **9**, 1761 (2017)
3. Andersson, P., Malmkvist, L.: Green IT Balanced Scorecard: a Model Developed for the Swedish Environment. Jönköping International Business School (2012)
4. Applied Trust a via West Company: Greening your IT infrastructure (2018). https://www.appliedtrust.com/resources/green-it/greening-your-it-infrastructure. Accessed 31 July 2018
5. Dedrick, J.: Green IS: concepts and issues for information. Commun. Assoc. Inf. Syst. **27** (11), 173–184 (2010)
6. Murugesan, S.: Harnessing Green IT: principles and practices. IT Prof. **10**(1), 24–33 (2008)
7. Molla, A.: Organizational Motivations for Green IT: Exploring Green IT Matrix and Motivation Models, AiSel, Asia (2009)
8. Accenture: Data Centre Energy Forecast Final Report (2008). www.accenture.com. Accessed 14 Feb 2018
9. Klavic, P., et al.: Changing computing paradigms towards power efficiency. Philos. Trans. R. Soc. A **372**(2018), 1–13 (2014)

10. Bermejo, B., Juiz, C., Guerrero, C.: On the linearity of performance and energy at virtual machine consolidation: the CiS2 index for CPU workload in server saturation. In: IEEE High-Performance Computing and Communications (HPCC-2018) (2018)
11. Padala, P., et al.: Performance Evaluation of Virtualization Technologies for Server Consolidation. Hewlett-Packard Development Company (2008)
12. Mueen, U., Azizah, A.: Server consolidation: an approach to make data centers energy efficient & green. Int. J. Sci. Eng. Res. **1**(1), 1–7 (2010)
13. Bermejo, B., et al.: Improving the energy efficiency in cloud computing data centers through resource allocation techniques. In: Research Advances in Cloud Computing. Springer, Singapore (2017)
14. Dabbagh, M., et al.: Efficient datacenter resource utilization through cloud resource over commitment. Memory **40**(50), 1–6 (2015)
15. Jain, A., et al.: International Conference on Energy Efficient Computing-Green Cloud Computing (ICEETS). IEEE (2013)
16. Deng, W., et al.: Harnessing renewable energy in cloud data centers: opportunities and challenges. IEEE Netw. **28**(1), 48–55 (2014)
17. Yuventi, J., Mehdizadeh, R.: A critical analysis of power usage effectiveness and its use in communicating data center energy consumption. Energy Build. **64**, 90–94 (2013)
18. Beloglazov, A., Buyya, R.: Adaptive threshold-based approach for energy-efficient consolidation of virtual machines in cloud data centers. In: MGC@ Middleware, p. 4 (2010)
19. Azevedo, D., et al.: Carbon Usage Effectiveness (CUE): a green grid data center sustainability metric. In: The Green Grid, White Paper, vol. 32 (2010)
20. Meng, X., Pappas, V., Zhang, L.: Improving the scalability of data center networks with traffic-aware virtual machine placement. In: Proceedings IEEE INFOCOM, San Diego (2010)
21. Owen, M., Shields, R.: Using enterprise architecture to develop a blueprint for improving your energy efficiency and reducing your carbon footprint (2008)
22. Deutsche Bank AG: Green IT: energy efficiency at work, 12 June 2018. https://www.db.com/cr/en/concrete-green-it.htm. Accessed 31 July 2018
23. Juiz, C.: New engagement model of IT governance and IT management for the communication of the IT value at enterprises. In: Communications in Computer and Information Science, vol. 194, pp. 129–143 (2011)
24. Juiz, C., Colomo-Palacios, R., Gómez, B.: Cascading ISO/IEC 38500 based balanced score cards to improve board accountability. In: Conference on ENTERprise Information Systems, Lisbon (2018)
25. Kaplan, R., Norton, D.: Linking the balanced scorecard to strategy. Calif. Manag. Rev. **39**(1), 53–79 (1999)
26. Sandkuhl, K., Seigerroth, U.: Method engineering in information systems analysis and design: a balanced scorecard approach for method improvement. Softw. Syst. Model. 1–25 (2018). https://doi.org/10.1007/s10270-018-0692-3
27. Juiz, C., Guerrero, C., Lera, I.: Implementing good governance principles for the public sector in information technology governance frameworks. Open J. Account. **3**, 9–27 (2014)
28. Gminder, C., Bieker, T.: Managing corporate social responsibility by using the sustainability-balanced scorecard. In: Proceedings of the 10th International Conference of the Greening of Industry Network, Goteborg (2002)
29. Marr, B.: What is a Modern Balanced ScoreCard? Management White Paper from Advanced Performance Institute (2010)

Conceptual Model for Software as a Service (SaaS) Enterprise Resource Planning (ERP) Systems Adoption in Small and Medium Sized Enterprises (SMEs) Using the Technology-Organization-Environment (T-O-E) Framework

Jose Valdebenito$^{(\boxtimes)}$ and Aldo Quelopana

Universidad Católica del Norte, 1270709 Antofagasta, Chile
jvp017@alumnos.ucn.cl, aldo.quelopana@ucn.cl

Abstract. Enterprise Resource Planning (ERP) systems under a Software as a Service (SaaS) model are being established as a viable solution for Small and Medium-Sized Enterprises (SMEs). These systems are a great opportunity for SMEs because they obtain the benefits of an ERP system without maintenance and investment costs associated with on-premise models. Since SaaS ERP systems success involves their adoption, it is important to know the factors influencing their adoption in the context of SMEs. Based on a literature review, a conceptual adoption model is proposed using the Technological, Organizational, and Environmental (TOE) framework consisting of the following factors: perceived value, security concerns, configurability and customization, organizational readiness, top management support, competitive pressure, and vendor qualities.

Keywords: SaaS · ERP · TOE · Technology adoption

1 Introduction

Enterprise Resource Planning (ERP) systems are an important element of modern IT enterprises structure because they help them to obtain competitive advantage in a digitally powered environment [1], by improving business processes and increasing organizational value [2]. ERP systems implementation is expected to render benefits to enterprises of all sizes. However, there are still several challenges in this regard because the success of ERP implementation projects has been quite poor lately [3–5]. This has caused Small and Medium-Sized Enterprises (SMEs) to be skeptical for implementing these systems [6].

Most studies related to ERP implementation failures have been conducted on on-premise systems (systems hosted and maintained by an enterprise which

© Springer Nature Switzerland AG 2019
Á. Rocha et al. (Eds.): ICITS 2019, AISC 918, pp. 143–152, 2019.
https://doi.org/10.1007/978-3-030-11890-7_15

demands the service) (e.g., [3–5]). This on-premise model, still prevailing in modern organizations, has well-known disadvantages such as high initial investment costs and a lot of time used for maintenance [7]. These disadvantages are expected to be significantly improved by Cloud Computing (CC) support [8], particularly by the SaaS model, which is becoming an effective way of providing business applications for companies of all sizes, particularly SMEs [6,9].

On the other hand, the technology adoption is a field in constantly developing. There are several theories that try to explain this phenomenon, for example Technology Acceptance Model (TAM) [10], Diffusion of Innovation (DOI) [11] and Technology, Organization, Environment (TOE) framework [12]. To a great extent, research on ERP systems adoption utilizes TAM constructs as a basis for conceptual and empirical studies [13]. Currently, there are few studies focusing on SaaS ERP systems adoption; however, there is a great interest in this field of study among researchers [6].

If ERP systems success involves the adoption of this technology [13], then, it is important to know the factors influencing ERP cloud systems adoption under SaaS model, in the context of SMEs. This study, therefore, aims to formulate a conceptual model, using the technological, organizational, and environmental contexts of TOE framework.

2   Literature Review

2.1   ERP SaaS en PyMEs

Traditionally, ERP resources (including data and database servers, among others) are hosted and kept internally by organizations. This is known as on-premise ERP. This on-premise approach has certain inherent disadvantages such as a high initial investment and an associated maintenance cost [8].

At present, SaaS-based solutions change the value frontier and may provide the same level of value at a lower price or a higher value at the same price [14]. This includes general cost minimization, the ability to provide real-time service to customers, easy access to global innovations, and scalability. So, more SMEs are beginning to go away from on-premise solutions [15].

SaaS ERP are less prevalent than other systems under SaaS model; however, they are gaining impulse among small-sized enterprises [6]. In this type of service, the ERP and its associated data are administered on a centralized basis by the supplier while customers accessing them through a web navigator [16], without requiring the physical installation of the system in local computers, neither storing data in local servers [8].

2.2   Technology Adoption

Technology adoption models and theories have been proposed by researchers to track the rhythm of dissemination of every innovation [17]. Among them are the ones mentioned in the Introduction (TAM, TOE, and DOI). Authors (see

[6,18,19]) notice that TAM focus on individuals, while TOE and DOI focus on firms. According to Gangwar *et al.* [20], the TOE framework is valid, robust, and most dominant in studies related to enterprise technology adoption. According to Awa *et al.* and Oliveira *et al.* [17,18], this framework has gained a great support from similar studies, over TAM, TRA, and TPB, becoming a generalized theory to study technology adoption [21]. On the other hand, several studies use TOE to understand ERP systems adoption [17,22–26], cloud systems adoption [27–33], and cloud ERP systems [6,34–37]. For these reasons, contexts provided by the TOE framework would be used to develop a conceptual model of factors influencing technology adoption.

2.3 The TOE Framework

The TOE framework was developed in 1990 by Tornatzky and Fleischer to study several IT and IS products on enterprises [12]. It is the first of its kind and proposes a set of generic factors explaining and predicting adoption probability [17]. TOE provides a holistic view of technology adoption by users, its implementation, its impact on the activities of the value chain and post-adoption dissemination among enterprises [38].

This framework identifies three aspects which influence the process of adopting and implementing a technology: technological, organizational, and environmental contexts.

The technological context describes the internal and external technologies of a firm. This includes practices and internal equipment, as well as a set of external technologies, such as perceived usefulness, technical and organizational compatibility, complexity, and learning curve [17].

The organizational context captures descriptive measures related to the form, such as the business scope, top management support, organizational culture, complexity of management structure, human capital quality and size [18].

The environmental context is related to those operational facilitators and inhibitors with which the company manages its business [12].

3   Proposed Model

There are many factors studied using TOE, for example, IT infrastructure, perceived compatibility, technology readiness, trading partner pressure, competitive pressure, firm size, regulatory support, and mimetic pressure (e.g., [17,27,29,32,39]). These factors may or may not be important for the particular technology of a study, which does not ensure that its effect on a different technology innovation is the same. For this reason, researchers use TOE constructs and varied studies on related fields to create robust adoption models (e.g., [6,17,27,29,32,34]).

3.1   Technological Context

Perceived Value: Perceived value, also called relative advantage, refers to the extent to which an innovation is perceived as an improvement, as compared with the idea it replaces [11, 36].

According to Lee [40], it is more likely that companies adopt an innovation when they perceive that this innovation brings a relative advantage to the business In studies related to cloud ERP (e.g., [34, 36, 37]), this factor is considered significant for predicting technology adoption. However, no direct evidence of the use of perceived value was found in the literature related to SaaS ERP. In studies on CC [27, 29, 32, 38, 41–44] and on-premise ERP systems [17, 24, 45, 46], perceived values have been considered a relevant and sometimes dominant factor.

Security Concerns: Security is a potential issue for SMEs [6]. The security includes topics such as perceived cloud systems security vulnerability, data privacy [36], external hacking, and sharing infrastructure with different vendor's customers [47].

This factor called security concerns can be defined as the extent to which a SaaS ERP system is perceived as insecure for storing and exchanging data and the execution of other business transactions [36]. Faasen *et al.* [47] suggest that security concerns have caused doubts as to SaaS viability to house this type of systems. Security concerns have been found to be a factor to consider when studying the adoption. Evidence can be found in studies related to cloud adoption [32, 48], cloud ERP [34, 36, 37], and those on the SaaS service model for ERP systems [47, 49].

Configurability and Customization: Configurability and customization are related to assistance provided to the customer so as to adapt the software to his individual requirements, i.e., unique organizational structure, legal framework, reports, formatting, processes [50], and the creation of something new associated with software features [6].

If business processes are considered to cause SMEs to be unique, differentiating them from competitors at the same or higher level, and, that SaaS systems offer less flexibility and a minimum customization option as compared with on-premise systems, then, according to [51], lack of configurability and customization affects SaaS ERP systems adoption negatively. The above-mentioned is not considered only a matter of system flexibility, but also of cost increase [49], due to internal process changes or lack of critical functionality for the business.

The factor related to configurability and customization of the system has not been studied so deeply as perceived value or security concerns. In relation to papers concerning SaaS ERP, studies [6, 47, 49] examine this factor, indicating what must be considered when studying adoption.

3.2 Organizational Context

Organizational Readiness: Organizational readiness is the measure of the financial and technological resources available in an organization, which may be used for adopting a SaaS ERP system [36]. Organizational readiness is measured by two sub-factors: financial readiness and technological readiness [36,38]. Financial readiness is an indicator of the possibility of an organization to fund the implementation and subsequent costs of a system. Technological readiness is a measure of IT sophistication in terms of use and management.

Although SaaS ERP systems reduce IT needs [52], certain resources are still needed to access the system and configure the application [6]. CC services may become part of the activities of the value chain, only if the enterprise has the necessary infrastructure, technical competencies, and financial support [27,38]. In addition, empirical studies [6,22,24,29,36,38,46] show that organizational readiness is a technology adoption factor.

It is important to note that some studies [6,22,24,29] only focus on the technological sub-factor of organizational readiness.

Top Management Support: Top management support is important. Managers must assign the necessary resources for adoption, minimize resistance, create a positive environment and attitude toward innovation, integrate services, and ensure a long-term view [29,34,36,38].

Top management support involves dedicating time to the IT program of the organization, review plans, follow up outcomes, and help to solve management problems related to IT integration with business processes [53]. In the case of SaaS, top management plays a fundamental role since this type of technology involves the integration of resources and process engineering [27].

As stated by Low *et al.* [27], without top management support, SMEs are less likely to adopt new technologies. Managers play the role of improving organizational performance, break the perceived performance gap, and exploit business opportunities with the use of a new technology [38]. Researchers have found that this factor has a positive influence on technology adoption [24,27,29,34,36–38,41–44,46,48,54,55].

3.3 Environmental Context

Competitive Pressure: Competitive pressure may be defined as the level of pressure an firm experiences due to competitors from the same industry [27, 29,34]. This competence influences IT adoption positively, particularly when a technology is a strategic need to compete in the market [46].

According to Kinuthia [36], cloud ERP systems offers a strategic tool, allowing the first companies in adopting this technology to obtain a considerable advantage [38], receiving the benefits of greater operational efficiency, better market visibility, and more precise real-time data [29]. So, the adoption of a SaaS ERP system as a strategic tool may help an SME to become an advantageous entity in the market [34]. On the basis of the above, many studies (e.g., [17,27,32,34,36–38,45]) have mentioned this factor is important for adoption and innovation.

Vendor Qualities: The vendor qualities of the SaaS ERP system are related to several factors concerning suppliers, among them are reputation, reliability, support, and the co-creation of value by the supplier and the firm [6,36,41,44,47]. The study [6] indicates that the software supplier reputation, his willingness to work with the customer throughout the implementation stage using his own employees as consultants, and considering improvements required by firms, are some of the factors influencing adoption. Whereas the work [36] deals with supplier's support. This refers to availability toward the customer (enterprise) in aspects such as training in systems use and technical support for implementation and use. The author points out that suppliers have the chance to show their abilities and the system scope by means of free training sessions.

In [41], supplier computing support is dealt with, considering it as actions that may influence the probability that an innovation is adopted. In [47], the authors found that SMEs show concern about the trust given by suppliers when implementing mission-critical software solutions. Authors in [44] state that supplier's credibility plays an important role to decide CC service adoption.

3.4   Adoption Factors

Finally, the literature indicates then that the factors that may influence SaaS ERP systems adoption by SMEs are: perceived value, security concerns, configurability and customization, organizational readiness, top management

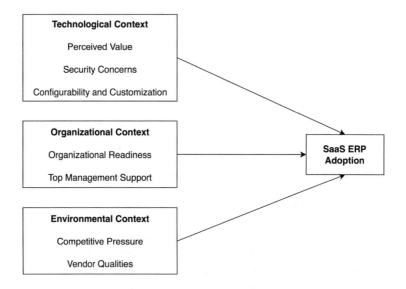

Fig. 1. Conceptual model for SaaS ERP adoption in SMEs.

Table 1. Summary of the factors obtained from the literature.

Contexts	Factors	References
Technological	Perceived value	$[17,24,27,29,32,34,36\text{–}38,41\text{–}46]$
	Security concerns	$[32,34,36,37,47\text{–}49]$
	Configurability and customization	$[6,47,49]$
Organizational	Organizational readiness	$[6,22,24,29,36,38,46]$
	Top management support	$[24,27,29,34,36\text{–}38,41\text{–}44,46,48,54,55]$
Environmental	Competitive pressure	$[17,27,32,34,36\text{–}38,45]$
	Vendor qualities	$[6,36,41,44,47]$

support, competitive pressure, and vendor's qualities (1). Figure 1 shows the model schematization obtained from the literature (Table 1).

4   Conclusions

The objective of this study is to generate a SaaS ERP adoption model for SMEs. A literature review about studies on CC adoption, ERP systems, and ERP cloud systems was conducted, putting emphasis on SaaS ERP studies. TOE framework was used as a basis, proposing a model consisting of seven factors obtained from the literature.

Factors with empirical validity or those not frequently studied in the past were considered for the construction of this model. So, the perceived value was found to have been deeply studied in the literature in the TOE technological context, having great empiric validity. Despite this fact, perceived value has not been considered in studies on SaaS ERP systems adoption yet. Therefore, it is regarded as a factor that may affect adoption. In the same way, security concerns and configurability and customization show validity in similar studies, particularly those studying cloud systems. In the organizational context, the factors considered as part of the model are organizational readiness, referring to SMEs economic and technological preparation when incorporating a new technology, and the factor related to top management support. In the environmental context, competitive pressure is favored by a great number of studies (while others do not favor it) as a factor prevailing in the adoption process. In addition, vendor's qualities are an issue that has aroused researchers' interest lately.

Finally, this study is a contribution to the field of technology adoption, providing more data about SaaS ERP systems adoption and opening up the field for future research using the adoption model proposed. So far, the model has not been validated, therefore, it is proposed to follow an experimental qualitative approach to go deeper into knowledge about factors influencing adoption, followed by a quantitative study giving validity to the model proposed.

References

1. Lewandowski, J., Salako, A.O., Garcia-Perez, A.: SaaS enterprise resource planning systems: challenges of their adoption in SMEs. In: 2013 IEEE 10th International Conference on e-Business Engineering (ICEBE), pp. 56–61. IEEE (2013)
2. Yan Huang, S., Huang, S.M., Wu, T.H., Lin, W.K.: Process efficiency of the enterprise resource planning adoption. Ind. Manag. Data Syst. **109**(8), 1085–1100 (2009)
3. Chen, C.C., Law, C.C., Yang, S.C.: Managing ERP implementation failure: a project management perspective. IEEE Trans. Eng. Manage. **56**(1), 157–170 (2009)
4. Mahmud, I., Ramayah, T., Kurnia, S.: To use or not to use: modelling end user grumbling as user resistance in pre-implementation stage of enterprise resource planning system. Inf. Syst. (2017)
5. Saade, R.G., Nijher, H.: Critical success factors in enterprise resource planning implementation: a review of case studies. J. Enterp. Inf. Manage. **29**(1), 72–96 (2016)
6. Seethamraju, R.: Adoption of software as a service (SaaS) enterprise resource planning (ERP) systems in small and medium sized enterprises (SMEs). Inf. Syst. Front. **17**(3), 475–492 (2015)
7. Mangiuc, D.M.: Enterprise 2.0-is the market ready? Account. Manage. Inf. Syst./Contabilitate si Informatica de Gestiune **10**(4) (2011)
8. Peng, G.C.A., Gala, C.: Cloud ERP: a new dilemma to modern organisations? J. Comput. Inf. Syst. **54**(4), 22–30 (2014)
9. Haselmann, T., Vossen, G.: Software-as-a-service in small and medium enterprises: an empirical attitude assessment. In: International Conference on Web Information Systems Engineering, pp. 43–56. Springer (2011)
10. Davis, F.D.: A technology acceptance model for empirically testing new end-user information systems: theory and results. Ph.D. thesis, Massachusetts Institute of Technology (1985)
11. Rogers, E.M.: The Diffusion of Innovation, 5th Edition (2003)
12. Tornatzky, L., Fleischer, M.: The Processes of Technological Innovation. Lexington Books, Lexington (1990)
13. Valdebenito, J., Quelopana, A.: Understanding the landscape of research in enterprise resource planning (ERP) systems adoption, pp. 35–39 (2018)
14. Lenart, A.: ERP in the cloud–benefits and challenges. In: Research in Systems Analysis and Design: Models and Methods, pp. 39–50 (2011)
15. Venkatachalam, N., Fielt, E., Rosemann, M., Mathews, S.: Small and medium enterprises sourcing software as a service-a dynamic perspective on is capabilities. Assoc. Inf. Sys. **7**, 15–2012 (2012)
16. Mell, P., Grance, T., et al.: The NIST Definition of Cloud Computing (2011)
17. Awa, H.O., Ojiabo, O.U.: A model of adoption determinants of ERP within T-O-E framework. Inf. Technol. People **29**(4), 901–930 (2016)
18. Oliveira, T., Martins, M.F.: Literature review of information technology adoption models at firm level. Electron. J. Inf. Syst. Eval. **14**(1), 110–121 (2011)
19. Rondan-Cataluña, F.J., Arenas-Gaitán, J., Ramírez-Correa, P.E.: A comparison of the different versions of popular technology acceptance models: a non-linear perspective. Kybernetes **44**(5), 788–805 (2015)
20. Gangwar, H., Date, H., Raoot, A.: Review on IT adoption: insights from recent technologies. J. Enterp. Inf. Manag. **27**(4), 488–502 (2014)
21. Zhu, K., Kraemer, K.L., Dedrick, J.: Information technology payoff in e-business environments: an international perspective on value creation of e-business in the financial services industry. J. Manag. Inf. Syst. **21**(1), 17–54 (2004)

22. Pan, M.J., Jang, W.Y.: Determinants of the adoption of enterprise resource planning within the technology-organization-environment framework: Taiwan's communications industry. J. Comput. Inf. Syst. **48**(3), 94–102 (2008)
23. Ilin, V., Ivetic, J., Simic, D.: Understanding the determinants of e-business adoption in ERP-enabled firms and non-ERP-enabled firms: a case study of the Western Balkan Peninsula. Technol. Forecast. Soc. Chang. **125**, 206–223 (2017)
24. Haberli Jr., C., Oliveira, T., Yanaze, M.: Understanding the determinants of adoption of enterprise resource planning (ERP) technology within the agri-food context: the case of the Midwest of Brazil. Int. Food Agribusiness Manag. Rev. **20**(5), 729–746 (2017)
25. Awa, H.O., Uko, J.P., Ukoha, O.: An empirical study of some critical adoption factors of ERP software. Int. J. Hum.-Comput. Inter. **33**(8), 609–622 (2017)
26. Melão, N., Loureiro, J.: ERP in the education sector: evidence from Portuguese non-higher education institutions. In: World Conference on Information Systems and Technologies, pp. 592–602. Springer (2017)
27. Low, C., Chen, Y., Wu, M.: Understanding the determinants of cloud computing adoption. Ind. Manag. Data Syst. **111**(7), 1006–1023 (2011)
28. Alkhalil, A., Sahandi, R., John, D.: An exploration of the determinants for decision to migrate existing resources to cloud computing using an integrated TOE-DOI model. J. Cloud Comput. **6**(1), 2 (2017)
29. Oliveira, T., Thomas, M., Espadanal, M.: Assessing the determinants of cloud computing adoption: an analysis of the manufacturing and services sectors. Inf. Manag. **51**(5), 497–510 (2014)
30. Masana, N., Muriithi, G.M.: Investigating TOE factors affecting the adoption of a cloud-based EMR system in the free-state, South Africa. In: International Conference on Emerging Technologies for Developing Countries, pp. 233–238. Springer (2017)
31. Priyadarshinee, P., Raut, R.D., Jha, M.K., Gardas, B.B.: Understanding and predicting the determinants of cloud computing adoption: a two staged hybrid SEM-neural networks approach. Comput. Hum. Behav. **76**, 341–362 (2017)
32. Hsu, C.L., Lin, J.C.C.: Factors affecting the adoption of cloud services in enterprises. Inf. Syst. e-Bus. Manag. **14**(4), 791–822 (2016)
33. Şener, U., Gökalp, E., Eren, P.E.: Cloud-based enterprise information systems: determinants of adoption in the context of organizations. In: International Conference on Information and Software Technologies, pp. 53–66. Springer (2016)
34. Salum, K.H., Rozan, M.Z.A.: Conceptual model for cloud ERP adoption for SMEs. J. Theor. Appl. Inf. Technol. **95**(4), 743 (2017)
35. Bhatia, S.S., Gupta, V.: Principles and practices for the implementation of cloud based ERP in SMEs. In: MATEC Web of Conferences, vol. 57. EDP Sciences (2016)
36. Kinuthia, J.N.: Technological, organizational, and environmental factors affecting the adoption of cloud enterprise resource planning (ERP) systems. Ph.D. thesis, Eastern Michigan University (2014)
37. AlBar, A.M., Hoque, M.R.: Determinants of cloud ERP adoption in Saudi Arabia: an empirical study. In: 2015 International Conference on Cloud Computing (ICCC), pp. 1–4. IEEE (2015)
38. Gangwar, H., Date, H., Ramaswamy, R.: Understanding determinants of cloud computing adoption using an integrated TAM-TOE model. J. Enterp. Inf. Manag. **28**(1), 107–130 (2015)
39. Awa, H.O., Ojiabo, O.U., Orokor, L.E.: Integrated technology-organization-environment (TOE) taxonomies for technology adoption. J. Enterp. Inf. Manag. **30**(6), 893–921 (2017)

40. Lee, J.: Discriminant analysis of technology adoption behavior: a case of internet technologies in small businesses. J. Comput. Inf. Syst. **44**(4), 57–66 (2004)
41. Alshamaila, Y., Papagiannidis, S., Li, F.: Cloud computing adoption by SMEs in the north east of England: a multi-perspective framework. J. Enterp. Inf. Manage. **26**(3), 250–275 (2013)
42. Martins, R., Oliveira, T., Thomas, M.A.: An empirical analysis to assess the determinants of saas diffusion in firms. Comput. Hum. Behav. **62**, 19–33 (2016)
43. Borgman, H.P., Bahli, B., Heier, H., Schewski, F.: Cloudrise: exploring cloud computing adoption and governance with the TOE framework. In: 2013 46th Hawaii International Conference on System Sciences (HICSS), pp. 4425–4435. IEEE (2013)
44. Lal, P., Bharadwaj, S.S.: Understanding the impact of cloud-based services adoption on organizational flexibility: an exploratory study. J. Enterp. Inf. Manage. **29**(4), 566–588 (2016)
45. Awa, H.O., Ukoha, O., Emecheta, B.C.: Using TOE theoretical framework to study the adoption of ERP solution. Cogent Bus. Manage. **3**(1), 1196571 (2016)
46. Ramdani, B., Kawalek, P., Lorenzo, O.: Predicting SMEs' adoption of enterprise systems. J. Enterp. Inf. Manage. **22**(1/2), 10–24 (2009)
47. Faasen, J., Seymour, L.F., Schuler, J.: SaaS ERP adoption intent: explaining the South African SME perspective. In: Enterprise Information Systems of the Future, pp. 35–47. Springer (2013)
48. Lian, J.W., Yen, D.C., Wang, Y.T.: An exploratory study to understand the critical factors affecting the decision to adopt cloud computing in Taiwan hospital. Int. J. Inf. Manage. **34**(1), 28–36 (2014)
49. Lechesa, M., Seymour, L., Schuler, J.: ERP software as service (SaaS): factors affecting adoption in South Africa. In: Re-Conceptualizing Enterprise Information Systems, pp. 152–167. Springer (2012)
50. Mcsi, M.: Configurability in SaaS (software as a service) applications. In: Proceedings of 2nd Annual Conference on India Software Engineering Conference, pp. 19–26 (2009)
51. Guo, P.: A survey of software as a service delivery paradigm. In: TKK T-110.5190 Seminar on Internetworking (2009)
52. DeSisto, R.: Software as a Service: Uncertainties Revealed. Gartner Research (2009)
53. Young, R., Jordan, E.: Top management support: mantra or necessity? Int. J. Project Manage. **26**(7), 713–725 (2008)
54. Yang, Z., Sun, J., Zhang, Y., Wang, Y.: Understanding SaaS adoption from the perspective of organizational users: a tripod readiness model. Comput. Hum. Behav. **45**, 254–264 (2015)
55. van de Weerd, I., Mangula, I.S., Brinkkemper, S.: Adoption of software as a service in Indonesia: examining the influence of organizational factors. Inf. Manage. **53**(7), 915–928 (2016)

Sharing Device Resources in Heterogeneous CPS Using Unique Identifiers with Multi-site Systems Environments

Diego Sánchez-de-Rivera[1](✉), Borja Bordel[1], Álvaro Sánchez-Picot[1],
Diego Martín[1], Ramón Alcarria[2], and Tomás Robles[1]

[1] Department of Telematics Systems Engineering,
Universidad Politécnica de Madrid, Madrid, Spain
{diegosanchez,bbordel,asanchez,dmartin,
trobles}@dit.upm.es
[2] Department of Geospatial Engineering,
Universidad Politécnica de Madrid, Madrid, Spain
ramon.alcarria@upm.es

Abstract. Cyber-physical systems environments offer value-added infrastructure by incorporating heterogeneous capabilities that provide a variety of data and processes over diverse and independent final devices. This can often leads to a complex setup that underuse the global capacities of the devices by preserving it in a single service process. By creating several identifiers for each service using specific devices, we can reuse multiple devices for several high layer individual services and processes. This paper proposes an architecture model and a division based on the different levels that are required to add the reutilization of the resources. This includes service replication schemes and device identification processes to the system. In addition, a first approximation of the methodology for providing unique identifiers is presented, in order to allow sharing device resources between different services.

Keywords: Cyber-physical systems · IoT · Service provision ·
Resource sharing

1 Introduction

Cyber-physical systems (CPS) are generally composed of heterogeneous devices that must be integrated to perform a common function that is in general a part of a certain service or process. An important part of all CPS in order to work properly is defining the reliability requirements of the specific information request that allow a system to controlling which devices are selected and verifying the correct behavior of the data received [1]. In an actual deployment, devices would also have a different state at each moment of the execution, being able to be affected by external components of the system and that it is necessary to have well controlled and tested.

The identification of functions, execution controls and execution optimizations in heterogeneous environments involve several modules that takes place in the process. Such as, the state of the defined capabilities in each final device, identify if its state is

© Springer Nature Switzerland AG 2019
Á. Rocha et al. (Eds.): ICITS 2019, AISC 918, pp. 153–164, 2019.
https://doi.org/10.1007/978-3-030-11890-7_16

the appropriate one in the execution context or the reusability factor possible in that task, in order to optimize the general utilization of the platform. In this article, the authors first define an additional division layer of the execution environments by the incorporation of the called "sites". After, a characterization process in which not only the devices, but also the execution service can be identified to share resources between several services of the system, by using unique identifiers for each entity is explained. The authors propose the use of identifiers, besides to identify the device itself, to know the status of the device in the face of a service execution [2].

One major drawback of using a predefined method to implement an algorithm to identify available entities in a certain system is the inability to predict future modification of the deployed algorithm. This is the basic approach of actual studies, where identifications are stored in a constant updated dictionary. We complemented the identification data with additional metadata to provide the more detail, such as the necessary information of the executed service. This dictionary, allows with the correct algorithms, to reduce the load of the active services by reutilizing some of the required data delivered by the final devices. Recent studies expose the limitations of cyber-physical devices when they are integrated to a running environment and cannot identify similar behaviors that could comply with the defined quality of service of the system [3].

The incorporation of power devices help the distribution of the management chain along the system and play an important key managing all other components [4]. This is why, in previous works, the problem was focusing majorly in the complex devices is the inability of controlling the lightest resources of our system, and nowadays these devices can perform very complex task that can overrides the capabilities of the system major components.

This paper study the actual architecture designs that have an impact in incorporating reutilization of the resources, as well as service replication schemes, and device identification [5]. Then a proposed model is presented in addition of a methodology for providing unique identifiers to different process. The document is structured as follows. Section 2 study related work and summarize reference models relevant in our approach. Section 3 presents the proposed architecture. Section 4 explains the service identification methodology along the user to device path. Section 5 presents the validation experiment of the system in a real world deployment and Sect. 6 summarizes the conclusions as well as the planned future works.

2   State of the Art

To provide a complete background of the current state of the art related to service provision in a heterogeneous system, some considerations and clarifications are required. Our vision of a service-based environment has at least two main procedures, which a reliable system need to perform in order to acquire a ready state.

The first stage in a new deployed system is the configuration process that leads to acknowledge the information of every entity, device or module. This installation phase can be manual or automatic, and in cases of large deployments can be challenging [9]. This is why a consistent installation phase is needed in order to obtain a truly automated environment.

The second stage takes into account the evolution over time of the system. In a dynamic system, the devices can suffer a variety of different events and some of them could indeed change the status or modify the system capabilities over time, without requiring a restart of the system or stopping it. It has to be noted that the discussed layer is not the only implementation layer that needs to be addressed, but in this paper, these procedures are the brought to the focus and variations over current research proposed.

Basic operations within a newly deployed system can include adding, removing and replacing devices. Each of those will require a predefined workflow in the adaptation process. While it can be a manual process, automated environments often relay the auto-configuration process to a superior layer, in charge of gather and save the capabilities in a persistent database [10]. This iterative process is very resource consuming and a fine tune is necessary in the entire chain in order to identify available resources and match them with the service requested from the user.

Recent research such as [6] presents a CPS framework with middleware integration that supports a dynamical architecture with artificial intelligence (AI) planning. Distributed approach is used in order to interface with multiple CPS or IoT sites. The process of maintain the required relationships between various resources located in a sub-network environment is analyzed and a process of "mix and match" are enabled.

Other approaches, that recently have seen increased their potential with the actual computational power of the newer devices are focus on integrate semantic models into the service identification process [7]. By separating three concepts, physical entity, location and CPS service they are capable of identify in terms of context the current execution progress and common information. The main drawback of this implementation can be the high computational time that it is necessary to process the ontology related information, as opposed to our proposal.

Regarding the second phase and trying to adopt a common architecture for a planning and execute services in a large and complex systems, [8] propose an outline architecture focused in tie service goals with changes in the domain model, using the characteristics of the real-life systems. This architecture can be seen as the starting point that can be enhanced with different mechanisms to auto-configure and characterize the global system.

3 Proposed Architecture

CPS are composed by several devices that often are in an immutable state or stablished in a certain manner that it is complicated to be altered. In that case, a device model can be defined in the planning phase of the CPS. This strategy involve a predefined model that can be pre-configured or not with the device, declaring its functions, capabilities and requirements in a regular mode. On other consideration, devices can be added to the system and removed dynamically, by decision of the administrator or by external factors that prevent the normal operation of the device, such as connectivity problems, battery discharge or even device malfunction [11].

In this paper, we add the concept of CPS subsystems, that are compound by a set of different CPS devices such as sensors and actuators that work in a relatively close environment, geographically or virtually and can be controlled by a unique so called gateway module.

Multi CPS subsystems paradigm approach can be done in several ways, and implement a universal method to provide the seamless integration must cover every aspect of the device to the user chain. Thus, it is important to develop a modular integration to allow future incorporations and modifications. In a primary approximation, we can set a primitive model categorization to characterize the different function types.

Our infrastructure differentiates several abstraction levels. It is so because the system integrates a complex service provision framework and doing this separation allows a granular identification of requisites and optimal strategy definition. In a real world CPS we can find several standalone devices acting as a computational resources that provides information and relays it to a more powerful and interconnected elements, which process it and generate useful information. On our infrastructure, considering the overview presented in Fig. 1, three levels are differentiated in order to obtain the general architecture layout.

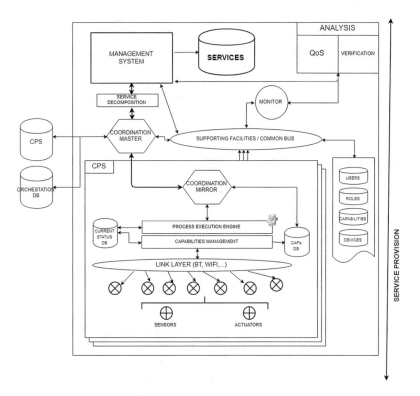

Fig. 1. Infrastructure overview

3.1   Service Provision

This is the primarily management level. At this level, all the service definition and rule provision is made in order to require low-level entities to provide the necessary data to obtain a result. Actors involve at this level are the knowledge experts that configure the rest of the infrastructure. A master entity that rule the entire system is needed in order to maintain control over the rest of the devices and gateways. In this paper, service provision level is defined as a transparent layer that is present and the service plans generated by this abstraction layer are relayed to the gateway interconnection to provide the necessary detail of the required devices. Thus, we identify several functions that this level contains.

Management System: The global entity that controls the available services incorporated to the system. As its main function is to generate and report the results of the running operations the user interface interact directly with it and provides the external data and parameters required to configure the system as a whole. Inside will reside for example the interfaces connecting the architecture with business logic models, applications or extensions that could be required in a controlled deployment.

Services Repository: The general database that stores the services, those previously identified at a precise time in the global system that has been mark as available. The database is accessed by the management system as it is populated by the services acknowledged in the lower levels.

Service Decomposition: In order to interact with lower levels and to be able to distribute and divide the services into the capabilities required from the final devices, a service decomposition block is placed as the main access enabling the actual communication and data relaying from and to the elements. These tasks are important in the data path, as they need to be merged and selected to fulfill upper layer requirements. Decomposition module also interacts with the services repository database, as it needs to know the properties of the service to allow the gather of the required capabilities.

3.2 Gateway Interconnection

This level aims to provide a context-aware domain in which the devices can be deployed to the system. The context that a certain gateway generates is isolated from other contexts but can be integrated with them in the service provision level.

This level functional model as represented in the Fig. 2 performs several coordination mechanism to interact with the management level by passing off the relevant information to the upper layer, and expecting the service and capabilities request data because of the anticipated process done by the superior level. This service plan contributes to the gateways by requesting the concrete capabilities and results that the devices connected are be able to provide.

Fig. 2. Coordination layer

Coordination Master: To enable the ability of interact with several devices subsystems, the proposed architecture is served by the incorporation of a coordination module and several coordination slaves. As the system can be increased in complexity, it is necessary to granulate the data along the device-service path. The coordination master takes the responsibility of distribute, queue, discard and prioritize available resources and capabilities, and it does that by relaying certain mechanism to the next explained module.

Coordinator Slave: It will be replicated in every existing subsystem and will be in a close relationship with the master. Their functions are beyond the basic gateway of a certain set of devices because of the incorporation of a process execution engine that performs delegated tasks from the master, as can be seen in Fig. 3.

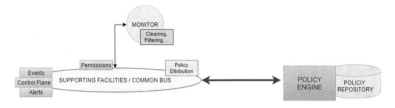

Fig. 3. Supporting facilities/Common bus

Supporting Facilities/Common Bus: To support the subsystem interconnection with the common infrastructure, we propose the addition a supplementary entity modeled as a ubiquity communication mechanism available for every module of this and the superior layers. The communication is based in loosed coupling and high scalability protocols such as MQTT to provide permanent availability and reliability of the global system. The incorporation of this module allows the control data to be decoupled from the main data path and will provide the chance of integrate several services such as monitoring and even analysis or verification process without interfering the data path [12].

3.3   Data Provision

The lowest level defined in the proposed infrastructure is dedicated to the end user devices connected to the platform. These devices are often from different manufactures that are not compatible with each other's and therefore could be difficult to interconnect the different services or capabilities that the devices offer to the system [13]. The methods of connecting each device to its correspondent gateway is relayed to the actual connection technology or in the general case, the connection is transparent to the service platform but there is some components that are common to all de devices providing actual data.

Process Execution Engine: Just after the coordination mirror takes into account the required resources that a certain devices (or a set of them) have, it is time of relay the charge of monitoring the execution process to this module. This is where the final communication with the devices or actuators is made, and the first entry gate of the raw information into the data path. The process execution engine will watch the execution of the required data gathering requested from the management system, as the services involved, and the actual capabilities used in the devices connected.

Capabilities Management: To help the previous module with the device interaction process, a supplementary service will be used in order to provide different database interaction at the device level. This service can provide device status, current capabilities, quality of service information and additional data that could be useful at this layer level.

4   From Devices to Services

In our work, we consider the gateway element of a sample CPS subsystem as more than a simple function module as explained previously. By using the incremented processing capabilities of the gateway device, the system can relay certain functions to the middle layer and release the service creation from the granular management of the devices. This helps the creation of more complex infrastructure as more gateway systems can be incorporated at the middle layer.

This is why, besides the behavior of the service interconnection procedures, the middle layer also acts as a trigger mechanism for the final device in which the actual data provision or is made. Due to the fact the devices can be part of several services at a time it is needed an entity capable of indicate them the most vital task.

Thus, the gateway layer defined in this paper performs two main tasks. The first procedure is being in charge of homogenize the connection interface by offering standard methods to interact with connected devices. These methods must be available in each device and they has to be supported by the hardware and implemented by the software.

Second task assigned to the gateway interconnection level takes into account once a successful addition to the system has been done, and implies the collection of every capability that the connected devices had incorporate and generate a unified system status to resume the state of the context at that time.

At this level, and in a top-down definition, the gateway gives the responsibility of the resource accommodation to the process execution engine and the capabilities management modules, which will have in their data base components a new service is assigned to it, and this is permanently updated. When a service execution need the resources of a known device, it triggers a new request message for the gateway in charge of the required device. The gateway now controls and manage the device by requesting the preferred behavior and dealing with the concurrent services that could be executing at the same time and using the same resources in the final device.

Gateway level uses the information provided by the upper layer to ease the device management procedures and incorporate execution privileges by replicating a reduced subset (only devices in it range) of the identifiers database, and this is why we propose the addition of unique identifiers for each service, which is detailed next.

To formalize the described method, authors define an execution environment E that shapes a Service Provision platform composed by different types of devices. In the model, a device D_i is represented by a collection of several m_n parameters in the following way (1):

$$D_i = \{id_i, m_1, \ldots, m_n\} \tag{1}$$

The device is identified by its unique identification id_i and a set of parameters in the form of key and value, selected and preconfigured in the device. These parameters, that are a subset of the D_i set, can be also identified from the standard messages that default device provide, thus in order to start detecting them only is necessary a first scan or connection with the link layer of choice.

The execution environment then can be expressed as a set of i devices currently connected.

When a device is connected to an execution environment that forms part of a global system management group previously configured, the capabilities management engine as well as the process execution engine are aware of the inclusion and start relaying the D_i set to the management system.

As stated before, the device i can be part of an isolated sub-system, so the information must flow through the common bus in order to be considered by the upper layer. In addition, the device model D could incorporate a "location" information to the parameter list and the process execution engine of each sub-system can decide if the execution is assigned for that device.

4.1 Incorporating Unique Identifiers

Capabilities offered by final devices in a cyber-physical system can be abstracted from the service layer by identifying services that the user can require. This manner, the presence of a certain device in a complex system can be seen as a part of a certain service that can be used and therefore, offered to the upper layer.

In our defined infrastructure, services are formed by grouping founded devices and capabilities and offering these to the management layer. Moreover, the management layer can request the status of a certain device or information about the reliability of the capabilities involved in an operation. The problem is that with several contexts running at the same time, it causes deviations in the status of the other contexts in an inevitable manner.

To overcome this problem, we propose the incorporation of unique identifiers to services along with devices identification. We define an identifier as a unique identification string or number which involves all the layers that a certain service or capability needs to be identify in a process context. This will be present in addition to the descriptor D_i, which is generated in a previous step as explained.

The identification string will be composed, as before, as a matrix of arrays, using declarative and exclusive elements that define an entity in our system. Device identification and Parameters can be extracted from D_i, as it contains the related information specific to the device. This is where the management layer can decide whether to reutilize the capabilities identified in the information model in another requested service, this process is executed into the resource engine.

With every new created service, a new collection is produced in the management system module, by grouping the selected elements. That is needed to control the amount of resources used in a certain time, and with that, the system can start to test out if any of the previous services already running can provide the same required results of the new service. The service then activates the shared resource engine and in a seamless actuation, the provision of the data is realized without the need of subsequent data request from the final devices.

Algorithm 1. – Resource matching process
for each Gateway *in* Service
  check Gateway presence
  for each Device *in Service*
   if Device exists *in Gateway*
    check Device
    for each Params[Key] *in* Device
     if Params [key] **exists in** *ExecutingParams*
      Params [key] = *ExecutingParams[key]
    else
     add Params[key] to ExecutingParams[]
    end for each
   end if
  end for each
end for each

It is assumed that the management module is aware of every service created, no matter the final user that creates the service so that the share of the resources between contexts is possible. Related to this, a clarification of the shared data boundaries is worth it: the user have not the exclusive exploitation, but only the device data will be shared, this is, the usage that the certain service will made of the raw data is kept private as the business layer of each service is preserved to a single user.

The service identification will be exchanged along with the capabilities used to permit the sharing of the results generated by the management engine and the set is replicated also in the gateway responsible of at least one device implied in the service. This way, the system can user the gateway entity to perform service selection task at the middle layer. In order to perform the share of the resources, a first test algorithm has been deployed to the management system. Algorithm 1 perform for every created service a iteration between current running services to discern which resources can be reused and avoiding repetitive requests.

5   Experimental Example

An evaluation has been designed by the authors to prove the real performance and validity of the proposed system in a real environment. A platform based on the Samsung Artik ecosystem [14] has been deployed in a real world situation and different scenarios has been proved in order to prove the validation of the system in a cyber-physical system implementation. The infrastructure has been used by different people profiles and tested in several cases.

This experiment is designed to validate the proposal of the architecture, and experience the real behaviors of the system, and compared with the lack of a unified management based on unique identifiers. The results of the real world deployment allows the authors to simulate a complex situation in which several devices are be added and then evaluate the performance of the system with different metrics.

The first component is using the low-power dedicated architecture model Artik 020, as it is specially designed to provide a small form device for IoT and CPS applications. This version has a Bluetooth LE wireless connection that allow a long battery life. For the gateway instead, we are using a more powerful processor which runs with a Linux operating system. This device needs to be powered from an external source, as the battery option is not usable for a long life operation.

5.1   Architecture Validation

To perform this validation, two gateways were deployed and four final devices connected, three located in the first gateway range and the other device discoverable by the second gateway. Each gateway is placed in a separated room, connected with the internet with wired Ethernet.

At first, the gateway tries to connect to every Bluetooth device found and an array of resources made in order to send it to the defined management system. Two types of resources are available in this test: a Temperature sensor and a Passive Infrared Sensor (PIR). Both of them available through the Bluetooth low energy characteristics and gathered in a simple network scan. The location string is literally coded in them and not modifiable in the current deployment. This allows to determinate the geographical situation of the sensor devices.

After gateways are started and the connection to the upper layer stablished, Management system boots and runs in an external server. Right now only offer the minimum functionality in order to validate the global system. The final user can create a service into the system by selecting the desired data sources and program simple rules with the data received. In this first experiment, only simple algebraic operations are allowed as well as conditional expressions. This allows the creation of rules in order to test the environment and validate it. In addition, a maximum time between new data request is specified in order to determine the minimum time that a resource can be requested again.

Figure 4 shows the processing time that certain procedures take to complete with the specified services created. System validation is done within the first service added and executed successfully. Then, we repeat measurements while adding more services to see how the system performs in relation with the workload. The created services were using only a combination of the sensors available in the connected devices. All services are created with a fixed maximum time of 10 s between each iteration. The new device procedure is triggered by the disconnecting or connecting procedure of a device, so each time that an addition or removal happen, the routine is initiated. Averages time are presented.

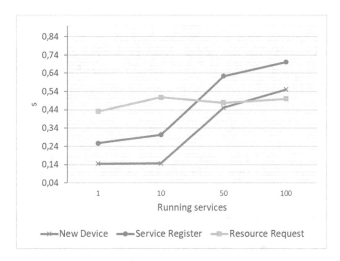

Fig. 4. Time measurements

As we can see in the results, it noticeably how the time of processing new devices and new services increase as more services are added to the running state. This increase is due because of the increment of the service matrix, as it include more shared resource and the algorithm needs to regenerate the assigned resource for each service.

On the contrary, the resource request processing time, that is, the time needed to obtain the sensor data from the final device maintains a similar performance in each measure. As the services request the required sensor data, the management system uses shared data acquired through the unique identification identities to interact with the gateway only if the maximum time has been completed, and sharing the data received between every service waiting. This allows a better resource utilization and low busy time as we can see in the results.

6   Conclusions

Cyber-physical systems are now greatly extended in a variety of situations; it has been even more since the IoT paradigm started to grow. The majority of the methodologies that the internet of things is using were started to being developed with CPS environments. In this sense, the use multi-site environments for a common service objective is one of the models that can be needed for a large deployments, and an architecture of a multi-site systems is one of the options that can be settled in that case.

Although the proposed model is best suited for a particular scenario, the work tries to abstract the necessary modules for a successful working environment with a scalable solution and ready for heterogeneous devices. In this paper, a proposal of an architecture model for multi-site the cyber-physical systems is presented as well as a division in different layer levels and definition of the main modules is described.

Acknowledgments. These results were supported by UPM's 'Programa Propio'. Additionally, the research leading to these results has received funding from the Ministry of Economy and Competitiveness through SEMOLA project (TEC2015-68284-R) and from the Autonomous Region of Madrid through MOSIAGIL-CM project (grant P2013/ICE-3019, co-funded by EU Structural Funds FSE and FEDER).

References

1. Harrison, R., McLeod, C.S., Tavola, G., Taisch, M., Colombo, A.W., Karnouskos, S., Delsing, J., et al.: Next generation of engineering methods and tools for SOA-based large-scale and distributed process applications. In: Industrial Cloud-Based Cyber-Physical Systems, pp. 137–165. Springer, Cham (2014)
2. Jammes, F., Karnouskos, S., Bony, B., Nappey, P., Colombo, A.W., Delsing, J., Bangemann, T., et al.: Promising technologies for SOA-based industrial automation systems. In: Industrial Cloud-Based Cyber-Physical Systems, pp. 89–109. Springer, Cham (2014)
3. Wu, C.L., Chun-Feng, L., Li-Chen, F.: Service-oriented smart-home architecture based on OSGi and mobile-agent technology. IEEE Trans. Syst. Man Cybern. Part C (Appl. Rev.) **37**(2), 193–205 (2007)
4. Hoang, D.D., Paik, H.Y., Kim, C.K.: Service-oriented middleware architectures for cyber-physical systems. Int. J. Comput. Sci. Netw. Secur. **12**(1), 79–87 (2012)
5. Alcarria, R., Robles, T., Morales, A., López-de-Ipiña, D., Aguilera, U.: Enabling flexible and continuous capability invocation in mobile Prosumer environments. Sensors **12**(7), 8930–8954 (2012)
6. Le, D.H., Narendra, N., Truong, H.L.: HINC-harmonizing diverse resource information across IoT, network functions, and clouds. In: IEEE 4th International Conference on Future Internet of Things and Cloud (FiCloud), pp. 317–324 (2016)
7. Sun, Y., Yang, G., Zhou, X.: A novel ontology-based service model for cyber physical system. In: IEEE 5th International Conference on Computer Science and Network Technology (ICCSNT), pp. 125–131 (2016)
8. Feljan, A.V., Mohalik, S.K., Jayaraman, M.B., Badrinath, R.: SOA-PE: a service-oriented architecture for planning and execution in cyber-physical systems. In: IEEE International Conference on Smart Sensors and Systems (IC-SSS), pp. 1–6 (2015)
9. Gurgen, L., Gunalp, O., Benazzouz, Y., Gallissot, M.: Self-aware cyber-physical systems and applications in smart buildings and cities. In: Proceedings of the Conference on Design, Automation and Test in Europe, EDA Consortium, pp. 1149–1154 (2013)
10. Lee, J., Bagheri, B., Kao, H.A.: A cyber-physical systems architecture for industry 4.0-based manufacturing systems. Manuf. Lett. **3**, 18–23 (2015)
11. Mayer, S., Verborgh, R., Kovatsch, M., Mattern, F.: Smart configuration of smart environments. IEEE Trans. Autom. Sci. Eng. **13**(3), 1247–1255 (2016)
12. Erradi, A., Maheshwari, P., Tosic, V.: Policy-driven middleware for self-adaptation of web services compositions. In: ACM/IFIP/USENIX International Conference on Distributed Systems Platforms and Open Distributed Processing, pp. 62–80. Springer, Heidelberg (2006)
13. Colombo, A.W., Karnouskos, S., Mendes, J.M.: Factory of the future: a service-oriented system of modular, dynamic reconfigurable and collaborative systems. In: Artificial Intelligence Techniques for Networked Manufacturing Enterprises Management, pp. 459–481. Springer, London (2010)
14. Samsung ARTIK Internet of Things (IoT) Platform. https://www.artik.io/. Accessed 31 May 2018

Digitalization Changing Work: Employees' View on the Benefits and Hindrances

Jussi Okkonen[⊠], Vilma Vuori, and Miikka Palvalin

Tampere University, 33014 Tampere, Finland
jussi.okkonen@uta.fi,
{vilma.vuori,miikka.palvalin}@tut.fi

Abstract. The Digitalization of work may enable better use of knowledge, which is expected to result in enhanced productivity and efficiency. Different organizations adopt digital ways of working, products, services and processes in different ways and paces. The extent and pace of capitalizing digitalization's potential in organizations has much to do with human attitude, ability to learn and the possibilities provided to use it in work. When looking digitalization from the employee viewpoint, it seems that the great expectations are not easily fulfilled. In fact, the effects of digitalization may be negative. This paper discusses the expectations and fears employees have towards digitalization changing their work. The issue is approached by using a qualitative data gathered along a survey. The aim was to find out if they see the effects digitalization has on their work as positive, negative, or something else, and which areas are highlighted.

Keywords: Digitalization · Digitalization of work · Benefits · Hindrances · Employees

1 Introduction

Digitalization has been understood to be a key driver of globalization for a long time. Organizations can operate without the restraints of time and presence as technological developments enable easier, faster and more affordable interaction among people. Consequently, digitalization has enabled entirely new business models and value creation mechanisms [1–3]. In fact, digitalization has rearranged the economics and business environments of organizations and simultaneously it has remodeled the ways people work.

Digitalization of work may enable better use of knowledge [4, 5] which is expected to result in enhanced productivity [6–10] and efficiency [11–13]. However, when looking digitalization from the employee's viewpoint, it seems that the great expectations are not easily fulfilled. In fact, the effects of digitalization seem to be twofold: By bringing about ever more information systems, applications, user interfaces and operating systems to enhance productivity and efficiency of work, digitalization has led to increasing information load, hectic pace of work, multitasking, and interruptions [14]. Studies confirm that users can experience ICT as demanding and stressful [15–17]. Another rather negative result of digitalization is potential weakening of social ties

© Springer Nature Switzerland AG 2019
Á. Rocha et al. (Eds.): ICITS 2019, AISC 918, pp. 165–176, 2019.
https://doi.org/10.1007/978-3-030-11890-7_17

and reducing social inclusion: by increased use of ICT people tend to have less face-to-face contacts [18] and in work context this may lead to weakening sense of community, and consequently issues with trust and motivation. Consequences of inadequate information systems, such as decreased job satisfaction and engagement with the organization [19], can negatively affect work quality and productivity [14].

Digitalization of work is a multilevel phenomenon ranging from, for example, using email instead of posted mail, utilizing videoconferences to enable remote face-to-face meetings, tapping into web platforms to sell products and services online, or outsourcing and automating routine tasks dealing with large data masses to software robots operating with artificial intelligence. Different organizations adopt digital ways of working, products, services and processes in different manner and pace. Issues with adopting digital ways of working may include change resistance among the employees, customers' and partners', general attitude towards digitalization, and the gap between current use of digital tools and the target level. Digitalization will surely change work in almost every organization at some point of time, but the extent and pace of capitalizing its potential has much to do with human attitude, ability to learn and the possibilities provided to use it in work context.

This paper discusses the expectations and fears that employees working in Finnish organizations have towards digitalization changing their work. The issue is approached by first conducting a literature review on the benefits and hindrances that digitalization may bring to work. Then, using an empirical case study, employees' attitudes towards these issues are clarified. The aim is to find out if they see the effects digitalization has on their work as positive, negative, or something else, and which areas are possibly highlighted.

This study aims to answer the following research questions:

(1) What are the expected benefits and feared hindrances regarding work performance caused by digitalization of work?
(2) How do the employees experience these benefits and hindrances in their everyday work?

The paper is constructed as follows: First, a brief theoretical discussion on the expected benefits and potential harmful effects of digitalization of work are presented. After that, the methodology and data gathering choices regarding the case study are described, followed by a presentation of the results of the empirical study. Finally, the conclusions and implications of the paper are provided along with suggestions for further research.

2   Two-Fold Effects of Digitalisation of Work

Digitalization is expected to enhance work performance by promoting productivity and efficiency as well as enabling better use of knowledge. Information and knowledge are the focal resources of contemporary organizations, and therefore, their effective flow is crucial for their performance: Effective knowledge flows may promote job satisfaction [20, 21], which, again, is linked to productivity [22, 23]. The expectation is that digitalization enhances knowledge flows by providing effective tools to act as mediums and enablers of work. Productivity advances hand in hand with the speed by which information flows through the process [24]. The better and more effective the knowledge flows within an organization are, the quicker employees are able to plan and perform their tasks [25]. However, even if effectively managed knowledge flow improves productivity [26, 27], poorly managed knowledge flow may cause information overload leading to reduction in productivity on both the individual and organizational levels [28].

Organizations often seek to supplement their resources and knowledge base by networking. Mobility and asynchrony, i.e. independence of time and place, are prerequisites for a modern networked organization, which again call for efficient digital tools. These tools enable and enhance co-creation, as they provide the means to, for example, simultaneously create and edit documents or arranging teleconferences, enabling asynchronous and multi-spatial work and making it more efficient. Efficiency is the optimal allocation of resources, i.e., any resource is available, but used only if needed [29]. However, even if a tool is understood to enhance efficiency, people tend to hold on to their old work conditions and habits because they often experience the demand for the constant integration of new technology as stressful [30]. In addition, personal preferences, organizational conventions, and the socio-technological environment might steer one toward using a certain tool.

In general, digital tools are designed to enable work, yet technology can also pose barriers that hinder the flow of knowledge from source to recipient (e.g. [31]) According to Riege [32], technology-related knowledge barriers include, among others, unrealistic expectations for information systems' performance and insufficient training and support regarding information systems. Technological shortcomings, such as poor usability or malfunctions, of a tool also negatively affect its expected support of work and, in fact, induce strain [17, 33].

Regardless of all the benefits and support these tools are designed to bring to work, they also seem to cause negative symptoms and disturbances of well-being (e.g. [34, 35]), which are often referred to as technostress [36, 37]. Digital tools provide better resources by making more information and knowledge available more easily. However, this may simultaneously cause information overload, as the human ability to process information has not increased [38].

The sense of losing control due to the abundant amount of information brings on inefficiency, procrastination and stress [39]. To benefit from digitalization, it is essential to identify and manage the information load in knowledge-related work. For individuals to better cope with the demands of work, there is a need to develop methods suited for different processes and conventions. Although digitalization may provide more freedom, independence and autonomy to work by enabling flexibility, mobility and asynchrony, it consequently creates an "always on" -mode in which work penetrates leisure time. The expectation of availability and the implicit pressure to reply immediately leads to time-management challenges and affects workers' well-being if they feel they do not have enough time to recharge between working hours [40–43]. The relation between the characteristics of ubiquitous, immersive, digital work environments and performance is a relevant theme for research to understand and, consequently, provide the means to diminish its supposed harmful effects.

As discussed on previous research on the phenomena we summarize that the factors in digitalized work environment enabling performance are Autonomy, Asychrony, Co-creation, Efficient and fast knowledge flow, Independency, and Mobility. Correspondingly, the factors constraining performance are identified as "Always on" -mode, Information overload, Procrastination, Stress and well-being, Technological shortcomings, and Time management challenges [44]. Next, using empirical data, we aim to explore the presence of these enabling and constraining factors in the daily work of employees.

3   Research Methodology

3.1   Data

Research was carried out in Finland between 2015–2018 and the data was collected using online survey. The topic was how working environment affects work and work performance. The survey included 49 Likert scale variables and four open-ended questions. In this, paper the focus is on those open-ended questions. Answering all the questions, including the open-ended, was optional. A questionnaire was sent to the participants by email, and they typically had about two weeks' time to respond. The response rates varied from 33% to 89%. The respondents were mainly from public organizations and some were from public corporations (which were previously public organizations). All the organizations were planning or went through large changes in their workplace, so they needed overviews on how their employees were experiencing their work environments, individual work practices, well-being and productivity. The organizations also planned to use or had used their own results to measure the impacts of the changes. The research data was collected for the organizations' own purposes to assist the change and the participants were informed that the data would be used for scientific purposes as well. The data includes responses before the change from 28 organizations and from 8 organizations also after the change. The total amount of respondents was 8568, and 5332 of them had responded to at least one of the four open-ended questions. All the respondents were considered as knowledge workers on different levels from assistants and junior level experts to project managers and to top management. All the respondents were Finnish-speaking.

3.2   Data Analysis

The quantitative analysis of the data was based on simple heuristic analysis by categorizing different factors brought about to categories of performance enablers or performance constraints. The factors presented as conclusion in Sect. 2 were used as subcategories. Each entry was read through and categorized. Based on researcher judgement each entry was also assessed according to criticality or weight as minor, moderate or critical. In analysis, minor enabler is something that is "nice to have". Moderate enabler is something that "should be". And critical enabler is something "must have". The constraints were treated accordingly their effect, i.e. minor, moderate or critical.

After classification of entries simple frequencies of enablers and restraints and distributions to minor, moderate and critical was calculated. Further analysis is based on these two items, i.e. how often it is mentioned and setting weight on how critical the factor is. Analysis is based on qualitative analysis, since classification is based on single researcher work. To some extent, it would be possible to apply quantitative analysis too, yet it would require classification at least by one other researcher, preferably several. Then it would be possible to use for example chi square test to test the significance of differences between the categories. More thorough assessment would also require additional testing for reliability of the classification by Kripperdorf's alpha. In this paper, qualitative analysis is the main method. The following section presents result, visualization of the distributions and analysis.

4 Results

Overall, 9843 entries of 10684 open-ended entries were related to factors that enable or restrain knowledge work in digital work environment. The other entries were more related to miscellaneous productivity issues, yet those had no relation to sociotechnical work environments. Most entries were about general managerial issue and the organizatizing of work in general. The entries contained relatively rich information on work environment and a total of 11834 entries on factors relating sociotechnical work setting were found.

Respondents entered 5235 factors that enable their work. Factors related to mobility had 289 entries, factor related to asynchrony had 1380 entries, factor related to co-creation had 122 entries, factors related to efficient and fast knowledge flow had 895 entries, factors related to independency had 1005 entries, and factors related to mobility had 1544 entries. Overall the emphasis was on enablers that supported spatial dispersion and asynchronic work on people's own working pace. Figure 1 summarizes enabling factors and the distributions to different importance categories.

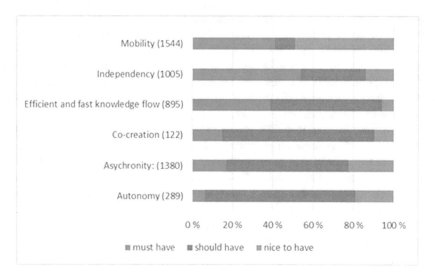

Fig. 1. Summary of number of entries per enabling factor and distribution of importance by enabling factors

Autonomy was considered mostly to have moderate effect on work and overall it was not major issue per se as it was mentioned 289 times. Respondents acknowledged their dependency on coworkers and customers. Only seldom they had full autonomy in their work. Asychronity had 1380 entries and it was seen as the second key enabler. In this dataset asynchrony had effect on work performance, since people could work by different digital platforms with their peers. Digitalization boosted asynchronic working and in general it was seen beneficial. Asynchorny has moderate effect as performance enabler as most of entries set on should have category. Co-creation had least effect as work performance enabler as there was only 122 entries. There might be several explanations, yet most evident is that even working in teams people considered themselves to work alone. However, in many responses people appreciated "relay" working, i.e. working with peers during iteration. On the other hand many people described themselves as single expert on their filed and thus there is no need to constantly work in groups.

Efficient and fast knowledge flow was mentioned 895 times, yet many saw it as a critical enabler. This is due to nature of the work respondents were doing. Making decisions or planning is highly dependent on the available information from different sources. Moreover, as people were spatially dispersed in their organizations the importance of correct and on-time information has a great effect as an enabler. Independency as an enabler had 1005 entries and it was seen important since most of the entries were in the "must have" category. This is mostly due to nature of the knowledge work in general, but mostly due to nature of knowledge work in these organizations. People work on their own core tasks and independency provided by digitalization gives them control on their work flow. As the work is not dictated by a certain process, but it is more organized the people executing it the effect on work performance is remarkable. Moreover, independency is also issue of better work-leisure fit, since people have more options on arranging their work.

Mobility had most entries as it was mentioned 1544 times. The different aspects of mobility in work enable performance since people have freedom of choice to conduct their task in different places, e.g. working distant, working in transportation, working at customer site, fieldwork, etc. This sets people free of spatial restrictions and most of the mobility is supported by digitalization as people have distant access to information, they have sufficient tools, infrastructure is accessible and not bound to certain location. Moreover, the performance effect is also due better utilization of time as people can use excess time or take action immediately.

Factors considered as restraints gained roughly 25% more entries totaling 6599 of them. "Always on" work communication mode or constant connectivity was mentioned 611 times. Most critical and most mentioned performance restraint was information overload as it gained 1994 entries. Procrastination was considered a minor restraint since it was mentioned in 421 entries. Stress and well-being was mentioned 1478 times. The third most important restraint category was technological shortcomings with 1389 entries. Time management challenges was mentioned 677 times. Figure 2 summarizes the restraining factors and distribution according to criticality.

"Always on" -mode is not critical to work performance, yet many respondents recognized it as a restraint. In many cases people felt that they were obliged to be available even if they wanted to work on an individual project, be on break, or even

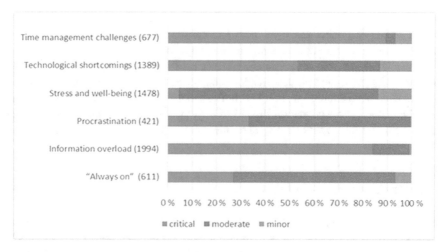

Fig. 2. Summary of number of entries per restraining factor and distribution of importance by enabling factors

during their time off the work. In several entries constant connectivity was related to the urge to be quick and productive.

Information overload was seen as the most critical issue restraining work performance. It has several direct effects as well as indirect ones. Respondents brought about in their entries that overload is due to poor digital skills, insufficient IT infrastructure, poor applications and services, poor explicit rules of information conventions, and poor habits. Information overload can be therefore seen as a cause, a symptom, and an effect of poor information ergonomics.

Procrastination by respondent or by the others seems to be critical and has an effect on performance. In digital environment procrastination is of conducted via email by sending insufficient or incomplete information. On the other hand, people described it also to be a working strategy, yet a harmful to performance though. However, procrastination did not gain so many entries. Stress and well-being was considered the second most critical restraint. In the entries the well-being at work was related to control of own work, getting things done, or general working conditions. In the entries well-being was something that could be perceived and also talked out. On the other hand, stress and well-being are more likely to be mere symptoms while the root cause is something else. However, very many people claimed that stress is causing poor performance.

Technological shortcomings have a direct effect on work as working infrastructure is a prerequisite for productivity. There were entries on how old, outdated, or impractical technology slows working or even totally restraints it. However, the number of entries was relatively low as digitally set work environment is very vulnerable in sense of technology. This somehow reflects technological fatalism, i.e. people can live with shortcomings in digital work environment if it is not totally out of order. Time management was seen as a critical issue in digital knowledge work. Respondents emphasized several issues about how digital environment, especially email, digital working platforms, customer information systems, and digital communication in general affect the work flow and cause distractions. The prioritization of work is done by digital environment, not the people executing the tasks. As presented above digitalization has an outstanding effect on knowledge work. It supports and leverages it by giving individuals freedom and access. On the other hand, digital work environments have enabled working without spatial restraints. As described above the technological advances have their flip side too, yet it is mostly an issue of how to adapt work environment, conventions and habits to better meet expectations and requirements.

5   Discussions and Conclusions

The purpose of this paper was to analyze and compare different factors that either enable or restrain knowledge work in different work settings, and further, to discuss the effect of these identified factors on work performance in knowledge work. In the theoretical part of the study enabling factors i.e. autonomy, asychrony, co-creation, efficient and fast knowledge flow, independency, and mobility were set for key factors. Correspondingly, the factors constraining performance were as "always on" -mode, information overload, procrastination, stress and well-being, technological shortcomings, and time management challenges were discussed as restraints of knowledge work. They have either a positive or a negative effect on the performance of knowledge work actions.

It was seen that there are many similar elements that support or hinder the knowledge work in different organizations, moreover many items entered have a flip side as well. A factor that someones considers to be an enabler may be a restraint for someone else. The enablers were typically the factors that facilitate the professionals' core work, such as information acquisition, refining and dissemination. Communication was considered important, especially asynchronic communication with peers and clientele. In addition, the enablers that were mentioned by the informants helped the knowledge workers to cope with their work pressure, handle the stress and regulate the work flow. Indeed, the enablers were strongly related to the wellbeing at work as well as sense of control and productivity. The core tasks of the knowledge workers consisted of knowledge work actions, such as information acquisition, combining, creation, dissemination, and communication. The wellbeing at the work and coping with the heavy workload have an indirect effect on the knowledge work actions and further, in work performance. The hindrances were related with the professionals' core tasks as well as wellbeing through the work and balancing work and leisure.

It seems that the knowledge work enablers aid more the core tasks of the people while the restraints are more commonly related to the wellbeing, stress and pressure at work. Factors affecting the core work and wellbeing at work were linked with one another to some extent. For many respondents, the factors causing stress and pressure had an indirect impact on their core work. If the core tasks are made difficult or disturbed, it can increase the pressure at work. On the other hand, having suitable sociotechnical work environment in terms of infrastructure, well-functioning work conventions, self-regulated working habits and sufficient skills supports productive working.

The study raised a lot of factors that can act as enablers or restraints of knowledge work. These elements similarly affect productivity as well as the well-being at work either positively or negatively. Key implications regarding knowledge work are about performance. There seems to be three main categories of performance. The first category is individual performance supported by individual knowledge, skills, and working habits. The second performance category is about organization supported by explicit operation procedures and socially constructed conventions. The third performance category is about social capital in sense of recognition of peers and key stakeholders in working domain and working with them. There categories are not exhaustive, yet as conclusion of enablers and restraints they bring about the key issues related to knowledge work in general. These implications have limited direct transferability to knowledge work.

The results implicate that more attention should be paid on managerial practices in organizing knowledge work. Planning sociotechnical work environment is the key, since asynchrony and spatial dispersion of work require more working via digital platforms or using communication tools. This is in close relation to ergonomics of knowledge work, i.e. how infrastructure is set, how working is explicitly instructed, and how working conventions dictate workflow and conducting the tasks. As brought about only part of issues could be planned, since conventions are dependent on the whole realm of work as well as habits are result of individual development.

The results presented in this paper are based on single researcher qualitative analysis. The validity of the factors is validated in previous research. The classification of the entries is valid as it is based on single researcher interpretation. Researcher triangulation and testing the uniformity by Krippendorfs alpha could increase the reliability of classification. As the data is now presented can be considered somewhat reliable and therefore the results are valid with certain restrictions. Restrictions come merely from the domain, since the data is not representative in general but exhaustive on public sector and publicly owned companies.

With these findings, this paper contributes primarily to the research on sociotechnical work environments. First, it is about setting knowledge work digital environment and how to prioritize different performance factors or assess the importance of those. Second, it contributes to research on information ergonomics as it elaborates how technology, people, and individual are actually forming complex activity system based on tacit, implicit and explicit rules, norms, conventions, and habits that shape activity called work. Moreover, as results underline information overload as a major issue in knowledge work organization also managerial implications can be drawn from it.

Further research on the domain should concentrate on making the classification finer in order it to better explain different interconnections between enabling and restraining factors. Moreover, the data could be better utilized through revision on interpretation by several researcher. By doing so actually more sophisticated analysis could be applied on data and go beyond simple heuristicsms!

References

1. Sassen, S.: Globalization and Its Discontents: Essays on the New Mobility of People and Money. New Press, New York (1998)
2. Neumeier, A., Wolf, T., Oesterle, S.: The Manifold Fruits of Digitalization: Determining the Literal Value Behind. 13. Internationale Tagung Wirtschaftinformatik WI (2017)
3. Porta, M., House, B., Buckley, L., Blitz, A.: Value 2.0: eight new rules for creating and capturing value from innovative technologies. Strategy Leadersh. **36**(4), 10–18 (2008)
4. Parida, V., Sjödin, D.R., Lenka, S., Wincent, J.: Developing global service innovation capabilities: how global manufacturers address the challenges of market heterogeneity. Res. Technol. Manag. **58**(5), 35–44 (2015)
5. Grudin, J.: Enterprise knowledge management and emerging technologies. In: Proceedings of the 39th Annual Hawaii International Conference on System Sciences, p. 57. IEEE Computer Society, Washington, DC (2006)
6. Shujahat, M., Sousa, M.J., Hussain, S., Nawaz, F., Wang, M., Umer, M.: Translating the impact of knowledge management processes into knowledge-based innovation: the neglected and mediating role of knowledge-worker productivity. J. Bus. Res. **94**, 442–450 (2017)
7. Michaelis, B., Wagner, J.D., Schweizer, L.: Knowledge as a key in the relationship between high-performance work systems and workforce productivity. J. Bus. Res. **68**(5), 1035–1044 (2015)
8. Chou, Y.-C., Chuang, H.H.C., Shao, B.B.M.: The impacts of information technology on total factor productivity: a look at externalities and innovations. Int. J. Prod. Econ. **158**, 290–299 (2014)
9. Ferreira, A., Du Plessis, T.: Effect of online social networking on employee productivity. S. Afr. J. Inf. Manag. **11**(1), 1–16 (2009)
10. Tuomi, I.: Economic productivity in the knowledge society: a critical review of productivity theory and the impacts of ICT. First Monday **9**(7) (2004)
11. Haas, M.R., Hansen, M.T.: Different knowledge, different benefits: toward a productivity perspective on knowledge sharing in organizations. Strateg. Manag. J. **28**(11), 1133–1153 (2007)
12. Porten, M., Heppelmann, J.: How Smart, Connected Products Are Transforming Companies. HBR **93**, 96–114 (2015)
13. Voth, D.: Why enterprise portals are the next big thing. Learn. Train. Innov. **3**(9), 24–29 (2002)
14. Franssila, H., Okkonen, J., Savolainen, R.: Developing measures for information ergonomics in knowledge work. Ergonomics **59**(3), 435–448 (2015)
15. Bordi, L., Okkonen, J., Mäkiniemi, J.P., Heikkilä-Tammi, K.: Employee-developed ways to enhance information ergonomics. In: Proceedings of the 21st International Academic Mindtrek Conference, pp. 90–96. ACM (2017)
16. Salanova, M., Llorens, S., Cifre, E.: The dark side of technologies: technostress among users of information and communication technologies. Int. J. Psychol. **48**(3), 422–436 (2013)
17. Tarafdar, M., Tu, Q., Ragu-Nathan, T.S., Ragu-Nathan, B.S.: Crossing to the dark side: examining creators, outcomes, and inhibitors of technostress. Commun. ACM **54**(9), 113–120 (2011)
18. Chen, W.: Internet use, online communication, and ties in Americans' networks. Soc. Sci. Comput. Rev. **31**(4), 404–423 (2013)

19. Ragu-Nathan, T.S., Tarafdar, M., Ragu-Nathan, B.S., Tu, Q.: The consequences of technostress for end users in organizations: conceptual development and empirical validation. Inf. Syst. Res. **19**(4), 417–433 (2008)
20. Bontis, N., Richards, D., Serenko, A.: Improving service delivery: investigating the role of information sharing, job characteristics, and employee satisfaction. Learn. Organ. **18**(3), 239–250 (2011)
21. Palvalin, M., Vuori, V., Helander, N.: Knowledge transfer and work productivity. In: Spender, J.C., Schiuma, G., Gavrilova, T. (eds.) Proceedings of 12th International Forum on Knowledge Asset Dynamics, St. Petersburgh, Russia, pp. 1120–1134 (2017)
22. Laschinger, H.K.S., Finegan, J., Shamian, J.: The impact of workplace empowerment, organizational trust on staff nurses' work satisfaction and organizational commitment. Adv. Health Care Manag. **3**, 59–85 (2002)
23. Miller, K.I., Monge, P.R.: Participation, satisfaction, and productivity: a meta-analytic review. Acad. Manag. J. **29**(4), 727–753 (1986)
24. Schmenner, R.W.: Service businesses and productivity. Decis. Sci. **35**(3), 333–347 (2004)
25. Wu, F., Huberman, B.A., Adamic, L.A., Tyler, J.R.: Information flow in social groups. Phys. A: Stat. Mech. Appl. **337**(1–2), 327–335 (2004)
26. Dyer, J., Nobeoka, K.: Creating and managing a high-performance knowledge-sharing network: the toyota case. Strateg. Manag. J. **21**(3 Special Issue), 345–367 (2000)
27. Titus, S., Bröchner, J.: Managing information flow in construction supply chains. Constr. Innov. **5**(2), 71–82 (2005)
28. Ben-Arieh, D., Pollatscheck, M.A.: Analysis of information flow in hierarchical organizations. Int. J. Prod. Res. **40**(15), 3561–3573 (2002)
29. Okkonen, J.: The use of performance measurement in knowledge work context. e-Business Research Center, Research Reports, vol. 9. Tampere University of Technology (TUT) and University of Tampere (UTA), Tampere (2004)
30. Tarafdar, M., Tu, Q., Ragu-Nathan, T.S., Ragu-Nathan, B.S.: Crossing to the dark side: examining creators, outcomes, and inhibitors of technostress. Commun. ACM **54**(9), 113–120 (2011)
31. Paulin, D., Suneson, K.: Knowledge transfer, knowledge sharing and knowledge barriers–three blurry terms in KM. Electron. J. Knowl. Manag. **10**(1), 81–91 (2012)
32. Riege, A.: Three-dozen knowledge-sharing barriers managers must consider. J. Knowl. Manag. **9**(3), 18–35 (2005)
33. Day, A., Scott, N., Kelloway, E.K.: Information and communication technology: implications for job stress and employee well-being. In: Perrewé, P.L., Ganster, D.C. (eds.) New Developments in Theoretical and Conceptual Approaches to Job Stress, pp. 317–350, Emerald Group Publishing (2010)
34. Mark, G., Iqbal, S.T., Czerwinski, M., Johns, P.: Bored mondays and focused afternoons: the rhythm of attention and online activity in the workplace. In: Proceedings of the 32nd Annual ACM Conference on Human Factors in Computing Systems, pp. 3025–3034. ACM, Toronto (2014)
35. Mark, G., Voida, S., Cardello, A.: A pace not dictated by electrons: an empirical study of work without email. In: Proceedings from SIGCHI 2012: Conference on Human Factors in Computing Systems, pp. 555–564. ACM, Austin (2012)
36. Gaudioso, F., Turel, O., Galimberti, C.: Explaining work exhaustion from a coping theory perspective: roles of techno-stressors and technology-specific coping strategies. Stud. Health Technol. Inform. **219**, 14–20 (2017)
37. Maier, C., Laumer, S., Weinert, C., Weitzel, T.: The effects of technostress and switching stress on discontinued use of social networking services: a study of facebook use. Inf. Syst. J. **25**(3), 275–308 (2015)

38. Woods, D.D., Patterson, E.S., Roth, E.M.: Can we ever escape from data overload? A cognitive systems diagnosis. Cogn. Technol. Work **4**(1), 22–36 (2002)
39. Mark, G., Iqbal, S., Czerwinski, M., Johns, P., Sano, A.: Email duration, batching and self-interruption: patterns of email use on productivity and stress. In: Proceedings of the 2016 CHI Conference on Human Factors in Computing Systems, pp. 1717–1728. ACM, San Jose (2016)
40. Barber, L.K., Santuzzi, A.M.: Please respond ASAP: workplace telepressure and employee recovery. J. Occup. Health Psychol. **20**(2), 172 (2015)
41. Barley, S.R., Meyerson, D.E., Grodal, S.: Email as a source and symbol of stress. Organ. Sci. **22**(4), 887–906 (2011)
42. Brown, R., Duck, J., Jimmieson, N.: E-mail in the workplace: the role of stress appraisals and normative response pressure in the relationship between e-mail stressors and employee strain. Int. J. Stress Manag. **21**(4), 325–347 (2014)
43. Wajcman, J., Rose, E.: Constant connectivity: rethinking interruptions at work. Organ. Stud. **32**(7), 941–961 (2011)
44. Vuori, V., Helander, N., Okkonen, J.: Digitalization in Knowledge Work: The Dream of Enhanced Performance. Cognition Technology & Work (2018). https://doi.org/10.1007/s10111-018-0501-3

Autonomous Cycles of Collaborative Processes for Integration Based on Industry 4.0

Cindy-Pamela Lopez[1(✉)], Marco Santórum[1], and Jose Aguilar[2]

[1] Departamento de Informática y Ciencias de la Computación,
Escuela Politécnica Nacional, Quito, Ecuador
{cindy.lopez,marco.santorum}@epn.edu.ec
[2] CEMISID, Universidad de Los Andes, Merida, Venezuela
aguilar@ula.ve

Abstract. The industrial process continues its evolution based on the growing technological advances, which provide improvements in production and the satisfaction of environmental needs. The more recent evolution is known as the Industry 4.0 paradigm. In this context, organizations have seen the need to create digital ecosystems and alliances as a competitive strategy. However, there are many aspects to overcome in order to achieve an effective collaboration between organizations with different functions. Some of them are about how to establish collaborative processes, how to identify the possibilities of the contribution of each company, and how to establish functions, responsibilities, and the optimal coordination between them. In light of this situation, we propose a collaborative model for integrating organizations, based on Autonomous Cycles of Data Analysis Tasks, which are self-adaptive to satisfy the changing customer's needs. All this will be made possible by making intensive use of "Everything mining", and new technologies.

Keywords: Business processes · Industry 4.0 ·
Collaborative processes · Virtual organizations · Autonomic computing

1 Introduction

The technological advances of recent years have impacted many areas of the human beings, such as education (virtual learning), culture (multimedia and virtual libraries), among others. These advances have also impacted the industry, generating the transition towards a new industrial revolution based on knowledge, innovation, data management, and industrial integration [1,2].

1.1 Industry 4.0

In general, the industrial evolution has been marked by scientific discoveries that, at certain moments, revolutionized the production of goods. We find well-defined periods characterized by the development of the steam engine, the introduction of electricity, mass production, automation, and the use of robots for repetitive tasks [3].

Milestones have impacted the production processes and proved to be generators of breakthrough stages and disruptive development, with respect to what was known up until that moment. They generated the reduction of time and steps in the processes, which determined greater production capacity, and therefore, more goods produced at lower manufacturing costs [4,5]. Industry 4.0 is the last historical moment of disruption. Basically, it is the introduction of new technologies in industrial processes, contributing to the integration of organizations in order to solve problems and requirements, to improve the efficiency, and to reduce the production times and costs [4,6].

A fundamental aspect in the Industry 4.0 is the Internet, which produces changes in the way how goods and services are offered [7–9]. It has gone from direct trade to online commerce, thus facilitating access to information and communication, both by businesses and consumers. That implies to process all of the immense amounts of data (Big Data) and to analyze all of the data produced by organizations and social networks, to turn it into valuable information (Data Mining), to connect physical devices, through sensors that interact with the cloud (IoT and Cloud Computing), among some of the innovative aspects that appear in the Industry 4.0. On the other hand, despite the paradigm of Industry 4.0, the way in which companies operate and are organized, causes a delay in responding to customer demand [8–10]. This lag translates into logistic bottlenecks, such as the accumulation of inventories, the dissatisfaction of customers, etc. In this context of technological progress, and to reduce the organizational gap, the concept of virtual organizations has appeared, a new form of business organization [11,12].

1.2   Problem

Despite all the technology that exists, achieving a virtual organization based on effective collaboration to solve the changing customer needs and to avoid waste in the manufacturing processes, is a difficult task because it requires analyzing the contribution possibilities of each organization, establishing the functions and responsibilities, establishing the collaborative processes, and coordinating this collaboration [13,14]. As each organization is different, an intelligent autonomous process will be needed to solve this problem. In this context, this work propose an automated model that allows the process of collaboration and integration between organizations in an agile, reliable, and real-time way. For that, which has been proposed in [15,16] is a type of autonomous intelligent supervision that allows reaching strategic objectives around a given problem. The autonomic cycles integrate a set of data analysis tasks, which autonomously and collectively work to achieve the strategic objectives pursued by these cycles. Each task interacts with the others, and has a specific role in the cycle: observing the process, analyzing and interpreting what happens in it, and making decisions about the process that allow reaching the objective for which the cycle was designed autonomic cycles of data analysis task. In this work, we seek to build a virtual organization, using the "autonomous cycle" concept, in which are defined collaborative processes between different organizations, in order to reach common goals. The paper is organized in the next way. Initially, the method used for the literature review is explained. In section three, the related works that were studied as the basis of this proposal are described, and a discussion of the results is made. Finally, in section four is presented the model of autonomous cycles of collaboration between organizations for the Industry 4.0 paradigm.

1.3 Methodology for the Literature Review

The literature review is based on the methodology proposed by Kitchenham [17] based on three stages: Planning of the review, Development of the review, and Publication of results.

1.4 Research Question

The main question in our context is: How to create a general automated environment in which organizations can collaborate fluidly through their processes under Industry 4.0 paradigm?

2 Related Works

The collaboration must consider several contexts, such as social, political, suppliers and investors, and additionally, the changing requirements of the clients [14]. According to the works studied in this context, organizations have relied on the new technologies that are part of the Industry 4.0 paradigm [4,18–20]. In general, modern business models have seen the need to associate in order to create ecosystems of organizations, which is a necessary environment for organizations to collaborate with each other, according to the needs and requirements of the customers [14,19]. To analyze the different works in this context, we will organize them in three domains:

- **Inter-organizational collaboration or business integration:** the goal of these works is proposed different ways of industrial integration.
- **Automation in collaborative manufacturing:** the main goal of these works is to study the automation of the collaborative process in the context of the industry.
- **"Everything Mining" and Big Data in order to form virtual organizations:** these works introduce the utilization of the "Everything Mining" and Big Data paradigms in order to define virtual organizations.

2.1 Inter-organizational Collaboration or Business Integration

The BMW automotive manufacturing plant in Leipzig, Germany, uses the Internet of Things (IoT), has augmented reality, the cloud, vertical and horizontal mechanism, and predictive systems, among other things, which is an example of an Industry 4.0 plant [5]. Fully automated and intelligent in all its manufacturing processes, whose workers only use mobile control tablets to access the data. The communication automation made it possible to minimize the margin of error and bring production to its maximum performance, by adding preventive monitoring, which means that the plant has a very high productivity per machine because of the ability to use predictive maintenance to reduce unplanned downtime, which, in turn, results in an optimal production performance [5]. To achieve this, the key factor was to create the "value chain" of information that included all those interested in the data of production and marketing processes, from suppliers to customers. In this way, it takes the statistics of production outside the borders of the factory, creating collaborative spaces of horizontal integration [21].

The works [4, 18–20] focus on external collaboration as the key to the success of a business. It is seen that the success or failure of an organization depends on its collaboration with other industries and companies [14]. Particularly, they propose a design that allows a strong collaboration between several organizations using a unique model that allows implementing several strategies. On the other hand, while it is true that companies are adopting different technologies to reduce the gap in the face of trends in the preferences of customers, suppliers, and business partners, it is no less true that today's collaborative environments have various limitations [8,9,22]; since there is no automation that allows effective collaboration in real time between organizations. In particular, there is no way to identify a problem before it happens and there are no effective ways to predict new requirements and needs of customers to satisfy [2,8,10].

2.2 Automation in Collaborative Manufacturing

A first work is the collaborative network of manufacture in the cloud called C2NET, for its acronym in English (Cloud Collaborative Manufacturing Networks), which is a project of Industry 4.0 of the European automotive manufacturing, within the "H2020 European Program." The automotive companies were forced to adapt to the changes to remain competitive in the market. Therefore, among other objectives, they have pursued the reduction to the minimum of the costs of orders and inventories, optimizing the requisition of materials in a joint and coordinated way [23]. For this, they have created a virtual collaboration space in a the cloud, which allows the planners of the factories to maintain the raw material and input requirements with specific algorithms, for making decisions that validate each other through the network. This process implies one on line responding when the requirements are received, accepting, and sending the planner the acceptance, so that the requisition plan is coordinated [23]. The communication of the programs occurs in real time. They also applied a vertical integration for the manufacture of their specific product, which forced the reorganization of both the physical space of production and the administrative processes. In the case of dealing with different factories integrated into a joint product, and whose materials came from other factories, a horizontal integration was applied [23,24]. In this case, the inventory levels for each one are combined with the production lines, and they share information in real time, in a complex system that would be impossible only with human participation.

In the electronic industry, technologies and road maps change frequently and to accommodate that it is necessary a flexible integration framework with inherent intelligence. For that purpose, an algorithm needs to be designed, which provides an automation intelligence allowing the industry to cope with such changes [25]. Different models of automation have been introduced in the past to handle this problem; however, after years of experimentation on these models, it has been found that the discrete automation model provides the best features including, flexibility and integration-friendly. Cloud computing also presents itself as a viable tool for smart manufacturing by utilizing Multi-Agent Systems (MAS). It exposes the manufacturing abilities as agents, the products are considered as manufacturing resources and provided as a service to end users [26]. Approaching the automation procedure holistically has proven to be a failure in most industries, as it lacks flexibility and is too intricate to be adaptable to different environments. Having said that, we need to ensure that the solutions we present also need to look at things in a broad perspective, and should be able to handle unforeseen happenings [25].

2.3 "Everything Mining" and Big Data in Order to Form Virtual Organizations

Collaborative spaces or business ecosystems are similar concepts, in which the main objective is to shorten the periods to attend to the changing customer's needs [19,23,24]. In this context, there are different proposals using concepts and technologies belonging to the Industry 4.0 paradigm, in which the different associated companies implement systems to automate their actions during the production process, creating an intelligent environment. This reduces the need for storage space for raw materials and supplies, unnecessary transportation, reduces the number of personnel that has to participate, and the probability of errors and delays. But within such a large volume of information, which is reliable and which is not? The problem arises in the selection of the necessary information for each organization. The volume of data in the network is gigantic, almost infinite. Specifically, Data Mining (DM) is revolutionizing the industry, since it presented a reliable alternative to the problem of data selection and analysis. Big Data Analysis involves collection and analysis of a huge amount of data from different sources to predict the behavior of a system. It has become fundamental in Industry 4.0 as it helps in identifying patterns, production models and many other things that prove to be quite beneficial in the long term. Many tools are being used to analyze huge amounts of data easier and more feasible; these tools include text mining, data mining, process, and many others. While these tools might be computationally expensive, they optimize the overall efficiency of the process by improving the costs and time to develop products, and provide statistics that help in decision-making.

Having said that, in order to deal with the different integration problems that arise, 'Everything Mining' has proven to be the best solution in recent times; it is a very broad and extensive concept that includes Big Data Mining, Process Mining, Things Mining and many more, to get a better understanding of an entity. It can be used to extract information that can later be used to manage the production process autonomously, while reducing the production costs, as well as optimizing the use of raw materials. Currently, works are being done to create smart factory frameworks where autonomous decision and distributed cooperation between agents are characterized by a self-managing multi-agent system, which uses big data analysis for feedback and coordination. Many intelligent coordination and negotiation mechanics for this system have also been introduced, which have been tested and improved to prevent deadlocks through simulations [27].

3   Proposal

This state-of-the-art poses challenges such as: in what way could automation be achieved that enables the effective construction of an immediate industrial collaborative network? What processes should be automated and how could interfaces be built between peers? What would be the key performance indicators that measure the operation of both the automation and the collaborative network? These are some of the main questions that motivate the present research. If the automation of certain integrating processes is feasible, then an organizational model must be created that formalizes this possibility. Automation would not be just another process for organizations; it would be a process of processes, with external ramifications that would include intelligent and autonomous coordination between organizations.

It has been seen in previous works that the collaborative problems are being solved through the application of the Industry 4.0 paradigm. However, no collaboration models were found to formalize these collaborative processes. Because of that, it is proposed to formalize these models to integrate organizations, forming a flexible virtual space that allows rapid response to the changing requirements of clients, considering that each organization, or a group of organizations, have independent systems, which, once interconnected, must perform a complete production process in an automated way.

3.1   Autonomous Cycles of Data Analysis Tasks

In light of the problem described, the automation of the collaborative processes between companies could be a process framed in current technological trends around of the paradigm of Industry 4.0 [15,16,28]. The diagram below (See Fig. 1) illustrates, in some detail, the cycle that we propose in this paper. The first step would include identifying the organizations that would be working together, along with understanding the needs of their customers. The second step would be the 'Everything Mining' process, where we would undercover all possible information regarding these organizations, by employing methods like Data Mining, Text Mining, Things Mining, among others. This step is further divided into three steps, which include observation, analysis and decision-making [16]. The third step is the key step in which the participating members will be identified, the objectives of the collaboration will be decided, and lastly, the contribution expected from each member will be stated. The last step would be to integrate all of these to create a virtual organization.

The proposed model of automation will seek to facilitate collaborative processes between organizations, designing Autonomous Cycles based on the knowledge gained from the mining processes. The main step is in knowing the organizations, which will be integrated, by mining processes over all of their data, in terms of the context and processes, in order to design a collaborative model for the creation of virtual organizations (VO) to response the changing demands of the customers [15, 16, 28].

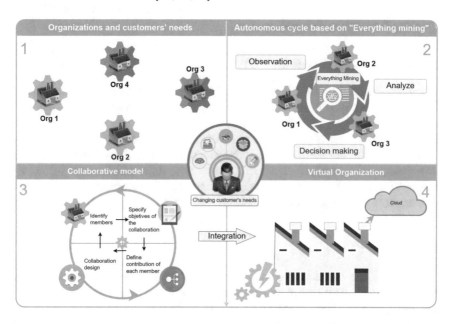

Fig. 1. Autonomous cycles of collaborative processes for integration based on Industry 4.0

3.2   Collaborative Model

The third step is the step in which the participating members will be identified, the objectives of the collaboration will be decided, and finally, the contribution expected from each member will be stated. In order to create a collaboration model and aim an effective integration are proposed three Autonomic Cycles:

- **Designing an initial collaborative plan:** This cycle involves the second and third step in Fig. 1. Firstly, in this stage the actors and goals are defined. Additionally, the elements of the context are analyzed applying data mining tasks, in order to find patterns and predict functionalities, to design a collaboration plan.
- **Supervision of the collaborative process:** One time the collaboration plan is designed, the process need being observed in runtime, in order to detect changes in goals, functionalities or actors, of the business ecosystem. This is a permanent process where the autonomous cycle applies data analysis tasks to detect issues in the collaboration process, or predict its futher behavior.
- **Improvement of the collaborative plan:** Based on the outcomes of the previous autonomic cycle, maybe will be necessary to improve or change the collaborative plan. In this way, the collaborative plan must be flexible and slef-adaptive, and this autonomic cycle determines the improvements in order to keep a correct collaborative plan.

3.3   Virtual Organization (VO)

The last step would be to integrate all to create a virtual organization (VO). A VO will be an ecosystem of autonomous organizations, shaped to aim a specific business opportunity. The VO allows the integration of the organizations with the goal of using their abilities or assets. This business ecosystem will be upheld by data from different sources, like the business processes, work in-process, work frameworks, and administrative structures. These data will be used in order to that the VOs achieve their business goals, such that the autonomous cycles based on "Everything Mining" will be the engine to get it.

4   Conclusion

In the light of the discussion above, the autonomous integration model that we proposed for organizations aims to solve the problems detected in collaborative processes, including their inability to generalize to new situations and to handle unforeseen circumstances or scenarios. The dynamic and flexible nature of the proposed model makes it a perfect candidate to improvise for new conditions. In addition to that, with the help of autonomous cycles, the collaborative processes will become self-adjusting or self-adapting according to the needs and requirements of the organization and its customers. To implement this, the four steps mentioned above will be employed; new technologies will be included especially Cloud Computing, Big Data, IoT and Machine Learning. These technologies would also prove to be one of the strong points for this model, as they add to its efficiency and make it outperform other existing models. Furthermore, they will also assist in making different stages of the proposed model a reality.

For instance, Big Data and Machine Learning will assist in data gathering and analysis to find trends that would help in decision making for the organizations, as well as to optimize different processes like storage of raw materials, transportation, and reduce production costs. The cost-benefit analysis of this model would highlight that while there would an increase in the cost as compared to other models, due to the use of latest technology, especially data analysis using machine learning algorithms, which would be computationally expensive, the overall increase in the work efficiency due to this model would cover for it.

References

1. Eberhard, B., Podio, M., Pérez, A., Radovica, E., Avotina, L., Peiseniece, L., Sendon, M.C., Lozano, A.G., Solé-pla, J.: Smart work: the transformation of the labour market due to the fourth industrial revolution. Int. J. Bus. Econ. Sci. Appl. Res. **10**, 47–66 (2017)
2. Mo, Q., Dai, F., Zhu, R., Da, J., Lin, L., Li, T.: A distributed business process collaboration architecture based on entropy in cloud computing. In: International Conference on Cloud Computing, vol. 1, pp. 126–134 (2014)
3. Costache, A., Popa, C., Dobrescu, T., Cotet, C.: The gap between the knowledge of virtual enterprise actor and knowledge demand of Industry 4.0. In: 28th Daaam International Symposium on Intelligent Manufacturing and Automation, pp. 0743–0749, February 2017
4. Baicu, A.V.: Methods of assessment and training of a company towards the enterprise 4.0. In: 28th Daaam International Symposium on Intelligent Manufacturing and Automation, pp. 1065–1073 (2017)
5. Wee, D., Kelly, R., Cattel, J., Breunig, M.: Industry 4.0 - how to navigate digitization of the manufacturing sector. Technical report (2015)
6. Bullinger, H.J., Neuhüttler, J., Nägele, R., Woyke, I.: Collaborative development of business models in smart service ecosystems. In: 2017 Portland International Conference on Management of Engineering and Technology (PICMET) (2017)
7. Papa, M., Kaselautzke, D., Radinger, T., Stuja, K.: Development of a safety Industry 4.0 production environment. In: 28th Daaam International Symposium on Intelligent Manufacturing and Automation, Austria, pp. 981–987 (2017)
8. Trantopoulos, K., Krogh, G.V., Wallin, M.W., Woerter, M.: External knowledge and information technology. MisQuarterly **41**(1), 287–300 (2017)
9. Roja, A., Nastase, M., Valimareanu, I.M.: Collaborative networks and strategic axes. Fundamental pillars of the development of technology entrepreneurial ecosystems. Rev. Int. Comp. Manag. **15**(5), 579 (2014)
10. Nikolic, B., Ignjatic, J., Suzic, N., Stevanov, B., Rikalovic, A.: Predictive manufacturing systems in Industry 4.0: trends, benefits and challenges. In: 28th DAAAM International Symposium on Intelligent Manufacturing and Automation, pp. 796–803 (2017)
11. Baldassari, P., Roux, J.: Industry 4.0: preparing for the future of work (2017)
12. Ivanov, D., Dolgui, A., Sokolov, B., Werner, F., Ivanova, M.: A dynamic model and an algorithm for short-term supply chain scheduling in the smart factory Industry 4.0. Int. J. Prod. Res. **54**(2), 386–402 (2016)
13. Kanth, L.: Automation development through robot cells in lean manufacturing system. Airo Int. Res. J. **XV**, 63012 (2018)

14. André, S., Elgh, F.: Creating an ability to respond to changing requirements by systematic modelling of design assets and processes. In: 2017 IEEE International Conference on Industrial Engineering and Engineering Management (IEEM), pp. 196–200 (2017)

15. Aguilar, J., Cordero, J., Buendía, O.: Specification of the autonomic cycles of learning analytic tasks for a smart classroom. J. Educ. Comput. Res. **38** (2017)

16. Aguilar, J., Buendia, O., Moreno, K., Mosquera, D.: Autonomous cycle of data analysis tasks for learning processes. In: Communications in Computer and Information Science, vol. 658 (2016)

17. Kitchenham, B.A.: Systematic review in software engineering - where we are and where we should be going. In: EAST 2012 Proceedings of the 2nd International Workshop on Evidential Assessment of Software Technologies (2012)

18. Schmidt, R., Möhring, M., Härting, R.C., Reichstein, C., Neumaier, P., Jozinović, P.: Industry 4.0 - potentials for creating smart products: empirical research results. Lecture Notes in Business Information Processing, vol. 208, pp. 16–27 (2015)

19. Kidanu, S., Cardinale, Y., Tekli, G., Chbeir, R.: A multimedia-oriented digital ecosystem: a new collaborative environment. In: 2015 IEEE/ACIS 14th International Conference on Computer and Information Science (ICIS) (2015)

20. Lee, E.A.: Cyber physical systems: design challenges. Technical report (2008)

21. Robla-Gomez, S., Becerra, V.M., Llata, J.R., Gonzalez-Sarabia, E., Torre-Ferrero, C., Perez-Oria, J.: Working together: a review on safe human-robot collaboration in industrial environments. IEEE Access **5**, 26754–26773 (2017)

22. Charbonnaud, P., Zbib, N., Archimede, B.: Interoperability service utility model and its simulation for improving the business process collaboration. In: Proceedings of the I-ESA Conferences on Enterprise Interoperability V, London. Springer (2012)

23. Andres, B., Sanchis, R., Poler, R., Saari, L.: Collaborative calculation of the materials requirement planning in the automotive industry. In: Proceedings of 2017 International Conference on Engineering, Technology and Innovation: Engineering, Technology and Innovation Management Beyond 2020: New Challenges, New Approaches, ICE/ITMC 2017, January 2018, pp. 496–503 (2018)

24. Friedl, A.: Meeting Industrie 4.0 challenges with S-BPM. In: S-BPM One 2018 Proceedings of the 10th International Conference on Subject-Oriented Business Process Management, April, pp. 1–6 (2018)

25. Quinones, H., Yang, Y.: Intelligent automation; key factor for high mix manufacturing. In: 2017 International Conference on Electronics Packaging, ICEP 2017, pp. 308–311 (2017)

26. Alexakos, C., Kalogeras, A.: Exposing MES functionalities as enabler for cloud manufacturing. In: Proceedings of IEEE International Workshop on Factory Communication Systems, WFCS (2017)

27. Wang, S., Wan, J., Zhang, D., Li, D., Zhang, C.: Towards smart factory for Industry 4.0: a self-organized multi-agent system with big data based feedback and coordination. Comput. Netw. **101**, 158–168 (2016)

28. Vizcarrondo, J., Aguilar, J., Exposito, E., Subia, A.: MAPE-K as a service-oriented architecture. IEEE Lat. Am. Trans. **15**(6), 1163–1175 (2017)

Customer Experience Analytics in Insurance: Trajectory, Service Interaction and Contextual Data

Gilles Beaudon$^{(\boxtimes)}$ and Eddie Soulier

Tech-CICO Team, Université de Technologie de Troyes (UTT),
10300 Troyes, France
{gilles.beaudon, eddie.soulier}@utt.fr

Abstract. The insurance market and particularly French Health-Insurers are affected by changes. That require from traditional actors to transform their Customer Experience management. Our paper, which relies on the third French mutual health-insurer, explores Customer Experience Analytics issues. We observed these issues throughout workshops focused on designing Customer Experience. Our observations show analytics based on Customer Journey Application which make use of decontextualized interactions. We approach the fact these analytics must apply the concept of trajectory as primary focus for user engagement. In order to articulate information system and human activity trajectories, we develop the "Contextualizing Artifact". It is grounded on the Context-System-Trajectory theory (CST). That new theory is mandatory to grasp Customer Experience beyond its marketing dimension. As first step of our artifact development we explain how to improve Customer Journey application with a combination of contextual dataset and classification techniques. This proposal relies on Service Interaction pattern (NISPARO) and provides new qualitative analytics.

Keywords: Customer experience management ·
Customer journey application · Information system trajectory · Context theory ·
Service interactions modeling · Classification techniques ·
Failure mode and effects analysis

1 Introduction: From Customer Journey Statistics to Customer Experience Insights

Our paper relies on the case of a mutual health-insurer, the third one in France. The French health-insurance market is completely transforming. The number of mutual health-insurers dropped from 1158 in 2006 to 421 in 2017 [1]. Many factors are involved in this process. Regulatory constraints upset market's rules. More and more aggressive competitors enter in this market. Lastly, the need for customer personalization grows up. Thus, offering relevant and omnichannel customer experiences is becoming both a necessity and a strategy for mutual health-insurers. Nevertheless, insurers have to transform their organizational structure to offer omnichannel

© Springer Nature Switzerland AG 2019
Á. Rocha et al. (Eds.): ICITS 2019, AISC 918, pp. 187–198, 2019.
https://doi.org/10.1007/978-3-030-11890-7_19

personalized services and to improve their Customer Experience understanding through customer journey insights.

1.1 Problem Statement

About the Customer Experience, [2] depicted the experiential views of consumer behavior through three concepts - "fantasies", "feelings and "fun". [2] define those concepts as "the fascinating and endlessly complex result of a multifaceted interaction between organism and environment". Then, [3] emphasize the "experience economy" as the furthest step of the evolution of the economic value. [3] claim that "an experience occurs when a company intentionally uses services as the stage, and goods as props, to engage individual customers in a way that creates a memorable event." Finally, [4] synthetize multiple authors' definitions of customer experience in: "the interaction between a person and a consumer object within a given situation. The interaction is both a process and a result. Interaction leads to an event coproduced by a customer and an enterprise which could become pleasant, memorable and sense-making. The customer may benefit from this interaction and could enthusiastically promote it via word-of-mouth and may wish to repeat it in future". Regarding these definitions, our main hypothesis here is the misalignment of the following components: organization, business model, service relation, data and information system. Within Customer Experience domain, insurers' practices are supported by statistics originated from Customer Journey Applications. According to health-insurer experts, Customer Journey is a relevant construct to analyze Customer Experience. However, that practices are precisely the point which emphasizes our misalignment hypothesis. Current statistics are just utilized as quantification of the enterprise' touchpoints and do not encompass the whole interactions.

Accordingly, this paper's research topic is to define how to improve Customer Experience Analytics through Customer Journey Applications including multifaceted interaction, customer engagement and situation. In the first section, we will analyze issues of Customer Experience Analytics within mutual health-insurer practices. The second section will introduce a new theory for Customer Experience management. That theory settles two concepts - trajectory and service interaction- and our Contextualizing Artifact relying on NISPARO pattern. The latter tries to reduce Customer Experience Analytics issues. Third section will discuss about our solution, which is based on the use of Contextual data and classification techniques to reach desired insights. Finally, this article will conclude with perspectives and work limits.

1.2 Related Works

The closely related works on Customer Experience is extremely extended. Therefore, we choose to focus this section on the specific aspect of our work which pays attention to enrich Customer Journey Application with Big Data and its techniques. To the best of our knowledge, concerning the understanding of the omnichannel Customer Journey [5] is mainly realized through mining sequential patterns [6] using sequence data [7, 8]. The Event Sequence technique is frequently utilized either for visualizations [9, 10] or for analyses (trends, IT adoption) in healthcare [11]. The problem is to use these works in the analytics field combined with Big Data. On the other hand, related work exists to categorize, evaluate and optimize a customer experience such as US Patents [12] but it does not imply contextual data. Another interesting work of [13] tries to identify Customer Journey by the means of clustering and temporal patterns techniques. But it neither offers analytics nor utilizes Big Data. Thus, depending on our researches, related work exists on different parts of our problem but none of them offer a complete vision on the linkage Customer Experience Analytics, Customer Journey Application and Big Data.

2 Current Customer Experience Analytics

For four months (November 2017–February 2018) we conducted fifteen exploratory interviews. Interviewees came from various insurer business units – Product Development, Marketing & Sales, Policy Administration, Customer Management and IT – and different levels of hierarchy – strategic, management, executive and experts. Each interview lasted one hour and a half. All have been transcribed. That process have confirmed and specified our hypothesis:

- Customer Experience is a strategic aspect of business transformation and have relevant digital dimension;
- Customer Journey is the construct used by marketing and customer relationship executives to understand Customer Experience;
- However, one of the major problems is to consider this phenomenon beyond its marketing dimension, especially when it concerns linking customer segments and journeys, grasped outside of dynamics and situated engagement or interaction notions.

In conjunction with these, we studied health-insurers experts' practices that deal with Customer Experience analysis using Customer Journey tools. In the first instance, they established one clear separation between *Customer Experience, Customer Journey and Customer Process* (Fig. 1).

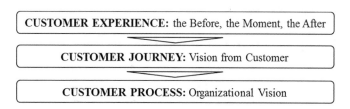

Fig. 1. Customer experience management definitions

Customer Experience defines all interactions lived by a person and fostered by a brand from the customers' point of views. These could be before, at the moment, or after the interaction. Customer Journey is composed of all the steps – interaction moments – occurring when the customer endured a relation with the brand, including selling, using or relational exchanges. Customer Process is all invisible activities to the customer that companies performed their business process and IT tools.

We challenged these specifics definitions with [14]. A similar distinction is made between customer experience and customer journey. Nevertheless [14] claimed: "customer journey is the enterprise prescribed vision of the customer trajectory within a chronologically given touchpoint organization". Thus, the creation of an artifact answering the complexity of customer experience in digital environment is a relevant approach. This artifact aims at designing cross-channel customer journey which develop managers' capabilities and help them reduce the complexity of customer experience management.

Throughout attended workshops of Customer Journeys Conception, we identified executives' practices and IT tools used. Marketing executives try to categorize and quantify journey touchpoints whether as "enchanting" or "irritating" relying on the customers' interactions. They make that categorization by using an IT tool: DataKili©. Results were not harnessed mostly because of the restricted vision of interaction concept. Neither organizational response nor customer satisfaction was addressed while using that tool. Thus, Customer Experience Analytics where not optimized enough design proper Customer Journeys.

To summarize, we identified two issues regarding Customer Experience Analytics based on customer journey. Number one: the limits of the Customer Journey tool as interactions quantification. Depending on experts, Customer Journey tool is an admitted mean to provide Customer Experience Analytics. Nevertheless, insights have to be situated within a customer trajectory. Number two: the data utilized. Executives' practices exploit unique anonym customer identification, types of interaction, channels and motives stored in a relational database. There are no contextual data that help defining interactions. Building on that issues, in the next section we define our artifact to grasp Customer Experience complexity based on Design Science Methodology [15] – the "Contextualizing Artifact" – which is grounded on engagement [16, 17] and trajectory [18] concepts, together entangled through a consistent "context" apparatus.

3   Context-System-Trajectory Theory and Service Interactions Modeling

3.1   CST Theory and Contextualizing Artifact

Since November 2017, coupled with our works on Customer Experience Analytics, we developed a new theory to grasp Customer Experience beyond its marketing dimension. We have given it the name of Context-System-Trajectory theory (CST). Within, our Contextualizing Artifact is grounded (Fig. 2).

In accordance with our previous works and publications, Contextualizing Artifact rests on a double hypothesis:

- customer experience information system and consistency activity journeys constitute one single process, according to imbrication perspective [19], the "intra-action" theory [20], and the entanglement and information system sociomateriality [21–23];

- the System/Trajectory pattern makes Context the joining element between System and Trajectory. So, our contextualized artifact could be positioned within the triplet Context/System/Trajectory concept also call CST theory. Context could be understood through elements which act directly on the experiential trajectory' building.

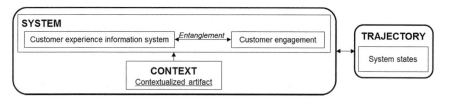

Fig. 2. Context-System-Trajectory (CST) theory

Contextualizing Artifact represents a complete model. The complexity of this construct implies to shape it step by step. Therefore according to Design Science Methodology [15] we build a method to answer our problem. That attempts to address Customer Experience Analytics issues capturing context surrounding Customer Journey. In order to forge this method we focused on the trajectory concept. Trajectories represent different states of a "system" (or object) that we observe. What is observed here is the Customer Experience Information System entangled with Customer Engagement within journeys. And the Contextualizing Artifact is the experimental apparatus used to characterize the state of the system. According to [24] analyzing trajectory implies characterizing ingredients. The ingredients are "context elements which act on project's trajectory" [24, p. 85]. Thus trajectory and context are conceptually linked. Trajectory concept is at the crossroad of interactionism [18], contextualism [25] and processualism [26, 27]. Trajectory's notions have been defined as follow by [24]:

- context: is the set of elements present in a situation;
- ingredient: refers to context's element identified by the researcher or the analyst as acting on specific project or system trajectory;
- engine: stands for ingredients movement and assembling generative mechanism during a trajectory;
- sequence: is the trajectory temporal segment which articulates an ingredient set following a significant disposition;
- turning point: the trajectory temporal segment characterized by an intense ingredients disposition reconfiguration leading to a change of trajectory orientation.

Those concepts allow analyzing regular movements within a trajectory as well as its brutal reorientation or turning point. If each component could be theorized according to various different perspectives, we could do the same for analysis methods of project trajectory within organization [24]. We consider these definitions as a starting point making trajectory concept relevant as "system states". Thus to improve customer journey software and finally customer experience analytics we focus on the "System" component of CST theory. This component consists in two main sub-components that have to be considered as entangled: Customer Experience Information System and

Customer Engagement. These components' main function is to define customer journeys aligned on health-insurers company strategy.

Our analyses of the System implementations revealed three configurations. The first one, separate systems are a major issue clearly identified. They provide predefined journey within one channel but no vision of the whole customer journey. The second one, the system called Customer Engagement Hub (CEH). CEH is a new interaction retention system that health-insurers try to implement. According to Gartner's (2017) definition, CEH "is an architectural framework that ties multiple systems together to optimally engage the customer". Adding that, "a CEH allows personalized, contextual customer engagement, whether through a human, artificial agent, or sensors, across all interaction channels. It reaches and connects all departments, allowing, for example, the synchronization of marketing, sales and customer service processes" [28]. Last configuration, omnichannel context-awareness information systems (smart contracts, collaborative insurance, micro-insurance) are actually too abstract ideas for health-insurance actors.

Therefore, to deal with Customer Experience Analytics issues we decided to use CEH recent approaches. As mentioned in Sect. 2, anonym customer identification, types of interaction, channels and motives are stored in CEH currently. These do not enable the capability neither to take into account real-world customer' situations nor the ubiquitous and pervasive computing systems they now daily use inside these situations. Accordingly, we propose to analyze system states through Service Interaction concept. This concept implies customer and employee having to work together in order to achieve a mutual understanding to get the task or service done. It is noted in [29]: "interactive service work is inherently characterized by uncertainty concerning the outcome or result of the interaction" [29, p. 6]. Thus, they are vivid and performative human interactions, which take place between prospects or clients, and, front-line or back-office service workers. So in order to address CEH current limitations and identify relevant contextual data, service interaction is appropriate because of its compatibility with trajectory concept. Consequently we tried to settle a computable service interaction model as a pattern for event mining.

3.2   Service Interaction Modeling

We propose to use context theory guiding our investigation. Context is the available information in a model representing real-world situation whereas situations are represented as a meta-level concept over context [30]. Thus, the "context" of the Contextualizing Artifact - articulation of Trajectory and Service Interaction - is distributed information from situations and their interactions, at a service systems level of abstraction.

It is why we propose to acquire the context of our Contextualizing Artifact – service interaction situation – from an expanded version of the Interact-Serve-Propose-Agree-Realize (ISPAR) model of service systems episodes. Within, a series of activities are jointly undertaken by at least two service systems and produce ten possible outcomes [31]. For example, in ISPAR model: realization of the proposed and agreed service is the desired outcome of a service system (outcome R). But, a proposal may not be successfully communicated or understood by other service systems (-P), and so, the interaction may be aborted. Secondly, a proposal may be communicated, but activities between the service systems may not lead to an agreement (-A), and so the service interaction may also be aborted. A case of particular interest is when an interaction (I) between service systems is not a service interaction (-S), but nevertheless the interaction may be welcomed by both service systems. Here, welcomed (W) non-service interactions should not be minimized, they often lay the foundation for future service interactions that may co-create great value.

Despite its relevance to represent service interaction, ISPAR model have to be adapted for our work. That change is needed to help Customer Experience experts analyze customers' real-world situations characterization. Therefore, we propose NISPARO model as an extension of ISPAR model and as main component for our Contextualizing Artifact. With the Event (E), we are capable to understand how an interaction and maybe a service interaction are triggered in real-world situation. With the Outcome (O), we should be able to provide analytics on activities that customers have achieved (or not achieved, –O) through the realization of the service interaction.

To summarize, we propose the use of an adapted service interaction pattern named NISPARO to build the first component of our Contextualizing Artifact. That component aims to enable the exploitation of new datasets (Contextual Data). These datasets characterize interactions at a service system level – service interaction – relying on CEH systems. It is our proposed model to enrich Customer Journey applications supporting Customer Experience Analytics. That model integrates multifaceted interactions, customer engagement, and situations. In Sect. 4, we will discuss about our model integration in Customer Experience Information System including new datasets and possible techniques.

4 Improve Customer Experience Analytics: Contextual Data and Classification Techniques

4.1 Customer Experience Information System

In order to explain how new datasets will be utilized to compute NISPARO with Big Data techniques; we have to discuss about the integration of our model within Customer Experience Information System. The ecosystem of that information system (Fig. 3) currently exists in the mutual-health insurer studied. White elements are those existing whereas grey ones are added or modified regarding our work. Here the goal is to support the business process named 'Customer Experience Analysis'. That process is accomplished by marketing executives (Business Process Component: Customer Experience Analysis). The application 'Customer Journey Analytics' supports this process and its intelligence tasks (Application component: Customer Journey Analytics). 'Customer Journey Analytics' application consists in two components, the first one exists and allows visualizing Customer Journeys (Application Component: Journey Visualization). The second one, will be created within our work (Application Component: Service Interaction Classification) and discussed in detail in Sect. 4.2. 'Customer Journey Analytics' application depends on an existing enterprise data management system named 'Customer Engagement Hub' (Application Component:

Customer Engagement Hub, CEH). It is also made up of two sub-components named datasets. 'Interaction Dataset' (Application Component: Interaction dataset) stores all the interactions of customers in the form of a quadruplet: Person, Interaction type, Interaction Channel, Interaction Motive. Each interaction has unique identification and a timestamp. We will add a new component to Customer Engagement Hub: 'Contextual Dataset' (Application Component: Contextual Dataset). It stores new data type needed by Big-Data techniques to classify interaction as an interaction service status regarding NISPARO pattern. In order to respect customer privacy and European security regulation we have strictly anonymized all utilized data.

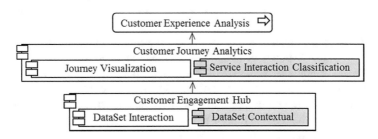

Fig. 3. Customer Experience Information System

4.2   Contextual Dataset Component

Firstly we will describe the Contextual Dataset application component. It will contain situations' contextual data from realized interactions; those will be associated to service interactions status. As explained this dataset will complete Interaction Dataset which is customer-centric. In the latter, through hashed customer identification, we should access all interactions of a customer (motive/channel/timestamp) and some of its anonymous properties (age group, state). The last data type available within Interaction Dataset is health-contract (product type, guarantees) information. All of these data are crucial for our works but they are not sufficient.

Accordingly, we aim to constitute the Contextual Dataset to extend data type accessible. That dataset is 'interaction-centric' and provides interactions' contexts. In computer sciences, and especially in the context-aware computing field, a context is, according to [32], "any information that can be used to characterize the situation of an entity". An entity is a person, place or object that is considered relevant to the interaction between a user and an application, including the user and applications themselves, and by extension, the environment the user and applications are embedded in. Thus, considering this definition we will constitute our Contextual Dataset with:

- contextual data of the interaction situation: interaction localization, device and near point of interest;
- contextual data concerning service interaction: front-line workers role, software used and achieved tasks;
- optional data: customer contents on product, brand or touchpoints, weather record, epidemiological record…

We will make the use of above data to achieve the discovery of an algorithm which aims to classify interactions as interactions services nodes using Big-Data techniques. In the next section, we will present the way in which to discover our algorithm by describing our Service Interaction Classification component.

4.3 Service Interaction Classification Component

The Service Interaction Classification component proposes to characterize a customer interaction with a service interaction status in the form of a NISPARO's final node. To make it possible, each interaction stored within customer engagement hub will receive a "current status of a service interaction". This will be possible by using new data type described in Sect. 4.2. Final result will be interactions' categorization as "I" (or "-I"), "S" (or "-S"), "P" (or "-P"), "A" (or "-A"), "R" (or "-R").

These categories constitute relevant aspects of the Customer Experience. According this, we do not clearly solve the question of the "N" (New event) and the "O" (Outcome) of NISPARO pattern. Conclusion section discuss about these qualitative aspects as the subject of future works.

To make the categorization possible we have to use learning classification techniques from Big-Data domain. We start with a complete dataset of pre-classified service interaction on NISPARO nodes in order to define those of interaction data or contextual data are the most relevant to determine an interaction service status. Principal component analysis technique combined with a classification algorithm (Random Forest, Naives Bayes or Support Vector Machine) will be our first approach to determine relevant data. Nevertheless we are fully aware that it will be insufficient. In order to be confident with our service interaction classification we have to add into our predictive algorithm the information of previous and next interactions status. It is an important aspect because of the trajectory concept. That point will be the second one discussed in conclusion. After all, we claim that combining this robust and well-known approach with new data type will allow innovation within Customer Experience Analytics based on Customer Journey application. Next session will discuss these news insights.

4.4 New Customer Experience Analytics

We saw in Sect. 2 that one of the main issues of Customer Experience Analytics within health-insurer companies is the lack of interaction contextualization. We proposed an approach to solve this using service interaction pattern named NISPARO. It allows classifying interactions by a combination of new data type (Contextual data) and Big-Data techniques. Following NISPARO status (nodes) we claim that it is possible, for each interaction, to deliver new qualitative and situated analytics and so, new Customer Experience insights. We use [33] who links phases and situations with ISPAR failures mode and effect analysis. Resuming and extending this work we can propose six service interaction failures:

1. No interaction (-I): insurer captures a relevant event but it is not convert into interaction;

2. No Service Interaction (-S): customer and insurer did not succeed in starting a work collaboration (misses phone calls for example);
3. No Proposal (-P): insurer has no response to the customer need;
4. No Agreement (-A): there was no possibility to get mutual understanding between insurer and customer regarding service proposal;
5. No Realization (-R): a problem happened to customer, insurer or both, which avoid the service realization;
6. No Outcome (-O): according to customer there is no particular achievement enabled thanks to interaction service realization.

These analyses depending on the combination of NISPARO status and failures mode and effect analysis will allow publishing quantitative analytics. Furthermore, will be capable of build a requesting capability relying on that qualitative aspects. For example our system will be able to answer request such as "Explore customer journey in which there is the most of failed flu vaccination service realization in February 2018 in Troyes after a customer demand". Thus, we provide news types of Customer Experience Analytics based on Customer Journey. Contribution of Big-Data techniques exploiting interactions' contextual data will help customer experience executives to fully grasp customer journey and co-design more appropriate experiences.

5   Conclusion

In this article we propose a solution dealing with Customer Experience Analytics issues based on Customer Journey application. Issues revealed directly come from existing conception of Customer Journey application. It focusses on touchpoints; do not take into account of multifaceted interaction, customer engagement and situation. We proved the need of a new theory to grasp Customer Experience: CST Theory (Context/System/Trajectory). Grounded in CST theory, we presented our Contextualizing Artifact to solve previous issues. As a complex artifact it should be composed of multiple IT artifacts. Thus we choose to focus at first on System states – Trajectory - component which is understood as the Context observing System. We proposed to compute it with an adapted service interaction pattern named NISPARO. We explained how we intend to build it by means of contextual dataset exploited with Big Data techniques. Finally we revealed new analytics provided by our solution. These aim at achieving customer journey contextual analysis and help to design relevant Customer Experience. As it still an exploratory work, we will have to prove all our assertions.

5.1   Limits

In parallel of developing our solution to prove its effectiveness for marketing executives; we are confronted to three challenges. The first one is time management. Time management is a major constraint of this work. Indeed, classify interaction as service interaction status based on NISPARO requires a precise timestamping. Yet, Datakili Proof of Concept has shown that manually recorded interactions realized by front-line workers were not done with regularity. It may introduce a bias to the results.

The second one, we will be the need of business experts. The latter have to be available in order to pre-qualify our datasets. Without them we will not be able to discover the classification algorithm. Neither availability of them is guaranteed nor the acquisition of new vocabulary introduced by NISPARO pattern. The last one, as we are using Design Science paradigm we may have an organizational limit. We must prove that new analytics are used by customer experience executives and improve their process. Thus, we have to involve them at the beginning of the realization to make sure of their engagement.

5.2 Perspectives

Our research has long term perspectives. The next step will consist in linking service interactions. Based on these graphs we may be able to develop new visualization type allowing representing "customer living story where they experience insurance product". Our final objective will try to resolve entanglement between customer experience information system and customer engagement.

References

1. Perrin, G., Mutuelles: La concentration du secteur se poursuit, L'Argus de l'assurance, April 2017
2. Holbrook, M.B., Hirschman, E.C.: The experiential aspects of consumption: consumer fantasies, feelings and fun. J. Consum. Res. **9**(2), 132–140 (1982)
3. Pine II, B.J., Gilmore, J.H.: Welcome to the experience economy. Harvard Bus. Rev. **76**, 97–105 (1998)
4. Flacandji, M.: Du souvenir de l'expérience à la relation à l'enseigne: une exploration théorique et méthodologique dans le domaine du commerce de détail, Thesis, Gestion et management, Université de Bourgogne, Dijon (2015)
5. Barwitz, N., Maas, P.: Understanding the Omnichannel customer journey: determinants of interaction choice. J. Interact. Mark. **43**, 116–133 (2018)
6. Huang, Y., Zhang, L., Zhang, P.: A framework for mining sequential patterns from spatio-temporal event data sets. IEEE Trans. Knowl. Data Eng. **20**(4), 433–448 (2008)
7. Gotz, D., Stavropoulos, H.: DecisionFlow: visual analytics for high-dimensional temporal event sequence data. IEEE Trans. Vis. Comput. Graph. **20**(12), 1783–1792 (2014)
8. Li, H., Sheopuri, A., Yi, J., Yu, Q.: Feature learning on customer journey using categorical sequence data. US20170293919A1, October 2017
9. Mathisen, A., Grønbæk, K.: Clear visual separation of temporal event sequences. In: IEEE Symposium on Visualization in Data Science, October 2017
10. Wongsuphasawat, K., Guerra Gómez, J.A., Plaisant, C., Wang, T.D., Taieb-Maimon, M., Shneiderman, B.: LifeFlow: visualizing an overview of event sequence. In: SIGCHI Conference on Human Factors in Computing Systems, New York, pp. 1747–1756 (2011)
11. Spaulding, T.J., Furukawa, M.F., Raghu, T.S., Vinze, A.: Event sequence modeling of IT adoption in healthcare. Decis. Support Syst. **55**(2), 428–437 (2013)
12. Winters, G., Elshof, J., Shmelev, A.: Process and system to categorize, evaluate and optimize a customer experience. US20170308917A1, 26 October 2017

13. Cordewener, M.H.H.: Customer journey identification through temporal patterns and Markov clustering—Eindhoven University of Technology research portal, Eindhoven University of Technology (2016)

14. Moschetti-Jacob, F.: Création d'un artefact modulaire d'aide à la conception de parcours client cross-canal visant à développer les capacités des managers des entreprises du secteur du commerce. Thesis, Paris-Dauphine (2016)

15. Hevner, A.R., March, S.T., Park, J., Ram, S.: Design science in information systems research. MIS Q. 28(1), 75–105 (2004)

16. Becker, H.S.: Notes on the concept of commitment. Am. J. Soc. 66(1), 32–40 (1960)

17. Thévenot, L.: Pragmatic regimes governing the engagement with the world. In: Knorr-Cetina, K., Schatzki, T., Savigny Eike, V. (eds.) The Practice Turn in Contemporary Theory (2001)

18. Strauss, A.L.: La trame de la négociation: sociologie qualitative et interactionnisme. Éditions L'Harmattan, Paris (1992)

19. Leonardi, P.M., Nardi, B.A., Kallinikos, J.: Materiality and Organizing: Social Interaction in a Technological World, 1st edn. Oxford University Press, Oxford (2012)

20. Barad, K.M.: Meeting the Universe Halfway: Quantum Physics and the Entanglement of Matter and Meaning. Duke University Press, Durham (2007)

21. Orlikowski, W.J.: Sociomaterial practices: exploring technology at work. Organ. Stud. 28(9), 1435–1448 (2007)

22. Orlikowski, W.J.: The sociomateriality of organisational life: considering technology in management research. Camb. J. Econ. 34(1), 125–141 (2010)

23. Orlikowski, W.J., Scott, S.: The entangling of technology and work in organizations. Working Paper Series, Department of Management Information Systems and Innovation Group, London School of Economics, January 2008

24. Oiry, E., et al.: Propositions pour un cadre théorique unifié et une méthodologie d'analyse des trajectoires des projets dans les organisations. Manag. Avenir 36(6), 84 (2010)

25. Pettigrew, A.M.: What is a processual analysis? Scand. J. Manag. 13(4), 337–348 (1997)

26. Van de Ven, A.H., Poole, M.S.: Explaining development and change in organizations. Acad. Manag. Rev. 20(3), 510–540 (1995)

27. Abbott, A.D.: Time Matters: On Theory and Method. University of Chicago Press, Chicago (2001)

28. Gartner, Inc. https://www.gartner.com/it-glossary/customer-engagement-hub

29. Schneider, A.: Handling the Clash Between Production & Consumption: A Situated View on Front-line Service Workers' Competencies in Interactive Service, Empirische Personal- und Organisationsforschung, vol. 55. Rainer Hampp (2016)

30. Akman, V., Surav, M.: The use of situation theory in context modeling. Comput. Intell. 13(3), 427–438 (1997)

31. Spohrer, J., Vargo, S.L., Caswell, N., Maglio, P.P.: The service system is the basic abstraction of service science. In: Proceedings of 41st Annual International Conference on System Sciences, Hawaii, p. 104 (2008)

32. Dey, A.K.: Understanding and using context. Pers. Ubiquit. Comput. 5(1), 4–7 (2001)

33. Beckett, R.: Service ecosystems supporting high reliability assets. Systems 5(4), 32 (2017)

ITSIM: Methodology for Improving It Services. Case Study CNEL EP-Manabi

Patricia Quiroz-Palma[1,2]([⊠]), Angela Suárez-Alarcón[3]([⊠]),
Alex Santamaría-Philco[1,2]([⊠]), Willian Zamora[1,2]([⊠]),
Viviana Garcia[2]([⊠]), and Elsa Vera-Burgos[2]([⊠])

[1] Universitat Politécnica de Valencia, Valencia, Spain
{patquipa,asantamaria}@dsic.upv.es,
wilzame@posgrado.upv.es
[2] Universidad Laica Eloy Alfaro de Manabí, Manta, Ecuador
{viviana.garcia,elsa.vera}@uleam.edu.ec
[3] Ministerio de Acuacultura y Pesca, Manta, Ecuador
Angela.suarez@acuaculturaypesca.gob.ec

Abstract. Currently, organizations that have many clients and users are available to access the available information technologies (IT). Technology and information services should be standardized and documented to ensure their quality. For continuous improvement, the IT Service Improvement Methodology (ITSIM) is proposed to support the organization, delivering efficient IT services, based on standards, policies and best practices IT management practices. A case study of the implementation of the methodology was made. ITSIM in the IT department of the National Electricity Corporation of Manabí, starting with the evaluation of the maturity of the domains: Acquire and implement, and Deliver and support. The processes and good practices proposed by the ITSIM methodology were applied. Resulting in the improvement in the quality of IT services and the satisfaction of the users of the organization.

Keywords: IT management · Best practices · Maturity models · CMMI · COBIT · ITIL

1 Introduction

At present, the Information Systems (SI) as well as the Information Technologies (IT) facilitate the operation of the companies [1]. Its use allows improving and automating operational processes, providing a platform of information necessary for decision making and achieving competitive advantages.

Information Technologies have been conceptualized as the "integration and convergence of computing and microelectronics, telecommunications and data processing technology" [2]. Where its main components are: information, equipment, factor human, infrastructure, software and information exchange mechanisms, policy elements and regulations, as well as financial resources. Organizations have a greater dependence on Information Technologies.

© Springer Nature Switzerland AG 2019
Á. Rocha et al. (Eds.): ICITS 2019, AISC 918, pp. 199–209, 2019.
https://doi.org/10.1007/978-3-030-11890-7_20

The departments of information systems in organizations have traditionally been seen as a support area for the business, without taking into account the quality of the processes and their constant evaluation to measure their profitability, effectiveness and the quality of the service offered to the entire organization. In an environment where periods of availability of services are increasingly broad, customer requirements are high, business is changing rapidly, it is important that information systems are properly organized and aligned with the business strategy.

Currently, there are methodologies that are based on the quality of service and effective and efficient development of processes that cover the most important activities of organizations in their information systems and information technology. This paper proposes the methodology for two IT processes, with the aim of implementing it in other processes in the future, and the following goals are pursued:

- Define the level of maturity of the IT domains: Acquire and Implement, and Deliver and Support, applying the domains of Control OBjectives for Information and Related Technology (COBIT) [7] y Capability Maturity Model Integration (CMMI) [10].
- Define the improvements, through good practices that will be applied according to the level of maturity obtained in the processes, applying the best practices proposed by ITIL [5].
- Implementation of best practices through the creation of policies, standards and procedures.

This article is organized by 5 sections. Section 2 reviews IT management models, process reengineering and business management through good practices. Section 3 presents the IT services methodology (ITSIM). Section 4 describes the implementation of the ITSIM methodology through a case study, step by step, from measuring the level of maturity of the processes, the implementation of the ITSIM methodology and the results obtained. In Sect. 5, the conclusions and future work derived from this investigation.

2   IT Management, Process Reengineering and Best Business Practices

2.1   IT Management

IT Management is a process-based management discipline that aims to align IT services with the needs of the organization, in addition to providing a certain order to management activities. Among the existing models, the use of: ITSM, COBIT, ISO/IEC 20000 and CMMI [12]. Below, the models is described.

2.1.1   ITSM

It is a technology through ITIL tools and ITSM software, it allows us to acquire a deeper knowledge of our clients, facilitating the acquisition of new ones, the loyalty of existing ones and the development of new business lines [3]. This set of ITSM solutions and ITIL tools allows for a more flexible IT environment, with greater availability and adaptable to the demands of users and the organization itself, offering the possibility of flexible integration with Customer Relationship Management (CRM) [4]. The Information Technology Infrastructure Library (ITIL) is a widely accepted industry framework. It adopts a focus on the operational development process, excellent IT support service and service delivery processes [5].

2.1.2   COBIT

COBIT is an IT governance framework that allows managers to bridge the gap between control demands, technical issues and business risks. COBIT allows the development of clear policies and "best practices" for the control of IT in all areas of organizations [6]. COBIT determines, with the support of the main international technical standards, a set of best practices for security, quality, efficiency and IT that are necessary to align IT with the business, identify risks, deliver value to the business, manage resources and measure performance, the achievement of goals and the level of maturity of the processes of the organization [7].

2.1.3   ISO/IEC 20000

The series ISO/IEC 20000 - Service Management standardized and published by the International Organization for Standardization (ISO) and International Electrotechnical Commission (IEC), is an ISO 20000 certificate that demonstrates that an IT organization is oriented to the needs of the clients, provides quality services and uses resources economically [8]. This certificate is always an advantage over the competition.

2.1.4   CMMI

Capability Maturity Model Integration (CMMI) is a model for the improvement and evaluation of processes for the development, maintenance and operation of software systems [9]. CMMI is widely used in organizations to determine the level of maturity of the processes to know the real situation and evaluate the performance and quality of the services they provide to their internal and external clients [10, 15].

2.2   Process Reengineering

Understanding what is a process and how it is an integral part of companies and institutions, whatever their nature, it is then possible to reach a definition of process reengineering. Hammer and Champy (1990) define process reengineering as "the fundamental re-conception and the radical redesign of business processes to achieve dramatic improvements in performance measures such as cost, quality, service and speed" [11]. Therefore, it is a new fundamental conception and a holistic vision of an organization. Process reengineering is radical to some extent, since it seeks to get to the root of things, it is not only about improving processes, but mainly, it seeks to reinvent them, to create competitive advantages, based on advances technological.

2.3   Best Practices Business Management

The "Best Practices Business Management" comprise a series of practical, easy-to-apply measures that an entrepreneur can put into practice to increase productivity, lower costs, reduce the environmental impact of production, improve the production process, as well as raise safety at work. Therefore, it is an instrument for cost management, environmental management and to initiate organizational changes [14]. Only paying attention to these three elements achieves a triple gain (economic, environmental and organizational) and establish the basis for a continuous and successful process of improvement in the company [16], as shown in Fig. 1.

Fig. 1. Benefits of best practices business.

3   IT Services Methodology (ITSIM)

For the creation of the methodology of processes for the improvement of IT, based on the methodologies of existing processes and best practices for business management, we consider the use of:

- ITIL as best practices and reference framework for IT Service Management.
- COBIT 5 as a frame of reference to facilitate the design and use of own models oriented to the Government and IT Management.
- ISO/IEC 20000 as a certifiable standard for IT Service Management.
- CMMI-SVC as reference model and body of knowledge for the Universal Management of Services.

From each of these models obtain the best aligned to the strategic planning of the organization, to create the Methodology ITSIM, the same that focused on two critical domains of IT:

- Acquire and implement
  - Acquire and maintain application software.
  - Develop and maintain IT resources.
- Deliver and support.
  - Define levels of service.
  - Manage problems and incidents.

In Fig. 2, the structure of the ITSIM Methodology for the domains Acquire and implement, and Deliver and give support is defined. The organization, users and clients, generate a great number of requirements, new applications or changes in the information systems that are in production. These requests are handled by the service desk, which redirects them to incident management and problem management. In the incident management the request is attended and if the first support line solves it, it registers it in the knowledge base. If you cannot solve it, you derive it to problem management. In the management of problems, the specialized team determines whether it is a maintenance or a new computer solution for version management and configuration management. All these services communicate with incident management for the registration of the solution in the knowledge base.

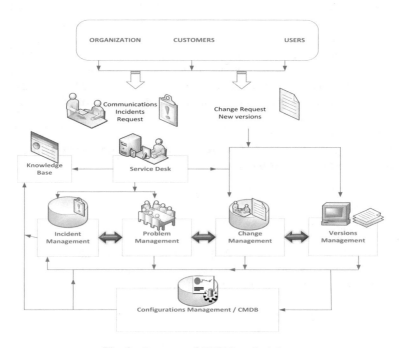

Fig. 2. Structure of ITSIM methodology

4 Case Study

4.1 Organization

The National Corporation of Electricity Public Company CNEL EP (https://www.cnelep.gob.ec) and its Manabi Business Unit, is responsible for the distribution and commercialization of electricity in the Province of Manabí. CNEL EP Manabí offers electric distribution service to a total of 1.25 million subscribers, covering 30% of the Ecuadorian customer market. It is an organization that has many human resources at its disposal, therefore it is important to consider that the security of information within the company is of vital importance and where you can identify the main processes and services of information technologies to effectively meet the requirements of internal and external customers of the organization.

4.2   Process Maturity Level

IT procedures in the organization are developed internally, and strategic decisions are made at the level of each project. Generally, all requirements are treated with high priority and distributed according to the areas or working groups. IT management has a significant impact when the information it provides is reliable and efficient, which helps us grow IT and is used to achieve a competitive advantage.

However, it is necessary to know and reach an optimum degree of maturity, so it is necessary to make the structure of a methodology with the guidelines of COBIT [12] and ITIL [13], which provide us with a necessary orientation on the structure of the process, and that clearly defines the IT activities in the organization, as a generic model of processes in domains, considering mainly the domains of *Acquire and Implement*, *Deliver and Support*, considered as the main needs of the organization. The measurement of performance is important for the development of IT and to show its results. It helps to understand, manage risks and maximize the benefits of IT.

4.2.1   Acquire and Implement

The domain *Acquire and Implement* is in a degree of maturity CMMI [10]: *Managed and Measurable* for the following reasons:

- There is an established methodology for the development and maintenance of applications.
- Dependence on the knowledge of key people.
- There are documented policies and procedures.
- High costs in the changes of the Acquired Systems without source code.
- Sporadic availability problems due to unexpected hardware failures.
- A Contingency Plan has not been defined in the existing systems.
- Integrity of data.
- Appropriate security levels.
- Quality control of the Systems.
- The requirements are clearly defined.

4.2.2   Deliver and Support

The domain *Deliver and Support* is at a CMMI maturity level [10]: *Managed and Measurable* for the following reasons:

- Acceptable downtime.
- Users are informed of the follow-up of the incidents.
- There is incident prioritization.
- Common incidents consume less time because there is a knowledge base.
- Management indicators to monitor incidents.
- There is an incident management system.
- Procedures of the Incident System are documented.

4.3   ITSIM Implementation

ITSIM implementation, should jointly consider the functions and common areas within the IT organization, allowing a structured adoption; for this it is necessary to begin by grouping the processes of the methodology so that the implementation itself is facilitated, the acceptance and soon reach of its benefits.

4.3.1   Acquire and Implement

IT solutions need to be materialized and identified through their development or acquisition of systems, and they must be integrated into business processes. Every system needs changes and maintenance to ensure the correct operation and continuous operation of business requirements.

In Fig. 3, the process flow of the domain acquiring and implementing the ITSIM methodology is detailed, from the reception of the requirement made by the clients (internal and external) of the organization, which is planned to define whether the IT solution will be developed, acquire or maintain systems, taking into account the business rules, as well as the identification of those involved in the project and the allocation of responsibilities, as a result of this stage you have communication plans, risk analysis and schedule of the computer solution. In the monitoring and follow-up stage, changes, development, testing, quality control and start-up of a computer solution are managed. If quality control determines that changes are needed, the application monitoring and monitoring cycle is generated again.

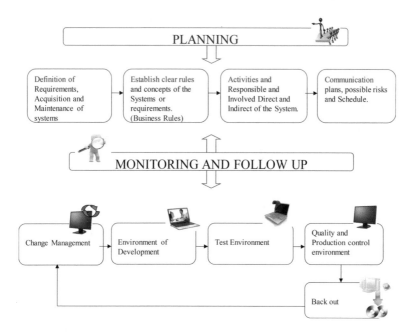

Fig. 3. Process flow of the acquire and implement domain, ITSIM methodology.

4.3.2   Deliver and Support

The IT departments of the organizations generate and provide the IT services and a group of IT clients, internal (users) and external (customers and suppliers) that demand these services and expect their timely and quality delivery. Relationships and communications between IT departments and IT clients must be channeled through a system that guarantees the optimization of the delivery and service support processes through the consolidation of IT Service Management.

Figure 4 defines the process flow of *Deliver and support* domain, this processes can be generated by the client, the organization or users through the service desk. According to the type of incident or request, it is derived to the management of incidents. Incident management solves it, if you cannot solve it, scale the incident to problem management or change management, respectively. Problem management diagnose the problem and solve it. Change management validates the changes made to solve the incident. The delivery management is responsible for testing and quality control of the solution. Configuration management performs the start-up and storage of the IT solution in the knowledge base. The domains *Acquire and implement*, and *Deliver and support* offered by the IT area were aligned to the processes described by the ITSIM methodology, adapting the procedures, formats and reports described in the previous flows.

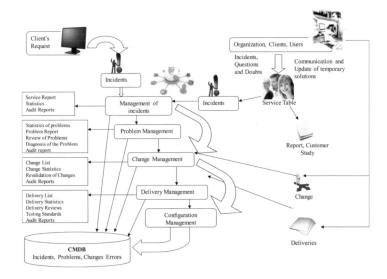

Fig. 4. Process flow of deliver and give support domain, ITSIM methodology.

4.4   Results

The implementation of good practices in IT processes provides benefits such as cost reduction, reduction of environmental impact and improvements within the organization [14]. In Tables 1 and 2 the comparison of the results obtained in the previous evaluation and after the implementation of the ITSIM methodology is carried out.

In Table 1, the activities of the Acquire and implement domain are compared, it can be seen that the planning of the systems, the procedures manuals and the documentation of each stage of development were complied with. The systems are modified prior to a change agreement with the user, all the personnel are informed of the changes made in the computer solution. In each project, the responsibilities of those involved are assigned from the beginning.

Table 1. Domain: acquire and implement

Activity	Previous Process	New Process
System planning	No	Yes
Procedures manual	No	Yes
Development documentation	End of the project	Yes
Risk estimation and monitoring	No	Forms
Changes according analysis	No	Yes
Periodic job meetings	Si	Yes
Communication to all staff	Only involved	Communication plan
Responsible for the project involve	Undefined follows	From the beginning

In Table 2, the activities of the Deliver and support domain are compared. The restoration of the service can be solved from the first level of support, the solutions of the incidents are recorded in a knowledge base, the communication with the users is improved and the established procedures and policies are applied. Finally, reports of incidents and the evaluation by the user of the provided service are obtained.

With the implementation of the ITSIM methodology, the procedures were standardized, documented and implemented, to obtain reports on each activity of the affected processes, as well as the user's assessment for continuous improvement in the quality of the IT services in the organization.

Table 2. Domain: deliver and support

Activity	Previous Process	New Process
Online information about incidents	Yes	Yes
Immediate restoration of a service	Technical support (30–60 min)	Yes (5–10 min)
Incident registration in an application	Yes	Yes
Knowledge base	No	Incident registration
Constant communication of changes	Sometimes	Yes
Policies and procedures to follow	Not applied	Yes
Report of indices	No	Yes
Improvements in the service	No	Yes
Evaluation of the service provided	No	Yes

5   Conclusions and Future Works

With the implementation of the ITSIM methodology it was possible to organize and improve the IT processes, offering better IT services, with a better level of quality applying new policies and established standards. It was essential to determine the degree of maturity of the processes to achieve improvements implemented the ITSIM methodology. The procedures were defined and documented, the documents of each activity were standardized using clear formats for all those involved in each procedure. Every solution was communicated, documented and registered, speeding up the solution of recurring incidents and obtaining timely reports for decision making. Finally, it was possible to demonstrate the improvement in the quality of the IT services, satisfaction of the final users of the organization and a more optimized level of maturity of the processes.

The benefits obtained by implementing this methodology are:

- Focus on business objectives and needs, improving cooperation and communication between customers and the IT area.
- Help organizations compare themselves with the competition and implement best practices of control objectives and related technology.
- Organizations generate trust and credibility towards their internal clients.
- It allows organizations to comply with regulatory requirements.

As future work, it is proposed to extend the ITSIM methodology to other IT domains, as well as the implementation of the ITSIM methodology to other organizations that need to improve the level of maturity of their IT services.

References

1. Antonio Muñoz Cañavate: Information systems in companies. http://www.hipertext.net
2. Information society. http://www.sociedadelainformacion.com
3. Gupta, R., Prasad, K.H., Mohania, M.: Automating ITSM incident management process. In: International Conference on Autonomic Computing, Chicago, pp. 141–150 (2008)
4. Jiangming, H., Ruijie, G., Taohua, O.: Notice of retraction - research into the crucial factors of successful enterprise CRM - a case of Haier CRM system. In: 2010 International Conference on E-Business and E-Government, Guangzhou, pp. 3123–3128 (2010)
5. Eikebrokk, T.R., Iden, J.: ITIL implementation: the role of ITIL software and project quality. In: 2012 23rd International Workshop on Database and Expert Systems Applications, Vienna, pp. 60–64 (2012)
6. ISACA. A Business Framework form the Governance and Management Enterprise IT (2012)
7. Harefa, K.R.P., Legowo, N.: The governance measurement of information system using framework COBIT 5 in automotive company. In: 2017 International Conference on Applied Computer and Communication Technologies (ComCom), Jakarta, pp. 1–6 (2017)
8. IEEE Standard - Adoption of ISO/IEC 20000-1:2011: Information technology – Service management – Part 1: Service management system requirements, in IEEE Std 20000-1-2013, pp. 1–48, 3 June 2013

9. Dong, S., Ren, A., Wang, X.: Application of organizational process asset library in high maturity process improvement. In: 2016 21st International Conference on Engineering of Complex Computer Systems (ICECCS), Dubai, pp. 223–226 (2016)
10. Sahin, E., Kaynak, I.K., Sencan, M.Ü.: A pilot study: opportunities for improving software quality via application of CMMI measurement and analysis. In: Joint Conference of the 23rd International Workshop on Software Measurement and the 8th International Conference on Software Process and Product Measurement, Ankara, pp. 243–246 (2013)
11. Institute of Industrial Engineers, Más allá de la Reingeniería, p. 4. CECSA, México (1995)
12. Sahibudin, S., Sharifi, M., Ayat, M.: Combining ITIL, COBIT and ISO/IEC 27002 in order to design a comprehensive IT framework in organizations. In: Second Asia International Conference on Modelling and Simulation (AMS), Kuala Lumpur (2008)
13. Yao, Z., Wang, X.: An ITIL based ITSM practice: a case study of steel manufacturing enterprise. In: 2010 7th International Conference on Service Systems and Service Management, Tokyo, pp. 1–5 (2010)
14. Business Management. http://degestionempresarial.blogspot.com/
15. Jackelen, G., Jackelen, M.: When standards and best practices are ignored. In; Proceedings 4th IEEE International Software Engineering Standards Symposium and Forum (ISESS 1999). Best Software Practices for the Internet Age, Curitiba, Brazil, pp. 111–115 (1999)
16. Ehsan, N., Malik, O.A., Shabbir, F., Mirza, E., Bhatti, M.W.: Comparative study for PMBOK & CMMI frameworks and identifying possibilities for integrating ITIL for addressing needs of IT service industry. In: IEEE International Conference on Management of Innovation and Technology, Singapore, pp. 113–116 (2010)

Automation of the Barter Exchange Management in Ecuador Applying Google V3 API for Geolocation

José Antonio Quiña-Mera[1,2]([✉]),
Efrain Rumiñahui Saransig-Perugachi[1], Diego Javier Trejo-España[1,3],
Miguel Edmundo Naranjo-Toro[1,4],
and Cathy Pamela Guevara-Vega[1,2]

[1] Computer Systems and Software Engineering Career,
Universidad Técnica del Norte, Ibarra, Ecuador
{aquina, ersaransig, djtrejo, menaranjo,
cguevara}@utn.edu.ec
[2] Software Engineering, Universidad de las Fuerzas Armadas -ESPE- Ecuador,
Sangolquí, Ecuador
[3] Computer Management, Pontificia Universidad Católica del Ecuador,
Ibarra, Ecuador
[4] Educational Research, Universidad Central del Ecuador, Quito, Ecuador

Abstract. This article presents the result of the automation of the barter management process in Ecuador. Bartering maintains, promotes and fosters the exchange of products and services without the use of money as an ancestral practice of the peoples and communities of the country. This study is a quantitative, exploratory, documentary, descriptive and field research. The results show that 92% of the targeted population welcome the automation of the process and 98% support the use of technology with regard to these regional ancestral barter practices. Based on these figures, we developed the software "Barter Management" and we embedded it into the Supertienda Ecuador platform. It was developed with the SCRUM agile methodology and with Google Api v3 for georeferencing. Finally, 10 acceptance tests were conducted to validate the software functionality.

Keywords: Barter · Google V3 API · Georeferencing ·
Software development · Acceptance tests

1 Introduction

Barter is an ancestral activity which started during the Neolithic period about 10 thousand years ago. It was particularly used by populations who were dedicated to agriculture and livestock. It is considered as one of the first means of trade between people. It was founded by Adam Smith (1776) and it consisted in exchanging surplus goods, i.e. products a person possessed but did not need, against some other goods the trader was interested in. The other trader also had surplus goods and was interested in getting another product [1].

© Springer Nature Switzerland AG 2019
Á. Rocha et al. (Eds.): ICITS 2019, AISC 918, pp. 210–219, 2019.
https://doi.org/10.1007/978-3-030-11890-7_21

According to Agustín Lencina de Leó, the advantages of barter include: Exchanging goods and services without the need of money, saving wealth in the monetary capital of a company, maximizing business or person' finances, increasing productivity, encouraging commercial relationships with companies from other industries and developing commercial channels [2].

On the other hand, barter has the following disadvantages. It is not always easy, or even possible, to meet someone who needs the product or service for which it is exchanged for. Even though parties rely on market price some ambiguity remains about the monetary value of the exchanged goods or services which makes decisions difficult. Barter does not allow the involvement of an intermediary, unless he is willing to work without getting anything in exchange, it means that the transaction must exclusively be dealt between both interested parties [3].

Currently, bartering remains in some Latin American countries such as Argentina, Bolivia and Ecuador specifically in Pimampiro canton (Imbabura province) which has become a national and international benchmark since 1990 [4]. For the 12,951 inhabitants in the communities and towns around Pimampiro, bartering is considered as a major commercial activity. About 4,000 people participate in a yearly fair which take place in the Holy Week - between the second and third week of March. During this event which lasts two days, in the evenings, people celebrate the festivity and prepare the traditional Ecuadorian "Fanesca" dish. Every year, the Universidad Técnica del Norte (UTN), representing the academic, along with the Superintendencia de Control del Poder del Mercado de Ecuador support this event strengthening and preserving these ancestral activities. For this reason, the "Trueque del Sol" is recognized as Intangible Culture Heritage of Ecuador since December 2017 (Agreement No. 137) [4].

The purpose of the Organic Law for the Regulation and Control of Market Power is to prevent, correct, eliminate and sanction the abuse of economic operators with market power, as long as the prevention, prohibition and sanction of collusive agreements and other restrictive practices. It controls and regulates economic concentration operations, considering the proper use and application of Information and Communication Technologies ICT [5–7].

The Economic Promotion of the City Council of Elche in Spain, under a collaborative agreement with city four associations of small businesses, helps local businesses to modernize their technological practices. Such initiative was inspired by a 2016 international study on e-Commerce showing that electronic commerce turnover in Spain increased by 20.3% to reach 5.948 million Euros in the second quarter of 2016, according to the latest e-commerce data from the Comisión Nacional de los Mercados y la Competencia – CNMC [8–10].

The problem identified by this research is that people involved in the barter activity in Pimampiro, especially small producers in the area (80% of the population) have to wait a specific time of the year to barter and they get limited products as they can only trade with people who physically attend this event.

The objective of this research work was to establish a proposal to automate the barter process in Pimampiro canton through the development of a software integrated into the technological platform Supertienda Ecuador of the Superintendence of Control of Market Power of Ecuador to facilitate participants interaction throughout the country [11].

Additionally, a study by Lanas and Medina shows barter in Pimampiro canton as a form of solidarity economy, its conclusions highlight that agriculture, livestock, handcrafted textile manufacturing and trade which have been and continue to be the main economic activities of Imbabura province have adopted over time different forms of production [12].

This paper is organized as follows: the Sect. 1 is the Introduction that describes the problem, background and the objective, Sect. 2 the Development that have materials and methods. It is divided in three phases (design of the research, design of the proposal and development of the proposal), Sect. 3 Results, Sect. 4 Discussion and Sect. 5 Conclusions.

2 Development

2.1 Materials and Methods

The research methodology followed three phases: the research design, the proposal design, and the proposal development. See Fig. 1.

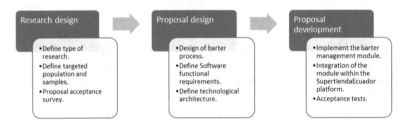

Fig. 1. Research methodology.

Phase 1: Research Design
Type of Research
This study is a quantitative, documentary, descriptive and field research.

Population and Sample
The population - or universe - is made of 4,000 people from the cantons of Imbabura, Carchi, Pichincha, Sucumbíos and southern Colombia. These people are involved in farming activity and they participate in the "Trueque del Sol". The sample is made of 571 people, with a 5% margin of error and a 99% confidence level. The survey consists of a technique of primary research method to obtain the level of acceptance of the proposal. See Eq. 1.

$$tm = \frac{\frac{z^2 \times p(1-p)}{e^2}}{1 + \left(\frac{z^2 \times p(1-p)}{e^2 N}\right)} \tag{1}$$

Where:

tm = sample size
N = population size = 4.000
e = margin of error = 0.05
z = z score based on a 99% confidence level = 2.58
p = proportion of individuals within the population meeting the study characteristic = 0.5

Survey to Determine the Level of Acceptance of the Proposal
The survey consisted of 17 questions. It was validated by 3 experts in the area of Software Engineering and 3 experts in the area of Tourism and Heritage of the UTN. They assessed each question relevance, its link with the objectives, and its wording and clarity. The final survey included 12 questions. In order to verify the reliability and validity of the survey, it was submitted to 50 people as a pilot test. These people were a randomly chosen subset of the sample. The processing and analysis of the data was done using Microsoft Excel 2016 and the SPSS social research statistical package. By crossing variables, it is determined the results.

Phase 2: Proposal Design
Design of the Barter Process
The barter process was designed by 2 experts in the area of Tourism and Heritage of the Superintendence of Market Power Control, 1 expert in the area of Software Engineering and 1 expert in the area of Industrial Engineering of the UTN, they developed and verified the process using a retrospective validation based on past experiences thanks to the documentation obtained during the field visit. See Figs. 2 and 3.

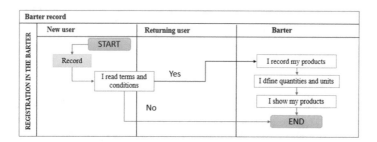

Fig. 2. Barter record process.

Functional Requirements of the Software
According to the IEEE 830 template, the functional requirements for software development were made up of 10 User Stories (US) established by: priority, risk, estimation and acceptance tests. They focused on the roles of client and administrator as stakeholders. The user stories were recorded in a stack called Product Backlog, to be developed by the programmer. See Table 1.

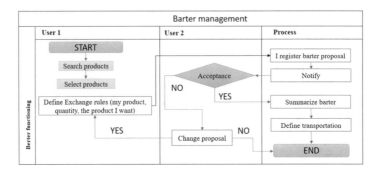

Fig. 3. Barter operation process.

Table 1. Software product list.

ID	Priority	User histories	Estimation (Hours)	Acceptation test
HU01	High	Authentification Module	8	2
HU02	High	Users' Record Module	8	2
HU03	High	Products Categories Record Module	8	2
HU04	High	Products Units Record Module	8	2
HU05	High	Products Record Module	8	2
HU06	High	Messages and Notification Module	8	3
HU07	High	Manage and Summarize Barter Module	40	2
HU08	High	Communication Module (chatting)	8	1
HU09	High	Statistical Reports Release	8	1
HU10	High	Maps with georeferenced data	8	2

Definition of Technological Architecture

The technological architecture was designed by 2 experts from the IT department of the Market Power Control Superintendence. It consists of a Model-View-controller (MVC) architecture, with Asp.Net programming language, Bootstrap framework and the use of Api v3 Google for georeferencing.

Phase 3: Proposal Development

Construction of the Barter Management Module

The construction of the barter management module was done using the agile software development methodology SCRUM (See Fig. 4). Six Sprints were applied in 128 h of development based on the list stack of products defined in phase 2, it is having 24 h and 16 h in the sprint 3 and sprint 4 (design and programming) respectively. The development of each sprint complied with the phases of planning, design, construction, testing of the Sprint backlog (user stories to be performed in the sprint). The increase of the potentially deliverable product was validated in the review meeting. Finally, the continuous improvement of work covered was discussed during the retrospective meeting.

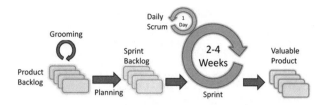

Fig. 4. SCRUM life cycle [13].

Integration within the "Supertienda Ecuador" Technological Platform
The integration between the SETP and the Barter management module was done at the level of the SETP's database objects where the same connection component is shared with the data repository. The main challenges were the standardization of nomenclatures in the definition of objects such as tables, views and stored procedures and the assignment of permissions to the user to connect to the database. See Fig. 5.

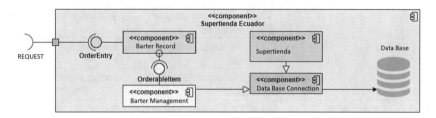

Fig. 5. Integration of barter module within the Supertienda Ecuador platform.

Acceptance Tests
After developing the user stories that appear in Table 1, those in charge of administering the *SETP* formally validated the functionalities of the Barter Management Module through User Acceptance Testing (UAT) conformed with the following acceptance criteria: Business processes of systems That have already been integrated (PNI), Operational processes (OPR), User procedures (UPR), Forms (FRM), Reports (RPT), Configuration data (CND) [18], the results of the tests are shown in Table 2. was made prior to the final delivery of the software.

Table 2. Software acceptance tests.

Barter system management				
ID	Users history	Tasks	Actual Results	Expected results
H01	Safety and Database	Database design	Successful	Successful
		Coding and Design (CD) of system Login	Successful	Successful
H02	User Register Module	CD of User Register module	Successful	Successful
H03			Successful	Successful

(continued)

Table 2. (*continued*)

Barter system management				
ID	Users history	Tasks	Actual Results	Expected results
	Product Category Register Module	CD of Product Category Register Module		
H04	Product Unit Register Module	CD of Product Unit Register Module	Successful	Successful
H05	Product Register Module	CD of Product Register Module	Successful	Successful
H06	Message and Notification Module	CD of Message and Notification Module	Successful	Successful
H07	Manage y Implement Barter Module	CD of "Offer Product" Submodule	Successful	Successful
		CD of "Check Existing Offers" submodule	Successful	Successful
		CD of "Check my products offers" submodule	Successful	Successful
		CD of "Barter Counter-offer" Submodule	Successful	Successful
		CD of "Transport selection" Submodule	Successful	Successful
H08	Communication Module (Chatting)	CD of the Communication Module	Successful	Successful
H09	Release of Statistical Report	CD of Statistical Report Release Module	Successful	Successful
H10	Maps with georeferencial data	CD of the Maps with georeferencial data Module	Successful	Successful

3   Results

The survey shows that 92% of the study targeted sample accept the automation of the barter process through the development of barter management software module that can be used every year in the ancestral activity "Trueque del Sol". It also shows 98% acceptance in the integration and use of technology in activities of cultural and ancestral nature in Ecuador. It is applicated with the acceptance tests of the "Barter Management Module".

The proposal of this research work was recognized and awarded in January 2018 by the Superintendence of Control and Market Power of Ecuador, for promoting an effective culture of competence in economic operators and consumers by supporting the national production of MiPymes, Artisans and the Popular and Solidarity Economy for the benefit of the country's society.

As future work, the developed proposal is to be implemented every year during the event "Trueque del Sol" to perform concept tests, verify and validate the functionality, the efficiency and effectiveness of the software.

4 Discussion

In Ecuador, the present technological project is the first one that supports the management of barter and that is also integrated into the Portal Supertienda Ecuador, this data is endorsed by the Superintendency of Control and Power of Markets of Ecuador; therefore, it is not possible to establish comparisons with respect to other solutions, however, in recent years some solutions related to the present research have been developed and implemented. In the following section, we explain how they relate to the current contribution with their own specificities.

The most recent project is a virtual barter system for mobile devices which was carried out by the University of the Armed Forces in 2016 [14]. However, it doesn't focus on any social issue, rather it tries to demonstrate the viability of technological tools and development methodologies of mobile applications in general and the software produced is not intended or related to any social sector of the country. On the other hand, the present work aims to support the reality of Pimampiro canton and it is in line with market regulation policies currently promoted by this government through the Superintendence of Control and Power of Markets of Ecuador. Indeed, as discussed above, the proposed software is embedded within the Supertienda Ecuador Portal and as such it is certified, recognized and endorsed.

Another similar work was carried out in 2015 at the University of the Americas within the framework of a master thesis in business administration [15]. It's about a startup 2.0 facilitating the exchange of goods without using money in Quito. It offers two technological tools for free: (a) a web portal and (b) an application for smartphones. However, as a commercial business, the company makes profit from: (a) a set of optional services enhancing transactions security and easiness, and (b) the sale of advertising space both on the web and on the mobile App.

The startup targets the population aged between 18 and 23 within the top half of the economic spectrum. This work is supported by a sound market research to offer services providing transaction security and convenience. By nature, this contrasts with the fundamental and universal concept of barter which is exchanging goods and/or services without the use of money [1]. This ancestral and millenary practice is supposed to support the solidarity economy in which there is no monetary value to avoid the consequences of unbalanced wealth distribution. In addition, similarly as money, which is not involved in transactions, intermediaries are not either. They reduce the value of the goods to be exchanged -unless intermediaries work for free or add value to the transaction [16].

This research work could be considered as a reference or baseline for other related projects, including in other cities of the country. For example, in the Bicentennial Park of the Metropolitan District of Quito, a project has been presented to design a barter spot dedicated to the exchange of knowledge and marketing [17]. The proponents point out that, the growing consumerism brings along a series of social and environmental issues and it is necessary to offer consumption alternatives with appropriate infras-tructures. In this sense, our work can contribute in Quito, especially if such initiative is implemented through the Superintendency of Control and Power of Markets of Ecuador.

5 Conclusions

The study and application of Google's API v3 was successfully implemented within the georeferencial maps module and within the barter management system such as it quickly facilitates the recovery of objects geographical coordinates.

The system of management of the barter works well and the times of delivery are respected thanks to the sound choice which were made for the base architecture of the technological platform of the Superstore Ecuador of the Superintendence of Market Control.

The development of the computer modules proposed in the barter management system allowed the barter process to be carried out at a regional and national level through the exchange of products and services, promoting and fostering this ancestral tradition.

The use of the agile SCRUM methodology in the development of the project made it possible to comply in a timely manner with the functional requirements defined by the client, achieving 100% software acceptance in production.

The 10 acceptance tests performed on the barter management system complied successfully with the functional requirements planned at the beginning of the proposal design.

References

1. Ferraro, E.: Trueque, intercambio y valor: un acercamiento antropológico. In: Íconos. Revista de Ciencias Sociales., vol. 14, pp. 150–152 (2002)
2. Hintze, S.: Trueque y economía solidaria (2003)
3. Krause, M.: Las limitaciones del trueque, pp. 107–120 (2003)
4. S. d. c. d. p. d. m. SCPM.: Supertienda Ecuador (2017). http://www.supertiendaecuador.gob.ec/Preguntas.aspx
5. SCPM Repositorio: Misión y Visión de la Superintendencia de Control del Poder de Mercado Ecuador (2016). http://scpm.gob.ec/es/site-map/articles/80-nosotros
6. LORCPM: Ley Orgánica de Regulación y Control del Poder de Mercado. Quito (2011)
7. Useit: E-commerce y su importancia (2018). https://www.useit.es/es/blog/e-commerce-y-su-importancia
8. Tantacon. Tecnología y Comunicación Grupo Onetec.: La Importancia del Comercio Electrónico y la necesidad de las empresas de disponer de Tienda Online, Parte primera: Tanta Comunicación, agencia de marketing digital (2017). https://tantacom.com/la-importancia-del-comercio-electronico-y-la-necesidad-de-las-empresas-de-disponer-de-tienda-online-parte-primera

9. Rodríguez Durán, D.O.: Creación de un motion graphics que describa los beneficios del trueque como una herramienta alternativa para adquirir bienes, servicios y conocimientos. Quito: Universidad de las Américas (2016)
10. Elche Followers: El Ayuntamiento comienza el proyecto de tiendas online para comercios locales (2016). http://www.elchefollowers.es/es/diario-follower/87-la-importancia-del-comercio-electronico-para-los-comercios-locales
11. Rodríguez, T.: La Importancia del Comercio Electrónico y la necesidad de las empresas de disponer de Tienda Online, Parte primera: Tanta Comunicación, agencia de marketing digital (2016). https://tantacom.com/la-importancia-del-comercio-electronico-y-la-necesidad-de-las-empresas-de-disponer-de-tienda-online-parte-primera
12. Lanas Medina, E.: El trueque una forma de economía solidaria presenta en la historia de Pimampiro. Sarance, vol. 26, pp. 13–28 (2010)
13. Sequal Solutions: Por qué utilizar Scrum (2017). http://sequal.com.mx/component/content/article/35-articulos-de-interes/347-ipor-que-utilizar-scrum.html
14. Villarruel Duque, P.P., Naranjo Torres, C.I.: Desarrollo de un sistema de trueque virtual para dispositivos móviles Android, utilizando software libre basado en lenguajes de programación interpretados. Universidad de las Fuerzas Armadas ESPE. Carrera de Ingeniería de Sistemas e Informática (2016)
15. Enríquez Tito, L.G.: Proyecto de viabilidad para la creación de una empresa 2.0 Que facilite el intercambio de bienes sin dinero (Trueques) en la ciudad de Quito. Universidad de las Américas (2015)
16. Salas Zapata, J.L.: Interacción comunicativa en el proceso social del trueque, un diálogo entre lo moderno y lo ancestral. Estudio de caso: mercado central de la parroquia de Cusubamba (2014)
17. Jaramillo Viteri, K.E., Evelyn, K.: Proyecto arquitectónico de diseño de un punto de trueque para el intercambio de conocimientos y comercialización en el parque Bicentenario de Quito. Quito: UCE (2015)
18. Ellingwood, J.: An Introduction to Continuous Integration, Delivery, and Deployment. DigitalOcean, New York (2017)

Using Model-Based Testing to Reduce Test Automation Technical Debt: An Industrial Experience Report

Thomas Huertas$^{(\boxtimes)}$, Christian Quesada-López$^{(\boxtimes)}$, and Alexandra Martínez$^{(\boxtimes)}$

Universidad de Costa Rica, San José, Costa Rica
{thomas.huertas,cristian.quesadalopez,alexandra.martinez}@ucr.ac.cr

Abstract. Technical debt is the metaphor used to describe the effect of incomplete or immature software artifacts that bring short-term benefits to projects, but may have to be paid later with interest. Software testing cost is proven to be high due to the time (and resource)-consuming activities involved. Test automation is a strategy that can potentially reduce this cost and provide savings to the software development process. The lack or poor implementation of a test automation approach derives in test automation debt. The goal of this paper is to report our experience using a model-based testing (MBT) approach on two industrial legacy applications and assess its impact on test automation debt reduction. We selected two legacy systems exhibiting high test automation debt, then used a MBT tool to model the systems and automatically generate test cases. We finally assessed the impact of this approach on the test automation technical debt by analyzing the code coverage attained by the tests and by surveying development team perceptions. Our results show that test automation debt was reduced by adding a suite of automated tests and reaching more than 75% of code coverage. Moreover, the development team agrees in that MBT could help reduce other types of technical debt present in legacy systems, such as documentation debt and design debt. Although our results are promising, more studies are needed to validate our findings.

Keywords: Test automation debt · Model-based testing · Legacy systems

1 Introduction

The technical debt (TD) metaphor was first used by Ward Cunningham in 1992, as "not quite right code which we postpone making it right" [3]. More recently, the concept has been extended to describe internal software development tasks chosen to be delayed, causing some software artifacts to remain incomplete or immature, which poses a risk for future problems [1]. Different artifacts produced during the software development process have been identified as sources of technical debt, hence, several strategies have been used to find all types of

© Springer Nature Switzerland AG 2019
Á. Rocha et al. (Eds.): ICITS 2019, AISC 918, pp. 220–229, 2019.
https://doi.org/10.1007/978-3-030-11890-7_22

technical debt that might have a negative impact on software projects [1]. Most efforts on identifying and managing technical debt have focused on the code. Nevertheless, debt can be hidden in other artifacts or parts of the project, such as test automation, which may also bring a significant negative impact to the project [1].

This paper proposes the use of model-based testing (MBT) as an approach to reduce the test automation debt (TAD), potentially impacting other TD categories as well. The study was conducted in a local branch of a renowned multinational IT company, specifically within a development team whose portfolio has 80% of legacy applications. This team has been making efforts to reduce the technical debt over the last years, but those efforts have addressed code and architecture debt only, leaving behind artifacts like automated testing. By adopting a MBT approach, the company seeks to find an alternative way to reduce the TAD present in their legacy systems.

The remainder of the work is structured as follows. Section 2 offers some background on technical debt and model-based testing. Section 3 summarizes relevant previous works in the area. Section 4 describes the design and context of our study. Section 5 shows and discusses the results. Section 6 presents the conclusions and outlines future work.

2   Background

Recent studies have shown that although *technical debt* is often associated mostly with code quality, it covers a much wider set of areas [5]. Alves et al. [1] made a systematic mapping study to characterize the types of TD, classifying them in categories such as architecture debt, documentation debt, test automation debt, requirements debt, people debt and others, which also have a correspondence with the stages of the traditional software development cycle. They also refer to *test debt* as the issues found in testing activities that can affect the quality of those activities. As a subcategory, they refer to *test automation debt* as the work involved in automating tests of previously developed functionality, to support continuous integration and faster development cycles.

Model-based testing is an advanced test approach that automatically generates test cases from a model of the system under test (SUT) [2]. The use of MBT is an alternative to practices of test design that tend to be hard to reproduce, poor documented, and dependant on engineers' expertise (or lack of it) [8]. The goal of MBT is to improve the quality and efficiency of test design and test implementation activities [8]. Building a comprehensive model based on the project's test objectives and providing this model as test design specification, is sufficient to automatically generate the test cases from it.

A *test requirement* is an item or event of a system that could be verified by one or more test cases [2]. A single software requirement should map to at least one test requirement [9]. Each test requirement states the conditions on how to enforce the intended behavior and how the SUT is expected to behave [9].

3   Related Work

The relationship between TD and testing has been studied by Holvitie et al. [4], who show that at least 4 out of 10 TD instances have to do with testing (the rest are related to implementation). Such testing inadequacies were reported to be related to *internal legacy*, which is the lack of quality in the architecture and implementation of software projects in the organization along time.

On the other hand, Wiklund et al. [10] studied TD in test automation, and they found no clear and comprehensive guidelines on how to design, implement and maintain an automated test execution system that enables keeping the accumulated TD on an acceptable level while providing benefit to the organization that uses the system.

Trumler et al. [7] studied the relationship between TD and code coverage. They describe that even though TD cannot be completely eluded, it can be at least partially avoided by the establishment of a high code coverage that enables a low defect rate. They mention how automated testing and code coverage can have a positive impact reducing technical debt while combined with a good definition and understanding of the software requirements.

Masri et al. [6] review and compare coverage criteria used in software testing. According to them, a coverage criterion needs to define a set of *test requirements* to be satisfied by the test suite, and then compute the percentage of satisfied requirements, yielding a metric of potential adequacy of the test suite.

The use of MBT in legacy applications was studied by Wendland et al. [9]. They employed a MBT process for safeguarding the correct migration of a legacy system to its modern implementation. They conducted a case study where MBT and UML were combined to model the SUT. They performed reverse engineering of the requirements through meetings with expert users, informal reviews of business rules, and walkthroughs of the legacy application. These requirements were used to derive test requirements, define the appropriate level of model abstraction, and decouple testing and development sides. They reported that no significant changes to the MBT methodology and tooling were needed.

Our work is partially based on the work of Xu [11], which served as reference for the MBT tool we used and experiment design. We also took some ideas from Wendland et al. [9] on adapting MBT to legacy systems, and from Alves et al. [1] on types of TD.

4   Case Study Design

The goal of this study was to apply model-based testing to existing legacy systems within an IT organization, and assess the impact on their TAD. Our research question was: *What impact does the use of model-based testing have on the TAD of a legacy system?*

4.1   The Context

The study was conducted in the context of a development team within the Information Technology Department of a multinational company. This team maintains a portfolio of legacy applications for the company, and consists of a business analyst, a project manager, and five software developers. The team worked in conjunction with the authors of this study, providing details of the SUTs, documentation, data needed to carry out the investigation and feedback on every stage of the process.

4.2   The Legacy Systems Under Test

Two legacy systems were selected from the application portfolio of the development team to be used in this study. They were chosen based on their level of technical debt (previously identified by the development team). None of the selected legacy systems had any test automation, hence had a high TAD (among other types of TD). Both systems were web applications hosted in the same infrastructure of the IT department, and interacted with their own relational database hosted in a dedicated database server. The first SUT, *LRC*, is a transactional website that allows users to create different types of requests, and has been in production for 10 years. It is composed of three modules, that provide the different operations available for the requests. The second SUT, *DSL*, was released 6 years ago, and has had minor updates during the last years. This is also a transactional website that allows the users to create jobs –an internal business concept– and queue them in the business process. This SUT is also composed of three modules for the jobs operations.

4.3   The MBT Tool

MISTA (previously named ISTA) is an academic model-based testing tool that allows the generation of automated test cases on different programming languages. It uses high-level Petri nets for specifying test models, so that complete tests can be generated automatically [11]. MISTA was selected as the MBT tool to use in this case study mainly because (1) it was freely available for download, (2) it was able to generate functional tests compatible with Selenium and with the infrastructure of the development team, and (3) it allowed the creation of the model through a Graphic User Interface (GUI). This study was limited to the capabilities of MISTA for modeling and mapping web applications as well as for generating functional tests for Selenium.

4.4   The Survey for the Development Team

We designed a survey for the development team, where participants were asked about their specific role in the team (developer, QA, project manager, business analyst) as well as their understanding of and working experience with technical debt remediation, if they had any. The survey also asked participants to identify the level of impact that using MBT and MISTA had on the TAD of the SUTs. Additionally, participants were asked to identify other types of technical debt that could potentially be impacted by the use of MBT. These types were based on the work of Alves et al. [1]. The survey was reviewed by two of the researchers and validated with three IT professionals, who checked that the questions were clear and understandable.

4.5   Procedure and Data Collection

Next we describe the procedure followed to use MBT on the SUTs and measure its impact on the TAD of the SUTs. Figure 1 shows a summary diagram of this procedure.

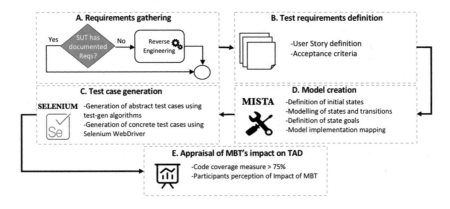

Fig. 1. Procedure for using MBT over legacy applications and assessing its impact.

A. Requirements Gathering. To model the SUTs, requirements were obtained from the documentation available for each system. For *DSL*, there were user stories stored in a Microsoft Team Foundation Server, and requirements specifications (in use case format) in a text document. A user guide and help page were also used as complementary information of the SUT features. In the case of *LRC*, it lacked documentation of system requirements, thus, a reverse-engineering process was performed. Such process was mainly based on the user help's wiki as well as walkthroughs of the system and business knowledge from the developers and project manager.

B. Test Requirements Definition. In this study, test requirements were obtained by consolidating all the information gathered from the different sources in the previous step, into a standard format. We created a spreadsheet with the following information per test requirement: title, details (in user story format), and acceptance criteria, including specific information that would help the model creation step.

C. Model Creation. To model the SUTs, we first defined the initial states, goal states and transitions. Then we created the PrT Net models in MISTA. After that, a *model-implementation mapping* (MIM) was done, setting the definition of asserts and required UI actions. All commands used were compliant with Selenium. Finally, the models were compiled and simulated with MISTA, in order to check that all the paths and goals were correct.

D. Test Case Generation. The generation of test sequences (i.e., abstract test cases) was performed using the available algorithms in MISTA. Next, test code generation was performed according to the MIM. Afterwards, test cases were translated from HTML Selenium to C#.NET, to make them more manageable inside a Microsoft Visual Studio solution, and to be able to execute them in different browsers.

E. Appraisal of MBT's Impact on TAD. TAD reduction was assessed in two ways: (1) by using a coverage criteria and (2) by surveying the participants on their perception of TAD reduction in the SUTs due to MBT. The following coverage criteria were used were: (*i*) at least 75% code coverage must be obtained for each SUT (derived from the code coverage goal of the development team), (*ii*) at least one test case should cover each test requirement, and (*iii*) the acceptance criteria of every test requirement must be met by the corresponding test case. A profiler tool called OpenCover was used to measure the percentage of code coverage attained by the tests. The survey asked members of the development team for their perception on how the use of MBT could reduce the SUT's TAD and other types of TD.

5 Results

We present here the results of our study, first showing what we obtained during the application of MBT to the SUTs and then what we found regarding the impact of MBT on the TAD of the SUTs.

5.1 Application of MBT to Legacy Systems

Built Models. Table 1 shows for each SUT module: the number of test requirements, the number of models built, and other model details such as the total number of transitions, total number of states, and total number of goal states. Each SUT contained three modules, with similarities among them, which facilitated the modeling task. The average modeling time depended on the number of states in each module.

Generated Tests. Table 2 shows the number of test cases generated per test generation algorithm for each SUT, and the code coverage achieved in each case. For *LRC*, the 'Reachability graph coverage' and 'Transition coverage' algorithms

Table 1. Number and characteristics of models generated per SUT module.

SUT name	Module	Number of test requirements	Number of models	Total transitions	Total states	Total goal states	Avg. modeling time (hours)
LRC	New request	2	1	2	3	2	1
	View requests	3	3	4	7	4	2
	Modify request status	12	1	12	6	2	2
	Total	17	5	18	16	8	5
DSL	New job	6	2	4	8	4	2
	View jobs	3	1	6	6	4	2
	Modify job status	3	1	8	5	2	1
	Total	12	4	18	19	10	5

generated not only the same number of tests but the same test cases. The 'State coverage' algorithm generated 10 test cases, but a quick manual analysis of the models shows that it only takes 9 scenarios to do a full state coverage. (The extra test is valid but unnecessary.) Using the 'Goal coverage' algorithm, the number of test cases equals the total number of goal states. For *DSL*, the 'Reachability graph coverage' and 'Transition coverage' algorithms also generated same number of test cases. The 'State coverage' algorithm generated 8 test cases, and a manual revision shows that all are required for full state coverage of this SUT. Using the 'Goal coverage' algorithm, the number of test cases equals the total number of goal states.

Table 2. Number of test cases and coverage achieved by each algorithm per SUT.

SUT name	Test generation algorithm	Number of test cases	Code coverage achieved
LRC	Reachability graph coverage	14	82%
	Transition coverage	14	82%
	State coverage	10	79%
	Goal coverage	8	76%
DSL	Reachability graph coverage	9	78%
	Transition coverage	9	78%
	State coverage	8	78%
	Goal coverage	10	78%

5.2   Impact of MBT on the TAD of Legacy Systems

Impact on Coverage. We present the results in terms of the three coverage criteria previously defined. First, all test generation algorithms produced a test suite that exceeded the 75% code coverage criterion. For *LRC*, some test suites obtained up to 82% code coverage, using the 'Reachability graph coverage' and 'Transition coverage' generation algorithms. For *DSL*, all test suites obtained a code coverage of 78%, regardless of the generation algorithm. Second, all test generation algorithms produced at least one test case that covered each test requirement of the SUTs. This was manually verified. As a single test case could cover more than one test requirement, the total number of test cases did not necessarily match the total number of test requirements. Third, all test generation algorithms produced test suites that met every test requirement's acceptance criteria. This was partially automated with asserts but also manually verified.

Perceived Impact on TAD. All participants perceived that the greatest impact of MBT was on TAD, rather than any other TD type. The reason being that none of the SUTs originally had automated tests, and with the MBT tool, it was possible to generate a suite of automated tests that surpassed the team's code coverage goal. Other categories of TD were also reported to be impacted by the use of MBT on the SUTs, particularly, documentation and people debts. For 'documentation debt', 4 (out of 7) participants (58%) perceived MBT could have a high impact, while 3 participants (42%) perceived it could have a medium impact. For 'people debt', 5 participants (71%) reported MBT could have a medium impact, while 2 (29%) perceived it could have a low impact. To support such choices, participants mentioned the existence of a SUT model and a set of test requirements. The more documentation (in the form of a model or test requirements), the more knowledge that is preserved in an independent source, therefore reducing documentation and people debt risks. For 'code debt', 5 out of 7 participants (71%) perceived MBT had no impact at all over the SUT, while 2 participants (29%) perceived a low impact. None of the participants perceived impact of MBT on 'service', 'versioning' or 'build' debts.

5.3   Discussion

The MBT approach was applied to both SUTs, but for *LRC* we first needed to reverse-engineer the SUT requirements, in preparation for the modeling step, where test requirements were the basis for creating the model. In the case of *DSL*, it was convenient to have requirements available, needing less preparation work before modeling the SUT.

Some observations beyond the scope of the research question are mentioned next. The first observation is that MISTA can be adapted to a real software development context, thanks to its compatibility with Selenium. Nevertheless, being an academic tool, it lacks documentation and support, which renders it not quite fit for the industry. A second observation is that modeling the SUT heavily relies on the skills of the person who creates the model. A third observation is that some other types of TD can also be reduced by using MBT, as indicated by the participants. A last observation is the steep learning curve on how to use the Selenium IDE with different browsers. This also raises concerns about the difficulty of using it on non-legacy applications, for which there is not a clear GUI defined by the time the MBT models are built.

Another lesson learned is to ensure a thorough process of requirements gathering and tests requirements definition, to ease the modeling process. Also, given that MISTA works based on state goals, it would make sense to define those goals since the test requirements definition stage, to enhance the resulting model. A *template model* could be thought for systems with similar modules and flows. For example, both of our SUTs have create, modify and view modules. Even though they have conceptual differences and different business rules, their models look very similar. Thus, a template model would make the modeling process faster.

6   Conclusions

In this paper, we discussed how to implement a model-based testing (MBT) approach as an alternative way to reduce the test automation debt of legacy systems. The chosen SUTs exhibited a high level of test automation debt. We used an MBT tool called MISTA to automate the generation of test cases. The study was conducted in the context of a development team that maintains a portfolio of legacy applications, and has seven members. Finally, we analyzed code coverage metrics and surveyed the team members about the impact of MBT on TAD reduction.

Our results show that the SUTs' TAD could be reduced by the use of MBT. All MISTA's test generation algorithms yielded test suites that covered all test requirements. The perception of the impact on use of MBT to reduce different types of TD in legacy systems was high for TAD and medium for documentation and people debts.

We consider MBT as a promising approach to improve test automation generation and reduce the TAD of legacy applications. MISTA is a good alternative to explore the use of MBT, mainly because it is free and open source. However, better and modern MBR tools can contribute on the spreading and consolidation of this technique in the industry. We believe that sharing these experiences in adopting model-based testing approaches for TD remediation in legacy applications is valuable for both researchers and practitioners.

Although our results are promising, similar studies or replications should be conducted in different contexts, to gather more empirical evidence on the use of MBT as a means to reduce the TAD of legacy systems. It would be interesting to expand this research in the future by exploring other types of legacy systems, MBT tools, roles, and organizations. Another interesting area of future work is how to improve the modeling and code coverage, possibly by refining the test requirements derivation step and the creation of model templates that can potentially enhance the methodology proposed in this study.

References

1. Alves, N.S.R., et al.: Identification and management of technical debt: a systematic mapping study. Inf. Softw. Technol. **70**(Supplement C), 100–121 (2016). ISSN 0950-5849
2. Christmann, S., et al.: ISTQB® Foundation Level Certified Model-Based Tester Syllabus. International Software Testing Qualifications Board Foundation Level Working Group (2015)

3. Cunningham, W.: The WyCash portfolio management system. In: Addendum to the Proceedings on Object-Oriented Programming Systems, Languages, and Applications (Addendum), OOPSLA 1992, pp. 29–30. ACM, Vancouver (1992)
4. Holvitie, J., Leppänen, V., Hyrynsalmi, S.: Technical debt and the effect of agile software development practices on it - an industry practitioner survey. In: 2014 Sixth International Workshop on Managing Technical Debt, pp. 35–42 (2014)
5. de Jesus, J.S., de Melo, A.C.V.: Technical debt and the software project characteristics. A repository-based exploratory analysis. In: 2017 IEEE 19th Conference on Business Informatics (CBI), vol. 01, pp. 444–453 (2017)
6. Masri, W., Zaraket, F.A.: Coverage-based software testing: beyond basic test requirements. Adv. Comput. **103**, 79–142 (2016)
7. Trumler, W., Paulisch, F.: How "Specification by Example" and test-driven development help to avoid technial debt. In: 2016 IEEE 8th International Workshop on Managing Technical Debt (MTD), pp. 1– 8 (2016)
8. Utting, M., Pretschner, A., Legeard, B.: A taxonomy of model-based testing approaches. Softw. Test. Verif. Reliab. **22**(5), 297–312 (2012)
9. Wendland, M.-F., et al.: Model-based testing in legacy software modernization: an experience report. In: Proceedings of the 2013 International Workshop on Joining AcadeMiA and Industry Contributions to Testing Automation, JAMAICA 2013, pp. 35–40. ACM, Lugano (2013)
10. Wiklund, K., et al.: Technical Debt in Test Automation, April 2012
11. Xu, D.: A tool for automated test code generation from high-level petri nets. In: Kristensen, L.M., Petrucci, L. (eds.) Applications and Theory of Petri Nets. PETRI NETS 2011, pp. 308–317. Springer, Heidelberg (2011)

A Software Platform for Processes-Based Cost Analysis in the Assembly Industry

Erik Sigcha[1] [iD], Eliezer Colina-Morles[2] [iD], Villie Morocho[1] [iD],
and Lorena Siguenza-Guzman[1(✉)] [iD]

[1] Department of Computer Sciences, Faculty of Engineering,
University of Cuenca, Cuenca, Ecuador
{erik.sigchaq,villie.morocho,
lorena.siguenza}@ucuenca.edu.ec
[2] Research Department, University of Cuenca, Cuenca, Ecuador
eliezer.colina@ucuenca.edu.ec

Abstract. Processes and resources management are important current discussions related to decision making in the industrial field. This fact motivates companies to search for management models to improve their processes and services continuously. In order to achieve this purpose, approaches such as Business Process Management (BPM) and Time-Driven Activity-based Costing (TDABC) are used as bases for models design. This article describes the validation process of a software platform constructed using Business Process Model and Notation (BPMN) and TDABC paradigms aimed at analyzing processes costs in assembly companies. This work contemplates a description of the methodologies applied, functionalities implemented and validations steps performed. The platform also serves to generate process diagnosis in assembly companies prior to full BPM implementation.

Keywords: BPMN · TDABC · Process · Cost · Management

1 Introduction

The high competitiveness level that currently exists in the assembly industry moves companies to concentrate efforts in efficiently managing their resources, seeking a direct impact in achieving objectives such as minimizing costs and maximizing profits. Thus, strategic management models or methods to optimize the use of their resources are especially important. Two essential aspects, which are part of the strategic management of resources to improve services and process, are the process and cost management. Among the techniques applied to perform process management, BPM comprises a set of principles, methods, and tools whose objectives are to create process-oriented organizations and to optimize their operation through the management of activities carried out internally [1]. Regarding cost management in companies, accounting is an essential mean to determine economic and financial states. In this area, accounting has been in constant evolution, contributing with several costing techniques

© Springer Nature Switzerland AG 2019
Á. Rocha et al. (Eds.): ICITS 2019, AISC 918, pp. 230–241, 2019.
https://doi.org/10.1007/978-3-030-11890-7_23

over time. TDABC is a costing system developed in 2004 which appropriately reflects the diversity of processes in companies, and provides accurate information about their costs [2]. TDABC calculations are based on two parameters: the unit cost of resource used and the estimated time needed to perform an activity [2].

Nevertheless, despite the two methods being complementary, only a few published studies are related to the simultaneous implementation of BPMN and TDABC regarding software tools aimed at analyzing the state of processes and costs in assembly companies [3]. One example is the development of a module called TD-ABC-D, presented in [4] and [5], oriented to costs management in university libraries. However, this software focuses on processes within service institutions rather than manufacturing companies. A study reported in [6] introduces the analysis and design of a software platform for processes and costs management but does not deepen on the TDABC costing approach. As a continuation of the before mentioned work, [7] presents the methodologies applied during the implementation of the platform for calculating process costs.

In the same context, this article describes the software architecture, features, and functionalities of a software platform for processes-based cost analysis in assembly companies. It also reviews the way this platform was tested and validated to analyze processes status within a real assembly company. This company, classified as a middle size enterprise, has more than 100 employees and assembles televisions with a production around of 40 K units/year. The remainder of this document is as follows; Sect. 2 summarises the previous work. The software platform and its functionalities are described in Sect. 3. Section 4 contains a detailed description of the validation steps performed on the platform. Finally, Sect. 5 is reserved for conclusions and discussion of future work.

2 Related Work

Information about the analysis and design phases of this work can be found in [6], where the BPMN standard was used for process modeling. To estimate process costs, TDABC was established as a base methodology for performing calculations. Figure 1 presents the steps executed to estimate costs per process or sub-process within an assembly company [8].

2.1 Conceptual Data Model

The conceptual data model, illustrated in Fig. 2, is a schema of entities whose contents and relationships are stored in the platform's database and provides data about elements participating in the costing process. The *Processes* entity allows having a referential map containing the main processes performed in an assembly company. *Sub-processes* are grouped in higher-level processes, and permit obtaining costs according to the referential processes map of an assembly company [8], which conventionally classify as operational, strategic and support processes [9]. Finally, the *Resources* entity con-

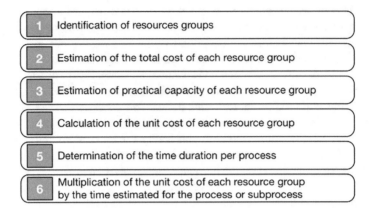

Fig. 1. Steps for estimating process cost using TDABC [8]

tains information about all the resources owned by a company. Among the required information are hours of availability, description, costs, practical capacity and type of resource (human, technological or material). This entity relates to the resource groups' identification step in the TDABC methodology, so accounting data of resources expenses is required as input for this data structure.

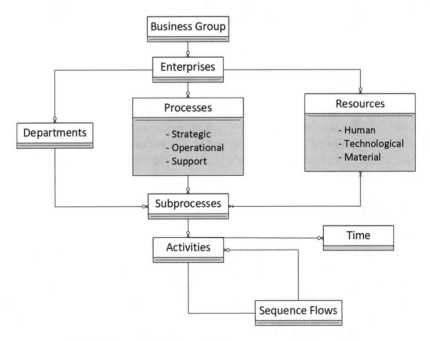

Fig. 2. Conceptual data model for process and costs analysis

3 Platform for Processes-Based Cost Analysis

The platform for processes-based costs analysis was developed as a web application using the Java programming language to support management and analysis of processes and costs in assembly companies. The software architecture, functionalities implemented and costs estimation functions are detailed next.

3.1 Software Architecture

The software architecture, depicted in Fig. 3, is based on the model-view-controller, MVC, architecture pattern and presents the information flow processed by the platform. The *Data Sources* block presents the methods used to obtain information about the company from process files and models, and the resource groups identified, which are essential information sources to perform TDABC. In the *Storage* block, the data entry in a database is shown. The database was constructed using the conceptual model depicted Fig. 2. The *Processing* block presents the elements of the MVC architecture, which is in charge of managing and processing the information contained in the database. Within the processing block, the Model sub-block contains structures to manipulate data using a programming language object. The Controller sub-block is responsible for executing the CRUD (an acronym for "Create, Read, Update and Delete") operations on the database. This sub-block records and offers the functionalities to perform the calculations and the platform's logic. Among these functionalities, BPMN is the structure employed to obtain the time of each sub-process diagram, and the TDABC functions are used to estimate costs per process. The View sub-block provides structures for the user interfaces and other functionalities, such as HTML pages, JavaScript code and style sheets, which allows the user to visualize information in a well-structured manner. In the *Data Visualization* block, the information is displayed to the user through interfaces and the data obtained in the Processing block. This allows the user to visualize an analysis of the process information, process representations with the BPMN standard and times and costs by processes in a company.

Data sources analysis and structuring had been manual tasks since the information had to be collected through interviews, observations and analysis of physical documentation to be formatted later as inputs required by the software. This step can be automated by capturing resources and activities information from accounting software and technological equipment within the assembly companies, as suggested in Fig. 3.

3.2 Platform Functionalities

The functionalities of the platform for processes-based costs analysis are the following:

Process Data Management. These functions and interfaces allow the user to create, update, visualize and delete information of processes, sub-processes, and activities. The main feature of this set of functions is using a diagram editor, based on the BPMN 2.0 standard, for process information input. Figure 4 shows a screenshot of the User Interface developed to manage process data using a BPMN diagram editor.

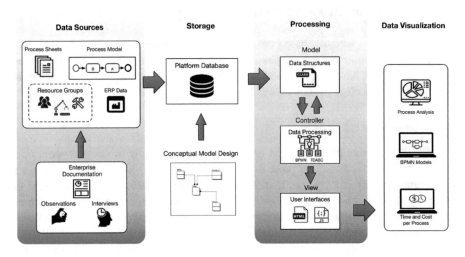

Fig. 3. Software architecture platform for process and cost analysis in the assembly industry.

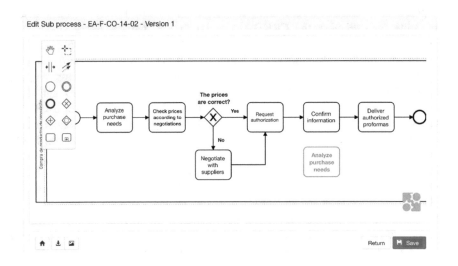

Fig. 4. Screenshot of the process edition UI

Business Data Management. Includes a set of forms and functions that allows entering, modifying, viewing and deleting information from entities belonging to the company (i.e., departments, resources or staff). To accelerate the implementation of these functionalities, the Java Server Faces (JSF) framework was used. JSF allows automatically generating interfaces and programming functions to perform the CRUD operations on database records. Figure 5 contains a screenshot of the user interfaces implemented to manage information of resources.

Process and Costs Analysis. This group of functionalities allows users to obtain results of cost estimates and processes duration using the TDABC system as

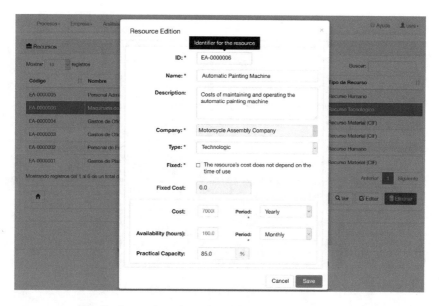

Fig. 5. Screenshot of user interfaces to edit resources data.

information reports. The reports generated by the platform are costs and times estimations by process and sub-process, costs by department, costs per production order, costs per product and generation of sub-process sheet. Figure 6 shows a screenshot of the costs and times estimation user interface.

Configuration. Constitutes a set of features to manage the general parameters of the software configuration. These include access control to the platform, user management functions, language configuration, currency and default parameters configurations.

3.3 Cost Estimation Function

Costs estimation function allows calculating the cost per process based on its activities times and resources used to perform it. This function is based upon the TDABC methodology presented in [8]. The required steps to estimate the cost per process are briefly described next. The first step to estimate process costs is to identify all expenses related to the resources used to perform the process. For this, accounting information is taken from balance sheets, income statements or accounts. Generally, these are total values for the whole company and are not grouped by resources as required by TDABC. Hence, these expense values must be classified in a way that may be assignable to company processes. For this, a classification, by cost centers, has been proposed, which was defined based on the main processes carried out in an assembly company according to a generalized process map; e.g., management, human resource management, assembly/production or sales. Once the resource groups have been identified, steps 2, 3 and 4 of the TDABC methodology are executed (see Fig. 1), which results in the unit cost ([\$/min]) for each resource group identified. Figure 7 shows the sequence of steps used to estimate costs by processes using the TDABC costing.

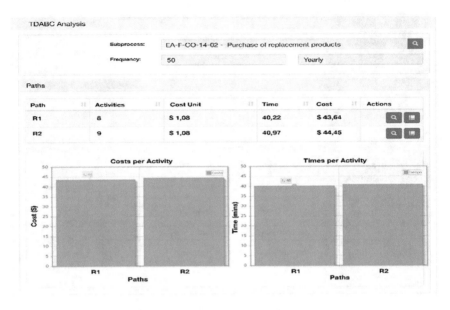

Fig. 6. Screenshot of the costs estimation user interface.

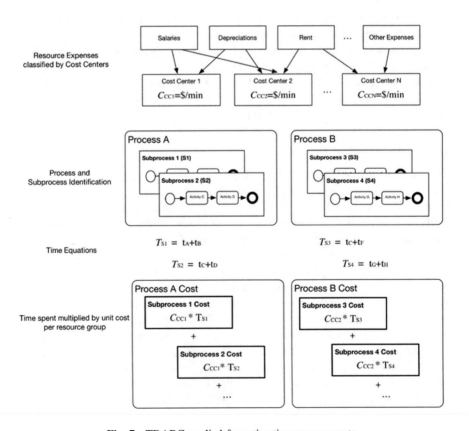

Fig. 7. TDABC applied for estimating process costs

To estimate process costs, a process identification phase is necessary to conduct through manual analysis using interviews, observations, and documentation. As a result, models and process files are obtained according to an ISO process map [9]. Once the company processes have been identified, the next step is generating time equations for each process. Since the BPMN standard has been used, a function that allows identifying activity sequences for executing a process was developed [7]. This function gives the execution time of each process, which is the objective of time equations, resulting in the second element required by TDABC. The final step in estimating process costs is to use the value of the cost unit of the resource groups and to multiply it by the time calculated with the time equations. This result in costs information for each process, sub-process and activities identified.

4   Software Validation

The platform validation considers three main aspects: data entry validations, testing of costs estimation functions and testing of process costs estimations.

4.1   Data Entry Validation

As an initial validation stage, a first partial validation was performed by entering information of 20 process sheets, collected during the process identification phase, into the platform using the functionalities of the process data management. This first data entry allows correcting some programming errors and determining the set of BPMN symbols to be implemented in the diagram editor, which are detailed in [7]. As a second stage, a new validation that comprises entry of the complete set of process models of the case study company was performed. This set contains approximately 200 process models that were entered in the platform database. This part of the validation allowed to verify if all process models were supported by the platform and could be stored using the platform database structure. Because of this validation, it was possible to test the BPMN diagram editor functionalities and to have a repository of process models in the database.

4.2   Testing of the Resources Cost Estimation Functions

These tests had the goal of verifying if the calculations performed to estimate the unit cost per resource were correctly done. To perform these tests, the JUnit[1] framework was used since it allows testing programming functions, by specifying input data, expected results and comparing these results. The function tested was the one regarding the unit cost estimation per resource; as this function executes the first four steps of the TDABC methodology (see Fig. 1). This function requires as input the resource data (name, cost per period, hours of availability per period and practical capacity percentage) and produces as output the resource cost per minute (unit cost). Data used to

---

[1] JUnit, https://junit.org/junit5.

perform this test comprises information of 21 resources records and were gathered from income statements and balance sheets from the case study company. An accountant expert processed the data, by using the TDABC methodology to obtain the expected unit costs, which were compared with the ones obtained from the automatized platform. Initially, small differences in the results were found due to the number of decimal digits used; but when rounding to six decimal digits, the costs estimation results were equal to the expected results.

Testing of Process Cost Estimation. The process cost estimation test verifies the correctness of costs values obtained in the platform. For this purpose, 36 process models belonging to an assembly line of a study case company were input into the software database. The study case is a middle size enterprise dedicated to the assembly of televisions. The 36 selected process models represent some of the phases required to assemble each one of the eight TV models produced by the company. Data from these process models were analyzed manually by an accountant expert to obtain estimated reference costs using the TDABC methodology [8]. Next, the cost per product functionality was used to calculate automatically the product costs grouping the process models by product. This allowed generating cost tables according to the processes involved in the production of each of the eight TV models in the assembly line. Table 1 shows an example of a cost table obtained in the platform, which includes the sub-processes and activities required to assemble a TV model, and their respective cost and times. Finally, a comparative table of platform results and expected results was developed. Table 2 shows times and costs automatically estimated by the platform, referential times and costs calculated manually (Ref. Time and Ref. Cost) and the differences between these values for all eight TV models analyzed. From Table 2, the difference values in most cases are less than 0.01%, which is justified regarding the number of decimal digits used in the platform, and in the reference calculations.

Table 1. Example of cost table by product.

Sub-process/Task	Time (min)	Cost ($)
Sub-process - Assembly TV Model 6		
Task – Set-Up	0,6095	0,177974
Task – Parts assembly	2,7323	0,797832
Task – Place accessories	3,0825	0,90009
Task – Connect electric system	0,7835	0,228782
Total:	7,2078	2,106048
Sub-process – Keyboards assembly TV Model 6		
Task – Receive Raw Material	0,03	0,00876
Task – Assemble accessories	2,76	0,80592
Task – Test	2	0,584
Task – Store	0,08	0,02336
Total:	4,87	1,422966

(*continued*)

Table 1. (*continued*)

Sub-process/Task	Time (min)	Cost ($)
Sub-process – Holster accessories TV Model 6		
Task – Receive remote control	0,0028	0,000818
Task – Holster	0,0152	0,004438
Task – Seal	0,0028	0,000818
Total:	0,0208	0,006078
Sub-process – Quality Control TV Model 6		
Task – Test functions and specifications	1,5015	0,438438
Task – Check product	0,5612	0,16387
Task – Fill out quality control sheet	0,2525	0,07373
Total:	2,3152	0,676479
Sub-process – Packing TV Model 6		
Task – Place packing carton and plastics	1,3567	0,396156
Task – Move to storage	0,6527	0,190588
Total:	2,0094	0,587127
TOTAL	**16,423**	**4,798**

Table 2. Comparison results of the cost estimation test

Product	Time	Cost	Ref. Time	Ref. Cost	Time Diff.	Time Diff. %	Cost Diff.	Cost Diff. %
TV Model 1	6,474500	1,891785	6,474	1,891	−0,000500	−0,000077	−0,000785	−0,000415
TV Model 2	8,434000	2,464331	8,381	2,448	−0,053000	−0,006284	−0,016331	−0,006627
TV Model 3	15,699167	4,587141	15,699	4,588	−0,000167	−0,000011	0,000859	0,000187
TV Model 4	20,320000	5,937303	20,311	5,934	−0,009000	−0,000443	−0,003303	−0,000556
TV Model 5	9,757167	2,850947	9,757	2,851	−0,000167	−0,000017	0,000053	0,000018
TV Model 6	16,423167	4,798687	16,423	4,798	−0,000167	−0,000010	−0,000687	−0,000143
TV Model 7	13,877667	4,054917	13,877	4,055	−0,000667	−0,000048	0,000083	0,000021
TV Model 8	25,663000	7,498474	25,664	7,499	0,001000	0,000039	0,000526	0,000070
Average					−0,007833	−0,000856	−0,002448	−0,000931

5   Conclusions and Future Work

This article describes a software tool and its validation for process analysis in assembly companies. The platform presented aims to provide support in obtaining an initial diagnosis, prior to the full implementation of Business Process Management or Enterprise Resource Planning, ERP, systems. The software functionalities design was inspired by the TDABC and BPMN methodologies, which are established approaches successfully applied in academic and business fields.

This work also presented a conceptual data model conceived for getting cost estimations and for modeling business processes workflows in assembly companies, based on the BPMN 2.0 standard. In addition, the conceptual data model allows obtaining data reports on costs and processes associated with departments and production lines. The software architecture showed the information analyzed and structured to implement a software platform. The architecture explained how the inputs were obtained, how they were stored and the type of analysis performed. Finally, algorithms to estimate costs and times per process have been developed with the aim of integrating TDABC and BPMN, and a set of functionalities for obtaining an initial diagnosis of the complete process, including information management for process models. These algorithms were tested and validated to verify the correctness of calculations and costs estimations, by comparing their results with referential costs per process provided by accountant experts. The resulting software, relying on open source tools, offers the possibility of taking advantage of TDABC and BPMN to determine a company's state based on its processes and resources.

Further work is considered to continue validation tasks of the platform by analyzing the processes in two more case studies. This might lead to some adjustments in the information format required for new processes and resources, and to verify the schema to store new data. This first version of the platform has some limitations, which may be solved by integrating a BPM engine or an ERP system as data sources for the platform. Therefore, it is proposed to update the architecture and data models to include this type of systems as other information sources. An additional improvement would be expanding the platform functionalities until it becomes a tool that provides full support for the automation of the BPM life-cycle [10].

Acknowledgments. This work is part of the research project "Modelo de Gestión para la Optimización de Procesos y Costos en la Industria de Ensamblaje" supported by the Research Department of the University of Cuenca (DIUC). The authors gratefully acknowledge the contributions and feedback provided by the IMAGINE Project team.

References

1. van der Aalst, W.M.P., Rosa, M.L., Santoro, F.M.: Business process management. Bus. Inf. Syst. Eng. **58**, 1–6 (2016)
2. Kaplan, R.S., Anderson, S.R.: Time-Driven Activity-Based Costing: A Simpler and More Powerful Path to Higher Profits. Harvard Business School Press, Boston (2007)
3. Siguenza-Guzman, L., Van den Abbeele, A., Vandewalle, J., Verhaaren, H., Cattrysse, D.: Recent evolutions in costing systems: a literature review of time-driven activity-based costing. ReBEL Rev. Bus. Econ. Lit. **58**, 34–64 (2013)
4. Cabrera Encalada, P., Ordoñez Parra, C.: Desarrollo de un módulo TDABC, aplicado al Centro de Documentación Regional Juan Bautista Vázquez, Tesis de Pregrado, Universidad de Cuenca (2012). http://dspace.ucuenca.edu.ec/handle/123456789/646
5. Siguenza-Guzman, L., Cabrera, P., Cattrysse, D.: TD-ABC-D: time-driven activity-based costing software for libraries. In: 80th IFLA General Conference and Assembly, Lyon, France (2014)
6. Merchán, E., Sigcha, E., Morocho, V., Cabrera, P., Siguenza-Guzmán, L.: Análisis y diseño de un software de gestión de procesos y costos en empresas de ensamblaje. Maskana **9**, 79–88 (2018)
7. Sigcha, E., Morocho, V., Siguenza-Guzman, L.: Towards the implementation of a software platform based on BPMN and TDABC for strategic management. CCIS, vol. 895 (2018)

8. Everaert, P., Bruggeman, W., Sarens, G., Anderson, S.R., Levant, Y.: Cost modeling in logistics using time-driven ABC: experiences from a wholesaler. Int. J. Phys. Distrib. Logist. Manag. **38**, 172–191 (2008)

9. Andrade Serrano, E., Elizalde Lima, B.: Levantamiento de procesos de ensamblaje de televisores para la empresa Suramericana de motores Motsur Cia. Ltda, Tesis de Pregrado, Universidad de Cuenca (2018). http://dspace.ucuenca.edu.ec/handle/123456789/29718

10. Dumas, M., La Rosa, M., Mendling, J., Reijers, H.A.: Fundamentals of Business Process Management. Springer, Heidelberg (2013)

Cloud-Oriented Packaging and Delivery

Claudio Navarro[1(⊠)] and Carlos Cares[2]

[1] TIDE S.A., Temuco, Chile
claudio.navarro@tide.cl
[2] University of La Frontera, Temuco, Chile
carlos.cares@ceisufro.cl

Abstract. Web Engineering concerns the software development of web applications. A particular type of web systems are those for providing cloud services. In these systems the packaging and delivery sub-process has complexities associated to multiple server configurations and continuing operation of existing services. In this paper we show a detailed procedure for packaging and delivery resulting from a three-phase qualitative approach: first, a case study for eliciting an existing process in a small software house, second an expert judgment approach for improving the existing procedure which its main recommendation was to include containers technology, and, third, a focus group for consolidating a unified view of the improved process.

Keywords: Software engineering · Web engineering ·
Packaging and delivery · Containers · Cloud computing

1 Introduction

Web engineering is a subtopic into Software Engineering which concerns the application of software engineering principles to web systems development. It has been widely supported as a different study area due to its particular technologies, quality factors, software process stages and challenges [8,9]. Among these challenges are availability, scalability, growing complexity, multi-technology legacy systems, and also the need to keep the alignment of web systems to its supporting business [4]. Most of these challenges have provoked the specialization and outsourcing of web development projects translating the demand for on time and on budget web systems from inside out of the organization boundaries [1,2]. However, in this scenario, the organization keeps the control of its own data centers and software systems, therefore, in all these cases, it is still a need to deal with the packaging and delivery stage. Therefore, it appears relevant to have a defined and supported process for this stage [13].

Moreover, cloud computing enables ubiquitous, convenient, on-demand network access to shared resources that should be provisioned with minimal management efforts. It has different service models (e.g. SaaS) and deployment models [7]. This type of web system imposes additional challenges on performance, robustness and, hence, a different focus on packaging and delivery.

© Springer Nature Switzerland AG 2019
Á. Rocha et al. (Eds.): ICITS 2019, AISC 918, pp. 242–251, 2019.
https://doi.org/10.1007/978-3-030-11890-7_24

In terms of background, as far as we know today, there are very few studies providing empirical support for dealing with packaging and delivery activities into web engineering projects and cloud computing services. Fortunately, there is a recent paper/essay a systematic literature review on packaging and delivery which shows a lack of methodological proposals for packaging and delivery and, at the same time, also a lack of supporting recommendations and suggestions on empirical research methods both, qualitative and quantitative [10].

In terms of proper empirical methods for software engineering research, Design Science is a contemporary proposal suitable for the production of useful artifacts into the software process. It has a cyclical structure proposing a knowledge generation activity each time, in order to improve the existing version of the desired artifact [11].

In this paper we report a design science approach to the design of a grounded packaging and delivery subprocess for a cloud computing supporting system under the service model of software as a service and under the deployment model of public cloud. We have conducted three research cycles of this methodology approach. In the first cycle we elicited an existing packaging and delivery subprocess, which has the advantage of supporting a 29110 certification process. In the second research we conducted an expert judgment research cycle receiving separated contributions for improvement from senior engineers. The gathered recommendations were expressed in a second version of the packaging and delivery subprocess. Finally, as a third research cycle, a focus group was conducted for generating a common, feasible and efficient perspective which became the third version of the subprocess.

We have organized the rest of the content as follows: In Sect. 2 we present the research design and in Sect. 3 we report the main methodological issues of each research cycle and also their main results. A relevant part of design science is to define the problem, therefore the related work that conducted us to the definition is reported in this section. In Sect. 4 we compare and analyze the results of each research cycle and we outline the future work, mainly in order to provide quantitative support to this proposal. Finally, in Sect. 5 we summarize the study under the perspective of reached research goals but also its current limitations and pending goals.

2   Methodology: A Design-Science Research Design

Design Science is a research methodology that allows conducting investigation projects under an iterative process of which the aim is the generation of useful artifacts that help some production process [11], in our case, the software process. It starts with the definition of the problem, which implies a review of the available knowledge. A second stage requires the definition of the objectives of the artifact to research. Using available theories, a first version of the artifact should be designed and developed, the next stage is demonstration, there are different ways to do that, case studies, proofs of concept, experiments, even structured interviews to evaluate feasibility and efficacy has been shown as sample of demonstrations. With this information, the evaluation should be produced and, eventually, this research outcome may be communicated. If the evaluation shows new improvement opportunities then a new research cycle may start again, either, redefining research objectives or re-designing the artifact on the light of the results.

The research design considered three cycles: the first cycle had the aim of obtaining a first version of the packaging and delivery subprocess. Therefore, we elicited an existing packaging and delivery procedure and, using this design, we conducted a case study in order to get a refined process of packaging and delivery. Even though it was already communicated in [10]. In the second cycle we include the refined stages in a new version of the process, i.e. it is redesigned, and then we conducted an expert judgment approach for evaluating this improved version. Finally, in the third cycle, we redesigned the process again with this information and we reevaluated the result again, this time, using a focus group approach. In Fig. 1 we illustrate both, the general research design approach, in the top part of the figure, and, in the bottom part pf the figure, we illustrate the three described research cycles. In the next section we summarize the most relevant results of each stage.

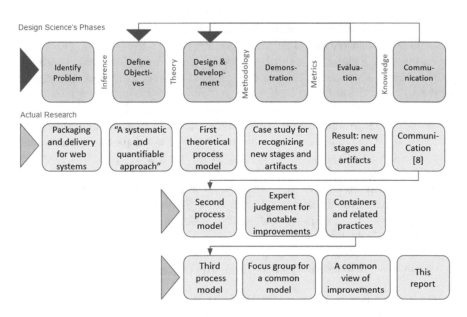

Fig. 1. Design-science research design for packaging and delivery subprocess

3   Results from the Design-Science Research Approach

In this section we summarize the definition of the problem, the objectives of the research and the three research cycles already described in the research design.

3.1   Problem Definition and Research Objectives

A recent study on packaging and delivery in web systems reports a systematic mapping review for the case of packaging and delivery of web systems [10]. In Fig. 2 we reproduce an outcome of this study. The values in the X axe is the research approach of each study which corresponds to: conceptual analyses (CA), concept implementation or proofs of concepts (CI), case studies (CS), laboratory experiment on software (LS), and simulation (SI). The contributions on methodology means packaging and delivery solutions involving or implying a process and/or refined tasks into this subprocess. The contribution is technical if the proposed solutions involve models or technological integrations which may help in some point of the subprocess. Finally, the contribution is in some tool if the subprocess is supported and driven by this tool.

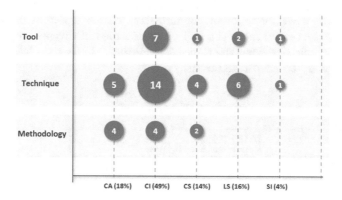

Fig. 2. Recent result on packaging and delivery from [10]

Neither of these articles propose a packaging and delivery process at a refined level (stages, inputs and outputs), there are no mentions of web systems implementing cloud computing services.

3.2   Case Study: A Refined Process of Packaging and Delivery

In this section we refer to the obtained process in the complete case study reported in [10]. The main product corresponds to the block "Second process model" in Fig. 1. The case was conducted in a Chilean SME which mainly works on custom-made software projects. A particular aim to obtain a representation compliant to ISO 9000 and ISO 29110 in order to keep its certifications.

The information was collected from a real but undocumented practice of packaging and delivery, through participant interviews, process tracking and semi-open surveys. Once the procedure was presented to participants, a test was applied in order to measure the level of agreement of the procedure representation. The related cronbach alpha results in 0.88 which is considered a trustworthy agreement of this measure [3].

The obtained representation describes how the source code, configuration files and databases, flows through the development environment to the final operation platform. The involved roles are the project manager, the technical leader and the operation's engineer of the client organization.

3.3 Expert Judgement for Notable Improvements

Expert judgment has been highlighted as a valid methodology for identifying trends, evaluating their usefulness and transferring results to specific contexts [14], resulting in an adequate methodology to apply for an improvement of the procedure recognizing enablers and obstacles.

Below are proposals for improvements, barriers and facilitating factors for the subprocess of web software packaging and delivery, obtained by applying this technique with the collaboration of a group of experts (see Table 1), from the SME software development industry, with a profile of more than 10 years of experience and more than 10 projects aimed at diverse sectors such as Health, Electronic Government, Education, Agro-food, Tourism, Mobile games, Retail/groceries, Transport and Forestry.

Table 1. Experts' profiles

Id	Main role	Experience			Education
		Years	Projects	Domains	
Expert 1	Sponsor	+10	+10	Electronic Government, Education, Agri-food, Mobile games, Retail/groceries	Civil engineer on industrial and software engineering, MSc on Computer Science and PhD
Expert 2	Project Manager	+10	+10	Electronic Government, Education	Technical Engineer on Software
Expert 3	Sponsor	+10	+10	Health, Electronic Government, Education	Civil engineer on industrial and software engineering, MSc on Computer Science
Expert 4	Senior developer	+10	+10	Electronic Government, Education, Agri-food, Tourism, Transportation, Forest	Technical engineer on software

We proposed a set of improvements, from the expert recommendations, to the packaging and delivery sub-process focused on the achievable targets for the target SME, which contemplate the incorporation of container technology (specifically Docker containers) and the complementary use of repositories to control sources of the infrastructure description scripts. This way was possible to simplify the procedures, reducing technical risks associated to the execution of manual operations and avoid any differences between the development, pre-operation and operation environments.

The Fig. 3 shows the improved general subprocess and their respective activities which implements the specific mentioned improvements.

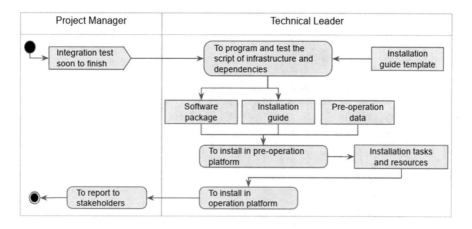

Fig. 3. General process of packaging and delivery

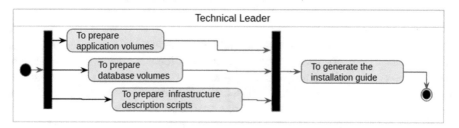

Fig. 4. Subprocess of incomes' preparation (To program and test the script of infrastructure and dependencies)

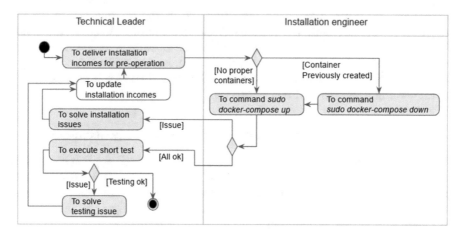

Fig. 5. Subprocess of pre-operation's installation

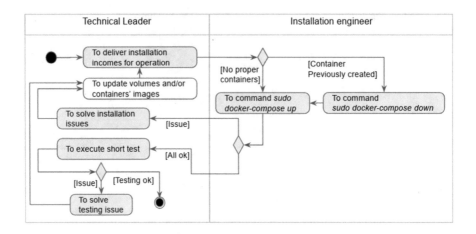

Fig. 6. Subprocess of operation's installation

The improved subprocess represented Fig. 4, shows the simplification of the tasks necessary for process incomes' preparation, focused now on to prepare virtual infrastructure elements (containers) and the installation guide, both stored on source code repositories. The Figs. 5 and 6 shows the simplification of the tasks necessary for the installation in pre-operation and operation environment. In this improved version, the tasks that must be carried out by the Installation Manager are reduced to the execution of a simple command to register (and/or download) the environment prepared by means of the infrastructure descriptor script. The execution of abbreviated tests was also considerably reduced instead of the more extensive integration tests required in the original version of the procedure. Although container technology ensures the portability of the virtualized infrastructure, the procedures consider the possibility of eventual issues during the installation that might require maneuvers related more to tasks of infrastructure management and container technology than to the particular web software being packaged and delivered.

3.4 Focus Group for a Common Perspective of Improvements

In summary, the focus group method is an empirical research approach, which is faster and more profitable, to obtain qualitative information and feedback from participants [6]. In order to generate a common, feasible and efficient perspective regarding the improved packaging and software delivery subprocess proposed in the previous cycle, we carried out a focus group in which web engineering professionals participated and it resulted in the production of useful and sufficient information to generate relevant improvements.

Defining the Research Problem. Through a case study, a version of the packaging and delivery software subprocess was validated with a good level of agreement regarding the actual practice. Improvements were proposed based on recommendations resulting from the application of expert judgment, generating a new version of the subprocess, incorporating virtualization technology at the operating system level (Dockers containers) and repositories with version control for infrastructure definitions. It was necessary to obtain feedback and information to analyze the proposed improvements to promote a common and feasible perspective in this regard.

Selecting the Participants. We used two main criteria to define the participants in the focus group: (i) Professionals with experience in Web engineering and (ii) those who participate in relevant roles in the sub-process, such as project managers or technical leaders. Some authors recommend 6 to 10 participants [12] and others do not rule out groups between 3 and 12 [5, p. 425]. To privilege the homogeneity and confidence of the group in a way that effectively produces contribution and collaboration (synergy), we conducted the session with 3 participants, all of them previously interviewed, prepared and instructed at the beginning of the session to encourage their participation and take advantage of the available time.

Planning and Conducting the Focus Group Session. We executed a focus group session, backing up with notes and audio recordings. The session began with the presentation of the participants, experience and role in Web projects. The objectives of the research and the proposed improvements for the packaging and software delivery subprocess were presented. The participants were then invited to freely express their opinions regarding the sub-process, activities and proposed actions. The session took a total time of 1:10 h. One of the researchers made the role of moderator and also the second researcher intervened in a complementary way to guide, motivate the discussion and/or point out possible agreements or disagreements. The discussion was semi-structured, with a well defined central research topic but giving enough space to extend the discussion to unscheduled concerns.

Analysis. The results of the focus group were obtained from the analysis of records and backup of audio recording. These results allowed a deeper understanding of the problems surrounding the software packaging and delivery subprocess. The most relevant potential improvements discussed were related to facilitating the preparation of infrastructure description scripts based on templates that should be defined and available from earlier stages of development and optimized for different architectures or types of software. Another improvement discussed was related to the software tests that were reduced to a minimum or abbreviated set of verifications, because the new technology minimizes the possibility of functional failures in virtualized and identical environments. Therefore, it was estimated that in the new subprocess the resources for tests could be dedicated to non-functional tests (response times, server resources, bandwidth, etc.), poorly addressed in the original subprocess.

4   Results Comparison and Analysis

In the first cycle of the design science approach, we proposed a refined process based on a case study [10] for the process of packaging and delivery of Web software, in a Chilean SME which mainly works on custom-made software projects, that has ISO 9000 and ISO 29110 certificates. However, the success of this process does not seem guaranteed due to numerous manual operations required over an infrastructure whose environments of development, pre-operation and operation are not always identical.

In the second cycle, the recommendations we got through expert judgment, provided important information to overcome these disadvantages by incorporating virtualization technology (Dockers containers), to guarantee the immutability of the infrastructure between environments and to minimize any incidents or issues, because packaged Web software already includes all the elements required for its proper operation. The process obtained takes account of these changes, reducing significantly the manual operations.

In the third cycle, through the focus group method, it was possible to achieve a common, feasible and efficient perspective of the proposed improvements, confirming the preliminary feasibility of its implementation and the opportunity to allocate resources for non-functional tests, previously dedicated to trivial and repetitive tasks, which will no longer be necessary.

In order to validate the proposed model, it will be necessary, in future work, to implement the proposed subprocess under real conditions by defining the metrics, indicators and data collection points, which allow, through analysis, to understand the real impact of the improvements.

5   Conclusions

In this article we have referred to the problem of packaging and delivery of web-based systems, in particular when these web systems implement cloud computing services under the model of software as a service and the deployment model of public cloud. The aim has been to generate a grounded procedure of packaging and delivery by using a proper empirical method.

By using a documented research methodology, as design-science research, we have reported three research cycles by using very known qualitative methods each time in order to produce an improved version of the packaging and delivery subprocess: a case study, an expert judgment research and, finally a focus group. This way an improved and detailed, and recognized useful process was obtained.

Our next challenge aims to gather enough empirical evidence of the suggested practices in one year, which means to collect data from 5 to 10 projects. Also, we are starting to work on the simulation over software production subprocess, therefore, we have also planned to use simulation as demonstration in another design-science cycle.

References

1. Airaj, M.: Enable cloud DevOps approach for industry and higher education. Concurr. Comput. Pract. Exp. **29**(5), e3937 (2017)
2. Austel, P., Chen, H., et al.: Continuous delivery of composite solutions: a case for collaborative software defined PaaS environments. In: Proceedings of the 2nd International Workshop on Software-Defined Ecosystems, pp. 3–6. ACM (2015)
3. Cronbach, L.J.: Coefficient alpha and the internal structure of tests. Psychometrika **16**(3), 297–334 (1951)
4. Gohil, K., Alapati, N., Joglekar, S.: Towards behavior driven operations (BDOps). In: Proceedings of the 3rd International Conference on Advances in Recent Technologies in Communication and Computing, pp. 262–264. IET (2011)
5. Hernández Sampieri, R., Fernández Collado, C., Baptista Lucio, P.: Metodología de la investigación. McGraw-Hill, México (2010)
6. Kontio, J., Lehtola, L., Bragge, J.: Using the focus group method in software engineering: obtaining practitioner and user experiences. In: Proceedings of the IEEE International Symposium on Empirical Software Engineering, ISESE 2004, pp. 271–280 (2004)
7. Mell, P., Grance, T., et al.: The NIST definition of cloud computing. NIST (2011)
8. Mendes, E., Mosley, N., Counsell, S.: The need for web engineering: an introduction. In: Murugesan, S., Deshpande, Y. (eds.) Web Engineering 2000, pp. 1–27. Springer. LNCS 2016 (2001)
9. Murugesan, S., Deshpande, Y., Hansen, S., Ginige, A.: Web engineering: a new discipline for development of web-based systems. In: Web Engineering, pp. 3–13. Springer (2001)
10. Navarro, C., Cares, C.: A real approach on web systems packaging and delivery. In: Procedings of the IEEE International Conference on Automatica (ICA) and the XXIII Congress of the Chilean Association of Automatic Control (ACCA), Concepción, 20–24 October, Chile (2018)
11. Peffers, K., Tuunanen, T., Rothenberger, M.A., Chatterjee, S.: A design science research methodology for information systems research. J. Manag. Inf. Syst. **24**(3), 45–77 (2007)
12. Rabiee, F.: Focus-group interview and data analysis. Proc. Nutr. Soc. **63**(4), 655–660 (2004)
13. Rathod, N., Surve, A.: Test orchestration a framework for continuous integration and continuous deployment. In: Proceedings of the International Conference on Pervasive Computing (ICPC), pp. 1–5 (2015)
14. Sjoberg, D.I., Dyba, T., Jorgensen, M.: The future of empirical methods in software engineering research. In: Future of Software Engineering, FOSE 2007, pp. 358–378 (2007)

Software and Systems Modeling

IoTV: Merging DTV and MDE Technologies on the Internet of Things

Darwin Alulema[1,2(✉)], Javier Criado[2], and Luis Iribarne[2]

[1] Universidad de las Fuerzas Armadas ESPE, Sangolquí, Ecuador
doalulema@sespe.edu.ec
[2] Applied Computing Group, University of Almería, Almería, Spain
{javi.criado,luis.iribarne}@ual.es

Abstract. Nowadays we live in a digital world of continuous changes in which digital platforms are more popular and new services have emerged, such as Netflix for video, or Amazon for retail purchases. Also, traditional companies such as telephony have ventured into the IoT (Internet of Things) with products of home automation, security, e-Health, among others. However, other platforms such as DTV (Digital Television) are not widely used, even when infrastructure is deployed. This infrastructure can be used in applications that take advantage of its large interface and presence in most homes, to transmit information through the television signal and receive information through the Internet. For this reason, in this paper we present a proposal based on MDE (Model-Driven Engineering) to facilitate developers the building of applications for DTV in IoT environments, using a DSL (Domain Specific Language) and a code generation engine.

Keywords: Model-Driven Engineering (MDE) ·
Domain Specific Language (DSL) · Internet of Things (IoT) ·
Digital Television (DTV)

1 Introduction

Today the Internet of Things (IoT) is transforming the way people communicate, collaborate and coordinate their daily lives. This change occurs due to the increase of devices connected to the Internet, which offers an ecosystem of technology integration, which allows: (a) extract data from everyday life, (b) analyze data in a virtual environment, and (c) give a value aggregated through analysis and algorithms that allow decision-making and quick responses [19].

The IoT has allowed the integration of software and hardware, to offer new services such as Smart Home, Smart City, Smart Agriculture and Wearables [10]. However, due to the great diversity of platforms this union is not an easy task. To solve this problem, projects such as Fiware, Nimbits or Amazon Web Services have proposed platforms for integration. These proposals do not cover all cases, because there are many devices from different manufacturers and with different

Á. Rocha et al. (Eds.): ICITS 2019, AISC 918, pp. 255–264, 2019.
https://doi.org/10.1007/978-3-030-11890-7_25

features. For this reason, a technology that could facilitate the development of applications is Model-Driven Engineering (MDE) [20]. MDE allows developers to standardize and automate the software development process. In this way, it is possible to continue expanding the systems to cover a large part of platforms by using models and transformations for the specification and generation of semiautomatic or automatic applications [4].

One of the platforms of greater coverage is the TV, which has incorporated new features that make it ideal for applications that wish to reach many people in a short time. However, it has not achieved the same acceptance as other technologies such as Smart TV, IPTV or video on demand. Especially because the convergence of hardware and software is directed to future smart cities, which will need to group several technologies, and where the TV can have a lot of potential to control household objects or to keep people informed [11].

TV is still valid with the new DTV (Digital Television) standards, such as Hybricast, HbbTV 2.0, ATSC 3.0 and Ginga [17]. These standards allow better audio and video quality, better coverage and interactivity. However, interactivity through television has not yet been fully deployed. Due to the difference between the number of TV sets and Internet connections. Even when the prediction of connected devices increases [13].

As it has been commented, the IoT is in a continuous advance, incorporating new technologies and expanding its field of action. To continue this expansion it is necessary to have a strong and flexible environment for the development of applications [5]. In this document we propose a methodology for the construction of IoT applications for DTV. This proposal has been called as IoTV (in a reference to the integration of the initials IoT and DTV). For which, a DSL (Domain Specific Language) has been developed, as well as a process of semiautomatic generation of applications. For this purpose Eclipse Modeling Framework (EMF) [6], Sirius [16] and Acceleo [1] have been used in the proposal.

The rest of the article is structured as follows. Section 2 reviews some related works. Section 3 presents the proposal for the development of TV applications (the IoTV methodology). Section 4 presents a test scenario that uses the proposed methodology. Finally, Sect. 5 outlives the main conclusions and identifies some future work.

2   Related Work

The scope of the IoT continues growing and integrating more platforms, which hinders the development of new applications. In this way, MDE allows developers to solve specific problems of integration, by considering and studying a general solution independent of the technology. For instance, in [5] the authors propose three development patterns for the provision of cloud services for mobile devices by using a metamodel as a structure to encapsulate and manage the resources of the applications. Although the proposal uses an MDE-based approach, it does not consider the DTV in IoT systems.

The authors in [15] focus their work on smart home systems, through standardized interfaces. Based on a model-driven perspective in the domain of Building Automation Systems (BAS), several types of text artifacts are generated for the OBIX standard, with access to communication technologies such as BACnet, KNX, EnOcean or (wireless) M-Bus. The main difference with our proposal is the graphic editor that we have developed and the use of digital television.

In the work accomplished in [7], the authors propose a graphical DSL tool for IoT systems, in which the sensors (accelerometers, GPS, pressure, light, temperature, gravity or proximity) can be interconnected in an IoT system through JSON objects. The platform allows the code to be developed on multiple platforms (Java for Android, C for Arduino, C# for Windows Phone or Objective-C for iPhone). The main difference with our proposal is the generation of code, which in our work will be done for DTV.

Another work is that proposed in [8], in which the authors focus on their work the hardware domain for IoT modeling the structure of the communication scheme for the Contiki operating system. For the generation of UDP applications, semiautomatically. In the proposal, a DSL and model transformations are developed to generate code. The main difference with our proposal is in the platform for which the code is generated, which in our case is the DTV.

On the other hand, in the work [14] the authors propose the use of the IoT in the military field, to monitor and visualize in real time the state of health during combat missions. For this purpose, they propose a methodology for the development of the systems, to obtain a conceptual model. The main difference with our proposal is that we have implemented the methodology and scope is broader than the military.

The authors in [18] analyze the applicability of Smart TV in Internet of Things (IoT) environments. The possible roles of TV (Information storage, visualization device, interaction point, data processor and data source) for the IoT infrastructure and the characteristics that a platform must have to act in the respective role are considered. The main difference with our proposal is the implementation of a tool that allows the development of applications for DTV.

Many of the investigations focus on the domain of hardware, however our proposal expands the domain by incorporating DTV. In addition a DSL is developed to facilitate the design process, and semiautomatic code is generated through M2T (Model-to-Text) transformations.

3   IoTV Methodology

This section describes the proposed methodology for the development of applications for IoTV according to MDE. The proposal requires six processes divided into two stages: one for specification and another for development.

In the specification stage, the DSL is designed and a M2T transformation process is implemented. This stage consists of three processes that are illustrated in Fig. 1. On the Engineer role side, the step #1 generates the metamodel, which defines the abstract syntax of the language according to the characteristics of the DTV and IoT applications. Then you can perform steps #2 in two different ways: (a) generate the M2T transformation engine, to generate the source code; or (b) develop the graphical editor, which corresponds to the graphical representation of the DSL, according to the metamodel created in the step #1.

In the development stage (Developer side), the application is designed and the source code is generated. This stage consists of three consecutive processes, illustrated in Fig. 1. The first one corresponds to the function established in the step #3, where a model is built using the graphical editor defined in the specification stage. This graphical editor is then used to describe the application scenario and generate the specific model of the scenario. This is a semi-automatic step and it requires the developer to configure the properties, such as the type of interface, the connections between the components and the names of the media files. After that, there is established the step #4, which is responsible for applying the model-to-text (M2T) transformation by means of the model generated in the step #3 and the M2T engine. This process automatically create the source code of the DTV. Finally, the functionality included in the step #5, generates the IoTV application, for which the developer manually incorporates into the project the code generated in the step #4 and the multimedia resources that the application will use.

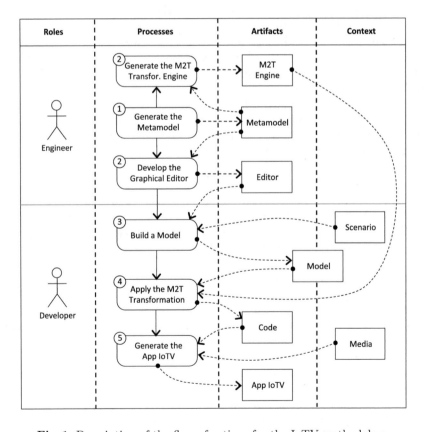

Fig. 1. Description of the flow of actions for the IoTV methodology.

3.1 A Metamodel Merging IoT and DTV

This subsection describes the metamodel for the development of applications for IoTV following an approach based on MDE, using the EMF (Eclipse Modeling Framework) [6]. The metamodel proposed in Fig. 2 allows us the development of applications with local and remote interactivity. This metamodel describes the interface of a DTV application, considering its four characteristic buttons, and allowing the access to multimedia resources or information provided by a web service communicated with an IoT system.

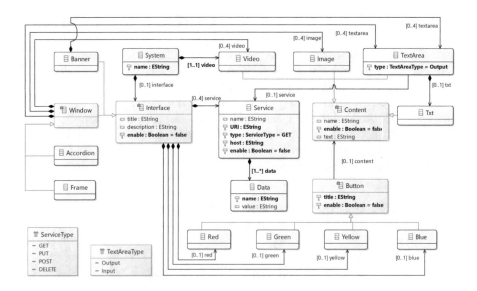

Fig. 2. The IoTV metamodel.

There is a wide variety of DTV applications, but for this work we have chosen three that can be seen in programs that can be watched on any TV channels [2]:

(a) **Banner**, vertically sorts the buttons on the left side of the screen and displays the consulted text at the bottom of the screen. This interface is mostly observed in news networks to display short messages;

(b) **Accordion**, vertically sorts the buttons on the right side of the screen and displays the invoked content in a field. This interface is mostly seen in store applications to expand information, and

(c) **Frame**, uses the 75% of the screen by reducing the video signal of the TV in a small field and orders in the available area information of the application and all the resources called by the buttons. This interface is mostly seen in applications that require more information as in cooking programs.

On the other hand, the visual elements to which the applications access are: **Video**, a TV signal that imbues the application that is transmitted, and short videos, which are multimedia resources called by the buttons; `Text`, exists the possibility of obtaining information locally (when it is inside the same data carousel), or remote (when it is consulted from a web service); `Button`, the traditional component elements of the DTV (red, green, yellow and blue) for interaction with users; `Image`, the image files that are displayed when called by the buttons; and `Service`, that represents the mechanism for the consumption of web services and temporary storage of this information on TV.

3.2   Graphical Editor

For the development of IoTV applications a graphical editor has been created based on the previous IoTV DSL. Figure 3b shows a piece of the editor and its configuration file. For the development of the editor, Sirius [16] was used to represent the metamodel of Fig. 2. Classes and their relationships specify which objects will be displayed and how they will be displayed. In Fig. 3a, the VSM (Viewpoint Specification Model) is observed. In the viewpoint, the representations are configured using a diagram, which contains: Nodes (video), Containers (banner, accordion and frame), Relations bases edge (area to service, button to area, button to image and button to video), and Sections for tools (interface, buttons, media and Connections).

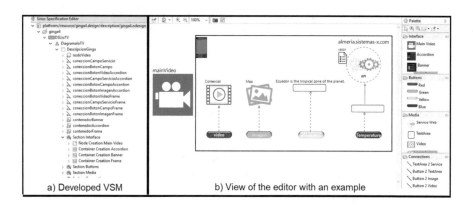

Fig. 3. Screenshot of the tool for developing the graphical editor.

Figure 3b shows the editor in two sections: Canvas, to present and edit the diagram, and the Palette of tools that includes the tools to draw and edit the visual representation. The Palette is divided into four sections. The `Interface` section allows you to choose between the three interfaces designed and the main video. `Buttons` refer to the four buttons of the DTV. `Media` for selecting the multimedia resources that will be called by the buttons. Finally the `Connections` section indicates the relationships between the elements present in the canvas.

3.3   MDE Engine

In the IoTV methodology, after the definition of the metamodel, one of the steps that is carried out is the generation of code. For achieving this purpose, a model-to-text transformation engine must be implemented. In our case, the Acceleo [1] tool has been used, which is an implementation of the MOF Model-to-Text Transformation Language standard (Mof2Text or MOFM2T) for the generation of code [3].

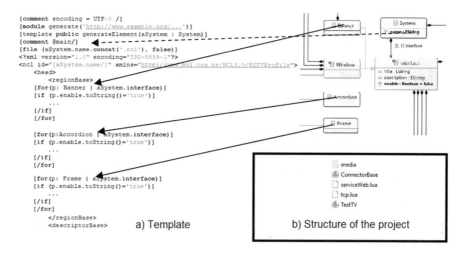

Fig. 4. Generate the M2T transformation engine.

For the generation of code, the Ginga standard has been used because it is included in DTV and IPTV (Internet Protocol Television) [9]. According to the structure of the applications for DTV with Ginga, the transformation process will create the NCL files for the media objects and their transitions, and the LUA files for information processing. Figure 4a, presents in a summary perspective a fragment of the transformation that is made according to the metamodel. In this case, an instance aSystem of the scenario model is created for the construction of the NCL script and the regionBase of the interface is built.

The transformation process generates four source code files and one resource folder, as shown in Fig. 4b. Regarding the four source code files, two of them are NCL: ConnectorBase, which is a library with all the connectors used in the project; and TestTV, which is the main file created with the value of the name attribute of the System class. The other two files are LUA: serviceWeb, the implementation for the consumption of the web service that generates temporary files with the information in the resources folder; and tcp, the library for the Internet access. Finally, the media resource folder contains all the multimedia resources (video, image and text).

4   A Case Study Scenario

To test the operation of our proposal, a scenario has been designed that can be illustrated in Fig. 5. This scenario has two actors: (a) the Content provider, has an indoor wireless weather station (Netatmo NHC-EC) [12], provides an App for DTV called `WeatherApp` that is issued by the carousel of the television signal and the web service that manages the data obtained from the temperature and CO_2; and (b) the Viewer, has a television set with Middleware Ginga, a modem for Internet access and remote control. The viewer from his home, when he receives the DTV signal, can interact locally or remotely with the application.

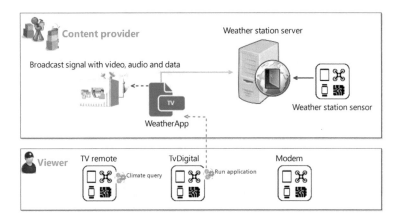

Fig. 5. Diagram of the architecture of an indoor climate service.

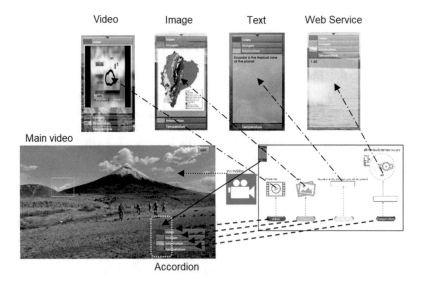

Fig. 6. Execution of the test case.

For the test case, the interface of the DTV application is `Accordion`, as shows Fig. 6. This interface has the four buttons: (a) Red, accesses to a video; (b) Green, accesses to an image; (c) Yellow, accesses to an informative text; (d) Blue, consumes the web service to display the temperature. With the model designed in the graphical editor, the transformation engine is used to generate the source code and a folder structure, as shows Fig. 4b. Finally the developer completes the application incorporating the multimedia resources in the `media` folder.

5   Conclusions and Future Work

This paper proposes a methodology to develop IoT applications integrated with the DTV, using model engineering for the generation of code. This proposal aims to facilitate the creation of applications, by not having to delve into specific aspects of programming languages.

A graphical editor is built as the initial concrete syntax of a domain specific language (DSL) in the methodology. The DSL is the main artifact of the proposed methodology and it is used for M2T transformation in the DTV application (the application is deployed in the TV and broadcasted in the data carousel of the TV signal). The methodology includes a manual step before generate the code application, which involves the developer to build the model corresponding to each scenario by using the graphical editor and filling some specific properties (e.g., the url of the web services).

As future work some second lines are still open: (a) extend the methodology for the automatic generation of code for the mobile and web platforms, (b) expand the metamodel to incorporate greater versatility in the design of the DTV interfaces, and (c) automate the entire process for the generation of applications.

Acknowledgments. This work has been funded by the EU ERDF and the Spanish Ministry MINECO under the AEI Projects TIN2013-41576-R and TIN2017-83964-R.

References

1. Acceleo — The Eclipse Foundation. https://www.eclipse.org/acceleo/. Accessed 22 Sept 2018
2. Alves, G., Barbosa, R., Kulesza, R., Filho, G.: A software testing process for Ginga products. In: Applications and Usability of Interactive TV. CCIS, vol. 689, pp. 61–73. Springer (2016)
3. Benouda, H., Azizi, M., Moussaoui, M., Esbai, R.: Automatic code generation within MDA approach for cross-platform mobiles apps. In: First International Conference on Embedded and Distributed Systems (EDiS), pp. 1–5. IEEE (2017)
4. Bruneliere, H., Burger, E., Cabot, J., Wimmer, M.: A feature-based survey of model view approaches. Softw. Syst. Model. **17**, 1–22 (2017)
5. Cai, H., Gu, Y., Vasilakos, A., Xu, B., Zhou, J.: Model-driven development patterns for mobile services in cloud of things. IEEE Trans. Cloud Comput. **6**(3), 771–784 (2016)

6. Eclipse Modeling Project: The Eclipse Foundation. https://www.eclipse.org/modeling/emf/. Accessed 22 Sept 2018
7. García, C., Espada, J., Núñez-Valdez, E., Garcíaz, V.: Midgar: domain-specific language to generate smart objects for an Internet of Things platform. In: International Conference on Innovative Mobile and Internet Services in Ubiquitous Computing, IMIS 2014, pp. 352–357. IEEE (2014)
8. Gomes, T., et al.: A modeling domain-specific language for IoT-enabled operating systems. In: Conference of the IEEE Industrial Electronics Society, IECON 2017, pp. 3945–3950. IEEE (2017)
9. ITU: H.761, Recommendation ITU-T, pp. 1–154. ITU (2014)
10. Kraijak, S., Tuwanut, P.: A survey on IoT architectures, protocols, applications, security, privacy, real-world. In: International Conference on Wireless Communications, Networking and Mobile Computing, WICOM - 2015, pp. 1–6. IEEE (2015)
11. Kumar, N., Goel, S., Mallick, P.: Smart cities in india, features, policies, current status, and challenges. In: International Conference on Technologies for Smart-City Energy Security and Power, ICSESP 2018, pp. 1–6. IEEE (2018)
12. Netatmo Connect. https://dev.netatmo.com/en-US/resources/technical/introduction. Accessed 1 Aug 2018
13. Nordrum, A.: Popular Internet of Things Forecast of 50 Billion Devices by 2020, IEEE Spectrum. https://spectrum.ieee.org/tech-talk/telecom/internet/popular-internet-of-things-forecast-of-50-billion-devices-by-2020. Accessed 19 Sept 2018
14. Reyes, R., Vaca, H., Paredes, M., Montoya, L., Aguilar, W.: Milnova: an approach to the IoT solution based on model-driven engineering for the military health monitoring. In: Proceedings of 2017 CHILEAN Conference on Electrical, Electronics Engineering, Information and Communication Technologies, CHILECON 2017, pp. 1–5. IEEE (2017)
15. Schachinger, D., Kastner, K.: Model-driven integration of building automation systems into Web service gateways. In: Proceedings of the IEEE World Conference on Factory Communication Systems, WFCS 2015, pp. 1–6. IEEE (2015)
16. Sirius - The Easiest Way to Get Your Own Modeling Tool. https://www.eclipse.org/sirius/. Accessed 1 Aug 2018
17. Sotelo, R., Joskowicz, J., Rondan, N.: Experiences on hybrid television and augmented reality on ISDB-T. In: IEEE International Symposium on Broadband Multimedia Systems and Broadcasting, pp. 1–7. IEEE (2017)
18. Yusufov, M., Kornilov, I.: Roles of smart TV in IoT-environments: a survey. In: Proceedings of the 13th Conference of open Innovations Association, FRUT 2013, pp. 163–168. IEEE (2013)
19. Zheng, D., Carter, W.: Leveraging the Internet of Things for a More Efficient and Effective Military. Center for Strategic and International Studies, pp. 1–52 (2015)
20. Zolotas, C., Diamantopoulos, T., Chatzidimitriou, K., Symeonidis, A.: From requirements to source code: a model-driven engineering approach for RESTful web services. Autom. Softw. Eng. **24**, 791–838 (2017)

Usability Quality Aspects Embedded in the Business Model

Juan Carlos Moreno[✉], Marcelo Martín Marciszack[✉],
and Mario Alberto Groppo[✉]

Centro de Investigación, Desarrollo y Transferencia de Sistemas de Información
(CIDS), Universidad Tecnológica Nacional – Facultad Regional Córdoba,
Maestro Lopez esq. Cruz Roja Argentina S/N, Ciudad Universitaria,
Córdoba, República Argentina
jmoreno33@gmail.com, marciszack@gmail.com,
sistemas@groppo.com.ar

Abstract. This work is a proposal for the identification Non Functional Usability Requirements Specification, at an early stage. The procedure starts in a business processes modeling domain, where essential specifications are captured. Then they are mapped to a structure called Baseline Requirements using the concept of paradigm transformation from Model Driven Software Design. In this context, usability specifications are introduced inside the scenarios using transformations, and take part in these scenarios by using Lexicon Extended Language (LEL). The use of LEL allows to describe the usability specifications and the construction of a dictionary, which will later be needed for integrating the usability specifications in the software programs developed. This also provides the possibility to identify, define and maintain the traceability of usability specifications defined at early stages.

Keywords: Usability · Business processes model · Scenarios ·
Lexicon Extended Language · Model transformation

1 Introduction

Developing applications efficiently has been one of the main worries in Information Systems Engineering, without omitting to consider the quality of those applications. This could be a consequence of several causes, such as: the advance of nanotechnology in microprocessor development leading to continuously smaller devices, the great technological progress in telecommunication networks and the expansion of Internet, increasing the need of web applications all over the world. Its impact can be seen not only in interactive applications through WEB 2.0 [1], but also in Social Networks [2]. Functional Requirements are usually given more relevance, independently from the software development methodology used. But product quality not only depends on these latter (FR), but also on Non Functional Requirements (NFR). What has been exposed raises the question if it is feasible to develop a methodological proposal that allows the introduction of usability requirements at early stages in software development, always starting from Software Engineering point of view.

© Springer Nature Switzerland AG 2019
Á. Rocha et al. (Eds.): ICITS 2019, AISC 918, pp. 265–273, 2019.
https://doi.org/10.1007/978-3-030-11890-7_26

Thus, the main goal was to develop a systematized procedure that allows, introducing usability aspects at early stages of software development life cycle, the analyst to rely on usability specifications before the software development is finished.

This work is structured as follows: the first part is an introduction of the state of the art, elements and concepts used to develop the methodological proposal are introduced later. After that, the process that allows to carry out the proposal and the tools used are explained, including an example and some observations on the whole experience. Finally, a conclusion on the final proposal is elaborated.

2 State of the Art

Information systems development in a web platform is a process that requires great knowledge of methodologies, security and diverse technologies, in order to build a useful and correct application for any user. In this way, the goal of Systems Engineering, besides the construction of functionally correct applications, is to construct quality applications through different methods and principles [3]. Often, when building an application or program, the focus tends to be on aspects related to the architecture, persistence and functionality of processes related to functional requirements (FR), not taking into account non-functional requirements (NFR). Additionally, the software quality is frequently measured based in common sense and developers experiences [4].

One of the relevant Non-Functional Requirements, in web application development, is Usability. The concept of usability has been defined by several International Organizations, which establishes rules on Quality Standards (ISO, IEEE). Such rules name Usability as a software attribute, and is related with the quality of such software.

In ISO/IEC 9126-1 [5], usability is considered as a software quality parameter and is one of the relevant characteristics of software. Usability is defined as "the capacity in which a software product can be understood, learnt and used by certain users under certain conditions in a specific use context". It considers external, internal and in use quality of a software product [6]. Usability is decomposed in sub-attributes, such as ease of learning, comprehension, operativeness and usability compliance [7].

ISO 25000 (Square) [8] considers Usability as a quality aspect under two different points of view: one is software, as a product itself, and the other would be the use capacity, from user perspective in a specific context. The goal is to provide certain criteria to help the analyst to build an integral and usable software product. This implies taking into account certain features in the construction of a product, such as: ease of understanding, learn curve, ease of use, help, technical support, attractiveness and compliance to rules.

It can be seen that Usability concept is evaluated from different perspectives in the different definitions mentioned, considering it as a quality feature inherent and intrinsic to software. Thus, the study of Usability has to be considered in the different stages of the software construction life cycle. Even if it is taken into account, it is important to remember that Usability is considered at a final stage in software construction. At this stage, any modification affects the system architecture, since interfaces are already designed and the cost of any modification is very high [9, 10]. One of the possible solutions to this problem is to include the analysis and introduction of Usability criteria at early stages, during the requirements elicitation stage.

There are a variety of methodologies for software construction, but a new paradigm of development, called Model Driven Software Development (MDA) has awoken special interest in the last decade, considering the features and advantages that provides for software architecture. Model Driven Software Development (MDA) has standardized the model transformation stages, to develop and build systems that are consistent to the original designed model, created during the requirements elicitation stage. In this transformations process, the traceability of requirements gains momentum since it is necessary to measure the magnitude of the impact of changes, in an update or system modification, and, at the same time, be able to introduce such changes in an automatic and immediate manner. In this proposal, essential requirements will arise from the business model, capturing both functional and non-functional requirements, and applying transformations that will allow the analyst to understand the domain of the problem. In this context and with the scenarios designed, Usability Non Functional Requirements will be worked through the analysis of their attributes.

3   Proposal Description

Elements and concepts used in the development of the proposal method are described in this section. It is described as a process, carried out in two stages.

3.1   First Stage

The first stage consists in the development of a business model, using BPMN [11], considering the following statements in the design of the model:

- Usability specifications, that are modeled, will form part in the non-functional requirements set to be satisfied. BPMN has a stereotype called "business rule", used for modelling this type of activities.
- Business rules are defined one time and are applied to every activity where non-functional requirements are detected. Therefore, an activity with the stereotype "Business Rule" will be generated for every usability specification and will then be associated with the activities where that requirement is detected. Later, these Business Rules will be mapped to a structure called Requirements Baseline [12, 13]. The Requirements Baseline structure uses scenarios to model the behavior of the system and Lexicon Extended Language (LEL) to represent the domain of the system. The activities of "Business Rule" type will be transformed to LEL entries and scenario restrictions respectively.
- Every activity will be mapped as scenarios, with the exception of those that the analyst defines with "Manual" stereotype, for being considered as non-automatable.
- Activities of "Business Rule" type that are associated with an activity defined with "Manual" stereotype, will neither be mapped as LEL entries nor scenario restrictions.

A model exchange file with XPDL format [14] will be later generated with the modelling tool, containing all model definitions.

3.2   Second Stage

The second stage consists in making the necessary transformations to introduce every definition generated in the previous stage and contained in the XPDL file, in the Requirements Baseline structure.

Baseline Mentor Workbench [15] (BMW) tool is used as mainframe, adding the following functionalities:

- "Restrictions" element is incorporated in scenario definition, containing the association to LEL entries, related to usability specifications.
- Necessary rules for scenario generation and LEL entries are created from the definitions captures in the XPDL file.
- Scenarios and LEL entries have to be manually completed. Scenario descriptions will be used in the system user interfaces generation. The usability requirements for the user interface to be built will be obtained from the captured restrictions and completed in every scenario.

Process and activities carried out in each stage can be seen in Fig. 1.

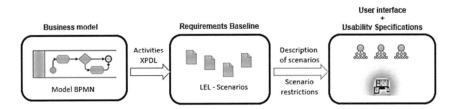

Fig. 1. Usability requirements elicitation process schema, starting from the BPMN model.

4   Procedure Applied Example

A students system, modeled in BPMN, will be used as an example domain to analyze this proposal, more specifically the "Manage Teachers" process indicated in Fig. 2.

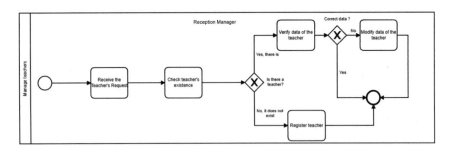

Fig. 2. "Manage Teachers" BPMN model example for a student's system.

4.1 Usability Criteria Specification

Usability specifications will be modeled as activities using the "Business Rule" stereotype, and will be associated to the activity that has to satisfy the specifications, as shown in Fig. 3.

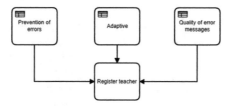

Fig. 3. Usability criteria specification starting from the business model.

Every activity will be mapped to the Requirements Baseline structure, with the exception of those selected with the "Manual" stereotype.

Activities selected as "Business Rule" that are associated with "Manual" type activities will not be mapped.

The complete result of the process can be seen in Fig. 4.

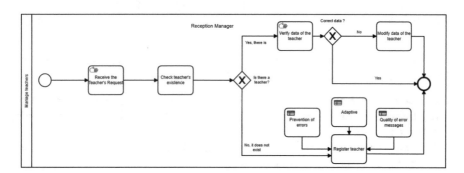

Fig. 4. Result after applying the usability criteria specification starting from BPMN model process.

Once the BPMN model is complete, the model has to be exported from the tool in a XPDL file. This file will be introduced in the BMW tool, and the needed functionalities will be added, in order to process the XPDL file with the BPMN model definitions.

The processing of the XPDL file will consist in the creation of a scenario for every activity, except in those of "Business Rule" type, which will be introduced inside of LEL as a symbol. LEL symbols are described using a notion corresponding to the meaning of the symbol, and an impact that indicates the effects of the symbol in the system. Each symbol has to be classified according to its function in subject, object, verb or state, and will have different notions and impacts depending on the classification in which they are [17, 18], as indicated in Table 1.

Table 1. Symbol definition heuristics.

Symbol	Notion	Impact
Subject	Describes who is the subject	Registers actions carried out by the subject
Object	Defines the object and identifies other terms which have some relation to the object	Describes actions that may be applied to the object
Verb	Describes who executes the action, when it happens and which are the procedures involved	Describes the restrictions on the action, which actions are started by the environment and the new situations that appear as a result of this action
State	Describes what it means and which actions can be executed as a consequence of this state	Describes other related situations and actions

- Verbs represent the actions carried out in the system. Actions are applied to objects or subjects.
- Subjects are the ones that carry out the actions indicated in verbs.
- Objects represent passive elements that receive the actions indicated in verbs, executed by subjects.
- States are used to describe specific conditions of objects or subjects.
- Usability specifications can be classified inside objects category, since they will be applied or evaluated in a specific moment through an action started by a subject.

LEL symbols corresponding to usability specifications will be described in a way that they comply to two rules simultaneously [16, 17]:

- **Circularity Principle:** limiting language in function of the domain through the maximization of LEL symbols, which is accomplished using symbols already described inside the LEL in the definition of notion and impact.
- **Minimal vocabulary Principle:** where the task is to minimize the use of symbols external to the application domain.

This will allow to maintain a data dictionary with all the definitions and the hierarchy of usability specifications that need to be satisfied.

Finally, the analyst will have to complete the description of the scenarios in a similar way as he would do with UML Use Cases [19]. The following will have to be described for each scenario:

- Title: needed to identify the scenario.
- Goal: main goal of the scenario, has to be coherent to the title.
- Context: it is used to describe the initial state, place and moment of execution of the scenario.
- Resources: LEL symbols of object type available for the execution of the scenario.
- Actors: LEL symbols of subject type that carry out actions in the scenario.
- Episodes: They represent the set of actions carried out by actors to execute the scenarios. An episode can appear in different scenarios.

"Check Teacher Existence", "Register Teacher" and "Modify Teacher" scenarios will be the ones derived directly from activities that do not have "Manual" stereotype.

LEL entries corresponding to usability specification will be automatically mapped inside the scenario under a new element called "Restrictions".

"Register Teacher" scenario can be seen in Fig. 5, with its corresponding restrictions mapped to the BPMN model. Remaining descriptions must be made by the analyst.

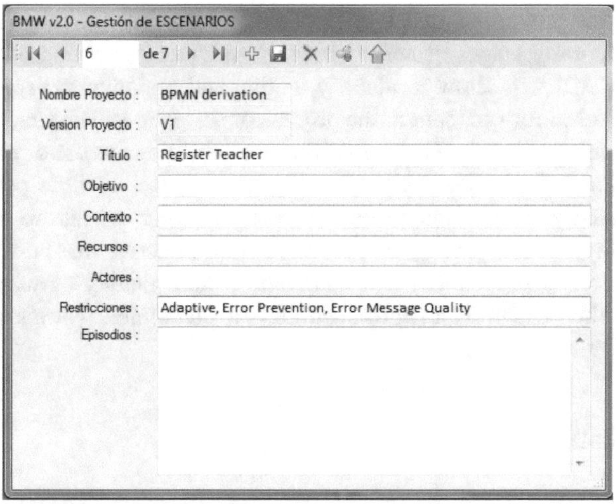

Fig. 5. Usability criteria introduced in the scenario management tool.

The resulting scenarios are later used to generate the user interfaces of the system to be built.

This process will allow to identify and define usability specifications at early stages in the development process. On the other hand, the use of LEL allows to generate and maintain a complete data dictionary, both in the definitions and in the hierarchies of usability specifications.

Finally, specifications defined in the BPMN modelling process that, after being mapped to LEL as its vocabulary, form a dictionary of data, will allow the analyst to maintain traceability of specifications from the start of BPMN modelling to the scenarios used for user interface generation.

5 Results

It is possible to capture requirements at early stages of modelling, starting with the business model. Usability specifications could be modelled inside BPMN, which were later introduced through transformations as LEL symbols in the Object category, taking into account circularity and minimal vocabulary principles. Additionally, the automatic mapping of activities can be done, by using XPDL files, for creation and definition of scenarios, allowing to associate them with usability specifications defined in the LEL vocabulary, in the scenario "Restrictions". This methodology also provides the possibility to generate a LEL dictionary of data complete with the definitions of usability specifications of transformed models.

6   Discussion

The obtained results confirm that it is possible to carry out the mapping of usability specifications at early stages of software development by applying a combination of methodologies, that will allow to identify, define and maintain the specifications that will have to be taken into account in the process of software generation, that include the user interfaces for the system. There is space for a debate on the use of patterns not only for the definitions of interfaces with usability aspects, but also the possibility to use metrics associated with such interfaces in an early manner in order to measure quality aspects (usability). These prototypes generated from patterns, that comply to usability requirements already predefined, will guarantee certain quality at an early stage and could provide the possibility to introduce certain predefined metrics to evaluate the presence of usability early on.

7   Conclusions

This process allow to identify, define, maintain and improve the traceability of usability specifications at early stages of software development, allowing to know, from the business model, the usability specifications that must be satisfied by user interfaces of the system to be built. The proposed methodology uses Scenarios and Lexicon Extended Language, and by means of transformations it allows to formulate Conceptual Models with Usability criteria specified at an early stage. Obtained results will be used for Pattern Study, starting from Business Models, for the design and construction of Conceptual Models.

References

1. Martins, J., Gonçalves, R., Pereira, J., Cota, M.P.: Iberia 2.0: a way to leverage Web 2.0 in Organizations. In: 2011 7th Information Systems and Technologies (CISTI), Iberican Conference (2011)
2. Martins, J., Gonçalves, R., Pereira, J., Oliveira, T., Cota, M.P.: Social networks sites adoption at firm level: a literature review. In: 2014 8th Information Systems and Technologies (CISTI), Iberican Conference (2014)
3. Pressman, R.S.: What a tangled wed we weave. IEEE Softw. **17**, 18–21 (2000)
4. Abrahao, S., Condori-Fernandez, N., Olsina, L., Pastor, O.: Defining and validating metrics for navigational models. In: Ninth International Proceedings of the Software Metrics Symposium, Australia, pp. 200–210. IEEE (2003)

5. Norma ISO/IEC ISO9126-1: Software Engineering -Product Quality - Part 1, Quality Model (2001)
6. Bevan, N.: Quality and usability: a new framework. Achiev. Softw. Prod. Qual. (1997)
7. Piattini, M.G., Garcia, F.O., Caballero, I.: Calidad de Sistemas Informáticos. México (2007). ISBN 978-970-15-1267-8
8. ISO/IEC 25000: Software Engineering - Software Product Quality Requirements and Evaluation (SQuaRE)
9. Bass, L., John, B.: Linking usability to software architecture patterns through general scenarios. J. Syst. Softw. **66**, 187–197 (2003)
10. Folmer, E., Bosh, J.: Architecting for usability: a survey. J. Syst. Softw. **70**, 61–78 (2004)
11. Object Management Group. Business Process Model and Notation (BPMN). Agosto 2011. http://www.omg.org/spec/BPMN/
12. Leite J.C.S.P., Rossi, G., et al.: Enhancing a requirements baseline with scenarios. In: Proceedings of RE 1997: International Symposium on Requirements Engineering, IEEE, January 1997
13. Leite J.C.S.P, Albuquerque Oliveira, A P.: A client oriented requirements baseline. In: Proceedings of RE 1995: Second IEEE International Symposium on Requirements Engineering, Inglaterra, March 1995
14. Xpdl.org: Welcome to XPDL.org. (2016). http://www.xpdl.org/index.html
15. Antonelli, R.: Traceability en la elicitación y especificación de requerimientos. Hdl.handle. net. (2016). http://hdl.handle.net/10915/4061
16. Leite, J.C.S.P.: Eliciting requirements using a natural language based approach: the case of the meeting scheduler problem, March 1993
17. Hadad, G., et al.: Informe Técnico: "Léxico Extendido del Lenguaje y Escenarios del Sistema Nacional para la Obtención de Pasaportes". Proyecto de Investigación, Departamento de Investigación, Universidad de Belgrano, Buenos Aires (1996)
18. Leonardi, C., Leite, J.C.S., Gustavo, R.: Una estrategia de Modelado Conceptual de Objetos, basada en Modelos de requisitos en lenguaje natural. Tesis de Maestría Universidad Nacional de la Plata. http://postgrado.info.unlp.edu.ar/Carrera/Magister/Ingenieria%20de% 20Software/Tesis/Leonardi.pdf
19. OMG: Unified Modelling Language: Superstructure Version 2.0, July 2005. http://www. omg.org/

Self-configuring Intelligent Water Drops Algorithm for Software Project Scheduling Problem

Broderick Crawford[1]($\boxtimes$), Ricardo Soto[1], Gino Astorga[1,2], José Lemus[1],
and Agustín Salas-Fernández[3]

[1] Pontificia Universidad Católica de Valparaíso, Valparaíso, Chile
{broderick.crawford,ricardo.soto}@pucv.cl,
jose.lemus.r@mail.pucv.cl
[2] Universidad de Valparaíso, Valparaíso, Chile
gino.astorga@uv.cl
[3] Universidad Tecnológica de Chile INACAP, Santiago, Chile
jsalasf@inacap.cl

Abstract. At present a large number exists of metaheuristics that can support some process in the industry; however, there is a great difficulty to be overcome before use, that is the adjustment of the parameters that they use. It is already known the significant impact that they have on their behavior the correct choice of their values. Given the importance that has the proper adjustment of the parameters, our work presents a self-adjusting alternative for a constructive metaheuristic called Intelligent Water Drops. To evaluate our proposal we solve the Software Project Scheduling Problem, obtaining very similar results and one case superior to the version with manual adjustment.

Keywords: Intelligent water drops · Project management ·
Software Project Scheduling Problem

1 Introduction

There is currently a diversity of metaheuristics, however all of them have a common point that is the difficulty of finding a good parameters configuration, in such a way as to obtain the best possible operation of the algorithm. For this reason, this is an area of interesting study for researchers. There are authors as [1] indicate that large percentage (90%) of the time of development of a metaheuristic is devoted to setting parameters. In general the parameters configuration we can classify it under two groups, on the one hand the off-line configuration, that is to find the appropriate values before that the algorithm begins to operate and on the other hand we have the online configuration, where the parameters are adjusted during the execution of the algorithm. Our work focuses on online configuration where we propose a scheme that allows you to

© Springer Nature Switzerland AG 2019
Á. Rocha et al. (Eds.): ICITS 2019, AISC 918, pp. 274–283, 2019.
https://doi.org/10.1007/978-3-030-11890-7_27

find appropriate values to a parameter of constructive metaheuristic Intelligent Water Drops (IWD) which has static and dynamic parameters. Our work is inspired by the concept of Autonomous Search (AS) where an interface with the end user is ideally reflected to provide a problem and this returns a solution with the minimum possible interaction [9]. Where there are some works where applies AS, for example Soto in [15] uses it to for generating good enumeration strategy blends in constraint programming and Crawford in [5] makes replacements on the progress of heuristics of under performance by other more promising.

The remainder of this work is organized in the following way. The algorithm used is explained in Sect. 2. In the Sect. 3, presents the problem used to validate our proposal. Section 4 presents our proposal for the adaptation of parameters for IWD. In Sect. 6 an analysis is carried out statistical results to validate our proposal. And finally, in Sect. 7 explains some conclusions about our results and the future works in this topic.

2   Intelligent Water Drops Metaheuristic

In 2007 was presented Intelligent Water drops by Shah-Hosseini in [13] it is based on the behavior of the water drops when flow through a river where they seek the optimal path by moving from one point to another. It is a population algorithm where the drops move in a discreet search space. The drops have two main properties which are the velocity and the amount of soil they collect. With the velocity we can determine the amount of soil collected which is incorporated into the drop.

The selection of a path of the water drop depends on the amount of soil because it is more attractive the area with less resistance. The behavior of the selection of the path is given by the following probability:

$$p_i^k(j) = \frac{f(soil(i,j))}{\sum_{\forall l \notin VC^k} f(soil(i,l))} \qquad (1)$$

where $f(soil(i,j))$ is calculated according to the Eq. 2 that corresponds to the inverse value of soil between the nodes i and j. On the other hand we used a small positive value for ε with the intention of avoiding the division for zero.

$$f(soil(i,j)) = \frac{1}{\varepsilon + g(soil(i,j))} \qquad (2)$$

where $g(soil(i,j))$ allows to obtain a positive value of the soil between the points i and j, this is shown in the Eq. 3.

$$g(soil(i,j)) \begin{cases} soil(i,j) \ if \ \min_{l \notin vc(IWD)} (soil(i,l)) \geq 0 \\ soil(i,j) - \min_{l \notin vc(IWD))} (soil(i,l)) \ else \end{cases} \qquad (3)$$

Each time a drop moves from i to j its velocity changes, which can be obtained with the Eq. 4.

$$vel^k(t+1) = vel^k(t) + \frac{a_v}{b_v + c_v \cdot soil(i,j)} \qquad (4)$$

where a_v, a_v and a_v correspond to constants; $soil(i,j)$ corresponds to the soil incorporated to the drop during its movement. This is calculated with the Eq. 5.

$$\Delta soil(i,j) = \frac{a_s}{b_s + c_s \cdot time(i,j : vel^k(t+1))} \tag{5}$$

where a_s, b_s and c_s correspond to constants; $time(i,j : vel^k(t+1))$ is the time that the drop takes to move between these two points being calculated according to the Eq. 6.

$$time(i,j : vel^k(t+1)) = \frac{HUD(i,j)}{vel^k(t+1)} \tag{6}$$

where HUD corresponds to the heuristic used for the problem.

To update the soil by where it passes the Drop we used the Eq. 7.

$$soil(i,j) = (1 - \rho) \cdot soil(i,j) - \rho \cdot \Delta soil(i,j) \tag{7}$$

where ρ is a value between $[0,1]$.

Each travel of a drop represents a solution, being chosen only the best solution of each iteration which is later used to update the general soil, using the following Eq. 8.

$$soil(i,j) = (1 - \rho_{IWD}) \cdot soil(i,j) + \rho_{IWD} \cdot soil(i,j)^{IB} \cdot \frac{1}{q(T^{IB})} \tag{8}$$

where ρ it is a positive value between $[0,1]$ and $q(T^{IB})$ the fitness of the best drop.

The following evaluates whether the new solution found is better than the global solution, then it is replaced by the new solution according to the Eq. 9.

$$T^{TB} = \begin{cases} T^{IB} & \text{if } q(T^{TB}) > q(T^{IB}), \\ T^{TB} & \text{else} \end{cases} \tag{9}$$

IWD has been used to solve different optimization problems such as workflow scheduling in cloud computing environment [8], supply chain optimization problem [10], capacitated vehicle routing problem [16], feature selection optimization problem [3], optimal location and sizing of distributed generation unit [12] among other uses.

3   Software Project Scheduling Problem

The SPSP is a combinatorial NP-HARD problem [18] that this present the software industry. It consists in carrying out a proper allocation of human resources to the different activities that compose a software project considering the skills and salary of each employee [2,17]. The problem has the following components:

– Tasks comprising the project. $T = \{t_1, \ldots, t_{|T|}\}$, where $|T|$ is the number of tasks to carry out the project.

- Skills needed for each task. $S = \{S_1, \ldots, S_{|S|}\}$, where $|S|$ is the number of skills the entire project.
- Employee skills. Is a subset of S corresponding to all the necessary skills to complete a task.

This problem has the following restrictions:

- Each task must be assigned to at least one employee.

$$\sum_{i=1}^{|E|} m_{ij} > 0 \qquad \forall j \in \{1, \ldots T\} \tag{10}$$

- The employees assigned to a task have all the necessary skills for this task,

$$t_j^{sk} \subseteq \cup \quad e_i^{sk} \quad \forall j \quad \{1, \ldots T\} \tag{11}$$

where t_j^{sk} represents the necessary skills for the task j and e_i^{sk} corresponds to the employee's skills i.

A possible solution to this problem can be the proposal for Crawford in [4, 6, 7] where a matrix ExT is used that contains the degree of dedication of each employee to a task.

4   Self-configuring the Intelligent Water Drops Algorithm

The adjustment of parameters can be a great disadvantage when selecting a metaheuristic, especially for the bioinspired that are characterized by having many parameters. In this section we explain our proposal of self-adaptation for a parameter od IWD, that is to say this is adjusted during the execution of the algorithm.

For the analysis 10 executions of the algorithm were realized in its standard version for the instance composed of 10 tasks, 5 employees and 5 skills. First it was used $Soil_{init}^k$ Leaving the parameter $Velocity_{init}^k$ fixed and later vice versa. The result can be visualized in the Fig. 1 where it is observed that $Velocity_{init}^k$ and $Soil_{init}^k$, have an impact on the behavior of the algorithm. In the case of the velocity it is observed that the best behavior takes place when it has value 50 while for the soil is presented for the value 1500.

In our work is performed the self-configuring the parameter of the velocity for the problem SPSP i.e. the user will not need to give an initial value. The general proposal is presented in Fig. 2, Where after the travel of all the drops of each iteration it assesses whether the best fitness found in the Iteration is different from the previous iteration if this is not so is increased by 1 a counter otherwise it is continued with the search of a new solution. Each time that increases the counter is compared to a value α which represents the quantity of times for that it is necessary to wait so that our proposal should assign a new value at the

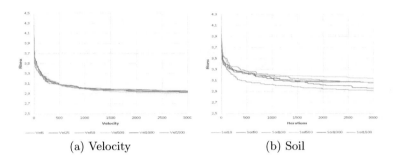

(a) Velocity   (b) Soil

Fig. 1. Influence of the velocity and soil in instance 5e-10t-5s.

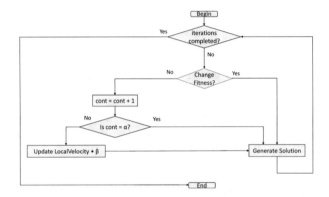

Fig. 2. Flowchart of the proposal of adaptation.

speed and with it the drops can get out of this stagnation of the search space. The new value of the velocity is given by the following Eq. 12:

$$\beta = Velocity_{init} \cdot nTask \cdot nEmployee \cdot nSkill \cdot \epsilon \tag{12}$$

where $Velocity_{init}$ is the speed of the current drop, nTask the number of tasks of the instance, nEmployee the number of employees of the instance, nSkill the number of skills of the instance and ϵ is a random value between $[0, 1]$.

The use of nTask, nEmployee and NSkill used in the Eq. 12 allows our proposal to incorporate external information into the algorithm And with this associate the characteristics of each instance to resolve with the metaheuristic. On the other hand the value α provides internal information indicating when to apply the velocity setting.

5   Results of Experiments

In this section we present the results of experiments carried out using our proposal. For this the algorithm was built using Java language and executed in a computer with the following features: Processor, Intel(R) Core(TM) i7-6700

CPU @ 3.40 GHz 3.41 GHz; Installed Memory (RAM), 16.0 GB; Operating System, Windows10 Pro; Type System, 64 bits, processor x64.

With the purpose to evaluate our work is used 4 instances similar to those employed in [7].

The values used for the parameters were the following: Amount of drops, 600; Number of iterations, 600; constant Soil $a_s = 1$, $b_s = 0.01$, $c_s = 1$; constant velocity $a_v = 1$, $b_v = 0.01$, $c_v = 1$, $Soil_{init}^k = 1500$.

Every instance was executed 30 times so much for the standard version like the version with autonomous search. The results are shown in Table 1 and the charts of convergence in the Figs. 3 and 4:

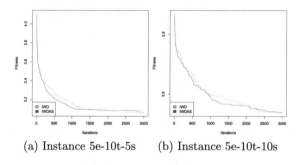

(a) Instance 5e-10t-5s   (b) Instance 5e-10t-10s

Fig. 3. Convergence instances 5e-10t-5s and 5e-10t-10s.

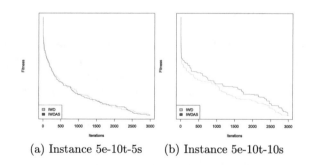

(a) Instance 5e-10t-5s   (b) Instance 5e-10t-10s

Fig. 4. Convergence instances 5e-10t-5s and 5e-10t-10s.

where RPD represents the relative percentage deviation and corresponds to the distance of our result with the best found. To calculate the RPD we use the Eq. 13.

$$RPD = \frac{Fitness - BestFitness}{BestFitness} \cdot 100 \tag{13}$$

where *Fitness* corresponds to the Fitness obtained by the algorithm and *BestFitness* the best result of all the algorithms used.

Table 1. Comparison to other techniques.

Instance	Algorithms	Fitness	RPD
5 employees	$ACO - HC$	**2.7750**	-
10 tasks	ACS	3.5149	26.6
5 skills	IWD_{STD}	3.1073	11.9
	IWD_{AS}	3.1446	13.3
5 employees	$ACO - HC$	**3.3449**	-
10 tasks	ACS	3.4049	1.7
10 skills	IWD_{STD}	3.6253	8.3
	IWD_{AS}	3.6603	9.4
10 employees	$ACO - HC$	**2.0967**	-
10 tasks	ACS	2.5773	22.9
5 skills	IWD_{STD}	2.5629	22.2
	IWD_{AS}	2.5313	20.7
10 employees	$ACO - HC$	**2.2660**	-
10 tasks	ACS	2.6440	16.6
10 skills	IWD_{STD}	2.8881	27.45
	IWD_{AS}	2.8992	27.9

6   Statistic Analysis

In this section we present a statistical analysis of the results obtained for $IWDSTD$ and $IWDAS$. For this analysis we apply the test of Kolmogorov-Smirnov-Lilliefors [14] to determine the independence of the data and Wilcoxons Signed Rank [11] is used to check the superiority of a strategy of resolution compared with the other. We use a significance level of 0.05 for both test.

The hypothesis considered for Kolmogorov-Smirnov-Lilliefors test is:

H_0 = The data follow a normal distribution.
H_1 = The data does not follow a normal distribution.

And given that the p-value obtained is <0.05 the hypothesis is rejected, then given that the data does not follow a normal distribution we apply the test not parametric called Wilcoxon-Mann-Whitney in order to verify that our proposal is better than the standard version. The Hypothesis used for this second test is:

H_0 = Standard IWD algorithm $\geq$ IWD algorithm with AS
H_1 = Standard IWD algorithm $<$ IWD algorithm with AS

Values were obtained > than 0.05 for p-value for this reason the hypothesis of this second test is accepted (Table 2).

The violin graphics shown in Figs. 5 and 6 are ratified the best results obtained.

Table 2. Results statistical analysis.

H_0	Technique	IWD_{STD}	IWD_{AS}
5 employees 10 tasks 5 skills	IWD_{STD}	-	0.07396821
	IWD_{AS}	0.92799973	-
5 employees 10 tasks 10 skills	IWD_{STD}	-	0.1310427
	IWD_{AS}	0.87193435	-
10 employees 10 tasks 5 skills	IWD_{STD}	-	0.8940872
	IWD_{AS}	0.108522035	-
10 employees 10 tasks 10 skills	IWD_{STD}	-	0.5278549
	IWD_{AS}	0.47770962	-

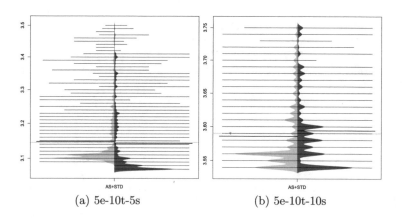

(a) 5e-10t-5s   (b) 5e-10t-10s

Fig. 5. Instances 5e-10t-5s and 5e-10t-10s.

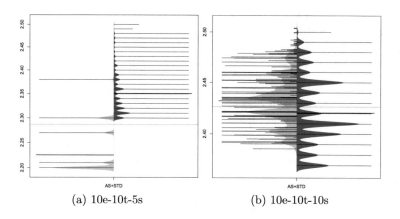

(a) 10e-10t-5s   (b) 10e-10t-10s

Fig. 6. Instances 10e-10t-5s and 10e-10t-10s.

7   Conclusions and Future Work

In this work, it was proposed the self-configuring of the velocity parameter of IWD metaheuristic that influences the behaviour of the algorithm by having a direct relation between velocity and soil removed. Although it was not possible to determine statistically that the version with self-configuring is better than the standard version the results were very similar which makes them encouraging. In the four instances used in this work, in the that contains 10 employees, 10 tasks and 5 skills was superior to the standard version and also to Ant Colony System. Moreover, in the instance with 5 employees, 10 tasks and 5 skills was superior to Ant Colony System. Given the results obtained, it is proposed as a future work the self-configuring of the soil parameter incorporating another type of internal information as can be the quality of the solution as also consider another condition of end of the algorithm.

Acknowledgements. Broderick Crawford is supported by grant CONICYT/ FONDECYT/REGULAR 1171243 and Ricardo Soto is supported by Grant CONI-CYT/FONDECYT/REGULAR/1160455, Gino Astorga is supported by Postgraduate Grant, Pontificia Universidad Catolica de Valparaíso, 2015 and José Lemus is supported by INF-PUCV 2018.

References

1. Adenso-Diaz, B., Laguna, M.: Fine-tuning of algorithms using fractional experimental designs and local search. Oper. Res. **54**(1), 99–114 (2006)
2. Alba, E., Chicano, J.F.: Software project management with GAs. Inf. Sci. **177**(11), 2380–2401 (2007)
3. Alijla, B.O., Lim, C.P., Wong, L.P., Khader, A.T., Al-Betar, M.A.: An ensemble of intelligent water drop algorithm for feature selection optimization problem. Appl. Soft Comput. **65**, 531–541 (2018)
4. Crawford, B., Soto, R., Astorga, G., Castro, C., Paredes, F., Misra, S., Rubio, J.M.: Solving the software project scheduling problem using intelligent water drops. Tehnički vjesnik **25**(2), 350–357 (2018)
5. Crawford, B., Soto, R., Castro, C., Monfroy, E.: Extensible CP-based autonomous search. In: International Conference on Human-Computer Interaction, pp. 561–565. Springer (2011)
6. Crawford, B., Soto, R., Johnson, F., Misra, S., Paredes, F., Olguín, E.: Software project scheduling using the hyper-cube ant colony optimization algorithm. Tech. Gaz. **22**(5), 1171–1178 (2015)
7. Crawford, B., Soto, R., Johnson, F., Monfroy, E., Paredes, F.: A max-min ant system algorithm to solve the software project scheduling problem. Expert Syst. Appl. **41**(15), 6634–6645 (2014)
8. Elsherbiny, S., Eldaydamony, E., Alrahmawy, M., Reyad, A.E.: An extended intelligent water drops algorithm for workflow scheduling in cloud computing environment. Egypt. Inform. J. **19**(1), 33–55 (2018)
9. Hamadi, Y., Monfroy, E., Saubion, F.: What is autonomous search? In: Hybrid Optimization, pp. 357–391. Springer (2011)

10. Kayvanfar, V., Husseini, S.M., Karimi, B., Sajadieh, M.S.: Bi-objective intelligent water drops algorithm to a practical multi-echelon supply chain optimization problem. J. Manufact. Syst. **44**, 93–114 (2017)
11. Mann, H.B., Whitney, D.R.: On a test of whether one of two random variables is stochastically larger than the other. Ann. Math. Stat. **18**(1), 50–60 (1947)
12. Prabha, D.R., Jayabarathi, T., Umamageswari, R., Saranya, S.: Optimal location and sizing of distributed generation unit using intelligent water drop algorithm. Sustain. Energy Technol. Assess. **11**, 106–113 (2015)
13. Shah-Hosseini, H.: Intelligent water drops algorithm: a new optimization method for solving the multiple knapsack problem. Int. J. Intell. Comput. Cybern. **1**(2), 193–212 (2008)
14. Shapiro, S.S., Wilk, M.B.: An analysis of variance test for normality (complete samples). Biometrika **52**(3/4), 591–611 (1965)
15. Soto, R., Crawford, B., Monfroy, E., Bustos, V.: Using autonomous search for generating good enumeration strategy blends in constraint programming. In: International Conference on Computational Science and Its Applications, pp. 607–617. Springer (2012)
16. Teymourian, E., Kayvanfar, V., Komaki, G.M., Zandieh, M.: Enhanced intelligent water drops and cuckoo search algorithms for solving the capacitated vehicle routing problem. Inf. Sci. **334**, 354–378 (2016)
17. Vega-Velázquez, M.Á., García-Nájera, A., Cervantes, H.: A survey on the software project scheduling problem. Int. J. Prod. Econ. **202**, 145–161 (2018)
18. Xiao, J., Ao, X.T., Tang, Y.: Solving software project scheduling problems with ant colony optimization. Comput. Oper. Res. **40**(1), 33–46 (2013)

Multivariate Discrimination Model for TNT and Gunpowder Using an Electronic Nose Prototype: A Proof of Concept

Ana V. Guaman$^{(\boxtimes)}$, Patricio Lopez, and Julio Torres-Tello

Universidad de las Fuerzas Armadas – ESPE,
Av. General Rumiñahui S/N, Sangolquí, Ecuador
{avguaman, jwtorres}@espe.edu.ec

Abstract. In this work a proof of concept for discriminating explosive substances is presented, where a discrimination model for the classification of TNT and gunpowder is developed. An electronic nose was designed for sensing volatile organic compounds present in TNT and gunpowder, and a model that combines Principal Component Analysis and Fisher Discriminant Analysis was built for enhancing class discrimination. The model was tested in two scenarios: discriminating among the two explosive substances and one non-explosive, and discriminating between explosives and non-explosives, obtaining better results in the second case. In order to test model confidence a permutation test was used proving an accuracy of 67% with a p-value <0.01 for the first scenario, and an accuracy of 86.6% for the second one. These results make us think that by enhancing the prototype characteristics in both hardware and software, we would be able to achieve better results.

Keywords: Electronic nose · Explosive discrimination model ·
Classification model · Permutation test

1   Introduction

It is easy to find information on how the terrorist attacks of September 11, 2001 have had impacts on the security issues both in the USA and worldwide. However, there have been many other terrorist activities in the past years around the world, such as bombings of buses in Israel and the Philippines, trains in Spain, and passenger jets in Russia [1]. In the case of Ecuador, the Constitution and laws guarantee human security through policies and integrated actions such as the control of weapons that has been one of the main policies of the Government. Statistics show that the rates of homicides and murders with firearms between 2009 and 2015 have been reduced; although there have been a few incidents during the last year [2] and the seizures remain high in the country. Possession and trafficking are still common in Ecuador and the National Police has identified many national and international routes and modes of transport in the illicit traffic of arms and ammunition [3].

The illegal trafficking of explosives through conventional means represents therefore a real challenge to civil security for future years, and law enforcement entities

Á. Rocha et al. (Eds.): ICITS 2019, AISC 918, pp. 284–293, 2019.
https://doi.org/10.1007/978-3-030-11890-7_28

throughout the world face the problem of detecting hidden explosive devices in luggage, cargo, mail, vehicles and on people. Human and canine inspection of large commercial payloads at borders are not a viable solution to detect explosives, so that it is becoming compulsory to develop integrated inspection systems by means of automated methods [1, 4]. Those methods are based in a variety of solutions, such as nuclear [4, 5], chemical gas sensors [6], MOS sensors [1, 7], quartz crystal microbalances [8, 9], ion mobility spectrometry (IMS), chemiluminescence, electron capture detection (ECD), electrochemistry, olfaction, etc. Even if currently there are various advanced methods of explosive detection, the simple ones are usually also very effective. The most known example of such simple technique is the electronic nose, which is an analytical instrument that consists of an array of sensors, a sampling circuit and a pattern classifier algorithm, capable of recognizing odors, either simple or complex [10].

One of the implementations of electronic noses found in literature is very similar to the one shown in the present paper. That portable device is composed of the multi-sensors chamber with three technologies of explosive vapors sensors and a laptop that controls the device with a Labview interface, processing all the data in real time [9]. However, an explosive model development is scarcely developed. Therefore, in this work, it is presented a discrimination model for identifying two different classes of explosives commonly used in military application, which are usually objects of illicit trafficking. The data was analyzed using an electronic nose which uses MOS or also called MOX (metal oxide) chemical sensors and the signal processing took place in an off-line fashion.

2   Materials and Methods

2.1   Electronic Nose Prototype

Electronic Nose (e-nose) is an analytical device for sensing volatile organic compounds (VOC) and can be used for producing either quantitative or qualitative analysis [11]. E-nose devices attempt to mimic the biological olfactory system by means of technical components; therefore it has to be composed by several functional blocks in order to reproduce biological activity, as it can be seen at the top of the Fig. 1. The first block is constituted by the pneumatic subsystem which allows VOC to flow from a substance to the sensors, and it is recommended that it includes a sampling mechanism that fosters substance volatilization. The second block of the system is the sensing mechanism, where chemical sensors are placed, which are the ones that react to the chemical compounds present in the substance acquired by the pneumatic subsystem. The third block corresponds to the acquisition subsystem that obtains information from sensors when excited and transduces it in some kind of measurable magnitude; for instance volts. Finally, the fourth and main block is the one that mimics brain functionality, which means that it interprets the information contained in the electric signal, performing either discrimination, identification or quantification of the compounds present in the substance.

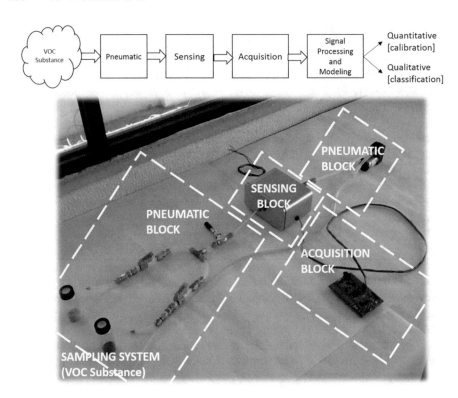

Fig. 1. Block diagram of a generic electronic nose and the implemented prototype used in this work.

The electronic nose used in this work has been already presented [12], however it is worthwhile giving a brief explanation such as follows: At the bottom of Fig. 1, it can be seen a picture of the e-nose prototype used for this work, which is constituted by four main blocks: (i) The pneumatic block has the main task of pumping in the air from the substance of interest into the sensing block. A sampling system is part of the pneumatic block and is composed by two vials where one contains the explosive substance and the other one is empty (contains only air). The VOC is extracted using headspace methodology [13]. Using electro valves the system pumps the volatile compounds from air through the empty vial (named blank) or volatile compounds related to explosive samples. The blank samples were used for both cleaning the entire system and to get the sensors to recover to their baseline. (ii) A sensing chamber which contains an array of six MOX chemical sensors from the FIGARO family (TGS822, TGS2610, TGS826, TGS825). The main purpose of the sensing chamber is to guarantee the flow of air to be evenly spread over all the sensors. (iii) The acquisition block is implemented on an STM32F407VGT6 board used as A/D converter due to the easy integration with Matlab, useful to build the HMI interface and to monitor the performance of the sensors. (iv) The final block not shown in Fig. 1, which is the one performing signal processing and modeling, was realized in an off-line fashion and the identification of explosive substances was performed on a desktop computer using Matlab R14.

2.2   Experimental Setup

Before getting into the actual tasks related to the work described in this paper, an initial experiment was carried out for extracting the VOC from explosive substances. Two explosive substances (TNT and gunpowder of military use) were analyzed in their raw solid state and then when mixed with other doping substances such as acetone and ethanol. A qualitative analysis was done in terms of sensor discovery response for each situation, determining their maximum voltage during excitation; and as a result ethanol was chosen as the dopant used for the following experiments.

In order to validate the discrimination model implemented in this work, three substances were analyzed: two explosives (TNT and gunpowder of military use) and one non explosive (Ethanol). Due to the limited access to explosive samples, 10 experiments were carried out for each substance, where 1 g of Trinitrotoluene 2, 4, 6 and 1 g of gunpowder in solid state were placed into vials, solved both of them in 1 ml of ethanol. In addition, 1 ml of ethanol was analyzed in order to understand the discrimination capabilities of the model between explosive and non-explosive substances.

The experimental procedure was done in the following manner: At the beginning of each day, the system was turned on during five minutes for pumping air through the pneumatic mechanism and sensing chamber, to guarantee that the system was cleaned. Thirty experiments (ten for each substance, and three per day) were carried out in a random sequence for ensuring not to create a memory effect during the analysis.

Every experiment lasted thirty minutes and four and a half seconds, where three cycles of cleaning and exposure of the samples was done. One cycle consisted of a cleaning period that lasted ten minutes and exposure to the sample which lasted one second and a half.

2.3   Signal Processing and Discrimination Model

Each sensor provides a response vector of 240 features per experiment, therefore it was necessary to concatenate the response of the six sensors in order to obtain a single vector for each experiment. After performing the concatenation procedure, a matrix of 30 experiments by 1440 features was used for building the discrimination model. A drift correction is always needed in order to diminish the drift phenomena that usually occurs in MOX chemical sensors such as the ones used in the e-nose implemented for this project. Therefore, a baseline correction operation was performed, which was done by subtracting the initial value of the transient response out of each sensor original response, and compensating any kind of additive drift effect of the sensor [14]. Furthermore, a peak alignment was done for correcting experimental errors mainly caused by temporal shifts when the sample was introduced in the system.

The discrimination model used was the so called Principal Discriminant Analysis (PDA), which is a combination of Principal Component Analysis (PCA) [15] and Fisher's Linear Discriminant Analysis (LDA) [16], originally proposed by Wang et al.

[17]. LDA is a supervised method that can be used either as classification or dimensionality reduction. An advantage of this method is that it produces a maximum class-separability reducing the data dimension to the number of classes minus one. However, LDA suffers the so called curse of dimensionality and can get overfitting results when there are more features than samples [18]. In this sense, PCA can be used as reduction of dimensionality algorithm previous to the application of LDA in order to avoid overfitting results. PCA is an unsupervised method, which reduces dimensionality, creating a new subspace, maximizing data variance. The selection of the number of principal components was based on the explained variance of the original features that captures more than of 70% of data variance. In this case, raw data of 30 × 1440 elements was reduced down to 30 × 5 (5 being the number of principal components) using PCA. This new subspace, in turn, is reduced to the final model to a dimension of 30 × 2 (three classes minus one).

In order to test the accuracy and performance of the model it was used a k Nearest Neighbor (kNN) classifier. kNN is a supervised algorithm based on how labeled samples in a model are distributed. A new sample is assigned to a specific class k nearest training sample [19]. The Euclidean distance is measured between the new class and the training classes, and the distances are ordered from small to large, then the new sample receives the class of the nearest k neighbors by majority vote; where k should be an odd number in order to avoid discrepancies and the range number should be between one to the maximum number of samples of one class. In this work, the value of k should be between one to nine, and k equal to 5 was selected to guarantee a classifier with a good generalizing.

In addition, a variation of the leave-one-out cross validation was used to test the generalization capabilities of the model [20]. The variation was called leave-one-day-out, meaning leaving one sample of each class as validation and the remaining data as training. Taking into account the small experimental samples used in this work, a permutation test was implemented. Permutation test [21] is a statistic test for determining if a classification was done by chance, in this case for establishing the robustness of the proposed model. The null hypothesis means that there is a close relationship between the data and the labels which cannot be learned reliably in the training step [22]; thus means the classification rate was done by chance. In order to reject the null hypothesis, labels are randomly changed and a new model is built getting as result a new classification rate. The classification rate done by changing labels has to be dissimilar to classification rate of the original labels. Equation (1) shows how permutation test probability is calculated.

$$p = (|\{D'E\ D : e(f, D') \leq e(f, D)\}| + 1)/\pi + 1 \tag{1}$$

$$D = \{x_i, y_i\}_{i=1}^{n} \tag{2}$$

$$D' = \{x_i, \pi(y)_i\}_{i=1}^{n} \tag{3}$$

Where, D is the original data set and π is the permutation of n elements (see Eq. 2); and D' is a randomized version of D which is obtained after applying the permutation over the labels [23] (see Eq. 3). In this work, π was set up to one hundred and for each iteration the cross validation methodology was run.

3   Results

3.1   Explosive Sample Analysis

Figure 2 depicts the response of a single sensor when TNT, acetone and ethanol were analyzed; and mixtures of TNT with acetone and ethanol. When TNT is analyzed in its raw state it is clearly observable that chemical sensors do not respond well, mainly due to the low vapor pressures of explosives. In general, trace detection of explosives involves applying different analytical techniques in order to extract enough information to be analyzed. In chemical sensing it is important to foster VOC generation, which implies doping the substance of interest with another in order to enhance the volatilization of the compounds. This phenomenon can be seen in Fig. 2 in which TNT being doped with either acetone or ethanol clearly shows a dynamic sensor response different from the ones obtained with pure acetone or ethanol. It can be seen that a highest peak is obtained when is TNT is doped with acetone than with ethanol. However, the recovery time when acetone is used as dopant is slow and a drift on the baseline is observed at the end of the analysis. For this reason, ethanol was used for doping explosive substances. Gunpowder presents a similar behavior and was doped using ethanol as it was already explained in section two.

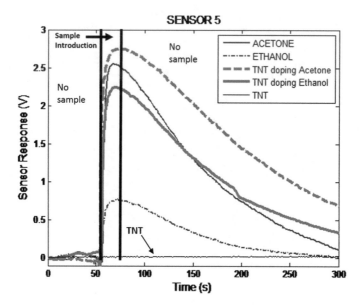

Fig. 2. A chemical sensor responses to TNT with and without doping, showing also a peak alignment for correcting temporal shifts.

Furthermore, Fig. 2 depicts the behavior of chemical sensors response where there are three clearly distinguishable stages. Initially there is a period of time when no samples are introduced yet; then comes the sensor excitation when the sample is introduced causing a maximum peak in the response of the sensor. Finally, a third stage happens when the sensor response decays mainly due to the no presence of compounds, which implies a cleaning period. The six chemical sensors of the array have the same behavior for a single experiment and this cycle is repeated three times consecutively, as explained before.

However, these slightly differences might be enhanced when using multivariate algorithms like PDA as proposed in this work. Additionally, huge dimensionality can be appreciated and taking into account the small number of experiments used in this work, it is very important to implement a robust methodology that avoids overfitting results.

3.2   Explosive Discrimination and Classification

A Principal Component Discriminant Analysis (PDA) model is presented in Fig. 3. It is straightforward to see that non-explosive samples (ethanol) are separated from the explosive ones. However, it is not feasible to interpret a separation between TNT and gunpowder. Another thing to notice is that TNT samples are more clustered than the gunpowder ones; we speculate that it happens because gunpowder is unstable and generates this kind of dispersion.

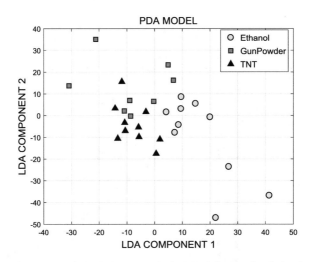

Fig. 3. Principal Component Discriminant Analysis (PDA) for the discrimination of explosives.

After applying leave-one-day-out cross validation methodology, a confusion matrix was obtained which results are shown in Table 1. The classification rate of the model is 67% and it is worth to note that most of the confusion in the classification is caused by misclassification between explosives. It is important to point out that just one out of 10 samples of ethanol were misclassified, meaning that the prototype is very specific for the discrimination.

Table 1. Confusion matrix of PDA model after performing cross validation strategy

Confusion matrix		Predicted class			Total
		TNT	GP	Ethanol	
Real class	TNT	**7**	2	1	10
	GP	4	**4**	2	10
	Ethanol	0	1	**9**	10
Total		11	7	12	**30**

Since the classification rate is relatively limited when distinguishing among the three substances and it is in fact close to a classification by chance, a permutation test was done. Figure 4 shows a histogram of 100 times where labels where randomly changed, and it can be seen that the error rate was between 40 and 90%. In thick line it is indicated the model error rate which is far from errors by chance. Applying Eq. 1, the p-value obtained was less than 0.01, which means that the model error or classification rate model was not done by chance. Therefore, the model results are robust enough to interpret that the prototype is able to distinguish explosive samples.

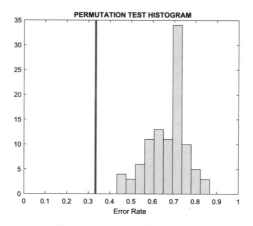

Fig. 4. Histogram of error rate after applying permutation test. The black thick line represents the error of the model which is 33%.

A similar analysis was done using two classes in the classification model (explosive vs non-explosive) and a matrix of confusion was obtained, which is shown in Table 2. An accuracy of 86.6%, specificity of 90% and sensitivity of 95% were obtained. Clearly, the response of the prototype is better when two different types of substances are being discriminated. This result implies that the e-nose prototype used here together with the implemented model is able to distinguish in a proper way between explosive and non-explosive samples.

Table 2. Classification rate results identifying explosive vs non explosive substances

Confusion matrix	Explosive	Non explosive	Total
Explosive	17	3	20
Non explosive	1	9	10
Total	18	12	30

4   Conclusions

In this work a proof of concept for distinguishing explosive substances was presented. In addition, a discrimination model for explosive substances was built, using an e-nose prototype consisting of an array of six chemical sensors. The model resulted in an accuracy of 67% when three different substances (two explosives and one non-explosive) were discriminated with a p-value <0.01. From the results it was evident that the prototype is not robust enough for distinguishing different explosives from the same chemical families. Nevertheless, when the system was set up to classify between non-explosive and explosive compounds, the performance of the prototype increased noticeably. This results make us think that by enhancing the prototype characteristics in both hardware and software, we would be able to achieve better results.

In addition, this work presents a methodology for building models and validating results, which should be used in a continuous manner in signal and interpretation processing analysis, due to its results showing capabilities for avoiding overfitting of the results. This could be used also for generating discrimination models that are more general.

Acknowledgments. This work was founded by Universidad de las Fuerzas Armadas ESPE under the project 2016-pic-009.

References

1. Gui, Y., Xie, C., Xu, J., Wang, G.: Detection and discrimination of low concentration explosives using MOS nanoparticle sensors. J. Hazard. Mater. **164**, 1030–1035 (2009)
2. Redacción el Comercio: Gobierno confirma secuestro de equipo periodístico de EL COMERCIO en Mataje (2018). http://www.elcomercio.com/actualidad/mataje-secuestro-equipoperiodistas-elcomercio-ecuador.html
3. Castillo Egüez, J.L.: Armas de fuego y políticas públicas (Ecuador 2.009–2.015) (Opinión) (2015)
4. Nebbia, G., Pesente, S., Lunardon, M., Moretto, S., Viesti, G., Cinausero, M., Barbui, M., Fioretto, E., Filippini, V., Sudac, D., Nađ, K., Blagus, S., Valković, V.: Detection of hidden explosives in different scenarios with the use of nuclear probes. Nucl. Phys. A **752**, 649–658 (2005)

5. Aleksandrov, V.D., Bogolubov, E.P., Bochkarev, O.V., Korytko, L.A., Nazarov, V.I., Polkanov, Y.G., Ryzhkov, V.I., Khasaev, T.O.: Application of neutron generators for high explosives, toxic agents and fissile material detection. Appl. Radiat. Isot. **63**, 537–543 (2005)

6. Barthet, C., Montméat, P., Eloy, N., Prené, P.: Detection of explosives vapours using a multi-quartz crystal microbalance system. Procedia Eng. **5**, 472–475 (2010)

7. Brudzewski, K., Osowski, S., Pawlowski, W.: Metal oxide sensor arrays for detection of explosives at sub-parts-per million concentration levels by the differential electronic nose. Sens. Actuators B: Chem. **161**, 528–533 (2012)

8. Guillemot, M., Dayber, F., Montméat, P., Barthet, C., Prené, P.: Detection of explosives vapours on quartz crystal microbalances: generation of very low-concentrated vapours for sensors calibration. Procedia Chem. **1**, 967–970 (2009)

9. Rousier, R., Bouat, S., Bordy, T., Grateau, H., Darboux, M., Hue, J., Gaillard, G., Besnard, S., Veignal, F., Montméat, P., Lebrun, G., Larue, A.: T-REX: a portable device to detect and identify explosives vapors. Procedia Eng. **47**, 390–393 (2012)

10. Stetter, J.R., Strathmann, S., McEntegart, C., Decastro, M., Penrose, W.R.: New sensor arrays and sampling systems for a modular electronic nose. Sens. Actuators B: Chem. **69**, 410–419 (2000)

11. Röck, F., Barsan, N., Weimar, U.: Electronic nose: current status and future trends. Chem. Rev. **108**, 705–725 (2008)

12. Lopez, P., Trivino, R., Calderon, D., Arcentales, A., Guaman, A.V.: Electronic nose prototype for explosive detection. Presented at the 2017 CHILEAN Conference on Electrical, Electronics Engineering, Information and Communication Technologies (CHILECON), Chile, October 2017

13. Banerjee, R., Tudu, B., Bandyopadhyay, R., Bhattacharyya, N.: A review on combined odor and taste sensor systems. J. Food Eng. **190**, 10–21 (2016)

14. Di Carlo, S., Falasconi, M.: Drift correction methods for gas chemical sensors in artificial olfaction systems: techniques and challenges. In: Advances in Chemical Sensors, p. 326. InTech, Rijeka (2012)

15. Jollife, I.T.: Principal Component Analysis. Springer, New York (2002)

16. Duda, R.O., Hart, P.E., Stork, D.G.: Pattern Classification (Pt. 1). Wiley, New York (2001)

17. Wang, M., Perera, A., Gutierrez-Osuna, R.: Principal discriminants analysis for small-sample-size problems: application to chemical sensing. 2004 Presented at the Sensors. IEEE (2004)

18. Luo, D., Ding, C., Huang, H.: Linear discriminant analysis: new formulation and overfit analysis. In: Proceedings of the Twenty-Fifth AAAI Conference on Artificial Intelligence, pp. 417–422 (2011)

19. Gupta, V., Mittal, M.: KNN and PCA classifier with Autoregressive modelling during different ECG signal interpretation. Procedia Comput. Sci. **125**, 18–24 (2018)

20. Arlot, S., Celisse, A.: A survey of cross-validation procedures for model selection. Stat. Surv. **4**, 40–79 (2010)

21. Good, P.I.: Permutation, Parametric, and Bootstrap Tests of Hypotheses. Springer-Verlag, New York (2005)

22. Mukherjee, S., Golland, P., Panchenko, D.: Permutation Tests for Classification. Massachusetts Institute of Technology, Cambridge, #2003-019 (2003)

23. Ojala, M.: Permutation tests for studying classifier performance. J. Mach. Learn. Res. **11**, 1833–1863 (2010)

Evaluating Model-Based Testing
in an Industrial Project:
An Experience Report

Rebeca Obando Vásquez[(✉)], Christian Quesada-López,
and Alexandra Martínez

Universidad de Costa Rica, San José, Costa Rica
{yohana.obando,cristian.quesadalopez,alexandra.martinez}@ucr.ac.cr

Abstract. Model-based testing (MBT) is an approach that automates
the design and generation of test cases based on a model that represents
the system under test. MBT can reduce the cost of software testing and
improve the quality of systems in the industry. The goal of this study is
to evaluate the use of MBT in an industrial project with the purpose of
analyzing its efficiency, efficacy and acceptance by software engineers. A
case study was conducted where six software engineers modeled one module of a system, and then generated and executed the test cases using an
MBT tool. Our results show that participants were able to model at least
four functional requirements each, in a period of 20 to 60 min, reaching a
code coverage between 39% and 59% of the system module. We discussed
relevant findings about the completeness of the models and common mistakes made during the modeling and concretization phases. Regarding
the acceptance of MBT by participants, our results suggest that while
they saw value in the MBT approach, they were not satisfied with the
tool used (MISTA), because it did not support key industry needs.

Keywords: Model-based testing · Automation · Industry ·
Experience report

1 Introduction

Assuring the quality of a delivered product is a key part of industrial software
development. Different types of testing techniques are used for product quality
assurance, including functional tests, whose goal is to detect deviations (non-
conformities) of the system under test (SUT) from its functional specifications or
requirements [1]. Model-based testing (MBT) is a software testing approach that
automates the creation of test cases based on a model of the system under test [2],
and is normally used for functional testing. MBT has five phases: modeling the
SUT from the requirements, choosing a test selection criteria, generating abstract
test cases, concretizing the test cases, and executing these test cases [3,4].

Many MBT approaches lack empirical evaluations or have not been transferred to industrial contexts [4]. This was our motivation for conducting an

© Springer Nature Switzerland AG 2019
Á. Rocha et al. (Eds.): ICITS 2019, AISC 918, pp. 294–303, 2019.
https://doi.org/10.1007/978-3-030-11890-7_29

industrial case study, where the use of MBT by a group of software engineers on a real system was empirically evaluated. In this study we used MISTA, a model-based testing tool that provides support for all five MBT phases [3]. Our work can shed some light on how MBT can be leveraged by the industry providing some evidence about the efficiency, efficacy and acceptance properties of MBT from the perspective of industry practitioners.

The rest of the paper is organized as follows. Section 2 summarizes relevant related work. Section 3 describes the design of the study. Section 4 presents our results, discussion and lessons learned. Section 5 outlines our conclusions and future work.

2   Related Work

A few studies that apply MBT to specific niches were found in the literature [5–7]. In particular, Weißleder and Schlingloff [5] describe the process to implement MBT in embedded applications using a new tool. Their study compares the efficiency and effort of generating test cases with MBT versus manually. Similar measures were reported by Ergun et al. [6] and Herpel et al. [7], where the time to model and the number of generated test cases were used to show the benefits of applying MBT. Code and branch coverage have also been used as metrics of efficiency when comparing MBT to manual testing [7].

Weißleder and Schlingloff [5] held an workshop with requirements experts and software engineers, in order to get feedback about the advantages of MBT and the need for training on model creation and on the use of MBT. These issues were also discussed by Ergun et al. [6], Gebizli et al. [8] and Schulze et al. [9]. Also, Herpel et al. [7] mentioned that a good understanding of the SUT's requirements was important to obtain a complete and correct model of the SUT. Moreover, Schulze et al. [9] suggest that the design of the model is a factor associated to the adoption process, due the time and training/experience required to complete the model.

According to Bettiga and Lamberti [10], the technology adoption process has two parts: the cognitive and the affective one. The cognitive part can be measured using techniques like the Technology Acceptance Model (TAM) [11], which assesses the adoption of new technology from three perceptions: usefulness, ease of use, and intention to use. The affective part can be measured by sensors or other mechanisms that are able to detect the physical and sensorial reaction (feelings) of users while interacting with the system [10].

Our study combines some of the efficiency, efficacy, and acceptance metrics from previous studies [5–8, 11], and defines a process for using MBT on a software project of the industry, which could be used in similar contexts to help promote industry adoption of MBT.

3   Case of Study Design

This case study was conducted according to Wohlin's protocol [12] and Runeson's guidelines [13] for case studies. The goal of this study was to evaluate the use of

MBT in an industrial context in terms of its efficiency, efficacy, and acceptance by the software engineers. To pursue this goal, we defined the following research questions:

RQ1: What is the efficiency and efficacy of the test cases generated using MBT?
RQ2: What level of acceptance does MBT have among software engineers?

3.1 Context

This study was conducted in a Fortune 100 company located in *country name*, which has several development teams dedicated to support both internal and external needs of the company.

Subjects. The participants were software engineers from different departments who work on designing and developing new solutions for internal projects. The subjects were selected based on convenience, according to their priorities, available time and interest in learning a new testing approach. The company allows their employees to devote some work time to learn about new technologies that can be useful in their daily work.

SUT. The object of the study was a web application that manages requests for internal use, which was released to production one year before the study and was under maintenance at the time of the study. We will refer to this application as APP1 for confidentiality reasons. APP1 has four modules but due to time restrictions of the subjects, only one of its modules was used as SUT in this study.

3.2 Process and Data Collection

The process followed in the study is shown in Fig. 1: the first step was the creation and calibration of the instruments, described in Subsect. 3.3. Steps 2 to 7 were followed by the participants. Participants received training on MBT and MISTA, completed a demographic survey, modeled the SUT, generated the test cases and ran them. The steps 8 to 10 were performed by the researchers of the study, who collected or computed the metrics and did the corresponding data analysis and interpretation.

3.3 Instruments

Next we describe each of the instruments used in this study.

- **Demographic survey:** it gathered information about the subjects: age, background, work experience, role, projects, and prior knowledge of the SUT.
- **MBT tool:** MISTA [14] was the MBT tool selected for this study. It allows the automation of the test case generation and execution processes using high level petri nets. MISTA allows the generation of abstract test cases using different coverage criteria, the concretization of test cases using its

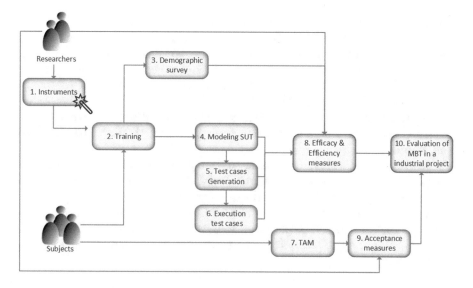

Fig. 1. Process followed in the study.

model-implementation description (MIM), and the execution of these test cases on the SUT. Given that the SUT is a web application, we chose HTML as language, reachability tree as the coverage criteria for test generation, and Selenium [15] as the tool for execution of functional test cases.

- **Training material:** it includes a presentation on MBT, on how to use MISTA, and a practical example of a model created with MISTA, together with links to MISTA tutorials.
- **Acceptance survey:** a survey that aims to assess the acceptance of MBT by the subjects. It has two sections: one with closed questions based on TAM [11] and another with open questions about the experience they went through, and will help explain the results obtained from the first section.
- **Observation notes:** during the sessions where the researchers interacted with the subjects, one researcher took notes about interaction and actions of the subjects.

3.4   Evaluation Criteria

The metrics used to measure the efficacy, efficiency and acceptance of MBT are based on [5, 11] and are described next.

- **Efficiency metrics:** we measured the modeling time (time it took each subject to create the model of the SUT) as well as the concretization time (time each subject took to convert –concretize– the abstract test cases generated from the model into executable test cases that can be run in Selenium). Both times were taken in minutes.

- **Efficacy metrics:** we measured the coverage percentage (percent of decision branches in the control structure of the source code that were executed while running the test cases created by the MBT tool [16]) and the number of requirements modeled (quantity of requirements each subject was able to model, from a total of five). For code coverage, we used a tool called OpenCover, that runs on the APP1 server and counts the number of visited branches. The number of requirements modeled was estimated by the researchers of the study, using their expertise in APP1. Both measures were in an ordinal scale.
- **Acceptance metrics:** we measured the perceived usefulness (degree to which the subjects believe that MBT with MISTA will increase their productivity or efficiency), perceived ease of use (degree to which the subjects believe that MBT with MISTA would be free of effort), and intention to use (degree to which the subjects think they will use MBT with MISTA in the future), according to TAM [11]. All these metrics are ordinal on a 5-point Likert scale.

4 Results and Analysis

4.1 Demographic Analysis of Participants

Here we describe the characteristics of our six subjects. All subjects were between 23 and 33 years old, and had completed a career in computer science or related areas. For subjects reported having a role of developers while two reported being software engineers. All subjects worked on software development and maintenance projects along with software validation and verification tasks. In terms of their general work experience, one subject had less than 3 years, four had between 3 and 6 years, and one had more than 10 years. Additionally, all subjects had 6 or less years working for their organization. Regarding their experience in modeling, four subjects reported to have less than 3 years of experience, one reported to have 6 to 10 years, and one reported to have more than 10 years. Moreover, five of the subjects stated that they had partially studied the material sent by email prior to the training session. One subject reported to know APP1. Finally, none of the subjects had knowledge of MBT or MISTA.

4.2 RQ1: Efficiency and Efficacy of Generated Test Cases

Efficiency. Table 1 shows the time it took each participant to model the SUT and concretize the abstract test cases. We observe that participants spent between 20 and 60 min creating the model. We found that the participant with prior knowledge of APP1 (P3) was able to complete the model (with all its requirements) faster. Participants P4 and P5 also achieved short modeling times, but neither of them finished modeling all the requirements (as seen in Table 2). The participant who took the longest time to model the SUT (P6), was the only one who included validation of entry data as part of his model. We found no

Table 1. Modeling and concretization time (in minutes) per subject.

Subject	Modeling time	Concretization time
P1	33	32
P2	45	29
P3	20	29
P4	23	31
P5	23	37
P6	60	33
Avg	34	31

Table 2. Number of requirements modeled, test cases generated, and coverage percentage achieved, per subject.

Subject	Requirements	Test cases	Coverage
P1	5	9	49%
P2	5	1	39%
P3	5	4	54%
P4	4	3	59%
P5	4	1	44%
P6	5	8	54%
Avg	4.7	4.3	49.8%

pattern relating modeling time to prior study of the material. On the other hand, participants' work experience was not considered a factor that affected the modeling time since they only modeled 'positive' scenarios: the subjects expressed that they focused on understanding how to use MBT rather than on developing a complete model, due to the time restrictions. From Table 1 we can also see that the average concretization time (31 min) is similar to the average modeling time (34 min). According to the participants, concretizing the generated test cases was not difficult (it was mere mechanical work) thanks to the previous training on MISTA and to the example model used as guideline.

Efficacy. Table 2 shows the number of requirements modeled, the number of test cases generated, and the coverage achieved, per participant. Using OpenCover, we found that APP1 had 732 branches in total. Participants achieved a coverage ranging between 39% and 59%. Such low percentages could be explained by the fact that most participants only focused on the 'happy path' when modeling. Participant P4 obtained the highest coverage percentage (59%) despite missing one requirement in his model. An analysis of P4's model shows that commands were added during the concretization stage to validate data entry, hence the higher coverage percentage. Participant P3, who had knowledge about APP1, modeled all the requirements and attained the second highest percentage coverage (54%). Likewise, participant P6 obtained a coverage percentage of 54% and modeled all the requirements, even adding data entry validations to the model. After analyzing P3's and P6's models, we discovered that they either did not contain a necessary state or did not add the corresponding commands in the MIM to simulate the behavior. This led us to an interesting finding: participants who modeled all the requirements did not necessarily achieved higher coverage than those who modeled less requirements.

We can also observe from Table 2 that participants generated between 1 and 9 test cases. Participant P2 obtained the lowest coverage percentage (39%) because his model only generated one test case and the corresponding concretized test case had some incorrect commands (possibly due to errors with the MIM). After analyzing of the number of test cases versus the coverage percentage achieved, we found that a higher quantity of test cases does not always guarantee higher coverage. For example, it is possible that a single test case with many steps covers (exercises) as many branches as several (shorter) test cases.

4.3   RQ2: Level of Acceptance of MBT Among Participants

Perceived Usefulness. Participants deemed MBT useful, with an average rating of 4 (out of 5) across the 'usefulness' questions. All subjects reported a positive perception of "Using MBT with MISTA would improve my job performance". Regarding the "Using MBT with MISTA would increase my productivity" and "Using MBT with MISTA would improve the quality of the work I do" statements, five subjects reported a positive perception and one was neutral. Besides, for the "Using MBT with MISTA would enhance my effectiveness on the job" and "Overall, I would find MBT with MISTA useful in my job" statements, four subjects reported a positive perception and two were neutral. On the other hand, three participants reported a neutral perception of the "MBT with MISTA would enable me to accomplish tasks more quickly" statement. In this case, responses to open questions revealed that participants believed MBT is a useful approach but MISTA is far from being an ideal tool. Furthermore, one participant reported a negative perception of "Using MBT with MISTA would make it easier to do my job", but we believe this is because his job was mainly related to database design and maintenance (deducted from his responses to open questions), which explains why he thinks that MBT would not make his job easier.

Perceived Ease of Use. Participants gave MBT an average rating of 3.5 (out of 5) across the 'ease of use' questions. Four subjects reported a positive perception and two were neutral on the following two statements: "Overall, I would find it MBT with MISTA easy to use" and "My interaction with MBT with MISTA would be clear and understandable". Also, two participants reported a positive perception of "I would find it easy to get MBT with MISTA to do what I want it to do" and four were neutral. For the claim "I would find MBT with MISTA flexible to interact with", we had two positive, two neutral, and two negative responses. On the other hand, the "I would find it takes a lot of effort to become skillful at using MBT with MISTA" statement, had one positive perception, two neutral and three negative perceptions. Interestingly, all but one (neutral) of the subjects were positive towards the statement "Learning to operate MBT with MISTA would be easy for me". Answers to open questions revealed that the subjects do not recommend any change in the process used to learn MBT, but would like to spend more time studying the theory of MBT and seeing more examples of models.

Intention to Use. Participants gave MBT an average rating of 3 (out of 5) across the 'intention to use' questions. Overall, our results show that the subjects liked the MBT approach but were not satisfied with the MBT tool used: MISTA. This is evidenced by the four positive, one neutral, and one negative responses to the statement "I plan to use an MBT approach in the next 6 months". And also by the four neutral and two negative responses to the "I plan to use MISTA in the next 6 months" statement. Open questions revealed that the subjects wanted to find and compare other MBT tools. Also, they agreed that MISTA was not powerful enough to use as an industry tool, but was useful for understanding the MBT process.

4.4   Lessons Learned

A pilot study can be helpful to remove potential noise factors and develop new instruments or refine existing ones. In our case, we performed a pilot study to validate the process and data collection defined in Sect. 3.2. Given that some participants from the pilot expressed that the modeling tasks were complex and difficult, we recommend to account for and control subjects' frustration at each stage of the MBT process. To control for that in this study, we (*i*) conducted a demo during the training session, (*ii*) clearly specified the SUT's requirements, (*iii*) had a *rescue* plan for each stage of the MBT process (essentially, we produced the artifact subjects were expected to complete, and if they did not finish it, we gave them ours so that they could continue the process), and (*iv*) did not allow model comparisons among subjects.

Creating a correct and complete the model is a crucial step in the MBT process, since it directly affects the quality of the generated test suite. The subjects reported modeling the SUT as the hardest stage of the MBT process. They recommended doing more research on MBT tools (to hopefully find better tools that facilitate model creation), and to delegate the creation of the model to other roles (instead of software developers) that are more familiar with modeling.

In summary, the two benefits of MBT reported by the subjects were: that the model itself is an easy way to visualize and understand the SUT, and that the MBT process could generate test cases faster than with the manual process. On the down side, the main problems indicated by the subjects were related to the modeling stage and the MBT tool. Particularly, they pointed out the following shortcomings of MISTA: it does not use the latest version of Selenium, MIM is not user-friendly, and MISTA had a poor user experience. The lesson learned here is that the choice of MBT tool is of utmost importance and can have a large impact on the results of the study.

5   Conclusions and Future Work

This paper reports on an industrial experience where a group of software engineers used model-based testing in a software project, with the aim of evaluating its efficiency, efficacy and acceptance. In this study, we used MISTA as the MBT tool, which allows the design of a model using high level petri nets, as well as the generation of test cases and their concretization with Selenium. A web application recently developed by the company was selected as the system under test. There were six participants in the study, who were software engineers from different departments within the company. They first received training on MBT and MISTA, and then carried out the entire MBT process over the SUT, from model design to test case execution. Metrics were collected during this process to assess the efficiency and efficacy of MBT as a testing approach, and its acceptance by the participants as a plausible technique to use in the near future.

Our results suggest that MBT is an efficient technique given that both average modeling time and average concretization time were short: 34 an 31 min, respectively. However, we noted that most subjetcs only modeled the 'happy path' scenarios, forgetting to include validation of data entry. In fact, the only subject who included data entry validation in his model took the longest time to model (60 min). This means that our results could be underestimating the modeling time, and that the modeling time depends on the scope of what is being modeled. Also, modeling time seems to be affected by subject's prior knowledge of the system being modeled, since the subject who finished modeling faster was the one with knowledge of the SUT. With respect to the concretization time, we found that there was not much variance between subjects, and this could be attributed to the fact that concretizing test cases is simple mechanical work that does not require a lot of effort, and therefore should not be significantly affected by the subject's experience.

The results also suggest that MBT is relatively effective given its average coverage of 49.8% with just 4.3 test cases (on average), for an average number of 4.7 (out of 5) requirements modeled. The relatively low coverage percentage attained (between 39% to 59%) could be explained by the fact that many requirements were incorrectly modeled or concretized. Furthermore, the quantity of test cases and the attained coverage do not directly correspond to the quality of the model generated.

Regarding the acceptance of MTB by the participants of the study, our results show that participants had a positive perception of the usefulness of MBT. Their perception of 'ease of use' was barely positive, with some subjects expressing concerns about the limitations of MISTA, including its poor user experience and outdatedness. Lastly, the subjects perception of 'intention of use' suggest that while they were interested in using MBT in the future, they wanted to search and compare other MBT tools, since MISTA did not meet their industry needs.

Given that some subjects considered modeling the SUT as the most difficult part of the MBT process, and some did not have much experience in modeling, we think it would be interesting, as future work, to replicate this study using subjects from different roles such as quality or test engineers, or even business analysts, who might have more developed modeling skills and a broader knowledge of the SUT.

References

1. Utting, M., Pretschner, A., Legeard, B.: A taxonomy of model-based testing approaches. Softw. Test. Verification Reliab. **22**, 297–312 (2012)
2. Biswas, S., Kaiser, M.S., Mamun, S.: Applying ant colony optimization in software testing to generate prioritized optimal path and test data. In: 2015 International Conference on Electrical Engineering and Information Communication Technology (ICEEICT), pp. 1–6 (2015)
3. Utting, M., Legeard, B.: Practical Model-Based Testing: A Tools Approach (2006)
4. Villalobos-Arias, L., Quesada-López, C., Martinez, A.: A tertiary study on model-based testing areas, tools and challenges: preliminary results (2018)
5. Weißleder, S., Schlingloff, H.: An evaluation of model-based testing in embedded applications. In: IEEE International Conference on Software Testing, Verification, and Validation (2014)
6. Ergun, B., Gebizli, C., Sozer, H.: Format: a tool for adapting test models based on feature models. In: IEEE 41st Annual Computer Software and Applications Conference (2017)
7. Herpel, H.J., et al.: Model based testing of satellite on-board software — an industrial use case. In: Aerospace Conference (2016)
8. Gebizli, C.S., Sözer, H.: Automated refinement of models for model-based testing using exploratory testing. Softw. Qual. J. **25**(3), 979–1005 (2017)
9. Schulze, C., et al.: Model generation to support model-based testing applied on the NASA DAT web-application - an experience report (2015)
10. Bettiga, D., Lamberti, L.: Exploring the adoption process of personal technologies: a cognitive-affective approach. J. High Technol. Manag. Res. **28**(2), 179–187 (2017)
11. Davis, F.D.: Perceived usefulness, perceived ease of use, and user acceptance of information technology. MIS Q. **13**, 319–340 (1989)
12. Wohlin, C., et al.: Experimentation in Software Engineering (2012)
13. Runeson, P., et al.: Case Study Research in Software Engineering: Guidelines and Examples (2012)
14. Xu, D.: A tool for automated test code generation from high-level petri nets (2011)
15. SeleniumHQ: SeleniumHQ. https://www.seleniumhq.org. Accessed 07 Mar 2018
16. Masri, W., Zaraket, F.A.: Coverage-based software testing: beyond basic test requirements. In: Advances in Computers, chap. 4, pp. 79–142 (2016)

Software Systems, Architectures, Applications and Tools

Classification of Software Defects Triggers: A Case Study of School Resource Management System

Nico Hillah[(⊠)]

DESI, University of Lausanne, Lausanne, Switzerland
nico.hillah@unil.ch

Abstract. In this work, we identify trigger factors of software defects that are responsible for severe defects. We conducted a case study on a system by classifying 842 defects according to their trigger factors and then identified the level of severity they have on this system. Knowing these types of triggers helps software maintenance teams improving the management of software defects by reducing the cost of maintaining the system, consequently the cost of software projects.

Keywords: Software defect triggers · Software severity · Defects classification

1 Introduction

IEEE standard 1044-2009 [1] defines Software defects (SDs) as *"An imperfection or deficiency in a work product where that work product does not meet its requirements or specifications and needs to be either repaired or replaced."* [1].

The classification of SDs helps the maintenance teams reducing the cost of correcting software bugs, detecting defective modules, and having efficient resource planning. Various studies have proposed and evaluated different approaches to collect and to analyze these SDs [1–4]. Other studies target the source of these defects by providing schemas and frameworks to help identifying these sources [5–8]. For our project, we have retained the EVOLIS framework [7] to identify the trigger factors that are at the source of the SDs.

To be able to know which factors among these trigger factors have a more severe impact on the system, we conducted a case study on a school resources management system. In fact, we studied the SDs of this system by identifying their trigger factors based on the EVOLIS framework [7] and then identified their severity weight on the system. The question we address in this paper is *"how to identify problematic SDs trigger groups using their severity weight?"*.

The paper will proceed as follows; first, we will define the software defect and its classification approaches. Second, we will present the case study, the classification results, and our contribution.

© Springer Nature Switzerland AG 2019
Á. Rocha et al. (Eds.): ICITS 2019, AISC 918, pp. 307–316, 2019.
https://doi.org/10.1007/978-3-030-11890-7_30

2 Related Works

2.1 The EVOLIS Framework

For our first classification project, we chose the EVOLIS framework [7]. This framework proposes a technique to classify SDs according to the factors that trigger them. "EVOLIS can be caused by a large variety of factors: bugs that need to be fixed, users that wish to have new functionalities, new market opportunities that require new software features, performance standards that the system must reach, technical changes in the environment with which the system must interact, obsolescence of applications and so on" [7]. EVOLIS identifies four main groups of factors that trigger SDs: (1) IS/users fit triggers (U.F) that are defined as any defect related to the user interface, the user documentation, and aptitude to use the system. (2) The technology triggers (TCH) are related to defects that concern the software as well as the hardware platforms as information system components. (3) According to the authors, the IS architecture triggers (ACH) concern *"different types of integration evolution, namely an evolution of integration among components of the system, among business functionalities, or an integration with systems outside of the company."* [7], and finally (4) the business-IS alignment triggers (B.IS) that *"address the co-alignment between business and information systems"* [7].

2.2 Software Defects

Previous research studies have proposed different approaches and schemas to classify SDs: the best-known schemas are (1) The Orthogonal Defect Classification (ODC) of IBM [6], the root cause analysis [5], (2) the HP Defect origins, types, and modes [8], and standards like the IEEE standard 1044-2009 [1]. In the same context, they also apply to data mining methods such as the Naïve Bayes Model [9], Clustering [10] or the regression model [11] to classify SDs. IEEE standard 1044-2009 also proposed a SDs classification approach based on their severity, priority, and origins. The IEEE standard 1044-2009 is the classification approach we retain for our second classification project. In fact, our objective is to classify SDs according to their trigger factors and their severity [1].

The approach of IEEE standard 1044-2009 proposes a simple and complete definition of the SDs severity types. Moreover, this severity attribute is one of the most used attributes in SDs classification in practice [12]. Our second SDs classification project is based on this attribute. The main advantage of choosing the severity attribute is the possibility for managers to identify which defects to correct first [12]. The IEEE's standard defines this attribute as *"The highest failure impact that the defect could (or did) cause, as determined by (from the perspective of) the organization responsible for software engineering."* [1]. There are five values of severity. They are classified by the most significant to the least significant ones in terms of the impact they have on the system (see Table 1).

Table 1. Severity values [1]

Attribute	Value	Definition
Severity	Blocking	Testing is inhibited or suspended pending correction or identification of suitable workaround
	Critical	Essential operations are unavoidably disrupted, safety is jeopardized, and security is compromised
	Major	Essential operations are affected but can proceed
	Minor	Nonessential operations are disrupted
	Inconsequential	No significant impact on operations

3 Methodology

3.1 Case Presentation

In order to identify the trigger factors that generate defects with the highest severity impact on the system, we conducted a case study of a school resources management system that we will name system B. The software development method used to develop system B is the scrum agile method [13]. A government institution owns it. The system is used for the human and material resources management of public schools. We classified 842 SDs of system B. The collection of SDs covers a period of 23 months from June 2014 to May 2016. The bug repository tool used by this organization is Jira [14]. The first version of the system B had been deployed at the beginning of 2013.

Each of these defects has the following information: the identity of the failure reporter, the date of reporting and solving the software defect (SD). In addition, each SD contains its description, the person who reported the case and the person who treated it. We only take into consideration the description characteristic in classifying these SDs.

3.2 Classification Method

Overall, we did three main classifications. First, we analyzed the SDs of system B by classifying them with the EVOLIS framework [7] (see Table 2). Second, we classified the same SDs according to the defect severity attribute of IEEE 1044-2009 standards [1] (see Table 3). Third, we combined both classifications. In detail, our method of classifying these SDs consists of four main steps:

Table 2. Classification of System B's SDs based on their trigger factors (EVOLIS)

Years	IS architecture	Business-IS alignment	Technology	IS/users fit	Total
2014	61	11	118	57	247
2015	87	22	251	90	450
2016	23	10	78	34	145
Total	171	43	447	181	842

Table 3. Classification of System B's SDs based on their severity

Year	Blocking	Critical	Major	Minor	Inconsequential	Total
2014	31	57	71	85	3	247
2015	57	112	135	145	1	450
2016	21	19	33	71	1	145
Total	109	188	239	301	5	842

In the first step, we collected SDs of system B from the Jira repository [14].

In the second step, we took each SD and identified its trigger factor or source based on its description. At this stage, we used the EVOLIS framework. We named this step "EVOLIS Classification".

In the third step (Severity Classification), we took again each SD and evaluated its severity impact (cost) on the system. This classification is done based on our severity-weighting model (see Table 4).

In the final step, we took each EVOLIS-Severity couple and ranked them according to the level of damage they may have on system operations (EVOLIS-severity classification).

4   Discussion and Contribution

4.1   Discussion

In this section, we analyzed the classification results threefold: the results of EVOLIS classification, followed by the results of severity classification, and finally, we combined and analyzed both results together.

First, the EVOLIS classification showed us that the top three groups of factors that trigger SDs are respectively the technology factors with 447 SDs, followed by the IS/Users factors with 181 SDs, and then the factors related to the IS architecture with 171 SDs. In the last position is the business-IS alignment with 43 SDs (see Table 2).

Second, the severity classification showed us that the top three high types of SDs are respectively the Minor type of SDs with 301 SDs, followed by the Major type with 239, and the Critical type with 188 SDs. The Blocking types are fourth with 109 SDs, and, in the last position, we find the Inconsequential SDs type with only five SDs (see Table 3).

We combined the two results in order to identify the groups of triggering factors that cause severe SDs (see Fig. 1). Doing so, we realized that limiting the results only to the number of SDs for each severity level group raises an ambiguity. In fact, counting only the number of SDs per trigger factor group does not give us the clear response on which trigger factors are responsible for severe SDs. E.g., how can we determine if five Blocking SDs have affected a system more than eight Critical SDs? In order to clear this ambiguity, we have associated a weighting factor to each level of severity according to their impact on the system (see Table 4).

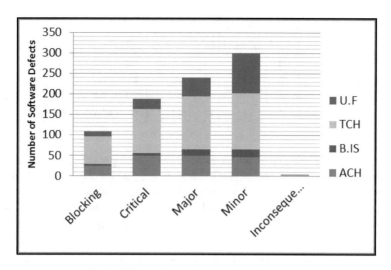

Fig. 1. Trigger factors by severity of system B

Table 4. The weighting factors for the severity level.

Severity level	Weighting factor
Blocking	40%
Critical	30%
Major	20%
Minor	8%
Inconsequential	2%

In addition, based on their definition, we separated the severity level into two groups: the first group we called "severe SDs" and the second group we named "no severe SDs". The severe SDs are any SDs that have an impact preventing the system to be operational. This group of SDs usually causes financial loss or any considerable resource loss to the system owner and to the system users. They are Blocking, Critical, and Major severity types.

The "no severe SDs" are any SDs that have an impact level that do not affect the system's operation: they are Minor and Inconsequential severity type. Thus, for the purpose of this study, we only considered the first group of severity level to be authentic severe SDs.

We then calculated the weighted score (W) for each trigger factor group based on the severity weight (see Table 5).

Table 5. Severe SDs weighted score of system B

	Weight	ACH	W-ACH	B.IS	W-B.IS	TCH	W-TCH	U.F	W-U.F
Blocking	0.4	25	*10*	4	*1.6*	68	*27.2*	12	*4.8*
Critical	0.3	49	*14.7*	6	*1.8*	109	*32.7*	24	*7.2*
Major	0.2	50	*10*	15	*3*	129	*25.8*	45	*9*
Total	0.9	124	*34.7*	25	*6.4*	306	*85.7*	81	*21*

Integrating both, results allow us to identify SD triggers that are causing more severe impact to the system. These results show that the technology trigger factors with the highest weighted score of 85.7, are responsible for most of the severe SDs followed by the IS architecture factors, with a weighted score of 34.7. Then come the IS/users fit triggers with a 21, and finally the business-IS alignment, with 6.4 (see Fig. 2).

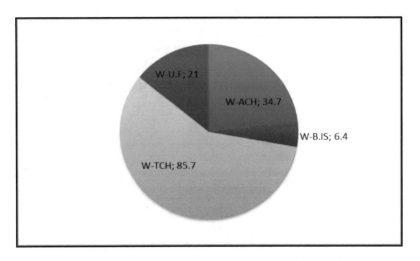

Fig. 2. Trigger factors by weighted severity of system B

4.2 Similar Case Study: Classifications of SDs of System A

Similarly, in our previous publication we conducted the same study on another system we named system A [15]. In this section, we will present this second case and compare its results to the one of system B.

Table 6. Severe weighted score of system A

Severe SD	Weight	B.IS	W-B.IS	ACH	W-ARC	TCH	W-TCH	U.F	W-U.F
Blocking	0.4	24	9.6	17	6.8	19	7.6	15	6
Critical	0.3	21	6.3	12	3.6	30	9	21	6.3
Major	0.2	66	13.2	99	19.8	87	17.4	70	14
Total	0.9	111	**29.1**	128	**30.2**	136	**34**	106	**26.3**

System A is a school management system and it belongs to an educational institute. Its purpose is to help schools in managing the grades of their students. More than 1500 teachers use this system for managing more than 90000 student grades. The first version of the system A had been released mid-2012. We classified in total 665 SDs of this system. The collection of SDs covered a period of 16 months, from January 2015 to April 2016. System A has released nine versions over this period [15]. The integrated classifications' results we obtained from this study are as follows (see Table 6):

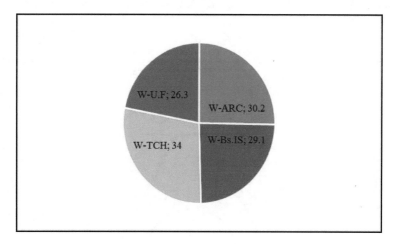

Fig. 3. Trigger factors by weighted severity of system A

The results of system A showed that the technology trigger factors, with the highest weighted score 34, are responsible for most of the severe SDs followed by the IS architecture factors, with a weighted score of 30.2. Then come the business-IS alignment, with a 29.1, and, finally the IS/users fit triggers, with 26.3 (see Fig. 3). We can also notice that there is a considerable gap between the number of SDs of the first two groups of triggers (IS architecture and technology) and the last two of them (business-IS alignment and IS/users fit). In the next section, we will present our contributions.

4.3   Contribution

The contribution of this paper is twofold:

First, to the question, *"how to identify problematic SDs trigger groups using their severity weight?"* we proposed our 4-steps method.

In order to make possible for other practitioners and other researchers to conduct and possibly observe similar results on their systems we summarized our method as follows:

- Step 1. Data collection: this step consists of collecting SDs of the system to study.
- Step 2. Identification of each SD triggering factor: in this step, we classify the SDs based on the EVOLIS framework in order to identify their trigger factors Our (EVOLIS classification).

- Step 3. Weighting of the severity level of each SD on the system. Here we classify the same SDs based on the severity attribute of IEEE standards (Severity classification).
- Step 4. Integrate results of steps 2 and 3: At this level, we classified the SDs based on both EVOLIS and IEEE 1044-2009 severity attribute in order to identify SDs having high severe impacts on the studied systems. This step is our EVOLIS and Severity classification.

We summarized these steps in Fig. 4. We must also point out that step 2 and 3 are interchangeable.

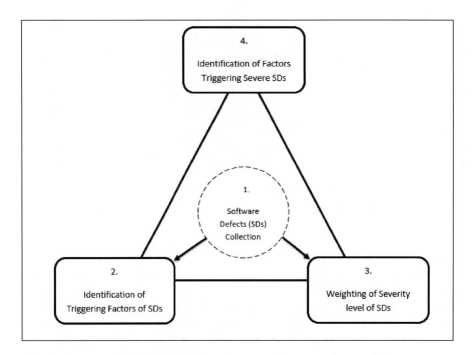

Fig. 4. 4-steps method to identify trigger factors causing most of severe SDs to a system

Second, to the question of which SD factors trigger most of severe SDs, we found that the following:

In the leading position are the technology trigger factors with 29% of the total weighted score for system A and 58% for system B. They are followed by the architecture trigger factors with 25% (30.8 weighted score) for system A and 24% for system B. We can see that in both cases, the same type of SD trigger factors occupies the first and the second position. In contrary, the third position is occupied by the business-IS alignment with 24% for system A while the IS/users fit occupied the same rank with 14% for system B. Finally, the business-IS alignment with 4% occupied the last place for system B, and IS/user fit with 22% occupied this position for system A.

5   Conclusion

With this case study, we have shown that severe SDs are mostly caused by technology and architecture trigger factors. We did so by classifying SDs according to four groups of trigger factors and matched them with the SDs severity level. The obtained results show us that the technology triggers are leading with 85.7 weighted scores. In the second position are the IS architecture triggers which come with 34.7 weighted scores, and are then followed by the IS/users fit triggers with 21. The last position is occupied by the business-IS alignment triggers with 6.4 weighted scores of the total SDs analyzed. As contribution, we also presented a method in order to identify problematic SDs trigger groups using their severity weight.

The results obtained from this study will help software teams to reallocate the resources of maintaining systems and help them to prioritize certain categories of SDs, thus reduce the cost of maintaining software systems.

In our future work, we will analyze other software systems and compare their results to the ones we obtained in this study. We will also investigate if there is any correlation between the first couple of trigger factor groups (IS architecture and technology) and the last couple of trigger factor groups (business-IS alignment and IS/users fit).

References

1. 1044-2009 IEEE Standard Classification for Software Anomalies (2009)
2. Grady, R.B.: Software failure analysis for high-return process improvement decisions. Hewlett Packard J. **47**, 15–24 (1996)
3. Grottke, M., Trivedi, K.S.: A Classification of software faults. J. Reliab. Eng. Assoc. Jpn. **27** (7), 425–438 (2005)
4. Duda, R.O., Hart, P.E., Stork, D.G.: Pattern Classification. Wiley, Hoboken (2000)
5. Leszak, M., Perry, D.E., Stoll, D.: Classification and evaluation of defects in a project retrospective. J. Syst. Softw. **61**, 173–187 (2002)
6. Chillarege, R., et al.: Orthogonal defect classification-a concept for in-process measurements. IEEE Trans. Softw. Eng. **18**(11), 943–956 (1992)
7. Métrailler, A., Estier, T.: EVOLIS framework: a method to study information systems evolution records. In: 2014 47th Hawaii International Conference on System Sciences (HICSS), pp. 3798–3807 (2014)
8. Huber, J.T.: A Comparison of IBM's Orthogonal Defect Classification to Hewlett Packard's Defect Origins, Types, and Modes. Hewlett Packard Company (1999)
9. Murphy, G., Cubranic, D.: Automatic bug triage using text categorization. In: Proceedings of the Sixteenth International Conference on Software Engineering & Knowledge Engineering (2004)
10. Dickinson, W., Leon, D., Fodgurski, A.: Finding failures by cluster analysis of execution profiles. In: Proceedings of the 23rd International Conference on Software Engineering, ICSE 2001, Toronto, Ontario, Canada, pp. 339–348 (2001)
11. Rajbahadur, G.K., Wang, S., Kamei, Y., Hassan, A.E.: The impact of using regression models to build defect classifiers. In: Proceedings of the 14th International Conference on Mining Software Repositories, pp. 135–145 (2017)

12. Wagner, S.: Defect classification and defect types revisited. In: Proceedings of the 2008 Workshop on Defects in Large Software Systems, pp. 39–40 (2008)
13. Hossain, E., Babar, M.A., Paik, H.: Using Scrum in Global Software Development: A Systematic Literature Review, pp. 175–184 (2009)
14. Atlassian: Jira | Logiciel de suivi des tickets et des projets. Atlassian. https://fr.atlassian.com/software/jira. Accessed 06 Apr 2018
15. Hillah, N.: Severe software defects trigger factors: a case study of a school management system. In: Antipova, T., Rocha, A. (eds.) Digital Science, vol. 850, pp. 389–396. Springer, Cham (2019)

Functional Requirement Management Automation and the Impact on Software Projects: Case Study in Ecuador

Cathy Pamela Guevara-Vega[1,2(✉)], Eric Daniel Guzmán-Chamorro[5],
Vicente Alexander Guevara-Vega[1,2],
Andrea Verenice Basantes Andrade[3,4],
and José Antonio Quiña-Mera[1,2]

[1] Applied Sciences Engineering Faculty,
Universidad Técnica del Norte, Ibarra, Ecuador
{cguevara, alexguevara, aquina}@utn.edu.ec
[2] Software Engineering, Universidad de las Fuerzas Armadas - ESPE,
Latacunga, Ecuador
[3] Education Science and Technology Faculty,
Universidad Técnica del Norte, Ibarra, Ecuador
avbasantes@utn.edu.ec
[4] Technology for Management and Teacher Practice,
Pontificia Universidad Católica del Ecuador, Ibarra, Ecuador
[5] Posgrado, Universidad Técnica del Norte, Ibarra, Ecuador
edguzmanc@utn.edu.ec

Abstract. This study intends to show the importance to improve and automate the functional requirement management through the implementation of a system and measure out the impact of it in software projects. The goal in some cases is to release the product in less time possible reducing the schedule in the main phase like the functional requirement elicitation. There is a blended approach quantitative and qualitative, exploratory, documentary work as well as field work. In this study some performance indicators were applied to measure the impact of the established system had, and it was based on the equations suggested by authors Uwizeyemungu & Raymond about how important is to apply software management of functional requirements. The system's impact measurement outcome was 4.10 over a scale of 5, a moderately significant outcome interpreted by 82% of user's acceptability percent value.

Keywords: Functional requirements · Requirement management · Software Engineering

1 Introduction

Worldwide software development supports and promotes the improvement of industrial, educational, cultural, social processes in addition to public health, food autonomy, the arts, and historical heritage among other mechanisms. Therefore, software development is immersed into the global productive matrix transformation in synergy with

Á. Rocha et al. (Eds.): ICITS 2019, AISC 918, pp. 317–324, 2019.
https://doi.org/10.1007/978-3-030-11890-7_31

(SDO) Sustainable Development Objectives that transform the world and the mission of United Nations Educational Scientific and Cultural Organization (UNESCO): "To build peace in the world's men and women's minds through the fulfilment of certain sustainable development agendas up until the year 2030 in the areas of Education, Natural Sciences, General Culture, Communication and Information". In this context, adding the contribution of science, technology and innovation, will make it possible to meet actual humanity needs [1].

Process automation through building software in Ecuador is fundamental for the country's economic. The Planning and Development National Commission (Senplades) in Ecuador in its 2017–2021 Development Planning project known as "A Whole Life", through the 5.6 regulation aims to promote research, knowledge, training, development, technological transferring, innovation, social venture and intellectual property protection to stimulate a change in the productive matrix by creating an association between the public and productive sectors and universities [2].

Organizations involved in software production have extensive work teams whose intended roles are to fulfil each and every software development's working cycle phase in contrast to other organizations which claim to develop software as their business core however, they just create such processes internally inside their data technology departments.

This kind of organizations does not deal with this type of requirement management adequately since they have limited personnel thus one person can do the role of computer programmer, designer and even the role of test user.

Femmer mentions that understanding quality measuring continues to be a significant problem [3], that causes sweeping changes within systems as these they are at a test phase or an implementation phase. Generally, such changes are the result of undefined or incomplete requirements even by the disregarded existence of requirements clearly by a client. An unbeneficial factor that must be considered is that often requirements obtained by a client have been established without the expertise of a specialized software development technical team [4].

Bernárdez mentions in her study called "An empirical approximation to a metric-based heuristics implementation for requirements for requirement validation", that system failure is distributed in 56% of the engineering requirement phase, 27% in the design phase 7% in codification and finally 10% in the remaining software development life cycle phases [5]. Such requirements should be precise, complete and consistent categorized by relevance, verifiability, alterability as well as traceability in order to help developers build software in the most optimum way [4].

López Echeverry, Cabrera, Valencia Ayala show that quality control software development is executed by the programmer [6], there is no quality assurance team to handle this aspect thus the chances of getting a quality system are highly limited. The Institute of Electrical and Electronics Engineers (IEEE) establishes standards that consist of good practice recommendations for software development, highlighting the following requirement standards: IEEE 830, ISO/IEC/IEEE 29148, PMBOK and SWEBOK [7]. Such standards particularly help stakeholders to develop software that meets a user's functionality and usability needs.

The IEEE 830 standard specifies recommended requirement specification software approach (ERS) as well as content and features a good ERS must have therefore, it is used as the basis for requirement traceability improvement [8].

The ISO/IEC/IEEE 29148 standard determines processes to be implemented at the requirement engineering systems and products software including services provided throughout its life cycle, consequently setting standards that may be applied to the requirements processes outlined in ISO/IEC 12207 apart from specifying data elements required in regards of the requirement establishing process [9].

The 54 years old Financial Cooperative Atuntaqui was this case study's financial institution placed at the Forefront of Ecuador's Savings and Credit Union System, it has nine offices located at the Northern Provinces of Imbabura and Pichincha made up by 85000 active business partners and 180 employees. At the initial phase of the software requirement management that took place at the institution's IT Systems Department, it was observed that the client requested and registered the information by using the requirement application form without the software development technical team's assistance, as a result the planning and timely software delivery issues in addition for creating a large number of changes to the initial requirements throughout the software's development life cycle.

The aim of this research work was to measure the impact that the implementation of a functional requirement management system has on software project development so that IT professionals and developers consider the importance the analysis and requirement phase has since maintenance cost and the fulfilling of clients' needs relies heavily on this phase.

2 Methods and Materials

Firs let's keep in mind that this research is comprised by a qualitative/quantitative, exploratory, documentary and field work blended approach, so the documentary investigation covered the first phase and data gathering was carried out starting on May 2016. Several indexed data bases such as Scopus, SCimago, Ebsco and ProQuest were accessed virtually from the North Technical University (UTN University's library) accessing Google Scholar which provided 31 references. The study was conducted at the IT Systems Department at the Savings and Credit Union's home office in the city of Atuntaqui in Ecuador. The target study group was composed by seven department managers in these areas: IT Systems, Operations, Finance, Securities, Business, Human Talent and Risk Management in addition to three software development team members, totaling ten employees. Three surveys with nine questions each one was applied to the target study group. These surveys were evaluated by three Software Engineering experts from North Technical University who also assessed the objective's relation to the drafting and understanding of each question. The first survey dealt with the lifting of software a requirement process, whereas the second survey identified the preferred standard for the lifting of software requirements. The third survey measured the level of satisfaction by the proposal put forth by this study. Furthermore, a 10-question interview also validated by two experts from North Technical University was conducted to the Systems Department Head, to the development administrator and two programmers totaling four a people group. Next, the interview determined issues that may have arisen during software development processes as well as cause and effect mechanisms including system flaws furthermore, likely improvement procedures were established that shaped a functional requirements management. Both interviews and surveys were conducted personally. Data analysis and processing was made using Microsoft Excel 2016 including a social research statistical package SPSS. Subsequently, the variables crossing process determined the input for the proposal's development process.

The following steps were taken to measure the level of impact the proposal implementation had, see Fig. 1:

Fig. 1. Steps to measure the impact of the implementation of the functional requirements management system of the institution.

Step 1. Identify the initial software functional requirement management process.

Step 2. Five software projects were selected and considered based on the large number of changes made to the first requirements showing 60.49% alterations as seen in Table 1, confirming cause and effect processes prompted by delays in deliveries in software projects.

Table 1. Changes to the initial requirements for the software project.

Name	Requirement total	Alterations to initial requirements	Percentage
Project A	22	15	6.18%
Project B	18	9	50%
Project C	25	12	48%
Project D	10	9	90%
Project E	6	4	66.66%
TOTAL	**81**	**49**	**60.49%**

Step 3. A process improvement proposal for the functional requirement software management process was introduced through the formulation of a conceptual baseline based on the IEEE 830 and ISO/IEC/IEEE 29148 standards. The interviews and surveys applied to the study target group defined the main cause for the large number of changes to the eliciting of initial requirements without the assistance of an IT expert team hence, the importance of improving the requirement management process by including the creation of an expert team. See Fig. 2.

Step 4. A functional requirement system process is implemented based on the established proposal.

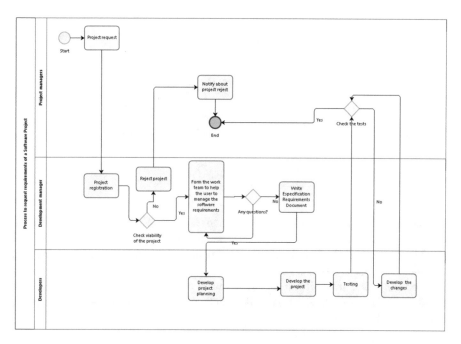

Fig. 2. Software management requirement project for the improvement process Financial Cooperative Atuntaqui.

Step 5. An impact measurement of the technological solution implementation was made through the formulas suggested by the following authors: Uwizeyemungu & Raymond [10]. See Eqs. 1 and 2.

Equation 1. Measurement involving the importance of the technological solution's implementation.

$$Impact\ of\ Effects = \sum_{i=1}^{n}(a_i * b_i) \tag{1}$$

Variable a corresponds to the technological implementation importance indicator within the institution's functional requirement management process, measured by the following performance indicators: (1) not important (2) somehow important (3) moderately important (4) and (5) very important.

Variable b corresponds to the functionality scope achieved by the technological implementation phase during execution time at a maximum value of 3 in the value scale, representing (0) none (1) weak (2) moderate and (3) strong.

Variable B corresponds to the highest value of b in order to reach the best software functionality.

Equation 2. Measuring the maximum range of importance of the implementation of the technological solution.

$$Impact \ of \ Variationss = \sum_{i=1}^{n}(a_i * B_i) \tag{2}$$

Equation 3. Technological implementation's global impact.

$$Global \ Impact = 5 * \frac{Impact \ of \ Effects}{Impact \ of \ Variations} \tag{3}$$

3   Results

To establish a formal specification requirement document that would be used in the study's proposal, two lifting requirement standards were implemented, the IEEE 830 and the ISO/IEC/IEEE 29148. They were evaluated by four IT experts from the institution through six indicators focused on precision, user experience, amount of effort, cost estimate and software finally software requirement validation. It was found indeed the IEEE 830 standard was adequately adapted to the institution's working environment. See Table 2.

Table 2. Feature comparison between IEEE-830 y ISO/IEC/IEEE-29148

Indicators	IEEE 830-1998	ISO/IEC/IEEE 29148-2011
Assists software clients to express their needs more accurately	Yes	Yes
Assists developers understand what clients want in a more precise way	Yes	Yes
Reduces the amount of effort for the development of the ERS document	Yes	No
Provides an assessment base for real cost estimates	Yes	No
Provides basis for requirement validation	Yes	Yes

Based on Table 1, on the survey's one outcome and on the interview reaction, it was resolved to create a team that specializes in software requirement eliciting so it works together with a final user.

For the optimization and effectiveness of the requirement management process the automation of this procedure was done through the implementation of a web system to a SCRUM methodology having a development language PHP, using a MySQL data base over a client-server architecture. Such system manages and generates the requirement specification document relying on the IEEE 830 standard apart from performing periodic project progress reviews through several graphic reports.

To measure the impact the system's implementation had on the requirement management process, Eqs. 1, 2 and 3 suggested by the authors Uwizeyemungu and

Raymond [10] were applied. The equations estimate was calculated by values assigned to performance indicators which in turn, were defined by the IT personnel interviews outcome thus the level of impact was finally determined. See Table 3.

Table 3. Equations 1, 2 outcomes/standardized impact level and over five-point scale.

Formula	Outcome
(1) $\sum_{i=1}^{n} (a_i * b_i)$	182
(2) $\sum_{i=1}^{n} (a_i * B_i)$	222
(1)/(2)	0.82
Result based in five points	**4.10**

Impact outcome is 4.10 over 5 which, expressed in percentages it comes to 82% resulting the acceptance for a new requirement management system suitable for the institution. See Table 3.

4 Discussion

The main objective of this study is to verify the impact the optimization and automation of functional requirement management processes have on software development

Those involved in this Research were seven institution department managers apart from three software development area members. The remaining personnel was not included because they do not require software development from the IT department. Should these employees need a task performed through this department, they certainly obtain it through their superiors.

From the first surveys outcome, 99% of respondents were not satisfied with the process used to inquire software development requirements since department heads made these practices without a team of experts' advice resulting in an obsolete document for data entering.

After searching similar tasks performed by diverse data bases like Springer, a few articles were found with the "software requirements management" search link having a degree of similarity to a completed project by Heikkilä, Paasivaara, Lasssenius, Damian and Engblom titled "Managing requirements flow from strategy to release in large-scale agile development: a case study at Ericsson". This study deals with defining more efficient processes for requirement management resulting in reduced development time, higher flexibility levels, planning efficacy, a highly motivated software developer, in addition to better communication processes [11].

In light of the development of software development tasks alike, it may be concluded that the most satisfactory requirement management not only helps the development team but also helps to improve the institution's image and credibility for services rendered to its clients [12].

Among the difficulties that were encountered in carrying out the work, was the fear to technology changes, in the same way, some inconveniences were the privacy of internal processes and several failures of collaboration on the part of some members of the staff inside the institution.

5 Conclusion

The requirement analysis phase is essential for the software's development life cycle emphasizing improvements to software development processes promoting the prevention of sudden setbacks as well as negative effects to the company's productivity.

The use and application of IT tools, the use of standardized documents and having trained IT personnel on requirement management issues may add quality value to a company's software development processes.

Once the value obtained from the prototype impact is defined together with the kind of use provided and the subsequent addition of components are likely to increase acceptance percentages in the near future.

References

1. UNESCO: United Nations Educational, Scientific and Cultural organization UNESCO and Sustainable Development Goals (2017). https://en.unesco.org/sdgs
2. SENPLADES: The National Secretary of Planning and Development (Secretaría Nacional de Planificación y Desarrollo) Ecuador (2017)
3. Femmer, H.: Reviewing natural language requirements with requirements smells – a research proposal, pp. 1–8. Research Gate (2013)
4. Safwat, A., Senousy, M.B.: Addressing challenges of ultra large-scale system on requirements engineering. Procedia Comput. Sci. **65**, 442–450 (2015)
5. Bernárdez Jiménez, B.: An empirical approach to the development of heuristics based on metrics for verification of requirements (Una Aproximación Empírica al Desarrollo de Heurísticas basadas en Métricas para Verificación de Requisitos), Sevilla, pp. 5–9, 49–62 (2004)
6. López Echeverry, A.M., Cabrera, C., Valencia Ayala, L.E.: Quality software introduction (Introducción a la calidad del software), p. 328. Scientia et Technica (2008)
7. Bourque, P., Fairley, R.: SWEBOK v3.0, New Jersey. IEEE Computer Society (2014)
8. IEEE 830: IEEE recommended practice for software requirements specifications, New York. IEEE (1998)
9. ISO/IEC/IEEE 29148: Systems and software engineering — life cycle processes — requirements engineering. IEEE, New York (2011)
10. Uwizeyemungu, S., Raymond, L.: Impact of an ERP system's capabilities upon the realisation of its business value: a resource-based perspective, pp. 73–87. Springer (2012)
11. Heikkilä, V., Paasivaara, M., Lasssenius, C., Damian, D., Engblom, Ch.: Managing the requirements flow from strategy to release in large-scale agile development: a case study at Ericsson. Empir. Softw. Eng. **22**, 2892–2936 (2017). https://doi.org/10.1007/s10664-016-9491-z. 10 January 2017, pp. 442–450 (2015)
12. Arpinen, T., Hämäläinen, T.D., Hännikäinen, M.: Meta-model and UML profile for requirements management of software and embedded systems. EURASIP J. Embed. Syst. **2011**, 14 (2011). https://doi.org/10.1155/2011/592168. Article ID 592168. Hindawi Publishing Corporation

RGAM: An Architecture-Based Approach to Self-management

Daniela Micucci[✉], Marco Mobilio, and Francesco Tisato

DISCo, University of Milano - Bicocca, Milan, Italy
{daniela.micucci,marco.mobilio,francesco.tisato}@unimib.it

Abstract. Software systems must self-adapt to changes in their execution environment and in the user requirements and usages. An architectural solution enables tackling more effectively problems related to self-adaptation. This paper presents an architecture-based approach to self-adaptation that relies on architectural reflection and on closed-loop principles. The approach has been developed and tested on a real scenario related to the management of emergencies in case of floods.

Keywords: Self-adaptation · Self-management ·
Architectural reflection

1 Introduction

Systems are more and more complex, large in size, and are required to adapt to changes [17]. Smart environments [2,3,18], cloud systems [4,15], and mobile applications [12–14], are a few examples of systems that need to adapt to changes in the user needs and the environments in which they are running.

Self-management has been recognized as a solution to build systems that are scalable, that support dynamic configurations, and that are flexible to the changes [10]. To achieve self-adaptive systems in a cost-effective way, a separation of concerns is required between the system and the aspects related to self-adaptation [5]. An architectural approach may support run-time software adaptation as it provides the adequate level of abstraction to face the challenges that self-adaptation poses [10]. Moreover, reflection may help in realizing a separation of concerns between system building ("base-level programming") and system configuration and adaptation ("meta-level programming") [6], thus it can be considered an enabling technology to self-adaptation [1].

Closed-loop mechanisms underly several approaches to self-adaption [16], as they adjust the target system to changes during its execution. These changes may stem from the software (e.g., failure) or the context (e.g., the user changes the device). Closed loops should constitute an architectural solution to self-adaptation, thus becoming first-class entities [17]. In this direction, several frameworks and conceptual models have been proposed. The IBM's MAPE-K [9]; the adaptation management by Oreizy et al. [11]; Rainbow by Garlan et al. [7]; Kinesthetics eXtreme by Kaiser et al. [8], just to mention a few.

© Springer Nature Switzerland AG 2019
Á. Rocha et al. (Eds.): ICITS 2019, AISC 918, pp. 325–334, 2019.
https://doi.org/10.1007/978-3-030-11890-7_32

In this paper we propose *Reflective Graph Abstract Machine* (RGAM), an *architectural approach* based on *architectural reflection* and supporting *closed-loop mechanisms*, that enables the *adaptive* execution of a distributed computation allowing the dynamic configuration and deployment of its elements. The architecture has been designed according to the real needs of an emergency management system, which is characterized by an highly dynamic running context.

2   RGAM Architecture

RGAM (Reflective Graph Abstract Machine) architecture is the union of two well distinguished but co-existing sub-architectures: the *base-architecture* that supports distributed computations, and the *meta-architecture* that supports their dynamic configuration and deployment (see Fig. 1). A *meta-application* applys domain-related strategies to configure the system by relying on the mechanisms made available by the meta-architecture. Observation mechanisms allow *domain* and *introspective* information to flow towards the meta-application, and control mechanisms allow *contracts* to flow towards the base-architecture. Domain information is related to the domain of the computation (e.g., river levels, forecasts), introspective information refers to the status of the computations (e.g., available bandwidth, occupied RAM), and contracts are requests for computation reconfigurations. The meta-application proposes a contract, the meta-architecture verifies its satisfiability, then the base-architecture makes it operative.

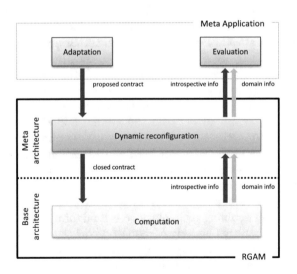

Fig. 1. The overall picture

2.1   Base-Architecture

The base-architecture defines an abstract base-machine supporting the execution of application domain distributed computations It includes *component, port,* and *connector,* which are reified as concrete elements at run-time.

A *component* is an architectural element that encapsulates computations and interacts with other components through its *ports* only. It is a unit of independent deployment and is characterized by a *status* that can be observed at the end of its computation. A *port* is a communication endpoint owned by a component. It reifies as architectural element the general concept of component interface. Ports model one-way data flows; therefore, input and output ports are distinguished. A *connector* is an architectural element that supports an information flow between component output ports and component input ports. A connector is an intrinsically distributed element consisting of a collection of *pins* that are bound to output ports (source pins) and to input port (sink pins). Pins are responsible for information transfer from output ports to inputs ports.

The base-architecture supports a set of *execution mechanisms* that allow a component to get/put data on its ports and, symmetrically, allow a connector to put/get data on its pins and to transfer data among pins. The base-architecture includes *interpreters* (be they hardware or software), which are capable of executing the internal behavior specified by a component.

The base-architecture supports a set of *reflective mechanisms* allowing the base-level system to be observed and controlled by meta-level activities: *reification mechanisms* that allow components and connectors to be reified as core-images on a computing node and made visible to the proper interpreter, and *introspective mechanisms* that support the alignment of introspective and domain data to/from base/meta architecture.

2.2   Meta-architecture

The meta-architecture supports the dynamic configuration and deployment of base-architecture elements to be executed in the context of the base-architecture. Configuration and deployment are "meta-problems", in that their domain is the run-time management of the elements of the base-architecture. This requires to view base-elements as *first-class objects*. This is achieved by representing base-elements via *meta-information* that is manipulated by *meta-activities* to configure and deploy base-elements. Meta-information includes: *computational graph meta-representation, ICT infrastructure meta-representation, deployed computational graph meta-representation,* and *computational graph schema.*

The *computational graph meta-representation* describes the base-elements that constitute the actual running computational graph as sketched in Fig. 2.

Both meta-components and meta-pins maintain meta-information (*MetaImplementation*) related to the concrete platform-dependent implementation reifying base-components and pins respectively.

Quality plays a crucial role as it is intended as any kind of information useful to take decisions about deployment issues. Operating systems, supported

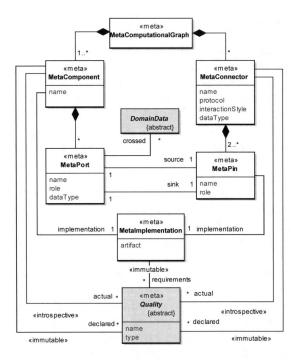

Fig. 2. Computational graph meta-representation

languages, number of cores, execution time, CPU usage, precision, jitter, delay, are only a few examples. Qualities strongly depend on the application domain, therefore it has been considered useful to define a general scheme that establishes how qualities can be defined, rather than provide a comprehensive list of qualities' features. In this way, the same element (e.g., a meta-implementation) may exhibit different qualities' features depending on the application domain. Some qualities are intended as intrinsic and immutable characteristic of an element («immutable» stereotype). For example, a meta-component declares a certain accuracy, precision, or power consumption; a meta-implementation declares to require a specific operating system or a specific run-time support. Other qualities dynamically change over time due to the underlying environment and the status of the computation. For example, the throughput of a meta-connector depends on the load of the underlying communication channels; the execution speed of a meta-pin or of a meta-component may depend from the actual overload of the computing node hosting them. Such kind of qualities have been stereotyped «introspective» because they embody introspective knowledge about the qualities of the base-elements.

Immutable qualities characterize *MetaImplementation* by specifying the resources required by the computing node candidate for hosting the implementation (e.g., supported languages, operating systems, free memory, network stack, and so on). Both *MetaComponent* and *MetaConnector* declare (i) a set of qualities whose values are immutable and do not directly depend from the specific base-implementation that is actually running (e.g., the accuracy of the computation), and (ii) a set of observed (i.e., introspective) qualities (e.g., execution time, throughput, used memory) whose values reflect the actual situation of the computation. Finally *MetaPorts* collect data (*DomainData*) that have been produced and consumed by base-components through base-ports.

The *ICT infrastructure meta-representation* describes the technological infrastructure in terms of computing nodes, communication links, and related qualities (see Fig. 3). The meta-representation of the ICT infrastructure is exploited by the meta-application strategies to drive the deployment of the base-elements (components and connectors via pins). A *MetaComputingNode* is the meta-representation of a base-computing node. It specifies a set of immutable qualities provided by the base-computing node (e.g., supported languages, operating systems, total memory), and a set of introspective qualities (e.g., free memory, CPU usage, battery charge). A *MetaCommunicationLink* is the meta-representation of a base-communication link. Likewise a meta-computing node, it specifies a set of immutable qualities provided by the base-communication link (e.g., supported protocols) and a set of introspective qualities (e.g., jitter).

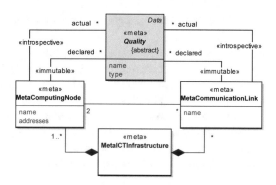

Fig. 3. ICT infrastructure meta-representation

The *deployed computational graph meta-representation* maintains the meta-information related to the actual deployment of the base-elements: it refines computational graph meta-representation by associating meta-components, meta-pins, and meta-connectors (meta representations of the base elements) to meta-ICT infrastructure elements (see Fig. 4). To make the diagram in Fig. 4 more readable, qualities are dropped (they are the same of Figs. 2 and 3).

The *computational graph schema* describes how the system could be configured in terms of components and connectors typologies (in Fig. 5, *LogicalComponent* and *LogicalConnector*). For each logical component and logical connector, the computational graph schema specifies which elements (in Fig. 5, *ComponentSchema* and *ConnectorSchema*) are available and their characteristics in terms of quality provided. They differ with respect to the quality provided, but

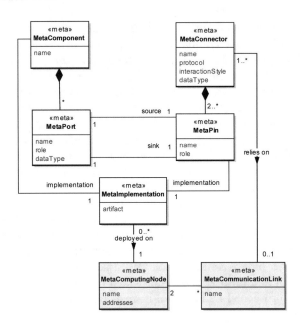

Fig. 4. Deployed computational graph meta-representation

they are equivalent with respect to the application domain functionality provided. Finally, for each component schema and pin schema, the computational graph schema specifies which different implementations are available and their characteristics. Implementations differ from the technological viewpoint (e.g., Java vs. C++), but they are fully equivalent from any other viewpoint.

The meta-application selects and asks to execute the configuration that best fits the actual situation by relying on *instantiation, un-deployment, allocation, deallocation, reification, alignment,* and *introspection* meta activities.

Instantiation activity deals with the instantiation of the computational graph schema (or portions) according to the observation of both introspective and contextualized information (e.g., users, technological infrastructure characteristics). In detail, the activity includes the selection of the realizations to be put in execution (i.e., component-schema, port-schema, connector-schema, and pin-schema) and their instantiation in terms of meta-components, meta-ports, meta-connectors, and meta-pin. This turns into the definition of the meta-computational graph representation (or, when reconfiguring, in an subset). Moreover, the activity includes the selection of the computing nodes that should host the computations (i.e., the implementation of the corresponding base-elements). This results in the definition of a "proposed" contract that specifies a proposed deployed computational graph meta-representation: it subsumes the information exploited to allocate components to computing nodes.

The *un-deployment* activities deals with the un-deployment of base-elements. The meta-application, still relying on introspective and domain information, asks for the de-allocation of base-elements from computing nodes. At the aim,

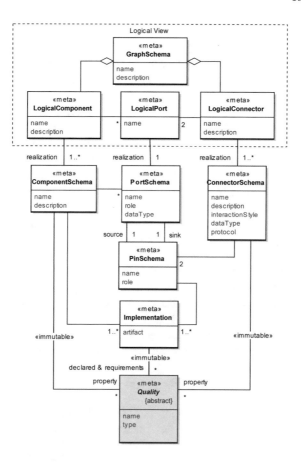

Fig. 5. Computational graph schema

the meta-application specifies which base-elements to remove still exploiting the contract meta-object. The "proposed" contract is then delivered to the meta-architecture to be managed by its allocation or deallocation activity.

The *allocation* activity receives from the meta-application a "proposed" contract dealing with reification of new base-elements. It is in charge of verifying the consistency between implementation requirements and node properties, and to build and deliver the "closed" contract to the *reification* activity. The "closed" contract concerns deployment, thus it contains all the information needed to reify the new base-elements.

The *deallocation* activity receives from the meta-application a "proposed" contract dealing with the deallocation of existing base-elements. It is in charge of verifying the correctness of the deallocation request (meta-elements/nodes correspondences and aligned update status of the computations), and to build and deliver the "closed" contract to the *reification* activity. In this case, the "closed" contract contains the information useful to un-deploy base-elements, that is, only the names of the corresponding meta-elements.

The *reification* activity reifies base-elements on target nodes by exploiting the *reification mechanism* provided by the base-architecture; therefore it is a base-activity, which is highlighted here to clarify the overall information flow. The reification activity receives a "closed" contract and exploits the information it carries to reify (or to destroy) on the target node concrete implementations of components, ports, and pins and to make them available to the interpreter.

The *introspection* activity aims at acquiring both the introspective information dealing with the system and the ICT infrastructure, and the domain information produced by the base-components. Likewise reification, the *introspection* activity relies on the introspective mechanisms provided by the base-architecture, and thus, it is a base-activity and is highlighted here to clarify the overall information flow.

Finally, the *alignment* activity maintains updated the meta-representations of both the computational graph and the ICT infrastructure. To support introspection activity, *knowledge pin*s have been introduced as special base-level architectural elements that are bound to ports like pins.

Allocation/deallocation and reification activities realize the *down-stream connection* of the causal connection, that is, the capability of reifying base elements on computing nodes. Introspection and alignment activities realize the *down-stream connection* of the causal connection, that is, the capability of aligning both *domain* and *introspective information* between the base and the meta level.

3   Discussion and Final Remarks

Both base and meta architectures support computations: the base architecture supports computations on *domain data*, whereas the meta architecture supports computations on *meta-representations* of the base elements. Base and meta architectures are largely autonomous and can be supported by different platforms.

A control loop architectural solution underlies RGAM (see Fig. 6): a meta-application exploits the knowledge of the computational graph schema, of both domain and introspective information about the running computational graph and the ICT infrastructure, and of the actual deployed computational graph to make decisions about reconfigurations. The meta-architecture provides all the mechanisms the meta-application can exploit to drive the computations and the base-architecture is in charge of actuating the selected computational graph.

There is a clean separation between base- and meta-issues: base-level mechanisms supporting the execution of distributed computations are carefully separated from mechanisms aimed at supporting meta-level activities. Moreover, the base-architecture does not embed any policy, because policies should be raised up to the meta-level in order to exploit both domain and introspective information. Both *components and connectors are concrete first-class objects* that can be explicitly observed and controlled at execution time by meta-level activities. This enables dynamic reconfiguration by promoting *interconnection independence*.

The classical separation between communication and computation layers vanishes. Components encapsulate information transformation, whereas connectors encapsulate information transfer. A distributed computation is viewed as a

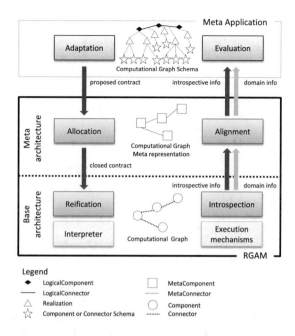

Fig. 6. The closed-loop in RGAM

pipe of components and connectors that are configured, deployed, and activated according to programming-level strategies in order to get an optimal quality of service. Thus, the base-architecture is light, because it only provides elementary and well-partitioned mechanisms aimed at supporting dynamic reconfiguration.

Two separated run-time supports reify base- and meta-concepts: the *run-time support on computational node* (RTonComp), which reifies the base-architecture, and the *run-time support on management node* (RTonMan), which reifies the meta-architecture and provides the interface the meta-application uses to interact with the system. A Java version of the RTonMan and both a Java and an Android version of the RTonComp have been developed. A standard library has been provided including a set of domain data, of qualities, and of pins.

The architecture has been tested in a emergency management system. In particular, the case study deals with the management of emergency in case of flooding and regards the evacuation of buildings. The definition of the actuation strategies have been derived from an in-depth analysis of some contingency plans of Italian municipalities. Our experimentation includes simulated components (e.g., the meteo forecaster and the river level sensor) as the aim was to verify the validity of the approach. The components are implemented in Java and Android. Finally, we have defined a meta-application that dynamically changes the computational graph schema, in terms of both running components and their allocation. We plan to complete the case study by substituting the simulated components and by providing other different implementations for them.

References

1. Andersson, J., de Lemos, R., Malek, S., Weyns, D.: Reflecting on self-adaptive software systems. In: Proceedings of the Workshop on Software Engineering for Adaptive and Self-Managing Systems (SEAMS) (2009)
2. Bernini, D., Micucci, D., Tisato, F.: A platform for interoperability via multiple spatial views in open smart spaces. In: Proceedings of the IEEE Symposium on Computers and Communications (ISSCC) (2010)
3. Bernini, D., Micucci, D., Tisato, F.: Space integration services: a platform for space-aware communication. In: Proceedings of the 6th International Wireless Communications and Mobile Computing Conference (IWCMC) (2010)
4. Chen, T., Bahsoon, R., Yao, X.: A survey and taxonomy of self-aware and self-adaptive cloud autoscaling systems. ACM Comput. Surv. **51**(3), 61:1–61:40 (2018)
5. Cheng, S.W., Garlan, D.: Stitch: a language for architecture-based self-adaptation. J. Syst. Softw. **85**(12), 2860–2875 (2012)
6. Dowling, J., Cahill, V.: The k-component architecture meta-model for self-adaptive software. In: Proceedings of the International Conference on Metalevel Architectures and Separation of Crosscutting Concerns (2001)
7. Garlan, D., Schmerl, B., Cheng, S.W.: Software architecture-based self-adaptation. In: Autonomic Computing and Networking, pp. 31–55 (2009)
8. Kaiser, G., Parekh, J., Gross, P., Valetto, G.: Kinesthetics eXtreme: an external infrastructure for monitoring distributed legacy systems. In: Proceedings of the Autonomic Computing Workshop (2003)
9. Kephart, J., Chess, D.: The vision of autonomic computing. Computer **36**(1), 41–50 (2003)
10. Kramer, J., Magee, J.: Self-managed systems: an architectural challenge. In: Proceedings of the Future of Software Engineering (FOSE) (2007)
11. Oreizy, P., et al.: An architecture-based approach to self-adaptive software. IEEE Intell. Syst. Appl. **14**(3), 54–62 (1999)
12. Riganelli, O., Micucci, D., Mariani, L.: Healing data loss problems in android apps. In: Proceedings of the International Workshop on Software Faults (IWSF), Co-located with ISSRE (2016)
13. Riganelli, O., Micucci, D., Mariani, L., Falcone, Y.: Verifying policy enforcers. In: Lecture Notes in Computer Science, LNCS, vol. 10548 (2017)
14. Riganelli, O., Micucci, D., Mariani, L.: Policy enforcement with proactive libraries. In: Proceedings of the 12th International Symposium on Software Engineering for Adaptive and Self-Managing Systems (SEAMS) (2017)
15. Shatnawi, A., Orrù, M., Mobilio, M., Riganelli, O., Mariani, L.: Cloudhealth: a model-driven approach to watch the health of cloud services. In: Proceedings of the 1st Workshop on Software Health (SoHeal) Co-located with ICSE (2018)
16. Shaw, M.: Beyond objects: a software design paradigm based on process control. SIGSOFT Softw. Eng. Notes **20**(1), 2738 (1995)
17. Souza, V.: A requirements-based approach for the design of adaptive systems. In: Proceedings of the International Conference on Software Engineering (ICSE) (2012)
18. Tisato, F., Simone, C., Bernini, D., Locatelli, M.P., Micucci, D.: Grounding ecologies on multiple spaces. Pervasive Mobile Comput. **8**(4), 575–596 (2012)

Energy Consumption for Anti-virus Applications in Android OS

Elsa Vera-Burgos[1(✉)], Willian Zamora[1,2(✉)], Homero Mendoza-Rodriguez[1(✉)],
Alex Santamaría-Philco[1,2(✉)], Denise Vera-Navarrete[1(✉)],
and Patricia Quiroz-Palma[1,2(✉)]

[1] Universidad Laica Eloy Alfaro de Manabí, Manta, Ecuador
{elsa.vera,homero.mendoza,dennisse.vera}@uleam.edu.ec
[2] Universitat Politecnica de Valencia, Valencia, Spain
wilzame@posgrado.upv.es, {asantamaria,patquipa}@dsic.upv.es

Abstract. The present investigation aims to carry out a comparative study on the energy consumption associated with the applications of anti-virus for smartphones running with the Android operating system. The characteristics and attributes of the devices used in this study are provided, with the details of the functionality offered by the different anti-virus applications. A methodology is proposed that includes the development of an application that through a service performs periodic measurements of the remaining percentage of the battery and the voltage demanded by the applications; allowing to estimate the variations of the voltage generated by anti-virus applications and their energy impact on the battery. The experimental results show that in general, anti-virus applications have a high power consumption with power levels ranging from 6 to 16 mW when the application is active, although the different anti-virus solutions are also verified.

Keywords: Energy consumption · Mobile devices · Anti-virus · Android · Smartphone

1 Introduction

Smartphones or mobile devices have attractive services and features of computing that compete with the performance of personal computers; however, they consume a lot of energy [1]. Most of the mobile devices use rechargeable electrochemical ion batteries lithium (Li-ion) that are short-lived, especially when the connection to data networks is kept active, and other applications and services are used continuously [2]. As a result, the duration of battery charging has become a problem of availability for the user, because the battery will not last long enough; therefore, battery life is one of the top priorities of the manufacturers of mobile devices [1].

Some tools measure the electrical consumption in different mobile devices; for example, the DOZE tool is included in the Android platform [3], that monitor the

© Springer Nature Switzerland AG 2019
Á. Rocha et al. (Eds.): ICITS 2019, AISC 918, pp. 335–345, 2019.
https://doi.org/10.1007/978-3-030-11890-7_33

percentage consumption, at the level of components and applications. Also, in Google Play one can find apps that model the power consumed by the principal components such as the CPU, communication interfaces, mobile screen, GPS and other applications used. When referring to these type of applications, the antivirus [4], appears, as the primary measure to protect the device against malware attacks, since this constitutes a constant and real threat that seriously affects the user, emotionally or financially. This scenario turns the antivirus into an indispensable application in a mobile device and, therefore, deserve to be studied especially concerning the subject of energy consumption [4], even more, if we consider that they have not evolved to the point of being efficient like the versions that protect personal computers.

This contribution is summarized as follows:

- A methodology was proposed that allows to monitor the energy consumption of antivirus applications on mobile devices with Android operating systems.
- An application was developed that runs in the background and allows to register the values relative to voltage, battery percentage and capacity.
- The results obtained were evaluated and compared through statistical graphs.

The rest of this document is organized as follows: in Sect. 2 we present some related works; in Sect. 3, we describe the different components that make up the proposal; in Sect. 4, we describe the evaluation methodology; in Sect. 5 the experimental results of the evaluation are shown and in Sect. 6 the conclusions and future works are presented.

2   Related Works

Battery life is an essential factor in the development and deployment of applications and services on mobile devices, so the user has to be informed of the available energy percentage, to provide it with a convenient use, considering that the execution of some applications demand higher battery consumption and the user can decide whether to use them or not.

Due to the need to know how much energy a mobile device consumes, several investigations have emerged, such as the one made by Manet et al. [5], in which they analyze the power consumption by the IEEE 802.11 interface in different modes of operation, demonstrating how the transmission traffic rate discharge the battery.

In other investigations [6,7] y [8], most of the energy consumed in mobile devices are attributed to communications; to the GPS and the display [9]; to the applications that execute processes in the background [2]; to Bluetooth, Wi-Fi, cellular radio, data network and even the lighting generated by the screen [2,6,10, 11]. In general, if the applications do not use the hardware prudently, it increases the energy consumption, for example, the frequency of waking the device in the background and simultaneous transfers of data through the Internet [10].

The research reviewed reflects several types of problems, some of the models designed to get the data and evaluate the power consumed of the various components of a smartphone are taken with electronic devices, able to obtain these values in real time, but these only work on devices with the same type of technology [8].

Regarding the developers of mobile applications, they must recognize the energy needs of their applications [12], because in a certain way each use that runs on a mobile device contributes differently to battery consumption [13] and antivirus programs cannot be an exception.

Finally, in the present bibliographical review works were found [14,15], which evaluate the energy efficiency of specific antivirus programs for the Android platform. Precisely, they measure energy consumption during various operations like the process of scanning the application after installation, the full scanning of the device, and scanning of the SD card. Our proposal differs from the previous bibliographical works since we measure the general impact of different antivirus programs on the power consumption in the battery of a smartphone, and additionally we propose a base methodology for a said proposal.

3   Description of the Proposal

The proposal of the methodology consists of four components that are shown in Fig. 1. The main component is the service that runs in the background, developed in Android Studio [16]. The second component is the reading of the energy consumption for the mobile device in its different states: Suspended, Active and Inactive. In the third component, the different antivirus applications that must

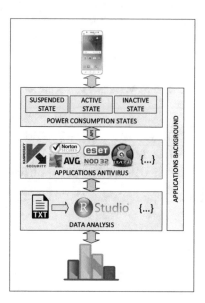

Fig. 1. Components of the proposed methodology.

be installed and initialized for their respective evaluation. Finally, the component that performs the statistical extraction and analysis of the results.

3.1 Power Management in Android

The Android operating system performs a power management associated with the methods of Linux, APM (Advanced Power Management) and ACPI (Advanced Configuration and Power Interface); however, it employs a more aggressive policy management of energy saving, with the premise that the CPU does not consume energy if there are no applications or services that need energy, therefore it has its own power management extension called PowerManager, which provides low level drivers in order to manage the peripherals supported.

A wakelock [17] is a function of the PowerManager service, which allows to control the energy state of the device dynamically [6], the applications and components have to create and acquire wakelocks to keep the assets active, if there is no active wakelock, the CPU shuts down and goes to a low consumption status.

There are three built-in states in the PowerManager state machine to the energy management model (i) SLEEP, (ii) NOTIFICATION and (iii) AWAKE. When an application acquires a complete wakelock, it can produce an event by screen activity or keyboard and the device will be maintained or changed to AWAKE status; If the waiting time passes or the on/off button is pressed the transition to the NOTIFICATION state occurs. While a partial wakelock is acquired, the device will keep in NOTIFICATION status if all partial wakelocks are released, the device goes to the SLEEP state, if in this mode all the resources are activated, the transition to the AWAKE state occurs.

Table 1 shows which wakelocks are available and used by Android to reduce the power consumption in a mobile device.

Table 1. Wakelocks in Android.

Flag value	CPU	Screen.	Keyboard.
PARTIAL_WAKE_LOCK	ON	OFF	OFF
SCREEN_DIM_WAKE_LOCK	ON	DIM	OFF
SCREEN_BRIGHT_WAKE_LOCK	ON	BRIGHT	OFF
FULL_WAKE_LOCK	ON	BRIGHT	BRIGHT

3.2 Antivirus Applications

In this component you will find the different antivirus that have been considered for evaluations. Five of the twenty-five security products for Android, in its most current versions, better ranked in a study conducted in November 2015 were taken as reference [18]. Table 2 shows the characteristics of the selected antivirus applications.

Table 2. Characteristics of antivirus applications.

Función	Avast	Eset	Kaspersky	McAfee	Norton
Malware detection	Yes	Yes	Yes	Yes	Yes
Anti-theft service	Yes	Yes	Yes	Yes	Yes
Call blocking	Yes	Yes	Yes	Yes	Yes
Filter messages	Yes	Yes	Yes	No	No
Secure browsing	Yes	Yes	Yes	Yes	Yes
Parental control	No	No	No	No	No
Backup copy	No	No	No	Yes	Yes
Data coding	No	No	No	No	Yes

3.3   Background Service

The measures of energy consumption of the mobile device have been taken by software, taking advantage of the capabilities provided by the Android operating system, through an application (app) developed to monitor the energetic demand. The application records at all times the remaining percentage of battery, capacity and voltage, in order to calculate the power consumed.

The source code of the application is composed of two Java classes:

– monitor.java.- Here you define the "activity" that starts or ends the service.
– service.java.- Defines the service and operations to be performed. When the service is loaded for the first time, the ACTION_BATTERY_CHANGED event occurs, which receives the voltage, capacity and the remaining percentage of energy in the battery.

To obtain the numerical value of the state of the battery, the BatteryManager class [19], which allows you to obtain battery data, such as capacity in mAh, voltage, technology, temperature, its charging state, among other information regarding the battery has been used. As mentioned before, the application registers the capacity, the voltage and the remaining percentage of the battery. The capacity is recorded in the variable EXTRA_LEVEL and the voltage in the variable EXTRA_VOLTAGE, these two variables are used to calculate power consumption.

When the application is run for the first time, a service is initialized and, no user interface is needed, and as long as variations in battery discharge does not occur, there will be no CPU consumption or any other resource of the device.

3.4   Data Analysis

In this component the stored data is extracted from the internal or external memory of the smartphone for statistical analysis in the graphic tool R [20], generating various graphics: (i) Box and Whisker plots to show the voltage, (ii) Bar graphs, to show the remaining energy of the battery, and (iii) Line graphs

used for comparing the discharge of the battery both with the execution of
the different antivirus applications without scanning the device, and with the
smartphone in suspended status. These results are shown in Sect. 5.

4   Test Methodology

The evaluation of the energy consumption has been done installing the application developed on the device; so that it is possible to obtain the consumption
value while the device is on but does not execute any task or process, in this
way a reference measure is obtained in order to observe how much energy consumption increases with the use of the antivirus application. To obtain exact
values, we must ensure that all data network interfaces are inactive and that
the only application running is the one developed on the device. In this way, we
will determine the state where the energy consumption is lowest. On the other
hand, to calculate the power consumed when antivirus applications are used,
two states have been defined:

- No Scanning: When antivirus applications are installed and with its active
  services.
- Scanning: When antivirus applications are constantly scanning files and other
  applications of the device; it is expected that in this state, it will increase the
  power consumed.

Figure 2 shows the flow diagram of the monitoring application. The mobile
device used for the different evaluations was a Samsung Galaxy J5 with Android

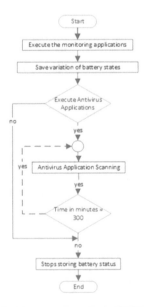

Fig. 2. Flow diagram of the monitoring application.

5.1.1 Lollipop, operating on version 3.10.49- 787809 of the Linux kernel. The developed application was installed, and three samples of five consecutive hours for each evaluation were made. The samples were recorded in ten-second intervals. In the antivirus applications for the case of scanning the folders and applications of the mobile device, we used a 32 GB MicroSD memory loaded with the detailed information as (photographs, videos, documents, music, and among others).

5 Results

The results are described below according to the proposal and methodology described in Sects. 3 and 4 respectively. The active, inactive and suspended states were evaluated. Because the battery is not exactly a linear device [6], exact power values are not always obtained, therefore, to determine the average consumption of an antivirus application you get several power values and the average of them is calculated.

5.1 Tests of Battery Discharge in Suspended State

In the tests performed, the device's battery is initially charged to 100%, where the monitoring application begins the registration of all the variations presented by the battery until the five hours is up. The discharge of the battery was evaluated when the device did not have any installed antivirus applications. This measure serves as a reference to verify when the consumption increased for each antivirus application. Figure 3 shows the discharge of the device's battery in suspended state (black line). It can be seen that battery life is prolonged without the use and installation of antivirus applications, since we can observe that it is still around 69% remaining capacity after the five hour mark is up.

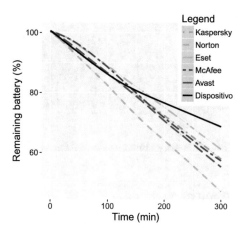

Fig. 3. Battery discharge by having the antivirus applications installed but not scanning the device, and with the smartphone in suspended state.

5.2   Anti-virus Applications in Suspended and No Scan Mode

After having the initial battery discharge data without antivirus applications installed on the device, reaching a remaining capacity of 69%, the evaluation was carried out with installed antivirus applications, running in the background but not scanning files. Figure 3 shows the results of these tests and reflects that the antivirus applications quickly reduce the battery life of the mobile device, Norton being the application that consumes the most energy leaving the battery in a 44% remaining capacity and Eset, the lowest power consumption in this state.

5.3   Antivirus Applications in Active and Scanning Mode

To verify the discharge time of the device's battery in Active mode, we proceeded to evaluate the antivirus applications in complete Scanning mode of the different folders in search of some type of virus or malware.

Figure 4 shows the voltage variation of the different antivirus applications, registering significant ranges between Norton and Kaspersky. Table 3 shows the remaining capacity of the battery according to the scan time of each antivirus application. The variations of the mobile device's voltage in the inactive and suspended states associated with the applications. In addition, McAfee as the most energy-consuming antivirus application, leaving the battery with a remaining capacity level of 96%, while the lowest-consuming antivirus application was Avast.

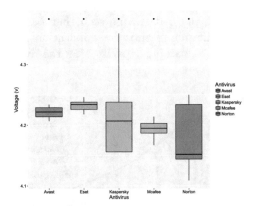

Fig. 4. Analysis of energy consumption of antivirus applications.

5.4   Power Measurement

To evaluate the energy consumption level of the mobile device, we collect data from the voltage and capacity readings of the device to calculate the power consumed by the device. In Li-ion batteries of mobile devices, the voltage changes

Table 3. Summary antivirus application in active state.

Scan time (s)	Antivirus	% Remaining battery
1080 seg	MacAfee	96%
1680 seg	Kaspersky	97%
450 seg	Eset	98%
430 seg	Norton	99%
300 seg	Avast	100%

during the discharge, allowing them to consider the power consumed. As the voltage varies, the capacity level of the battery decreases. This means that the power consumed can be calculated from the changes in voltage and the capacity level of the battery cell. The formula to calculate the energy consumed during a specific time interval is the following:

$$P = \frac{C_1 * V_1 - C_2 * V_2}{t_1 - t_2}[6] \tag{1}$$

Where, P refers to the power consumed, the variables $(t_1 - t_2)$, refers to the time interval, the values of the remaining capacity level are represented with C_1 y C_2 expressed in mAh and with V_1 and V_2 as reference to the voltage value. The actual available capacity of the mobile device must be taken into account due to deterioration and with the passage of time this can decrease significantly.

Table 4 shows the values of power consumed in each one of the states explained above and for each one of the evaluated antivirus applications.

Table 4. Power consumption by antivirus applications.

	Antivirus (mW)				
	Avast	Eset	Kaspersky	Norton	McAfee
Active	6.45	8.33	14.94	11.45	16.33
Suspended	15.13	11.74	14.19	18.10	13.44
Inactive	1.01	1.20	1.08	2.07	1.30

6   Conclusion

In this paper, a methodology was proposed that allows to monitor the energy consumption of mobile devices with Android operating systems, associated with antivirus applications. Application developed allows recording the levels of energy consumption produced by the above antivirus applications selected. In the tests performed, it was verified that, in two of the three states (Suspended, Active and Inactive), the Norton antivirus application is the one that showed the highest battery consumption. On the other hand, the antivirus apps that presented lower energy consumption were Eset and Avast. The results show that there is a significant decrease in energy, based on this it can be deduced that security providers do not develop antivirus applications with moderate energy consumption profiles. It should be noted that, in this evaluation, the performance of these antivirus apps was not measured. For future work, it is intended to improve the application for better data collection, as well as evaluate other parameters concerning the benefits of the antivirus applications through different devices brands and models.

References

1. Perrucci, G.P., Fitzek, F.H., Sasso, G., Kellerer, W., Widmer, J.: On the impact of 2G and 3G network usage for mobile phones battery life. In: 2009 European Wireless Conference, pp. 255–259. IEEE (2009). https://doi.org/10.1109/EW.2009.5357972
2. Martins, M., Cappos, J., Fonseca, R., Martins, M., Cappos, J., Fonseca, R.: Selectively Taming Background Android Apps to Improve Battery Lifetime
3. Optimizing for Doze and App Standby. http://developer.android.com/training/monitoring-device-state/doze-standby.html
4. Pieterse, H., Olivier, M.S.: Security steps for smartphone users, pp. 1–6 (2013). https://doi.org/10.1109/ISSA.2013.6641036
5. Manet, L.E., Anne, F., Jean-yves, M., Serge, F.A.: Energy Consumption Models for Ad-Hoc Mobile Terminals
6. Estudio del consumo de energía. In: Univeridad Carlos III de Madrid, pp. 54–55 (2012)
7. Bytheway, I., Grimwood, D.J., Jayatilaka, D.: Wavefunctions derived from experiment. III. Topological analysis of crystal fragments., Acta crystallographica. Section A, Foundations of crystallography 58(Pt 3), pp. 232–243 (2002)
8. Malik, M.Y.: Power Consumption Analysis of a Modern Smartphone, p. 11 (2012). arXiv:1212.1896
9. Carroll, A., Heiser, G.: An analysis of power consumption in a smartphone. In: USENIXATC 2010 Proceedings of the 2010 USENIX Conference on USENIX Annual Technical Conference, p. 21 (2010)
10. Datta, S.K., Bonnet, C., Nikaein, N.: Minimizing energy expenditure in smartdevices (ICT), pp. 712–717 (2013). https://doi.org/10.1109/CICT.2013.6558187
11. Vallina-Rodriguez, N., Hui, P., Crowcroft, J., Rice, A.: Exhausting battery statistics, p. 9, February 2010. https://doi.org/10.1145/1851322.1851327
12. Anand, A., Manikopoulos, C., Jones, Q., Borcea, C.: A Quantitative Analysis of Power Consumption for Location-Aware Applications on Smart Phones, pp. 1986–1991 (2007). https://doi.org/10.1109/ISIE.2007.4374912
13. Shye, A., Scholbrock, B., Memik, G.: Into the wild, p. 168 (2009). https://doi.org/10.1145/1669112.1669135
14. Bickford, J., Park, F., Varshavsky, A., Park, F.: Security versus Energy Tradeoffs in Host-Based Mobile Malware Detection
15. Polakis, I., Diamantaris, M., Petsas, T.: Powerslave : Analyzing the Energy Consumption of Mobile Antivirus Software, vol. 3, pp. 165–184 (2015). https://doi.org/10.1007/978-3-319-20550-2

16. Download Android Studio and SDK Tools. http://developer.android.com
17. Datta, S.K., Bonnet, C., Nikaein, N.: Android power management: current and future trends. In: 2012 The First IEEE Workshop on Enabling Technologies for Smartphone and Internet of Things (ETSIoT), December 2015, pp. 48–53. IEEE (2012). https://doi.org/10.1109/ETSIoT.2012.6311253
18. AV-TEST- The Independent IT-Security Institute (2015).https://www.av-test. org/es/antivirus/moviles/android/noviembre-2015
19. Android Developer BatteryManager. http://developer.android.com/intl/ reference/android/os/BatteryManager.html
20. R-Foundation, R project available (2015).https://www.r-project.org

Spatial Data Infrastructure as the Core for Activating Early Alerts Using EWBS and Interactive Applications in Digital Terrestrial Television

Villie Morocho, Rosario Achig$^{(\boxtimes)}$, Fabian Santander,
and Sebastian Bautista

Department of Computer Science,
University of Cuenca, Cuenca C.P. 01.02.03, Ecuador
{villie.morocho, rosario.achig, fabian.santander,
sebastian.bautista}@ucuenca.edu.ec

Abstract. Integrated Services Digital Broadcasting Terrestrial (ISDB-T) is the standard adopted in Ecuador since 2010. There is a lack of proposals for taking advantage of communication alternatives with citizens, which the new digital technology will bring. The speed for alerting the community about dangers associated with natural hazards is an essential issue. There are many devices to alert population but none of them has the level of dissemination within households as the television does. This paper presents a software solution for early alert using the UCuenca-SDI (Spatial Data Infrastructure) core to activate an Emergency Warning Broadcasting System (EWBS) protocol in a Digital TV. The solution has two components: (a) A schema to integrate the trigger module into the SDI, where the activation code is generated for EWBS. (b) An interactive app software developed over GINGA, where citizens can access instructions about actions, such as maps and text messages, established by activating emergency organizations to evacuate or to make decisions concerning a particular event.

Keywords: Spatial Data Infrastructure · Digital TV · EWBS · ISDB-T · Ginga

1 Introduction

Landslides are one of the most destructive geological processes that affect humans, causing thousands of deaths and property damage for tens of billions of dollars each year [1]. A slip is the movement downhill of a soil or mass of rocks that occurs predominantly on well-defined breaking surfaces or on relatively narrow zones of intense shear [2]. The effects of landslides are devastating for population and infrastructure. Therefore, it is important to prevent and to alert population of possible events, specifically, in this case of mass landslides. The capacity and speed to evacuate people and infrastructure protection by agencies with responsibility for security and emergency response are crucial tasks during this type of event. In Ecuador, the Risk Management Secretary (SGR) is responsible for alerting the population about natural disasters occurrence.

© Springer Nature Switzerland AG 2019
Á. Rocha et al. (Eds.): ICITS 2019, AISC 918, pp. 346–355, 2019.
https://doi.org/10.1007/978-3-030-11890-7_34

This research contributes to improving the capacity to alert digital television technology users, and takes into account several facts: (a) analog blackout is close to occur; (b) national television coverage (TV penetration is greater than internet) even in vulnerable areas, (c) speed to transmit an alert.

Social media contributes with useful data for searching and locating sub-events of particular disasters, and may also be helpful in activities related to relieving and rescuing during natural hazards; despite they are not conceived for real time alerts, this data may be of great aid in particular types of natural hazards [3]. On the other hand, maps based on GIS technology are especially designed for regional planning, and are used to determine areas where local landslide hazards may exist, but do not allow for population alerts either [4]. In [5], it is reported an useful tool to visualize natural hazards, where a specific digital elevation model (DEM) data is created as a basis, and relevant map layers of information are added and displayed as overlays on the terrain, allowing interactivity with user, providing an important improvement in disaster preparedness tools.

This paper includes a brief introduction of technology related to DTT and EWBS; the proposed architecture is showed; and the software application with real data tests are considered. The last section of the paper is devoted to conclusions and prospect of future work.

1.1 Digital Television

Digital Terrestrial Television (DTT) corresponds to TV transmission signals using digital rather than conventional analog methods. The main advantages of DTT are better video, image and sound quality, greater number of channels and interactivity with the viewer [6]. Interactivity is the ability to offer additional content to television programs, the user can see information associated with audiovisual content, channel programming, participate in contests, voting, buy products or services, and even participate in television programs, using the remote control. Interactivity is possible due to applications that complement the programming [7].

Ecuador officially adopted the ISDB-T standard in March 2010, called ISDB-Tb due to its Brazilian adaptations; this way, Ecuador aligned with a standard adopted in most Latin American countries [8]. The change from analog to digital television in Ecuador will start in 2018; meanwhile there will be a progressive transition starting in big cities. The Ministry of Telecommunications and the Information Society (MINTEL) has conceived that at least 90% of the population must be prepared to receive the digital signal in their homes. Currently, 577 television stations operate in the country, of which 30 transmit digital signals with a temporary concession to provide DTT service, in cities such as Quito, Guayaquil, Cuenca, Santo Domingo, Manta, Latacunga and Ambato [9].

1.2 DTT to Transmit Emergency Alerts

EWBS protocol is used to send warning signals over digital television. For example, issuing emergency bulletins to alert the population about an impending disaster [10]. The countries of Central and South America that adopted the ISDB-T standard have shown interest in the EWBS system, creating great expectations and carrying out activities with technical support from Japan [11].

To send a warning signal, it is necessary to identify the areas potentially affected by some type of disaster. For this, the Digital Broadcasting Experts Group (DiBEG) has assigned an encoding to specific areas of Central and South America [12]. For Ecuador, the areas classified and codified are cantons (country subdivision). Encodings are transmitted-interpreted by broadcasting stations, and issued to the corresponding areas.

1.3   Codes Generation for EWS Protocol Through UCuenca-SDI

Geographic information systems are increasing in number, and have been introduced in different applications, providing territory information from geographical, urban, rural and environmental viewpoints [13]. To send a warning signal, area codes are consulted and subsequently are entered in the transmission package. This demands an important time in an alert situation. This paper proposes a map viewer that allows concentrating codes representing each canton, through a dynamic selection of interest areas, relying on some advantages offered by an SDI, such as layers, WMS and WFS services.

2   Outline of the Proposed Solution

This paper presents the design and implementation of a system for generating early warning codes with the collaboration of the DTT Working Group of the Army Forces University (ESPE). The transmission/reception tests were performed using signal channels from the ESPE DTT Laboratory. Figure 1 depicts a diagram solution for alerts transmissions through DTT, whose detailed explanation was reported in [14].

Initially, the SGR or other authorized institution generates the information to be transmitted through the DTT signal. Once an alert is activated, its information includes canton code, type/subtype of vulnerability and level of alert.

This information is received by the DTT transmission system through a web service and it is delivered to the viewer through the EWBS emergency alert protocol. When the event is triggered, the interactive application is sent to the user from the application server. The interaction can be local or remote; in case of remote interactivity, the return channel will be used via Internet to update information from the SDI Server.

SDI is designed for generating alert codes and interactivity, thus, there are various information sources from which SDI has access. The UCuenca-SDI Core contains information for user review such as evacuation routes, meeting points, emergency plans, among many other maps. Different from the connectivity provided by other channels, such as cellular networks, the SDI interaction using a DTT system would even allow connectivity in cases of natural disasters. Additionally, it is a channel with a better chance of recovery and coverture.

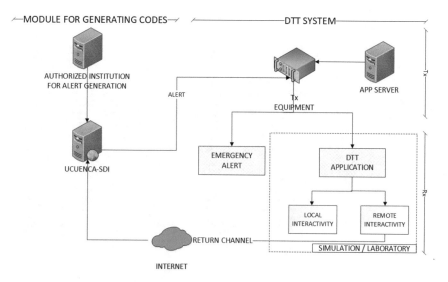

Fig. 1. Diagram of sending alerts and interactive applications on DTT.

3   Module for Generating EWBS Through the UCuenca-SDI

Through a map viewer, this module allows graphic selection of cantons where an emergency occurs, so that this information can be sent to the alert system using the EWBS protocol.

3.1   Preparing Information

An Ecuador canton layer has been set for canton selection on a map. The National Institute of Statistics and Census (INEC) generated it, and it is available in the UCuenca-SDI services. In this layer, the attribute of the coding proposed by the DiBEG is not available. Table 1 includes the corresponding canton codes contained in the layer, used for sending warning signals.

This information is stored in a spatial database PostGIS and published in the map server GeoServer. Therefore, it is available through the WMS or WFS services to be used by different GIS tools, owned UCuenca-SDI and others.

3.2   Module Architecture

A new module was implemented taking advantage from the architecture design of UCuenca-SDI [15] for integrating different tools and the information required to have a multidisciplinary environment work. As showed in Fig. 2, the layers of spatial database are obtained by geoserver, after that, the SDI consumes those layers through WOS/WMS services of geoserver to select the cantons and obtain their codes, then, they are stored in a relational database, to be consumed through a WEB service or downloaded in a JSON file.

Table 1. Canton codes.

DPA_ANIO	DPA_CANTON	DPA_DESCAN	DPA_PROVIN	DPA_DESPRO	DPA_BINARY
2011	1316	24 DE MAYO	13	MANABI	000000000001
2011	2202	AGUARICO	22	ORELLANA	000000000010
2011	0602	ALAUSI	06	CHIMBORAZO	000000000011
2011	0902	ALFREDO BAQUE...	09	GUAYAS	000000000100
2011	1801	AMBATO	18	TUNGURAHUA	000000000101
2011	1002	ANTONIO ANTE	10	IMBABURA	000000000110
2011	1604	ARAJUNO	16	PASTAZA	000000000111
2011	1503	ARCHIDONA	15	NAPO	000000001000
2011	0702	ARENILLAS	07	EL ORO	000000001001
2011	0806	ATACAMES	08	ESMERALDAS	000000001010

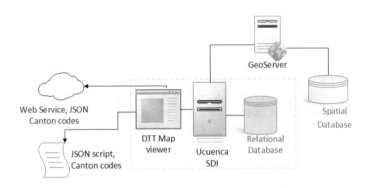

Fig. 2. Architecture of the DTT map viewer module

3.3   DTT Map Viewer

Once different tools and services were integrated, a map viewer was developed for expanding towards geographic analysis. This application provides the necessary support for a script generation that covers canton coding identified through the analysis and geographical interaction to send alerts. An authentication is necessary to have access to different functionalities offered by this viewer. This is because information about events is exclusive of the entities authorized to generate alerts.

DTT viewer offers a simplified operation, through the manipulation of specific tools, such us:

- Graphing panel, with task of selection of cantons and assignment of styles.
- Option for registration and generation of information.

Once the interest areas have been selected, the layer is saved and the script covering all the canton information is generated. The system captures the map with all the existing geographic information, and then additional parameters such as title, type and description of the event (possible vulnerability events) are entered.

This image is stored in a specific directory, and the parameters entered together with the information of the selected canton in the map are stored in the UCuenca-SDI database. Then, the event history is generated and it is consumed through a web service, which facilitates a script in the JSON format. This script can be downloaded through the UCuenca-SDI or simply be employed by different applications.

The DTT map viewer, showed in Fig. 3, is designed to add additional layers from the same server or from alternate sources through web map services. This way, it is possible to have information to enrich the perspective of the selected areas. The captured image will be utilized by the DTT application to be presented to the user.

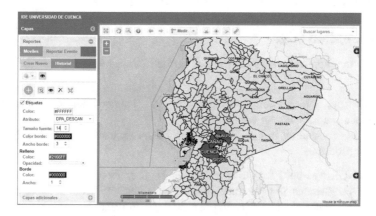

Fig. 3. Map viewer for selecting cantons

4   Interactive Application

In addition to sending an emergency alert, the system broadcasts interactive contents to support the emergency (through the television channel), which are transmitted once the official entity (SGR) activates the alert. It was developed using the GINGA-NCL-LUA programing language.

This interactive application aims to provide the population with useful and updated information for helping decisions making. The application implementation phases are presented with detail next.

4.1   Analysis and Design Stage

For the execution of this stage, the DEVTVD-ÁGIL methodology was applied for the development of TVDi software and its uses was verified in different types of interactive applications. This methodology is being developed by the DTT Group of the University of Cuenca, Ecuador. The templates of the requirements specification and software design documents were used.

Additionally, the templates generated automatically by the Multi-Platform Tool developed by ESPE were analyzed [16] in order to select the ones to use in this project.

The SGR information was structured for designing interactive content. The sources were coordinated with SGR.

The main options of interactive application are:

1. Current situation. Is there an alert level?
2. What are landslides? Contains information about landslides.
3. General recommendations. Consists of information about what to do before, during and after an event.
4. Recommendations in slope areas. Includes information regarding what to do in slope areas.
5. Maps. Shows a gallery of maps about refuges and safe places.

4.2   Development Stage

The interactive application was developed in the Eclipse platform, with the following plugins:

- NCL-Eclipse, which allows coding in the NCL language and correct syntax and relationship errors among components.
- NCL-Lua, for coding in the Lua scripting language, useful for performing mathematical processes.

    Ginga-NCL was used for the execution of NCL applications.

    Figure 4 illustrates the interface of the interactive application. Note that if the receiver has a return channel over Internet, updated information can be obtained from the UCuenca-SDI server.

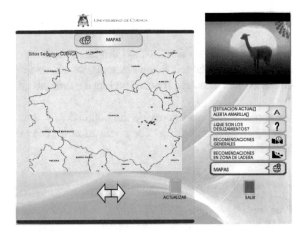

Fig. 4. Interactive application interface

In addition, the Digital TV Management Web module was implemented as a component of the UCuenca-SDI, which allows managing the content of the interactive application, selecting the alert level and downloading the application in a .zip format. Figure 5 illustrates the interface of the management web module.

Fig. 5. Management web module

4.3 Testing Stage

Tests were performed in two environments: simulation (VMWare, Fedora Linux) and Tx/Rx equipment (DTT Laboratory).

A comparison between two versions of interactive application, executed in DTT Laboratory is showed in Table 2. It is important to emphasize that loading time changes according to the brand and firmware of the receptor. Charge times from 22 to 33 s were measured.

Table 2. Comparison versions

Previous version	Current version
There is not focus color	The focus color was set to purple for highlighting options
The application has help button	The help button was removed, because of the application must be intuitive
Three maps can be loaded in gallery	"N" maps can be loaded in Gallery
"Update button" was not implemented	"Update button" enable download of updated maps if user have return channel
The video of programming was not visible because of problems in layers	The problem was solved
Return channel was not used	The average charging time with return channel is one second for image

5   Conclusions

Ecuador is a country where the effects by natural hazards have been devastating, with a high potential for danger to people. This paper has presented an approach for early warning by means of DTT using EWBS, which has been tested by controlled experiments at a laboratory level. The software architecture centered in a SDI core has represented the main contribution for the proposed solution.

The SDI core is used as source for generating codes to be transmitted to a channel station, and for the agency that manage emergencies. This way, the population will be able to receive real information in seconds, provided by a broadcast system, with a higher penetration level as compared to the internet service.

The relevance of this work is integrating in the same platform, accessible via internet with security levels, crucial information for the population that is managed by the official agency for dealing with emergency matters.

In a year time has been programmed the analogue transmission switch-off, and therefore, every effort in order to have applications with real interactivity is a trend in Ecuador. In this sense, the proposed EWBS might become a viable alternative for saving people lives, infrastructures and so on.

This work have contributed to the techno-scientific development of a scheme for interactivity and integration of DTT with the SDI. This work also contribute to include the obligatory nature of the EWBS protocol within the ISDB-T standard for Digital TV in Ecuador.

Acknowledgment. This project was financed by the DIUC funds under the XV project contest, with the name "Geoprocessing SDI as support for vulnerability analysis and territorial planning" 2017–2018. This project had the collaboration of DTT working group and laboratory of ESPE, through Dr. Gonzalo Olmedo. Special thanks to Ing. Ricardo Peñaherrera and Ing. Leonardo Espinosa of the SGR for their support regarding early warnings and DTT.

References

1. Brabb, E.E., Harrod, B.L.: Landslides: extent and economic significance. In: Proceedings of the 28th International Geological Congress: Symposium on Landslides (1989)
2. Cruden, D., Varnes, D.: Landslide types and processes. Transp. Res. Board Spec. Rep. **247**, 36–75 (1996)
3. Abhik, D., Toshniwal, D.: Sub-event detection during natural hazards using features of social media data. In: 22nd International Conference on World Wide Web, New York, NY, USA, pp. 783–788 (2013)
4. Lee, S., Poudyal, C.P.: Landslide hazard mapping using geospatial models. presentado en 2nd International Conference on Computing for Geospatial Research & Applications, New York, NY, USA, Article no. 34, 5 p. (2011)
5. Woolard, F., Bolger, M.: Using unity for immersive natural hazards visualization. presentado en SIGGRAPH Asia 2015 Posters, New York, NY, USA, Article no 34, 1 p. (2015)
6. Ministerio de Telecomunicaciones: ¿Qué es la TDT?, TDT - Televisión Digital Terrestre. https://tdtecuador.mintel.gob.ec/que-es-la-tdt/. Accedido 18 May 2018

7. Gobierno de España: Televisión digital - Interactividad. http://www.televisiondigital.gob.es/TelevisionDigital/tecnologias/Interactividad/Paginas/interactividad.aspx. Accedido 18 May 2018
8. Ministerio de Telecomunicaciones: Antecedentes, TDT - Televisión Digital Terrestre. https://tdtecuador.mintel.gob.ec/antecedentes-tdt/. Accedido 18 May 2018
9. Ministerio de Telecomunicaciones: Ciudades con cobertura TDT, TDT - Televisión Digital Terrestre. https://tdtecuador.mintel.gob.ec/normativas-para-concesionarios-de-senal-abierta/. Accedido 18 May 2018
10. Villacrés Jiménez, D.P.: Implementación de un sistema piloto de transmisión de alerta de emergencia sobre la televisión digital terrestre en el ECUADOR, May 2013
11. ITU Association of Japan: Estandarización del Sistema de Alertas de Emergencia EWBS en América Central y Sudamérica (2013)
12. ISDB-T: ISDB-T Harmonization document part 3: Emergency Warning Broadcast System EWBS, 30 November 2015
13. Sheina, S., Shumeev, V., Matveyko, R., Babenko, L., Khamavova, A., Kartamysheva, A.: GIS and territorial planning. In: Proceedings of the International Conference on Computing in Civil and Building Engineering 2010, vol. 30, pp. 1–12 (2010)
14. Morocho, V., Achig, R., Bautista, S.: Esquema de activación de alertas tempranas en Televisiòn Digital generadas a través de una IDE. Revista Geoespacial **14**(2) (2017)
15. Morocho, V., Santander, F.: De los rígidos códigos internos de una IDE a un ambiente gráfico de administración y gestión: IDE Ucuenca V3.5. Revista Geoespacial **12**, 45–56 (2015)
16. Pillajo, C.A., Ochoa, J.S., Acosta, F.R., Olmedo, G.F.: Herramienta multiplataforma para generación automática de aplicaciones interactivas Ginga-NCL basado en plantillas. Maskay **6**(1), 8–12 (2016)

Performance Evaluation of Apache Zookeeper Services in Distributed Systems

Renato Toasa[1,2(✉)], Clay Aldas[2(✉)], Pablo Recalde[1(✉)], and Rosario Coral[1(✉)]

[1] Departamento de Ciencias de la Ingeniería,
Universidad Tecnológica Israel, Quito, Ecuador
{rtoasa, precalde, rcoral}@uisrael.edu.ec
[2] Universidad Técnica de Ambato, Ambato, Ecuador
clayaldas@uta.edu.ec

Abstract. Zookeeper was designed to be a robust service, it exposes a simple API, inspired by the filesystem API, that allows to implement common coordination tasks, such as electing a master server, consensus, managing group membership and managing metadata. This work makes a basic implementation and evaluation of Zookeeper and simulation with some of the services offered to evaluate them and discover how distributed applications can benefit. Obtaining as a result that Zookeeper offers a stable, simple and high performance coordination service that provides the necessary tools to manage distributed applications without worrying about the loss of information.

Keywords: Zookeeper · Coordination · Consensus · Leader election

1 Introduction

The distributed coordination in the cloud is a very important aspect in the current services that are offered through the internet. The main advances and challenges to be solved within the distributed coordination in cloud computing are: the consensus and the process of choosing the process leader [1].

An approach to coordination is the development of services for each of the different coordination needs. For example, Amazon Simple Queue Service focuses specifically on queues. Other services have been developed specifically for leader choice and configuration [2].

Applications distributed on a large scale require different forms of coordination to execute their services. The configuration is one of the most basic forms of coordination. In its simplest form, the configuration is just a list of operating parameters for system processes, while more sophisticated systems have dynamic configuration parameters [2].

Consensus and leader choice are common in distributed systems, often these processes need to know what other processes are running and what these processes are in charge of [3]. Consensus allows the processes to reach a common decision based on certain values, and the election of a leader is a process in which a unique process is chosen to take a certain role and lead a certain activity commissioned by a process [4].

© Springer Nature Switzerland AG 2019
Á. Rocha et al. (Eds.): ICITS 2019, AISC 918, pp. 356–364, 2019.
https://doi.org/10.1007/978-3-030-11890-7_35

There are several projects that work with these services such as: Apache Zookeeper, Chubby and Doozer. Apache Zookeper was chosen since its implementation is relatively simple and will allow to achieve the objectives of this work quickly [5]. Zookeeper is a high performance coordination service to build distributed applications, and among the main services is the consensus and choice of leader, Zookeeper offers an API for its implementation [6].

The algorithms that intervene in this processes are complex to understand and execute, therefore you must resort to a tool that facilitates the work, this is where Apache Zookeeper intervenes that will allow to implement and evaluate these processes in order to understand their operation and define in what type of systems can be included to improve their functioning.

The rest of the paper is organized as follows. Section 2 describes the literature review. Section 3 presents the zookeeper implementation, while the application to evaluate is shown in Sect. 4. Next, in Sect. 5 tests and results are detailed, and in Sect. 6 conclusions are presented.

2   Literature Review

The evaluation of performance in different systems and computer processes is very necessary to verify that the operation is correct and meets its objective, allowing to compare different alternatives, determine the impact of a new element or characteristic in the system, measure relative performance between Different systems, there are several investigations that work in the evaluation of performance such as: evaluation of database [7], evaluation of sensors [8], evaluation of communication protocols [9].

Currently there are large companies that use Zookeeper and its services to manage part of its critical processes, among these large companies are: Yahoo that uses Zoo-Keeper for tasks such as: leader choice, collision detection, and metadata storage, facebook use Zookeeper for: the storage of data records through multiple machines, for failover and for service detection [1]. In addition, there are several investigations that work with zookeeper and are detailed below.

Work [3] describes Zookeeper as a service for coordinating processes of distributed applications. ZooKeeper aims to provide a simple and high performance kernel for building more complex coordination primitives at the client. It incorporates elements from group messaging, shared registers, and distributed lock services in a replicated, centralized service.

It is also used Zookeeper to build distributed applications implementing open APIs that enables developers to apply their own powerful coordination primitives. There is a case study about Zookeeper implementations in the payment process where synchronization and coordination is not meet due to dual server implementation, where orders are placed through a store application [10].

The research [11] presents an alternative protocol Instant Commit, which is write delay optimized. We implemented the adaptation logic and Instant Commit in Zoo-Keeper and show its potential with a workload-aware adaptation trigger and propose to switch ZooKeeper's underlying consensus protocol at runtime while preserving all guarantees for the overlying applications.

In the same way as this research, there are others that evaluate the functionalities offered by Zookeeper, [12] use the tool "Modbat" that generates test cases for concurrent client sessions, and processes results from synchronous and asynchronous callbacks, this work has detected multiple previously unknown defects in ZooKeeper. While [13], present a Dynamic Service Discovery System with a Priority Load Balance Strategy. This system not only integrates Zookeeper and tradition web project written with Spring and SpringMVC to achieve the dynamic service discovery function, which feeds the basic needs of the Priority Load Balance Strategy and in [14] is evaluate the use of ZooKeeper in a distributed stream computing system called System S to provide a resilient name service, dynamic configuration management, and system state management. The evaluation shed light on the advantages of using ZooKeeper in these contexts as well as its limitations.

The proposed work realizes a basic implementation of Zookeeper and an evaluation to consensus and choice of leader that offers zookeeper, in this way it is possible to analyze the operation of these processes in the distributed systems and to identify how reliable they are through, functional tests and data loss.

3   Zookeeper Implementation

ZooKeeper does not expose primitives directly. Instead, it exposes a file system-like API comprised of a small set of calls that enables applications to implement their own primitives. We typically use recipes to denote these implementations of primitives. Recipes include ZooKeeper operations that manipulate small data nodes, called znodes, that are organized hierarchically as a tree, just like in a file system (see Fig. 1) [2].

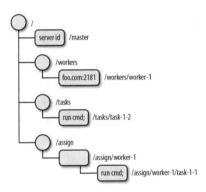

Fig. 1. Zookeeper structure

To get started, is necessary to download the ZooKeeper distribution of the main page, and unzip this files [6]. To set up ZooKeeper locally or in multiple servers in standalone mode and create a session, it uses the *zkServer* and *zkCli* tools that come with the ZooKeeper distribution under bin/.

Before executing any request against a ZooKeeper, the client must establish a session with the service. The concept of sessions is very important and quite critical for the operation of ZooKeeper. All operations a client submits to ZooKeeper are associated to a session. Sessions offer order guarantees, which means that requests in a session are executed in FIFO (first in, first out) order. Typically, a client has only a single session open, so its requests are all executed in FIFO order [1].

The Zookeeper API has the possibility to make direct calls to the system, Every time Zookeeper is started, a session is created in the system, maintaining an active process until the Zookeeper execution is finished, if a server is disconnected, the client is automatically routed to another server and starts a session with it (see Fig. 2).

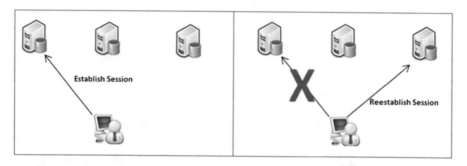

Fig. 2. Zookeeper sesions

To start zookeeper the following instruction is used:

```
bin / zkServer.sh start
```

4 Application to Evaluate

In this work we use three virtual servers; first start 3 instances of zookeeper, each instance is a server. For the servers to be communicated between, it is necessary to establish some data for the configuration, for this we open the file *zoo.cfg* that is in the *conf* folder of Zookeeper, and the following code is written.

```
tickTime=2000
initLimi t=10
syncLimit=5
dataDir=./ data
clientPort=2181
server.1=127.0.0.1:2222:2223
server.2=127.0.0.1:3333:3334
server.3=127.0.0.1:4444:4445
```

The zkServer file is used to start the servers (see Fig. 3).

Fig. 3. Zookeeper instances

Once this is done, the Zookeeper service is available, now it is going to be configured the client that is going to connect to the service. The connection lists all the machines that are active: ports of the active machines, in this case the connection string is: 127.0.0.1:2181, 127.0.0.1:2182, 127.0.0.1:2183 which is what was created in the configuration files of the Zookeeper servers.

To connect the client with the servers is use the *zkClient* File (see Fig. 4).

Fig. 4. Zookeeper client

The clients are connected in a random order to the servers in the connection chain. This allows ZooKeeper to achieve simple load balancing. However, customers are not allowed to specify a preference for a server to connect.

Now a master-slave structure is created with Zookeeper's Znodes, The Master-Slave model involves 3 Znodes:

The /workers znode is the parent znode to all znodes representing a worker available in the system. If a worker becomes unavailable, its znode should be removed from /workers.

The /tasks znode is the parent of all tasks created and waiting for workers to execute them. Clients of the master-worker application add new znodes as children of /tasks to represent new tasks and wait for znodes representing the status of the task.

The /assign znode is the parent of all znodes representing an assignment of a task to a worker. When a master assigns a task to a worker, it adds a child znode to /assign.

The Fig. 5 show the creation of this Znodes.

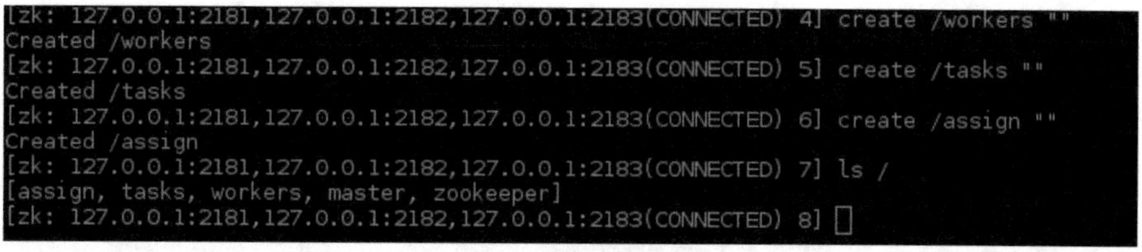

```
[zk: 127.0.0.1:2181,127.0.0.1:2182,127.0.0.1:2183(CONNECTED) 4] create /workers ""
Created /workers
[zk: 127.0.0.1:2181,127.0.0.1:2182,127.0.0.1:2183(CONNECTED) 5] create /tasks ""
Created /tasks
[zk: 127.0.0.1:2181,127.0.0.1:2182,127.0.0.1:2183(CONNECTED) 6] create /assign ""
Created /assign
[zk: 127.0.0.1:2181,127.0.0.1:2182,127.0.0.1:2183(CONNECTED) 7] ls /
[assign, tasks, workers, master, zookeeper]
[zk: 127.0.0.1:2181,127.0.0.1:2182,127.0.0.1:2183(CONNECTED) 8] []
```

Fig. 5. Znodes creation

The Client adds tasks (see Fig. 6) in the system, no matter what the task really performs so the client is supposed to ask the master-worker system execute a *cmd* command with the task to be performed.

```
[zk: 127.0.0.1:2181,127.0.0.1:2182,127.0.0.1:2183(CONNECTED) 7] create -s /tasks/task- "cmd"
Created /tasks/task-0000000000
[zk: 127.0.0.1:2181,127.0.0.1:2182,127.0.0.1:2183(CONNECTED) 8] []
```

Fig. 6. Task creation

The client sends several requests to the zookeeper servers, with the objective of verifying the load balancing in these, how the consensus of the servers is made and which process takes on the role of Leader. All servers reach consensus and decide which server will take care of the workload of the server that failed.

5 Test and Results

An application was programmed in Java, which performs the monitoring of the Zookeeper servers and the requests of the clients (see Fig. 7).

A summary of the data obtained in the tests that were carried out are shown in Table 1.

This data was obtained from the output files of each server, information was also obtained on CPU consumption, Memory and it was identified as being minimal, and it does not affect server performance. The SAP tool was used to obtain CPU consumption values and these values are plotted with GNUPLOT (See Fig. 8).

Of the three previous scenarios there are a minimum consumption of CPU. In this way the effectiveness of Zookeeper is checked.

Fig. 7. Developed application

Table 1. Summary of evaluation

Case	N' requests	N' clients	N' servers	Time	Action
1	100	100	5	2 min	0 servers fail
2	500	100	5	5 min	1 server failed The next server assumed the workload
3	1000	100	5	5 min	1 server failed The requests are distributed in the 4 servers

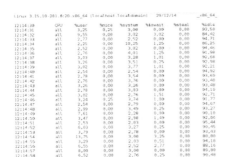
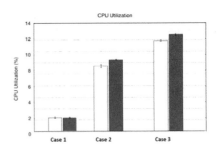

Fig. 8. CPU utilization by case

6 Conclusions

It was found that Zookeeper is a stable, simple and high performance coordination service that provides the necessary tools to manage distributed applications because if one server fails, another one ready to replace it, this is due to the process in which all the servers they reach a consensus and choose a leading server that assumes the load of the server that fails.

With the programmed simulation it is determined that ZooKeeper reaches values performance of several operations and requests per second for the workloads, without affecting server performance.

Zookeeper offers a great modularity which allows using both the API (Application Programming Interface) complete or using separately some of its characteristics to optimize some process in operation or already structured.

References

1. Guachi, T., Mauricio, R.: Análisis de los procesos de decisión común en sistemas distribuidos tolerantes a fallos a través de Zookeeper (2015)
2. Junqueira, F., Reed, B.: ZooKeeper: Distributed Process Coordination (2013)
3. Hunt, P., Konar, M., Junqueira, F.P., Reed, B.: ZooKeeper: wait-free coordination for internet-scale systems. In: USENIX Annual Technical Conference, vol. 8, p. 11 (2010)
4. Chaudhuri, S.: More choices allow more faults: set consensus problems in totally asynchronous systems. Inf. Comput. **105**(1), 132–158 (1993)
5. chubby & zookeeper: different consistency level | Xudifsd chubby & zookeeper: different consistency level | Rational life. http://xudifsd.org/blog/2016/06/chubby-zookeeper-different-consistency-level/. Accessed 03 Sept 2018
6. Apache ZooKeeper - Home. https://zookeeper.apache.org/. Accessed 03 Sept 2018
7. Flores, A., Ramírez, S., Toasa, R., Vargas, J., Urvina, R., Lavin, J.M.: Performance evaluation of NoSQL and SQL queries in response time for the E-government. In: 5th International Conference on eDemocracy eGovernment, pp. 257–262 (2018)
8. Vargas, J., Mariño, C., Aldas, C., Morales, L., Toasa, R.: Kinect sensor performance for Windows V2 through graphical processing. In: 10th International Conference on Machine Learning and Computing, ICMLC 2018, pp. 263–268 (2018)
9. Silva, C., Toasa, R., Martinez, H.D., Veloz, J., Gallardo, C.: Secure push notification service based on MQTT protocol for mobile platforms. In: XII Jornadas Iberoamericanas de Ingenieria de Software e Ingenieria del Conocimiento 2017, JIISIC 2017 - Held Jointly with the Ecuadorian Conference on Software Engineering, CEIS 2017 and the Conference on Software Engineering Applied to Control and Autom (2017)
10. Goel, L.B., Majumdar, R.: Handling mutual exclusion in a distributed application through Zookeeper. In: 2015 International Conference on Advances in Computer Engineering and Applications, pp. 457–460 (2015)
11. Frommgen, A., Haas, S., Pfannemuller, M., Kohldehofe, B.: Switching ZooKeeper's consensus protocol at runtime. In: 2017 IEEE International Conference on Autonomic Computing (ICAC), pp. 81–82 (2017)
12. Artho, C., et al.: Model-based API testing of apache ZooKeeper. In: 2017 IEEE International Conference on Software Testing, Verification and Validation (ICST), pp. 288–298 (2017)

13. Song, M., Luo, G., Haihong, E.: A service discovery system based on zookeeper with priority load balance strategy. In: 2016 IEEE International Conference on Network Infrastructure and Digital Content (IC-NIDC), pp. 117–119 (2016)
14. Pham, C.M., Dogaru, V., Wagle, R., Venkatramani, C., Kalbarczyk, Z., Iyer, R.: An evaluation of zookeeper for high availability in system S. In: Proceedings of the 5th ACM/SPEC International Conference on Performance Engineering - ICPE 2014, pp. 209–217 (2014)

Framework for Supporting JavaScript-Based Mobile Agents

Carlos A. Silva[1,2(✉)], Carlos Grilo[3(✉)], Jorge Veloz[2(✉)],
and Nuno Costa[3(✉)]

[1] School of Technology and Management, Polytechnic Institute of Leiria,
R. Gen. Norton de Matos, 2411-901 Leiria, Portugal
2162317@my.ipleiria.pt
[2] Facultad de Ciencias Informáticas, Universidad Técnica de Manabí,
Av. José María Urbina y Che Guevara, 130104 Portoviejo, Ecuador
jorge.veloz@fci.edu.ec
[3] School of Technology and Management, CIIC, Polytechnic Institute of Leiria,
R. Gen. Norton de Matos, 2411-901 Leiria, Portugal
{carlos.grilo,nuno.costa}@ipleiria.pt

Abstract. The emergence both of World Wide Web and distributed systems solutions allowed many communication paradigms to appear, being the client-server the most common today. Here, we present an agent-based platform for the mobile agents computing paradigm. The platform components have been designed to allow the development, execution, tracking and the ability to move JavaScript mobile agents through the local network and Internet. All components are based on the JavaScript programming language in order to reach desktop and mobile operating systems, such as Android and iOS. This initiative arose as a way of dealing with problems raised by the considerable amount of existing Java based mobile agents platforms, which force the installation of the Java Virtual Machine on the devices, making complicated their execution in operating systems like macOS and others non-Java friendly operating systems, including mobile ones.

Keywords: Agent-based platform · Mobile agents paradigms · JavaScript · Java Virtual Machine · Mobile operating system

1 Introduction

With the appearance of the World Wide Web, some computational paradigms emerged to optimize the use of resources, both computational and network resources. For example, client/server, remote evaluation (REV), code on demand and mobile agents paradigms appeared as solutions for clients that have bandwidth limitation and others [1]. The client-server model is commonly used to develop applications executing the process on the server side, lightening the load in the client (web browsers or client applications). For example, obtaining personal information, instant messaging services (push messaging), Global Positioning System (GPS) navigation services, and others follow the client/server paradigm. In the last couple of years, the approach to reduce the

© Springer Nature Switzerland AG 2019
Á. Rocha et al. (Eds.): ICITS 2019, AISC 918, pp. 365–375, 2019.
https://doi.org/10.1007/978-3-030-11890-7_36

burden of processes on the client's side has been resumed again with the appearance of platforms for application development executed on the server side. For example, developing web applications using the Java programming language through the Vaadin framework or web applications developed with JavaScript using tools like Node.js runtime with Angular development framework or React framework, and others.

In the mobile agents paradigm approach, the agent is moved (code and data stored) and executed, maintaining their state among all the hosts of the network. As opposed to systems that only allow the exchange of nonexecutable data, systems incorporating mobile agents can achieve significant gains in performance and functionalities [2].

The solution proposed in this paper consists in a set of software components for an agent-based platform that allow the execution of mobile agents in desktop operating systems such as Windows, Linux and macOS, including operating systems of mobile devices, e.g. Android and iOS. One of the developed components is a framework that allows to create, execute, monitor and track all the agents running in a network. The platform uses JavaScript as programming language because it can be executed in any operating system where the developed software components are installed and executed.

Throughout this document, the state of the art is presented and discussed in Sect. 2, while Sect. 3 describes the solution's functional specification. Section 4 presents the implementation in real devices. Section 5 describes the evaluation environment and test's results. Finally, conclusions and future work are presented in Sect. 6.

2   State of the Art

In the past decade, many mobile agents platforms have been developed. Some of the well-known mobile agents platforms include Mole [3], Aglets [4], Concordia [5], D'Agents [6], Ara [7], TACOMA [8], JADE [9], Ajanta [10], Tryllian's agent development kit [11], Fipa-os [12], Grasshopper [13], and JACK Intelligent Agent [14].

An important aspect to take into consideration when selecting a Middleware/ Framework for the development of systems based on mobile agents is the agent programming language, type of license that the development and implementation platform supported.

Table 1 shows license type, agent programming language support, license type and the last update date of the most well-known mobile agents middleware/frameworks. As can be seen, most of them are developed using the Java language. Java applications can be executed on any devices that have a running Java Virtual Machine (JVM). However, there are widely used operating systems as, for example, macOS and iOS, that are not enough compatible to run the JVM, including all applications developed in this environment.

The work presented here is a first step to overcome this weakness in the area with the goal of expanding the number of devices to be used in mobile agent-based solutions. A lot of agent-based systems have been proposed and multi-agent systems have been studied as solutions for current problems in different areas. For example, in [15], the authors developed a virtual butler that provides the interface between the elderly and a smart home infrastructure. The middleware platform used in this work was JADE [9] and the mobile agents were developed using the Java language. Additionally, the authors of [16] proposed a mobile agent-based solution for solving the energy sink-hole

Table 1. Mobile agents middleware/framework list.

Middleware/Framework	Agent programming language	License	Last updated
Concordia	Java	Unknown	July 2017
Aglets software development kit	Java	IBM Public License	July 2017
Java Agent DEvelopment framework	Java	LGPL Version 2	July 2015
JACK TM intelligent agent	Java	Proprietary	2002
Ajanta	Java	Proprietary	1999
Tacoma network agent	C, Tcl/Tk, and others	Qt Public License	1999
Tryllian's agent development kit	Java	Not available today	Not available

problem, aiming to extend the network life by reducing redundant data being passed to the nodes near to the sink and, thereby, reducing the load and saving battery life. The algorithm was implemented using aglets and the analytical results showed significant improvement in the network lifetime. Furthermore, in [17], an IEEE FIPA compliant mobile agents system has been developed and an agent-based real-time traffic detection and management system (ABRTTDMS) was designed.

In [18], the authors developed an energy-efficient, fault-tolerant approach for collaborative signal processing information (CSIP) among multiple sensor nodes using the mobile-agent-based computing model, with the aim of increasing energetic efficiency of the equipment and improve fault tolerance. Additionally, in [19], the authors proposed the use of mobile agents for the design of an infrastructure for the data integration in the network (DSN), to decrease the consumption of bandwidth. According to the evaluations done, the authors were able to conclude that the new infrastructure based on mobile agents in their most optimal performance can save more than 98% of the execution time, due the less data transfer time spent.

3   Solution's Functional Specification

Figure 1 shows the high level architecture of the proposed solution. The solution is composed by *mobile agent(s)*, *middleware*, *registry server* and *mobile agent framework*. *Mobile agents* are the programs to be executed, while the *middleware* is a software layer that supports execution of the agents and their mobility through devices. This means that each node that can be visited by a mobile agent must have the middleware layer installed. The *mobile agent framework* acts as a dashboard for all the stablished mobile agent environment, allowing programmers do write mobile agent's source code, compiling, deploying and keep tracking of the existing mobile agents. Finally, the *registry server* intermediates the communication between middlewares and mobile agent frameworks. The specified agent-based platform allows the agents execution, moving them through different local networks and through the Internet.

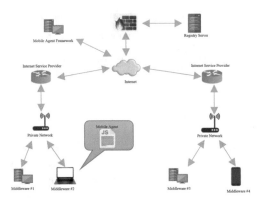

Fig. 1. High level agent-based platform system architecture.

3.1 Criteria Evaluation

For the specification and development of an agent-based platform, the following criteria were considered: (1) The programming language (PL) to build both agents and software components of the developed platform (middleware and mobile agent framework); (2) The communication protocol system for message passing between middlewares; (3) The communication architecture.

3.1.1 The Programming Language

The choice of the programming language took place at the beginning of the project and it was carried on according to the following requirements: (1) Be object-oriented (agents can be considered object structures); (2) Be robust (for error handling during importation and agent's execution); (3) Support for object serialization (convert code executed to a string of bytes); (4) Be flexible (simple variable declaration and function support); (5) Be compliant with traditional operating systems, including mobiles ones.

We have chosen the JavaScript programming language because it complies with all these requirements. One important aspect using JavaScript is that it is possible to develop native mobile applications using the React Native framework, while for desktop applications development it is possible to use the Node.js runtime. For the agent-based framework development, the Electron development framework can be used, allowing to build desktop applications using JavaScript.

3.1.2 The Communication Protocol System

The communication infrastructure is IP based and each middleware needs a network interface (ethernet or Wi-Fi adapter). The communication protocol used was TCP/IP v4 using a WebSocket API, which allows bidirectional and asynchronous communication. This means, that the WebSocket API is able to send both agents and messages (messages types specified with keywords using the JSON format) without the need to stop the current process in order to wait for a response.

3.1.3   Communication Architecture

The most suitable communication architectures available to build an agent-based platform are the peer-to-peer and client-server architectures. In the peer-to-peer architecture, all computers within the network have direct communication. On its turn, the client-server model consists of a central host in a network and several other computers as clients. In the initial phase of this project, prototypes were developed with the objective of evaluating the most appropriate communication architecture for such agent-based platform. Both prototypes are able to send an object from one host to another. The following results led us to conclude that the client-server model is the most appropriate for an agent-based platform:

- Thanks to the existence of a centralized host in the client-server model, the search of new middleware running within the network is not necessary (required in peer-to-peer model). The central host would be known by all hosts, diminishing the time of execution and lightening the middleware by removing these features;
- The devices with a middleware running do not need to be inside the same local network (required in peer-to-peer). The centralized host is configured with a public IP that allows access to any middleware from any place that has an Internet connection. Furthermore, clients can have the same (private) IP address, be it in the different network segment or in different local networks. This does not cause problems with the identification of the host clients as happens in the peer-to-peer model;
- In comparison with the peer-to-peer model, it is not necessary to perform the additional configuration in network devices, such as NAT or VPN. It is only required that the centralized host running the registry server allows communication in the assigned ports.

4   Implementation

The following sections describe each developed platform components concentrating on implementation aspects.

4.1   Registry Server

The registry server works as a routing server allowing to move agents through the Internet. It stores internally both middlewares and frameworks host list name. The middleware host list is shared with all hosts connected and the *registry server* has the ability to send asynchronous messages to all connected hosts about different events that occur as, for example, the host connection/disconnection to/from the network. In terms of implementation, the registry server is a software server implemented in Node.js using the JavaScript programming language with libraries showed in the next software layer diagram (see Fig. 2).

The registry server main aim is both allowing mobile agent movement and communication messages through the Internet. The registry works with two listening services (middleware and framework separately) that allow to send/receive messages and

Fig. 2. Registry server software stack.

then the suitable function is invoked according to the message type. For example, it notifies all connected frameworks about all agents tracking.

4.2 Middleware

The middleware is the core component and allows the execution and mobility of agents as it receives agents arriving to the host. The middleware has the ability to instantiate arriving agents and put them to run. Apart from the runtime environment provided by middleware, it also exposes an API that running mobile agents can invoke in in order to request the following actions: *MoveTo ([params])* - it allows the agent to move from one host to another; *GetAvailableHosts ([params])* - it returns the middlewares hostname list; *GetHostName([params])* - it returns the local middleware host name assigned by the registry server; *GetRandomHost ([params])* - it returns a random hostname from available host list.

The developed agent-based platform includes two different middleware versions using the JavaScript programming language with the same functions and capabilities: a version for desktop operating systems, called *desktop middleware* (for Windows, Linux and macOS operating systems) and another version for mobile operating systems, called *mobile middleware*, which supports Android and iOS operating systems.

The desktop middleware was implemented using Node.js runtime, while the mobile middleware version was implemented using the React Native framework, it allows the creation of the mobile middleware as a native application. Figure 3 below shows the complete software stack per host, including the running mobile agent.

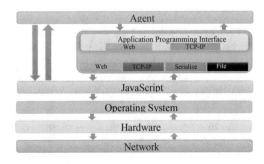

Fig. 3. The complete software stack per host.

4.3   Mobile Agents

A mobile agent is a set of tasks the user wants to execute in every middleware equipped host. In practice, a mobile agent is a JavaScript program that must be compliant with an established (software) interface/prototype presented next.

```
exports.runAgent = function() {
    //AGENT CODE
};
```

According to the above prototype, the random *hello world* mobile agent could be written as follows:

```
exports.runAgent = function() {
  this._name = "Scenario 002"; -> Agent Name assigned
  var agent = this;
  console.log("Hello, World!"); -> Print hello World
  var host;
  var api = this._api; -> API Importation
  setTimeout(function() {
    api.getRandomHost(function(err, result) { -> Call Middleware
function
      if (err) {console.log('Error: ' + err);}
      else {
          host = result;
          api.moveTo(agent, host, function(err, result) { -> Move
            if (err) {
            console.log('Error: ' + err);
            } else {
            console.log('\033[2J');
            }
        });
      }
    });
  },5000); -> Waiting 5 seconds for each movement.
};
```

4.4   Mobile Agent Framework

The mobile agent framework was developed also in JavaScript using the Electron Framework, it allows cross platform desktop apps built with JavaScript, HTML and CSS. Figure 4 illustrates the mobile agent framework software stack.

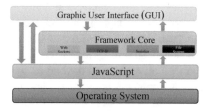

Fig. 4. Mobile agent framework stack.

The mobile agent framework provides a graphic interface that exposes six different windows: Window 1 is a text editor for mobile agents programming and it allows to send the agent to run on a specific host through a drop-down list, which contains a middleware hostname list; Window 2 shows all middlewares hostnames connected to the registry server and it allows information gathering about hardware and operating systems; Window 3 is a graphic interface for monitoring (tracking) the mobile agent movement between middlewares; Window 4 shows the middleware hostname list with running agents names (see Fig. 5); Window 5 shows the list of current connected frameworks, highlighting the IP address and allows gathering information about hosts; Windows 6 allows to configure IP address and port connection.

Fig. 5. Mobile agent framework – middleware connected information interface.

5   Test and Evaluation

The developed platform was tested in two different environments: LAN and WAN (see Fig. 6). The first one is a controlled local network and the second one includes the integration of two local networks and the Internet. Table 2 shows the devices list, operating systems versions and the used software versions (Node.js runtime and the React Native versions). Mobile devices only need to have installed the mobile application developed with React Native.

For each environment mentioned above three test-case scenarios were evaluated. In scenario 1, a mobile agent was developed with the aim of printing the message "Hello, World!" to the console on each device it moves to. The agent movement pattern is sequential, which means, it is executed on Middleware 1, Middleware 2… to Middleware *n*. However, in scenario 2 an agent was developed to print the "Hello, World!" message in the console, with a random movement pattern. There are three ways to develop an agent that moves in a random way: (a) through the *getAvailableHost ([params])* function, which returns a middleware hostname list, (b) using the *getRandomHost([params])* function, which returns a middleware hostname randomly; (c) using

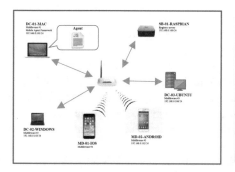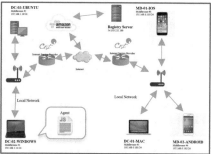

Fig. 6. LAN (left) and WAN (right) test environments.

Table 2. Devices used in the test environments.

Device information	Operating system	Version
MacBook Pro (Retina, 13-inch, Early 2015)	macOS High Sierra 10.13.4	Node.js v6.11.0
Dell system Inspiron N4110	Windows 10.0	Node.js v8.9.4
Intel desktop board	Ubuntu 16.04.2 LTS	Node.js 4.2.6
iPhone 6s	iOS 11.3.1 (15E302)	React native 0.46.11
Samsung S4 mini GT-I9192	Android 4.4.2	React native 0.46.11
Raspberry PI 3 model B V1.2 2015	2018-04-18-raspbian-stretch	Node.js 4.8.2

the reserved word "*random*" as destiny host name using the *MoveTo([params])* function. This last approach makes the task simpler and easier. Scenario 3 includes the test and evaluation of an agent with the aim of gathering the MAC address for each host it executes on, following a sequential movement pattern. The collected MAC addresses list is printed to the console when the agent returns to the starting host. All the test-case scenarios returned the expected results in all used devices. In the case of the third test-case scenario, it was necessary to include code to identify the underlying operating system, as the way of getting the MAC addresses is different for mobile platforms.

6   Conclusions and Future Work

There are already mobile agent-based platforms that allow the deployment of agent-based solutions, but they do not have enough support for mobile operating systems today. To accomplish the aim of this research an agent-based platform was developed for desktop and mobile devices using the JavaScript language. Furthermore, to facilitate the development, deployment and agent tracking, a framework was idealized and developed, and it can be seen as a dashboard that allows to visualize and gather information on both middlewares and frameworks running within the network.

The platform allows mobile agents to be moved in local networks and through the Internet. A suitable agent structure was also defined and can be easily handled as an *object* with the aim of being possible to be serialized, which provides the ability to move agents through the network and facilitates their manipulation by the middleware. The middleware offers support for Windows, Linux and macOS and mobile devices OS such as Android and iOS. The results obtained are promising for future research or applications development in this area.

Future work includes incorporating the capacity of agents to establish communication with other agents that are running on other hosts for information sharing. We also expect to endow the platform with the ability to run developed agents in other programming languages, such as LUA, TCL, among others.

Acknowledgments. Thanks to both the Secretaría de Educación Superior, Ciencia, Tecnología e Innovación (SENESCYT) and the Instituto de Fomento al Talento Humano of Ecuador, this research work would not have been possible without his financial support.

References

1. Lange, D.B.: Mobile objects and mobile agents: the future of distributed computing? In: European Conference on Object-Oriented Programming, pp. 1–12. Springer, Heidelberg (1998)
2. Knabe, F.C.: Language support for mobile agents. Doctoral dissertation, Carnegie Mellon University (1995)
3. Baumann, J., Hohl, F., Rothermel, K., Strasser, M., Theilmann, W.: MOLE: a mobile agent system. Softw.-Pract. Exp. **32**(6), 575–603 (2002)
4. Lange, D.B., Oshima, M.: Programming and Deploying Java Mobile Agents with Aglets. Addison-Wesley, Boston (1998)
5. Wong, D., Paciorek, N., Walsh, T., DiCelie, J., Young, M., Peet, B.: Concordia: an infrastructure for collaborating mobile agents. In: International Workshop on Mobile Agents, pp. 86–97. Springer, Heidelberg, April 1997
6. Gray, R.S., Cybenko, G., Kotz, D., Peterson, R.A., Rus, D.: D'Agents: applications and performance of a mobile–agent system. Softw.-Pract. Exp. **32**(6), 543–573 (2002)
7. Peine, H.: Application and programming experience with the ara mobile agent system. Softw.-Pract. Exp. **32**(6), 515–541 (2002)
8. Johnansen, D., Lauvset, K.J., van Renesse, R., Schneider, F.B., Sudmann, N.P., Jacobsen, K.: A TACOMA retrospective. Softw.-Pract. Exp. **32**(6), 605–619 (2002)
9. JADE - Java Agent Development Framework. http://jade.tilab.com/. Accessed Sept 2018
10. Tripathi, A.R., Karnik, N.M., Ahmed, T., Singh, R.D., Prakash, A., Kakani, V., Vora, M.K., Pathak, M.: Design of the Ajanta system for mobile agent programming. J. Syst. Softw. **62**(2), 123–140 (2002)
11. Tryllian's. http://www.tryllian.com/. Accessed Sept 2018
12. Emorphia. http://fipa-os.sourceforge.net/index.htm. Accessed Sept 2018
13. Baumer, C., Breugst, M., Choy, S., Magedanz, T.: Grasshopper - a universal agent platform based on OMG MASIF and FIPA standards. In: First International Workshop on Mobile Agents for Telecommunication Applications (MATA 1999), pp. 1–18 (1999)

14. JACK. http://www.aosgrp.com/. Accessed Sept 2018
15. Costa, N., Domingues, P., Fdez-Riverola, F., Pereira, A.: A mobile virtual butler to bridge the gap between users and ambient assisted living: a smart home case study. Sensors **14**(8), 14302–14329 (2014)
16. Yadav, M., Sethi, P., Juneja, D., Chauhan, N.: An agent-based solution to energy sink-hole problem in flat wireless sensor networks. In: Next-Generation Networks, pp. 255–262. Springer, Heidelberg (2018)
17. Chen, B., Cheng, H.H., Palen, J.: Integrating mobile agent technology with multi-agent systems for distributed traffic detection and management systems. Transp. Res. Part C: Emerg. Technol. **17**(1), 1–10 (2009)
18. Qi, H., Xu, Y., Wang, X.: Mobile-agent-based collaborative signal and information processing in sensor networks. Proc. IEEE **91**(8), 1172–1183 (2003)
19. Qi, H., Iyengar, S., Chakrabarty, K.: Multiresolution data integration using mobile agents in distributed sensor networks. IEEE Trans. Syst. Man Cybern. Part C (Appl. Rev.) **31**(3), 383–391 (2001)

How Edge Computing Transforms the Security of Cloud Computing

Cesar de la Torre[1]([✉]), Marco de la Torre[2], Juan Carlos Polo[1], and Fernando Galárraga[1]

[1] Universidad de Fuerzas Armadas - ESPE, Sangolquí, Ecuador
{cadelatorre,jcpolo,jfgalarraga}@espe.edu.ec
[2] Wroclaw University of Science and Technology, Wroclaw, Poland

Abstract. These days, network administrators are interested in delivering high performance network cores and edges, ensuring optimal visibility of services and simultaneously reducing to minimum the IT risk against network downtimes. The edge computing's distributed architecture is an alternative to reduce the number of attack vectors. Transform network cloud services in edge computing components of cloud is fundamental because it is the best fit for latency-sensitive microservices. Identity enforcement/validation, for example, is a latency-sensitive task that all SECaaS applications could run at the edge for a superior end user experience. The ability to run latency-sensitive microservices may spawn a new generation of apps that will be developed with end-user proximity/location as a crucial design element. As a beginning point, to start the development of these new technologies, it is important the design of new models of analysis.

Keywords: Cloud computing · Information security · Networking

1 Introduction

In the past, the principal ideas of IT security focused on perimeter defenses firewalls, proxies, and content filtering. The idea of cloud computing has dramatically changed the scenario from the last-mentioned concepts. Because of new ideas to implement corporate IT services over the Internet.

Cloud computing is a general term for an old idea that evolves delivering hosted services using remote rather than local computing power and storage, to run virtualized applications that users access via corporate WAN/LAN or Internet. The edge can help to filter and reduce the amount of data that needs to be sent to the cloud or data center for processing. As a consequence, it helps to reduce the overall amount of traffic load on the Internet, and from the security point of view it contributes to reduce an application's attack surface [1].

Clouds are classified in publics and privates. A public cloud sells services to anyone on the Internet and the administration is out of enterprise's control. Besides, a private cloud is a proprietary data center or environment that supplies hosted services to a limited number of people or devices in a personalized environment. The idea is that smaller attack surface makes it harder for hackers and bots to compromise or steal data. See Table 1.

© Springer Nature Switzerland AG 2019
Á. Rocha et al. (Eds.): ICITS 2019, AISC 918, pp. 376–385, 2019.
https://doi.org/10.1007/978-3-030-11890-7_37

Table 1. Deployment of security components [2].

Administration	Location	Service requirement
Internal	User office/desk	Antimalware, data leak protection, client agents
Internal	Remote office	Security management services, local services
Internal	Data centers (private cloud)	Consolidate servers, virtualized applications. Central security management consoles
External	Third-party (public cloud)	Recollection of enterprise events (public cloud), virtualized environments for event and data analysis. Infrastructure as a service (IaaS)
External	Third-party (public cloud)	Distribution of analysis results and good practices, platform as a services
External	Third-party (public cloud)	SECaaS (security as a services). Security instances and security technology on-demand

For security, flexibility and cost effectiveness, organizations are using today private and public services and resources. The principal goal of cloud computing is to offer an easy, scalable access to applications in real time at the lowest possible cost and enforcing the security paradigms. The importance of micro security services is to guarantee identity enforcement or validation. It allows the visibility of virtualized applications or access to data. For a facility of analysis, we assume that cloud security is the sum of latency-sensitive microservice components. These are deployed on: network security, application security, service security, user security, and so on.

In this analysis, we assume that service visibility should be interrupted as a result of the increase of number of attack vectors or excessive traffic volume. Those interrupt the service response in a real time, compromising the well-being of services. Therefore, for high service availability, it is essential to increase the bandwidth efficiency and an optimal synchronization of active micro security services, ensuring legitimate end-user traffic gets to its proper cloud environment on time.

1.1 The Challenge of Edge Computing

Edge Computing is a distributed information technology (IT) architecture in which client data is processed at the periphery of the network, as close to the originating source as possible. In this term, time-sensitive data in an edge computing architecture may be processed at the point of origin by an intelligent device or sent to a private cloud located in close geographical proximity to the client. Data that is less time sensitive is sent to the public cloud for historical analysis, big data analytics and long-term storage.

While we consider that in all path routes it is transmitted massive amounts of raw data over a network links, it is a tremendous load on network resources. So, it is more efficient to process data near its source and send only the data that has a value over the network to a remote data center. A major benefit of edge computing is that it improves time to action and reduces time down response to milliseconds; meanwhile it also conserves network resources. The concept of edge computing is not expected to replace cloud computing. Rather, edge computing is a distributed architecture which pretends to reduce the system vulnerabilities and deliver to a more intelligent edge client the expert tools necessaries to avoid malware infections and security exploits.

An important assumption to this analysis is that all network links after a correct deployment of agents (micro security services) will have enough capacity to satisfy all user requirements. For example, this occurs in data flow influence, packet delivery resiliency, and retransmissions. Also, deployed agents will have the capacity to maintain updated the inventory of data, provide necessary encryption tools and give indispensable heuristic to understand and stop security attacks [3].

2   Definitions

Virtualized applications in a cloud consist of three elements: production, transport and consumption. The production is related to the computing and storage process. Consumption is the work performed in client devices. Transport is the networking technology that binds production to consumption. In the next paragraphs, we assume that production and consumption are the result of agents, which have the ability to interconnect themselves.

When virtualized application processes (agents) validate a user, identity is liable to accept and respond the requirements of other agents it is denominate service visibility. This is the principal parameter to evaluate the optimal deployment of virtualized application in a cloud, see Fig. 1.

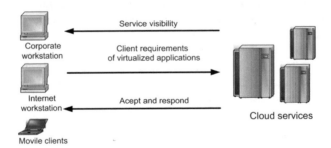

Fig. 1. Service visibility.

2.1   Edge Computing and Network Performance

Proactive management of micro security agents over LAN/WAN requires development of new components. It includes new programing languages oriented to analyze data, tools for security analysis, and tools analyzing network and application performance.

IT administrators to evaluate optimal network performance and security introduced by agents need to consider the following criteria: average improvement in network throughput, average improvement in application availability, average improvements in application response time, and minimum risk of user applications. Although, average improvement in other application response time is a critical element in network design, it is not considered in this work. Indeed, it could increase the network vulnerability and it is possible to offset this parameter through the optimization of the other three parameters. Consequently, the benefit of distributed agents to authenticate and validate end user identities and enforce API routing policies, ensuring only legitimate end-user traffic gets to its proper cloud environments is calculated as the minimization of threats in network flows. It is the reduction of total cost of transmission between nodes through discard of unsafe packets [8].

The gain of this work is to transform the cloud services throughout optimization of network and application that require of real time interaction and allows the possibility of hyper-personalization of web apps. While on the other hand, the distribution of micro security agents to critical links is a solution to the need for greater bandwidth and security in the end nodes.

$$\max_{flow} Benefit = \min_{networkTreaths} f(givenNetwork, userRisk, matrixFlow, networkThreats)$$

(1)

2.2 Principal Indicators for Edge Computing

The principal metrical applied to distribute edge computing agents are: risk in user applications, number of hits with unknown destination IP, and requirement of controlled access to VMs developed with end-user proximity/location design. So, the probability of micro security agent deployment depends of a set of variables $P = (p_1, \ldots, p_m)$, that can be calculated as function of the before mentioned indicators (independent variables).

$$p_i = f_i(applicationRisk, numberUnknowIP, accessVM)$$

(2)

Definition 1. Given a set of nodes in which client data is processed at the periphery of the network P, where $P = (p_1, \ldots, p_m)$ and m is the total number of those data processing nodes; exist a set of security policies O that require of validate en-user identities and enforce API routing policies, where $O = (o_1, \ldots, o_n)$ and n is the total number of security elements, the result of joints the both sets $P \cup O$ is a set of communication agents $A = (a_1, \ldots, a_m)$ which manage multiples clouds as a single platform [3].

2.3 Graphs and Notations Required

Indeed, many different families of graphs exist. We will analyze only the random graphs because they are the most common graphs found in practice and have a wide application for computer network analysis. In random graphs, arcs connect vertices without any structural regularity and the probability that an arc connect two vertices is independent of the vertices [4].

Since all the graphs are directed in this work, the network connections will be most probably bidirectional and the bandwidth in links are different for each way (asymmetric). Therefore, in real network computers the different forms how two agents in a network connect can be represented as a graph isomorphism.

Let direct graph $G = (V, E)$ consist of a finite non-empty set V of vertices and a binary relation E, i.e. a subset $E \subseteq VxV$. The elements E are called *edges or arcs*. An *edge* $(i, j) \in E$ is considered to be oriented from i *to* j. An undirected graph is a graph whose edges set E is symmetrical, i.e. $(i, j) \in E$ and $(j, i) \in E$. In next paragraphs the term graph will be used to refer to a direct graph, because consider that undirected graphs are just a particular case of direct graphs.

Given a graph $G = (V, E)$, E can be represented by an adjacency matrix $Adj(G) = A$ with size $|V|x|V|$, where $a_{ij} = 1$ if $(i, j) \in E$ and $a_{ij} = 0$ if $(i, j) \notin E$.

2.4 Network Definitions

It will be necessary to introduce some specific notation to be used in our network definitions.

Definition 2. Let each agent $a_i \in A$ represent one of two different types of process:

1. The traffic volume of services (processes) generated to validate end-user identities.
2. The traffic volume of API routing policies, necessary to maintain other application services.

The whole network is represented as a direct graph $G = (V, E)$, where $V = \{v_1, \ldots, v_n\}$ is the set of vertices, E is the set of edges $E \subseteq VxV$ and represents network connections. Links between vertices are considered directed for the sake of generality, because connections will be most probably bidirectional, and the bandwidths in links are different for each way (*asymmetric*).

Definition 3. Let E is a set of links connections between vertices V, such for all links between vertices (i, j) the following statement holds:

1. If exist a link between vertices (i, j), then $e_{ij} = 1$ and $e_{ij} \in E$.
2. If not exist a link between vertices (i, j), then $e_{ij} = 0$ and $e_{ij} \notin E$.

We consider a general wide area network, where *internal nodes are gateways* and *external (border) nodes are servers, clients or gateways to different subnets*. The agents are deployed in external network nodes V and exist for each ordered pair of communications agents (a_i, a_j) a static route path in the graph G.

Definition 4. Let a graph $G = (V, E)$ can be represented by adjacency matrix $Adj(G) = A$ with size $|V|x|V|$, then we assume for element $a_{ij} \in A$ is true the following statements:

1. The routes are statics in time.
2. The communication between agents is through the shortest path.
3. The time delay is 0 and not depended of hops and distances.
4. The capacity of each link is enough for any routed traffic, even if others communication process (a_k, a_l) are using the same link at the same time.
5. In each vertex v_i is deployed only one agent a_i.

The (*static*) route between two agents, see Definition 4, from *external vertex v_i to v_j* is denoted as *Route(a_i, a_j)*, where (a_i, a_j) represent the respective agents deployed in respective vertices, and includes at least one internal node or gateway *(router or NGFW)*. The internal nodes form a core network represented as a direct graph $G_c = (W_c, R_c)$, where $(W_c \subset V, R_c \subset E)$.

Definition 5. Let a graph $G_c = (V_c, E_c)$; represent a core network, the following statements are hold:

1. The set $R = \left(r_{lk;l \neq k}\right)_{l,k=1}^{l,k=n}$ represent a set of static routes between internal nodes l and k.
2. For each pair of agents $(a_i, a_j) \in A$ is only one static route $r_{lk} \in R$.
3. Because all links in our graph are directed, exist a routes $(r_{lk}, r_{kl}) \in R$.

Definition 6. A partition of a set S is a sequence $S = (S_1, \ldots, S_r)$ of disjoint nonempty subsets of S such that $S = \cup_{i=1}^{r} S_i$. The sets S_i are called the cells of S. The empty partition will be denoted by $\emptyset$.

Definition 7. Let $S = (S_1, \ldots, S_r)$ and $T = (T_1, \ldots, T_s)$ be partitions of two disjoint sets S and T, respectively. The concatenation of S and T, denoted SoT, is the partition $(S_1, \ldots, S_r, T_1, \ldots, T_s)$. Clearly, $\emptyset oS = S = So\emptyset$.

Usually, the partition of the vertex set according to the degree of each vertex is used as the starting point for partition of a graph [5].

Definition 8. Let $G = (V, R)$ be a graph. The degree partition of G, denoted DegreePartition(G), is a partition $V = (V_1, \ldots, V_r)$ of V such that for all $i, j \in \{1, \ldots, r\}$, $i < j$ implies $\text{Deg}(V_i, G) \succ \text{Deg}(V_j, G)$; where: $\succ$ implies the second one precedes the first one in lexicographic order.

Partitions may be further refined by two means. The first one is to classify the vertices in the cells of a partition considering the adjacency type they have with a certain pivot vertex in the graph considered. This way, cells may be split into up to four distinct cells. We call this process a vertex refinement. The second refinement classifies the vertices in the cells using their available degree in a given pivot set (cell). This leads to what we call a set refinement.

To build sequences of partitions in which an initial partition of a graph is taken (for example the degree partition) and subsequent partitions are generated, each from its previous one, by applying one of the refinements defined above. Vertex refinements are tagged as VERTEX (if the pivot set has only one vertex), SET (if a set refinement is possible with some pivot set), or BACKTRACK (when a vertex refinement is performed with a pivot set with more than one vertex).

Definition 9. Let $G = (V, R)$ be a graph. A sequence of partitions for graph G is a tuple (S, R, P), where $S = (S_0, \ldots, S_t)$, are the partitions themselves, $R = \left(R^0, \ldots, R^{t-1}\right)$ indicate the type of refinement applied at each step, and $P = (P^0, \ldots, P^{t-1})$ choose the pivot set used for each refinement step, such that all the following statements hold.

1. For all $i \in \{0, \ldots, t-1\}$, $R^i \in \{\text{VERTEX, SET, BACKTRACK}\}$, and $P^i \in \{1, \ldots, |Si|\}$.
2. For all $i \in \{0, \ldots, t-1\}$, let $S_i = \left(S_1^i, \ldots, S_{r_i}^i\right)$, $V^i = \cup_{j=1}^{r_i} S_j^i$, $S^{i+1} = (S_1^{i+1}, \ldots, S_{r_{i+1}}^{i+1})$.

Then:

a. For all $x \in \{1, \ldots, r_{i+1}\}$, it exists $y \in \{1, \ldots, r_i\}$, such that $S_x^{i+1} \subseteq S_y^i$ and HasLinks (S_y^i, V^i, G).

b. For all $x \in \{1, \ldots, r_{i+1} - 1\}$, $S_x^{i+1} \subseteq S_y^i$ implies $S_{x+1}^{i+1} \subseteq S_z^i$ where $y \leq z$.

c. $R^i = SET$ implies $S^{i+1} = SetRefinement(S^i, S_{p_i}^i, G_{V^i})$; where SetRefinement means that no exist a vertex refinement of S^{i+1}.

d. $R^i \neq SET$ implies $S = VertexRefinement(S^i, S_{p_i}^i, G_{V^i})$ for some $v \in S_{p_i}^i$.

3. Let $S^t = (S_1^t, \ldots, S_r^t)$, $V^t = \bigcup_{j=1}^r S_j^t$ then for all $x \in \{1, \ldots, r\}$, NumLinks $(S_x^t, V^t, G) = 0$ or $|S_x^t| = 1$.

Partition refinement has been traditionally performed by splitting cells according to the adjacencies their vertices have with all the cells in a partition, However, this can be quite costly. Therefore, we do things the other way around; i.e. we take cells, not to try to have them split, but to try to split other cells (or itself) using vertex or set refinement. This approach is much less costly in terms of time and, on the short term, also space, but, on the long term, it needs more space (though limited to $O(n^2)$ as required), and leads to the same stable (or equitable) partition. Furthermore, it is not necessary to consider singleton cells (cells with only one vertex) more than once. Hence, we discard singleton cells once they have been used for a vertex refinement. This reduces the complexity of the problem, and reduces the memory requirements of the algorithm [3].

3   The Formal Model

For each ordered pair of agent nodes $(a_i, a_j) \in A$, all the micro security agents will have the same authentication data base. This ensures that the legitimate end-user traffic gets to its proper cloud environment. Therefore, ratio $H_r(a_i, a_j)$ represent the fraction of total transmitted packets to be discarded by the micro security agents because represent a network traffic not considered into Definition 2, in consequence $H_r(a_i, a_j) < 1$. The hit detection ratio represent the probability of packets arriving to vertices, in which is installed a micro security agent, could be discarded and not forwarded to their destination [6].

The brute average traffic that goes from agent a_i to agent a_j is the amount of packets in the network that are originated in node a_i and have a node a_j as destination and complain with terms of Definitions 2 and 3. The traffic is measured in *packets/seconds*, and is denoted as $BruteTr(a_i, a_j)$. For instance, the net average traffic, denoted as $NetTr(a_i, a_j)$, that goes from agent a_i to agent a_j is:

$$NetTr(a_i, a_j) = BruteTr(a_i, a_j) - H_r(a_i, a_j) * BruteTr(a_i, a_j)$$

So:

$$NetTr(a_i, a_j) = BruteTr(a_i, a_j)(1 - H_r(a_i, a_j)) \tag{3}$$

Although, packets flowing in Internet differ in dimension, for this paper it is assumed that all packets have the same size. The constant C denotes the cost of transmitting a packet through a link r_{lk}. Thus, the cost of transmitting a packet through a *route path* will only depend on the length of the route $|Route(a_i, a_j)|$. Otherwise, routes will be at least of length 2 according to our premises.

$$a_i \quad w_1 \quad w_2 \qquad w_{X-1} \quad w_X \quad a_j$$

Fig. 2. Sample route from agent a_i to a_j.

Let us consider the *route* from agent a_i to a_j, see Fig. 2, let $Route\,(a_i, a_j) = (a_i, w_1, w_2, \ldots w_x, a_j)$, there exist x_{ij} hopes, where: $x_{ij} \in |W_c|$, number of gateways in the path from a_i to a_j, where is possible install a micro security agent, considered as part of Definition 5. Besides, the cost of transmitting the brute traffic demand from node a_i to a_j denoted as $CostTr(a_i, a_j)$ [7], should be:

$$\text{CostTr}\,(a_i, a_j) = BruteTr\,(a_i, a_j)\,(1 + x_{ij})\,C \tag{4}$$

For the next paragraphs it is assumed that all latency sensitive micro security agents to be installed are equals in sense of Definition 2 and its incidence is described inside a set of Definition 9. So, it is a latency sensitive task that all SECaaS application could run at the edge for a superior end-user experience. In addition, time required for doing updates is 0 in all elements. Consequently, the packet discards is desired to be done in the nearest Vertice considering end-user proximity/location as a crucial design element.

This means that traffic between nodes (a_i, a_j) is the traffic flowing on the network immediately after having passed the first micro security agent that is on the route, see Definition 1 *and dimension of subset S its optimal agree with Definition 9.*

$$\underset{\substack{i,j \in [1,n] \\ n=|A|}}{\forall}\ (a_i, a_j) \in AxA,\ \exists (a_i, w_1, \ldots . w_m, a_j) \subset Route(a_i, a_j)$$

$$\Rightarrow w_p \in W, p \in [1, m],\, BruteTr(a_i, a_p) \geq BruteTr(w_p, w_{p+1})$$
$$\Rightarrow NetTr(a_i, a_j) = BruteTr(w_p, w_{p+1})$$

Proof 1. However, if the micro security agent is located in the node nearest to the origin of packets *(w₁ the first hope in the route from a_i to a_j)*, then the transmission cost will have the lowest possible value, *see* Proof 1, because packets are discarded as quickly as possible and Eq. 5 will be the following:

$$\text{Cost}Tr(a_i, a_j) = x_{i,k} BruteTr(a_i, a_k) C + x_{k+1,j} NetTr(a_{k+1}, a_j) C \tag{5}$$

Equation 5. Where: $i < k < j$; and k represent the node were is installed an agent.

The expression $x_{k+1,j}NetTr(a_{k+1}, a_j)$ of Eq. 5 represents the yield benefit from the existence of micro security agents in route path from a_i to a_j. If devices were placed in last hop of path, the yield benefit would have been of only $(NetTr(a_{k+1}, a_j)H_r(a_i, a_k))$, see Eq. 3, it is $x_{k+1,j}$ times less than if agent were placed in the first hope in the path[10]. This illustrates how important it is to allocate the agents near the source of traffic. If the distance between the source and the agent's decrees, the benefits of discard packet are more evident, partitioning the network in segments to access distributed cloud services located in edge is overkill and defeats the benefits of enabling and using the edge. Note, for global network traffic the importance of locate the security devices in principal routes and optimize flows in the maximum number of route paths; because connections between agents are asymmetric, its importance calculate in separate ways the location of micro security agents that contribute to reduction of flows for Route(a_i, aj) and Route(a_j, a_i) [6].

4   Conclusions

In the next years it will appear a new generation of technologies which will demand of a identity validation closest to the source because critical services could not depend of long-distance routes associated to increase attack vectors and security exploits.

The distribution of cloud services to the edge network is fundamental to achieve a reduction of transmitted massive amount of raw data over a network, considering that is more efficient to process data near its source and send only the data that has a value over the network to a remote data center.

The edge computing has a principal role to introduce new generation services running in segmented cloud environments, which are fundamental to work for organizations that have a geographically dispersed infrastructure. In such scenario, the importance of access to intermediary data centers of high-performance represent the opportunity to improve the performance and the ability for a service to act upon perishable data in a fraction of a second.

References

1. NIST: Developing a framework to improve critical infrastructure cybersecurity, Submitted On 04 August 2013
2. Weinhardt, C., Anandasivam, A., Blau, B.: Cloud computing – a classification, business models, and research directions. BISE 1, 391–399 (2009)
3. De la Torre, A., De La Torre, M.: Network forensic analysis in the age of cloud computing. In: CEPOL Budapest, Conference (2016). https://www.cepol.europa.eu/sites/default/files/18-marco-de-la-torre.pdf
4. Amman, P., Wijesekera, D., Kaushik, S.: Scalable, graph-based network vulnerability analysis (2002)

5. De la Torre, C., Polo, J.C.: Cloud computing and network analysis. In: Borzemski, L., Świątek, J., Wilimowska, Z. (eds.) Information Systems Architecture and Technology: Proceedings of 39th International Conference on Information Systems Architecture and Technology – ISAT 2018. ISAT 2018. Advances in Intelligent Systems and Computing, vol 852. Springer, Cham (2018)
6. De la Torre, C.: Design of secure and cost-efficient networks to support cloud computing applications. In: ISAT 2011. Service Oriented Networked Systems, pp 280–289 (2011)
7. Cisco: Annual security report, pp. 10–28 (2016)
8. IDC: The role of virtual WAN optimization in the next generation datacenter, December 2012

Sketching by Cross-Surface Collaboration

Jorge Luis Pérez-Medina[1]([✉])[ID] and Jean Vanderdonckt[2][ID]

[1] Intelligent and Interactive Systems Lab (SI^2 Lab),
Universidad de Las Américas (UDLA), Quito, Ecuador
jorge.perez.medina@udla.edu.ec
[2] Université catholique de Louvain (UCL),
LouRIM - Place des Doyens, 1, 1348 Louvain-la-Neuve, Belgium
jean.vanderdonckt@uclouvain.be

Abstract. Human-Computer Interaction and agile practices in software engineering are not two separate domains, but rather agile is a work principle that is applied in software development, of which user interface design represents a significant part. Agile projects could require such an approach, which typically iterates on the user interface using low-fidelity prototypes. This paper motivates, presents, and assess capabilities of a software for collaborative sketching of user interfaces on multiple surfaces of interaction, ranging from mobile phones to wall screens. We proposes a Collaborative User Centered Design (CUCD) method for user interface prototyping to supporting cross-surface collaboration by sketching, enabling fast, flexible, intuitive and reusable prototype.

Keywords: Collaborative Design · Electronic sketching ·
Multi-surface · Prototyping · User interface design

1 Introduction

While software modeling tools are more often used during early steps of the software development lifecycle (SDLC) [16], including the User Interface (UI) [5], other techniques are followed such as: free-form approaches. Office tools, white boards and papers arranged on a wall, are frequently used [20]. Indeed, these can help to produce more and usable tools and bring together people who understand the activities and needs of developers and other stakeholders to produce User Interface Design (UID) throughout the SDLC.

Designers can explore the requirements using essential UI prototypes (UIP) [1]. It represents UI requirements in a technology-independent manner. Although essential UIP many stakeholders find them too abstract and instead prefer screen sketches and not surprisingly working software [1]. Designers will use the simplest tool which will get the job done. Whiteboard diagrams typically suffice for initial requirements models. The Agile Software development (ASD) encompasses a series of methods for software development. A survey [1] concludes that 92.7% of respondents indicated that agile teams were using *Whiteboard Sketching*

© Springer Nature Switzerland AG 2019
Á. Rocha et al. (Eds.): ICITS 2019, AISC 918, pp. 386–397, 2019.
https://doi.org/10.1007/978-3-030-11890-7_38

for requirements gathering. 85.5% of those teams found the effort worthwhile. Similarly, sketching activities were the most common primary approach to modeling [2]. Over 90% of respondents [3] indicated that their agile teams did some sort of up front requirements modeling. These results are aligned to the daily effort taken by User Experience (UX) designers during gathering requirements. It consists of using sketching in early of the UI design [9] as an informal and flexible manner for producing any models for requirements and analysis phases. Sketching can also be used by any stakeholders as it is a way of expression which does not require any advanced modelling skills. Sketches are therefore refined to achieve user validation.

We aim at providing support for collaborative and flexible modeling in early steps of a User Centered Design (UCD) process where stakeholders and final users can collaborate by using multi-surface provided by multi-device. It consists of using sketching in early of the UI design. Sketching can also be used by any stakeholders using any device as it is a way of expression which does not require any advanced modelling skills. Sketches are therefore refined to achieve user validation. Iterative UID tools, used in Human-Computer Interaction (HCI), generally support electronic sketching, giving more freedom to change sketches and more flexibility in creating and evaluating a design prototype [10]. Prototyping the UI can detect ergonomic issues that will arise even before the first line of code is produced. The prototype is also a valuable communication tool; the

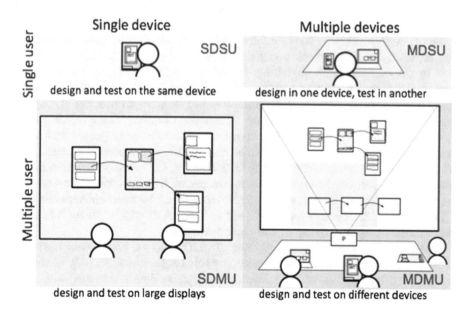

Fig. 1. Four physical configurations of a CUCD method considering multiple users and devices: SDSU = single device and single user; SDMU = single device and multiple users; MDSU = multiple devices and single user; MDSD = multiple devices and multiple users.

operation of a tool can be described to a non-computer user with greater ease. Prototyping is characterized by collaborative interactions, therefore a tool to support UID should ideally be flexible in order to accommodate different design situations, i.e., different activities, people from different areas using different media and devices.

We propose a combination of four possible configurations (Fig. 1) supporting collocated synchronous interactions identified in [15]. It considers the different users (such as, designers and users) and devices. This paper presents a flexible CUCD method based on sketching to support UI prototyping along a SDLC. It promotes flexible communication among stakeholders and end users in early steps of development when it is vital to understand the system requirements. Section 2 presents some related work including sketching and basic concepts of design by sketching. Section 3 describes the CUCD method. Section 4 introduces the Collaborative Sketching System (CSS) which supports the enactment of the CUCD method. An experimental report concludes this section. Finally, conclusions and some remarks for future work are presented.

2   Related Work

Sketches [6] are rapid drawings. Stakeholders and final users are able to carry out UID by sketching during collaborative sessions of brainstorming for requirements gathering. Sketching is useful as a design tool. It stimulates a re-interpretive cycle in the individual designer's idea generation process [12]. The prototyping activity is then complementary to the sketching activity in order to achieve understanding, especially if the artifact being designed has more dimensions than those used by the designer while sketching. Design is a hard challenging for designers. During the initial phases of this process designers have only a vague goal in mind. They does not have a clear understanding of what to produce. Design is essentially a problem of wicked nature[1]. On wicked problem the complexity can be due incomplete or contradictory knowledge, the number of people and opinions involved, the large economic burden, and the interconnected nature of these problems with other problems. In the HCI domain, UID shares the same aforementioned properties: the wicked nature, the use of *"cheap"* tools like sketching, the use of low-fidelity models such as prototypes, etc. The main difference is that UID is included in more general methods such as UCD which has specific roles with different responsibilities to the party involved.

Prototyping is one the core activities of UCD process. A review of past and current graphical user interface prototyping tools is presented in [19]. Recently, multi-device interaction where many different cross-device systems and tools have been investigated [14]. Among the aspects discussed are specific topics for the development of UIs between devices addressing problems and solutions suitable for non-technical users. In another context, we found many software

---

[1] Wicked nature is a term used for describe a social or cultural situation/problem that is difficult to solve because the problem is not understood until after the formulation of a solution.

development frameworks like [8], interactive applications and academic research supporting sketching activities [13,17,21]. [17] presents an extensive comparison of sketching tools considering UI design by sketching and Collaborative Design. FlexiSketch TEAM [21] allows multiple users to simultaneously use their own tables to quickly gather and validate requirements, for which UCD can be used. [4] concludes that paper prototypes have been found to be useful in practice in many more ways than just for usability testing. [13] puts forward the relationship between the agile practices and UI design. Finally, [7] shows the importance of using UCD as an early practice in the SDLC. The inclusion of the collaborative method within UCD and Agile is aligned with previous proposals, such as Communication-Centered design and Extreme Designing [18], where the focus is on the communication between designers and users.

3   A Collaborative User Centered Design Method Based on Sketching

The CUCD method is derived from the observation of a collection of user-centric methods concentrated on the construction of usable and ergonomic UIs (like Brainstorming, Paper Prototyping, etc.). Theses activities engages the user, its activities and their environment in all stages of an interactive application's analysis and design. It involves the activities of Drawing, Prototyping, Sharing/Testing and Discussing/Reflecting. In the **Drawing**, designers define the structure of the interface, producing some sketches depicting a UI in the form of screens, roughly one for each *"state"* of the UI. This activity is then iteratively performed between the elicitation of requirements and design activities when designers and users need to obtain the system specifications. Designers and end-users would sometimes be divided into subgroups. An aspect very important to achieve as it facilitates the collaborative sessions between stakeholders and end-users. **Prototyping** allows designers to define the **behavior** by assembling the different screens in a logical manner. In the **Sharing/Testing** normally, each subgroup sticks their produced prototype on a wall, arranged in a logical way, forming a sort of storyboard. An overall validation happens at that stage, where the navigation flows are tested. Very often, the end-users point to the drawings following the previously designed navigation, in a sort of *"Storytelling"*. The last stage is the **Discussing/Reflecting** where once the design is discussed and the possible problems are highlighted, the modifications are agreed. Then, designers and end-users proceeds to another iteration of drawing/prototyping/sharing until both parts are satisfied with the results.

The observations of UCD activities served as the fundamental concepts for devising the method and consequently the requirements for the tool:

1. *A software tool for design sessions.* After the design sessions, the designers would make a report that contains details about what was produced and decided. This is the first point in which a software tool could be of use, since it would register automatically everything that is produced on a session.

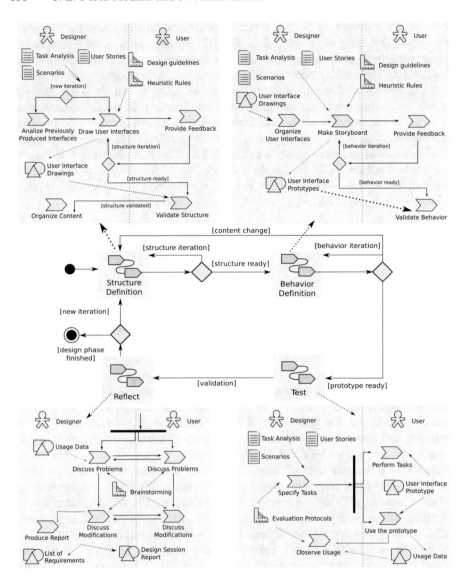

Fig. 2. The method workflow.

2. *Aparté sketching.* Often, some participants may feel uncomfortable about sketching in front of the rest of the group. For those participants, the designer usually gives a private sheet of paper, to be placed on the wall after the user feels his/her design view is good enough for discussion. This Aparté (aside) sketching activity would engage users as an active part of the solution design, and not only for validation.

3. *Different devices and fidelities.* Despite the fact the designers most part of time uses only pen and paper to conduct their design's sessions, they would frequently bring different media according to the users' needs. For instance, they would bring "pixel-perfect" (high-fidelity) software tools like Axure and Visio if they needed to design the final look of a system. The users also usually provided picture of interfaces previously done, in case of a redesign and often bring in their own devices or devices that are important for simulating the usage of the system in a real situation.

4. *Realistic prototypes.* Sometimes it's hard to imagine a design for a specific platform without seeing what it would look like on it. A lot of redesign could be avoided simply by drawing the interface on the very device it is indented to run. In other words, the physical domain would enable designers to make use of the device's properties that are often lost or ignored using just a paper simulation.

Based on the UCD activities and the fundamental concepts, the CUCD method is divided into four basic phases: *Structure Definition, Behavior Definition, Test Definition,* and *Reflect Definition.* The goal of devising such a method is to support the activities identified on the paper prototyping practice, keeping in mind the basics of Sketching and Prototyping as tools for helping designers and users to communicate, and also to include technological support. The workflow that links all the disciplines together is presented in Fig. 2.

The **Structure Definition Phase** allows Designers to define the structure of the interface, depicting a UI from screens either by sketching them or by composing interface elements according to the gathering requirements. All the elements of the UI are set according to the interaction needs of the user and to the functional specifications. In this step, which can be conducted while interviewing the user, the Designer concentrates on the content of the interface and on what is going to be shown. The activities begin with the definition of the structure by designers and users and iterate (**[structure iteration]/[structure ready]**) proceeding to the next step when both parts agree. Figure 2 details this step further, having the Designer starting the process by either Analyzing Previously Produced Interfaces and Drawing UI or simply by drawing them. For those activities, the Designer uses models such as User Stories, Task Analysis, Scenarios and Guidelines such as Design Guidelines and Heuristic Rules. The User should provide feedback about the produced interfaces until the structure is agreed and validated. Then, the Designer should organize the content in order to proceed to the next step.

In the **Behavior Definition Phase** Designers are focused on the practical usage of the system, or how the information will be shown, while the different screens are assembled in a sequential way (Fig. 2). The activities start with the Designer assembling the different UIs produced previously and then proceed to Make the Story-board which connects the interfaces in a logical manner, taking into consideration the models produced in the Structure Definition. The UI Prototypes are created in this activity for further validation by the User.

In the **Test Definition Phase** each subgroup puts their prototypes on the wall, arranged in a logical way, forming a sort of storyboard (Fig. 2). This step requires the active participation of the User, since this is the role that validates the prototype. The Designer specifies the task to be conducted, that can match the Scenarios, User Stories, Task Analysis, etc. to be evaluated, either by planning a user study with Evaluation Protocols. The User should perform the tasks while using the prototype, and the Designer has the role of observing the usage, from which usage data might be collected and evaluated.

In the **Reflect Definition Phase** Designers and Users can use brainstorming techniques to discuss possible problems and modifications. This activity starts by splitting the work between designers and users in parallel. Nevertheless both perform essentially the same activity, since they discuss problems and modifications. Designers have the responsibility to produce a report containing a Requirement List (regardless of whether the overall iteration is over or not). The list can be seen as a series of post-it notes stuck to the wall. In case the iteration is over, a design session report is produced, otherwise a new iteration can begin.

3.1 Discussion

Flexible software development encompasses a series of methods for software development that, in overall, consider software design as a problem of *wicked nature*. One of the highest priorities in flexible and ASD process is the development of runnable software that can be validated by both Stakeholders and Users. Generally, a prototype is used as a proof of concept. It can be considered as a first model towards a software development process. The holistic view in ASD is an iterative process of discovery, coding and validation, with small iterations. This vision does not consider that system requirements will all be provided correctly. Our method is a recommendation for rapid gathering and validating requirements, for which UCD can be used.

The CUCD method is presented independently of any technological support. However, offering a tool support for the method considering the collaboration among distributed teams requires also to take into account a multi-surface context where designers and final users they could working together using multiple devices to interact. In order to show how to include multiple devices or surfaces on the method, a combination of the configurations can be defined which considers the different users (or roles, which contains both designers and users) and devices. We can consider single and multiple users and also, single and multiple devices, thus giving a combination of 4 possible configurations of roles and devices (see Fig. 1). Each step of the method can use one of the 4 possible *"user X device"* combinations, being expressed by n^r, where n is the number of combinations and r is the steps of the method. In total, there are 256 possible combinations of single/multiple users, single/multiple devices and the phases Structure definition, Behavior definition, Test, Reflect.

4   The Collaborative Sketching System

The Collaborative Sketching System (CSS) is "a multi-surface collaborative tool for UI design that allows the sketching and simulation of UIs on many different devices". The tool was developed using a web technology - Google Web Toolkit (GWT). Therefore, it is accessible online and works on almost every modern web browser which makes it reachable from various platforms and interaction surfaces. The CSS offers a distributed multi-surface workspace. It provides users to interact around a UID method. Users can participate by using their personal device or different types of devices. These devices ranging from tiny to large interactive surfaces at the same time. Figure 3 illustrates a use of the CSS on the same session with a phone, a tablet and a tabletop computer. The main capabilities and features of the tool are: (1) Multi-surface and collaboration; (2) Infinite workspace and (3) Supports the four step of the CUCD method.

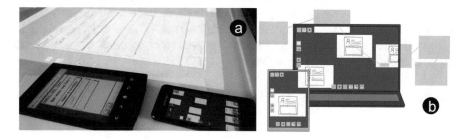

Fig. 3. (a) A collaborative sketching session on a phone, a tablet and a tabletop computer. (b) The infinite workspace concept of the collaborative sketching system.

The CSS allows the user to draw on virtual sheets of paper on a wall, that is represented as an infinite workspace (Fig. 3). Many devices can be used at the same time as they show parts of that workspace in an analogous way to a map software. Users can zoom in/out to show more or less details, sketch interfaces and "play" them as if they were real systems.

In the CSS the "Interface production" (Fig. 4) is conducted mainly by designers. It eventually includes users for providing feedback, validating the structure and optionally producing themselves the interface drawings. By activating, the Sketch mode, it is possible for the user to create "scenes" and sketch on them. The designer can add, remove and organize scenes spatially. The CSS includes an import feature that allows the possibility for the user to construct interfaces in different levels of fidelity: low (sketching), medium (wireframe mockup) and high (UI screenshot).

The "Prototype production or Behavior definition" (Fig. 4) is also conducted mainly by designers and eventually includes users for providing feedback, validating the behavior and optionally producing prototypes themselves. With the Prototype mode, the user is able to create interactive zones on top of the scenes (e.g. a "next" button) and connect these zones to another scene existing in the

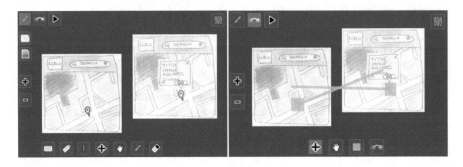

Fig. 4. The interface production (left) and the behavior definition (right) of the collaborative sketching system

defined structure (e.g. the scene that is targeted by the "next" button). The last feature consists in the test of the developed prototype in any device. By selecting the "Play" mode, designers can select the UI where he wants to start and the tool will show only the selected scene in full screen. The prototype might be navigated since the defined interactive zones will react to users input, showing the next scene when an action occurs.

4.1   Experiment

We have conducted a usability study with 20 participants having different backgrounds, (11 female or 55%), aged between 15 and 57 years old ($M = 34.25$, $SD = 10.26$). Participants were divided into 2 groups and instructed to sketch a system UI for children. An IBM Computer Satisfaction Usability (IBM CSUQ) Questionnaire [11] was completed by the participant to give their feedback about the CSS. Five additional questions were added to IBM CSUQ concerning the usage of animation pictures and UI navigation concerns: (20) "For me, animated images are not recommended"; (21) "I always know where I am and how to go where I want to"; (22) "Returns and outputs are explicit and visible"; (23) "The address of the current page allows me to return easily"; (24) "The caption or alternative text allows me to understand the meaning of the image"; (25) "The colors are chosen in order to leave the information readable".

Figure 5 depicts acceptable results of a user trial applied to 20 participants. Both results of the SysUse show that the groups in general have been appreciated. The results suggest that the CSS provides the support required for our CUCD method based on sketching. The InfoQual also suggests that participants from both groups have appreciated the quality of the information manage by the CSS. An aspect that we will to consider is the possibility to incorporate a history of annotations into the scenes useful when final users perform the simulations of the interfaces. The InterQual shows that the user interface of CSS is acceptable for the participants. However, we consider that the tool requires additional improvements for example: the possibility to manage versions of the same document and the possibility to testing the behaviour of the user interfaces without having to access the interface production. The OVERALL satisfaction suggests that CSS is positively perceived for the participants of the groups. The system overall usability suggests that the participants from both groups agree with the statements made in the survey. Table 1 shows cumulated results of the IBM CSUQ for the most significant questions questions according to the responses. Group 1 responded that the information provided in the CSS is easy to understand (Q13), while group 2 responded that the CSS is simple to use (Q2). Regarding the least voted questions Group 1 disagrees that the moving images are not recommended (Q20). Finally, Group 2 stated that they did not believe that they were productive quickly using CSS (Q8).

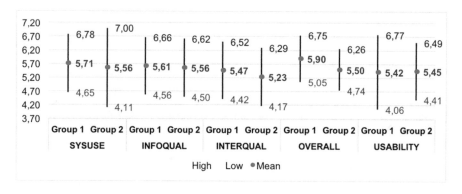

Fig. 5. The IBM CSUQ results.

Table 1. Cumulated results of the CSUQ. (1) The most suffrage question; (2) The less suffrage question; (3) The most critical participant; (4) The less critical participant

Question	Group 1	Group 2
(1)	**13** ($\mu = 6, 15; M = 6, 00; \sigma^2 = 0, 51; \sigma = 0, 67$)	**2** ($\mu = 7, 00; M = 7, 00; \sigma^2 = 0, 00; \sigma = 0, 00$)
(2)	**20** ($\mu = 4, 30; M = 4, 00; \sigma^2 = 1, 66; \sigma = 2, 05$)	**8** ($\mu = 3, 25; M = 3, 00; \sigma^2 = 0, 73; \sigma = 0, 91$)
(3)	**6** ($\mu = 4, 64; M = 4, 00; \sigma^2 = 0, 91; \sigma = 1, 04$)	**20** ($\mu = 4, 40; M = 5, 00; \sigma^2 = 1, 47; \sigma = 1, 85$)
(4)	**2** ($\mu = 6, 96; M = 7, 00; \sigma^2 = 0, 08; \sigma = 0, 20$)	**2** ($\mu = 6, 04; M = 6, 00; \sigma^2 = 0, 84; \sigma = 1, 02$)

5   Conclusions and Future Work

We have presented a CUCD method based on sketching to support early prototyping along flexible and ASD process. It is presented as independent of technological support, which is then added in order to support the assessed requirements. The approach is far from introducing new software development methodology that integrates UI engineering with the process, but is rather an attempt to include the steps of the proposed method into existing practices through UCD. The CSS, constructed to support the method, takes into consideration the contemporary technological support of multiple-surface. Its tools have the original property of functioning with multiple possible combinations of designers, users and devices at the same time. Future works will focus on describing how the UIs designed by sketches with our CUCD method can be linked with others activities that happen within a regular software development process. In this view, we consider performing several experimental studies which involve classical known SE methodologies to find new insights and improvements to our method.

References

1. Ambler, S.W.: Agile adoption rate survey results, March 2007
2. Ambler, S.W.: Modeling and documentation practices on it projects survey results, July 2008
3. Ambler, S.W.: 2013 agile project initiation survey results (2013)
4. Aston, J., Meszaros, G.: Adding usability testing to an agile project. In: Proceedings of Agile 2006 Conference (AGILE 2006) (2006)
5. Bodart, F., Hennebert, A.M., Leheureux, J.M., Provot, I., Vanderdonckt, J., Zucchinetti, G.: Key activities for a development methodology of interactive applications, pp. 109–134. Springer, London (1996)
6. Buxton, B.: Sketching User Experiences: Getting the Design Right and the Right Design. Morgan Kaufmann Publishers Inc., San Francisco (2007)
7. Fox, D., Sillito, J., Maurer, F.: Agile methods and user-centered design: how these two methodologies are being successfully integrated in industry. In: Agile 2008 Conference, pp. 63–72. IEEE (2008)
8. Gonzalez-Perez, C.: Filling the voids-from requirements to deployment with OPEN/Metis. In: ICSOFT (1), p. 19 (2010)
9. Landay, J.A., Myers, B.A.: Interactive sketching for the early stages of user interface design. In: Proceedings of the CHI 1995, pp. 43–50. ACM Press (1995)
10. Landay, J.A., Myers, B.A.: Sketching interfaces: toward more human interface design. Computer 34(3), 56–64 (2001)
11. Lewis, J.R.: IBM computer usability satisfaction questionnaires: psychometric evaluation and instructions for use. Int. J. HCI 7(1), 57–78 (1995)
12. Van der Lugt, R.: Functions of sketching in design idea generation meetings. In: 4th Conference on Creativity & Cognition, pp. 72–79. ACM (2002)
13. McInerney, P., Maurer, F.: UCD in agile projects: dream team or odd couple? Interactions 12(6), 19–23 (2005)
14. Nebeling, M., Kubitza, T., Paternò, F., Dong, T., Li, Y., Nichols, J.: End-user development of cross-device user interfaces. In: 8th ACM SIGCHI Symposium on Engineering Interactive Computing Systems, pp. 299–300. ACM (2016)
15. Olson, G.M., Olson, J.S.: Distance matters. HCI 15(2), 139–178 (2000)
16. Ossher, H., van der Hoek, A., Storey, M.A., Grundy, J., Bellamy, R.: Flexible modeling tools (FlexiTools 2010). In: 32nd ACM/IEEE International Conference on Software Engineering-Volume 2, pp. 441–442. ACM (2010)
17. Sangiorgi, U.B., Beuvens, F., Vanderdonckt, J.: User interface design by collaborative sketching. In: Proceedings of the Designing Interactive Systems Conference, pp. 378–387. ACM (2012)
18. da Silva, B.S., Aureliano, V.C.O., Barbosa, S.D.J.: Extreme designing: binding sketching to an interaction model in a streamlined HCI design approach. In: 7th Brazilian Symposium on Human Factors in CS, pp. 101–109. ACM (2006)

19. Silva, T.R., Hak, J.L., Winckler, M.: A review of milestones in the history of GUI prototyping tools. In: 15th IFIP TC. In: 13th International Conference on Human-Computer Interaction, vol. 22, p. 267. University of Bamberg Press (2015)
20. Whittle, J., Hutchinson, J., Rouncefield, M., Burden, H., Heldal, R.: Industrial adoption of model-driven engineering: are the tools really the problem? In: International Conference on MDE Languages and Systems, pp. 1–17. Springer (2013)
21. Wuest, D., Seyff, N., Glinz, M.: Sketching and notation creation with FlexiSketch team: evaluating a new means for collaborative requirements elicitation. In: 23rd International Requirements Engineering Conference, pp. 186–195. IEEE (2015)

Efficiency Analysis Between Free and Paid Hardware and Software in a Pneumatic Press

Pamela Espejo[1]($\boxtimes$), Vicente Hallo[1], Andrés Gordón[1],
Nancy Velasco[1], Darío Mendoza[1]($\boxtimes$), Verónica Gallo[2],
and Fernando Saá[3]

[1] Departamento de Energía y Mecánica,
Universidad de las Fuerzas Armadas ESPE, Sangolquí, Ecuador
{pmespejo,vdhallo,amgordon,ndvelasco,
djmendoza}@espe.edu.ec
[2] Facultad de eléctrica, Instituto Tecnológico Superior Guayaquil,
Ambato, Ecuador
vgallo@institutos.gob.ec
[3] Ingeniería Industrial, Universidad Tecnológica Indoamérica, Ambato, Ecuador
fernandosaa@uti.edu.ec

Abstract. The present work proposes the automation of a pneumatic press, based on free software and hardware. The project analyzes the behavior of free software versus proprietary technology. The project is based on creating a prototype of a distribution system. The proprietary technology consists in a Siemens S7-1200 programmable logic controller, which monitors the system with an OPC server and a (Human Machine Interface) user interface based on LabVIEW, M. In the other hand, the free technology consists of an M-DUINO controller based on Arduino monitors the system with a LAMP server, and hosts a website programmed in PHP; reducing implementation costs in control processes, without affecting the performance and quality in the industrial process. Research is done to determine the reliability of free software.

Keywords: Data acquisition · Automation · Free versus paid software · Efficiency software

1 Introduction

Today companies pursue to be competitive in the labor market, delivering their products on time and improving its quality. However, they are limited by the high cost in the automation of their production processes; because the current solutions involve proprietary hardware and software, directly affecting the price of the final product. Besides that, with the technological improvements and the collaboration of thousands of people, free hardware and software solutions have been developed for the industry, and high levels of reliability and security, achieving the reduction of implementation costs, thus allowing small businesses become competitive in low budget conditions and high quality in the product.

© Springer Nature Switzerland AG 2019
Á. Rocha et al. (Eds.): ICITS 2019, AISC 918, pp. 398–405, 2019.
https://doi.org/10.1007/978-3-030-11890-7_39

Currently there are works being done with free technology, allowing the user to modify both hardware and software. For example, after facing of the problems outlined above, Thiago Rodrigues has proposed the creation of a logical controller, both in hardware and software aspects, with architecture and standardized features that make it competitive, robust, reliable and easy to use; with open access technology [1]. In the work carried out by Alberto Medrano [2] in 2015, the importance of managing codes and free designs associated with electronic devices is explain to improve profitability in the control of oil plants in Venezuela [2].

In 2011, Barreño [2] published an article refering to the design of embedded systems for industrial control loops based on free software that analyzes the scarce implementation of free software in developing countries largely due to the fact that technology-based companies are very small, with low levels of production and use outdated design methodology. One of the main arguments to promote sustainable development with free hardware is the accelerated, innovative and active production of common goods.

Lazalde [4] states that free hardware is an ideal technology to promote spaces for citizen innovation with social impact, that is, laboratories such as FabLabs and makerspaces grouping together universities, companies, government and citizens in general.

Besantes [5] automated the hot water recirculation system using free hardware at IESS Latacunga Hospital, demonstrating that hardware and free software technology can be applicable to small and medium-sized companies in order to reduce implementation and start-up cost of automation systems.

Should be mentioned that the hybrid system involves free access technology for the creation of control and acquisition data interfaces (SCADA) based on java and OPC libraries [6], with which, the linking of both hardware proprietary and free access software can be used as supplements and generate robust systems with the same reliability as when using only proprietary hardware and software.

Mahato has presented a revision of web-PLC, for the timely incorporation of a WEB server within the PLC (Programmable Logic Controller), facilitating access to the information of the production process and allowing monitoring and even control from various locations [7].

With the industrial control and experimental physics control system (EPICS) and through MODBUS/TCP communications, a communication has been established between an AH500 PLC and a personal computer for the implementation of a real-time, web-based IOT (Internet of Things) application. web for signals monitoring acquired by the PLC, making necessary only a device with web browser [8].

In [9] is presented a significant contribution to link a system with the GSM (Global System for Mobile Communications) networks to send alert signals through text messages, which implies a starting point to the generation of information tables in mobile terminals.

This work proposes the implementation of a SCADA system, using free hardware and software.

2 Development

A standard production system has been proposed in the industrial field for the present investigation, defined by Fig. 1.

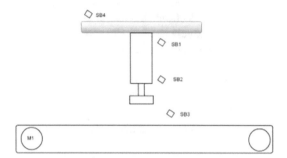

Fig. 1. System for analysis

The control of the pneumatic press has keypads for the ignition (ON) and shut down (OFF) of the system, and an emergency stop. Pressing the ON button, a heating element of the press is heated at 30°C (monitored by a closed-loop PID control and a temperature sensor) and turns on the M1 motor connected to a variable frequency drive (VDF) at a speed pre-established that takes a product by the conveyor belt to the press, which is retracted. The piston in reverse is monitored by the sensor SB1 and the advance by the sensor SB2. The SB3 sensor detects the product that is going to be pressed and lowers the press for 4 s. Then, the piston rises and the motor of the conveyor belt is activated to make way for the next product.

There is a regulator to manually control the pressure that enters the piston, being acquired by an SB4 sensor of absolute pressure, the normal pressure is 2 bars and when it exceeds 5 bars the alarm is turned on in the HMI. The SB3 sensor keeps count of the products that are pressed and controls the jams in the conveyor belt. It has 2 pilot lights that indicate whether the system is on or off, also alerts for jams, pressure and high and low temperature.

2.1 Hardware Requirements

The devices described in Table 1 have been used for the implementation of the production systems prototype; upon which the central analysis will be done.

2.2 Software Requirements

Both free software and proprietary software have been used for the comparative analysis of the performance of the implemented system and the creation of visualization interfaces for the industrial process. Table 2 lists the software in detail.

Table 1. Hardware requirements for the prototype.

Quantity	Description
1	Proximity sensors SMT8
3	Temperature sensors
1	Pressure sensor 539757
1	Inductive position sensor M18, outgoing assembly 548645
1	Three-phase motor AEG TYP AM 63 NY4
1	ElecTlock 5/2-way flip-flop with LED539778
1	Conveyor belt 16.5 × 40 cm
1	Piston 152888
1	PLC M-Duino
1	PLC S7-1200
10	Cables with banana type jacks
1	Micromaster 440 frequency inverter
1	Valve Shut-off valve with filter unit and egulator 40691
1	Distribution block (152896)
5	Channels
1	Multimeter
4	Screwdrivers

Table 2. Software requirements for the prototype.

Quantity	Description
1	Website PHP
1	Labview
1	OPC
1	LAMP SEVER

The prototype of the production system for the study can be seen in Fig. 2(a). The connections have been made based on the diagram of Fig. 2(b). This system will be connected later to each PLC to compare the operation.

Fig. 2. System construction.

2.3   Open-Source Software and Controller

The interface implemented using M-Duino PLC and a monitoring interface in Server is presented in Fig. 3.

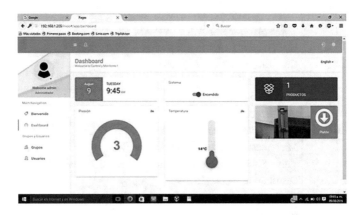

Fig. 3. Dashboard Implemented as a replacement to the conventional HMI.

3   Results

3.1   Analysis of System Variables Implemented in S7-1200 and Labview

As shown in Fig. 4, the proportional, integral and derivative control (PID) implemented to keep the temperature controlled at a value of 30 °C was successful, the reaction time and stabilization of the control system was optimal for the process to be developed from the start-up of the production system.

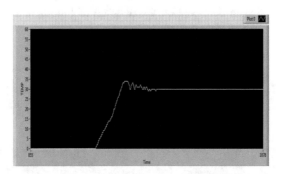

Fig. 4. PID control implemented in Labview.

On the other hand, the automated control variables of the system have also presented a rapid reaction (Fig. 5) from the start-up of the production process, with a minimum delay of milliseconds within the timeframe between the detection of environmental variables and the reaction of the actuators to generate the minimum delay and optimize the production times.

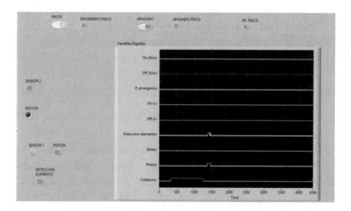

Fig. 5. Analysis of the control variables in SCADA implemented in LabVIEW.

3.2   Analysis of System Variables Implemented in M-Duino and LAMP Server

Figure 6 shows the reaction of the PID control implemented in the M-DUINO PLC, such response is similar to the one of the systems implemented in proprietary hardware and fulfills its task with the same reliability, demonstrating that it can be an effective substitute for control stages in the industry.

Fig. 6. Analysis of the control variables in SCADA implemented in LabVIEW.

For the analysis of the reaction of the control signals from the actuators, in Fig. 7 the control signals are presented in response to the actuators, this response is instantaneous due to the use of interrupts enabled in the controller.

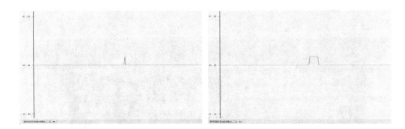

Fig. 7. Analysis of the control variables in SCADA implemented in M-DUINO.

After analyzing the case studies, having implemented control and data acquisition systems, positive results were obtained regarding the temperature control section and also in the control stages of all the devices and mechanisms. Such findings have demonstrated that both programmable logic controllers have a similar behavior and that free hardware devices and open access monitoring systems reduces the cost of implementation, becoming a reliable alternative applicable not only to small business.

4   Conclusions

The M-DUINO PLC offers the advantage of working on free software and hardware without additional expenses after purchase compared to traditional PLCs.

The creation of the visual environment and the acquisition of the variables data: ON/OFF status, temperature, pressure, presence and speed facilitate the information storage in a dynamic environment; the signal values sent are saved in real time, providing a stable control of them.

The Web application created shows the ON/OFF status of the press, Press temperature (°C), Pressure sensor (Bar), Product Count, Piston Status up or down, Date and Time. To access to the information, the user must be registered in the system, either as administrator or operator.

The graphical interface of the Web application is user-friendly allowing a comfortable iteration between the user and the process that is being controlled and monitored. The database created after the acquisition is compatible and easy to use from the application in the Website since it uses phpMyAdmin to manage data tables.

Such findings have demonstrated that both programmable logic controllers have a similar behavior and that free hardware devices and open access monitoring systems reduces the cost of implementation, becoming a reliable alternative applicable not only to small business.

References

1. Alves, T.R., Buratto, M., de Souza, F.M., Rodrigues, T.V.: Openplc: an open source alternative to automation. In: Global Humanitarian Technology Conference (GHTC). IEEE (2014)
2. Chourio, L.: Hardware libre y abierto, modelos de negocios para america latina y el caribe. In: ALTEC (2015)
3. Camargo, C.: Metodologia para la transferencia tecnologica en la industria electronica basada en software libre y hardware copyleft. In: Congreso Argentino de Sistemas Embebidos CASE 2012 (2012)
4. Lazalde, A.: Hardware libre recomendaciones para el fomento de la innovacion ciudadana. infraestructuras tecnicas abiertas, NA, pp. 1–2 (2015)
5. Talve, W.: Automatizacion de sistema de recirculacion de agua caliente utilizando hardware libre en el hospital IESS Latacunga. Ingeniero, UTA, Ed. UTA (2015)
6. Dutta, S., Sarkar, A., Samanta, K., Das, R., Ghosh, A.: Supervision of control valve characteristics using PLC and creation of HMI by SCADA. In: 2014 First International Conference on Automation, Control, Energy and Systems (ACES). IEEE (2014)
7. Mahato, B., Maity, T., Antony, J.: Embedded web PLC: a new advances in industrial control and automation. In: Second International Conference on Advances in Computing and Communication Engineering (2015)
8. Joshi, R., Jadav, H.M., Mali, A., Kulkarni, S.V.: IoT application for real-time monitor of PLC data using EPICS. In: International Conference on Internet of Things and Applications (IOTA) (2016)
9. Bhaskarwar, T.V., Giri, S.S., Jamakar, R.G.: Automation of shell and tube type heat exchanger with PLC and LabVIEW. In: International Conference on Industrial Instrumentation and Control (ICIC), May 2012
10. Industrial Shields: M-DUINO FAMILY, SPAIN Divina Pastora 13-15 Baixos 3
11. SIEMENS: S7 Controlador programable S7-1200, December 2009

Cyber-Physical Systems for Environment and People Monitoring in Large Facilities: A Study Case in Public Health

Borja Bordel$^{(\boxtimes)}$, Ramón Alcarria, Álvaro Sánchez-Picot,
and Diego Sánchez-de-Rivera

Universidad Politécnica de Madrid, Madrid, Spain
{bbordel,asanchez,diego.sanchez}@dit.upm.es,
ramon.alcarria@upm.es

Abstract. Traditionally, public health policies have been based on sociological studies where data are collected by experts, and citizens are informed through media such as television. Nevertheless, new digital society and innovative technological paradigms enable the realization of these activities in a much more efficient and automatic manner, allowing (besides) a personalized and ubiquitous interaction with people. Although different technological solutions could be employed for this purpose, Cyber-Physical Systems (CPS) is the most promising one. In fact, in the last ten years, many different cyber-physical applications for environment and people monitoring have been reported, which (we argue) could be employed to support and promote public health policies. Therefore, in this paper it is described a new Cyber-Physical System to promote healthy behaviors to reduce cardiovascular risks. The system monitors people's behavior (through a laser barrier and WiFi access points), and incentives positive and healthy activities by providing specific digital services. The proposed solution was simulated and implemented, deployed and validated using commercial components and statistical procedures, comparing the obtained result to which obtained from traditional mechanisms.

Keywords: Cyber-Physical Systems · Large facilities · Public health · Environment monitoring · People monitoring

1 Introduction

Cyber-Physical Systems (CPS) [1] are integrations of computational and physical processes, where embedded devices interact with the physical world creating feedback control loops. This new kind of systems may be employed in several different scenarios: from education to industrial systems [2]. The three-layer implementation architecture usually employed to build these solutions (see Fig. 1) [4], besides, also matches the requirements and design characteristics of other technologies which initially are not related to CPS, such as Enhanced Living Environments (ELE) [3] or Ambient Intelligence (AmI) [5].

In the particular case of these secondary technologies, such as AmI or ELE, the basic functionality to be supported by CPS is the sensitization of people and their

© Springer Nature Switzerland AG 2019
Á. Rocha et al. (Eds.): ICITS 2019, AISC 918, pp. 406–416, 2019.
https://doi.org/10.1007/978-3-030-11890-7_40

environment. In CPS, these functionalities are represented by the concept "human-in-the-loop" [6], which is a relevant part of the CPS conceptual map. This idea includes all pending challenges in relation to the integration of humans in the feedback control loops defined by CPS. Although some works have proposed that humans may take active roles as service providers in CPS [7], in cyber-physical application for AmI and ELE scenarios people are always passive agents whose behavior is sensitized and analyzed using certain knowledge extraction techniques (see Fig. 1).

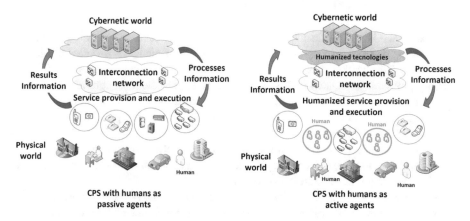

Fig. 1. Comparison between CPS where people are passive and active agents

With this approach, a new catalogue of applications may be covered using CPS. Among all them, probably, one of the most interesting fields is public health [8]. In fact, public health policies usually take complex sociological studies as input, where data are manually and personally collected by experts. Then, these data must be processed, and results and awareness campaigns are communicated to citizens using traditional media such as television. Although this methodology has been successful during the last years, new digital society and innovative technological paradigms, such as CPS, have enabled new more efficient and automatic manners to carry out and increase the impact of these activities. In particular, we argue, sensitizing capabilities and control loops could be employed to support and promote public health policies and interact with people in a more personalized and ubiquitous manner.

Therefore, in this paper, it is proposed a CPS to promote healthy behaviors to reduce cardiovascular risks, by encouraging physical exercise among inhabitants in large facilities such as office complexes or public bodies (universities, hospitals, etc.). The solution was based on laser barriers and WiFi access points, employed to sense the people's behavior. This information is analyzed, determining the impact of the public health policies under study. Besides, the impact of these policies may be increased by the provision of specific digital services (such as exclusive access to multimedia content). The solution was implemented using consumer electronic devices and materials, and was deployed in the School of Telecommunication at the Technical University of Madrid. A specific simulation tool was employed to evaluate the obtained improvements by using this new approach. Finally, statistical methods were employed to determine the performance of our proposal.

The rest of the paper is organized as follows: Sect. 2 describes the state of the art on CPS for environment and people monitoring. Section 3 describes the main proposal (functional architecture and hardware implementation), and Sect. 4 evaluates the performance of this solution, including simulations. Finally, Sect. 5 presents some conclusions.

2 State of the Art on Predictive Models

Proposals on Cyber-Physical Systems for environment and people monitoring are sparse. Most of them discuss, from a theoretical point of view, the problem of human-in-the-loop [6]. Works about the future of human-in-the-loop applications [9] and security problems associated to this approach [10, 11] are the most common proposals. With a more practical view, different technologies to reason about the human behavior may be found: robotic solutions [12], intelligent social agents [13], artificial intelligence frameworks [14], etc.

All these solutions, however, are focused on data processing and other very specific issues, which cannot be coordinated to create a global architecture. Then, vertical solutions are required to address this problem.

In this sense, some specific CPS for particular purposes have been reported. Cyber-Physical Systems for environmental monitoring [15], where IoT nodes are connected to a central server and some mobile devices to interact with users, may be found. Besides, crowd sensing solutions based on CPS have been reported [16]. In these solutions, a cloud platform is deployed to which all user interfaces (running in smartphones and other similar devices) are connected to send information about people. A very relevant group of solutions is those which are focused on medical purposes. Although no solution for public health has been reported, different proposals to monitor the state of people with health problems and control their medical treatment have been reported. Some of them are designed to be deployed in hospitals and similar environments [17] and other to be employed in the users' home [18].

The main problem of all these solutions is the use of personal devices (such as smartphone) as essential sensing instruments. Although these devices may be employed to enrich the system operation and data analysis, to obtain information about the human behavior in an unobtrusive manner other unmanaged components (such as sensitizing nodes) are needed.

3 Proposed CPS for Healthy Habit Promotion

In this section, the creation of a Cyber-Physical System to promote healthy behaviors among inhabitants in a public university is described.

Many different policies and actions to promote healthy behaviors among inhabitants in a large facility such as a public university may be designed. In this case, we have selected policies oriented to the cardiovascular public health, looking for users to increase the use of the traditional stairs, instead of using the elevators.

Traditionally, these policies have been supported by posters with motivational messages about the benefits in health obtained when using standard stairs; with experts evaluating the impact of these posters in situ. To, both, increase the impact of these motivational messages, and evaluate their successful rate in a more automatic manner, a cyber physical architecture is proposed (see Fig. 2a).

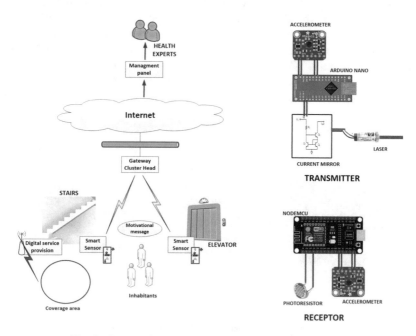

Fig. 2. Proposed CPS **(a)** Architecture **(b)** Implementation

As can be seen, the use of stairs and elevators is monitored by smart sensor nodes, based on laser technologies and algorithms for counting people [21]. All these nodes have a unique ID employed to send the obtained information to a central cluster head (or gateway) using wireless technologies. This gateway, then, sends this information to a management panel where health experts may consume the results. Besides, in the stairs area, a digital service is provided to people using this element (stairs). In this initial system, the proposed service to be deployed is a WiFi area, where people may freely access to special paid contents. Finally, around all the facility, motivational messages are placed to promote and inform users about this initiative and the benefits of healthy habits.

The proposed implementation consists of commercial hardware platforms for CPS development, such as Samsung Artik and NodeMCU (see Fig. 2b). Each sensor node includes two elements. The first element includes an Arduino microcontroller which controls a 5mW-power red laser though a current mirror based on bipolar junction transistors (BC107); and a MMA8451 accelerometer to ensure the laser is correctly placed (aligned with the receptor). If the accelerometer detects an inadequate movement or position, the microcontroller stops the light emission in the laser to avoid security problems. The second element is in charge of receiving the laser light. A nodeMCU microcontroller with an embedded WiFi System-on-Chip (ESP8266) is connected to another accelerometer (to ensure the correct device placement), and a photoresistor whose output signal is sampled. The basic behavior of this element is an infinite loop where the signal level across the photoresistor is evaluated. If signal is vanished during a short time, it is considered a person has crossed the laser barrier.

On the other hand, the cluster head is based on Samsung Artik 710 architecture: a single-board computer implementing Ubuntu architecture. This gateway executes a JavaScript web server (Node.js) which receives information from NodeMCU microcontrollers by means of HTTP PUT requests (see Fig. 3a). This same gateway is employed as service provision system (WiFi access point). Finally, motivational messages are deployed using standard posters (see Fig. 3b).

```
PUT /gateway.js  HTTP/1.1
Host: gateway.edu
Content-type: application/json
Content-length: 32

{position:OK, detectedPeople:1}
```

Fig. 3. Proposed CPS (**a**) PUT request from nodes to gateway (**b**) Motivational posters

4   Experiments and Discussion

The proposed experiments to promote the use of stairs using the designed CPS were carried out in the Technical University of Madrid, specifically in the School of Telecommunication. Two spaces with different configurations were considered. Table 1 presents the characteristics of both spaces (see Fig. 4).

Table 1. Spaces considered in the proposed experiments

Name	Elevator mean wait time (s)	Distance to the closest stairs (m)	Floor	Mean traffic volume (people/h)
B-building	57	20	0	149
C-building	20	5	2	53

The experiment consisted of four phases: (i) during the first phase no intervention was considered; (ii) in the second phase only motivational posters, as in traditional public health policies, were deployed; (iii) the third phase was characterized by the use

Fig. 4. Considered spaces **(a)** B-building **(b)** C-building

of the proposed CPS to automate the data capturing and increase the impact of the initial intervention; (iv) finally, during the fourth phase, to the CPS monitoring the human behavior it is added a service provision system (WiFi access point), with coverage in the stairs area.

During all these phases, the use rate of both the elevators and stairs was evaluated, comparing the obtained results for the different phases. As results are partially overlapped, a Mann-Whitney U test was conducted. The Mann-Whitney U test is a non-parametric test of the null hypothesis that two samples come from the same population against an alternative hypothesis, comparing the mean values of the two samples. It is used to evaluate if two different data populations are similar or different (higher or lower). The p-value indicates the significance level of Mann-Whitney U test.

The experiments were developed using two different techniques: a CPS simulator and a real deployment. Next subsections describe and discuss results for both evaluation techniques.

4.1   Experiments in Simulation Scenarios

In order to perform the proposed experiments using simulation techniques, it is employed the Hydra simulator [19], considering rules for human behavior extracted from previous evaluations about the use of stairs and elevators in large facilities [20]. Figure 5 shows the graphical interface employed to configure the scenarios.

Fig. 5.   Simulation interface

Table 2 shows the obtained results from the statistical analysis for the simulation experiment and both spaces, when considering the results obtained from different phases as reference. As can be seen, in B-building scenario, obtained results are globally better than in C-building scenario. In fact, as considered elevators in C-building scenario are in the second floor, more people use the stairs by default (first phase); so the potential improvement is lower.

Table 2. Results for the simulation scenario

Reference	B-building			C-building		
	Phase #2	Phase #3	Phase #4	Phase #2	Phase #3	Phase #4
Phase #1	**	**	***	*	**	**
Phase #2	–	*	**	–	NS	*
Phase #3	–	–	*	–	–	NS

*NS not significant; * significant at $p < 0.05$; ** significant at $p < 0.005$; ***significant at $p < 0.001$*

The most relevant improvements are obtained when using both the proposed CPS for people monitoring and the digital service provision system (phase #4) in comparison to people behavior when no intervention is performed (phase #1), or when only motivational messages are deployed (phase #2). Then, these results prove that the proposed CPS for supporting public health policies increases the impact of these policies in respect to traditional techniques.

4.2   Experiments in Real Deployments

To finally prove the validity of the proposed solution, and the correctness of the obtained simulation results, it is carried out an experimental validation in a real and relevant scenario (see Table 1). During fifteen days data were collected and stored. Figure 6 shows the different elements (hardware devices) which were deployed. Activities involving people in this work were always evaluated and approved by the ethics committee for research activities of the Universidad Politécnica de Madrid [22]. No action was taken to modify the native population, whose composition is (approximately): professors 10% (male: 70%, female: 30%), students 90% (male: 70%, female: 30%).

A key technological factor to be considered is the correctness of the proposed mechanism for counting people. To guarantee the validity of the obtained results from the proposed deployment (see Fig. 7) it is evaluated the error rate in people counting. To perform this operation, the number of detected people crossing the laser barrier and the corresponding real number were acquired, and a well-known algorithm based on "virtual gates" is employed [21]. Thus, error rate may be associated, almost completely, to the proposed laser barrier and cyber-physical infrastructure.

Figure 8 shows the error rate in the system, for different amounts of people crossing the laser barrier at the same time. As can be seen, the total aggregated error is below 5%. Besides, highest errors appear for biggest people groups, whose appearance probability is much lower (as we are considering the marginal probability for each

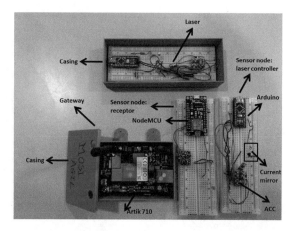

Fig. 6. Hardware device for experimental validation

Fig. 7. Real deployment

people is equal in all cases). Therefore, we are considering the proposed mechanism for counting people as a valid solution.

Once proved the validity of the proposed mechanism for counting people, the system operation is started. Measurements obtained by the system are stored in a specific database, embedded in the management portal. These measurements feed an R program which automatically calculates the Mann-Whitney U test. Results are showed in Table 3. As can be seen, although in this case the results in the B-building are still globally better than the results for the C-building, they are closer than obtained results in the previous section (simulation). This difference may be caused by the interest of users in the provided digital services, which moves them to use stairs even if being in the second floor. Besides, it must be taken into account that the large elevator wait time in the B-building scenario may also encourage users to use the stairs.

Anyway, as also showed in the simulation evaluation, these results prove that the proposed CPS for supporting public health policies increases the impact of these policies in respect to traditional techniques.

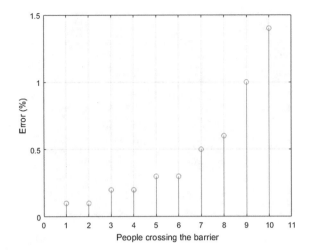

Fig. 8. Error rate

Table 3. Results for the real deployment and experimental validation

Reference	B-building			C-building		
	Phase #2	Phase #3	Phase #4	Phase #2	Phase #3	Phase #4
Phase #1	**	**	***	*	**	***
Phase #2	–	*	**	–	*	**
Phase #3	–	–	*	–	–	*

*NS not significant; * significant at p < 0.05; ** significant at p < 0.005; ***significant at p < 0.001*

5 Conclusions and Future Works

Cyber-Physical Systems (CPS) are integrations of computational and physical processes, where embedded devices interact with the physical world creating feedback control loops. This new kind of systems may be employed in several different scenarios: from education to industrial systems. In fact, in the last ten years, many different cyber-physical applications for environment and people monitoring have been reported, which (we argue) could be employed to support and promote public health policies. In this paper it is described a new Cyber-Physical System to promote healthy behaviors to reduce cardiovascular risks. The system monitors the people's behavior (through a laser barrier and WiFi access points), and incentives positive and healthy activities by providing specific digital services. The proposed validation supports that the proposed CPS for supporting public health policies increases the impact of these policies in respect to traditional techniques.

In future works the long-term impact of proposed technologies will be evaluated, to measure the tolerance development of users to these approaches. Besides, a more detailed analysis considering different social groups will be developed.

Acknowledgments. These results were supported by the Ministry of Economy and Competitiveness through SEMOLA project (TEC2015-68284-R) and from the Autonomous Region of Madrid through MOSI-AGIL-CM project (grant P2013/ICE-3019, co-funded by EU Structural Funds FSE and FEDER).

References

1. Bordel, B., Alcarria, R., Robles, T., Martín, D.: Cyber–physical systems: extending pervasive sensing from control theory to the Internet of Things. Pervasive Mob. Comput. **40**, 156–184 (2017)
2. Sánchez, B.B., Alcarria, R., de Rivera, D.S., Sánchez-Picot, A.: Enhancing process control in industry 4.0 scenarios using cyber-physical systems. JoWUA **7**(4), 41–64 (2016)
3. Bordel, B., Pérez-Jiménez, M., Sánchez-de-Rivera, D.: Recognition of activities of daily living in Enhanced Living Environments. IT CoNvergence PRActice (INPRA) **4**(4), 18–31 (2016)
4. Bordel, B., Alcarria, R., Pérez-Jiménez, M., Robles, T., Martín, D., de Rivera, D.S.: Building smart adaptable Cyber-Physical Systems: definitions, classification and elements. In: International Conference on Ubiquitous Computing and Ambient Intelligence pp. 144–149. Springer, Cham, December 2015
5. Sánchez-Picot, Á., Martín, D., de Rivera, D.S., Bordel, B., Robles, T.: Modeling and simulation of interactions among people and devices in ambient intelligence environments. In: 2016 30th International Conference on Advanced Information Networking and Applications Workshops (WAINA), pp. 784–789. IEEE, March 2016
6. Schirner, G., Erdogmus, D., Chowdhury, K., Padir, T.: The future of human-in-the-loop cyber-physical systems. Computer **46**(1), 36–45 (2013)
7. Bordel, B., Alcarria, R., Martín, D., Robles, T., de Rivera, D.S.: Self-configuration in humanized cyber-physical systems. J. Ambient Intell. Humaniz. Comput. **8**(4), 485–496 (2017)
8. Lupton, D.: M-health and health promotion: the digital cyborg and surveillance society. Soc. Theory Health **10**(3), 229–244 (2012)
9. Munir, S., Stankovic, J.A., Liang, C.J.M., Lin, S.: Cyber physical system challenges for human-in-the-loop control. In: Feedback Computing, June 2013
10. Cardenas, A.A., Amin, S., Sastry, S.: Secure control: towards survivable cyber-physical systems. In: 28th International Conference on Distributed Computing Systems Workshops, ICDCS 2008, pp. 495–500. IEEE, June 2008
11. Cardenas, A., Amin, S., Sinopoli, B., Giani, A., Perrig, A., Sastry, S.: Challenges for securing cyber physical systems. In: Workshop on Future Directions in Cyber-Physical Systems Security, vol. 5, July 2009
12. Leeper, A.E., Hsiao, K., Ciocarlie, M., Takayama, L., Gossow, D.: Strategies for human-in-the-loop robotic grasping. In: Proceedings of the Seventh Annual ACM/IEEE International Conference on Human-Robot Interaction, pp. 1–8. ACM, March 2012
13. Dautenhahn, K.: The art of designing socially intelligent agents: science, fiction, and the human in the loop. Appl. Artif. Intell. **12**(7–8), 573–617 (1998)
14. Cranor, L.F.: A framework for reasoning about the human in the loop (2008). https://www.usenix.org/legacy/event/upsec/tech/full_papers/cranor/cranor.pdf
15. Mois, G., Sanislav, T., Folea, S.C.: A cyber-physical system for environmental monitoring. IEEE Trans. Instrum. Meas. **65**(6), 1463–1471 (2016)

16. Hu, X., Chu, T.H., Chan, H.C., Leung, V.C.: Vita: a crowdsensing-oriented mobile cyber-physical system. IEEE Trans. Emerg. Top. Comput. **1**(1), 148–165 (2013)
17. Lee, I., Sokolsky, O.: Medical cyber physical systems. In: 2010 47th ACM/IEEE Design Automation Conference (DAC), pp. 743–748. IEEE, June 2010
18. Hossain, M.S.: Cloud-supported cyber–physical localization framework for patients monitoring. IEEE Syst. J. **11**(1), 118–127 (2017)
19. Sánchez-Picot, Á., Martín, D., Bordel, B., Sánchez-de-Rivera, D.: Aml environments simulations approach integrating social and network aspects: a case study. J. Ambient Intell. Smart Environ. **10**(4), 303–314 (2018)
20. MOSI-AGIL project. http://www.gsi.dit.upm.es/mosi/
21. Kopaczewski, K., Szczodrak, M., Czyżewski, A., Krawczyk, H.: Application of virtual gate for counting people participating in large public events. In: International Conference on Multimedia Communications, Services and Security, pp. 316–327. Springer, Heidelberg, May 2012
22. Ethics committee for research activities of the Universidad Politécnica de Madrid. http://www.upm.es/Investigacion/gestion_proyectos/ComiteEtica. Accessed 1 Nov 2018

T Wave Alternans Analysis in ECG Signal: A Survey of the Principal Approaches

Nancy Betancourt[1,2(✉)], Carlos Almeida[3(✉)], and Marco Flores-Calero[4,5(✉)]

[1] Departamento de Informática y Ciencias de la Computación,
Escuela Politécnica Nacional, Quito, Ecuador
nancy.betancourt@epn.edu.ec
[2] Departamento de Ciencias Exactas, Universidad de las Fuerzas Armadas - ESPE,
Quito, Ecuador
[3] Departamento de Matemática, Escuela Politécnica Nacional, Quito, Ecuador
carlos.almeidar@epn.edu.ec
[4] Department of Intelligent Systems, I&H Tech, Latacunga (Cotopaxi), Ecuador
[5] Universidad de las Fuerzas Armadas - ESPE, Quito, Ecuador
mjflores@espe.edu.ec

Abstract. The T wave alternans (TWA) is an important phenomenon not only within the clinical field but within the scientific and technological field, it has been considered an important, non-invasive, very promising indicator to stratify the risk of sudden cardiac death. Due to its microvolt amplitude and background noises, sophisticated signal processing techniques are required for its detection and estimation. In this paper we present a survey of the state of the art focusing to detect sudden cardiac death by analyzing the T wave on long-term ECG signals.

Keywords: ECG · Sudden cardiac death · T-wave analysis · T-wave alternans

1 Introduction

Sudden cardiac death (SCD) is defined as the natural death caused by cardiac disease occurring at most one hour after symptoms appear or within twenty four hours in a person without any precondition that seems fatal [1]. SCD is a leading cause of cardiovascular mortality [2–4] producing nine million deaths worldwide according to the World Health Organization [5]. It is estimated that 400,000 deaths per year are caused by SCD in the United States [6,7]. Only 1–2% of patients can survive when SCD occurs outside of a hospital [6]. Most patients are identified only after they have experienced severe hearth diseases such as ischemia, infarction, ventricular conduction abnormalities, or potassium abnormalities [7–9]. However, these patients represent 2% to 3% of the total SCD victims [3]. In Ecuador, heart ischemic diseases mortality is about 29.32%

© Springer Nature Switzerland AG 2019
Á. Rocha et al. (Eds.): ICITS 2019, AISC 918, pp. 417–426, 2019.
https://doi.org/10.1007/978-3-030-11890-7_41

of the total deaths recorded [10]. The incomprehensible nature of this devastating disease increases the urgency to develop new methods to estimate and predict this pathology, which leads to a more effective prevention [11].

The electrocardiogram (ECG) represents an important standard clinical procedure for the investigation of cardiac abnormalities [12]. The ECG reflects the electrical activity of the heart, which is composed of two phases called depolarization and repolarization (contraction and relaxation, respectively) [13]. Each beat corresponds to an electrical wave that crosses the different structures of the heart. These waves are knowing as P wave, Q waves, R and S (QRS complex) and T wave.

The T-wave represents ventricular repolarization in the ECG. The changes in the shape or amplitude of the T wave is known as T-wave alternan (TWA) [12, 14]. These magnitudes are typically in the order of microvolts [15] so it is difficult to detect the alternation. Therefore, the development of new effective algorithms for their detection is required. TWA is a heart rate dependent phenomenon that has proven to be a non-invasive indicator to stratify cardiac risks [6,16]. This phenomenon has been associated with malignant ventricular arrhythmia and SCD [17,18].

This paper presents the results of a literature review performed to analyze and evaluate relevant papers in field of TWA analysis and the SCD detection/estimation. The rest of the paper is divided in the following parts. Section two presents a general principles of ECG and T wave alternan analysis scheme. Section four shows the methodological review to TWA analysis in the case of detection and estimation. The results are presented in the last section.

2   General Principles

2.1   The Electrocardiogram

The electrocardiogram (ECG) is the record of the electrical activity of the heart. Each heartbeat is triggered by an electrical impulse that is normally generated in specialized cells in the right upper cavity of the heart (cells of the sinus node). An electrocardiogram records the time and strength of these signals as they move through the heart [12,13].

The ECG signal study is important because it allows the early diagnosis of the main cardiac diseases such as irregularities in heart rhythm (arrhythmias), coronary artery disease, structural problems in the heart chambers or a previous heart attack [12,17,18].

The ECG is also known as the "standard 12-lead electrocardiogram" because it gathers information from 12 different areas of the heart [19]. These images are created by electrodes that are placed on the skin of the chest and, sometimes, the extremities. Electric activity is recorded as waves in a graph, with different patterns corresponding to each electrical phase of the heartbeat.

Usually, the frequency range and the dynamic range of an ECG signal are between 0.05–100 Hz and 1–10 mV, respectively. The ECG signal is characterized by five peaks labelled by the letters (P, Q, R, S, T), (see Fig. 1).

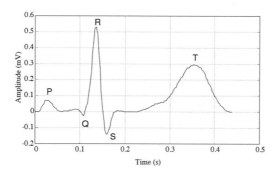

Fig. 1. ECG for one normal heartbeat showing typical amplitudes and time durations for the P, QRS, and T waves [20].

The first deviation observed in the signal is called the P wave and represents the activation (depolarization) of the atria. After the P wave, a series of waves is produced due to the activation of the ventricles (QRS complex). The ST-T complex, formed by the ST segment and the T wave, is observed, representing the repolarization of the ventricles. If there is a change in the shape or amplitude of the T wave, we speak of TWA. TWA consists of a variation of the morphology, amplitude or duration of the T wave that occurs continuously every two beats [12,21,22]. TWA is a heart rate dependent phenomenon that has been shown to be a non-invasive, reliable indicator for stratifying cardiac risk, malignant ventricular arrhythmias and SCD [16].

2.2 T-Wave Alternans Analysis

A TWA analysis can be divided into three stages (Fig. 2). The input of the system is the ECG signal. The output Ho indicates that the T wave alternan is present in the signal. In other hand, H1 represents the alternans absence.

Fig. 2. Procedure to develop TWA

1. **Preprocessing Stage.** The output of the preprocessing stage is a set X of ECG filtered segments. Once the ECG signal is captured it is important to prepare the signal to facilitate the subsequent TWA analysis. Alternan is a phenomenon associated with cardiac repolarization (ST-T complex) and it occurs beat by beat; then, it is then necessary to define a process of segmentation and alignment of the waves. Therefore, the detection of the QRS complex and the segmentation of the ST-T complexes are necessary tasks. Additionally, it is possible to use filter techniques to improve the quality of the signal [12,23,24].

2. **Features Extraction.** The objective of this stage is to reduce the number of data to be processed while maintaining the greatest amount of information of the TWA [4,14]. In their paper, [25] used continuous and discrete waveletes to feature extraction such as: Morphology analysis, fiducial points locating and time interval measurement. The output in this stage is a set Y that contain the signal without redundancies.

3. **TWA Analysis.** In this stage, the set Y is analysed permitting to detect the presence (Ho) or absence (H1) of alternans, after that its amplitude is estimated [13,26].

3   Methodological Review

Several methods have been developed in recent years to detect and quantify TWA. Some of the most widely employed methods in clinical practice as follow:

Modified Moving Average Method (MMA): In the article presented by [21], the authors showed a new approach to detect and estimate alternans in the T wave. The magnitude of TWA is obtained by means of the maximum absolute difference of averages of series of even and odd beats calculated in T waves or ST-T complexes.

Different approaches have been presented based on MMA for example: In their paper [23] propose a new method called template matched-filter based scheme for detection and estimation of t-wave alternans (TMFD), in this work a preprocessing stage for the MMA method to ensure an optimal alignment of the computed averages. In the work of [27], the authors propose an Enhanced modified moving average (EnMMA) the accuracy of the method was improved with an better aligned prior to distance calculation. In order to achieve the improvement, the authors added a preprocessing stage based on continuous dynamic time warping (DTW).

Spectral Methods (SM): SM was proposed by [28] in this method digitized ECG beats are aligned, and periodogram based power spectral evaluations are calculated for every sample in the segment of interest. The value of a added spectrum at 0.5 cpb is compared with the spectral noise level to decide if TWA is present. Different versions of the SM have been presented. In their paper [6] presents an improved spectral method that contains three component: enhanced spectral method (EnSM); spectral analysis of T-Slope variations (TSV); and singular value decomposition (SVD). Another version of SM is presented by [7] named non-negative matrix factorization (NMF)-Adaptive SM, the Adaptive SM have the advantages of both SM and Modified moving average. In [12] the paper presents a three step TWA detection and estimation strategy which consists of filtering the ECG signal using a variant of Kalman filter (KF), segmenting the ST-T wave based on ECG phase and applying the spectral method to detect TWA and estimate its value.

Complex Demodulation Method (CD): In this method, the beats are aligned, and TWA is showed in each series as a sinusoidal signal of frequency and variable amplitude and phase. TWA amplitude in each beat-to-beat series is projected by demodulation of the 0.5-cpb component and low-pass filtered to obtain a continuous beat-to-beat alternans measurement [29].

Correlation Method: A single cross correlation coefficient is computed for every ST-T complex against a representative for a heartbeat series. The single beat-to-beat series of coefficients is evaluated by a time-domain zero-crossing counter. If the correlation index alternants for some consecutives beats, a TWA episode is detected [30,31].

Karhunen-Loève Transform (KLT): The KL transform has been used for its ability to achieve the maximum compaction of energy in a few coefficients [32]. Two proposals have made use of this transform: In the first of the proposals, each ST-T complex is represented by the first four coefficients of its KL transform. Then each beat-to-beat series of KL coefficients is analyzed spectrally using a periodogram [14]

Capon Filtering Method (CF): In this method, the low-pass filter is replaced by the filter cap. It is the filter system that preserves the alternating component, minimizes the power of the signal at its output. The filter is based on the data and is obtained from the autocorrelation function of the input signal [33].

Statical Test Method (ST): This method is based on statistical tests: Student's t tests for independent and paired samples, applied to study if there are differences between the characteristics of the T wave between the odd and even beats and the Rayleigh periodicity test [34]. Recents methods have been developed, [17] presents a nonparametric adaptive surrogate test to assist in accurate detection of TWA, independent of the particular estimation algorithm being used. In their paper, [18] present a new class of algorithms, based on the Monte Carlo method, for the detection and quantitative measurement of alternans.

Laplacian Likelihood Ratio Method (LLR): The LLR method calculates the maximum likelihood estimation of the TWAs by assuming a Laplacian noise distribution, and applies the generalized likelihood ratio test to decide whether the TWAs are present or not in the ECG [22]. In [4], they propose the use of a multilead TWA analysis scheme that combines LLR method and periodic component analysis (πCA), an eigenvalue decomposition technique whose aim is to extract the most periodic sources of the signal.

Poincaré Mapping Method (PM): PMM are formed by plotting T wave magnitude of alternante beat. Semiperiodic signal such as TWA, appear as close clusters. TWA magnitude is the intercluster distance, [35].

4   Results of the State of the Art

We inform the reader that all the methods presented in the last section are considered important; however, in this document we will focus on the analysis of the results of the methods (SM and MMA) by the following reasons: According to the literature SM and MMA are the most used methods; These methods have been included in medical equipment such as CH2000 and Heartwave (Cambridge Heart Inc., Bedford, MA). The results are shown following the scheme (Fig. 2).

4.1   Preprocessing and Feature Extraction

The paper [23] presents TMFD approach, The QRS and T-wave peak detection is performed using the waveform locator available at Physionet. The comparative evaluation is carried out with three most common classical techniques in the case of stationary as well as non-stationary TWA (SM, MMA and CM). In the preprocessing stage, [27] presents an enhanced MMA method using Dynamic Time Warping (DTW) curve alignment. DTW is described as a method that can eliminate shift-related artifacts from measurements by correcting a sample vector of length J towards a reference of length I. This method performs well for different levels of TWA, noise, and phase shifts, but it is sensitive to the alignment of the T-waves.

The method proposed by [6] Enhanced modified moving average (EMMA), an digital filters were utilized to remove general arterial interference and to limit the ECG bandwidth between 1 Hz and 50 Hz. After that, Pan & Tompkins method is used to indicate R waves. With detected R points, T points can be located by cross-checking on the maximum points of ECG and all the zero-crossing points of dECG. In their research, the noise band is at range [0.42 0.46] cpb, and $P_0.5$ is the maximum value at range [0.47 0.5] cpb by considering potentially TWA frequency shifting. In their job [12] artificial generated Gaussian noise is added to the ECG recordings with SNR varying from −30 dB to 30 dB. Moreover to assess the performance in the presence of non-stationary noise, real muscle artifact (MA) and electromyography (EMG) noise were taken from MIT BIR noise stress database. Fiducial points are selected at the onset by employing mechanism for QRS and T-wave detection.

4.2   TWA Analysis

The approach [23] detect presence of TWA under the assumption of random Gaussian noise. The implementation of TMFD is discussed in the context of two different templates, median (TMFD-1) and mean (TMFD-2). The proposed method outperforms CM over the entire range of SNR between −15 dB and 35 dB. In Gaussian noise, performance of TMFD-1 is comparable with SM when detection probability is maximized and is 2 dB degraded for Laplacian noise. In better signal conditions (SNR = 25 dB), the bias of TMFD approaches SM for alternant magnitude >40 μ V in the Gaussian case and for magnitudes >20 μ V with real noises.

In [27], when there are phase shifts in the register and with some noise, EMMA outperforms MMA in all cases by 25%. The experiments under different baseline wandering conditions also demonstrated that EMMA is more robust than MMA.

[6] presents an improved spectral method successfully indicates 82.4% of people within SCD database at high risk, and 76.7% of people within normal database at low risk. In [7], the Receiver operating curves (ROC) is used to evaluate the results. In a ROC curve the true positive rate (Sensitivity) is plotted in function of the false positive rate (100-Specificity) for different cut-off points of a parameter. Each point on the ROC curve represents a sensitivity/specificity pair corresponding to a particular decision threshold. The area under the ROC curve (AUC) is a measure of how well a parameter can distinguish between two diagnostic groups (diseased/normal). In the method Adaptive SM, ROC were computed for each method with the area under the curve indicating relative TWA signal discrimination. NMF-Adaptive SM had the greatest area under the ROC (0.92) followed by the SM's Kscore of (0.77), SM without the Kscore (0.74), and MMA (0.70) ($p < 0.001$). By ROC curve analysis, TWA discrimination with NMF-Adaptive SM was superior to SM with k-score ($p < 0.001$). The estimation accuracy comparison in [12] is carried out in terms of relative bias (Rb), standard deviation and mean square error (MSE). Extended Kalman smoother (EKF) and Unscented KF (UKF) provide an advantage of 10 dB in achieving best Rb = O. UKS provides a higher Rb at low SNRs (-20 dB).

In SM, the stratification of risk is analyzed by means of the maximum alternating magnitude, this value is associated with a high level of risk of SCD if it is greater than or equal to 60 μ V during an ambulatory and routine test greater than or equal to 47 μ V after an episode of myocardial infarction. SM in a method that requires a stable heart rate of 105–110 beats per minute over a period of time, using a specialized exercise protocol, pharmacological agents or atrial pacing. Due to these restrictions, approximately 20–40% of the tests are classified as "indeterminate", either due to factors related to the patient such as the inability to reach the target heart rate, excessive ventricular ectopia, atrial fibrillation or technical problems as noise in the recording. MMA [21] is a time domain approach that consists of continuously estimating the average beat of even and odd beats calculated in T waves or ST-T complexes. Normally, the TWA level is reported every 10 or 15 s, which makes the MMA more versatile and more appropriate in ambulatory recordings. The MMA method can be applied in stress tests and ambulatory tests, however, the exact values of the thresholds have not yet been defined to calculate the maximum alternating magnitude, resulting in the classification of the tests being around 75% accuracy.

4.3   Databases

The different papers presented in the state of the art used databases that allowed to perform the experiments and validate the methods, these methods are presented in Table 1.

Table 1. Databases of ECG signals used in the state of the art for TWA

Article	Database	Method
[4]	Physionet TWA Database	LLR
[23]	MIT-BIH Arrhythmia Database, from MIT Physionet	TMFD
[27]	Simulated ECG signals: European STT database (0123, e0103, and e0105)	EnMMA
[31]	MIT-BIH Arrhythmia Database, from MIT Physionet	CM
[2]	Simulated ECG signals	SM
[6]	MIT/BIH Sudden Cardiac Death Holter Database (seventeen records with T waves presented)	SM
[12]	Simulated ECG signals	SM
[17]	NSRDB, CHFDB and SCDDB	ST
[18]	NE	ST
[21]	7-French USCI quadripolar catheter	MMA

5   Conclusions

Documents presented in the state of the art are focusing on the detection and quantification of the amplitude in the T-wave to determine the risk of SCD. According this MMA and SM are the most used methods. It can see that the accuracy improves but only in certain cases and under certains conditions, which complicates a comparison between methods since the same database or the same sample size is not used. On the other hand, the proposed methods are tested using either synthetically generated signals or using the physionet database. The reliability of current systems is still debatable because their results are not enough robust.

In this context is important to develop new methods to detect and quantify the alternation of the T wave, to surperate the results presented in the state of the art. For this, we will use the current computational power and new AI techniques, which will allow us to develop new efficient and lightweight algorithms for improving the efficiency in SCD detection. Also, these new algorithms may be part of the new eHealth systems.

References

1. Gimeno-Blanes, F.J., Blanco-Velasco, M., Barquero-Pérez, Ó., García-Alberola, A., Rojo-álvarez, J.L.: Sudden cardiac risk stratification with electrocardiographic indices - a review on computational processing, technology transfer, and scientific evidence. Front. Physiol. **7**, 1–17 (2016)
2. Narayan, S., Botteron, G., Smith, J.: T-wave alternans spectral magnitude is sensitive to electrocardiographic beat alignment strategy. **24**(ii), 593–596 (1997)

3. Pham, Q., Quan, K.J., Rosenbaum, D.S.: T-wave alternans: marker, mechanism, and methodology for predicting sudden cardiac death. J. Electrocardiol. **36**, 75–81 (2003)
4. Monasterio, V., Clifford, G.D., Laguna, P., Martí Nez, J.P.: A multilead scheme based on periodic component analysis for T-wave alternans analysis in the ECG. Ann. Biomed. Eng. **38**(8), 2532–2541 (2010)
5. World Health Organization: The top 10 causes of death (2014)
6. Shen, T.W., Tsao, Y.T.: An improved spectral method of detecting and quantifying T-wave alternans for SCD risk evaluation. Comput. Cardiol. **35**, 609–612 (2008)
7. Ghoraani, B., Krishnan, S., Selvaraj, R.J., Chauhan, V.S.: T wave alternans evaluation using adaptive time-frequency signal analysis and non-negative matrix factorization. Med. Eng. Phys. **33**(6), 700–711 (2011)
8. Valverde, E., Arini, P.: Study of T-wave spectral variance during acute myocardial ischemia, pp. 653–656 (2012)
9. Murukesan, L., Murugappan, M., Iqbal, M.: Sudden cardiac death prediction using ECG signal derivative (heart rate variability): a review, pp. 8–10 (2013)
10. INEC: Principales causas de mortalidad
11. Chugh, S.S.: Early identification of risk factors for sudden cardiac death. Nat. Rev. Cardiol. **7**(6), 318 (2010)
12. Irshad, A., Bakhshi, A.D., Bashir, S.: *Department of Electrical Engineering, College of Electrical and Mechanical Engineering, National University of Science and Technology, Islamabad, Pakistan. **Department of Electrical Engineering, University of Engineering and Technology, Lahore, Pakistan, pp. 222–227 (2015)
13. Blanco-Velasco, M., Cruz-Roldán, F., Godino-Llorente, J.I., Barner, K.E.: Nonlinear trend estimation of the ventricular repolarization segment for T-wave alternans detection. IEEE Trans. Biomed. Eng. **57**(10 PART 1), 2402–2412 (2010)
14. Martínez, J.P., Olmos, S.: Methodological principles of T wave alternans analysis: a unified framework. IEEE Trans. Biomed. Eng. **52**(4), 599–613 (2005)
15. Adam, D.R., Smith, J., Akselrod, S., Nyberg, S., Powell, A.O., Cohen, R.J.: Fluctuations in T-wave morphology and susceptibility to ventricular fibrillation. J. Electrocardiol. **17**, 209–218 (1984)
16. Haghjoo, M., Arya, A., Sadr-Ameli, M.A.: Microvolt T-wave alternans: a review of techniques, interpretation, utility, clinical studies, and future perspectives. Int. J. Cardiol. **109**(3), 293–306 (2006)
17. Nemati, S., Abdala, O., Monasterio, V., Yim-Yeh, S., Malhotra, A., Clifford, G.D.: A nonparametric surrogate-based test of significance for T-wave alternans detection. IEEE Trans. Biomed. Eng. **58**(5), 1356–1364 (2011)
18. Iravanian, S., Kanu, U.B., Christini, D.J.: A class of Monte-Carlo-based statistical algorithms for efficient detection of repolarization alternans. IEEE Trans. Biomed. Eng. **59**(7), 1882–1891 (2012)
19. Stroobandt, R.X., Barold, S.S., Sinnaeve, A.F.: ECG from basics to essentials: step by step (2016)
20. Tompkins, W.J.: Biomedical Digital Signal Processing: C-Language Examples and Laboratory Experiments for the IBM PC. Prentice Hall, Hauptbd (2000)
21. Nearing, B.D., Verrier, R.L.: Modified moving average analysis of T-wave alternans to predict ventricular fibrillation with high accuracy. J. Appl. Physiol. **92**(2), 541–549 (2002)
22. Martínez, J.P., Olmos, S.: A robust T wave alternans detector based on the GLRT for Laplacian noise distribution. In: Computers in Cardiology, pp. 677–680. IEEE (2002)

23. Bashir, S., Bakhshi, A.D., Maud, M.A.: A template matched-filter based scheme for detection and estimation of T-wave alternans. Biomed. Signal Process. Control **13**(1), 247–261 (2014)

24. Fujita, H., Acharya, U.R., Sudarshan, V.K., Ghista, D.N., Sree, S.V., Eugene, L.W.J., Koh, J.E.: Sudden Cardiac Death (SCD) prediction based on nonlinear heart rate variability features and SCD index. Appl. Soft Comput. J. **43**, 510–519 (2016)

25. Gualsaqui Miranda, M.V., Vizcaino Espinosa, I.P., Flores Calero, M.J.: ECG signal features extraction. Ecuador Technical Chapters Meeting (ETCM), pp. 1–6. IEEE (2016)

26. Goya-Esteban, R., Barquero-Perez, O., Blanco-Velasco, M., Caamano-Fernandez, A.J., Garcia-Alberola, A., Rojo-Alvarez, J.L.: Nonparametric signal processing validation in T-wave alternans detection and estimation. IEEE Trans. Biomed. Eng. **61**(4), 1328–1338 (2014)

27. Cuesta-Frau, D., Micó-Tormos, P., Aboy, M., Biagetti, M.O., Austin, D., Quinteiro, R.A.: Enhanced modified moving average analysis of T-wave alternans using a curve matching method: a simulation study. Med. Biol. Eng. Comput. **47**(3), 323–331 (2009)

28. Smith, J.M., Clancy, E.A., Valeri, C.R., Ruskin, J.N., Cohen, R.J.: Electrical alternans and cardiac electrical instability. Circulation **77**(1), 110–121 (1988)

29. Nearing, B.D., Huang, A.H., Verrier, R.L.: Dynamic tracking of cardiac vulnerability by complex demodulation of the T wave. Science **252**(5004), 437–440 (1991)

30. Burattini, L., Zareba, W., Couderc, J., Titlebaum, E., Moss, A.: Computer detection of non-stationary T wave alternans using a new correlation method. In: Computers in Cardiology, pp. 657–660. IEEE (1997)

31. Noohi, M., Sadr, A.: T wave detection by correlation method in the ECG signal. In: The 2nd International Conference on Computer and Automation Engineering (ICCAE), vol. 5, pp. 550–552 (2010)

32. Laguna, P., Ruiz, M., Moody, G., Mark, R.: Repolarization alternans detection using the KL transform and the beatquency spectrum. In: Computers in Cardiology, pp. 673–676. IEEE (1996)

33. Martinez, J.P., Olmos, S., Laguna, P.: Simulation study and performance evaluation of T-wave alternans detector. In: Proceedings of the 22nd Annual International Conference of the IEEE Engineering in Medicine and Biology Society, vol. 3, pp. 2291–2297. IEEE (2000)

34. Srikanth, T., Lin, D., Kanaan, N., Gu, H.: Presence of T wave alternans in the statistical context-a new approach to low amplitude alternans measurement. Comput. Cardiol. **29**, 681–684 (2002)

35. Strumillo, P., Ruta, J.: Poincare mapping for detecting abnormal dynamics of cardiac repolarization. IEEE Eng. Med. Biol. Mag. **21**(1), 62–65 (2002)

Smart University: A Review from the Educational and Technological View of Internet of Things

Dewar Rico-Bautista[1]($\boxtimes$), Yurley Medina-Cárdenas[2], and Cesar D. Guerrero[3]

[1] Systems and Informatics Department,
Universidad Francisco de Paula Santander Ocaña,
Algodonal Campus Vía Acolsure, 546551 Ocaña, Colombia
dwricob@ufpso.edu.co
[2] Integrated Management System Coordinator,
Universidad Francisco de Paula Santander Ocaña,
Algodonal Campus Vía Acolsure, 546551 Ocaña, Colombia
ycmedinac@ufpso.edu.co
[3] Center of Excellence and Appropriation on the Internet of Things,
Universidad Autónoma de Bucaramanga,
Avenida 42 No. 48 – 11, 680003 Bucaramanga, Colombia
cguerrer@unab.edu.co

Abstract. This article reviews from the Internet of Things the emerging concept of smart university. The purpose is to present the problem situation, justification, hypothesis and conceptual framework about the term and characteristics of smart universities published in the scientific literature, from which the role of the IoT is emphasized as a fundamental element in the conception and implementation of projects and initiatives that affect the successful development of these organizations.

Keywords: Internet of Things · Technologies · Smart campus · Smart university

1 Introduction

In Colombia, the National Development Plan [1] is based on three pillars, including education. It assumes education as the most powerful instrument of social equality and economic growth in the long term, with a vision aimed at closing gaps in access and quality to the education system. To consolidate education as one of the pillars, the plan includes the following transversal and regional strategies: *Strategic competitiveness and infrastructure; social mobility; countryside transformation; security, justice and democracy for the peace construction; good government and green growth.*

Similarly, the private competitiveness council launches its eleventh National Competitiveness Report [2], where it discloses certain figures from Colombia. In education, higher education quality in Colombia has an important gap to close. Only 14.9% of the higher education undergraduate programs and 14.5% of the higher

© Springer Nature Switzerland AG 2019
Á. Rocha et al. (Eds.): ICITS 2019, AISC 918, pp. 427–440, 2019.
https://doi.org/10.1007/978-3-030-11890-7_42

education institutions have high quality accreditation. The quality in higher education, taken from the Colombian National Accreditation Council model [3], implies the continuous effort of the institutions to fulfill in a responsible manner the specific demands of each of their mission functions: research, teaching and social projection; committing them to continuous improvement [4]. For the fulfillment and management of these mission processes, the university has inputs and resources: human (teachers, managers and administratives), physical (physical infrastructure), financial (transfers from the nation and territorial entities, own revenues for enrollment and academic rights and technology (IT infrastructure and information systems)) [5, 6]. These inputs of each university are measured from the ministry of education through different models and indicators, which ultimately represent their total management capacity, thus constituting an efficiency measure.

Technology becomes a fundamental element for the universities, without which it could not carry out any of its essential functions [7]. To improve the management of their resources, the universities design their strategic direction according to their own needs, but do not take into account the national requirements that demand their innovation and competitiveness in areas such as technological infrastructure. Their development plans consider technology as a support, but not as a strategic necessity for its operation. Its implementation is delegated to meet specific software and hardware needs, but the current capacity to contribute to the strategic aims of the universities is not evaluated.

Being the basic pillar of the development of the future communities, it is allowed for them to provide *intelligence* to all their areas and generate services that provide an improved quality of life to citizens and greater sustainability to cities and universities [8]. However, they should not be conceived only as a tactical element, they should not be planned in isolation, but should be part of the overall planning of the university [9].

This paper is divided mainly into (i) introduction, (ii) Architectures and technologies, (iii) Internet of things in the educational field, (iv) Smart Campus, (v) Smart University and finally, conclusions.

2   Architectures and Technologies

The world has changed, technology has changed the world. Technology and society are now inseparable fellows, they have merged to offer a new level of services. Information Technology (IT) is not just talking about computers, mobile devices, sensors, networks, it is to talk about new concepts and socio-technological paradigms [10].

To analyze, propose and develop infrastructure in an organization, for example in universities, the architectures and their technologies must be known [11, 12]. To this end, an overview is offered regarding the system architecture, the technologies, the implementation in smart campuses and smart universities [13], (see Fig. 1). In the case of layered architecture, the number of layers decides the architecture complexity. The five-layer architecture is the ideal architecture from the perspective of security and compatibility [14].

Diverse technologies support a generalized network infrastructure, highlighting the IoT development and its devices [15, 16]. These are classified into devices with limited resources and resource-rich devices. Resource-rich devices such as a smartphone, a standard personal computer, or a server have sufficient hardware, software, and memory

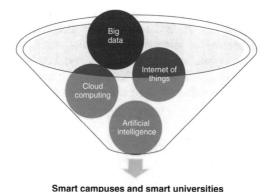

Smart campuses and smart universities

Fig. 1. Technologies that are implemented in smart campuses and smart universities. Source: Author

that support the TCP/IP protocol. Devices with resource limitations such as devices based on microcontrollers, sensors and actuators do not have sufficient hardware/ software capabilities that support the TCP/IP protocol [14].

In recent years, several published studies examined IoT from different aspects [17]. The work done by Atzori [18, 19] presented the different visions of IoT, generating a first approach to its understanding. In the review carried out by Al-Fuqaha et al. [20], protocols and possible applications of IoT are presented among with key challenges for future work including security and privacy issues [21–23]. Also, the work done by Andrea et al. [24] presented security vulnerabilities and IoT challenges from the point of view of applications, networks and physical systems. The work of Sha et al. [25], set out IoT problems and opportunities.

Botta et al. [26] considered the integration of *cloud computing* with IoT. Likewise, Wu and Zhao [27], proposed a new IoT infrastructure, namely WInternet, that meets several IoT requirements. Sheng et al. [28]; Palade [29]; Rodrigues and others [30], present *middleware* as a key technology in the realization of IoT systems, which is often described as a computer system designed to be the intermediary between devices and IoT applications. In the paper by Lee, Bae and Kim [31], they focus on the network layer, which for them is the most important part of the IoT environmentrealization. The idea of IoT, therefore, has evolved over time and changes will continue to occur in the coming years with the arrival and incorporation of new technologies, such as cloud computing, ICN, Big data, social networks, among others. [20, 21, 32–35].

3   Internet of Things in the Educational Field

Although a series of efforts have been made, most of the existing reviews and works have mainly focused on technological aspects of IoT. Specifically, there is no review available on applications in the field of education [36]. This requires a thorough IoT study to obtain a general understanding of the complex discipline of this research area and the challenges that emerge in its implementation [18]. Some of these challenges include security and privacy [37–39], availability, mobility, reliability, performance, interoperability, scalability, trust and management [40]. In terms of IoT applications, (see Fig. 2), there are several established in the needs of potential users, such as: smart cities [41]; smart energy and grid [42]; smart transportation and mobility; smart homes; smart buildings and infrastructure [43]; smart factories and manufacturing [44]; smart health; food and water monitoring and safety; smart networks [45]; and smart campuses [46].

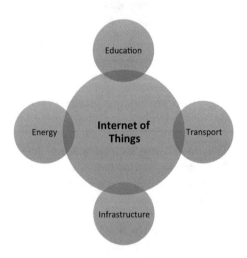

Fig. 2. Internet of things applications according to the review. Source: Author

Technology has played an important role in higher education connecting the three bodies: students, teachers and administrative staff [47]. The great impact in the field of education has not only changed traditional teaching practices, but has also produced changes in the infrastructure of educational institutions, generating the concept of smart campuses [48].

Using IoT as a tool to improve education and facilitate educational life, some of the works related to this topic are presented below. The work of Xu and Hela [49] examines the use of Cloud Computing and IoT to incorporate the structure of educational resources and provides an integration model. The work of Whitmore, Agarwal and Xu [50] confirms that, at this time, the research being done on IoT is largely focused on technology. Once the technology matures, research should be extended to the areas of management, operations, law, economics and sociology, among others. Another study [51] analyzes the impact of four different technologies, including IoT, Cloud Computing, data mining and Triple-Play in modern distance education. The research work [52], describes the application of IoT and Cloud Computing in education and also distinguishes between smart campus and digital campus. At work [53], an integrated architectural model was presented to develop an IoT system within an academic framework.

Atzori, Iera & Morabito [18] present an approach to evolution from its evolutionary nature. Three main stages in the evolution of the paradigm are identified, each one characterized by key enabling technologies, architectural reference solutions and available products. Nevertheless, what IoT is today and what it will be in the future is, undoubtedly, the result of the convergence in the primary evolutionary trajectory of all R&D experiences in several ICT domains, (see Fig. 3).

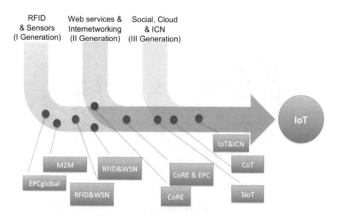

Fig. 3. IoT evolution. Source: Atzori, Iera, & Morabito [18]

4 Smart Campus

The term smart campus emerged from the concept of smart cities applying the principles of smart cities to the campus operation [54]. It is an emerging industry, with an infinity of solutions already adopted by several universities around the world. They are developed on the basis of digital campuses. Compared to traditional digital campuses, smart campuses provide services in a timely manner, reduce effort and reduce operating costs [55]. The smart campus implies that the institution will adopt advanced technologies to automatically control and supervise campus facilities and provide high quality services to the campus community, that is, students and staff [56]. This leads to increasing the campus efficiency and responsiveness, and to having better decision-making, space use and student experience [57].

A debate is opened on the current challenges and future orientations of research in the field of smart campuses enabled for IoT. In the first place, the adoption of IoT on campus would lead to the use of hundreds, if not thousands, of sensors, actuators and other objects, which can be a huge burden for manual configuration. Certainly, it is necessary to thoroughly investigate new forms of automated configuration of its devices [58]. Second, the variety of detection tasks and IoT applications in smart campuses imposes the use of heterogeneous objects. This heterogeneity presents a great obstacle to allow detection, so new interoperability standards are needed to integrate several devices in a single system. In addition, a standardized data format is needed for the description of the data generated by the IoT devices [59]. Third, it is expected that a large number of smart objects collaborate with each other to provide information about the environment, generating a huge volume of data. The handling of such large sensory data sets out a great challenge, since it requires advanced data fusion and optimization techniques. If data processing must be done in real time, high-speed communication links and powerful processing units will be needed [60, 61]. A higher education institution must ensure that both its computer team and its pedagogical approaches support the use of IoT [62].

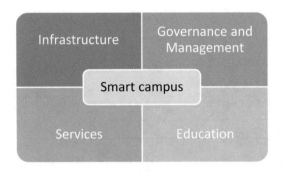

Fig. 4. Smart campus and its four thematic axes. Source: Cerdeira y Mendes [78]

A work that draws attention is the one presented by Cerdeira and Mendes [63], where the concept of smart campus is divided into four thematic axes, (see Fig. 4): Infrastructure; Governance and Management; Services and Education. Although a smart campus is not characterized only by the use of the technologies written in the infrastructure axis, as suggested by several works, it is notable that this axis serves as support for the other dimensions, mainly in the provision of data and connectivity, the governance and management of those institutions. The data generated by those services serve as input for other initiatives, mainly those aimed at the campus management, in activities aimed at making decisions. The implementation of smart campuses, despite being related to the great use of Information and Communication Technologies, must be aligned with the institutional strategy and the context in which the institution is inserted, with a view to measuring the value that will be delivered by each initiative. Otherwise, the availability of technological solutions may be doomed to failure, since it must be preceded by planning, which in turn must have the participation of all the involved parties, always looking for a win-win relationship through the balance between the interests [64].

5   Smart University

Currently, the term "smart university" has been incorporated in the literature. Universities can be considered smart universities, since they profitably use the available technologies to improve their performance and improve their graduates' quality [65, 66]. Despite this, there is still room for improvement and universities should become smarter universities [67]. In a smarter university, the best technological solutions encourage collaboration and cooperation among people [68, 69]. Smart universities and smart education are rapidly emerging and growing in areas that represent an integration of (1) smart systems, smart objects and smart environments, (2) smart technologies, various branches of computer science and computer engineering, (3) state-of-the-art smart systems, agents and educational software and/or hardware tools, and (4) innovative pedagogy, teaching strategies and learning methodologies based on advanced technologies [70, 71].

From the concept of smart university, an institution can be considered as smart if it has the following [72]:

- Special intellectual environment for the continuous improvement of the educational process subjects and objects competences that presupposes formal and informal education.
- A highly developed university technical "architecture" that contributes to the active application of smart technologies and devices in the educational process.
- Educational content designed on the basis of unified data and description regulations.
- Compatibility of software developed for different operating systems (so-called "seamless"), which guarantees equal access capabilities for all participants in the educational process from devices of any kind.
- A complete set of adaptive educational programs that ensure individualized education.
- An e-learning and m-learning well-developed system based on Web 2.0 tools, which provides broad access for students to educational content from around the world, regardless of time and place.
- A well-developed electronic university within a superior framework.
- Realization of a Quick Start concept for learning in the electronic university framework.
- Participation of practical experts in curricular design and effectiveness evaluation.
- Smart Learning; Learning-by-doing (Virtual labs); Adaptive Teaching; Automatic translation Systems (from/to English Language); Collaborative Learning; Gamification in Learning.

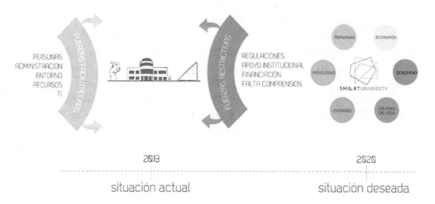

Fig. 5. Basic components of a smart university. Source: Maciá, Berná, Sánchez, Lozano & Fuster [8]

Maciá, Berná, Sánchez, Lozano & Fuster [8] states that the concept of Smart University, (see Fig. 5), maintains the same main objective of improving the quality of life of its community by applying globally, intensively and sustainably the IT under the principle of service to citizens. That is why it is a controlled environment, in which economic, sustainability, construction or development policies decide locally and do not depend on external factors, although they follow criteria similar to national or international policies and strategies [68, 73–75]. The current infrastructure allows to have globally-connected digital systems, able to support applications, services, platforms and sensors, all of them working together under a standardized ecosystem or framework of regulations and protocols.

6   Results and Discussion

The literature analysis revealed that the research that is being carried out on IoT focuses to a large extent on technology at this time. Research, as technology, smart campuses and universities mature (IoT and others), should be extended to areas such as management. The literature review produced some important findings that can focus efforts on project development [50]. These include:

- University technology is not well represented in management literature.
- The literature is dominated by research related to IoT technology.
- The coverage of business models driven by technology is scarce.
- Little work has been done on issues related to legal and governance frameworks.

Many researchers are trying to define a conceptual model for the smart university and identify its main characteristics, components, technologies and systems, reducing the role of traditional teaching/learning methods in universities. It is considered that a priority in the realization of smart university projects is the virtual classroom system, modern in terms of hardware and software, centered on the student. There is a great variety of architectures with smart devices, technologies and applications used in the educational environment that create even more difficulties in standardization. So far, there is no standard model with well-defined concepts and principles [74].

The *"smartification"* process, taken not as an objective, but as a way of life, a means, a continuous improvement process, implies a change of paradigm or model, by means of which it is intended to reinforce the concept of a more open university, and as a consequence adapt its management model to the new times by redesigning its relations with the public and private sectors, the relationship with the university community, its synergies and its transversal axes where all actors and their infrastructures must be coordinated for a common purpose: sustainability and quality of life [8, 57, 73, 76, 77]. For this reason, it is vital that any university can and must constitute a meeting point among different public and private economic agents, providing an environment where companies can discuss different options, receive scientific support and participate in innovative projects.

The Internet of Things (IoT) vision foresees a future Internet that incorporates physical objects that offer hosted functionality as IoT services. To facilitate seamless access and life cycle management of distributed and heterogeneous IoT resources, service-oriented computing and resource-oriented approaches have been widely used as promising technologies. However, a reference architecture that integrates IoT services in any of these two technologies remains an open research challenge [17].

7 Conclusions

Despite the importance and considerable time to adopt new technologies in universities [78, 79]; in countries such as Spain [7], Mexico [80] and Ecuador [81], there was no study that allowed to visualize the technology situation in these institutions [7, 78, 79, 81–83]. By generating such study, and its corresponding analysis, allowed a better use of available resources and the possibility of visualizing collaborative projects between them, thus favoring the investment optimization. In Colombia so far, there is no such study.

Given the concept definition and delimitation of *smart city* and *smart university*, the next step is to define a framework to evaluate the degree of smart development of those services [73, 84–86], taking into account for example, the AENOR Technical Standardization Committee 178 framework, whose objective is the definition of a series of metrics and indicators, both quantitative and qualitative, of a smart community under the following activities [8, 87, 88]:

- To define a simple model that allows, in a practical way, the evaluation of the degree of development of a smart service.
- To have a simple tool to make an assessment of the status of its services in relation to the goal of making them smart.
- To try to establish a model that is complementary to the works carried out within the framework of the aforementioned Technical Committee.

There are other global initiatives and international standards that seek the standardization of indicators; among them, the ISO 37120:2014 standard stands out, where it is sought to monitor and track the progress of the city services' performance and sustainability and the quality of life, starting point to be applied to a community as a University. This scenario, both in cities and universities, reveals that we are facing a new social paradigm that, due to its complexity and transversality, has evolved without a general reference model.

As there is a great variety of smart devices, technologies and applications used in the educational environment, difficulties in standardization are created. Most of the existing reviews and works so far have mostly focused on specific aspects of technology, architectures and infrastructure. Specifically, there is no available review on the applications in the sphere of management in the educational field, what opens a workspace since there is no standard model with well-defined concepts and principles [74]. This shows the need to *design models and architectures* that support this explosion of ideas, technologies and services for citizens [55].

Acknowledgements. Authors would like to acknowledge the cooperation of all partners within the Center of Excellence and Appropriation on the Internet of Things (*Centro de Excelencia y Apropiaciónen Internet de las Cosas, CEA-IoT*). Authors would also like to thank all institutions that supported this work: the Colombian Ministry for the Information and Communication Technologies (*Ministerio de Tecnologías de la Información y las Comunicaciones - MinTIC*), and the ColombianAdministrative Department of Science, Technology and Innovation(*Departamento Administrativo de Ciencia, Tecnología e Innovación - Colciencias*) through the National Trust for Funding Science, Technology and Innovation Francisco José de Caldas (*Fondo Nacional de Financiamiento para la Ciencia, la Tecnología y la Innovación Francisco José de Caldas*), under project ID: FP44842-502-2015.

References

1. C. de la república de Colombia: Plan nacional de desarrollo 2014-2018 (2015)
2. Consejo Privado de Competitividad: Informe nacional de competitividad 2017-2018, p. 271 (2017)
3. Generales, P., De Calidad, C.: Modelo de acreditación CNA (2006)
4. Enrique, L., Silva, O.: La calidad de la universidad. Más allá de toda ambigüedad, pp. 1–14 (1997)
5. Roa, A.: Hacia un modelo de aseguramiento de la calidad en la educación superior en Colombia: estándares básicos y acreditación de excelencia. Educ. Super. Calid. y acreditación. Alfa Omega Colomb. Bogotá, pp. 101–107 (2003)
6. Ministerio de Educación Nacional: Propuesta metodológica para la distribución de recursos Artículo 87 de la Ley 30 de 1992 Vigencia 2013, p. 6 (2013)
7. Gómez, J., Jimenez, T., Gumbau, J., Llorens, F.: UNIVERSITIC 2017 Análisis de las TIC en las Universidades Españolas (2017)
8. Maciá, F.: Smart University. Hacia una universidad más abierta, Primera (2017)
9. Fernández Martínez, A., Llorens Largo, F.: Gobierno de las TI para universidades (2016)
10. Mineraud, J., Mazhelis, O., Su, X., Tarkoma, S.: A gap analysis of Internet-of-Things platforms. Comput. Commun. **89–90**, 5–16 (2016)
11. Lin, J., Yu, W., Zhang, N., Yang, X., Zhang, H., Zhao, W.: A survey on internet of things: architecture, enabling technologies, security and privacy, and applications. IEEE Internet Things J. **4**(5), 1125–1142 (2017)
12. Marotta, M.A., Both, C.B., Rochol, J., Granville, L.Z., Tarouco, L.M.R.: Evaluating management architectures for Internet of Things devices. IFIP Wirel. Days, vol. 2015–Janua, no, January 2015
13. Zapata-Ros, M.: La universidad inteligente La transición de los LMS a los Sistemas Inteligentes de Aprendizaje en Educación Superior The transition from Learning Management Systems (LMS) to Smart Learning Systems (SLS) in Higher Education. RED. Rev. Educ. a Distancia. Núm, **57**(10), 31–41 (2018)
14. Ara, T., Gajkumar Shah, P., Prabhakar, M.: Internet of Things architecture and applications: a survey. Indian J. Sci. Technol. **9**(45), December 2016
15. Khan, R., Khan, S.U., Zaheer, R., Khan, R.: Future internet: the internet of things architecture, possible applications and key challenges. In: Proceedings - 10th International Conference on Frontiers of Information Technology, FIT 2012, pp. 257–260 (2012)
16. Rodríguez Molano, J.I., Montenegro marín, C.E., Cueva Lovelle, J.M., Molano, J., Marin, C., Cueva, J.: Introducción al Internet de las Cosas. Redes Ing. **6**(7), 53–59 (2015)
17. Dar, K., Taherkordi, A., Baraki, H., Eliassen, F., Geihs, K.: A resource oriented integration architecture for the Internet of Things: a business process perspective. Pervasive Mob. Comput. **20**, 145–159 (2015)

18. Atzori, L., Iera, A., Morabito, G.: Understanding the Internet of Things: definition, potentials, and societal role of a fast evolving paradigm. Ad Hoc Netw. **56**, 122–140 (2017)
19. Nitti, M., Atzori, L., Cvijikj, I.P.: Friendship selection in the social internet of things: Challenges and possible strategies. IEEE Internet Things J. **2**(3), 240–247 (2015)
20. Al-Fuqaha, A., Guizani, M., Mohammadi, M., Aledhari, M., Ayyash, M.: Internet of Things: a survey on enabling technologies, protocols, and applications. IEEE Commun. Surv. Tutorials **17**(4), 2347–2376 (2015)
21. Tan, L., Tan, L., Wang, N.: Future internet: the Internet of Things. In: 2010 3rd International Conference on Advanced Computer Theory and Engineering, pp. V5-376–V5-380 (2010)
22. Liu, R., Wang, J.: Internet of Things: application and prospect. In: MATEC Web Conference, vol. 100 (2017)
23. jun Lee, W.: When the future technology is now: paradoxical attitudes of consumer and evaluation of iot service. Int. J. Smart Home **10**(6), 115–126 (2016)
24. Chatzigiannakis, I., Vitaletti, A., Pyrgelis, A.: A privacy-preserving smart parking system using an IoT elliptic curve based security platform. Comput. Commun. **89**, 165–177 (2016)
25. Tandon, A., Shah, V., Gajjar, S.: Emerging applications perspective for Internet of Things. In: Proceedings of the Second International Conference on Information and Communication Technology for Competitive Strategies - ICTCS 2016, vol. 04–05, pp. 1–7 (2016)
26. Botta, A., De Donato, W., Persico, V., Pescap, A.: Integration of Cloud computing and Internet of Things: a survey. Futur. Gener. Comput. Syst. **56**, 684–700 (2016)
27. Xiong, N., Liu, R.W., Liang, M., Wu, D., Liu, Z., Wu, H.: Effective alternating direction optimization methods for sparsity-constrained blind image deblurring. Sensors **17**(1), E174 (2017)
28. Ngu, A.H., Gutierrez, M., Metsis, V., Nepal, S., Sheng, Q.Z.: IoT middleware: a survey on issues and enabling technologies. IEEE Internet Things J. **4**(1), 1–20 (2017)
29. Razzaque, M.A., Milojevic-Jevric, M., Palade, A., Clarke, S.: Middleware for Internet of Things: a survey. IEEE Internet Things J. **3**(1), 70–95 (2016)
30. da Cruz, M.A.A., Rodrigues, J.J.P.C., Al-Muhtadi, J., Korotaev, V., Albuquerque, V.H.C.: A Reference model for Internet of Things middleware. IEEE Internet Things J. **PP**(99), 1 (2018)
31. Lee, S.K., Bae, M., Kim, H.: Future of IoT networks: a survey. Appl. Sci. **7**(10), 1072 (2017)
32. Weyrich, M., Ebert, C.: Reference architectures for the Internet of Things. IEEE Softw. **33**(1), 112–116 (2016)
33. Gagliardi, D., Schina, L., Sarcinella, M.L., Mangialardi, G., Niglia, F., Corallo, A.: Information and communication technologies and public participation: interactive maps and value added for citizens. Gov. Inf. Q. **34**(1), 153–166 (2017)
34. European Technology Platform on Smart Systems Integration, Internet of Things in 2020 (2008)
35. Akyildiz, I.F., Nie, S., Lin, S.C., Chandrasekaran, M.: 5G roadmap: 10 key enabling technologies. Comput. Networks **106**, 17–48 (2016)
36. Gul, S., Asif, M., Ahmad, S., Yasir, M., Majid, M., Malik, M.S.A.: A Survey on role of Internet of Things in education. IJCSNS Int. J. Comput. Sci. Netw. Secur. **17**(5), 159–165 (2017)
37. Broadband Internet Technical Advisory Group (BITAG): Internet of Things (IoT) Security and Privacy Recommendations (2016)
38. Flauzac, O., Gonzalez, C., Nolot, F.: New security architecture for IoT network. Procedia Comput. Sci. **52**(1), 1028–1033 (2015)
39. Sánchez-Torres, B., Rodríguez-Rodríguez, J.A., Rico-Bautista, D.W., Guerrero, C.D.: Smart Campus: trends in cyber security and future development. Rev. Fac. Ing. **27**(47) (2018)

40. Li, S., Da Xu, L., Zhao, S.: The Internet of Things: a survey. Inf. Syst. Front. **17**(2), 243–259 (2015)
41. Góngora, G.P.M.: Revisión de literatura sobre ciudades inteligentes: una perspectiva centrada en las TIC. Ingeniare **19**(19), 137–149 (2016)
42. Bagheri, M., Movahed, S.H.: The effect of the Internet of Things (IoT) on education business model. In: 2016 12th International Conference on Signal-Image Technology & Internet-Based Systems (SITIS), pp. 435–441 (2016)
43. Aqeel-ur-Rehman, A.Z.A., Shaikh, Z.A.: Building a Smart University using RFID technology. In: 2008 International Conference on Computer Science and Software Engineering, vol. 5, pp. 641–644 (2008)
44. Gierej, S.: The framework of business model in the context of industrial Internet of Things. Procedia Eng. **182**, 206–212 (2017)
45. de Carvalho Silva, J., Rodrigues, J.J.P.C.: IoT network management: content and analysis (2017)
46. Abuarqoub, A., et al.: A survey on Internet of Things enabled smart campus applications. In: Proceedings of the International Conference on Future Networks and Distributed Systems, pp. 50:1–50:7 (2017)
47. Mohamed Soliman, E.: Experimental evaluation of Internet of Things in the educational environment. Int. J. Eng. Pedagog. **7**(3), 50–60 (2017)
48. Pandey, R., Verma, M.: Current Emerging Trends in IOT: a survey and future prospects, vol. 8, no. Iii, pp. 339–344
49. Xu, Y., Helal, A.: Scalable cloud-sensor architecture for the Internet of Things. IEEE Internet Things J. **3**(3), 285–298 (2016)
50. Whitmore, A., Agarwal, A., Da Xu, L.: The Internet of Things—a survey of topics and trends. Inf. Syst. Front. **17**(2), 261–274 (2015)
51. Cheng, X., Xue, X.: Construction of smart campus system based on cloud computing. In: Proceedings of 2016 6th International Conference on on Applied Science Engineering and Technology, vol. 77, no. Icaset, pp. 187–191 (2016)
52. Widya Sari, M., Wahyu Ciptadi, P., Hardyanto, P.: Study of Smart Campus Development Using Internet of Things Technology, vol. 190 (2017)
53. Malatji, E.M.: The development of a smart campus - African universities point of view. In: 2017 8th International Renewable Energy Congress, IREC 2017 (2017)
54. Kar, A., Gupta, M.P.: How to make a Smart Campus - Smart Campus Programme in IIT Delhi (2015)
55. Galego, D., Giovannella, C., Mealha, Ó.: Determination of the smartness of a University Campus: the case study of aveiro. Procedia Soc. Behav. Sci. **223**, 147–152 (2016)
56. Valks, B., et al.: Smart campus tools - adding value to the university campus by measuring space use real-time Article information: To cite this document: About Emerald www. emeraldinsight.com Smart campus tools - adding value to the university campus by measuring space us. J. Corp. Real Estate **20**(2), 103–116 (2017)
57. Ali, M., Majeed, A.: How Internet-of-Things (IoT) Making the University Campuses Smart?, pp. 646–648 (2018)
58. Arsan, T.: Smart systems: from design to implementation of embedded Smart Systems. In: 2016 HONET-ICT, pp. 59–64 (2016)
59. Tarouco, L., Boesing, I., Barone, D., Rosa, G.R.P.: Internet das Coisas na Educação trajetória para um campus inteligente. An. dos Work. do CBIE 2017, p. 1220 (2017)
60. Xiong, L.I.U.: A study on smart campus model in the Era of Big Data. Adv. Soc. Sci. Educ. Humanit. Res., vol. 87, no. Icemeet 2016, pp. 919–922 (2017)

61. Feng-Ling, W.: Research on the application of smart campus construction under the background of Big Data. In: 2nd International Conference on Computer, Network Security and Communication Engineering (CNSCE 2017), pp. 246–252 (2017)
62. Mujun, W.: Smart campus-based study on optimization model for the computer information processing technology in universities and colleges. Rev. la Fac. Ing. **32**(15), 524–529 (2017)
63. Cerdeira Ferreira, F.H., Mendes de Araujo, R.: Campus Inteligentes: Conceitos, aplicações, tecnologias e desafios. Relatórios Técnicos do DIA/UNIRIO **11**(1), 4–19 (2018)
64. Zhan, S.: The Reconstruction Strategy "Internet+" from the Perspective of Education. (2017)
65. Aldowah, H., Ul Rehman, S., Ghazal, S., Naufal Umar, I.: Internet of Things in higher education: a study on future learning. J. Phys. Conf. Ser. **892**, 012017 (2017)
66. Bueno-Delgado, M.V., Pavón-Marino, P., De-Gea-García, A., Dolón-García, A.: The Smart University Experience: an NFC-based ubiquitous environment. In: 2012 Sixth International Conference on Innovative Mobile and Internet Services in Ubiquitous Computing, pp. 799–804 (2012)
67. Shvetsova, O.A.: Smart education in high school: new perspectives in global world. In: 2017 International Conference "Quality Management, Transport and Information Security, Information Technologies" (IT&QM&IS), pp. 688–691 (2017)
68. Coccoli, M., Guercio, A., Maresca, P., Stanganelli, L.: Smarter universities: a vision for the fast changing digital era. J. Vis. Lang. Comput. **25**(6), 1003–1011 (2014)
69. Rico-Bautista, D., Parra-Valencia, J.A., Guerrero, C.D.: IOT: Una aproximacion desde ciudad inteligente a universidad inteligentE. Rev. Ingenio UFPSO **13**(1), 9–20 (2017)
70. Savov, T., Terzieva, V., Todorova, K., Kademova-Katzarova, P.: Contemporary technology support for education. In: CBU International Conference Proceedings, vol. 5, p. 802 (2017)
71. Rueda-Rueda, J., Manrique, J., Cabrera Cruz, J.: Internet de las Cosas en las Instituciones de Educación Superior (2017)
72. Semenova, N.V., Svyatkina, E.A., Pismak, T.G., Polezhaeva, Z.Y.: The realities of smart education in the contemporary Russian Universities. In: Proceedings of the International Conference on Electronic Governance and Open Society: Challenges in Eurasia, pp. 48–52 (2017)
73. Staskeviciute, I., Neverauskas, B.: The Intelligent University's Conceptual Model. Inz. Ekon. Econ. **4**, 53–58 (2008)
74. Banica, L., Burtescu, E., Enescu, F.: The impact of internet-of-things in higher education. Sci. Bull. Sci. **16**(1), 53–59 (2017)
75. Heinemann, C., Uskov, V.L.: Smart Universities, vol. 70 (2018)
76. Hipwell, S.: Developing smart campuses #x2014; A working model. In: 2014 International Conference on Intelligent Green Building and Smart Grid, pp. 1–6 (2014)
77. Green, J.: The Internet of Things Reference Model. Internet of Things World Forum, pp. 1–12 (2014)
78. C. U. ESPAÑOLAS: Analisis de las TIC en las Universidades Españolas (2015)
79. Gómez, J., Jimenez, T., Gumbau, J., Llorens, F.: UNIVERSITIC 2016 Análisis de las TIC en las Universidades Españolas, p. 150 (2016)
80. ANUIES, Estado actual de las Tecnologías de la Información y las Comunicaciones en las Instituciones de Educación Superior en México (2017)
81. Padilla, R., Cadena, S., Enríquez, R., Córdova, J., Llorens, F.: Estado De Las Tecnologías De La Información Y La Comunicación En Las Universidades Ecuatorianas (2017)
82. Antonio Fernández Martínez, F.L.L.: Universitic Latam 2014, no. 1 (2014)
83. Valls, J., Villers, R., Duque, G.: Estado Actual de las Tecnologías de la Información y las Comunicaciones en las Instituciones de Educación Superior en México (2016)
84. Ontiveros, E., Vizcaíno, D., López Sabaer, V.: Las ciudades del futuro: inteligentes, digitales y sostenibles futuro: inteligentes, digitales y sostenibles (2016)

85. Del, E., Une, D.: Norma Española Accesibilidad Universal en las Ciudades Inteligentes (2017)
86. Muñoz López, L., Proyecto, D., Antón Martínez, P., Fernández Ciez, S.: El Estudio y Guía Metodológica sobre Ciudades Inteligentes ha sido dirigido y coordinado por el equipo del ONTSI. Deloitte (2015)
87. Iberoamericano, O.: Manual Iberoamericano de Indicadores de Educación Superior: Manual de Lima, p. 88 (2016)
88. Ministerio de Modernización Innovación y Tecnología: La Importancia de un Modelo de Planificación Estratégica para el Desarrollo de Ciudades Inteligentes, p. 32 (2017)

Accessibility and Gamification Applied to Cognitive Training and Memory Improvement

Ana Carol Pontes de Franca[1]([⊠]), Arcângelo dos Santos Safanelli[2],
Léia Mayer Eyng[3], Rodrigo Diego Oliveira[4], Vânia Ribas Ulbricht[3],
and Villma Villarouco[1]

[1] Universidade Federal de Pernambuco, Recife, Brazil
acpsicologa@gmail.com, vvillarouco@gmail.com
[2] Programa de Pós-Graduação em Engenharia de Produção,
Universidade Federal de Santa Catarina, Florianópolis, Brazil
safanelli.arcangelo@gmail.com
[3] Programa de Pós-Graduação em Engenharia e Gestão do Conhecimento,
Universidade Federal de Santa Catarina, Florianópolis, Brazil
leiamayer@gmail.com, vrulbricht@gmail.com
[4] Universidade Federal do Parana, Curitiba, Brazil
rodrigo@rodrigodiego.com.br

Abstract. The increase in people's life expectancy, with the consequent projection of 1 billion elderly people in less than 10 years and of 2 billion people in 2050 –22% of the global population–, makes it necessary to develop assistive technologies that contribute to the permanence and the protagonism of the elderly in their social context. In parallel, it is observed that smartphones have brought an increase in the use of applications by the elderly. Considering this segment of users, it was verified the need to identify accessible applications focused on cognitive memory training, due to: (1) the progressive increase in the number of elderly people in the Brazilian population; (2) the decline of the cognitive functioning resulting from the natural process of aging; and (3) the lack of accessible and dynamic applications focused on cognitive training. In order to provide greater user engagement while performing tasks and to make the user experience more enjoyable, engaging and challenging, gamification has been employed in the design and development of applications. Based on this scenario, we sought to identify the main ideas regarding accessibility and gamification applied to cognitive training and the memory improvement of elderlies. For that, a Systematic Literature Review (SLR) was conducted, originated from the following question: "How to create an accessible gamified application for cognitive memory training of the elderly?". The state of the art and the few previous studies revealed the importance of researches on this subject.

Keywords: Gamification · Accessibility · Cognitive training · Elderly · Memory

The original version of this chapter was revised: The author's name has been corrected. The correction to this chapter is available at https://doi.org/10.1007/978-3-030-11890-7_91

1 Introduction

Scientific and technological development, combined with some economic and social advances, have been contributing to significant changes in nutrition, housing, sanitation and public health in Brazil. This process of improving living conditions has led to a reduction in mortality rates, a consequent increase in the life expectancy of Brazilians and a progressive increase in the number of elderly people.

This segment has been arousing the interest of mobile technology products and services developers: willing to spend and ensure better quality of life and greater autonomy, the elderly have been increasingly investing in the purchase of mobile devices.

As reported by IBGE [1] in the National Household Sample Survey conducted in 2014, 136.6 million Brazilians, corresponding to 77.9% of the population, had a mobile phone for personal use. Compared to the previous year, the use among elderly people showed an increase of 4% points, from 51.6% in 2013 to 55.6% in 2014.

With the popularization of smartphones, the trend is that mobile phone users over the age of 60 migrate to these devices and progressively start using applications. However, some challenges emerge from this context. Declines in the physical and cognitive functioning resulting from the process of senescence[1] make it common the progressive increase of multiple limitations in the elderly (e.g. memory, hearing, vision, motor coordination, communication, etc.).

In order to minimize these and other problems, accessibility and gamification have been employed in applications design and development to provide greater engagement for the elderly and to make the user experience more enjoyable, engaging and challenging. This employment makes the use of the application allows the interaction, locomotion and integration of the elderly with their family and community, building networks of knowledge sharing.

Addressing these demands and focusing on these users, a Systematic Literature Review (SRL) was conducted to make a state-of-the-art survey that could identify an accessible gamified application for cognitive memory training of elderly people, considering: (1) the progressive increase in the number of elderly people in the Brazilian population; (2) the decline of the cognitive functioning resulting from the natural process of aging; and (3) the lack of accessible and dynamic applications focused on cognitive training.

2 The Method

In order to better support the creation of accessible applications aimed at cognitive training and memory improvement of the elderly, the Systematic Literature Review (SLR) was performed according to the method presented by *Cochrane Collaboration*

---

[1] According to the physician Wilson Jacob Filho, senescence is characterized by the natural process of aging. It concerns all changes produced in a living organism due to its evolution over time. There are no related diseases. They are due to the physiological processes common to all elements of the species [2].

[3]. For this, an explicit and heuristic [4] algorithm was applied, consisting of seven steps:

1. Research question;
2. Search and selection of studies;
3. Critical evaluation of the studies;
4. Selection of data for analysis;
5. Analysis and presentation of data;
6. Interpretation of data;
7. Improvement and update.

2.1   Systematic Literature Review (SLR)

The Systematic Literature Review consists of a secondary exploratory study on a specific theme and purpose. According to Dresch *et al.* [5], "they are secondary studies used to map, find, critically evaluate, consolidate and aggregate the results of relevant primary studies on a specific research question or topic [...]" in order to provide the researcher with a comprehensive view and updated recent studies about areas and topics of interest [5].

The Systematic Literature Review conducted was aimed at mapping and analyzing studies related to the creation of gamified and accessible applications that help improve memory through cognitive training to be used by healthy elderly people. In other words, this study targets at elderly with cognitive decline of memory due to the natural aging process, who use smartphones and who are familiar with the technology.

The result will be applied in the foundation of the concepts that guide the subject, in the evaluation of its relevance for the scientific community, and in the forthcoming development of an application for the same purpose.

2.2   Research Question and Search Strings

In the initial moments of a study, to formulate the question is decisive for research success, since "poorly formulated questions lead to obscure decisions about what should or should not be included in the review" [6]. In order to better contribute to the creation of accessible gamified applications for the elderly with decline in cognitive functioning due to the natural process of aging, we tried to answer the following question: "How to create an accessible gamified application for cognitive memory training of the elderly?".

From the problem question, we extracted the keywords for the Systematic Review. Ten words were obtained from this process: five in English and five in Portuguese. The keywords represent the concepts and the main topics of the research. Once the keywords were defined, they were crossed, totaling twenty combinations to be searched. Ten combinations were obtained in Portuguese and ten in English, aiming to make the research comprehensive and focused on the proposed subject. The result is described in Table 1.

Table 1. Keywords and combinations.

Portuguese	English
Gamificação AND Idosos	Gamification AND Elderly
Acessibilidade AND Idosos	Accessibility AND Elderly
Treino Cognitivo AND Idosos	Cognitive Training AND Elderly
Memória AND Idosos	Memory AND Elderly
Gamificação And Acessibilidade	Gamification AND Accessibility
Gamificação AND Treino Cognitivo	Gamification AND Cognitive Training
Gamificação AND Memória	Gamification AND Memory
Acessibilidade AND Treino Cognitivo	Accessibility AND Cognitive Training
Acessibilidade AND Memória	Accessibility AND Memory
Treino Cognitivo AND Memória	Cognitive Training AND Memory

2.3   Critical Evaluation of the Studies

At this stage of the research, inclusion and exclusion criteria for the articles were applied directly to the platforms provided by the selected databases (Scopus, *Web of Science* and Scielo.org). The purpose of these criteria was to constrict the results and seek greater relevance. This was necessary because the thousands of results that arise from a search within the proposed subject would make the research too broad and unfeasible. Twenty word combinations were searched in each of the bases, and a total of 45,113 results were obtained. The total of non-constricted results per database can be seen in Table 2.

Table 2. Non-constricted results per database

Database	Results
Scopus	24.812
Web of Science	19.572
Scielo.org	729

For purposes of constriction and greater relevance for the research, the following criteria were adopted:

- Search by title;
- Articles only;
- Languages of the results: English and Portuguese;
- Last five years;
- Sort by relevance;

- Sample: 25% between 50 and 100 results;
- Sample: 10% above 100 results;
- Selection by title reading (relevance to the problem); and
- Selection by abstract reading (relevance to the subject).

Applying the search criteria (by title, articles only, published in the last five years, Portuguese and English languages, and sorted by relevance), a considerable constriction was achieved and 398 articles were obtained among the three databases surveyed. The sample criteria were applied for the combinations of keywords searched that presented results above fifty articles, constricting the search to 240 selected articles, of which 41% (99 articles) were originated from Scopus, 11% (26 articles) from Scielo. org, and 48% (115 articles) from Web of Science, as shown in Fig. 1.

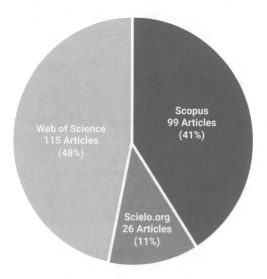

Fig. 1. Articles per database

For purposes of refinement, 75 duplicate articles were eliminated from the study. There were then 165 articles for the next step: selection of the data for analysis. The result of the constrictions for selecting the articles whose titles would be analyzed can be seen in Table 3.

Table 3. Constrictions for selecting the data to be analyzed.

Description	Results
Total articles found	45.113
Constricted articles with application of criteria	398
Articles constricted with sample criteria	240
Total duplicate items	75
Selected papers for title analysis	165

2.4   Selection of Data for Analysis

We selected 165 articles for reading and title analysis, which contained at least one of the research keywords. The titles of the selected articles were then read to verify whether or not they were qualified for the subject under study. Besides indicating the qualification, a brief justification of the selection or refusal of the article was made so that there was consensus among the parties. The data were organized in a spreadsheet, with columns containing: date of search, keyword, database, title and link to the article, justification, qualified (yes or no), abstract, and qualified abstract (yes or not).

The process of constriction occurred through reading these data and analyzing their relevance for the subject under study. The articles whose titles explicitly addressed the following terms were considered valid: 'treino cognitivo para idosos', 'gamificação', 'aplicativos' ou 'jogos computadorizados' ('cognitive training for the elderly', 'gamification', 'applications' or 'computerized games'). In this process, 139 articles were discarded, and 26 remained for reading and analysis of abstracts. Then the abstracts of the 26 articles were read, in order to evaluate their relevance to the subject studied. In this second constriction, 18 articles were discarded, and only eight were left for complete reading, as presented in Table 4.

Table 4. Articles extracted from the research ordered by the number of citations.

Nº	Article	Database	Citations
1	A Brain-Computer Interface Based Cognitive Training System for Healthy Elderly: A Randomized Control Pilot Study for Usability and Preliminary Efficacy	Web of Science	10
2	Cognitive and neural plasticity in older adults' prospective memory following training with the Virtual Week computer game	Scopus	5
3	Efficacy of a computerized cognitive training application for older adults with and without memory impairments	Scopus	5
4	Effects of Computer-assisted Cognitive Rehabilitation Training on the Cognition and Static Balance of the Elderly	Web of Science	4
5	Cognitive Training System for Dementia Prevention Using Memory Game Based on the Concept of Human-Agent Interaction	Scopus	2
6	Development of Computer-Aided Cognitive Training Program for Elderly and Its Effectiveness through a 6 Months Group Intervention Study	Web of Science	1
7	Engaging Elderly People in Telemedicine Through Gamification	Web of Science	1
8	Cognitive Function Training System Using Game-Based Design for Elderly Drivers	Scopus	0

2.5   Analysis and Presentation of Data

At the end of the SLR process, eight articles were selected and read in full. They served as a basis for the foundation of concepts related to cognitive training, memory improvement, gamification and accessibility for the elderly. Figure 2 shows the graphical representation of the process applied in the Systematic Literature Review.

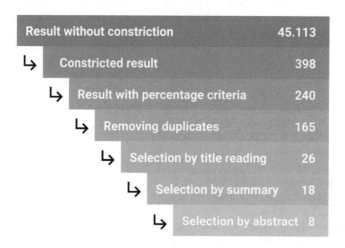

Fig. 2. Graphical representation of the SLR process.

2.6   Data Interpretation

The process of data interpretation is variable. Each researcher has his/her own mind-set and, in order to extract the concepts of the selected articles in a collaborative way, a table was created listing the name of the article, category (word representing concept), description or direct citation, author and year.

In Table 5 below, only the articles consulted, their authors, their year of publication and the corresponding category were listed:

Through this process, ten categories were created with their respective descriptions and source citations, allowing a comparative and relational analysis among the concepts extracted from the selected articles. The concepts extracted in this process were applied in the construction of the theoretical foundation, discussed in Sect. 3 of this article, which consists of "supporting theoretical aspects of research through the ideas of other authors" [17].

2.7   Improvement and Update

According to *Cochrane Collaboration* [3], the Systematic Literature Review is continuous and can be updated in line with researchers needs, due to new discoveries, criticisms and suggestions about the research, and mainly because of the daily inclusion of new articles in the databases.

Table 5. Concepts extracted from selected articles.

N°.	Article	Category	Author and year
1	A Brain-Computer Interface Based Cognitive Training System for Healthy Elderly: A Randomized Control Pilot Study for Usability and Preliminary Efficacy	Cognitive rehabilitation Elderly	Lee et al. (2013) [9]
2	Cognitive and neural plasticity in older adults' prospective memory following training with the Virtual Week computer game	Prospective memory Neural plasticity Cognitive plasticity Cognitive training	Rose et al. (2015) [10]
3	Efficacy of a computerized cognitive training application for older adults with and without memory impairments	Elderly Cognitive training Artifacts	Maseda et al. (2015) [11]
4	Effects of Computer-assisted Cognitive Rehabilitation Training on the Cognition and Static Balance of the Elderly	Cognitive rehabilitation	Lee et al. (2013) [12]
5	Cognitive Training System for Dementia Prevention Using Memory Game Based on the Concept of Human-Agent Interaction	Cognitive training Human-agent interaction	Kitakoshi et al. (2015) [13]
6	Development of Computer-Aided Cognitive Training Program for Elderly and Its Effectiveness through a 6 Months Group Intervention Study	Cognitive decline Motivation	Otsuka et al. (2015) [14]
7	Engaging Elderly People in Telemedicine Through Gamification	Telemedicine Gamification Artifacts	de Vette et al. (2015) [15]
8	Cognitive Function Training System Using Game-Based Design for Elderly Drivers	Gamification Cognitive decline Cognitive training Old man Motivation Artifact	Hiraoka; Wang; Kawakami (2016) [16]

Since the purpose of this SLR is to substantiate and identify the criteria for creating an accessible gamified application for the cognitive memory training of the elderly, new issues should be investigated and deepened for the ultimate goal to be achieved. At the end of the Review, the corpus was delimited in eight articles. In the next section, the main aspects of the theoretical foundation will be approached, based on the concepts obtained from the survey of the state of the art.

3   Discussion

The natural process of aging, also known as senescence, is characterized by the appearance of problems related to the decline of cognitive functioning, and memory is the most affected cognitive process [11].

In accordance with Rose et al. [10], from 50% to 80% of memory problems in everyday situations involve prospective memory. This memory is related to the ability to remember and perform intentioned and planned activities, and is critical to a successful and independent life.

For cognitive rehabilitation of the elderly, cognitive training is traditionally used [9, 12], as well as it can be employed in telemedicine [15]. Maseda et al. [11] conceptualize cognitive training as a set of systematic, non-pharmacological interventions that aim to improve, maintain or restore capacities through repeated and structured practice of cognitive activities.

Rose et al. [10] classify cognitive training programs as compensatory and restorative. The compensatory approach is characterized by teaching a specific strategy or technique that can be used to circumvent or compensate a specific cognitive deficit (e.g., to train a person to use a mnemonic strategy). The restorative approach focuses on repairing or improving the functioning of the neurocognitive processes involved in several domains of cognition (e.g., adaptive training for working memory).

Substantial performance gains related to prospective memory and evidence of neuroplasticity were found in people undergoing the virtual cognitive training program [10]. However, there is usually a shortage of studies that integrate cognitive training with subjective aspects. Maseda et al. [11] warn of the importance of introducing non-cognitive variables such as: well-being, quality of life and daily functioning.

For Otsuka et al. [14] the high motivation of the elderly can be maintained by adjusting the difficulty of the tasks. It is also necessary to provide some artifice to avoid boredom. In this sense, the inclusion of new stimuli or tasks could motivate the elderly to invest more effort and energy in learning. Therefore, gamification presents itself as an opportunity to reduce their feelings of fear and frustration in relation to technology. Hiraoka, Wang and Kawakami [16] understand gamification as the use of game elements in non-gaming systems in order to improve user engagement and experience.

Gamification has a significant potential for use by the elderly. The challenge lies in their lack of familiarity with games [15].

Finally, Vette et al. [15] emphasize the importance of universal design for meaningful gamification, focused on learning. In such terms, gamification would make cognitive training more engaging, fun, and challenging, so that gains obtained in virtual training can be observed in user's everyday life situations.

4   Conclussion

This article had as main purpose to identify how to create an accessible gamified application for cognitive memory training of the elderly, through the Systematic Review of Literature about the increase in the number of elderlies among Brazilian population, the decline in cognitive functioning due to the natural aging process (senescence) and the scarcity of accessible and dynamic applications focused on cognitive training.

Based on this survey, it was possible to identify that the elderly people have access to smartphones, but it does not enable them for the use of apps, since the multiple uses existing in current devices are unknown for many of them. Thus, in order to understand the target audience of this research, it is necessary to know the elderly, their use of the smartphone, the reasons that led them to adquire the device, the difficulties and facilities they find in the use.

With regard to the decline in cognitive functioning (senescence), which imposes multiple losses to the elderly -like reduction of their capacities of orientation, memory, visuospatial ability, and language-, it was validated that the most frequent is the memory loss, which consequently implies the reduction or loss of functional capacities and social isolation, leading to a possible stage of illness that may evolve to death. The search for alternatives to minimize the losses resulting from the process of social isolation, and the feeling of incapacity before oneself and the world, should be prioritized not only from an academic point of view, but also aiming at the effective applicability in society, guaranteeing the accessibility of the elderly group in its multiple social functions.

Finally, we validated the scarcity of accessible and dynamic applications focused on cognitive training for elderly. We noted the importance of investigations not restricted to telemedicine, such as the use in the Netherlands, but to the cognitive stimulation that employs gamification in a significant way, considering the characteristics of users, subjective and situational aspects, in order to adapt and integrate digital games to interests and needs of the elderly.

The gamified cognitive training should be created with the information obtained and the knowledge built up from the same elderly people inserted in the learning process, by means of insertion and effective permanence in the short, medium and long-term training, and in this way, they can contribute with practical, creative and dynamic solutions that provide better quality of life.

Nowadays, it is possible to verify the scarcity of studies on cognitive training and memory improvement, on mobile devices, with a dynamic approach, applied to the elderly group, that allow these people to use the game wherever they are, effectively, efficiently and safely, even if they have some type of limitation and/or disability.

For forthcoming works, it is proposed the development of researches and dynamic applications that explore cognitive training and memory improvement from a preventive approach, in order to obtain information and clarification about the effects of preventive training in a given period of time, applied to the elderly and with their effective participation.

References

1. IBGE (Instituto Brasileiro de Geografia e Estatística): Acesso à Internet e à Televisão e Posse de Telefone Móvel Celular para Uso Pessoal. Pesquisa Nacional de Amostra por Domicílios (2014). http://www.ibge.gov.br/home/estatistica/populacao/acessoainternet2014/default. shtm. Accessed 11 Apr 2017

2. SBGG (Sociedade Brasileira de Geriatria e Gerontologia): Senescência e senilidade, qual a diferença? http://www.sbgg-sp.com.br/pub/senescencia-e-senilidade-qual-a-diferenca/. Accessed 24 Apr 2017

3. Cochrane Collaboration (Eua) (Org.): Cochrane Handbook for Systematic Reviews of Interventions. http://training.cochrane.org/handbook. Accessed 15 Apr 2017

4. Berg, C.H., Flores, A., Fadel, L., Ulbricht, V.: Pessoas Cegas e Representação Espacial: Uma Revisão Sistemática de Literatura. Ergodesign & HCI, Rio de Janeiro (2015)

5. Dresh, A., Lacerda, D.P., Júnior, J.A.V.A.: Design Science Research: Método de Pesquisa para Avanço da Ciência e Tecnologia. Bookman Editora, Porto Alegre (2015)

6. Araujo, C.A.: Revisão Sistemática e Meta-Análise (2001). http://usinadepesquisa.com/ metodologia/wp-content/uploads/2010/08/meta1.pdf. Accessed 24 Feb 2017

7. Elsevier. Scopus. https://www.elsevier.com/americalatina/pt-br/scopus. Accessed 15 Apr 2017

8. Web of Science. https://images.webofknowledge.com/images/help/WOS/hp_database.html. Accessed 15 Apr 2017

9. Tih-Shih, L., Goh, S.J.A., Quek, Y.S., Phillips, R., Guan, C., et al.: (2013). http://doi.org/10. 1371/journal.pone.0079419. Accessed 04 Apr 2017

10. Rose, N., Rendell, P.G., Hering, A., Kliegel, M., Bidelman, G.M., Craik, F.I.M.: Cognitive and neural plasticity in older adults' prospective memory following training with the Virtual Week computer game. J. Front. Hum. Neurosci. 9 (2015). Accessed 04 Apr 2017 http:// journal.frontiersin.org/article/10.3389/fnhum.2015.00592/full

11. Ana, M., Calenti, J.C.M., López, L.L., Naveira, L.N.: Efficacy of a computerized cognitive training application for older adults with and without memory impairments. Aging Clin. Exp. Res. (2013). https://link.springer.com/article/10.1007%2Fs40520-013-0070-5. Accessed 04 Apr 2017

12. Lee, Y.M., Jang, C., Bak, I.H., Yoon, J.S.: Effects of computer-assisted cognitive rehabilitation training on the cognition and static balance of the elderly. J. Phys. Therapy Sci. http://doi.org/10.1589/jpts.25.1475. Accessed 04 Apr 2017

13. Daisuke, K., Ryo, H., Iwata, K., Suziki, M.: Cognitive training system for dementia prevention using memory game based on the concept of human-agent interaction (2015). https://fujipress.jp/jaciii/jc/jacii001900060727/. Accessed 04 Apr 2017

14. Otsuka, T., Tanemura, R., Noda, K., Nagao, T., Sakai, H., Luzo, Z.W.: Development of computer-aided cognitive training program for elderly and its effectiveness through a 6 months group intervention study. Curr. Alzheimer Res. 12(6), 553–562 (2015). https://www. ncbi.nlm.nih.gov/pubmed/26027812. Accessed 04 Apr 2017

15. de Vette, F., Tabak M., Dekker-van, W.M., Vollenbroek-Huten, M.: Engaging elderly people in telemedicine through gamification. JMIR Ser. Games. 3(2). https://www.ncbi.nlm. nih.gov/pubmed/26685287. Accessed 15 Apr 2017

16. Hiraoka, T., Wang, T., Kwakami, H.: Cognitive function training system using game-based design for elderly drivers. IFAC-PapersOnLine 49(19), 579–584 (2015). https:// sciencedirect.com/science/article/pii/S2405896316322157. Accessed 04 Apr 2017

17. Portal da Educação. Fundamentação Teórica. https://www.portaleducacao.com.br/conteudo/ artigos/pedagogia/fundamentacao-teorica/31156. Accessed 19 Apr 2017

Detection and Segmentation of Ecuadorian Deforested Tropical Areas Based on Color Mean and Deviation

Henry Cruz[1](✉), Juan Meneses[2], Wilbert Aguilar[1], and Gustavo Andrade-Miranda[3]

[1] Universidad de las Fuerzas Armadas-ESPE, Sangolquí, Ecuador
{hocruz,wgaguilar}@espe.edu.ec
[2] Research Center on Software Technologies and Multimedia Systems for Sustainability (CITSEM), Technical University of Madrid, Ctra. Valencia, Km. 7, 28031 Madrid, Spain
juan.meneses@upm.es
[3] Facultad de Ingeniería Industrial, Universidad de Guayaquil, Av. Las Aguas, Guayaquil, Ecuador
gxavier.andradem@ug.edu.ec

Abstract. This paper presents an evaluation on a novel statistical method applied to segment Ecuadorian deforested tropical areas; this is based on color average and deviation which is named Average and Deviation Segmentation Method (ADSM). In order to achieve this aim, the digital treatment of the images has been carried out, seeking to obtain the color characteristic of a region of interest. Later, a post-processing step based on active contours is used to delimit the deforested areas detected. The ADSM algorithm can use different color spaces (RGB, HSV, YCbCr) which make it application-independent. Additionally, it provides a rejection filter that allows reducing the false positive and ensuring more accurate detections. The experiments carried out are based on segmentation quality as well as the detection accuracy, obtaining true positive rates of 98.57%. Finally, despite the difficulty of the evaluation in this type of images, it has been possible to verify the accuracy of the proposed algorithm, helping to reduce problems in cases of partial occlusions in saturated images.

Keywords: Segmentation · Mean · Deviation · Deforestation · Evaluation

1 Introduction

Deforestation has a negative impact on the environment. For that reason, Ecuadorian institutions seek strategies to preserve the ecosystem, the biological resources, natural resources (tangible and intangible) protected areas and especially tropical forest areas [1]. Therefore, it is important to identify the deforested areas in order to control, prevent its extension and carry out plans to mitigate its impact on the environment [2]. The deforestation in tropical forest areas is analysed commonly via digital image processing of satellite images, radars, LIDAR systems, among others [3, 4]. These technologies may not be very accessible (high economic costs) which delays in

© Springer Nature Switzerland AG 2019
Á. Rocha et al. (Eds.): ICITS 2019, AISC 918, pp. 452–461, 2019.
https://doi.org/10.1007/978-3-030-11890-7_44

obtaining results in the short and medium term. In addition, they do not always provide a good observation of the deforested areas due to the presence of clouds and other atmospheric phenomena, especially in the cases of jungle areas [5]. Nowadays, other ways to detect deforested areas are available through the use of unmanned aerial vehicles (UAVs) and small unmanned aircraft systems (UASs), also called drones, which can be operated in areas of frequent cloud cover such as tropical jungle [6, 7]. For other hand, some methods use the colour information to get accurate segmentations about deforestation through colour indexes [8] or to detect fire forest [9] as well as cropped fields [10].

In Ecuador, the images of tropical forests are highly saturated and with great contrast, therefore, the colour is valuable information in this kind of application. In this work, it has been proposed an evaluation on statistical method to segment the deforested areas in the Ecuador tropical forests based on color mean and deviation through ADSM [11]. For the final delimitation of the deforested regions, an active contour algorithm is used. The paper is organized as follows. Section 2 explains briefly the ADSM methodology. Section 3 shows the results obtained using a database of images of the Ecuadorian tropical forests obtained from an Unmanned Aerial Vehicle system (UAV's) of the Ecuadorian Air Force Research and Development Center (CIDFAE). Section 4 presents the conclusions and future works.

2   Materials and Methods

This section provides a detailed explanation of the methodology carried out to detect deforested areas through ADSM. The detection is based on properly ordered steps, which are described below

- Selection of the pixels values of interest in a respective color space.
- Segmentation of the region of interest through ADSM.
- Automatic acquisition of multiple regions of interest.
- Final segmentation by active contours.

2.1   Database

The CIDFAE database has 200 images of the Ecuadorian tropical forest areas obtained from an Unmanned Aerial Vehicle system (UAV's) of the Ecuadorian Air Force Research and Development Center (CIDFAE) with an electro-optic camera. The images of the tropical forest have resolution of: RGB24 (320 x 240 pel), RGB24 (640 x 480 pel) and I240 (1280 x 1024 pel).

2.2   Segmentation of the ROI's Through ADSM

The process starts with the manual selection of the pixel values of interest (p_i) as shown in Fig. 1. Later, the ADSM segmentation is performed and the regions of interest are found (ROI's). In our case, these ROI's are found through the acquisition of several areas of interest (A_i) that are part of the deforested areas to be found.

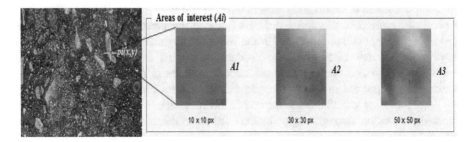

Fig. 1. Selection of p_i value and A_i computation.

The segmentation by ADSM includes two principal steps. The first one takes the color means information of p_i and the second one found out deviation parameters. To obtain the average color, we consider the information of the pixel of interest (p_i) within a set of three areas of interest (A_i) described in (1) and (2).

$$p_i \in A_1; \; pi \in A_2; \; pi \in A_3 \tag{1}$$

$$A_1 \; U \; A_2 \; U \; A_3 = \{p_1, p_2, p_3, \ldots \ldots .pn\} \tag{2}$$

These areas may contain all the interesting pixels with similar values in the RGB color space. The number of areas has been chosen based on observational and empirical evidence which shows that a greater number of areas generates not useful information. However, this implementation of areas will depend on the specific application-purpose that the reader wants to give to the ADSM method. The A_i are obtained from the p_i, generating windows pixels sizes (grouping of 10×10, 30×30, 50×50 pixels can be option). Subsequently, in each of the A_i a statistical average of the values of the components in RGB is computed.

To ensure that the values of p_i have the correct weight in the estimation of the color average. A weighting (W) is established for each of the A_i in such a way that the A_1 has greater statistical weight compared to A_2 and A_3, so the weights are $W_1 > W_2 \; W_3$. The total weight value, over all areas A_i, is computed through statistical mean and later a threshold called deviation is taken according method studied [11].

The deviation is based on the determination of a tolerance interval (TI) that allows to reduce the probability of error and it is defined as $TI = a, b$. For the RGB color space a and b were chosen as 10 and 20, respectively. Values greater than 20 produce over-segmentation meanwhile values lesser than 10 produce under-segmentation. In order to obtain the segmented image, the foreground is separated from the background based on the deviation margins. Figure 2 depicts the result using ADSM algorithm.

2.3   ADSM Based on HSV and YCbCr Color Spaces

The ADSM can be adapted in order to be used with HSV and YCbCr color spaces for the task to segment deforested areas. The procedure to detect deforested areas using HSV and YCbCr follows the same approach used for RGB. Firstly, the pixel of interest

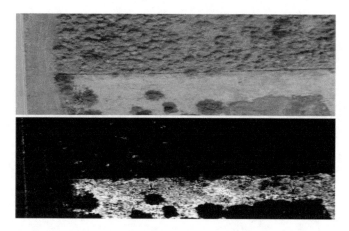

Fig. 2. Segmentation of a deforested area using ADSM method in RGB.

is chosen and the different A_i are computed by (1) and (2). The weighting relationship is obtained and the deviation value is calculated, the values obtained are calculated in HSV TI is in the range of a = 0.15 and b = 1 meanwhile YCbCr have values of $a = 3$ and $b = 223.5$. Figures 3 and 4 depicts the results obtained using HSV and YCbCr respectively.

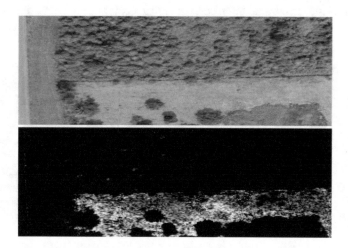

Fig. 3. Deforested area segmentation using ADSM method and HSV.

2.4   Deforested Area Delimitation by Active Contours

The active contours are thin elastic bands which are coupled appropriately to non-rigid and amorphous contours. The active contour model is controlled by two kinds of energies: external and internal. The external energy is generated by processing the image, producing a force that is used to drive the active contour towards features of

Fig. 4. Deforested area segmentation using ADSM method and YCbCr

interest. Whereas, the internal energy serves to impose a piecewise smoothness constraint. Therefore, the active contours are used to complement the detection of the deforested areas.

The snake model proposed by [12] is adapted to our task, allowing more precise delimitations. Figure 5 shows the results obtained after performing the snake algorithm. However, it is important to mention that it is not an indispensable tool and in real-time applications, it generates a high computational cost in relation to the number of interactions required. It was used in order to establish accuracy of the segments.

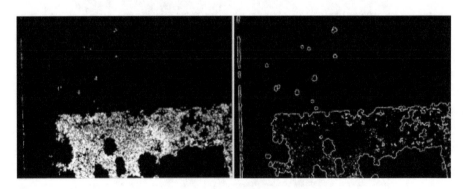

Fig. 5. Segmentation of a deforested area using ADSM and Snake.

2.5   ADSM Implementation

In order to validate the ADSM application either for RGB, HSV, and YCbCr color spaces, some experimental tests have been carried out using the database of the CIDFAE. The computer used for the experimental tests is an Intel (R) Core (TM) i7 3.4 GHz processor. A graphical interface developed in Matlab R2017b is implemented.

Through the interface, the detection and discrimination of the deforested areas can be visualized. This graphic interface includes a classifier of successes and failures. Through this tool, it was possible to perform image assessments.

3   Experimental Results

The experiments focused on proving two types of evidence: segmentation quality, and accuracy detection and these tests have been carried out within the research center CITSEM (Spain) and the CICTE (Ecuador).

3.1   Segmentation Quality

In order to assess the segmentation quality, different metrics have been computed. The first three computes the degree of overlap between two segmentations a manual segmentation (known as the expert) and automatic segmentation (known as the machine) [13]. To this group belongs the global constancy error (GCE), the local constancy error (LCE) and object-level consistency error (OCE). The second measures the similarity between clusterings (manual and automatic segmentation). One of its important properties is that is not based on labels, and thus can be used to evaluate clusterings as well as classifications. The metric implemented in this work was the Rand Index (RI). Lastly, the information variation metric (VoI), which is defined as the distance metric that exists between two segmentations through a conditional random average of image data, known as entropy was computed. The metrics presented are common used in image segmentation evaluation [14]. Figure 6 shows the results obtained with ADSM and the manual segmentation, and Table 1 summarize the percentage of accuracy obtained with the different metrics (GCE, LCE, OCE, RI, and VoI).

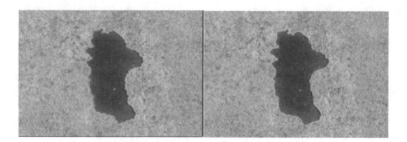

Fig. 6. ADSM versus expert segmentation

The images with dense vegetation are difficult to segment, due to the changes of tonalities, irregular non-symmetrical shapes, changes of illumination, etc. Therefore, the results obtained can be interpreted as quite accurate under the proposed method; LCE and GCE reach a 95.79% efficiency meanwhile OCE presents a quality of segmentation of 80%, the accuracy value is lower because the evaluation is stricter than

Table 1. Segmentation quality assessment with LCE, GCE, OCE, RI, and VoI

Metric	LCE	GCE	OCE	RI	VoI
Average	0,042	0,042	0,192	0,994	0,030
% Eff	95,79	95,79	80,80	99,46	96,98

others. RI presents a 99.46%. VoI parameter shows a very low error rate with a 96.98% of precision. The quality of the segmentation has been also evaluated with the three different color spaces in order to know the most convenient one for deforested areas problem. Table 2 depicts the average obtained for each color space considering the VoI. The intensification of the execution of this comparison is to ensure accurate detection of deforested areas and consequent decrease of false positives.

Table 2. VoI obtained when ADSM is combined with RGB, HSV or YCbCr.

Color space	RGB	HSV	YCbCr
VoI Average	0,0302	0,0267	0,2430
% (Efficiency)	96,98	97,33	75,70

3.2 Accuracy Detection

It is possible to determine the confusion matrix and compute the true positives detections (TP), false positives detections (FP), true negatives detections (TN), and false negative detections (FN). Figure 7 shows the TP, FP, TN, and FN with different tonalities. From the confusion matrix, the rate of true positives (TPR) and the rate of false positive (FPR) are computed. TPR allows defining the effectiveness in terms of correct detections. On the other hand, FPR allows estimating the number of detections that have been incorrect, such as in [8–11]. Table 3 summarizes the results of the detection evaluation of deforested areas A_i from all databases. From Table 3, it can be observed an average accuracy of 98.57%.

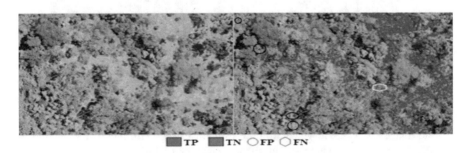

TP   TN ○FP ○FN

Fig. 7. Automatic ROI's detection represented by the TP, VN, TF and FN.

Table 3. Detection accuracy based on TP, TN, FP and FN.

Factor	% obtained
TP	98,57
TN	98,32
FP	1,69
FN	1,43
TPR	0,9857
FPR	0,0169
Total images evaluated	100

Additionally, Fig. 8 depicts a sample of 5 images segmented with the ADSM procedure. The images show that ADSM solves the problems of partial occlusions and cluttering in the deforested tropical forest areas.

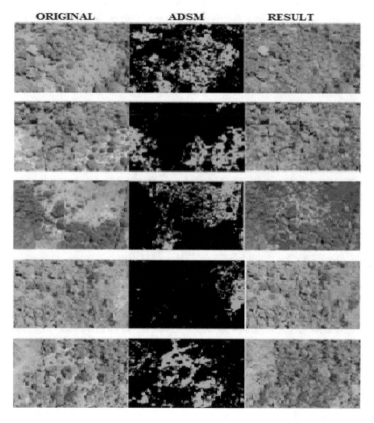

Fig. 8. Segmentation using ADSM of deforested tropical forest areas.

4 Conclusions and Future Works

In this article, has been evaluated the use of ADSM in the automatic acquisition of multiple objectives equal in color which has allowed the detection of deforested tropical areas. It has also shown the usefulness of the combination of ADSM with active contours in this type of images, especially considering the non-symmetry of them.

The results show high percentages of accuracy and quality in the segmentation. Additionally, it can be found that the color space with the best performance in this type of images is the HSV (97.33% of effectiveness). Finally, all these tasks have allowed us to show an application algorithm, which shows a 98.57% effectiveness rate of positive detections and no positive detections. An extension of the presented work is the automatic detection of objectives and develops an application to be used in real-time tracking systems.

Acknowledgments. The authors, gives thank to Technological Scientific Research Center of the Ecuadorian Army (CICTE) and the Research Center on Software Technologies and Multimedia Systems for Sustainability (CITSEM), for the collaboration obtained.

Henry Cruz Carrillo gives thanks Ecuadorian Air Force Research and Development Center (CIDFAE) for the collaboration obtained.

References

1. Southgate, D., Sierra, R., Brown, L.: The causes of tropical deforestation in Ecuador: a statistical analysis. World Dev. **19**(9), 1145–1151 (1991)
2. Sierra, R.: Traditional resource-use systems and tropical deforestation in a multi-ethnic region in North-west Ecuador. Environ. Conserv. **26**(2), 136–145 (1999)
3. Richards, J., Xiuping, J.: Remote Sensing Digital Image Analysis An Introduction, 4th edn. cap. 8, pp. 193–338. Springer (2005)
4. Aragão, L.E., Anderson, L.O., Fonseca, M.G., Rosan, T.M., Vedovato, L.B., Wagner, F.H., Barlow, J.: 21st Century drought-related fires counteract the decline of Amazon deforestation carbon emissions. Nature Commun. **9**(1), 536 (2018)
5. Baldeck, C.A., Asner, G.P., Martin, R.E., Anderson, C.B., Knapp, D.E., Kellner, J.R., Wright, S.J.: Operational tree species mapping in a diverse tropical forest with airborne imaging spectroscopy. PLoS One **10**(7), e0118403 (2015)
6. Dandois, J.P., Olano, M., Ellis, E.C.: Optimal altitude, overlap, and weather conditions for computer vision UAV estimates of forest structure. Remote Sens. **7**(10), 13895–13920 (2015)
7. Mohan, M., Silva, C.A., Klauberg, C., Jat, P., Catts, G., Cardil, A., Dia, M.: Individual tree detection from unmanned aerial vehicle (UAV) derived canopy height model in an open canopy mixed conifer forest. Forests **8**(9), 340 (2017)
8. Cruz, H., Eckert, M., Meneses, J., Martínez, J.F.: Precise real-time detection of nonforested areas with UAVs. IEEE Trans. Geosci. Remote. Sens. **55**(2), 632–644 (2017)
9. Cruz, H., Eckert, M., Meneses, J., Martínez, J.F.: Efficient forest fire detection index for application in unmanned aerial systems (UASs). Sensors **16**(6), 1–15 (2016). https://doi.org/10.3390/s16060893, (893)

10. Hassanein, M., Lari, Z., El-Sheimy, N.: A new vegetation segmentation approach for cropped fields based on threshold detection from hue histograms. Sensors **18**(4), 1–25 (2018). 1253
11. Cruz, H., Meneses, J., Andrade, G.: A real-time method to detect remotely a target based on color average and deviation. CCIS. Springer (895) (2018) (in Press)
12. Kass, M., Witkin, A., Terzopoulos, D.: Snakes: active contour models. Int. J. Comput. Vis. **1**(4), 321–331 (1988)
13. Martin, D., Fowlkes, C., Tal, D., et al.: A database of human segmented natural images and its applications to evaluating segmentation algorithms and measuring ecological statistics. In: Proceedings of International Conference Computer Vision, Vancouver, pp. 416–425 (2001)
14. Cruz, H., Eckert, M., Meneses, J.M., Martínez, J.F.: Fast evaluation of segmentation quality with parallel computing. Sci. Program. **2017**, 1–9 (2017)

Multimedia Systems and Applications

Novel Artist Identification Approach Through Digital Image Analysis Using Machine Learning and Merged Images

Peter Stanchev[1,2(✉)] and Michael Kolinski[1]

[1] Kettering University, Flint, MI, USA
{pstanche, koli8922}@kettering.edu
[2] IMI-BAS, Sofia, Bulgaria

Abstract. Artist identification has been an interesting task for centuries. Machine learning algorithms are used to solve this problem. An artist's profile is obtained through an artist's merged images enhanced with layer and transparency tools. The machine learning J48 algorithm is utilized for classification. Cohen's Kappa, F-measure, and Matthew's correlation coefficient statistics are applied to compare the results obtained.

Keywords: Image analysis · Machine learning · Artist identification

1 Introduction

Examining and analyzing a painting to determine whether an artist was the creator of it has been around since art was first collected. All art from paintings, to sculptures, needs to have an identified artist associated with it. Today, this is done using more advanced technological methods as well as the discerning eyes of connoisseurs. Hyperspectral imaging as well as machine learning tools are hot topics in the current field. Machine learning algorithms could be used for identifying the artist by creating an artist signature.

Different styles in art paintings are connected with techniques used on the one hand and aesthetic expression of the artist on the other. The process of forming an artist's style is a very complicated process, where current fashion painting styles, social background and personal character of the artist play a significant role. All these factors lead to forming some common trends in art movements and some specific features, which distinguish one movement from another, one artist's style from another, one artist period from another, etc. On the other hand, the theme of the paintings also stamps its specifics and can be taken into account. The compositions in different types of images (portraits, landscapes, town views, mythological and religious scenes, or everyday scenes) also set down some rules, aesthetically imposed for some period. When humans interpret images, they analyze image content. Computers are able to extract low level image features like color distribution, shapes and texture. Humans, on the other hand, have abilities that go beyond those of computers. Humans draw their own subjective conclusions. They place emphasis on different parts of images, identify objects and scenes stamping their subjective vision and experience. MPEG-7 descriptors over tiles were used to identify the artist in [1].

© Springer Nature Switzerland AG 2019
Á. Rocha et al. (Eds.): ICITS 2019, AISC 918, pp. 465–471, 2019.
https://doi.org/10.1007/978-3-030-11890-7_45

Hyperspectral imaging is a technique that collects information across the electro-magnetic spectrum of light in each pixel of an image. This has been used not only to identify the artist but even to discover potential paintings "behind" other paintings [2]. Images were found by looking at a specific wavelength of light not normally visible to human eyes. Hyperspectral image analysis is not even limited to just art. Agriculture, urban planning and even space exploration are spheres of application for hyperspectral image analysis [3].

Machine learning is the use of an algorithm that "learns" how to perform a task using decision structures and a set of data. It performs its task on artist identification by giving the computer a test set of art from various artists and repeatedly having it attempt to identify the artist. This is done until a reasonable accuracy is obtained. There are difficulties in doing this, as some of the tools used such as hyperspectral image description have what is known as the curse of dimensionality. This is an issue caused by the fact an artist may not have enough "attributed" paintings to be able to put forth an adequate dataset to the computer for it to be able to learn [3]. Data shows increasing the size of the training set does not necessarily increase accuracy as there simply may be too many features present [4]. The Support Vector Machines can achieve an accuracy of approximately 85% using training sets from 25 to 375 images [4]. There is a strong case that eventually computers may be able to identify an art piece, but not until the issues involved are resolved. Other errors in machine learning are attributed to a bias caused by under fitting linear equations and general variance in the computer's data [3].

Others look at the problem of artist identification by examining wavelets of the sound or light made by the artistic piece. The wavelets allow multiscale and multi-orientation image decompositions that can identify consistencies or inconsistencies within certain images [5]. Something as simple as brushstrokes can be examined with wavelets. Different researches have come up with different methods for using the wavelets for image identification. Penn State's approach uses orthonormal wavelets, providing a critically sampled representation of the data via quick sub band filtering [6]. Princeton also uses sub band filtering but abandons the critical sampling to obtain a greater orientation selectivity [6]. Small variations of wavelets such as ridge lets and contour lets are also of use for differentiating between images and can even potentially do better than wavelets in some cases with accuracy ranging from anywhere from 49.6% with ridge lets and 76.2% with contour lets [7]. With wavelets paintings can be examined in depth by scholars and art connoisseurs. Attempts to create a "signature" for an artist using a variety of their work as a data set ends up with a 71.5% accuracy rate [8].

Comparing digital images on the visible light spectrum does not appear to be an advanced technique. The interesting factor of the study is the idea of merging different images into one "signature" image and using that as a basis of comparison. This is different from simply finding an average of the RGB pixels on a set of images as the merging should create darker tiles in general as the colors move closer to black. Should a reasonable correlation be found this could be a great additional tool for hyperspectral analysis as there may also be patterns found by merging images together to create an infrared or ultraviolet "signature". One could perhaps even find patterns using wavelets to find common brush strokes or more precise colors. Computers may be able to have another tool or angle of analysis for determining whether an art piece is legitimate, though it would not solve the problems dimensionality presents.

2 Research Methodology

A total of 25 images per artist are used for the comparisons. The images are enhancing in GIMP 2.8 [10] by using the Layer and transparency tools. After that the images are processed in Weka 3.8.2 [9] using the package imageFilters version 1.0.3 to allow for comparing how Machine Learning algorithms work on different artists. There are two artists for each arff file provided to Weka which is formatted as follows:

```
@relation [ARTIST1]_vs_[ARTIST2]
@attribute filename string
@attribute class {[ARTIST1], [ARTIST2]}
@data
[NAME OF IMAGE FILE], [ARTIST1]
...
[NAME OF IMAGE FILE, [ARTIST2]
```

This gives the program each image file as well as the proper artist to test against with both artists being the dataset. The primary filter chosen from the package is the Color Layout Filter. After applying the directory to the images to Weka in preprocess, the program adds additional features for its machine learning algorithms to work off. The filename string is then removed from the dataset before moving over to the Classify tab. The standard classifier used to check the images is J48 algorithm using the default settings with no changes. This is an algorithm used to generate a decision tree which is used for classification.

Five sets of five images are used to create five additional images for each artist such as the images presented in Fig. 1. The images are chosen at random and layered in a random order but are not used twice. After opening each image in their own distinct layer, the opacity for each image is set to 50. No changes are made to the alignment or size of the images. The option to merge visible layers is used and expanded as necessary is chosen and the image is merged. The image is then exported as a .JPG at 100 qualities to prevent as much compression to the image as possible. Images like the one in Fig. 2 are produced and this is done until five images are acquired for each artist and data set. The image set is run in Weka with the same settings.

The data is collected on two different runs. The first run replaces five of the original images with the set of the five new merged images, where each of the five imaged replaced is in a different merged image. This is to see how close the algorithms detect images as originals. The second run simply adds on the images without removing any of the originals. This is done repeatedly with each artist to see if there is a noticeable trend that occurs.

Fig. 1. An example of the images from a set of Sandro Botticelli's work used in the experiment

Fig. 2. A merged image resulting from Sandro Botticelli's art images in Fig. 1.

3 Interpretation of the Results

The artists used in our experiments are Sandro Botticelli, Caspar Friedrich, Georges Braque, Gustav Klimt, Michelangelo Caravaggio, Francisco Goya, William Turner, and Raphael Urbino. Some of the results obtained are given in Table 1 and Fig. 3. The results were interesting to see, as the differences between Cohen's Kappa, F-Measure, and Matthew's Correlation Coefficient differed significantly between runs. The kappa statistic is a measure of how much chance takes place in the choice of the computer program's decisions. In general, a small kappa statistic means that calculated data is more closed to the real data. The F1-score or F-measure accounts for the ability of the algorithm's precision and recall of the identified images. Matthew's correlation coefficient is an additional measurement of the quality of the machine to sort true positives and true negatives. On average, adding the extra images to the artist dataset improved the scoring of this three statistics.

Some datasets resulted in lower statistics in all essential statistics and an increase in the mean absolute error when the merged images were added to the dataset. One noticeable artist, whose added and replaced runs always had a negative influence was the artist Gustav Klimt. This was not an isolated problem to the two artists and several of Friedrich's images from the Botticelli v Friedrich run, which was the most successful run, also had the issue of not having similar sizes yet resulted in a far more favorable run. There is a common factor among Friedrich's merged images. The dark center figure of the trees, cross, or woman in the merged images may give the algorithm a distinct point to latch onto as Botticelli's merged images have no trace of any dark center piece like Friedrich's images.

Ultimately the largest issue present toward the methodology is that the statistics such as correctly classified instances did not go up by a sizable enough margin, even on the more successful runs. The largest increase in accuracy was approximately nine percent. The more successful runs that did not result in a loss in accuracy resulted in a change of approximately four to five percent. Looking at the sets, if the data did not increase by an amount of approximately fifteen percent the algorithm potentially did not identify any additional original artist images. Since ten images in total were added to the set, it would be a desirable outcome to have an increase in accuracy of around 14 percent. In addition, it was noted that changing the order of the layering could ultimately change the composition of the merged images, though it should be a concern if the merged images become too large a portion of the tested dataset.

After the replacement runs it was noticed that the replacement runs were quite similar to the actual artist image only runs. Minor losses were had in accuracy, Cohen's kappa, and Matthew's correlation coefficient on average. The most significant of this was the Caravaggio v Goya run having a decrease in accuracy of approximately 11.7%. However, other runs resulted in a far more moderate difference such as the Raphael v Turner run losing only approximately 3% accuracy, or even the Raphael v Caravaggio increasing by approximately 3%. This means that the merged images can potentially replace some images and still run a similar profile to the actual artist set.

The more successful image sets which had a positive influence on stats like Cohen's kappa, Matthew's correlation coefficient and accuracy simply were not

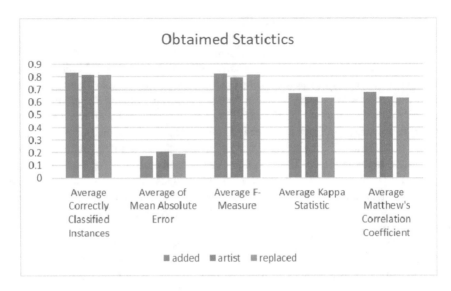

Fig. 3. Averages of statistics derived by several runs

Table 1. Data statistics obtain by the first run

Artists	Correctly classified instances (%)	Kappa statistic	Mean absolute error	Precision	Recall	F-measure	MCC
Botticelli v Braque	71.67%	0.43	0.29	0.72	0.72	0.72	0.43
Botticelli v Friedrich	78.33%	0.57	0.24	0.79	0.78	0.78	0.57
Botticelli v Klimt	73.33%	0.47	0.28	0.74	0.73	0.73	0.47
Braque v Friedrich	86.67%	0.73	0.14	0.87	0.87	0.87	0.74
Braque v Klimt	63.33%	0.27	0.37	0.64	0.63	0.63	0.27
Caravaggio v Goya	96.67%	0.93	0.07	0.97	0.97	0.67	0.93
Friedrich v Klimt	91.67%	0.83	0.09	0.92	0.92	0.92	0.83
Raphael v Caravaggio	81.67%	0.63	0.19	0.82	0.82	0.82	0.63
Raphael v Goya	90.91%	0.82	0.09	0.91	0.91	0.91	0.82
Raphael v Turner	76.67%	0.53	0.25	0.77	0.77	0.77	0.54
Turner v Caravaggio	89.09%	0.78	0.32	0.89	0.89	0.89	0.78
Turner v Goya	83.33%	0.67	0.17	0.83	0.83	0.83	0.67

sufficient to warrant using this method in its current form. The largest change in accuracy came from the Botticelli v Friedrich run at an increase of accuracy at 8.8% despite an increase in the size of the data set where the merged images make up ten out of seventy images or approximately 14.3%.

While the added run's data increases in accuracy and several other key statistics, it does not do so by a sizable amount or with any strong reliability. We can't see this currently being worth the time or effort put into making the merged images. Several runs resulted in a loss of accuracy and only confused further the algorithm.

4   Conclusions

It may be possible to increase the performance of the algorithm by changing the methodology of creating the images to be more in line with each other by aligning the image borders and adjusting the sizes of the art pieces to be closer to each other.

Another angle of attack towards refining the methodology would be changing the opacity of the images to increasingly opaque from the top to the bottom, allowing the ordering of layering to have less of an effect on the merge.

We will apply the developed methodology with a combination of semantic image characteristics such as contrast and harmony and hope to improve the results.

References

1. Ivanova, K., Stanchev, P., Vanhoof, K., Ein-Dor, P.: Semantic and abstraction content of art images. In: MCIS 2010 (2010). http://aisel.aisnet.org/mcis2010/42
2. Herens, E., Defeyt, C., Walter, P., Strivay, D.: Discovery of a woman portrait behind La Violoniste by Kees van Dongen through hyperspectral imaging. Herit. Sci. **5**(1) (2017). https://doi.org/10.1186/s40494-017-0127-4
3. Gewali, U., Monteiro, S., Saber, S.: Machine learning based hyperspectral image analysis: a survey. ISPRS J. Photogrammetry and Remote Sens. (2018). arXiv:1802.08701
4. Blessing, A., Wen, K.: Using Machine Learning for Identification of Art Paintings (2018)
5. Lyu, S., Rockmore, D., Farid, H.: A digital technique for art authentication. Proc. Natl. Acad. Sci. **101**(49), 17006–17010 (2004). https://doi.org/10.1073/pnas.0406398101
6. Johnson, C., et al.: Image processing for artist identification. IEEE Signal Process. Magaz. **25**(4), 37–48 (2008). https://doi.org/10.1109/msp.2008.923513
7. Ozparlak, L., Avcibas, I.: Differentiating between images using wavelet-based transforms: a comparative study. IEEE Trans. Inf. Forensics Secur. **6**(4), 1418–1431 (2011). https://doi.org/10.1109/tifs.2011.2162830
8. Shirali-Shahreza, S., Abolhassani, H., Shirali-Shahreza, M.: Fast and scalable system for automatic artist identification. IEEE Trans. Consum. Electron. **55**(3), 1731–1737 (2009). https://doi.org/10.1109/tce.2009.5278049
9. Durrant, B., Frank, E., Hunt, L., Holmes, G., Mayo, M.: Weka (Version 3.8) (Computer software). https://www.cs.waikato.ac.nz/ml/weka/
10. Natterer, M., Neumann, S.: GIMP (Version 2.8.22) (Computer software). https://www.gimp.org/. Accessed 14 Feb 2018

On Improving the QoS of Video Applications with H.264 over WPANs

Luis Cobo[1($\boxtimes$)], Carlos-Hernan Fajardo-Toro[1],
and Alejandro Quintero[2]

[1] University EAN, Calle 79 No. 11-45, Bogota, Colombia
{lacobo, chfajardo}@universidadean.edu.co
[2] École Polytechnique de Montréal, 2500, chemin de Polytechnique, Montréal,
QC H3T 1J4, Canada
alejandro.quintero@polymtl.ca

Abstract. WPANs are gaining in popularity as cheap, easy to deploy networks to collect real-time contextual information. Research in the area of WPANs has been focused mostly on energy efficiency with only a few papers on supporting video applications on these networks. In this work we will present a new method to improve the quality of video applications over WPANs. The proposed solution is based on a cross-layer architecture to collect information on the state of the network, and to dynamically modify the parameters of the H.264 coder at the source accordingly. We prove through extensive simulations on ns-2 that this adaptive implementation of the H.264 video coder improves the quality of the video at the receiver as compared to the regular static implementation.

Keywords: QoS · Video transmission · WPAN

1 Introduction

In recent years, wireless personal area networks (WPANs) utilization for collecting contextual information has witnessed a dramatic growth. These networks are cheap to deploy and often require minimal configuration. Two IEEE work groups (WGs) were defined to develop two types of WPANs, the 802.15.4 WG for low rate WPANs [1] and the 802.15.3 WG for high rate WPANs. The latter could not produce a normalized physical layer (PHY) due to disagreement between the two main industrial groups and was closed in 2005. Since, research effort has been put to support multimedia application on 802.15.4 WPANs. Nevertheless these efforts remain shy since mainstream research in WPANs is dedicated to other aspect such as power efficiency.

In this paper we will propose a cross-layer architecture to support video transmission over 802.15.4 WPANs. This architecture combines the application layer with a dynamically adjustable H.264/AVC video coder, with the transport layer with its modified UDP transport protocol with a simplified receiver reporter protocol based on real time transport control protocol (RTCP), and the ad hoc on demand distance vector (AODV) routing protocol at the networking layer. We will prove through extensive simulation that this architecture considerably improves the quality of the video at the receiver and the performance of the WPAN. Despite of we count, a few years ago, with

© Springer Nature Switzerland AG 2019
Á. Rocha et al. (Eds.): ICITS 2019, AISC 918, pp. 472–479, 2019.
https://doi.org/10.1007/978-3-030-11890-7_46

new video encoding standards, such as HEVC (High Efficiency Video Coding) presented in [9] and AV1 presented in [10], they are unsuitable for WPANs video transmission. These codecs are very good in video quality when we need to video in high rates and high resolution in a high throughput network, but in low rate network, their efficiency is not as good as H.264 [11].

The rest of this article is organized as follows: in the next section we briefly introduce WPANs and H.264/AVC and discuss related works. In Sect. 3 we present the proposed architecture. This section will be followed by the implementation of the solution and the simulation results. Finally, in Sect. 5 we conclude this work.

2 Related Work

The problem of multimedia transmission over 802.15.4 WPAN networks has been studied in [2]. In this article, the authors propose to implement a new field in the contention free period (CFP), the multimedia contention free period (MCFP) that is dedicated for the transmission of multimedia content. They also propose a new protocol, the cross-layer multimedia guaranteed time slot (CL-MGTS) which collects at the receiver cross-layer information from the application, MAC and physical layers locally and from the sender, and changes the appropriate parameters at the application (i.e. video data rate) and the MAC (i.e. gap value) layers. In [3], the authors propose to introduce a traffic-shaper buffer with its controller in order to control the bursty nature of video data. This controller is allowed to drop B frames of the MPEG4 group of pictures (GOP). Furthermore, the authors propose to add a rule based fuzzy (RBF) scheme to control the output rate from the video encoder and the arrival rate at the traffic-shaping buffer. In another work [4], the authors try to improve multipath multimedia transmission in wireless multimedia sensor networks (WMSN). They propose a new buffer structure in intermediate nodes formed of four queues, one for each frame type (I, P and B) and the last for non-video data with decreasing priority. Packet scheduling in these buffers follow a round robin scheme. Another work [5] is abed on queue priority. In this work, the authors propose a cross layer design and a three queues MAC layer buffer design to schedule MPEG frames based on their type. The authors of this article also propose to modify the time slot allocation mechanism used on TDMA to accommodate nodes with high priority data.

3 Proposed Architecture

Unlike wired networks where rigid transparency and independence between layers of the protocol stack yields the best results, cross-layer approach is a very promising approach in WPANs and has been the base of many architectures as the ones presented in the previous section. Figure 1a represents the cross-layer architecture that we propose to implement in the receiver node (generally the coordinator). Figure 1b represents the cross-layer architecture in the source (generally leaf nodes).

Following the architecture in Fig. 1a, on the transport layer we implement the UDP protocol that we modify slightly to include a sequence number and a time stamp in the header of each video packet. This information is usually included in the real time transport protocol (RTP) header which, in most network implementations, coexists

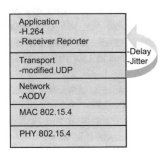
a. at the receiver

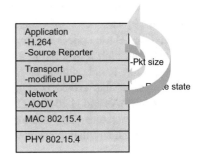
b. at the source

Fig. 1. Cross-layer architecture

with UDP for multimedia sessions. However, this protocol is too heavy to be implemented on WPANs, and replacements have been proposed as explained in [6]. In our case we found that these protocols add overhead that is irrelevant to our solution, and we decided that including the needed information in the UDP header is a better approach especially since network and higher layers protocols are not yet normalized for WPANs. Finally at the application layer, we of course have the H.264 decoder, and we add a receiver reporter protocol that we developed. This protocol collects information through the cross-layer architecture and takes a decision following the algorithm in Fig. 2a, which it than transfers to the source in a receiver report (RR).

H.264 start in Normal mode
 If RR received
  If transmission was stopped
   H.264 start in Normal mode
  Else, extract mode and last received frame
   H.264 switch mode
   H.264 restart from last received frame
   UDP switch packet size
 Else if no RR received || AODV no route
  H.264 stop
	If frame received
	 Increment counter
	 Last_frame = received_fid
	 Collect delay, jitter from UDP
	 If first frame || expected frame
	  If delay > thresh_D || jitter > thresh_J
	   switchDJ = TRUE
	  else switchDJ = FALSE
	 else if not expected frame
	  frame_loss += received_fid – expected
	  if delay > thresh_D || jitter > thresh_J && frame_loss > thresh_FL
	   switchU = TRUE
	  If delay > thresh_D || jitter > thresh_J
	   switchDJ = TRUE
	  else switchDJ = FALSE
	  if frame_loss > thresh_FL
	   switchL = TRUE
	  else switchL = FALSE
	 if switchU = TRUE
	  mode = mode – 1
	  if mode < 0; mode = 0
	  build RR (mode, last_frame) and send
	 if timer up
	  if switchL; mode = CR
	  if switchDJ; mode = mode – 1
	   if mode < 0; mode = 0
	  build RR (mode, last_frame) and send

a: at the source	b: at the destination

Fig. 2. Receiver reporter and mode switching algorithm

In order to minimize signalling overhead in the WPAN, we defined four functioning modes at the source. Now instead of sending in the RR the measured parameters of the network at the receiver, we use the parameters to take a decision at the receiver and send to the source only the recommended functioning mode and the sequence number of the last received frame. On the network layer we implement the AODV routing protocol which is a good protocol to collect information on the state of the route, i.e. route established/repaired, broken link, this information is shared with the application layer. The receiver than takes a decision following the algorithm presented in Fig. 2b and informs the modified UDP of the appropriate packet size to use following the functioning mode in effect as represented in Fig. 1b. The four functioning modes define a selection of parameters in the H.264 coder and the transport layer as follows:

Normal mode

- Profile: High
- Level: 1.1
- Entropy coding: CAVLC
- Quantization parameters (QP): QPI, QPP and QPB, 18, 20 and 22.5 respectively
- Group of pictures (GOP): 20
- Number of inter-P frames: 9 B frames
- Frames per second (FPS): 30
- UDP packet size: 512 Bytes

Congestion Prevention (CP) mode

- Profile: High
- Level: 1.1
- Entropy coding: CAVLC
- Quantization parameters (QP): QPI, QPP and QPB, 22, 24 and 27.5 respectively
- Group of pictures (GOP): 20
- Number of inter-P frames: 9 B frames
- Frames per second (FPS): 30
- UDP packet size: 256 Bytes

Prevention mode

- Profile: High
- Level: 1.1
- Entropy coding: CABAC
- Quantization parameters (QP): QPI, QPP and QPB, 22, 24 and 27.5 respectively
- Group of pictures (GOP): 10
- Number of inter-P frames: 4 B frames
- Frames per second (FPS): 30
- UDP packet size: 128 Bytes

Congestion Recovery (CR) mode

- Profile: High
- Level: 1

- Entropy coding: CAVLC
- Quantization parameters (QP): QPI, QPP and QPB, 24, 26 and 30 respectively
- Group of pictures (GOP): 10
- Number of inter-P frames: 4 B frames
- Frames per second (FPS): 30
- UDP packet size: 256 Bytes

4 Simulation Results

We implemented the proposed architecture on ns-2 where we added the mannasim module [7] for WPAN simulation. As a video traffic generator, we used myEvalVid [8]. We simulated a network formed of thirty wireless nodes moving randomly in a 500 × 500 m^2 and communicate at random time and for random periods of time with the network coordinator at a CBR of 100 kbps to simulate random link breakage and congestions. We use three video sequences: akiyo-cif, bridge-far-cif and coastguard-cif to study the performance of the proposed solution on different levels of movement in the video. And we define five simulation sets where the only thing that changes is the H.264 functioning mode: In the first set the coder functions in the normal mode; in the second the coder functions in the CP mode; in the third the coder functions in the prevention mode; in the fourth the coder functions in the CR mode; finally in the fifth we tested our solution having the H.264 change modes dynamically in response to the received RR. We study the feasibility of our solution and compare it to the case where each of the four predefined modes is statically used throughout the whole simulation time by following three parameters: the average peak signal noise ratio (PSNR) of the received video to study the impact on the quantitative QoS of the video application; the average delay and jitter to study the impact on the qualitative QoS of the video application; the packet loss ratio that along with the delay and jitter results allow to study the impact on the network performance.

Figure 3 shows the results for the average PSNR of each video sequence at the receiver for the different functioning modes of the source. These results show first that the akiyo-cif sequence is better suited for transmission over WPANs since the PSNR calculated for this sequence is best between the three (the average PSNR is 39 dB for akiyo-cif, 33 dB for bridge-far-cif, and 18 dB for coastguard-cif). Furthermore, the impact of the functioning mode at the coder is also most significant for this sequence where our solution yields the best result when compared to the static functioning modes. For the other two sequences, i.e. bridge-far-cif and coastguard-cif, the results for all functioning modes are almost identical. The same behavior can also be observed for the mean opinion score (MOS) results represented in Fig. 4. This parameter gives a better idea of the QoS of the received video when compared to the quality coded video at the source. Again, the akiyo sequence yields the best results between the three sequences and coastguard the worst. Furthermore, for the akiyo and the coastguard sequences, the adaptive mode gives the best results (4.769 and 1,316 respectively) when compared to the other functioning modes (the second best are the Normal mode with 4.651 and the CR mode with 1.254 respectively). For the bridge-far sequence, the adaptive mode is second with 3.95 after the CR mode with 4.145.

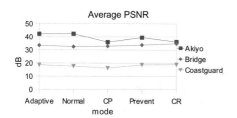

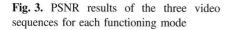

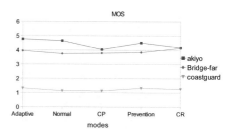

Fig. 3. PSNR results of the three video sequences for each functioning mode

Fig. 4. MOS results of the three video sequences for each functioning mode

Figures 5 and 6 show the results for the delay and jitter respectively of each video sequence at the receiver for different functioning modes at the source. In Fig. 4 we can see that for the akiyo-cif sequence, the adaptive mode gives a lower delay (91 ms) than all other functioning modes (second lowest is 185 for CR mode). For the bridge-far sequence on the other hand, only the normal mode yields a better result than the adaptive mode (296 ms vs. 474 ms respectively), all other modes induce higher delay in the network. Finally, for the coastguard-cif sequence, the delay is very stable for all modes varying between 45 ms and 85 ms.

The jitter results shown in Fig. 6 are very similar to the delay results explained previously. Indeed, we can see in this figure that while the adaptive mode yields better results than all other modes for the akiyo-cif sequence (10 ms for adaptive vs. 13 ms for the second best which is the CR mode), for the bridge-far-cif sequence, only the normal mode has a lower jitter (4.3 ms) than the adaptive mode (8.6 ms). The jitter for the coastguard-cif sequence is stable for all modes between 1 ms and 2.5 ms.

Finally Fig. 7 shows the result ratio of packets lost to packets sent. Here too the results are most conclusive for the Akiyo-cif sequence where we find a noticeable increase in the packet loss ration between the adaptive mode (15%) and all other modes (35% for the normal mode which is the second best). For the bridge-far-cif sequence on

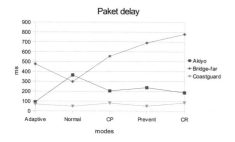

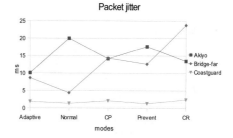

Fig. 5. Packet delay results of the three video sequences for each functioning mode

Fig. 6. Packet jitter results of the three video sequences for each functioning mode

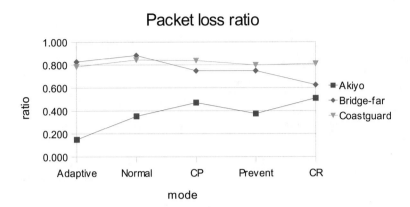

Fig. 7. packet loss ratio of the three video sequences for each functioning mode

the other hand, the results show the opposite behaviour. Indeed for this sequence the adaptive mode is second worst (82% right after the normal mode with 88% of loss), where the CR mode is best with 62% of loss. The results for the coastguard sequence are inconclusive as before and remain stable for all modes. These results also show what we had observed in the PSNR results: that video sequences with low movement and background change like the akiyo-cif sequence are better suited for transmission over WPANs than high movement sequences like the two other sequences.

5   Conclusion

In this work we proposed a new method to support video applications over WPANs. The solution we proposed implements an adaptable version of the H.264 coder which changes its coding parameters following four pre-defined functioning modes to adapt to the state of the network. The state of the network is identified at the reporter application we developed, using data collected from the through a cross-layer architecture. We implemented and simulated the proposed solution on ns-2 and we showed that this method improves the quality of the video application as well as its impact on the WPAN.

References

1. IEEE: Part 15.4: Wireless Medium Access Control (MAC) and Physical Layer (PHY) Spec- ifications for Low-Rate Wireless Personal Area Networks (WPANs). IEEE, New York (2006)
2. Garcia-Sanchez, A.J., Garcia-Sanchez, F., Garcia-Haro, J., Losilla, F.: A cross-layer solution for enabling real-time video transmission over IEEE 802.15.4 networks. Multimed. Tools Appl. **51**, 1069–1104 (2011)

3. Kazemian, H.B.: An intelligent video streaming technique in ZigBee wireless. In: 2009 IEEE International Conference on Fuzzy Systems (FUZZ-IEEE), 20–24 August 2009, Piscataway, NJ, USA, pp. 121–126 (2009)

4. Karimi, E., Akbari, B.: Improving video delivery over wireless multimedia sensor networks based on queue priority scheduling. In: 2011 7th International Conference on Wireless Communications, Networking and Mobile Computing (WiCOM), 23–25 September 2011, Piscataway, NJ, USA, pp. 1–4 (2011)

5. Ismail, N.S.N., Yunus, F., Ariffin, S.H.S., Shahidan, A.A., Rashid, R.A., Embong, W.M.A. E.W., Fisal, N., Yusof, S.K.S.: MPEG-4 video transmission using distributed TDMA MAC protocol over IEEE 802.15.4 wireless technology. In: 2011 Fourth International Conference on Modeling, Simulation and Applied Optimization (ICMSAO 2011), 19–21 April 2011, Piscataway, NJ, USA, pp. 1–6 (2011)

6. Akyildiz, I.F., Melodia, T., Chowdhury, K.R.: A survey on wireless multimedia sensor networks. Comput. Netw. **51**, 921–960 (2007)

7. MR Group. Mannasim Framework (2017). http://www.mannasim.dcc.ufmg.br/index.htm

8. Chih-Heng, K.: How to evaluate H.263/H.264/MPEG4 video transmission using the NS2 simulator? (2017). http://140.116.72.80/∼smallko/ns2/myevalvid2.htm

9. Sullivan, G.J., Ohm, J.R., Han, W.J., Wiegand, T.: Overview of the high efficiency video coding (HEVC) standard. IEEE Trans. Circ. Syst. Video Technol. **22**(12), 1649–1668 (2012)

10. Grange, A., Quillio, L.: AV1 Bitstream & Decoding Process Specification (2018). https://aomediacodec.github.io/av1-spec/av1-spec.pdf

11. Layek, M.A., Thai, N.Q., Hossain, M.A., Thu, N.T., Talukder, A., Chung, T.C., Huh, E.-N.: Performance analysis of H.264, H.265, VP9 and AV1 video encoders. In: 2017 19th Asia-Pacific Network Operations and Management Symposium (APNOMS), pp. 322–325. IEEE (2017)

Computer Networks, Mobility and Pervasive Systems

At a Glance: Indoor Positioning Systems Technologies and Their Applications Areas

Jaime Mier, Angel Jaramillo-Alcázar$^{(\boxtimes)}$, and José Julio Freire

Facultad de Ingenierías y Ciencias Aplicadas,
Universidad de Las Américas, Quito, Ecuador
{jaime.mier,angel.jaramillo,jose.freire}@udla.edu.ec

Abstract. The location of users by electronic devices is in great demand due to the facilities provided by the Internet of Things (IoT). In the search to make a correct decision, the present study has been written which aims to use the knowledge acquired from bibliographies, technical documents and empirical research to develop a comparative study of the existing technologies in terms of Indoor Positioning Systems (IPS) refers. The IPS allow to develop diverse applications among others with the control of accesses, the security in networks, the management of information, and optimization of the services of emergency, among others. Obtaining data in real time provides an important advantage over what clients or administrators need. This advantage can be well exploited in indoor environments. The correct choice of a connection technology for IPS offers innovative solutions to needs of users to improve their quality of life and provide an opportunity to use daily applications.

Keywords: IPS · AOA · RSSI · AAL · BLE · Analysis parameters · Accuracy · Precision

1 Introduction

Internet of things (IoT) has shown an opening (until now) without limits for the use of interconnection technologies between different devices. The opening of this technological model provides the opportunity for daily objects to adhere to the Internet to be recognized and managed by other equipment or specific systems.

The objective of this article it is to provide an analysis of the existing IPS technologies making a comparative study and a summary for each application area using analysis of parameters in order to provide a research solution for future studies dedicated to indoor positioning using wireless technologies. Indoor location aims to be a trend in terms of mobile technologies. The different information and communication services based on location, currently require a decisive innovation that facilitates catapulting new business models and at the same time making the communication infrastructure profitable [8]. Indoor Positioning Systems (IPS) are a set of hardware and software solutions that allow the location of objects or people wirelessly within buildings. The IPS are integrated systems that, wanting to be part of the convergence technologies, they have been proposed to be designed to work in synergy. These systems are responsible for collecting unprocessed data through sensors for later and,

© Springer Nature Switzerland AG 2019
Á. Rocha et al. (Eds.): ICITS 2019, AISC 918, pp. 483–493, 2019.
https://doi.org/10.1007/978-3-030-11890-7_47

with the use of software and application of location estimation methods, convert this information into representations of physical location that are understandable to the end user [14]. The use of different sensors or hot spots called beacons is common inside the building to locate devices and users. Beacons are active or passive devices (depending on the system in which they are applied) that emit a single signal that is interpreted by other devices to facilitate distance identification. These sensors provide an exceptionally accurate positional tracking [5].

The rest of this article is organized as follows. In Sect. 2, we present a description of features and operation of the diverse wireless communication technologies. In Sect. 3, we explain about the different application areas for IPS technologies. Next, in Sect. 4, we make a review of analysis parameters for each IPS technologies and systems. Going on the next point, in Sect. 5, we present a study about IPS solution systems currently existing. In Sect. 6, we proceed to expose a comparative analysis for each IPS systems. Finally, in Sect. 7, we conclude the research and show our future perspectives for these technologies.

2 Technologies for IPS

A. Wi-Fi: 802.1x protocol communication follows a centralized model, therefore, a network consists of one or more access points and a multitude of clients connected to them. Each access point of the network periodically emits a beacon to make its presence known to users, who can thus know at all times the available wireless networks. The networks issued by this protocol and their different variations work in the 2.4 GHz band between 14 overlapping channels (not all of them are valid in all countries) [9]. Wi-Fi-based IPS systems make use of the Received Signal Strength Indicator (RSSI), which is a reference scale measured in mill watts (mW) in order to measure the signal strength received by a device and triangular its location.

B. Based on Mobile Systems: The Global System for Mobile Communications (GSM) is an open digital cellular technology that transmit mobile voice and data services. GSM operates in the 900 MHz and 1.8 GHz bands in Europe and the 1.9 GHz and 850 MHz bands in the United States. GSM services transmit through the 850 MHz spectrum in Australia, Canada and many countries in Latin America. GSM land networks now cover more than 90% of the world's population. The location based on GSM is present in 80-85% of the current cell phones that work worldwide and consumes minimum energy [18]. Several scientific investigations have addressed the disadvantages of GSM localization, including time-based systems, the Angle of Arrival (AOA) which is an estimation method to determine the propagation of a radio frequency signal in cellular networks through RSSI [12].

C. Bluetooth: This standard is based on the master/slave operation mode. The term piconet is used to refer to the network formed by a device and all the devices that are within its range. The main features of Bluetooth technology are Bluetooth devices with low cost, low power consumption, small range, robustness and global use. Bluetooth provides a transfer speed of 1 Mbps and uses an unlicensed frequency band of 2.4 to

2.485 GHz, that is, it uses the ISM area (Industrial, Scientific and Medical) [14]. In addition, Bluetooth offers a radio connection with other systems. This needs to maintain a constant use of mobile terminals has allowed that, without leaving the transmission standard and providing a wide range of up to 100 m, the Bluetooth Low Energy (BLE) is necessary in wireless units [16]. BLE is characterized by a very small size, low cost, low power consumption of several years.

D. Radio Frequency Identification: RFID is a generic term used to describe a system that transmits the identity of an object or person wirelessly using radio waves. It is grouped under the broad category of automatic identification technologies. RFID tags are attached to all objects in a system from which owners want to store certain information, so that this information can be easily retrieved and used later. RFID tags consist of a microchip that can usually store up to 2 KB of data and a radio antenna. A reading device is used to retrieve information from the labels and, depending on the labels used, between 20 and 1,000 labels can be read every second. RFID systems use low, high, ultra-high or microwave frequencies. Active tags usually operate at 455 MHz, 2.45 GHz or 5.8 GHz, and passive tags use 124, 125, 135 kHz, 13.56 MHz, 860–960 MHz and 2.45 GHz [10]. A higher frequency provides a better range, but it is more difficult to control because the energy is sent over long distances and is more easily reflected.

E. Infrared: Infrared wireless communication makes use of the invisible spectrum of light only below red in the visible spectrum. This means that IR communication is blocked by obstacles that prevent the passage of light. Infrared can be used in outdoor communication with Gbps data rates, for example, to connect local area networks in different buildings. But, in such cases, the laser is used as an optical source, and this is not suitable for most indoor uses due to possible transmission interruptions and high operating costs [9]. The IR modules can be small, inexpensive and consume little Energy. Since IR signals cannot penetrate through the walls, it is suitable for sensitive communication because it will not be accessible outside the room or a building.

F. Ultrasound: These US-based or ultrasound-based location systems do not penetrate solid walls, and do not require line of sight between the labels and the detector. Ultrasound waves are mechanical waves and do not interfere with electromagnetic waves. The positioning with ultrasound is suitable for both proximity detection and multilateration, especially for applications where the accuracy of the room scale is sufficient, proximity detection with ultrasound is very effective as it does not penetrate the walls [3]. The frequencies normally used are 40–75 kHz, which allows accurate measurements of the distance between transmitter and receiver at intervals of up to 10 m. An advantage of indoor tracking by ultrasound is that it does not require line of sight, so objects that are hidden or located in drawers or filing cabinets can still be tracked.

G: Ultra Wideband: UWB technology provides a fairly accurate estimate of distances and even in conditions that are not favorable for communications. The ability to provide these estimates should be based on a good temporal resolution dedicated to UWB signals, their robustness against multi-path fades, also known as Rayleigh fades, and their ability to cross obstacles [16]. The use of this technology facilitates the location of targets and especially in densely populated environments and can be used to infer the location of a moving agent. According to the Federal Communications Commission of the United States, UWB systems use a signal that occupies a frequency spectrum greater than 20% of the frequency of a central carrier. That is, it has a bandwidth in the range of 500 kHz to 5 GHz However, according to the International Telecommunications Union (ITU) this window of operation could be extended to 10.6 GHz, but could affect several radio services [11]. UWB is a communication channel that distributes information over a wide portion of the frequency spectrum. This allows UWB transmitters to transmit large amounts of data while consuming little transmission power.

3   IPS Application Areas

The list of applications below demonstrates the omnipresent need for indoor positioning capability in our modern way of life. The most commercially relevant applications for the mass market are the so-called Location Based Services (LBS). Ambient Assistant Living (AAL) systems provide assistance to elderly people in their homes in their daily activities. The applications at home are medical monitoring such as vital signs monitoring, emergency detection and detection of falls, but also personalized service and entertainment systems, such as intelligent audio systems. In hospitals, monitoring the location of medical personnel in emergency situations has become increasingly important [13]. Medical applications in the hospital also include patients and equipment monitoring, for example, patient fall detection. Accurate positioning is required for robotic assistance during surgeries. Environmental monitoring is used to observe some phenomena such as heat, pressure, humidity, air pollution and deformation of objects and structures. A Wireless Sensor Network (WSN) consists of a small autonomous, economical and spatially distributed nodes with limited processing and computing resources and radios for wireless communication [4]. Indoor positioning capabilities provide important benefits in law enforcement, rescue services, and fire services. The police benefit from several relevant applications, such as instant detection of theft or robbery, detection of the location of trained police dogs to find explosives in a building, locate and recover stolen objects. Additionally, products for post-incident investigations, crime scene recovery, statistics and training, but also crime prevention. Mechanical engineering is developing into intelligent systems for automatic manufacturing. For many industrial applications, knowledge of the interior position is an essential functional element, such as robotics guidance, industrial robots, cooperation of robots, intelligent factories (for example, tool assistance systems in automobile assembly lines), monitoring automated and quality control. Indoor positioning capabilities can help find labeled maintenance tools and equipment dispersed throughout a plant in industrial production facilities. Positioning for cargo management systems in airports and ports and for rail traffic offers unprecedented opportunities to increase its efficiency. In Table 1, we present a resume of IPS technologies by application areas.

Table 1. IPS technologies by application areas

IPS application areas	IPS technologies						
	Wi-Fi	GSM	Bluetooth	RFID	Infrared	Ultrasound	Ultra wideband
Home and consumption	*	*	*	*	*	*	*
Health care	*		*	*			*
Environmental monitoring		*			*	*	*
Emergency services	*		*	*		*	
Industry	*		*	*	*	*	*
Logistics and automation	*		*	*			*

4   Analysis Parameters

The diversity of existing technologies and solutions for locating both objects and people in the interior means a problem. Choosing properly a technology to use is really complex since it must take into account the characteristics of each of them, their advantages and disadvantages, which will be useful when building an indoor location system.

A. Accuracy: It is the most obvious property to consider when evaluating and comparing in different positioning systems. Accuracy is a measure of the closeness of one or more positions to a location that is known and defined in terms of an absolute reference system. The accuracy can be given in kilometers, meters, centimeters, etc. Based on location estimation methods, it can be said that the accuracy of an IPS is the average Euclidean distance between the estimated position and the true position.

B. Precision: It is a measure given in percentage, and is based on a relative reference system. Precision is a repetitive measure, and tells us how often we can expect to obtain specific accuracy. For example, if 95% of the position readings of a system are within 10 cm from the true position, it is said to have an accuracy of 10 cm in time, or 10 cm to 95% confidence.

C. Security: The security and privacy of the users must be guaranteed when using any technology, and even more so when exchanging information among millions of interconnected objects. Security mechanisms must be implemented and maintained to protect data against intruders, theft and misuse. Information about the user's location is sensitive data that may present a problem.

D. Scalability: It is the ability of systems to react and adapt without losing quality, and to manage a continuous growth of operation in a fluid way. The scale of the number of users indicates that the number of units located per period of time. The geographical scale refers to the possible changes in the furniture infrastructure in which the IPS solutions are applied.

E. Capacity: For communication systems designed specifically to transmit indoor location data, capacity is essential and fundamental since simple processes can be interrupted due to the overflow of the transmission capacity, thus generating a defective quality of service.

F. Cost: The cost of an IPS can be measured by different dimensions: money, time, space and energy. The cost of installing and maintaining the system includes the cost required for the implementation and the expenses necessary to maintain the functionality of the system, while the cost of infrastructure components and positioning devices may include the costs of purchasing components and preparing them, as well as the space and energy needed to execute those components. Maintenance costs include expenses required to keep the system functional. Space costs involve the amount of infrastructure installation and the size of the hardware.

G. Topology: The network topology refer to how the devices are connected, both active and passive, and the form of communication. The topology directly affects the performance and capacity of the connected devices within the network and therefore a correct topology will optimize costs on the system.

5   IPS Technological Solutions

A. Active Badge: The original system can locate individual objectives in a building, monitoring its presence in different rooms. This is done by equipping rooms with one or more network sensors, which detect diffuse infrared transmissions emitted by active badges. The badges are used by individuals, and emit unique identifiers every 10 s or on demand. The range of the system is approximately 30 m, and line of sight is necessary. Conventional badges batteries last approximately one year with the time intervals between the set emissions in 10 s [1].

B. Active Bat: In this system, the people and objects that are located are equipped with wireless devices called bats, and receivers connected by a network cable installed in known and fixed positions in the ceiling. Bats measure around 8 x 4 x 2 cm, and extract energy from a single AA lithium cell that with low power characteristics has a life of about 1 year. The 3D positioning in Active Bat has an accuracy of 3 cm in approximately 95% of the time. The maximum position update rate in each radio cell is 150 updates per second [2].

C. Cricket: It was designed to avoid the inconvenience of user privacy inherent in the systems previously reviewed (Active Badge or Active Bat), in such a way that allows their devices to learn their location. Cricket devices can be manufactured economically and commercially for less than $ 10. Cricket can be as accurate as 1 to 3 cm in real deployments. The signals operate at 433 MHz, with the default transmission power level and antennas that provide a range of about 30 m indoors when there are no obstacles. The maximum range of ultrasound is 10.5 m when the listener and the beacon are one in front of the other and there are no obstacles between them [15].

D. NaviFloor: It is a system based on fiberglass reinforced sheets that contain integrated RFID passive tags. The NaviFloor labels are installed in 25 cm squares and whose NXP chips measure 45 x 45 mm; These receivers comply with the ISO 15693 standard and communicate in the 13.56 MHz band. They have a proven accuracy of up to 50 cm at 95% of the measurements. Requires an active sensor for approximately 2/3 of the room area of up to 28 m^2 [7].

E. Radar: This system was developed by Microsoft researchers. Most of the work in the implementation is to build this mapping, which can be constructed using an empirical radio map or with the radio propagation model. The precision is of is handled in the range of 2 to 3 m, approximately the size of a typical office room. The accuracy is approximately 4.3 m 50% of a unit time [3].

F. Parctab: This personal digital assistants (PDA) with several IR Diodes spaced around the box and also have a multidirectional receiver. The reach is approximately 6 m, and when placed on the roof in the middle of a room it offers a very good coverage. Diffuse emissions make that loss of signal (LOS) is not necessary, because tranceivers and PARCTABs can detect infrared light reflected from surfaces.

G. Locust Swarm: Locust uses wireless infrared devices that transmit their location information using infrastructure nodes. A locust covers an area of approximately 6 m in diameter, depending on the distance to the floor. The locust measure around 3×8 cm and are connected to a small panel of solar cells measuring approximately 15×15 cm.

H. IRIS-LPS: The system consists of a series of IR-emitting labels and a mounted stere stationary camera. The stereo camera consists of two USB cameras with 120° lenses mounted at 20 cm from each other. The cameras measure the angle of arrival of the light emitted from the labels. The label consists of an LED with a narrow angle of 20° and a range of more than 10 m.

I. Ekahau: Provides real-time and multi-tiered tracking of Wi-Fi devices or Wi-Fi tags. The average accuracy of up to 1 m in less than a second is achieved in the interior by using 5–7 access points, or an average accuracy of 2–3 m by using 3–5 access points. The system can place more than 100 devices per second on a typical desktop PC, and more with more powerful hardware such as a dedicated server [6].

J. Cordis Radio Eye: The system consists of a unit mounted on the roof. CRE is able to provide an accuracy of up to 50 cm in 50% of the sample time. The devices can be monitored with a coverage angle of 110°, and a compensation between coverage and accuracy is required, decreasing the coverage area.

K. Bluetooth Based Systems: The Topaz location system is a local area positioning software and a hardware system that calculates the local position of Bluetooth tags and other devices (for example, mobile phones, PDAs, etc.). It has an average positioning accuracy of up to 2–3 m (7–10.5 ft.). Zonith Bluetooth uses the frequency range of 2.4 to 2.485 GHz for an accuracy range of 0–25 m. The measurement accuracy of Zonith varies from 5 m to 99% of the measurements [19].

L. UWB Systems: UWB technology does not require direct visibility and is not affected by the existence of other communication devices. The cost of UWB equipment is low and it consumes less energy than other solutions. The Zebra's DartWand handles a frequency range of 6.35 GHz at 6.75 GHz with an accuracy of up to 30 cm in 99% of readings and allows to configure up to 100 DartTags for each module and is fully compatible with WLAN networks. This system has proven its correct use for solutions in areas such as: Transportation, Industry and Asset Control [17].

M. Dolphin: It is similar to Active Bat and Cricket in the emission of reference node of an RF signal and an ultrasonic signal simultaneously so that the surrounding nodes and in this way can be measured the TOA used for the distance measurement. The positioning is less than 5 cm, degrading between 10 and 15 cm in the nodes they use for the location.

N. Bristol IPS: It is an ultrasonic system that tries to reduce costs using a minimal infrastructure. It covers a room of approximately 4 × 7 m and you can implement precisions of 10 to 15 cm. The system is self-positioning and uses four ceiling-mounted transmitters connected to a transmitter module that contains the ultrasonic controllers, the microcontroller and the radio transmitter.

6   Comparative Analysis of IPS by Technology in Relation to Its Analysis Parameters

Next, Table 2 is shown that includes all the parameters previously analyzed on each technology of interior positioning systems.

Table 2. IPS by technology in relation to its analysis parameters

Systems	Technology	Parameters							
		Positioning algorithm	Accuracy	Precision	Scalability	Security	Cost	Topology
Active Badge	Infrared	Viterbi algorithm	1–2 m	1 m at 50%	1 Node up to 30 m	Low	Medium	Ad-Hoc
Active Bat	Ultrasound	Gaussian Processes	3–10 cm	3 cm at 95%	1 Node up to 8 × 8 m	Low	Low	Ad-Hoc
Cricket	RF & Ultrasound	*	10–15 cm	15 cm at 90%	1 Node up to 8 × 8 m	Low	Low	Ad-Hoc/Star
NaviFloor	RF	Probabilistic method	<50 cm	50 cm at 95%	2 Nodes up to 28 m	Medium	High	Grid
RADAR	WLAN	Viterbi algorithm	2–3 m	4.3 m at 50%	1 Node up to 200 m	High	High	Grid
PARCTAB	Infrared	Probabilistic method	1–2 m	1 m at 90%	1 Node up to 2 × 3 m	Low	Medium	Ad-Hoc
Locus Swarm	Infrared	*	2–15 cm	15 cm at 50%	1 Node up to 2 × 3 m	Medium	Medium	Ad-Hoc
IRIS-LPS	Infrared	PD based	<1 m	1 m at 50%	1 Node up to 100 tags	Medium	Medium	Ad-Hoc
Ekahau	WLAN	Probabilistic method	<1 m	1 m at 50%	3–5 Nodes each 50 m	High	High	Grid
Cordis Radio Eye	WLAN	LSM	<50 cm	50 cm at 50%	1 Node up to 200 m	High	High	Grid

(*continued*)

Table 2. (*continued*)

Systems	Technology	Parameters						
		Positioning algorithm	Accuracy	Precision	Scalability	Security	Cost	Topology
Topaz	Bluetooth	PD based	2–3 m	2 m at 95%	2 Nodes up to 2–15 m	High	Low	Star
Zonith	Bluetooth	PD based	0–5 m	5 m at 99%	1 Node each 25 m	High	Medium	Star
Ubisense	Ultra Wideband/GSM	LSM	15 cm	30 cm at 99%	2-4 Nodes each cell (10–200 m)	High	Low	Star
DartWand	Ultra Wideband/WLAN	TDoA	<30 cm	30 cm at 99%	1 Node up to 100 labels	High	Medium	Star
Dolphin	Ultrasound	Gaussian processes	5–10 cm	10 cm at 50%	1 Node up to 8 × 8 m	Medium	Low	Ad-Hoc
Bristol IPS	Ultrasound	Gaussian processes	10–15 cm	15 cm at 95%	1 Node up to 8 × 8 m	High	Low	Ad-Hoc

Table 3 shows a comparison between the solutions previously exposed and the areas of application with which they have the most affinity. It is possible to observe that most of these solutions are adapted to the needs of home consumption.

Table 3. IPS solutions in application areas

IPS systems	Application areas					
	Home and consumption	Health care	Environmental monitoring	Emergency services	Industry	Logistics & automation
Active Badge	*				*	*
Active Bat	*	*		*		*
Cricket	*	*				*
NaviFloor	*	*		*	*	*
RADAR		*	*	*	*	
PARCTAB	*				*	*
Locus Swarm			*		*	
IRIS-LPS			*			*
Ekahau	*	*			*	*
Cordis Radio Eye	*	*			*	*
Topaz	*	*		*	*	*
Zonith		*		*	*	*
Ubisense	*			*	*	*
DartWand	*				*	*
Dolphin		*		*		*
Bristol IPS		*		*		*

7 Conclusions

The accuracy provided by indoor positioning systems, regardless of the technology in question, may vary depending on the location estimation method used. The choice of an estimation method appropriate to the chosen IPS will allow the optimization of resources, since the number of readings will also depend on the number of beacons integrated into the network. The right antenna placement will provide a more accurate reading of the devices to be located regardless of the technology chosen. The indoor location has an important challenge to overcome and this is to deal with the obstacles in the rooms despite having an exact mapping of them. Technological solutions, such as those based on infrared, are not recommended in environments whose line of sight can be continually impeded. With aim to overcome these setbacks, RF-based systems such as NaviFloor are the right choice as it tracks at ground level. The diversity of different technological solutions for the internal positioning, shows how deeply interdisciplinary is the field of telecommunications and reflects that almost any signal, technique or sensor can be exploited for this purpose. In spite of abundance of approaches that exist to address the problem of indoor positioning, current solutions cannot face with the level of performance that significant applications require.

References

1. AT&T Laboratories Cambridge. The Active Badge System (2002). https://www.cl.cam.ac.uk/research/dtg/attarchive/ab.html
2. AT&T Laboratories Cambridge. The Bat Ultrasonic Location System (2002). https://www.cl.cam.ac.uk/research/dtg/attarchive/ab.html
3. Bahl, P., Padmanabhan, V.N.: Radar: an in-building RF-based user location and tracking system. In: Proceedings of the Nineteenth Annual Joint Conference of the IEEE Computer and Communications Societies, INFOCOM 2000. IEEE (2000)
4. Bensky, A.: Wireless Positioning: Technologies and Applications. Artech House, Boston (2008)
5. Brassart, E., Pegard, C., Localization, Mouaddib M.: Using infrared beacons. Robotica **18**, 153–161 (2000). https://doi.org/10.1017/S0263574799001927
6. Ekahau. Wireless Design: Ekahau Site Survey y Wi-Fi Planner (2018). https://www.ekahau.com/es
7. Future Shape. SensFloor: NaviFloor based positioning system (2017). https://future-shape.com/en/system/
8. Gartner Inc. At-a-Glance. Internet of Things: Connected means informed. Inc. Cisco Systems (2015)
9. Hamidović, H.: WLAN - Bežične lokalne računalne mreže (2009). https://goo.gl/Chc3LH
10. Hightower, J., Borriello, G.: Location Sensing Techniques. IEEE Computer. Technical Report for University of Washington (2001)
11. Tristant, P.: Ultra Wide Band (UWB) and Short-Range Devices (SRD) technologies. ITU (2009). https://goo.gl/i9cwsz
12. Caffery, J.J., Stuber, G.L.: Overview of radiolocation in CDMA cellular system. IEEE Commun. Magaz. **36**(4), 38–45 (1998)
13. Krilwich, B.: Indoor Location: Sensor Technologies for 2015 and Beyond. Grizzly Analytics (2015)

14. Mautz, R.: Indoor Positioning Technologies (2012)
15. M.I.T. Computer Science and Artificial Intelligence Laboratory. The Cricket Indoor Location System (2005). http://cricket.csail.mit.edu/
16. Svalastog, M.S.: Indoor Positioning-Technologies, Services and Architectures (2007)
17. Ubisense. UWB Sensor Systems (2018). https://goo.gl/zoD4JV
18. Otsason, V., Varshavsky, A., LaMarca, A., de Lara, E.: Accurate GSM indoor localization, vol. 3660, pp. 141–158. Springer (2005)
19. Zonith. ZONITH indoor positioning module: bluetooth positioning and Lone-Worker protection (2018). https://goo.gl/r2Ljwe

An Industry 4.0 Solution for the Detection of Dangerous Situations in Civil Work Scenarios

Borja Bordel[1]([✉]), Ramón Alcarria[1], Tomás Robles[1], and David González[2]

[1] Universidad Politécnica de Madrid, Madrid, Spain
{bbordel,trobles}@dit.upm.es, ramon.alcarria@upm.es
[2] Espacios Castellanos de Innovación, Toledo, Spain
david.gonzalez@ec-innova.es

Abstract. Industry 4.0 is envisioned to apply Cyber-Physical Systems to production systems, assembling lines and other industrial solutions. However, new and innovative approaches may be also designed to solve traditional problems in a more efficient and low-cost manner. In particular, one of the most recognized problems in industry is the prevention of occupational hazards. Many different specific scenarios and risks could be identified in industrial scenarios, but nowadays the industrial sector where (probably) more risks are present is civil works. The use of heavy equipment, where drivers have a limited visual field, is a potential problem for workers at ground level. In this paper it is proposed an Industry 4.0 solution for this situation, where workers are provided with a beaming element, whose signal is received in a central node placed in the machinery. Received signal are processed and analyzed to create alarms and other messages to the driver. The proposed solution is validated in a real scenario using real heavy equipment.

Keywords: Industry 4.0 · Security · Civil works · Wireless solutions · Signal processing

1 Introduction

Cyber-Physical Systems (CPS) [1] are integrations of computational and physical processes, where embedded devices interact with the physical world creating feedback control loops. As this paradigm is based on control theory, it may be applied to scenarios where control systems are required.

One of the most important applications in this group is industry. In fact, in 2014, in Germany, it was defined the concept of Industry 4.0 [2]. Industry 4.0, or fourth industrial revolution, refers the future industry where robots and other similar systems will be replaced by CPS. Physical processes and environments are sensed and collected data are processed to obtain some understanding about the world [3]; this information feeds control loops whose outputs are the actions to be taken by the computational elements to drive the physical world to the desired situation. This approach, on the other hand, may be employed to create a virtual representation of an environment or situation, on which a computational system can make decisions.

© Springer Nature Switzerland AG 2019
Á. Rocha et al. (Eds.): ICITS 2019, AISC 918, pp. 494–504, 2019.
https://doi.org/10.1007/978-3-030-11890-7_48

Industry 4.0 will reach all sectors, from agri-food industry to hostelry, and should address current and future challenges providing new and innovation solutions, with a higher efficient and lower cost. In particular, one of the most relevant challenges in industry is the occupational hazards prevention. This important social problem affects several scenarios, although the most warring is civil works.

In civil works, heavy machinery moves around a perimeter where workers at ground level are also operating. Moreover, drivers in this machinery have a very limited vision field, making difficult to identify risks. In this situation, several dangerous situations may occur including equipment that runs over workers (see Fig. 1). Although using current industrial solutions this problem may be difficultly addressed, through Industry 4.0 technologies it may be solved.

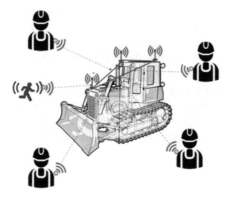

Fig. 1. Dangerous situations in civil work scenarios

Therefore, in this paper, it is proposed an Industry 4.0 solution for civil work scenarios, in order to prevent dangerous scenarios and occupation hazards. In this solution, workers will be provided with a beamer element using WiFi wireless technology, whose signal will be received and processed by a receptor in the machinery (see Fig. 1). The receptor includes a smart antenna and a personal device to display information. The solution is implemented using commercial technologies and deployed in real scenarios to evaluate its performance.

The rest of the paper is organized as follows: Sect. 2 describes the state on Industry 4.0 solutions. Section 3 describes the main proposal, including the architecture and implementation; and Sect. 4 evaluates the performance of this solution. Finally, Sect. 5 presents some conclusions.

2   State of the Art

Industry 4.0 is a very innovative idea which has received a lot of attention during the last years. Any different works about this topic have been reported.

Firstly, works on architecture for Industry 4.0 systems have been proposed. In general, these proposals are focused on the design of manufacturing solutions [4, 5] which are usually event-driven and considered five different layers (connection, control, cognition, cyber level and configuration). Other architectures are focused not on Industry 4.0 systems but on the CPS that support it [6].

Second, works on Industry 4.0, which probably are the most common. Different reports about the research opportunities in Industry 4.0 [7, 8], the idea of Industry 4.0 [9], and the social and economic implications of Industry 4.0 [10] may be found.

In respect to technological proposals, the concept of Smart factory is one of the most strongly related paradigms to Industry 4.0. Self-organized smart factories [11] are usually based on Internet-of-Things (IoT) solutions and energy saving techniques [12] to create efficient and autonomous production systems.

In this sense, in order to manage all information provided by IoT nodes in Industry 4.0 and Smart factories, different Big Data [13] for data analysis have been reported. These data may be also include information about workers, as new human-computer-interaction paradigms are being investigated in the context of Industry 4.0 [14].

Finally, works discussing the base technologies and mechanisms most appropriate to support Industry 4.0 systems have been reported. Techniques based on Lean [15] and virtualization [16] may be found.

The problem of all these proposals is that they are focused on traditional industrial systems, such as production plants and assembling lines. However, other sectors present also pending challenges to be addressed using Industry 4.0 principles. In this paper, we advance to this purpose by investigating in civil work scenarios.

3   Proposal

In this section, the design and implementation of the proposed solution is discussed. First the architecture and network design are described; later the system implementation is analyzed and finally a data processing algorithm to make acquired information valid to calculate alarms and other similar signals is reported.

3.1   Architecture

The proposed architecture is based on different overlapped star topology networks (see Fig. 2), with two clearly different functions: detecting the position of workers at ground level (yellow network in Fig. 2), and sending additional data to provide enhanced services to drivers and managers (red network in Fig. 2). The center of these networks is a smart antenna, in charge of preprocessing and adapting information to be acquired, analyzed and displayed by the management panel and the driver assistance system. Workers are provided with a beaming personal device which may include additional sensors to feed the network supporting enhanced services.

The system consists of a smart device (smartphone, tablet or other similar component) where a mobile application runs, receiving information from the smart antennas, processing it to calculate the selected alarms and displaying these alerts. In this system, moreover, managers may select the configuration parameters to personalize the alarms and the system behavior. This system, besides, periodically stores information about the workers' behavior in a remote database to calculate statistics. The

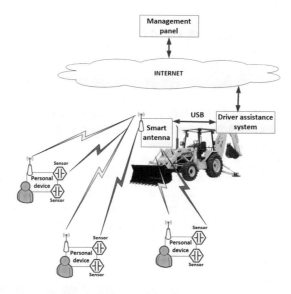

Fig. 2. Basic system architecture

management panel is a web application running in a Node.js server, which calculates statistics from data stored in the database. This instrument may be employed by managers to define occupational hazard prevention.

In the basic design, only one start network is employed to locate works in the scenario. However, techniques based on a unique receiving antenna only may support alarms calculated from approximated works' positions. In order to increase the system precision, three (or more) receiving antennas may be considered to enable the use or

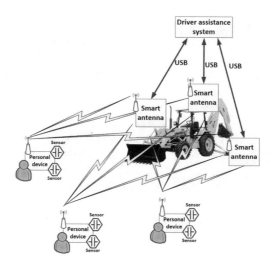

Fig. 3. Enhanced system architecture

trilateration solutions (see Fig. 3). Using the second network, employed to support enhanced services, GPS information could be collected to also improve the system precision.

3.2   System Implementation

To implement the proposed architecture in the previous section, it is employed the WiFi technology and hardware commercial components (Samsung Artik platform and NodeMCU microcontrollers).

The smart antennas are implemented using the Samsung Artik [17] platform. Specifically, Artik 530 devices installing Linux Ubuntu operating system are employed (see Fig. 4a). In these devices, the WiFi module is configured in dual mode to act as access point and client at the same time. Two programs to support each one of the two overlapped networks considered in the system architecture are designed and deployed.

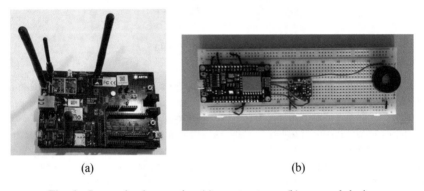

(a)   (b)

Fig. 4. System implementation (a) smart antenna (b) personal device

First, using the specific Artik programming environment and libraries and C language it is created a program to periodically scan available WiFi networks. Received information is processed, and each discovered network is described using a JSON object (see Fig. 5) which is communicated to the driver assistance subsystem through a USB interface.

```
{ID: {IDSystem:PERIMETER, IDGroup: Cuadrilla#1, IDPerson:
FFFFFFFFFFFF}, IDAnte:1, pow:8, rssi:-59, time:1496833248}

{ID: { IDSystem:PERIMETER, IDGroup: Cuadrilla#1,
IDPerson:FFFFFFFFFFFF}, IDAnte:1, pow:8, rssi:-59, time:1496833249}
```

Fig. 5. WiFi networks presented as JSON objects

Second, using a Node.js sever, it is created a web server to which workers' personal devices are connected to send information collected from the additional sensors they include. This server employs the WiFi module as access point and is programmed in JavaScript language. Configuration operations are performed using scripts in bash language.

On the other hand, personal devices for workers are based on NodeMCU micro-controllers, which implement ESP8266 WiFi System-on-Chip (see Fig. 4b). NodeMCU devices implements Arduino architecture, and may easily control two additional sensors included in the personal devices: a MMA8451digital accelerometer (implementing I2C interface) and an ADF-746 GPS receptor (implementing an UART port). These devices are configured to create a WiFi network which is detected and discovered by the smart antenna.

The driver assistance subsystem consists of an Android tablet and is connected through an USB interface to the smart antennas. A Java mobile application is employed to configure and display alarms.

Finally, the management panel is implemented in a Linux server using Node.js technologies and HTML files.

3.3   Data Processing Algorithms

When operating, the smart antennas receive information from personal devices and are able to discover the WiFi networks they generate. This information, however, cannot be directly employed by the driver assistance subsystem to create alarms, as (see Fig. 6) power signals have a very high noise level and may fluctuate.

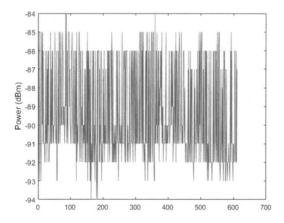

Fig. 6. Raw power signal in the proposed system

To soften and remove the noise and interferences from these power signal (and then) enable the driver assistance subsystem to calculate the configured alarms, the following algorithm is proposed (see Fig. 7).

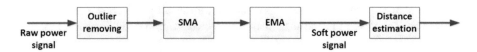

Fig. 7. Proposed processing algorithm

Firstly, signal is divided in sort periods containing around twenty (20) seconds of information. From all selected samples, 20% of them are removed: 10% of samples with highest value and 10% with lowest value. With this action, the signal dispersion is reduced, but fluctuations due to workers movements can be still detected. Then, using the remaining samples, it is calculated a simple moving average (SMA).

This SMA may still contain same slow but relevant interferences, so in order to create a smooth curve, it is applied an exponential moving average (EMA), where α parameter has been selected to be $\alpha = 0.08$ (1).

$$y[n] = \alpha \cdot x[n] + (1 - \alpha) \cdot y[n - 1] \tag{1}$$

With all these operations (see Fig. 8), it is obtained a smooth signal which may be employed to calculate alarms and other information. However, using only power information, it is not possible to obtain an accurate worker position estimation using the standard Friis law [18]. Besides, as farther a worker is, the calculated position has a higher error, following an exponential law. Therefore, to calculate the worker position, we are using a numerical method and we are only positioning workers in three relevant areas (see Fig. 9).

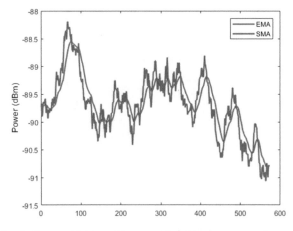

Fig. 8. Processed (smooth) power signal in the proposed system

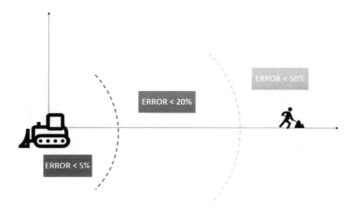

Fig. 9. Relevant areas to map the worker position

The proposed numerical method employs a power regression from the normalized signal obtained from divide all samples between the RSSI (Received Signal Strength Indicator) of the specific beaming system (2).

$$distance = A \cdot \left(\frac{power}{RSSI}\right)^B + C \qquad (2)$$

To calculate A, B and C parameters specific numerical methods must be employed. Parameters A and B are obtained from regression calculation methods, and parameter C must be calculated to minimize the mean square error.

The obtained results from this mathematical method are mapped into a three-zone circular crown (see Fig. 9):

- Red area contains predictions with an error lower than 5%, those which are obtained for the shortest distances.
- In the orange area predictions with an error below 20% are represented.
- Finally, in yellow area, predictions with an error below 50% are placed.

In that way, at least three different alarms may be configured.

4   Experimental Validation

A validation was designed in order to obtain and analyze experimental results. The system was deployed in a civil work scenario, including a bulldozer and four workers (see Fig. 10a). Proposed hardware elements were protected with a plastic casing (see Fig. 10b).

Workers were placed at different distances and provided with personal devices to detect their position. Distance estimations and alarms generated by the proposed solution were acquired and stored; and finally analyzed through statistical methods. To perform this evaluation, an initial setup process was carried out to calibrate the EMA for the suburban scenario where the experiment was developed.

Fig. 10. Experimental validation (a) scenario (b) encapsulated devices

Obtained results are presented in Table 1. As can be seen, the global mean error is below 10%, the usual limit to determine the correctness of low-precision prediction solutions (it is important to take into account that the proposed system is not a positioning system).

Table 1. Results: mean square error

Real distance (m)	Estimated distance (m)	Error (%)
20	22.59	12.96
25	28.47	13.89
30	29.75	0.84
35	26.66	23.83
40	41.17	2.92
Mean error (%)		9.07

Results show that the obtained predictions depend strongly on the quality of the setup process, and the measures obtained during this procedure. In this case, erroneous information for workers placed at 35 meters was introduced during configuration, and predictions for workers at this distance, then, present the higher error (above 20%).

Techniques to ensure the data quality [19], then will be then investigated and proposed in future works.

5   Conclusions

Industry 4.0 is envisioned to apply Cyber-Physical Systems to production systems, assembling lines and other industrial solutions. Many different specific scenarios and risks could be identified in industrial scenarios, but nowadays the industrial sector where (probably) more risks are present is civil works.

In this paper, it is proposed an Industry 4.0 solution for civil work scenarios, in order to prevent dangerous scenarios and occupation hazards. In this solution, workers will be provided with a beamer element using WiFi wireless technology, whose signal will be received and processed by a receptor in the machinery.

The proposed solution was evaluated in a real scenario using commercial hardware and communication technologies. The final mean error in the predictions is below 10%.

Acknowledgments. The research leading to these results has received funding from the Ministry of Economy and Competitiveness through SEMOLA project (TEC2015-68284-R), from the Centre for the Development of Industrial Technology (CDTI) through PERIMETER SECURITY project (ITC-20161228) and from the Autonomous Region of Madrid through MOSI-AGIL-CM project (grant P2013/ICE-3019, co-funded by EU Structural Funds FSE and FEDER).

References

1. Bordel, B., Alcarria, R., Robles, T., Martín, D.: Cyber–physical systems: extending pervasive sensing from control theory to the Internet of Things. Pervasive Mob. Comput. **40**, 156–184 (2017)
2. Sánchez, B.B., Alcarria, R., de Rivera, D.S., Sánchez-Picot, A.: Enhancing process control in Industry 4.0 scenarios using cyber-physical systems. JoWUA **7**(4), 41–64 (2016)
3. Bordel, B., de Rivera, D.S., Sánchez-Picot, Á., Robles, T.: Physical processes control in Industry 4.0-based systems: a focus on cyber-physical systems. In: Ubiquitous Computing and Ambient Intelligence, pp. 257–262. Springer, Cham (2016)
4. Lee, J., Bagheri, B., Kao, H.A.: A cyber-physical systems architecture for industry 4.0-based manufacturing systems. Manuf. Lett. **3**, 18–23 (2015)
5. Theorin, A., Bengtsson, K., Provost, J., Lieder, M., Johnsson, C., Lundholm, T., Lennartson, B.: An event-driven manufacturing information system architecture for Industry 4.0. Int. J. Prod. Res. **55**(5), 1297–1311 (2017)
6. Bagheri, B., Yang, S., Kao, H. A., Lee, J.: Cyber-physical systems architecture for self-aware machines in industry 4.0 environment. IFAC PapersOnLine **48**(3), 1622–1627 (2015)
7. Stock, T., Seliger, G.: Opportunities of sustainable manufacturing in industry 4.0. Procedia CIRP **40**, 536–541 (2016)
8. Weyer, S., Schmitt, M., Ohmer, M., Gorecky, D.: Towards Industry 4.0-Standardization as the crucial challenge for highly modular, multi-vendor production systems. IFAC Papersonline **48**(3), 579–584 (2015)
9. Lasi, H., Fettke, P., Kemper, H.G., Feld, T., Hoffmann, M.: Industry 4.0. Bus. Inf. Syst. Eng. **6**(4), 239–242 (2014)
10. Rüßmann, M., Lorenz, M., Gerbert, P., Waldner, M., Justus, J., Engel, P., Harnisch, M.: Industry 4.0: The Future of Productivity and Growth in Manufacturing Industries, vol. 9. Boston Consulting Group (2015)
11. Wang, S., Wan, J., Zhang, D., Li, D., Zhang, C.: Towards smart factory for industry 4.0: a self-organized multi-agent system with big data based feedback and coordination. Comput. Netw. **101**, 158–168 (2016)
12. Shrouf, F., Ordieres, J., Miragliotta, G.: Smart factories in Industry 4.0: a review of the concept and of energy management approached in production based on the Internet of Things paradigm. In: 2014 IEEE International Conference on Industrial Engineering and Engineering Management (IEEM), pp. 697–701. IEEE (2014)
13. Lee, J., Kao, H.A., Yang, S.: Service innovation and smart analytics for industry 4.0 and big data environment. Procedia CIRP **16**, 3–8 (2014)

14. Gorecky, D., Schmitt, M., Loskyll, M., Zühlke, D.: Human-machine-interaction in the Industry 4.0 era. In: 2014 12th IEEE International Conference on Industrial Informatics (INDIN), pp. 289–294. IEEE (2014)
15. Kolberg, D., Zühlke, D.: Lean automation enabled by industry 4.0 technologies. IFAC-PapersOnLine **48**(3), 1870–1875 (2015)
16. Wan, J., Tang, S., Shu, Z., Li, D., Wang, S., Imran, M., Vasilakos, A.V.: Software-defined industrial internet of things in the context of industry 4.0. IEEE Sensors J. **16**(20), 7373–7380 (2016)
17. Wootton, C.: Samsung ARTIK Reference: The Definitive Developers Guide. Apress, Berkeley (2016)
18. Schelkunoff, S.A., Friis, H.T.: Antennas: theory and practice, vol. 639. Wiley, New York (1952)
19. Bordel, B., Alcarria, R., Robles, T., Sánchez-Picot, Á.: Stochastic and information theory techniques to reduce large datasets and detect cyberattacks in Ambient Intelligence Environments. IEEE Access **6**, 34896–34910 (2018)

Intelligent and Decision Support Systems

A Fuzzy-Based Failure Modes and Effects Analysis (FMEA) in Smart Grids

Andrés A. Zúñiga$^{(\boxtimes)}$, João F. P. Fernandes, and P. J. Costa Branco

IDMEC, LAETA, Instituto Superior Técnico, Universidade de Lisboa,
Av. Rovisco Pais 1, 1049-001 Lisbon, Portugal
{andres.zuniga,joao.f.p.fernandes,
pbranco}@tecnico.ulisboa.pt

Abstract. A Smart Grid combines Information and Communication Technology (ICT's) with innovative power equipment in a fully autonomous and sustainable electrical grid. However, their growing and complex interdependence must be considered in a Smart Grid reliability analysis. Reliable data about its components is not available or even inaccurate most of the time to apply in a Smart Grid the classic method of Failure Modes and Effect Analysis (FMEA). For this reason, this work proposes to a fuzzy-based FMEA that is designed to compute the failure modes risk level in Smart Grids. The results achieved show its capability in improving not only the perception of risk, but also rank in a better way the impact of failure modes.

Keywords: FMEA · Fuzzy systems · Smart Grid · Risk analysis

1 Failure Modes and Effect Analysis FMEA

Failure Modes and Effects Analysis FMEA is a systematic technique for failure analysis, and can be used to define, identify, and eliminate known and/or potential failures, or process errors in a system, process, or service [1]. FMEA is a qualitative reliability analysis technique and is often used as first step in reliability studies [2].

In FMEA terms, a *failure* can be defined as the loss of the intended function of a device (or system) under specific conditions. A *failure mode* describes the way the failure occurs [3].

Each failure mode is characterized by three *risk factors*: the *severity S that* quantifies the effect of the failure mode, the *occurrence O* that represents the frequency a failure mode, and the *detection D* that represents how detectable the failure mode can be. Each of the three risk factors are usually expressed in *risk categories* represented by an integer number on a numerical scale, usually defined in an interval from 1to 10 as in references [2], or in interval from 1 to 5 as in reference [4]. Tables 1, 2 and 3 show some typical ratings for severity, occurrence and detection used in FMEA; these ratings were used in this work.

To assess how hazardous can a failure mode be, FMEA stablishes a *risk level* for each failure mode. This risk level is called as *Risk Priority Number*, *RPN*, and is

© Springer Nature Switzerland AG 2019
Á. Rocha et al. (Eds.): ICITS 2019, AISC 918, pp. 507–516, 2019.
https://doi.org/10.1007/978-3-030-11890-7_49

Table 1. Ratings for severity S used in classical FMEA.

Rating	Effect	Code	Severity of effect
5	Hazardous	EHA	Hazard consequence and the recovery of failure mode is very cost and time expensive
4	Very high	EVH	Energy supply reduction beyond acceptable limits but the repair time and costs of failure mode is acceptable
3	Major	EM	Energy supply reduction important but acceptable, corrective maintenance is needed
2	Low	EL	Low energy supply reduction, no maintenance is needed
1	Minor	EMI	Energy supply slightly affected

Table 2. Ratings for occurrence O used in classical FMEA.

Rating	Failure mode occurrence	Code	Occurrences/year
5	Frequent	OF	3.33×10^{-1}
4	Probable	OP	5.00×10^{-2}
3	Occasional	OO	5.00×10^{-3}
2	Very unlikely	OVU	5.00×10^{-4}
1	Remote	OR	5.00×10^{-5}

Table 3. Ratings for detection D used in classical FMEA.

Rating	Detection	Code	Criteria
5	Absolutely impossible	DAI	Monitoring does not detect the failure
4	Low	DL	Low chance for monitoring to detect a failure
3	Moderate	DM	Moderate chance for monitoring to detect a failure
2	High	DH	High chance for monitoring to detect a failure
1	Almost certain	DAC	Monitoring will almost certainly detect a failure

computed as shown in Eq. (1) by the simple multiplication of three risk factors S, O and D. According to Eq. (1), the *RPN* level can take values from 1 to 125.

$$RPN = (S)(O)(D) \qquad (1)$$

Failure modes can be ranked according its computed *RPN* value: the higher the *RPN* of a failure mode, the greater is the risk for a component and/or system. Failure modes are prioritized from the highest ranked *RPN*, with priority 1, to the higher one.

Failure modes with high priority usually require corrective actions. Such actions may include (but are not limited to) design change, material upgrade and, sometimes, revision of test plans [2]. *RPNs* should be recalculated after the corrections to verify whether the risks have been mitigated and also to check the effectiveness of the corrective precaution for each failure mode [2].

Although FMEA is considered as a powerful tool for qualitative reliability and risk analysis, the *RPN* indicator computation procedure has some disadvantages:

- *RPN* computation procedure does not consider differences between the three risk factors *O*, *S* and *D* (no weight associated to these risk factors);
- Although it is true that higher *RPN* is usually associated to more critical failure modes, this is not always true. For example, a failure mode with a *hazardous severity* rated as 5, in Table 1, a *low occurrence* rated as 1, in Table 2, and *moderate detection* rated as 3, in Table 3, has a *RPN* equal to 15. This *RPN* is lower than another failure mode with *low severity* rated as 2, *probable occurrence* rated as 4 and *low detection* rated as 4 and *RPN* equal to 32. It is evident that a failure mode's effect, represented by severity value, has a very important role in the criticality assessment of such failure mode;

Despite the aforementioned drawbacks, FMEA still is a powerful tool for evaluating the reliability and the risk of any type of system [2].

FMEA is widely used in several industrial and commercial applications as oil and gas, energy, mining, nuclear power, chemical process and, lately, also in healthcare, among others. Related to electrical power systems, its main applications are currently oriented to wind energy [4], photovoltaic generation [5], capacitors banks [6], substations [7], power transformers [8], distribution systems [9], and in smart grid to study the delivery of electricity to the clients [10].

In real applications, the criteria used to assess the three-risk factor (*O*, *S* and *D*) usually is expressed in natural language and contains a high level of imprecision, ambiguity and vagueness. Fuzzy systems can help modelling the human knowledge and expertise to perform numerical computation when linguistic terms are involved. By this reason, fuzzy systems are a valid option to improve the calculation process for *RPN* in FMEA analysis.

2 Fuzzy Systems Applications in FMEA

Fuzzy systems were widely applied in FMEA analysis. Works in references [11–13] show a review for these applications. In this fuzzy-based FMEA approach, fuzzy sets are used to represent the risk factors *S*, *O*, *D*, and *RPN*, called as *fuzzy* variables, and a fuzzy reasoning mechanism is used to compute the *RPN* value.

2.1 Fuzzy Rules

Because the risk factors *S*, *O* and *D* and *RPN* can be represented by fuzzy sets, the *RPN* computation process can be performed by using *fuzzy reasoning*, also known as *approximate reasoning* [14]. In fuzzy systems, approximate reasoning refers to a mode of reasoning in which the input-output relation of a system's variables is expressed as a collection of IF-THEN fuzzy rules, where the antecedents and consequents are represented as fuzzy sets [14]. For example, a fuzzy rule for a particular failure mode can be *IF (Severity is hazardous with warning) AND (Occurrence is remote) AND (Detection is high) THEN (RPN is very high)*.

2.2   Fuzzy Inference System

The Fuzzy Inference System *FIS* is a computational framework that formulates input/output mappings through fuzzy if-then rules and fuzzy reasoning. FIS consists of three conceptual components [14]: A *rule base*, which contains a selection of *if-then* rules; a *database*, which contains the membership functions used in the linguistic antecedent and consequent fuzzy rules; and, a *reasoning mechanism*, which performs the inference procedure upon the rules and given facts to derive a reasonable output or conclusion. Figure 1 shows a general FIS structure.

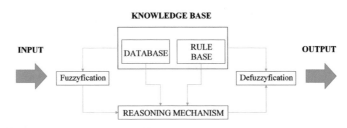

Fig. 1.   Structure of a fuzzy inference system (FIS).

The FIS types and their reasoning mechanisms are fully detailed in Fuzzy Systems literature [14]. The Mamdani fuzzy inference system was used for the proposed fuzzy-based FMEA presented in the next section.

3   Fuzzy-Based FMEA Application in Smart Grid

In the classical FMEA, the risk priority number *RPN* is computed by the direct multiplication of ratings for *S*, *O* and *D*. In risk assessment terms, it is possible to say that severity *S* and occurrence *O* represents the direct impact of a failure mode in the system.

Now, for the proposed fuzzy-based FMEA, it is possible to group the severity and occurrence to define, in a first stage, an intermediate risk factor called as impact, *I*, that results from fuzzy relation between *S* and *O*. In a second stage, the fuzzy *RPN*, denoted here as *FRPN*, is computed by using the impact, *I*, and the detection, *D*. This two-stages approach makes it easier the designation of each fuzzy rule.

Table 4 shows the five linguistic categories attributed to the impact I as being Catastrophic (IC), Very Hazardous (IVH), Moderately Hazardous (IMH), Low Hazard (ILH), and Minor Hazard (IMH).

Similarly, a set of five linguistic categories were define for the and Fuzzy Risk Priority Number FRPN as being a risk Extreme (RE), High (RH), Moderate (RM), Low (RL), and Minor (RMI).

Table 4. Linguistic categories for the impact *I* and the Fuzzy Risk Priority Number (*FRPN*).

IMPACT (I)		FRPN	
Category	Code	Category	Code
Catastrophic	IC	Extreme	RE
Very hazardous	IVH	High	RH
Moderately hazardous	IMH	Moderate	RM
Low hazard	ILH	Low	RL
Minor hazard	IMIH	Minor	RMI

3.1 Fuzzy Sets for Severity *S*, Occurrence *O* and Detection *D* Risk Factors

The five categories of the risk factors involved in the fuzzy *RPN* computation, as the occurrence *O*, severity *S*, detection *D*, impact *I* and *FRPN*, all previously listed in Tables 1, 2, 3 and 4, are represented by fuzzy sets. Occurrence *O*, severity *S* and detection *D* risk factors were defined each one with 5 ratings represented by integer values in the range between 1 and 5. By this reason, each risk factor was represented by a fuzzy set, one for each category, and uniformly distributed.

The impact *I* (product of *O* and *S*) can take values in the range between 1 to 25, thus defining its universe of discourse, and having five categories also represented each one by a fuzzy set, all uniformly distributed. At last, the fuzzy risk priority number *FRPN* (product of *I* and *D*) can take values in the range between 1 to 125, thus defining its universe of discourse, and also represented by 5 uniformly distributed triangular fuzzy sets along that range (Fig. 2).

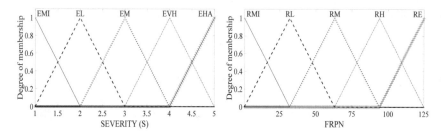

Fig. 2. Triangular fuzzy sets used to represent the severity and fuzzy risk FRPN variables, all uniformly distributed by each universe of discourse.

In this work, two types of membership functions are used to represent the categories for *S*, *O*, *D* and *FRPN*: a symmetrical triangular membership function and a symmetrical Gaussian membership function. As an example, Fig. 2 shows the representation of severity *S* and *FRPN* using their 5 symmetrical triangular membership functions, all uniformly distributed in each universe of discourse.

3.2   Fuzzy Rules Definition

A number of 25 fuzzy rules were defined for criticality matrixes associated to impact I and the *FRPN*. The impact I results from the combinations of severity S and occurrence O; because they have 5 fuzzy sets each, the impact is composed by 25 possible fuzzy rules as in Table 5a. Similarly, *FRPN* results from the combination of impact I and detection D, and then is composed by 25 possible fuzzy rules to compute *FRPN* as in Table 5b.

Table 5.   Criticality matrices containing the fuzzy rules for impact I (a) and *FRPN* (b).

a)

		OCCURRENCE				
		OF	OP	OO	OVU	OR
SEVERITY	EHA	IC	IC	IC	IC	IVH
	EVH	IC	IC	IVH	IVH	IVH
	EM	IVH	IVH	IMH	IMH	IMH
	EL	IVH	IMH	IMH	ILH	ILH
	EMI	IMH	IMH	ILH	IMIH	IMIH

b)

		IMPACT				
		IC	IVH	IMH	ILH	IMIH
DETECTION	DAI	RE	RE	RE	RE	RH
	DL	RE	RE	RH	RH	RH
	DM	RH	RH	RM	RM	RM
	DH	RH	RM	RM	RL	RL
	DAC	RM	RM	RL	RMI	RMI

The following rule is an example extracted from Table 5a (for impact I computation). First, the respective linguistic expression was first established. After, one "translated" the expression to its fuzzy representation.

- *Rule 1:* If the severity is hazardous and occurrence of the failure mode is frequent to occur then the impact will be catastrophic.
- *Fuzzy rule 1:* If (severity is EHA) AND (occurrence is OF) THEN (impact is IC).

The following rule is also an example extracted from Table 5b (for *FRPN* computation):

- *Rule 14:* If the detection of failure mode is moderate and the impact of the failure mode is low hazard then the fuzzy risk priority number is moderate.
- *Fuzzy rule 14:* If (detection is DM) and (impact is ILH) then (FRPN is RM).

3.3   Fuzzy Inference System

Two Mamdani's FIS were defined to compute the impact and the *FRPN:*

- One for computing the fuzzy risk priority number FRPN, that uses triangular membership functions, and help to simulate a linear transition in membership grades; and,
- A second FIS for to compute the FRPN but using Gaussian membership functions, that simulates a nonlinear transition in membership grades.

Each of the two Mandani FIS have two inputs, each of them represented by 5 fuzzy sets and a single output also represented by 5 fuzzy sets, as described in Sect. 3.1.

4 Results of Fuzzy-Based FMEA Application in a Smart Grid

The classical FMEA and the proposed fuzzy-based FMEA were applied to evaluate the risk priority number on eight components of a Smart Grid. The tests were performed considering the following assumptions:

(a) All fuzzy linguistic terms were represented by symmetrical membership functions, which maximum value is the rating for each category of S, O and D;
(b) Two shapes for the membership functions were used: triangular and Gaussian, and;
(c) Only single local effects were analysed. Therefore, the interdependences between different components were not considered for the current analysis;
(d) The failure modes were prioritized according to their higher to lower RPN, where the highest RPN has priority 1 and the lowest RPN has priority 8.

In a first test called as FMEA1, a classical FMEA was conducted and ratings for S, O and D were determined, then the RPN was computed using Eq. (1).

In the second test called FMEA2, the fuzzy-based FMEA was applied to compute a $FRPN$ value considering symmetric and triangular membership functions for both the input and output variables. The rating values for S, O and D shown in Table 5 were used as inputs to the fuzzy inference system, thus inferring the values for impact I and fuzzy risk priority number $FRPN1$.

In the third test called FMEA3, the fuzzy-based FMEA was applied to compute a $FRPN$ value considering now not triangular functions but symmetric Gaussian membership function for both input and output variables, continuing to be uniformly distributed. As for the second test, the rating values for S, O and D in Table 6 were used as inputs to the fuzzy inference system to obtain the impact I and the fuzzy risk priority number $FRPN2$.

Table 6 shows the results of the tests FMEA1 and FMEA2. It is possible to identify three failure modes (FM5, FM6 and FM7) with the *highest severity S* rank valued as 5 and categorized as *hazardous* in Table 1, and *low detection D* rank valued as 4 and categorized as *low* in Table 3 (remember that detection has a higher rank for worst detectability). However, since those failure modes have *lower* occurrence O ranked as 1 and categorized as *remote* Table 3, their RPN is valued as 20 and were prioritized as 5, 6 and 7. It is important to note that *hazardous severity S* implies that failure mode produce a hazard effect according to Table 1.

In addition, failure modes FM1, FM2 and FM3 have a severity S rank valued as 4 and categorized as *very high* in Table 1, that is, their severity rank is less than severity ranks for those failure modes FM5, FM6 and FM7 but it were prioritized as 1, 2 and 3. Nevertheless, FM1, FM2 and FM3 have the highest RPN because their occurrence O is higher than occurrence for FM4, FM5 and FM6.

Table 7 shows the results of the tests FMEA1 and FMEA3 and as for fuzzy-based FMEA using Gaussian membership functions, the fuzzy-based FMEA prioritize the failure modes in better way. It is important to note that FMEA3 prioritize the failure mode FM8 with the highest priority 1 although does not has the higher severity; similarly, FM2 is prioritized as 2 although does not has the highest severity. As results obtained using FMEA2, prioritization were improved for failure modes with worst severity FM5, FM6 and FM7, categorized as *hazardous* and valued as 5.

Table 6. FMEA and fuzzy-based triangular results.

Code	Failure mode	Component	S	O	D	Classic FMEA1		Fuzzy triangular FMEA2	
						RPN	Priority1	FRPN1	Priority2
FM1	Bearing-inner/outer race and ball faults	Wind turbine Gearbox	4	3	3	36	1	94	6
FM2	Cable break	Network link	4	2	4	32	2	115.3	5
FM3	Electrical fault	Power Transformer	4	2	3	24	3	94	7
FM4	Open/short circuit	PV panel bypass diode	3	2	4	24	4	94	8
FM5	Human error/sabotage	Human Machine Interface HMI	5	1	4	20	5	116	1
FM6	Tank damage	Power Transformer	5	1	4	20	6	116	2
FM7	Early operation	Protective Relay	5	1	4	20	7	116	3
FM8	Spurious opening and closure	Power circuit breaker	4	1	5	20	8	116	4

Table 7. FMEA and Fuzzy-based Gaussian results.

Code	Failure mode	Component	S	O	D	Classic FMEA1		Fuzzy Gaussian FMEA3	
						RPN	Priority1	FRPN2	Priority3
FM1	Bearing-inner/outer race and ball faults	Wind turbine Gearbox	4	3	3	36	1	85.2	7
FM2	Cable break	Network link	4	2	4	32	2	100.9	2
FM3	Electrical fault	Power Transformer	4	2	3	24	3	85.2	8
FM4	Open/short circuit	PV panel bypass diode	3	2	4	24	4	87.9	6
FM5	Human error/sabotage	Human Machine Interface HMI	5	1	4	20	5	100.8	3
FM6	Tank damage	Power Transformer	5	1	4	20	6	100.8	4
FM7	Early operation	Protective Relay	5	1	4	20	7	100.8	5
FM8	Spurious opening and closure	Power circuit breaker	4	1	5	20	8	116	1

5 Discussion

In both fuzzy-based FMEA tests, FMEA2 and FMEA3, the prioritization for failure modes with worst severity valued as 5 were improved when compared with prioritization assigned using the classical FMEA1. This fact is very important because classic FMEA prioritization is only based on the simple arithmetic multiplication of the three risk factor S, O and D and does not considers any weighing to distinguish the importance of these risk factor in the system overall risk assessment. Fuzzy inference system considers a weighted average method to compute the output, this characteristic of fuzzy systems allow to assign quantitatively more importance to higher valued risk factors O, S and D and therefore the RPN is better prioritized.

The main issue in above referenced works lies in that different combinations of O, S and D may produce the same RPN rating for different failure modes and, although their hidden risks or impacts on the system could be different, the classical FMEA cannot distinguish them [7, 10]. In classical FMEA, critical failure causes, sometimes with bigger RPN than other noncritical failure modes, can see their maintenance strategies being ignored, as concludes Baliea in [10].

Results obtained show that proposed fuzzy-based FMEA can improve the prioritization method of classical FMEA used in the smart grid environment, especially when the interdependences in the cyber-physical system are relevant.

6 Conclusions and Future Work

This paper summarizes our preliminary work concerning a more global objective of using fuzzy-based FMEA to analyse a Smart Grid. The paper focused on a basic application of fuzzy systems principles to improve the classic FMEA analysis for the main components on a Smart Grid. A comparison between the classical FMEA and fuzzy-based FMEA was conducted in order to verify the advantage of the fuzzy approach in relation to the first one.

Although classical FMEA is a powerful tool for the qualitative analysis of reliability, the procedure to evaluate the different modes of failure is very simple and causes errors in the prioritization of your level of risk.

Results obtained show that fuzzy-based FMEA proved to be very satisfactory in relation to improve the perception of risk and, more relevantly, improve the failure modes ranking and prioritization, when compared with the classical FMEA approach.

The prioritization is a very important issue in reliability analysis, especially in systems with too many failure modes similarly ranked, because prioritization is used to decide about the preventive maintenance actions and systems improving for detection and mitigation actions for hazard failures, which are very cost expensive.

Some points are being considered to improve the proposed analysis:

- Consider the interdependences between the different components failure modes and also their local and global effects on the Smart Grid;
- To analyse different membership function shapes for risk factor's categories and their sensitivity to the final risk grade;

- Tune the membership function to fulfill specific characteristics for each failure modes and to capture in the best way the expert knowledge and expertise about reliability and risk criteria;
- Redefine how a certain fuzzy rule is important considering the addition of weights for the antecedent in the fuzzy inference system.

Acknowledgements. This work has been partially supported by: Secretaría Nacional de Educación Superior, Ciencia, Tecnología e Innovación (SENESCYT) of the Ecuadorian Government, and also supported by national funds through the Fundação para a Ciência e a Tecnologia (FCT) of the Portuguese Government with references UID/EEA/50008/2013 and through IDMEC, under LAETA, project UID/EMS/50022/2013.

References

1. Teixeira, Â.P.: FMEA/FMECA. Class Notes - Systems Reliability and Maintainability. Instituto Superior Técnico, Universidade de Lisboa, Lisboa (2017)
2. Liu, H.-C.: FMEA Using Uncertainty Theories and MCDM Methods. Springer, Singapore (2016)
3. International Electrotechnical Commission: IEC 60812:2006 – Procedure for Failure Mode and Effects Analysis (FMEA), Geneva, Switzerland (2006)
4. Dinmohammadi, F., Shafiee, M.: A fuzzy-FMEA risk assessment approach for offshore wind turbines. Int. J. Prognost. Health Manage. 4(Sp2), 1–10 (2013)
5. Villarini, M., Cesarotti, V., Alfonsi, L., Introna, V.: Optimization of photovoltaic maintenance plan by means of a FMEA approach based on real data. Energy Convers. Manage. 152, 1–12 (2017)
6. Pourramazan, A., Saffari, S., Barghandan, A.: Study of Failure Mode and Effect Analysis (FMEA) on capacitor bank used in distribution power systems. Int. J. Innov. Res. Elect. Electron. Instr. Control Eng. 5(2), 113–118 (2007)
7. Araújo, W.: Metodologia fmea-fuzzy aplicada à gestão de indicadores de continuidade individuais de sistemas de distribuição de energia elétrica. Master Thesis, Universidade Federal de Santa Catarina, Florianópolis S.C., Brasil (2008)
8. Kaur, J., Singh, B.N.: Condition monitoring of power transformer using failure modes and effects analysis (FMEA). Int. J. Innov. Res. Sci. Eng. Technol. 6(9), 19108–19115 (2017)
9. Yssaad, B., Khiat, M., Chaker, A.: Maintenance optimization for equipment of power distribution system based on FMECA method. Acta Elecht. 53(3), 218–223 (2012)
10. Baleia, A.: Failure modes and effects analysis (FMEA) for smart electrical distribution systems. Master Thesis, University of Lisbon, Lisbon (2018)
11. Vinodh, S., Aravindraj, S., Narayanan, R.S., Yogeshwaran, N.: Fuzzy assessment of FMEA for rotary switch: case study. TQM J. 24(5), 461–475 (2012)
12. Bowles, J.B., Pelaez, C.E.: Fuzzy logic prioritization of failures in a system failure mode, effects and criticality analysis. Reliab. Eng. Syst. Saf. 50(2), 203–213 (1995)
13. Tay, K.M., Lim, C.P.: Fuzzy FMEA with a guided rules reduction system for prioritization of failures. Int. J. Qual. Reliab. Manage. 23(8), 1047–1066 (2006)
14. Jang, J.-S.R., Sun, T.-S.: Neuro-Fuzzy and Soft Computing: A Computational Approach to Learning and Machine Intelligence. Prentice Hall, Englewood Cliffs (1997)

A Sales Route Optimization Mobile Application Applying a Genetic Algorithm and the Google Maps Navigation System

Cristian Zambrano-Vega[1(✉)], Génesis Acosta[2], Jasmin Loor[2], Byron Suárez[2], Carla Jaramillo[2], and Byron Oviedo[1]

[1] Facultad de Ciencias de la Ingeniería, Universidad Técnica Estatal de Quevedo, Quevedo, Los Ríos, Ecuador
{czambrano,boviedo}@uteq.edu.ec
[2] Carrera de Ingeniería en Sistemas, Universidad Técnica Estatal de Quevedo, Quevedo, Los Ríos, Ecuador
http://www.uteq.edu.ec

Abstract. Nowadays, the Route Optimization Problem (ROP) is one of the most studied combinational optimization problems that researchers study. Although it is easy to define, its solution is hard. Therefore, it is one of the NP-hard problems in the research literature. It can be used to solve real-life problems such as route planning and scheduling, and transportation and logistics applications. Using the optimal tour results in efficient use of time and fuel. This paper aims to develop an Android Application that can provide optimal tour (shortest distance) to visit a set of clients. Genetic Algorithm is used to solves the problem and is implemented using the Google API and Android OS. The source code of the application is available at url https://github.com/Genethh/VentasExpress.

Keywords: Sales route optimization · Genetic Algorithms · Minimarkets · Distribution

1 Introduction

Companies are using more analytics to enable better sales force decisions, for example, the sales territory design, or the way in which the responsibility for accounts is assigned to salespeople or sales teams. The distribution of customer workload and opportunity across the sales force has a direct impact on salespeoples ability to meet customer needs, realize opportunities, and achieve sales goals [4].

Every day sales reps face difficult decisions. What clients to prioritize, what to say during a sales meeting, how to make maximize their schedules, just to name a few decisions. The best way to ensure that the Sales Reps meet more

Á. Rocha et al. (Eds.): ICITS 2019, AISC 918, pp. 517–527, 2019.
https://doi.org/10.1007/978-3-030-11890-7_50

customers, and minimize their mileage is Route Optimization. Routing optimization algorithms basically designs for the best routes to reduce travel cost, energy consumption and time. During this process, commercial territories are delimited in order to help agents to sell more efficiently. Delimitation is done by allocating a number of existing and potential clients to each distribution representative acting in a given area, usually, but not always, established on geographical criteria. Due to non-deterministic polynomial-time hard complexity, many route optimizations involved in real-world applications require too much computing effort (Fig. 1).

Fig. 1. Route optimization problem illustrated.

1.1   Logistics E-Commerce in Ecuador

According to the experts who visited Ecuador in the Ecommerce day on July 2, 2015, the local reality is about to be infected by the Ecommerce fever, companies that move away from the trend of selling online will simply be left behind in the market competition. Although Ecuador has already taken the first steps, there are still aspects that must be worked on, logistics is one of the main points because 6% of online purchases are lost due to the slow delivery, changes to implement to avoid the abandonment of the purchase are several, in mentioned:

- *All products have to reach the customer:* One of the advantages of online sales is the ease they provide users to purchase a number of products anywhere and receive them at home.
- *Failures in delivery generate dissatisfied customers:* In the case of receiving a product hit, out of time, or even wrong, the customer will blame directly to the store or distributor.
- *The logistic cost can determine the profitability of an online business:* Both the operating costs and the delivery values that customers must pay are important economic factors to take into account when setting up a business.

The contribution of this paper is propose a sales route optimization mobile application applying a genetic algorithm and the Google Maps navigation system, designed to optimize any route map, turning it into something that involves less time spent driving while taking in more customer visits. Our proposal provides a sales route optimization and a place to store the customer details.

The rest of this paper is organized as follows: First, the Route optimization problem definitions are described in Sect. 2. In Sect. 3 we detail the specification of Genetic Algorithms. The Genetic Algorithm implemented to solve the Route Optimization Problem is described in Sect. 4. The features and specifications of our proposal are detailed in Sect. 5 and finally, Sect. 6 outlines some concluding remarks and suggest some lines of future work.

2   Route Optimization Problem

The mathematical formulation of the Route Optimization Problem is defined as follow: Consider a set of customers $C = 1, 2, ..., N$ and a set of potential sellers $S = 1, 2, ..., |S|$ dispersed in a given region with geographical coordinates (*long*, *lat*). It is desired to design a sales route plan that defines a minimum number of sellers required to fulfill the customer's demand β. The sellers will attend the demand of customers during the weekdays $W = 1, 2, 3, 4, 5, 6$ denoted by index t in the scheduling plan per week. Finally, it is desired to get the optimal daily routing. The distance between two location is computed using the Haversine formula [4, 11], the procedure to calculate this Haversine Distance d is detailed as follow:

1. R = ← 6371 # Earth radius in kilometers
2. Δ lat ← radians(lat2 − lat1)
3. Δ long ← radians(long2 − long1)
4. $lat1$ ← radians(lat1)
5. $lat2$ ← radians(lat2)
6. a ← $sin(\frac{\Delta lat}{2})^2 + \cos(\text{lat1}) * \cos(\text{lat2}) * sin(\frac{\Delta long}{2})^2$

7. c ← $2 * \text{asin}(\sqrt{a})$
8. d ← R * c

Route Optimization is a solution for two of the most difficult computer science problems: the Traveling Salesman Problem (TSP) and the Vehicle Routing Problem (VRP).

2.1   Traveling Salesman Problem TSP

Nowadays, The traveling salesman problem (TSP) is one of the most studied combinatorial optimization problems that researchers study. It is one of the NP-hard problems in the research [2]. The main problem of the TSP is to find a tour of a given number (n) of cities with a minimum distance, where each city visited exactly once and returning to the starting city [1]. Mathematically, it can be defined as given a set of n cities, named $c_1, c_2, ..., c_n$, and permutations, $\delta_1, \delta_2, ..., \delta_n!$, the objective is to choose δ_i such that the sum of all Euclidean distances between cities in a tour is minimized [7].

TSP issue can be solved by enumerating $\frac{(n-1)!}{2}$, where n is the number of cities and then select the route with the shortest length. The position of the city with the number 4 city (a, b, c, d) as shown in Fig. 2 that have $\frac{(4-1)!}{2} = 3$, possible routes that can be passed. Finished, the shortest route is (a, c, b, d, a) or (a, d, b, c, a) with a route length of 32.

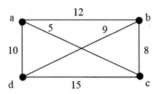

Fig. 2. Four city TSP.

2.2 Vehicle Routing Problem (VRP)

The vehicle routing problem has been one of the elementary problems in logistics ever since because of its wide use. Vehicle Routing Problem (VRP) can be described as follows. Suppose there are M vehicles each of which has a capacity of Q and N customers who must be served from a certain *depot(terminal station)*. The goods each customerasks for and the distance between them are known in advance. The vehicles start from the *depot(terminal station)*, supply the customers and go back to the depot. It is required that the route of the vehicles should be arranged appropriately so that the least number of vehicles is used and the shortest distance is covered.

The following conditions must be satisfied:

- The total demand of any vehicle route must not exceed the capacity of the vehicle.
- Any given customer is served by one, and only one vehicle.
- Customer delivery should be done efficiently and economically.

3   Genetic Algorithm

John Holland proposed Genetic Algorithm in 1975. In the field of artificial intelligence genetic algorithm is a search heuristic that mimics the process of natural evolution. Genetic Algorithm belongs to class of evolutionary algorithm. GA begin with various problem solution which are encoded into population, a fitness function is applied for evaluating the fitness of each individual, after that a new generation is created through the process of selection, crossover and mutation. After the termination of genetic algorithm, an optimal solution is obtained. If the termination condition is not satisfied then algorithm continues with new population [7]. The whole process of genetic algorithm is described in Algorithm 1 and its flowchart is illustrated by Fig. 3.

Result: Polutation of solutions P
t ← 0;
Initializate P(t);
Evaluate [P(t)];
while *not Termination Condition* **do**
    | t ← t+1;
    | P'(t) ← Selection [P(t-1)];
    | P"(t) ← Crossover [P'(t)];
    | P"'(t) ← Mutation [P"(t)];
    | P(t) ← Replacement(P(t-1),P"'(t));
    | Evaluate [P(t)];
end

Algorithm 1. Basic Genetic Algorithm

The components of the Genetic Algorithms are:

- **Representation:** Several representation methods have been proposed for represent a problem of TSP like a binary, matrix and path representation. The path representation is probably the most obvious representation of a tour on TSP [5,7].
- **Initialization:** In this step, an initial population is generated from a random selection of solutions (chromosomes). Genetic algorithm performance is influenced by the size of the population if the problem becomes more difficult then the size of the population should increase [7,9].
- **Evaluation:** After initial population, to improve the solutions each chromosome in the population is evaluated by using the fitness function. The fitness function we are using inverse of the objective function. In this way, we can find the best chromosome that corresponds to the shortest path.
- **Selection:** Process selection is choosing two parents from the population for further crossover process. Tournament selection is used for this section, because of its efficiency and simple implementation [7,8].
- **Crossover:** Crossover is the process that mimics mating between two chromosomes with the goal of producing offspring. A new offspring is generated via the order crossover (OX) of two parent chromosomes. OX is used to perform the path/permutation representation, because OX rather concerns circular permutations [6,7].
- **Mutation:** The function of mutation is to add the diversity of the population and prevent the chromosomes falling into the local minima [7,12]. The current implementation uses exchange mutation. Which randomly selects two genes of a chromosome and exchanges them. After creating new population by crossover and mutation, there is a big chance, that we will lose the best chromosome. Therefore elitism method (elite selection) is used. The main objective of elitism is to determine the best individuals and immediately copy them over to the next generation.
- **Termination condition:** Termination condition is the criteria used to stop the evolutionary process. When the evolutionary process has reached the maximum time period then stop.

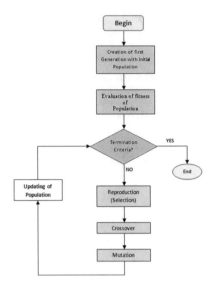

Fig. 3. Flowchart of genetic algorithm

4 Genetic Algorithm Applied to Route Optimization

In this section we detail the implementation of the Genetic Algorithm applied in our mobile application to obtain the best route to visit the clients.

4.1 Representation of the Solutions

The classical genetic algorithm paradigm deals with the solutions encoded as a literal string, called chromosomes. A chromosome is the representation of a single solution of the problem and requires additional encoding/decoding steps to be defined in the algorithm [10]. The solutions in our proposal, are represented as an array of destinations, which contains the GPS coordenates: Latitude and Longitude.

4.2 Fitness Function

A fitness function is a type of objective functions which summarizes the goodness of a solution with a single figure of merit. In our proposal The fitness function is calculated as the sum of the Haversine distance of all the destinations of one tour. The fitness function to evaluate the individuals (candidate tour) is described in the Eq. 1:

$$f(x) = \frac{1}{\sum_{i=1}^{n-1}(d(i, i+1)) + d(n, 1)} \tag{1}$$

Where n is the number of places to visit and d represents the Haversine distance between the destination i and destination j.

4.3   Procedure

The procedure of the genetic algorithm is described as follow:

1. Create a group of many random tours in what is called a *population*. This algorithm uses a random initial population that sort randomly all the places to visit.
2. This initial population is evaluated calculating the fitness of each tour.
3. A new population of individuals (tours) called *OffSpring* is created using the reproduction operators.
    (a) The selection operator *Binary Tournament* is used to identify chromosomes which will be used in reproduction and will survive in the next generation. Two individuals are selected randomly and, in this case, the tour that requires the lowest fitness is selected. Two shorter tours parents are chosen.
    (b) With the two individuals selected, a new individual is generated using the Uniform Order Crossover. Hopefully, this new child tour will be better than either parents. This is done to prevent all tours in the population from looking identical.
    (c) A swap mutation operator is used with intention to prevent getting stuck in the local optimum and increase a probability to find the global optimum. Two selected randomly destination of the tour are swapped. A small percentage of the time, the child tours are mutated.
4. The longer individuals (tours) of the current population are replaced with the shorter new tours, creating a new current population. The size of the population remains the same.
5. Repeat the third step until a maximum number of iterations (generations).
6. The best individual is selected, with the shortest fitness (lowest distance between destinations), this represents the quasi-optimum sales route. As the name implies, Genetic Algorithms mimic nature and evolution using the principles of Survival of the Fittest.

4.4   Google Maps

Google maps provide an intuitive and highly responsive mapping interface with aerial imagery and detailed street data. In addition, map controls and overlays can be added to the map so that users can have full control over map navigation. Map panning can also be performed by dragging the map via the mouse or by using "arrow" keys on a keyboard. Google maps can be customized according to application specific needs [3].

5   RouteOptApp Mobile Application

The system implemented will solve the Travelling Salesman problem to determine the optimum route on Google map using Google API and Genetic Algorithm in Android OS. The system starts from getting different Geo Locations

(Latitude and Longitude) on the map triggered by the user in String Format through Geo Coding. The Google Geo Coding API provides a direct way to access a Geo Coder via an HTTP request. Then, a distance matrix of these locations to be visited by the user is determined by parsing the JSON file based on pair of source and destination Geo Points.

5.1   Features

The mobile sales route optimization application that applies a genetic algorithm and the Google Maps navigation system has the following characteristics:

- It is developed for mobile devices with Android operating system, developed in the integrated development environment (IDE for its acronym in English Integrated Development Environment) Android Studio with API 21: Android 5.0 (Lollipop).
- The database is implemented in the MySQL database manager.
- Web services are developed in the PHP programming language using the NetBeans IDE.
- The Google Maps API SDK for Android is also used in the mobile application to view maps based on Google Map data specifically in native Android applications.
- It also accesses the user's location when registering a new client.

5.2   How it Works

The mobile application has three modules or users that perform different functions. The first user is the *Administrator* who is basically in charge of adding the Sellers and these at the same time can enter the Clients who are the agents that use the application to place their orders to the distributor of products.

The Administrator user performs the following functions: Contains the category section where you can perform the operations of viewing, adding and updating the categories to which a specific product belongs; the Product module in the same way allows us to carry out the operations indicated above to control the mass consumption items that the stores may request, in this option the category to which the product belongs is first considered in order to be able to add it; in the Vendors option, all the workers of the product distributor are listed, when creating a new vendor, the NIF ID is assigned as user and password so that they can enter their platform; the application gives each seller a bonus for each order delivered to customers, on the Bonuses tab you can see the bonuses that each seller receives for the work performed, the bonus will correspond to the percentage established by the administrator applied to the basic salary, this rate is defined in the Bonus Rate tab; The route tab is entered and the routes to be traversed by each vendor are assigned, the vendors are visualized where the vendor that wishes to assign a route is selected.

The user Seller performs the function of achieving the delivery of the order requested by the client on the established date, for this purpose it uses a genetic algorithm that uses the coordinates of the store to show the optimal route that the seller needs to get to his store. destination at the agreed time. In the Customers module you can add the stores or users of the application, you get the current location of the seller: longitude and latitude, assuming that it is in the store at the time of adding it; In the Orders tab the seller will take control of the orders of the clients that he must deliver, he will be able to change his status to "delivered", add products to the order and the most interesting to visualize the route he must travel through points marked on a map; The Routes tab lists the routes entered by the administrator, and at the same time shows the clients added for this route and the route on the map.

The third user, the Client, is the consumer of the application to whom it is desired to provide the best service. You can access your module through your login assigned by the vendor, perform the functions of updating the information about your store in the Profile tab and in the Order Tab you can place your order

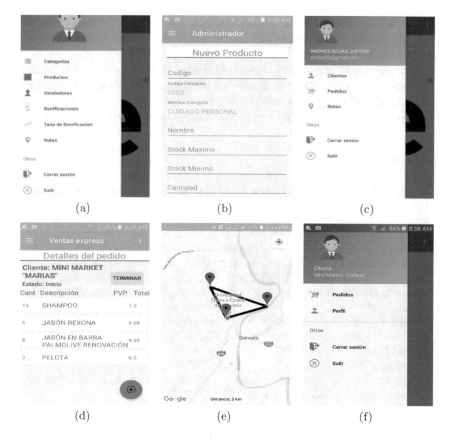

Fig. 4. Screenshots of the mobile application. Optimal distance obtained by implementing genetic algorithm optimal path shown by lines

at the desired time, view the ones you have made, see the detail of each one, add products to the order and take control of the status (Started or delivered).

5.3 Example of Use

The mobile application has three users, the Administrator screens are shown below. The administrator's menu has the tabs categories, products, seller, bonuses, bonus rate, routes, close session and exit (Fig. 4).

6 Conclusions

- The implementation of a geolocation mechanism served to obtain the quick location of the clients and in this way show their location in the route to be followed by the company's distributor.
- The use of a genetic algorithm made it possible to optimize the route to follow by the seller so that he can access the mini markets registered in the system, in this way to do it effectively when delivering the order, reducing mobilization costs.
- The development of a mobile application allows the visualization of the data of the route to be carried out by the distributor, obtaining customer information (mini markets) being registered with greater ease in the application in order to obtain a bonus according to the standards stipulated by the company.

References

1. Bryant, K., Benjamin, A.: Genetic algorithms and the traveling salesman problem, pp. 10–12. Department of Mathematics, Harvey Mudd College (2000)
2. Garey, M.R.: Computers and Intractability: A Guide to the Theory of NP-Completeness. W. H. Freeman & Co., New York (1997)
3. Helshani, L.: An android application for Google map navigation system, solving the travelling salesman problem, optimization throught genetic algorithm. In: Velencei, J. (ed.) Proceedings of FIKUSZ 2015, pp. 89–102. Faculty of Business and Management, Óbuda University, Keleti (2015). https://ideas.repec.org/h/pkk/sfyr15/89-102.html
4. Hervert-Escobar, L., Alexandrov, V.: Iterative projection approach for solving the territorial business sales optimization problem. Procedia Comput. Sci. **122**, 1069–1076 (2017)
5. Larranaga, P., Kuijpers, C.M.H., Murga, R.H., Inza, I., Dizdarevic, S.: Genetic algorithms for the travelling salesman problem: a review of representations and operators. Artif. Intell. Rev. **13**(2), 129–170 (1999)
6. Liu, R., Jiang, Z., Geng, N.: A hybrid genetic algorithm for the multi-depot open vehicle routing problem. OR spectr. **36**(2), 401–421 (2014)
7. Narwadi, T., Subiyanto: An application of traveling salesman problem using the improved genetic algorithm on android Google Maps. In: AIP Conference Proceedings, vol. 1818, p. 020035. AIP Publishing (2017)

8. Razali, N.M., Geraghty, J., et al.: Genetic algorithm performance with different selection strategies in solving TSP. In: Proceedings of the World Congress on Engineering, vol. 2, pp. 1134–1139. International Association of Engineers Hong Kong (2011)

9. Reese, A.: Random number generators in genetic algorithms for unconstrained and constrained optimization. Nonlinear Anal.: Theory Methods Appl. **71**(12), e679–e692 (2009)

10. Vaira, G., Kurasova, O.: Genetic algorithm for VRP with constraints based on feasible insertion. Informatica **25**(1), 155–184 (2014)

11. Veness, C.: Calculate distance and bearing between two latitude/longitude points using Haversine formula in javascript. Movable Type Scripts (2011)

12. Wang, Y.: The hybrid genetic algorithm with two local optimization strategies for traveling salesman problem. Comput. Ind. Eng. **70**, 124–133 (2014)

Contextual Analysis of Comments
in B2C Facebook Fan Pages
Based on the Levenshtein Algorithm

Danny Jácome, Freddy Tapia$^{(\boxtimes)}$, Jorge Edison Lascano,
and Walter Fuertes

Universidad de las Fuerzas Armadas ESPE, Sangolquí, Ecuador
{dajacome,fmtapia,jelascano,wmfuertes}@espe.edu.ec

Abstract. The present study proposes the implementation of an algorithm to determine the degree of reliability of Business to Consumer Facebook fan pages, to mitigate possible cheating, scams or fraud. We use data mining to filter information from comments expressed by online stores' customers. The experiment determines a word dictionary to find matches using a Natural Language Processing technique. The main contribution of this research is the analysis of the context of online stores followers and customers that use Facebook as a business tool, this analysis is based on comments as opposed to the classic way of counting the number of likes. This analysis discards wrong and miswritten comments.

Keywords: Social networks · Facebook · Data mining ·
Business to Consumer · Levenshtein algorithm · Natural Language Processing

1 Introduction

According to the National Institute of Statistics and Census (INEC by its abbreviation in Spanish), in Ecuador about 540 million dollars go every year through Electronic Commerce. However, the Chamber of Commerce of Guayaquil estimates that it is higher, 700 million. This because "there is a percentage of Ecuadorian population that uses credit cards issued abroad to buy internationally without paying taxes for the capital outflow", explained Leonardo Ottati [1].

Although, it is believed that e-Commerce is not safe, due to various vulnerabilities that users are exposed to, experts claim the opposite and ensure that buying online is even safer than shopping in on-site stores. In the worst case scenario of a robbery, usually, the company assumes the costs related to the transaction [2]. This makes social networks the right place for marketing their products.

In Ecuador, 97.9% of people over the age of 12 have a Facebook account [1]. Worldwide, Facebook owns the largest number of social networks users. Social networks help people keep in touch and find potential business partners [3] either manually or by smart ads. A Natural Language Processing technique (NLP) uses an algorithm that allows to look for similar words on the same topic [11].

The Graph API allows you to pass in and extract data on the Facebook platform. It is also a low-level HTTP-based API that is used to query data, publish new stories, manage

© Springer Nature Switzerland AG 2019
Á. Rocha et al. (Eds.): ICITS 2019, AISC 918, pp. 528–538, 2019.
https://doi.org/10.1007/978-3-030-11890-7_51

ads, upload photos and perform different tasks [4, 5]. Most uses of the Graph API require access tokens that an application generates at the beginning of each session.

The new trend of creating online stores for linking to consumers have generated small and medium-size enterprises with Fan Pages to massively offer products or services, and reach potential customers. Eventually, these pages are dedicated to sales and advertising, hence there is money related to scams, cheating and possible frauds.

There are two characteristics that a social and behavioral measurement instrument must meet on respect to "psychometric solidity": reliability and validity. They consist of evaluating the psyche of an individual and quantifying it in a numerical way [6]. Reliability, means that the acquired data is safe, reliable and identical, even if different processes are performed. There can be internal reliability, when two or more people involved in the project agree with the conclusions, and external reliability, when people outside the project perform tests at different times, reaching the same results.

Comments in social networks, allow for a large number of inconsistent results, because people write in a poor way for many reasons, such as: spelling errors, variation in pronunciation, lack of integrity in the data, use of abbreviations, vulgarisms and idioms. For this reason, it is necessary to use a reliable and efficient matching algorithm that guarantees the integrity of the results. In our case, the data comes from different Fan Pages of the following virtual stores: All Store, Aio Shop and Monkey Shop. Stores dedicated to clothing accessories sales, they have over 17 thousand followers.

This proposal handles internal reliability, guided by a phenomenological systemic, and hermeneutic orientation, reliability is oriented to an interpretative level of agreement between different phenomenon observers. 70% is considered a good level of reliability. For example, of 10 judges, there is consensus between 7 of them [7]. This study uses the Natural Language Processing (NLP) and the Levenshtein's distance algorithm. it aims to analyze the context by means of similarities or coincidences of words within the comments of Fan Page followers. The objective of the research is to implement an algorithm that finds similar words and phonetic variations.

This paper is organized as follows: Sect. 2 describes the related work. Section 3 describes different word matching algorithms, the PMBOK probability and impact matrix. Section 4 describes the experimental topology of the algorithm. Section 5 evaluates and discuss results. Finally, Sect. 6 shows conclusions and future work.

2   Related Work

Works based on distance algorithms are well accepted for context analysis. For example, Mosley [8], makes an analysis of Twitter insurance publications, as comments are in a range between 1 and 140 characters, this represents a maximum of 35 words per comment. They perform a data cleansing, and then changes the structure of the data from horizontal to vertical to be able to compare word by word against a set of standard words. Mosley uses Levenshtein's distance algorithm and the comparison rules that weigh a certain score depending on the type of correction made to a word.

Martel and Carranco [9], studied on the aggressiveness in social networks, based on a Facebook comments database, and using the R analysis tool, which has the Levenshtein's algorithm built-in package "stringdist". Bilenko [10] and Cohen [11] worked with a group of phonetic and character matching techniques. Based on tokens and hybrids, they tested and then compared them. They did not analyze error rates, i.e., typographical errors and the size of the data set. Hassanzadeh presents a general study of word matching techniques, he evaluates accuracy. Neither the typographical errors or the amount of data in each group for the tests were considered.

Peng [12] presents a comparative work on the performance of name matching techniques, they considered some factors that Hassanzadeh did not take into account, such as the error rate and the size of the data set. However, similarly to Hassanzadeh, Peng did not consider typos and first names.

After reviewing the existing research on the field, we can see that there is no unique and exclusive technique to solve all character matching tasks. Particularly when the data is in Spanish and words not belonging to the dictionary are used. Also the use of neologisms and Ecuadorian idioms is not considered, they can change all conclusive analysis depending on the region or population area. Hence, there is need of a method that consider possible typos, euphemisms, regional idioms, abbreviations, or simply words that are not in a dictionary.

3   Matching Algorithms

The problem of similar name matches (Qualifying Adjectives - QA), is solved by more than one technique. The accuracy in the recognition as in the search for Word Match, depends on the strengths and capabilities of the proposed technology and algorithms.

The problem of word matching, includes distance, spelling, phonetic and sound techniques. Table 1 shows a comparison of the different search algorithms for matching two terms, this will help us determine which algorithm is most suitable for our research.

Table 1. Comparison of matching search algorithms

PLN task	Technique	Method	Operation	Type of match	Algorithm
RNE Name Match Search	Spelling and Distance	Distance Calculation	Edition Substitution Elimination Transposition	Exact Partial	Guth Levenshtein
	Sounds and Phonetics	Reduction to phonetic code based on pronunciation	Substitution Rules Phonetics	Partial Phonetics	Soundex Soundx-sp Phonex Metaphone NYSIIS
	Composite	Distance Calculation Reduction to phonetic code based on pronunciation	Edition Substitution Rules Phonetics	Exact Partial Phonetics	Damereau Levenshtein TFID Editex Jaro-Winkler Ngrams Qgrams
RNE Names Segmentation	Pattern Recognition	Decision Trees Bayesian Networks Neural Networks	Classification	Exact Partial Similarity Synonym	ID3 Naive – Bayes Perceptrón Multicapa
	Pairing Patterns	Distance Calculation	Edition	Abbreviation	KMP/BM/BMH/BMS
				–	LZT8/LZW

3.1   Matching Techniques

There are several techniques that allow the comparison of texts from different sets or data sources, this is considered a relatively complex task. This task can be performed by techniques associated with pairing words by search and recognition. Two groups of techniques are considered: (1) phonetics; and (2) spelling and distance.

Coincidence of phonetic codes: phonetic correspondence techniques are established in the perception of sounds represented by a single letter or syllable. A value is assigned to a sequence of characters based on the sound they produce. For each word, a canonical form generates a code, by means of phonetic code matching algorithms the similarity of two character strings can be determined. Some techniques are: Soundex, Soundex Enhanced (Soundexp), Phonex, Metaphone and NYSIIS.

Comparison of spelling and edit distance: the unique difference of the characters coincides with words spelling. The words are compared by detecting typographical errors or coincidence in their entirety. These algorithms do not need phonetic transformations at all [12]. Some algorithms evaluated in this paper are: Distance from Levenshtein, Distance from Damerau-Levenshtein, Matching spelling from Guth, Ngrams, Jaro and Jaro Winkler.

3.2   RestFB Library

Graph API allows the extraction of comments that are published on social networks. This API is the link between our application and Facebook, thanks to the RestFB library that allows to obtain the information in tables (flat files) and then save them in our database.

It is a Facebook Graph API client, flexible and written in Java. It is open source software released under the MIT License. The RestFB API uses a method to obtain information and publish new elements on Facebook. With this library you get direct connection to the Graph API. Following is the Data extraction algorithm:

```
String postId = "323907354456822_807114659469420";
Connection<Comment> commentConnection =
  fbClient.fetchConnection(postId + "/comments",
  Comment.class, Parameter.with("limit", 10));
int personalLimit = 10000;
int n=1;
for (List<Comment> commentPage: commentConnection){
  for (Comment : commentPage){
    System.out.println(" N:" +n+" Id:
    "+comment.getId()      +"Comentario:"+
    comment.getMessage());
    personalLimit--;
      if (personalLimit == 0) {return;}
}
```

3.3   Dictionary of Words and Semantic Weight

People today have different ways of writing on social networks. They use idioms, or neologisms, often not defined in the Spanish Royal Academy (RAE), therefore, we created a dictionary of positive and negative words, the result from observation and analysis of three web pages: All Store, Aio Shop and Monkey Shop. We used the following algorithm, to break a sentence down into words to compare them later.

```
Algorithm to separate and compare words
   while (st.hasMoreTokens()) {
      s2 = st.nextToken();
      numTokens++;
      System.out.println("Palabra "+numTokens+": "+s2);
      ControladorLogin pdc = new ControladorLogin(s3,
            s2);
      pdc.calcularDistancias();
   }
```

3.4   Levenshtein Algorithm

This algorithm is based on alphabetical comparison techniques and uses edit distance metrics. A weight is assigned to each position compared between text strings, in terms of the number of insertions, deletion and substitution operations. Scenarios are required to edit and convert the first string to the second. The similarity is determined based on the minimum editing distance. The minimum distance distld (s, t) between the chains s and t is represented in Eq. 1.

$$\text{distld}(s, t) = \min \sum_{i=1}^{N} W_i(|x|, |y|) \tag{1}$$

Levenshtein not only compares strings of characters but also phonemes. In such a way that, s is the source phonetic transcription chain and t is the destination phonetic transcription chain. N is the number of transformations between s and t, i is the index that represents each operation of the set to obtain the Levenshtein distance. And so | x | and | y | are the phonemes involved in the edition. W is the minimum transformation weight to match the phonemes. The distance is symmetrical and it is always true that:

$$0 \leq \text{distld}(s, t) \leq \max(|s|, |t|), \text{y abs}(|s| - |t|) \leq \text{distld}(s, t) \tag{2}$$

The similarity function is given by Eq. 3:

$$\text{simld}(s, t) = 1, 0 - \left(\frac{\text{distld}(s, t)}{\max(|s|, |t|)}\right) \tag{3}$$

Levenshtein's editing distance is defined as "the cost of a better sequence of editing operations to convert a string s into t". It is a typical editing operations for insertion, deletion and substitution, to which a cost must be assigned to every one [13, 14].

```
program Inflation (Output)
private void calcularDistancias() {
  char[] cad1 = cadena1.toCharArray();
  char[] cad2 = cadena2.toCharArray();
  for(int i=0; i<=cad1.length; i++) distancia[i][0]=i;
  for(int j=0; j<=cad2.length; j++) distancia[0][j]=j;
  for(int i=1; i<=cad1.length; i++) {
    for(int j=1; j<=cad2.length; j++){
      int cambio=0;
      if(cad1[i-1]!=cad2[j-1]) cambio=1;
      distancia[i][j] = minimo(distancia[i-1][j]+1,
      distancia[i][j-1]+1,distancia[i-1][j-1]+cambio);
    }
  }
}
```

The algorithm was used many times to write in a hasty way because we misspell a word and this can completely change the meaning of a sentence. This algorithm is used to compare word by word from our dictionary with those gotten from the posts, accepting or rejecting the words as appropriate. Experts recommend that a word have a maximum distance of 4, because if it is higher, the word would completely change the meaning of the sentence. For our research and in order to apply it accordingly, we take a maximum distance of 3. Since we are more likely to make mistakes when writing in smart phones, therefore should be stricter.

4   Proposed Algorithm

The distance algorithms are used to obtain optimal algorithms for words recognition and similarities, these algorithms find a match of the data with a pre-established data dictionary. Comments and their variants will be used without misrepresentations. The phases of the proposed algorithm are described below:

Comment Loading. The comments downloaded from social networks are loaded in their natural form and only the blank spaces are discarded, also the line breaks are identified from each comment. This will help establish a format for each comment regardless of its size.

Comment Separation. Once the comments are loaded, we proceed to separate them into simple words, each word is stored in a variable, so we can perform different operations in each word.

Editing Distance. Levenshtein distance is applied, a coefficient of similarity expressed in percentages, indicates the distance calculated for each word with each of the words contained in the data dictionary. In this phase, we get integrity, accuracy and precision.

Evaluation. The records from previous phase are placed in descending order by the coefficient of similarity. Those matches with the highest coefficient value must be separated into a final data set. The rest of the low value matches are discarded. Duplicate matches with high or low values are analyzed to detect the presence of homonyms. These cases can be determined by means of a second restriction condition (e.g., place of birth). The final result is an exact match data set.

Quantification. We used PMBOK to give a quantitative value to words, so the comments are analyzed in their context, considering the frequency of each word and the incidence that it has for our research. To determine the level of accuracy of the Fan Page, a dictionary of words was used to compare with the comments obtained from the posts. Figure 1 shows the experimental topology, as well as the main tools that were necessary for the data extraction and analysis.

Fig. 1. Experimental topology

5   Evaluation of Results and Discussion

For evaluating results, the formula of false negatives was not taken into account, since by using the probability and impact matrix the probability of an erroneous term known as "false positives" can be regulated.

To evaluate the results, the formula of false negatives was not taken into account because their sensitivity and specificity are not evaluated. By using the probability and impact matrix, the probability of an erroneous term known as "false positives" can be regulated.

5.1   Semantic Word Weight

To start with the analysis, we considered the positive and negative words that are used by followers and customers to congratulate or claim for the service, quality and reliability of the products. This process is done manually taking into account an equal number of comments for each Fan Page, additionally, the meaning of each selected word is taken from the RAE and for those words that are not registered, definitions were entered by the researchers and collaborators. The probability of finding these words in the posts was analyzed, according to (4). Table 2 describes the probability of occurrence for each positive word.

$$\frac{N° veces\, palabra\, encontrada}{N° posts}. \tag{4}$$

Table 2. Comparison of matching search algorithms

Word	Aio Shop		All Store		Monkey Shop	
	No appearance	Probability	No appearance	Probability	No appearance	Probability
Excelente	7	0.175	0	0	0	0
Recomendado	4	0.1	0	0	0	0
Seriedad	8	0.2	0	0	0	0
Bueno	23	0.575	10	0.25	9	0.225
Seguro	6	0.15	2	0.05	6	0.15
Calidad	5	0.125	0	0	0	0
Gracias	9	0.225	2	0.05	1	0.025
Satisfecho	2	0.05	0	0	0	0
Confiable	0	0	0	0	0	0
Perfecto	2	0.05	0	0	0	0
Agradable	1	0.025	0	0	0	0
Mejor	5	0.125	1	0.025	2	0.05
Extraordinario	0	0	0	0	0	0
Honesto	0	0	0	0	0	0
Encanta	0	0	5	0.125	0	0
Chévere	0	0	2	0.05	0	0
Bacano	1	0.025	4	0.1	0	0
Sin comentarios	222		180		214	

In the same way, we proceed with the negative words. Then every word will be weighed with an impact value on a 0–1 range, where 0 is a low impact value and 1 is the maximum value. Depending on the meaning it may have and the experience of the experts, i.e., the administrators of the Fan Pages, as shown in Table 3.

With the probability of occurrence and the impact values. We determine the quantitative value of each Dictionary word, by multiplying the probability by the impact. Data shown in Table 4 will help conclude the validity and reliability.

Once the Dictionary has been determined, we can interpret the results for every Fan Page under study, it will be done by a Pareto diagram, to easily visualize the results

5.2 Visualization Through Pareto Diagram

We seek for the words mostly found in the different Fan Pages, in relation to the dictionary, the process was carried out individually for positive and negative words. For the first Aio Shop page the positive words that stood out in the comments were: Bueno, Excelente, Gracias, Calidad, Satisfecho y Agradabable Fig. 2. In the same way we proceeded to tabulate the negative words for Aio Shop Fan Page, Fig. 3. In All Sore Fan Page, the results on positive words were: Bueno, Encanta, Seguro y Mejor Fig. 4.

As for the negative words mostly often found in All Store Fan Page, see Fig. 5, one of the influential words is "Estafa" (scam). Featured on Monkey Shop Fan Page results were: Seguro, Mejor y Excelente. See Fig. 6. In the Monkey Shop Fan Shop, the negative values are shown in Fig. 7.

In a global analysis made to the three fan pages we can conclude that not all the information recorded in the comments, are useful.

<table>
<tr><td colspan="4">Table 3. Impact of words</td></tr>
<tr><td>Word</td><td>Impact</td><td>Word</td><td>Impact</td></tr>
<tr><td>Excelente</td><td>0.9</td><td>Pésimo</td><td>0.8</td></tr>
<tr><td>Recomendado</td><td>0.9</td><td>Estafa</td><td>0.9</td></tr>
<tr><td>Seriedad</td><td>0.8</td><td>Malo</td><td>0.01</td></tr>
<tr><td>Bueno</td><td>0.5</td><td>Mentirosos</td><td>0.9</td></tr>
<tr><td>Seguro</td><td>0.9</td><td>Farsante</td><td>0.9</td></tr>
<tr><td>Calidad</td><td>0.5</td><td>Desconsiderado</td><td>0.7</td></tr>
<tr><td>Gracias</td><td>0.5</td><td>Desconfianza</td><td>0.9</td></tr>
<tr><td>Satisfecho</td><td>0.9</td><td>Ineficiencia</td><td>0.7</td></tr>
<tr><td>Confiable</td><td>0.7</td><td>Irresponsable</td><td>0.9</td></tr>
<tr><td>Perfecto</td><td>0.5</td><td>Terrible</td><td>0.6</td></tr>
<tr><td>Agradable</td><td>0.5</td><td>Peor</td><td>0.01</td></tr>
<tr><td>Mejor</td><td>0.5</td><td>Bloquean</td><td>0.8</td></tr>
<tr><td>Extraordinario</td><td>0.5</td><td>Defraudado</td><td>0.8</td></tr>
<tr><td>Honesto</td><td>0.7</td><td>Decepción</td><td>0.8</td></tr>
<tr><td>Encanta</td><td>0.8</td><td>Deshonestos</td><td>0.8</td></tr>
<tr><td>Chévere</td><td>1.0</td><td>Engañosa</td><td>0.6</td></tr>
<tr><td>Bacano</td><td>1.0</td><td></td><td></td></tr>
</table>

<table>
<tr><td colspan="4">Table 4. Impact of words</td></tr>
<tr><td>Word</td><td>Impact</td><td>Word</td><td>Impact</td></tr>
<tr><td>Excelente</td><td>0.9</td><td>Pésimo</td><td>0.1575</td></tr>
<tr><td>Recomendado</td><td>0.9</td><td>Estafa</td><td>0.09</td></tr>
<tr><td>Seriedad</td><td>0.8</td><td>Malo</td><td>0.16</td></tr>
<tr><td>Bueno</td><td>0.5</td><td>Mentirosos</td><td>0.2875</td></tr>
<tr><td>Seguro</td><td>0.9</td><td>Farsante</td><td>0.135</td></tr>
<tr><td>Calidad</td><td>0.5</td><td>Desconsiderado</td><td>0.0625</td></tr>
<tr><td>Gracias</td><td>0.5</td><td>Desconfianza</td><td>0.1125</td></tr>
<tr><td>Satisfecho</td><td>0.9</td><td>Ineficiencia</td><td>0.045</td></tr>
<tr><td>Confiable</td><td>0.7</td><td>Irresponsable</td><td>0</td></tr>
<tr><td>Perfecto</td><td>0.5</td><td>Terrible</td><td>0.025</td></tr>
<tr><td>Agradable</td><td>0.5</td><td>Peor</td><td>0.0125</td></tr>
<tr><td>Mejor</td><td>0.5</td><td>Bloquean</td><td>0.0625</td></tr>
<tr><td>Extraordinario</td><td>0.5</td><td>Defraudado</td><td>0</td></tr>
<tr><td>Honesto</td><td>0.7</td><td>Decepción</td><td>0</td></tr>
<tr><td>Encanta</td><td>0.8</td><td>Deshonestos</td><td>0</td></tr>
<tr><td>Chévere</td><td>1.0</td><td>Engañosa</td><td>0</td></tr>
<tr><td>Bacano</td><td>1.0</td><td></td><td></td></tr>
</table>

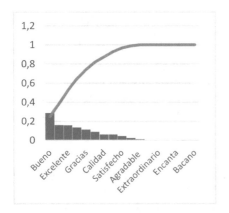

Fig. 2. Positive words for Aio Shop

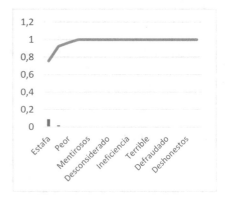

Fig. 3. Negative words for Aio Shop

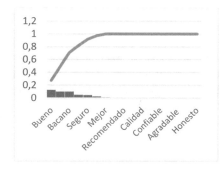

Fig. 4. Positive words for All Store

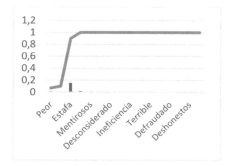

Fig. 5. Negative words for All Store

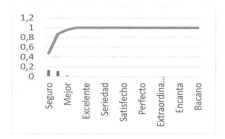

Fig. 6. Positive words for Monkey Shop

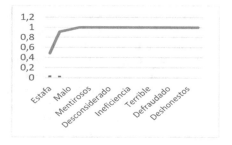

Fig. 7. Negative words for Monkey Shop

6   Conclusions and Future Work

Levenshtein's algorithm performs word comparison in an optimal way, based on words similarities. It can increase the probability to find miswritten words or typos. But it also increases the probability of finding erroneous coincidences with words that present a different grammar meaning.

Only 32.88% of the information found in the comments was considered significant, mainly due to the fact that the analyzed fan pages keep in reserve information regarding to price and products. Thus many comments are intended to request this information, instead of good (positive) or bad (negative) criticism towards products or sellers.

Using the probability and impact matrix can reduce Levenshtein's algorithm error, because according to the impact (weight) a word possesses, the word will be considered in the conclusive section.

Data mining uses too much temporal memory. In this sense Java IDEs are not efficient tools. Future work should consider the study and use of different tools and languages before the experiment setup. We could also eliminate certain words, punctuation marks, and special characters to reduce the analysis time.

To improve the process of comments comparison with each word in the dictionary, one can change the structure of the comments that are horizontally structured to a vertical direction, in this way the comparison is simpler because it is necessary to load all comments, but only one word at a time, this decreases the memory and time needed to carry out the process.

As future work, we propose the use of a much larger set of data. In this way, the predictions will be more accurate. We also intend to analyze different type of data, for example, cyber bullying comments. Additionally, we will compare the results from our algorithm with other public algorithms, e.g., a sentiment analysis tool.

References

1. INEC. Encuesta de Condiciones de vida realizada por el Instituto Nacional de Estadísticas y Censos. Quito (2014)
2. Ecommerce en Ecuador: Datos interesantes | elEmprendedor.ec, El Emprendedor, 05 July 2013
3. Haucap, J., Heimeshoff, U.: Google, Facebook, Amazon, eBay: is the internet driving competition or market monopolization? University of Düsseldorf, Düsseldorf Institute for Competition Economics (DICE), no. 83 (2013)
4. Overview - Graph API - Documentation, Facebook for Developers. https://developers.facebook.com/docs/graph-api/overview. Accessed 23 Sept 2018
5. Project Management Institute: Guía de los Fundamentos Para la Dirección de Proyectos, 5th edn. Project Management Institute (2014)
6. Virla, M.Q.: Confiabilidad y coeficiente Alpha de Cronbach, September 2018
7. Pruebas y evaluacion psicologicas: introduccion a las pruebas y a la medicion | ronald jay cohen | comprar libro 9789701057049. https://www.casadellibro.com/libro-pruebas-y-evaluacion-psicologicas-introduccion-a-las-pruebas-y-a-la-medicion/9789701057049/1094006. Accessed 23 Sept 2018
8. Mosley, R.: Social media analytics: data mining applied to insurance twitter posts. In: Casualty Actuarial Society E-Forum, Winter 2012, vol. 2 (2012)
9. Martel, W.: Determinación de niveles de agresividad en comentarios de la red social Facebook por medio de Minería de Texto (2016)
10. Bilenko, M., Mooney, R., Cohen, W., Ravikumar, P., Fienberg, S.: Adaptive name matching in information integration. IEEE Intell. Syst. 18(5), 16–23 (2003)
11. Cohen, W.W., Ravikumar, P., Fienberg, S.E.: A comparison of string distance metrics for name-matching tasks. In: Proceedings of the 2003 International Conference on Information Integration on the Web, Acapulco, Mexico, pp. 73–78 (2003)
12. Peng, T., Li, L., Kennedy, J.: A comparison of techniques for name matching. GSTF J. Comput. JoC 2(1), 55–61 (2018)
13. Miguélez, M.M.: Validez y confiabilidad en la metodología cualitativa. Paradigma 27(2), 7–33 (2016)
14. Cohen, W., Ravikumar, P., Fienberg, S.: A comparison of string metrics for matching names and records. In: KDD Workshop on Data Cleaning and Object Consolidation, vol. 3, pp. 73–78 (2003)

Stock Market Data Prediction Using Machine Learning Techniques

Edgar P. Torres P.[1(✉)], Myriam Hernández-Álvarez[1],
Edgar A. Torres Hernández[2], and Sang Guun Yoo[1] ⓘ

[1] Facultad de Ingeniería de Sistemas,
Escuela Politécnica Nacional, Quito, Ecuador
{edgar.torres,myriam.hernandez,sang.yoo}@epn.edu.ec
[2] Facultad de Ciencias Administrativas,
Pontificia Universidad Católica del Ecuador, Quito, Ecuador
etorresh777@gmail.com

Abstract. This paper studies the possibilities of making prediction of stock market prices using historical data and machine learning algorithms. We have experimented with stock market data of the Apple Inc. using random trees and multilayer perceptron algorithms to perform the predictions of closing prices. An accuracy analysis was also conducted to determine how useful can these types of supervised machine learning algorithms could be in the financial field. These types of studies could also be researched with data from the Ecuadorian stock market exchanges i.e. Bolsa de Valores de Quito (BVQ) and Bolsa de Valores de Guayaquil (BVG) to evaluate the effectiveness of the algorithms in less liquid markets and possibly help reduce inefficiency costs for market participants and stakeholders.

Keywords: Machine learning · Stock market · Artificial intelligence · Prediction

1 Introduction

This paper explores the possibility of using computers as tools to automatically process data gathered from financial activities and extract relevant information to achieve the goals of the participants involved in this field. The objective is to perform an automatic process of data using artificial intelligence techniques, in particular the application of machine learning algorithms. The final goal is to manage the extraction of quantitative data with relevant information from the stock market. For the purpose of this paper, we have used the historical data of opening price, closing price, highest price, lowest price, and volume traded of Apple Inc. stocks which has been gathered using Google Finance.

We think that the price of a stock incorporates all the relevant data related to the whole process of supply and demand of the stock market. This process of price discovery in the markets leaves traces in its historical prices.

© Springer Nature Switzerland AG 2019
Á. Rocha et al. (Eds.): ICITS 2019, AISC 918, pp. 539–547, 2019.
https://doi.org/10.1007/978-3-030-11890-7_52

This research is meant to be a first approximation to this subject by using machine learning techniques to process stock market data that are readily quantifiable. Furthermore, it is possible to perform predictions about the possible variations of Apple Inc.'s (NASDAQ: AAPL) stock through a historical price data set that includes different types of prices and their fluctuations while noting the impact of the traded volume for each market session. Finally, our goal is to evaluate how accurate these predictions are and measure its error.

The rest of the present paper is organized as follows. Section 2 explains some basic concepts and previous works related to the present paper. Then, Sect. 3 describes the details of the stock market price prediction activities. Finally, Sect. 4 delivers the conclusions of this work.

2   Background

2.1   Basic Concepts

Before proceeding with the stock market prices analysis, it's important to define specific concepts used in the present research:

- Data mining represents the use of automated techniques to analyze data to discover relationships between different variables in the data.
- Machine learning or automated learning is an Artificial Intelligence branch that develops techniques that allow computers to learn.
- Supervised Learning is a machine learning technique that generates a model from a training data set that is capable of predicting variables given input.
- Unsupervised Learning is a machine learning technique where models are formed by fitting it to the dataset used.
- Feed Forward Neural Network is an artificial neuronal network where connections between units don't form a cycle.
- WEKA is a collection of machine learning algorithms that can perform data mining operations.

2.2   Previous Works

Great effort has been placed to analyze financial data since the economic incentive is considerable for participants. Because of this reason, there are several researches about predictions and future tendencies of stock market prices [1]; some of them uses machine learning techniques such as Support Vector Machine (SVM) [3, 4] and others use text analysis to understand emotions and other quantifiable data from relevant financial information sources [1, 2, 4]. These types of techniques extract information regarding the participant's emotions and feelings about a particular subject, in this case, the stock market data. Generally speaking, it's possible to extract the data from multiple sources, like highs or lows in prices with a notable traded volume, but also financial statements, press releases, and other information from the company [2]. Another source for this type of information can be microblogs in social networks [3], and in particular stock tweets [3, 4], although this kind of analysis falls outside the scope of this article. In conclusion, there are many possibilities to trace the emotions and feelings of stock market participants. Finally, other approaches have tested this type of information from local data [2], expert data and other participants [3], and systematic processing of lexicographic, syntactic and natural language for sentimental analysis. [5].

2.3 Weka

Weka which means Waikato Environment for Knowledge Analysis is a free machine learning software developed at University of Waikato in New Zealand. Weka includes a collection of tools and algorithms for data analysis and prediction, and it is widely used for research [6, 7].

There is plenty of available information about WEKA's algorithms capabilities and usage for financial data through machine learning techniques. These sources are available in WEKA's official website and also other sites that offer tutorials on how to use its software, as well as papers researching the mathematical and technical aspects of WEKA's algorithms. [8–12].

WEKA's classifiers provides different models to predict nominal and numerical quantities. Among others, WEKA offers the following algorithms: Decision trees and lists, Classifiers based on instances, Support Vector Machines (SVM), Multilayer Perceptron, Logistical Regression, Bayesian Networks, and "Meta" Classifiers.

3 Stock Market Data Prediction Using Machine Learning Algorithms

3.1 Generalities

The objective of this research is to forecast or predict future closing prices of Apple Inc.'s (NASDAQ: AAPL) stock. To this end, we have used historical price data of the Open, Close, High, Low and Volume of the last 250 trading sessions. We obtained this information from Google Finance.

An important supposition for this research is that the different historical price types and their traded volume leave traces of the price discovery process. By extension, this method relies heavily on the emotions, feelings, and expectations of participants, some of which are often irrational and purely speculative. Hence, price jumps or crashes with heavy or light traded volume should have significance, and these subtle relations can be modeled by the application of machine learning techniques as previously mentioned. In this way, a prediction of the following closing prices can be made with a certain degree of acceptable accuracy. We've chosen to predict the variable Close because it's the last price at which all participants agreed for the financial security in the traded session.

Much has been written about daily price fluctuation in stock market prices, often suggesting that these movements resemble a Random Walk [13], or a Brownian motion [14], among others. Nevertheless, it is accepted in finance that stocks have had a historically positive Baseline Drift which can be attributed to prices incorporating the growth of earnings and equity publicly traded companies. In this aspect, since in the last 250 stock market sessions, Apple Inc. has released multiple quarterly reports, it stands to reason that these results should be reflected in its stock price. On the other hand, the day to day trading of the stocks will provide with different types of data that machine learning algorithms should be able to model. We expect that by using this information, we can reduce the algorithms exposure to random noise and improve its accuracy for forecasting future closing session prices.

3.2   Algorithm Selection and Universe to Be Analyzed

The present work has used the WEKA software to execute the machine learning algorithms in our stock market data sets. As mentioned previously, we have obtained the data from Google Finance which contains the Open, High, Low and Close prices of Apple Inc. stock and the volume of trades of the last 250 sessions. Regarding to the machine learning algorithms, we have used the following WEKA packages:

- weka.classifiers.trees.RandomTree
- weka.classifiers.functions.Multilayer Perceptron

We tested both of these algorithms on the same dataset, i.e., the historical price values of the last 250 trading sessions of Apple Inc.'s (NASDAQ: AAPL) stock.

As previously mentioned, we considered six attributes for the algorithms: (1) Date, (2) Open, (3) High, (4) Low, (5) Close, and (6) Volume.

- Date: Market session date
- Open: The opening price for the market session
- High: The highest price reached during the market session
- Low: The lowest price reached during the market session
- Close: The closing price for the market session
- Volume: The total number of trades performed on the stock during the market session

The attribute Close is the one to be forecasted and compared to its real data, so that the accuracy of the algorithms can be tested and measured for errors and over fitting. Figures 1 and 2 present the results of the executions performed using Random Tree and Multilayer Perceptron algorithms.

Algorithm: weka.classifiers.trees.RandomTree

=== Run information ===

Scheme: weka.classifiers.trees.RandomTree -K 0 -M 1.0 -V 0.001 -S 1
Relation: prediccion 2-sep
Instances: 251
Attributes: 6
             Date
             Open
             High
             Low
             Close
             Volume
Test mode: evaluate on training data

=== Classifier model (full training set) ===

RandomTree
==========

High < 104.4
| High < 98.94
| | Open < 94.72
| | | High < 94.07
...
| | | High >= 123.09
| | | | Date < 41.5 : 122.57 (1/0)
| | | | Date >= 41.5 : 122 (1/0)

Size of the tree : 225

Time taken to build model: 0 seconds

=== Predictions on training set ===
  inst#   actual predicted   error
     1   112.31   112.453   0.143
     2   110.15   110   -0.15
     3   112.57   112.453   -0.117
...
   250   106.73   106.73   -0
   251   ?   105.024   ?

=== Evaluation on training set ===

Time taken to test model on training data: 0.08 seconds

=== Summary ===
Correlation coefficient   0.9998
Mean absolute error   0.1151
Root mean squared error   0.16
Relative absolute error   1.6105 %
Root relative squared error   1.9553 %
Total Number of Instances   250
Ignored Class Unknown Instances   1

Fig. 1. Execution of Random Tree

Algorithm: weka.classifiers.functions.MultilayerPerceptron

=== Run information ===
Scheme: weka.classifiers.functions.MultilayerPerceptron -L 0.3 -M 0.2 -N 500 -V 0 -
S 0 -E 20 -H a
Relation: prediccion 2-sep
Instances: 251
Attributes: 6
    Date
    Open
    High
    Low
    Close
    Volume
Test mode: evaluate on training data

=== Classifier model (full training set) ===

Linear Node 0
    Inputs   Weights
    Threshold 0.16407389183567725
    Node 1   -0.7409666964660168
    Node 2   1.7171258338000799
    Node 3   -1.371651766795869
Sigmoid Node 1
    Inputs   Weights
    Threshold -1.225106292966126
    Attrib Date   0.01979747908695488
    Attrib Open 0.4749210034034923
    Attrib High -0.5455483243023302
    Attrib Low   -0.7448374160532633
    Attrib Volume   0.5265070532721522
Sigmoid Node 2
    Inputs   Weights
    Threshold -1.4617941741850469
    Attrib Date   0.006059752397296158
    Attrib Open -1.3715046053801243
    Attrib High 1.6670130380975303
    Attrib Low   1.4209018925134493
    Attrib Volume   0.11134270375207257
Sigmoid Node 3
    Inputs   Weights
    Threshold -1.3965598781124515
    Attrib Date   0.05273749507771454
    Attrib Open 0.7651119521019094
    Attrib High -1.1018676853352654
    Attrib Low   -1.5648480563559939
    Attrib Volume   -0.23675701241891794
Class
    Input
    Node 0

Time taken to build model: 0.1 seconds

=== Predictions on training set ===

  inst#   actual predicted   error
    1   112.31   111.523   -0.787
    2   110.15   111.099   0.949
...

  250   106.73   106.161   -0.569
  251   ?   107.005   ?

=== Evaluation on training set ===

Time taken to test model on training data: 0.15 seconds

=== Summary ===

Correlation coefficient   0.9976
Mean absolute error   0.4529
Root mean squared error   0.6058
Relative absolute error   6.3376 %
Root relative squared error   7.4024 %
Total Number of Instances   250
Ignored Class Unknown Instances   1

Fig. 2. Execution of Multilayer Perceptron algorithm

3.3   Results Evaluation and Analysis

Both executions of WEKA's algorithms fit the actual historical Price data (Correlation factor of 0.9998 for the first one and 0.9976 for the second with a maximum adjustment possible of 1.0) very closely and errors are tolerable (mean absolute error of 0.1151 for the first one and 0.4529 for the second). Hence, it is reasonable to conclude that the predictions using these two algorithms are suitable and acceptable for the application. Table 1 shows the details of the comparison of the results and Table 2 shows a portion of the historical data used for analysis.

Table 1. Comparison of algorithms

No.	Attributes	Algorithm 1	Algorithm 2
1	Correlation coefficient	0.9998	0.9976
2	Mean absolute error	0.1151	0.4529
3	Root mean squared error	0.16	0.6058
4	Relative absolute error	1.6105%	6.3376%
5	Root relative squared error	1.9553%	7.4024%
6	Total number of instances	250	250

Table 2. Examples of historical data

Date	Open	High	Low	Close	Volume
1	108.59	109.32	108.53	108.85	21257669
2	108.86	109.1	107.85	108.51	25820230
3	108.77	109.69	108.36	109.36	25368072
4	109.23	109.6	109.02	109.08	21984703
5	109.1	109.37	108.34	109.22	25355976
6	109.63	110.23	109.21	109.38	33794448
...	...	...	...	...	...
245	112.49	112.78	110.04	110.37	52906410
246	110.23	112.34	109.13	112.34	61520170
247	110.15	111.88	107.36	107.72	75988194
248	112.03	114.53	112	112.76	55962842
249	112.17	113.31	111.54	113.29	52896384
250	112.23	113.24	110.02	112.92	83265146

Table 3. Example of forecasted data

Date	Actual close	Predicted close	Error
251	105.17	105.024	0.146

Based on the previous analysis, it is possible to conclude that the first algorithm has a higher correlation coefficient and a lower error than the second one. Although, both algorithms present a good performance. Finally, Table 3 presents the forecast for the Close attribute for a test set using the Random Tree Algorithm (algorithm with better result).

4 Conclusions

The present work has shown how it is possible to perform forecasts and predictions of future stock market data using artificial intelligence techniques, specifically machine learning algorithms. The application of WEKA proved to be very valuable for this purpose and its use could be further researched in the financial field. This tool counts with many different algorithms which could be used for various types of economic data that could provide interesting insights to market participants and society. Additionally, in future works, it could be possible to quantify and analyze emotions and feelings expressed in a text through blogs, stock tweets, or other mediums, to increase the effectiveness of the predictions. Finally, we recommend that the Ecuadorian stock exchanges (BVQ and BVG) make this type of information (open, high, low, close, volume and otherwise data concerning the sentiment of market participants) public and readily available, through free online channels if possible. This transparency of information would facilitate the application machine learning algorithms and artificial intelligence to Ecuadorian financial securities to further research their application in other markets, which may help reduce costs of market inefficiencies. Ecuadorian government should pass laws to make transparent stock market information. This would constitute a valuable innovation for Ecuadorian stock market.

References

1. Schumacher, R.P., Chen, H.: Textual analysis of stock market prediction using breaking financial news: the AZFin text system. ACM Trans. Inf. Syst. **27**, 1–19 (2009)
2. Giannini, R.C., Irvine, P.J., Shu, T.: Do local investors know more? A direct examination of individual investors' information set. Working paper (2014)
3. Bar-Haim, R., Dinur, E., Feldman, R., Fresko, M., Goldstein, G.: Identifying and following expert investors in stock microblogs. In: Proceedings of the Conference on Empirical Methods in Natural Language Processing, pp. 1310–1319 (2011)
4. Plakandaras, V., Papadimitriou, T., Gogas, P., Diamantaras, K.: Market sentiment and exchange rate directional forecasting. Algorithmic Financ. **4**(1–2), 69–79 (2015)
5. Ruiz-Martínez, J.M., Valencia-García, R., García-Sánchez, F.: Semantic-based sentiment analysis in financial news. In Proceedings of the 1st International Workshop on Finance and Economics on the Semantic Web, pp. 38–51 (2012)
6. Kumar, A., Malik, H., Chandel, S.S.: Selection of most relevant input parameters using WEKA for artificial neural network based solar radiation prediction models. Renew. Sustain. Energy Rev. **31**, 509–519 (2014)

7. Kalmegh, S.: Analysis of WEKA data mining algorithm REPTree, Simple Cart and RandomTree for classification of indian new. Int. J. Innov. Sci. Eng. Technol. **2**(2), 438–446 (2015)
8. Teuvo, K.: Self-Organizing Maps. Springer Science & Business Media, Heidelberg (2001)
9. Russell, S., Norvig, P.: Artificial Intelligence: A Modern Approach. Prentice-Hall, Upper Saddle River (1995)
10. Aha, D.W.: Tolerating noisy, irrelevant, and novel attributes in instance based learning algorithms. Int. J. Man Mach. Stud. **36**(2), 267–287 (1992)
11. Wettschereck, D., Aha, D.W., Mohri, T.: A review and empirical evaluation of feature weighting methods for a class of lazy learning algorithms. Artif. Intell. Rev. **11**(1–5), 273–314 (1997)
12. Hornik, K., Buchta, C., Zeileis, A.: Open-source machine learning: R meets weka. Comput. Stat. **24**(2), 225–232 (2009)
13. Fama, E.F.: Random walks in sotck market prices. Financ. Anal. J. **51**(1), 75–80 (1995)
14. Osborne, M.F.M.: Brownian motion in the stock market. Oper. Res. **7**(2), 145–173 (1959)

A Comparison of Machine Learning Methods Applicable to Healthcare Claims Fraud Detection

Nnaemeka Obodoekwe and Dustin Terence van der Haar[(⊠)]

Academy of Computer Science and Software Engineering, University of Johannesburg, Cnr Kingsway and University Road, Johannesburg 2006, Gauteng, South Africa
nnaemekaobodo@gmail.com, dvanderhaar@uj.ac.za

Abstract. The healthcare industry has become a very important pillar in the modern society but has witnessed an increase in fraudulent activities. Traditional fraud detection methods have been used to detect potential fraud, but for certain cases they have been insufficient and time consuming. Data mining which has emerged as a very important process in knowledge discovery has been successfully applied in the health insurance claims fraud detection. We implemented a prototype that comprised different methods and a comparison of each of the methods was carried out to determine which method is most suited for the Medicare dataset. We found that while ensemble methods and neural net performed, the logistic regression and the naive bayes model did not perform well as depicted in the result.

Keywords: Healthcare · Fraud detection · Machine learning

1 Introduction

The health insurance industry, a pillar in the modern-day society, that serves the purpose of providing affordable healthcare to individuals. Healthcare has become a necessity for households, hence the cost of healthcare forms a part of household expenditure. The increased cost of healthcare has made it a luxury rather than a basic need [1]. One of the reasons for the increased cost of health insurance can be attributed to the money lost through fraud in the healthcare system [2].

Fraud has impacted several aspects of life and the healthcare industry is no exception. Fraud occurs in the medical billing process leading to loss of funds by the insurance company which leads to the insurance company charging higher premiums to make up for the lost funds. The impact of fraud in healthcare has other implications aside from the monetary implications such as health risks that can arise from the altering of a patient's record by the physician [3].

Traditional fraud detection methods such as rule based statistical methods have been applied to detect possible fraud in the healthcare claims process, but

© Springer Nature Switzerland AG 2019
Á. Rocha et al. (Eds.): ICITS 2019, AISC 918, pp. 548–557, 2019.
https://doi.org/10.1007/978-3-030-11890-7_53

these methods no longer suffice due to the large number of claims to be processed and the variety of patterns these fraudulent activities take. Machine learning methods have been applied to other fraud detection problems such as credit card fraud detection and has also been applied to fraud detection in healthcare claims [4].

In the research study, we unpack the problem of healthcare claims fraud detection as well as the impacts it has. We then analyse how machine learning has been applied to the healthcare fraud claims detection problem by discussing the similar systems. With an understanding of the current research being done in the area, we create a data mining model that takes an exploratory approach to solving the problem of healthcare claims fraud detection by comparing and analysing different methods. We implement these methods in a prototype and then analyse the results to see which methods performed best with the Medicare dataset.

2   Problem Background

Abraham Maslow states the physiological needs of any individual are the most basic innate human needs that need to be satisfied and has the highest priority [5]. Maintaining a healthy body condition, eating, basic security is tantamount to satisfying the safety needs of an individual. To have good health, one needs access to adequate and affordable healthcare. Unfortunately, the cost of healthcare has been on the rise, making healthcare more of a luxury than a basic need [1]. One of the factors that have contributed to the increased cost of healthcare is the impact of the funds lost to fraudsters through healthcare claims fraud.

Before going deeper into the depth of health insurance, a definition of health insurance is needed. Health insurance represents a contract that a person pays an agreed premium to an insurance provider for a designated healthcare cover. The health insurance industry involves the transfer of funds and has been affected by fraudulent activities perpetrated by individuals that seek to gain illegal access to these funds.

Health insurance waste in healthcare is most times unrelated to fraud as it mainly the provision of unnecessary health services. Health insurance waste can only be seen as fraud and abuse when the act is intentional. Waste can occur when services are over utilized and then results in unnecessary expenditure [6].

Health insurance abuse is the billings of practices that either directly or indirectly is not consistent with the goals of providing patients with services that are medically necessary and these practices meet professionally recognized standards as well as being fairly priced [6].

Health insurance fraud is purposely billing for services that were never performed and or supplies not provided, medically unnecessary services and altering claims to receive higher reimbursement than the service produced [6].

To tackle the problem of fraud in the medical billing process, health insurance companies make use of traditional rule-base models, but these models do not suffice anymore due to several factors such as the large volume of claims to be processed which makes the medical billing process prone to error, slow and sometimes inefficient. Machine learning methods can be used to improve the detection of possible healthcare claims fraud. The next section discusses the related works on the applications of machine learning methods in the detection of possible fraudulent healthcare claims.

3   Review of Related Work

Machine learning methods have been effective in automatically extracting patterns from data to derive knowledge which yields meaningful results such as detecting which submitted claims are likely fraudulent. The first work we consider is the Outlier-based health insurance fraud detection for US Medicaid data presented by Thornton et al. [7]. They made use of Medicaid data for dental services which is a healthcare provider in the US that caters to low income people. Their model made use of 3 different univariate machine learning methods which are the linear regression, time series plot as well as box plot. They also used a multivariate method through clustering to detect possible health insurance fraud. The dataset used contained a case study for 500 dentists and they successfully identified 17 activities that can be deemed fraudulent among the 360 records analysed.

We also reviewed "Graph Analytics for Healthcare Fraud Risk Estimation" by Branting et al. which made use of a graph to link providers, drug prescriptions and the procedures [8]. They used two algorithms where the first algorithm was used to calculate the similarity to predetermined fraudulent and non-fraudulent providers while the second algorithm calculates the estimated fraud risk through location of practitioners. They achieved an F-score of 0.919 and an impressive AUC of 0.960.

Bauder et al. also carried out several works in the area of detecting fraud in the health insurance process. One of the systems, a multivariate outlier detection in Medicare claims payments applying probabilistic programming methods [9]. They created a base for what the expected Medicare payments should look like for each type of provider. Outliers were then identified by comparing payment amounts with the normative case and the deviations are categorised as outliers.

4   Experimental Setup

To establish a way to conduct the research, we define a research methodology that form the guide to solving the problem at hand. The quantitative research approach was chosen as it allows for the statistical analysis of the data and maintains an objective standpoint.

The data that was used for the study is the Medicare payments data between 2012–2015. The dataset contained payments and utilization healthcare claims data as well as the details about the procedures rendered to individuals. The data from the List of Excluded Individual or Entities (LEIE) was used to create the ground truth labels.

5   Model

Applying a machine learning algorithm to derive knowledge is just a piece in the puzzle of creating an effective data mining model. The data mining model consists of different processes and each of these processes play a major role in deriving knowledge from the data. In this section, we unpack these individual processes as well as the methods that were used for each individual process.

5.1   The Data Collection Phase

Data is the raw material to be processed in a data mining system. Data can be structured, unstructured or semi-structured. The Medicare dataset used, was in a structured format. Both datasets were loaded and stored in a database for further analysis.

5.2   The Data Pre-processing and Transformation Phase

Normally, data in the real world is dirty, contains missing values and can be incorrect. Therefore, a lot of work to be done cleaning up the data to get it to the form that will be suitable for use. The Medicare data used was already pre-processed by the Centre for Medical Services (CMS). The work first task we performed was filtering the Medicare dataset for only non-prescription data. The Medicare dataset did not contain any label to be used to differentiate between fraudulent and non-fraudulent claims, therefore we used the LEIE dataset to flag the claims that were detected as fraudulent. We made use of fuzzy matching on the practitioner's first name, last name and ZIP code to link the Medicare payment data to a practitioner in the LEIE dataset, as there was no explicit join between the two datasets.

After labelling the dataset and identifying ground truth labels, the next task performed was indexing categorical string features. We based the initial selection of features on the work done by [10] using the same Medicare dataset as shown in Table 1. We also used feedback gotten from the model as calculated by the feature importance to adjust the features used for the model.

Finally, we applied a 70:30 split to the dataset. 70 % of the data was used for training the machine learning model while the remaining 30 % was used for testing the outcome of the model.

5.3   Machine Learning Application

Now that we have completed the pre-processing and transformation of the data, we apply machine learning algorithm to derive insights from the data. We used an exploratory approach in creating the machine learning model as several methods were used and results were collected on the performance of the different methods used.

Table 1. Description of medicare features.

Feature	Description
NPI	Unique provider identification number
last_name	Provider's last name
First_name	Provider's first name
Zip	Provider's 5-digit zip code
provider_type	Medical provider's specialty (or practice)
line_srvc_cnt	Number of procedures performed per provider
bene_unique_cnt	Number of distinct beneficiaries per day services
average_submitted_chrg_amt	Average of the charges that the provider submitted
average_medicare_payment_amt	Amount paid to the provider for services performed

1. The Bayesian classifier is a simple classifier based on statistical methods and it assigns probabilities to each member of the class in such a way that a given sample can be categorised into one class. It makes a strong assumption on the independence of features.
2. Random forest classifier, an ensemble learning method made up of a sequential construct of several decision trees in the training phase and outputting the class that is the mode of the classes (classification) or mean prediction (regression) of the individual trees.
3. Logistic regression is another machine learning method based on statistical principles. It is highly suited for predicting categorical features. In predicting a binary outcome, the logistic regression model makes use of the binomial logistic regression and for multiple outcomes it makes use of multinomial logistic regression.
4. Gradient Boosted tree classifier is yet another powerful classification method. The classification method uses ensembles of decision trees and applies the technique known as boosting to improve performance. The idea of boosting emanates from the attempt to combine weaker learners to become better learners. The gradient boosted tree classifier is a combination of a loss function, a weak learner and the additive function responsible for combining the weak learners and reducing the loss function.
5. The artificial neural network is a machine learning method based off the functioning of the neurons in the brain of a biological system. It is made up of a network of interconnected nodes. The nodes do not contain any computation but rather, they function as a group of linear functions. The nodes in the neural network are grouped into layers. The behaviour of each node is defined by an activation function.

6   Results

Once the data has been passed through to the machine learning process, we evaluate how well the model performed by using the following pre-determined benchmarks. We used the following metrics to evaluate the different machine learning models created: Weighted Precision, Weighted Recall, AUC, Test Error, Sensitivity, Specificity, F1. In this section, we unpack how each of the implementations performed against the defined metrics.

Table 2. Result for the performance metric of the different machine learning methods.

ML model	Weighted precision	Weighted recall	AUC	Test error	Sensitivity	Specificity	F1
Naive Bayes	73.7	82.1	47.0	17.9	11.6	99.4	74.6
GBT	**93.3**	**93.3**	**97.0**	**6.7**	**73**	98	**93**
Random forest	88.6	88.61	92	11.4	41.3	**98.8**	86.9
Logistic regression	63.5	82.3	63.5	17.7	45.2	93.0	74.3
Neural network	89	90	93.8	10	71.2	94.8	90

6.1   Naive Bayes

We implemented the Bayesian model using the Apache SparkML library using multinomial distribution of features and a smoothing of 1.0. The model performed poorly with a ROC curve of 47.0 as seen in the figure below, which entails that the predictions of the model was just as good as random guesses. The Naive Bayes model scored well in the other metrics and also recorded a low test error of 17.9%.

6.2   Logistic Regressions

The logistic regression model was a slight improvement to the Naive Bayes model. The logistic regression model was created with a regularization parameter of 0.3 and the elastic net parameter was set at 0.8 based on the recommendation of the SparkML documentation. The logistic regression model was run over several iterations and achieved the best result at 10 iterations. An improved AUC of 63.5% was achieved as seen in Fig. 2, which is better than the Bayesian model but not good enough for predictions. The model achieved an accuracy of 83.7% (Fig. 1).

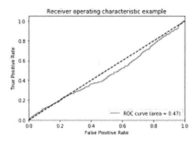

Fig. 1. ROC curve for the Naive Bayes classification model

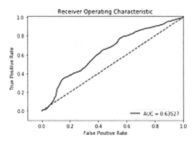

Fig. 2. ROC curve for the logistic regression classification model

6.3   Random Forest Classifier

The random forest classifier was made up of 10 decision trees and a maximum of one hundred bins. The random forest offered a slight improvement to the logistic regression with an AUC of 63.5%. The random forest classifier had a low specificity score of 41.3% and an accuracy of 89.6%. The AUC score of 92.0% makes the random forest classifier suitable for the Medicare dataset as the score implies a great improvement to random guess noticed in the previous models.

6.4   Gradient Boosted Tree Classifier

The gradient boosted tree classifier model presented a great improvement to the previous models. The model was run through 10 iterations. The gradient boosted tree classifier had a weighed precision and recall of 93.3% each. It had an accuracy of 93.3% and an F1 score of 93 %. It recorded a sensitivity score of 73% and a specificity of 98%. The most important improvement was the noticeable increase in the AUC score of 97.0%. Based on the results the gradient boosted tree classifier will be ideal for detecting possible fraud in healthcare claim using the Medicare data (Figs. 3 and 4).

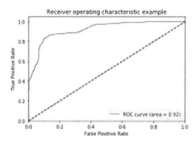

Fig. 3. ROC curve for the random forest classification model

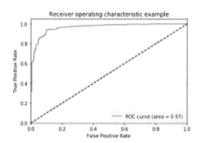

Fig. 4. ROC curve for the gradient boosted tree classification model

6.5 Artificial Neural Net

The artificial neural network slightly underperformed the gradient boosted tree classifier as can be seen in the metrics in Table 2. The model made use of the binary cross-entropy as the loss function. The model had a weighted precision and a weighted recall of 89% and 90% respectively. It also had an accuracy of 90% with specificity and sensitivity measuring 71.2% and 94.8% respectively. The F1 score was 90% and it also had an AUC of 93.8%.

7 Discussion

We have shown how the proposed model functions as well the results that were derived from the model. There are several limitations associated with the model which can act as a hindrance in the implementation of the model. The following limitations were evident through the implementation of the model.

7.1 Critique

Lack of Labelled Data. There was difficulty in obtaining openly available data with ground truth label. We made use of data from the LEIE database for ground truth label. The availability and use of a dataset that has been labelled initially would be a better validation of the model (Fig. 5).

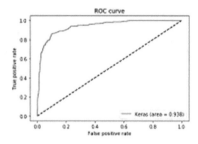

Fig. 5. ROC curve for the artificial neural network classification model

Sample Dataset. The data that was used in the model was only concerned with the healthcare provider. Although fraud mainly occurs through the healthcare provider, it doesn't imply that fraud cannot occur through the other entities in the health insurance ecosystem. The dataset we used constrain the research to only consider fraud from the medical practitioner.

7.2   Support

Applications of Alternative Machine Learning Methods. The choice of the machine learning methods to be used in a data mining system is important. The approach taken in the study is an exploratory approach, applying different types of machine learning methods and the assessing them against the benchmark. The use of multiple machine learning methods allowed for a better analysis of how each method performs with the Medicare dataset as well as the different advantages and disadvantages of each model.

Flexibility of Solution. The model is structured in a modular manner which allows for a flexible implementation of a healthcare claims fraud detection system. The modular implementation of the system means that the different components of the system can be easily replaced thereby reducing the burdens of future system upgrades.

8   Conclusion

The task of building a system that enables the identification of possible fraudulent healthcare claims has become very important as the demand for healthcare increases. The medical billing process, due to its complexities and the volume of claims to be processed has been exploited by fraudsters looking to gain illegal advantage from the system. Machine learning methods have enabled the identification of these fraudulent claims and have improved the effectiveness of the medical billing process.

The study explored the application of different machine learning methods to detect fraudulent claims in the medical billing process. The result generated from the application of these machine learning methods were collected. The analysis of the result showed that the ensemble methods and the artificial neural network performed best with the Medicare dataset.

Through the insights gained from the study, we were able to identify the strengths and weaknesses of the model. The Medicare dataset limited the system to only identify fraud that occurs through the medical practitioner as the Medicare payment data only contained data regarding the medical practitioner. Notwithstanding the limitations, the success of the model in identifying the fraudulent healthcare claims as well as the explorative approach taken to determine the which machine learning method makes this a viable solution.

References

1. Yu, H.: Impacts of rising health care costs on families with employment-based private insurance: a national analysis with state fixed effects. Health Serv. Res. **47**(5), 2012–2030 (2012)
2. Singh, A.: Fraud in insurance on rise. Technical report, Ernst & Young (2011)
3. Davis, L.E.: Growing health care fraud drastically affects all of us, October 2017
4. Rabiul, J., Nabeel, M., Ahsan, H., Sifat, M.: An evaluation of data processing solutions considering preprocessing and "special" features. In: 11th International Conference on Signal-Image Technology & Internet-Based Systems (2015)
5. McLeod, S.: Maslow's hierarchy of needs. Simply Psychol. **1** (2007)
6. Bush, J., Sandridge, L., Treadway, C., Vance, K., Coustasse, A.: Medicare fraud, waste and abuse. In: Business and Health Administration Association Annual Conference (2017)
7. Thornton, D., van Capelleveen, G., Poel, M., van Hillegersberg, J., Mueller, R.M.: Outlier-based health insurance fraud detection for U.S. medicaid data. In: 16th International Conference on Enterprise Information Systems (2014)
8. Branting, L.K., Reeder, F., Gold, J., Champney, T.: Graph analytics for healthcare fraud risk estimation. In: Advances in Social Networks Analysis and Mining (ASONAM) (2016)
9. Bauder, R.A., Khoshgoftaar, T.M.: A probabilistic programming approach for outlier detection in healthcare claims. 15th IEEE International Conference on Machine Learning and Applications (ICMLA) (2016)
10. Bauder, R.A., Khoshgoftaar, T.M.: Medicare fraud detection using machine learning methods. In: 16th IEEE International Conference on Machine Learning and Applications (2017)

Predicting Death and Morbidity in Perforated Peptic Ulcer

Hugo Peixoto[1], Lara Correia e Silva[2], Soraia Pereira[2], Tiago Jesus[2],
Vítor Lopes[3], António Abelha[1], and José Machado[1(✉)]

[1] Algoritmi Research Center, University of Minho, 4710 Braga, Portugal
{hpeixoto,abelha,jmac}@di.uminho.pt
[2] University of Minho, Campus de Gualtar, 4710 Braga, Portugal
{a74923,a74713,a73787}@alunos.uminho.pt
[3] Centro Hospitalar do Tâmega e Sousa, 4564 Penafiel, Portugal
72777@chts.min-saude.pt

Abstract. Peptic ulcers are defined as defects in the gastrointestinal mucosa that extend through the muscularis mucosae. Although not being the most common complication, perforations stand out as being the complication with the highest mortality rate. To predict the probability of mortality, several scoring systems based on clinical and biochemical parameters, such as the Boey and PULP scoring system have been developed. This article explores, using data mining in the medical data available, how the scoring systems perform when trying to predict mortality and patients' state complication. We also try to conclude, from the two scoring systems presented, which predicts better the situations described. Regarding the results, we concluded that the PULP scoring allows a better mortality prediction achieving, in this case, above 90% accuracy, however, the results may be inconclusive due to the lack of patients who died in the dataset used. Regarding the complications, we concluded that, on the other hand, the Boey system achieves better results leading to a better prediction when it comes to predicting patients' state complication.

Keywords: Data mining · Dataset · Scoring systems · PULP · Boey · Death · Health complications

1 Introduction

Peptic ulcers are defined as defects in the gastrointestinal mucosa that extend through the muscularis mucosae [1]. Complications of peptic ulcer disease include bleeding, perforation, penetration, and gastric outlet obstruction. Its incidence varies geographically – in developed countries, hemorrhage is the most common cause (up to 73%), followed by perforation (9%) and obstruction (3%). Although not being the most common complication, perforations stand out as being the complication with the highest mortality rate [2]. Different numbers present in countries categorized as developing countries, as a review from Nigeria demonstrates, with obstruction being the most common cause of complication (56%), followed by perforation (30%) and bleeding (10%) [3]. Pieces of evidence implicates that the primary causes of complicated peptic ulcer disease are H. pylori infection and nonsteroidal inflammatory drugs. This work is based on a dataset that evaluates several clinical and biochemical parameters in patients with the diagnosis of peptic ulcer perforation, in order to classificate them in two scoring systems – Peptic Ulcer Perforation (PULP) [5] and Boey [6] – previously established to determine a patient's prognosis with this pathology.

The main objective of this work is to establish a relation between the scoring systems (PULP and Boey) in patients with perforated peptic ulcer and the outcome (mortality and morbidity) [4]. In order to achieve all the goals proposed was necessary to follow a detailed process that allows you to analyze large amounts of data. This process is entitled "Data Mining". To easily understand the method used, first will be explored, superficially, the concept of Data Mining and the main features of all the process. Then, we will explore, briefly, some of the work already done in this area. Only after that, the Data Mining process will effectively start, passing through all the five phases that constitute the process. In the end, it will be done an extensive analysis of the results and drawn some conclusions relevant to the propose.

2   Methodologies, Materials and Methods - Data Mining

Data mining is the process of extracting knowledge that matters from massive amounts of data. The idea is to build computer programs that sift through databases automatically, seeking patterns. If found, these regularities will likely generalize to make accurate predictions or descriptions on future data. It is an interdisciplinary field with contributions from many areas, such as statistics, machine learning, information retrieval, pattern recognition and bioinformatics. The main techniques for data mining include classification and prediction, clustering, outlier detection, association rules, sequence analysis, time series analysis and text mining, and some new techniques such as social network analysis and sentiment analysis. It should also be noted that this process can be divided into two major categories, different from each other, the supervised and the unsupervised methods. The major difference between them is the final goal sought with of the application of the method. While the supervised search the achievement of a value, requiring a target attribute, the unsupervised search for relations between the data, such as affinities and intrinsic structures [7–9].

Six Data Mining techniques were used to induce the data mining models: JRip, OneR, NB, DT, J48 and K-Means. JRip is one of the basic and most popular algorithms. It implements a propositional rule learner, repeated incremental pruning to produce error reduction (RIPPER). It as a building state (formed by a growing phase and a pruning phase) and an optimization stage [10, 11]. In its turn, OneR (short for One Rule) is a class for building a OneR classifier: it is a simple classification algorithm that uses a one-level decision tree. In other words, it uses the minimum-error attribute for prediction, discretizing numeric attributes [12]. Then, the NB, stands for Naive Bayes, it is a class for a Naive Bayes classifier that uses estimator classes. Numeric estimator precision values are chosen based on analysis of the training data. The NB classifier learns very quickly and is easy to apply [13, 14].

Another model used was DT, that stands for Decision Table, and it is a class for building and using a simple decision table majority classifier. Decision tables are very easy to understand and one of the simplest hypothesis spaces possible [11]. J48, the fifth of the six models used, is a class for generating a pruned or unpruned C4.5 decision tree. At least, it was used K-Means model which is a cluster data using the k means algorithm. It can use either the Euclidean distance (default) or the Manhattan distance. If the Manhattan distance is used, then centroids are computed as the component-wise median rather than mean [15]. In order to implement those data mining models, it was necessary to follow a specific process, that delimits all the steps and decisions that must be taken. The decision fell on the CRISP-DM process, a Cross-Industry Process for Data Mining. The CRISP-DM methodology provides a structured approach to planning a data mining project and consists on a hierarchical process model that divides the process of data mining into six phases: Business Understanding, Data Understanding, Data Preparation, Modeling, Evaluation and Deployment [16–18]. This methodology was followed throughout this work. Therefore, in order to correctly perform each of the steps of CRISP-DM, it was necessary to resort to different tools and try to combine their results. The main software's used were WEKA and R. WEKA is a machine learning software that allows a specific analyzation of the data, based on a collection of machine learning algorithms for data mining tasks [12]. On the other hand, R is a free software environment for statistical computing and graphics that is supported by the R Foundation for Statistical Computing and based on R language, widely used for developing statistical software and data analysis [16].

3   Related Work

Data mining, as already mentioned, is a powerful tool that allows us to find certain patterns in a dataset for instance. As described in the paper "Principles of Data Mining" it allows discovering interesting aspects of the dataset, some patterns, that can be very useful [19]. The process, however, when applied in the healthcare area has some peculiarities, because not all the data is available to be treated since medical information can sometimes be sensitive and for ethical and legal aspects cannot be revealed. The paper "Uniqueness of medical data mining" addresses these differences in data mining when applied in medical data [20]. Despite this uniqueness, data mining is very useful, even in medical data. The paper "Data Mining in Healthcare and Biomedicine: A Survey of the Literature" describes how data mining can reveal knowledge hidden within all the medical data and how this knowledge can be very useful [21]. Already in the paper "Application of Data Mining Techniques to Healthcare Data" is described how data mining is useful in medical data and shows its advantages. It also shows the steps involved in data mining and successful applications of the process [22]. When it comes to the comparison of the results obtained by the scoring systems, the paper "Scoring systems for outcome prediction in patients with perforated peptic ulcer" is possible to visualize an evaluation of several score systems, in which is concluded that the most used system is the Boey system and that the PULP system, although promising, need to be validated externally before the implementation [23].

4   Data Mining Process

As told, previously, all the process of classification and analysis of the data provided was based on the CRISP-DM methodology. The process followed, and all the results obtained, are presented in this section.

4.1   Business Understanding

The death and the existence of complications in patients with ulcers is a reality, unfortunately. This leads to a necessity, the attempt to predict these adverse situations and try to protect the patient both of death and of the existence of complications in their clinical reports. This study aims to achieve a prediction of the two parameters in the study in order to focus all the resources available to the situations that need them the most.

4.2   Data Understanding

In order to achieve all the results to which we propose the study base has to be a well-achieved dataset that contains good and adequate data. This allows achieving reliable results that can be applied to real situations. The ulcer dataset, on which the whole study is based, has 69 instances and multiple attributes that allows achieving PULP and Boey scoring punctuation.

Table 1. Analysis of the statistical distribution of PULP/Boey parameters.

Punctuation System	Nominal Variable	Positive cases (%)
PULP	Hepatic cirrhosis	1.45
	Use of steroids	4.35
	Shock on Admission	17.39
	Drilling time up to diagnosis >24 hours	43.48
	Serum Creatnine	17.39
	ASA 2	62.32
	ASA 3	23.19
	ASA 4	10.10
	ASA 5	1.45
Boey	Severe co-morbidity	14.49
	Preoperative Shock	17.39
	Drilling time	44.93

Punctuation System	Numerical Variable	Mean	Standard Deviation	Minimum	Maximun
PULP/Boey	Age	55.24	17.65	21	90

The dataset contains, also, the PULP and Boey result as well as the class death and complicate in which can be found the result of the patient, if he died and got complicated. Through the analyze of the data is possible to visualize that 31.88% of them suffer of complications derived from the disease but only 5.8% of the patients died from it. In regard of gender distribution, it is possible to see that 63.66% of the patients are males, and 36.36% are females.

As for the classes in the study, is possible to analyze the distribution of the values for each attribute available. Since the dataset is based on "yes" or "no" information, almost all the attributes are nominal. Only one of the attributes is numerical, the age of the patient. Therefore, is possible to analyze the number of positive cases regarding each attribute, as shown in Table 1.

Once analyzed the attributes that allow to calculate the PULP score and Boey score and, since this is the main parameters in the prediction of death in real situations. So, is possible to verify the distribution of these parameters' values by all the dataset, as shown in Fig. 1.

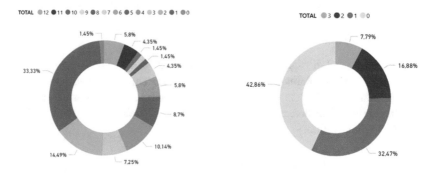

Fig. 1. Distribution of the PULP score (left) and the Boey score (right) by the dataset.

4.3   Data Preparation

One fundamental step of the CRISP-DM is the preparation of the data and the selection of all the important attributes in order to determine which are the most important ones. In the dataset used, it made no sense to eliminate any of the attributes, essentially because the most important ones are the PULP score and the Boey score. However, a dataset with this score and the class was too small, and in order to counteract this, all the attributes used to estimate this score are used. The dataset had, however, to be divided in order to study each of the systems, independently, so that, in the end, was possible to compare the accuracy results and conclude which of the punctuation systems is more assertive.

Focus our further study in the dataset that analysis the influence of PULP score on the prediction of death is possible to highlight some procedures that were followed. In addition to the several tests made using the original data, it was important to analyze the influence of some procedures in the final results. One of this was the oversampling. For such, a dataset with the triple of death cases, this is, instead of the 4 death cases, it has become 12 death cases, in a total of 75 cases. The other approach was proceeding to the age normalization. In the original dataset, the age has values between 21 and 90. So that, it can be normalized and tested in another different way, to be possible to conclude which is the best way to study this particular data, in this case.

Number of Clusters

In addition to the normalized dataset and the oversampling, the influence of clustering technique in the final results was also studied. This study begins with the determination of the ideal number of clusters that feat the dataset. For that is necessary to analyze the sum of squared errors returned in Clutterer output. For the case of PULP versus death, the ideal number is 6, since it is the minimum value achieved after a jump in them. However, analyzing the information, it was concluded that the expected number of clusters should be 2, because of the dimension of the target attribute, this is, because the target attribute is "Yes" or "No", it is more logical to split the results into two groups, one for the "Yes" and other for the "No". In this way, both possibilities will be studied and compared, in order to obtain the best results.

4.4   Modelling

The next step of the study is to induce the prepared data in WEKA, an open-source machine learning software tool. For these, it was necessary to choose an approach and identify a number of Data Mining Techniques, in order to obtain results.

The combination of different algorithms, scenarios and other parameters allows to obtain different models and set the best one. Each model depends on six parameters that can elaborate a huge number of different models and can be described as represented in Eq. 1.

$$DMMn = \{Af + Si + DMTy + SMc + DAb + TGj\} \tag{1}$$

Af = {Classification} Si = {Scenarios} = {S1}
DMTy = {Data Mining Techniques} = {JRip, OneR, NB, DT, J48, K-Means}
SMc = {Sampling Method} = {Cross-Validation, Percentage Split}
DAb = {Data Approach} = {With Oversampling, Without Oversampling, With Cluster, Without Cluster, With Prunning, \\Without Prunning, With Normalization, Without Normalization}
TGj = {Target} = {Death}

Analyzing all the information specified is possible to conclude that the data mining model will be:

DMM = {1 classification, 1 scenario, 6 techniques, 2 methods, 8 approaches', 1 target} which leads to a total of 96 models induced.

4.5   Evaluation

The evaluation of the models constructed was made through three different statistic metrics based on the confusion matrix resulting from the application of the model. Based on the value of True Positives (TP), False Positives (FP), True Negatives (TN) and False Negatives (FN), it is possible to calculate Accuracy, according to Eq. 2:

$$Accuracy = (TP + TN)/(TP + FN + TN + FN) \tag{2}$$

This way is possible to analyze all the results obtained and conclude which of the models and data approaches can achieve the best results. As referred there are several approaches to the data information that were studied in order to find the best way to predict death through PULP score, trying to understand which combination of factor can achieve a higher accuracy. Therefore, will be studied factors such as Number of Clusters and the influence that this can lead in accuracy results, as well as the influence that oversampling and normalization can represent in the result of accuracy (Table 2).

Table 2. Evaluation results obtained to death prediction through PULP score.

Technique	Method	Data approach	Accuracy (%)
NB	Cross-Validation	–	92.75
NB	Cross-Validation	Cluster (k = 2)	94.20
JRip	Percentage Split	Prunning and Cluster (k = 6)	86.96
JRip	Percentage Split	Prunning and Oversampling	96.15
OneR	Cross-Validation	Oversampling and Cluster (k = 2)	97.40
JRip	Cross-validation	Oversampling, Prunning and Cluster (k = 6)	94.81
NB	Cross-Validation	Normalized	92.75
NB	Cross-Validation	Normalized and Cluster (k = 6)	94.20
J48	Cross-Validation	Normalized, Prunning and Cluster (k = 7)	91.30

Analyzing the results is possible to withdraw the results to the three principal data approaches (oversampling, normalized and original) and compare their results, as we can see in Fig. 2.

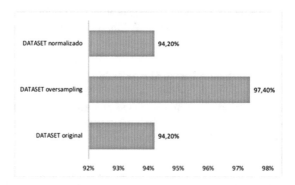

Fig. 2. Results obtained for the three principal data approaches.

Similarly, to the procedure made for obtaining the accuracy for death prediction using PULP score, the prediction of death using Boey score, and the prediction of complications using both, PULP and Boey score, were studied and the results compared. The results obtained can be found in Fig. 3.

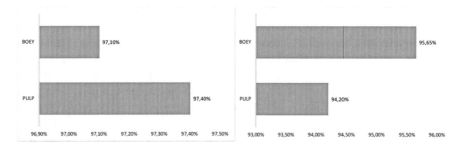

Fig. 3. Results obtained for predicting death (left) and complications (right) in patients with ulcers.

The final step of this study is the analysis of the correlation existing between all the attributes and the two targets studied. This will allow finding the most relevant attributes among all, this is, the ones that have a major influence in the final state of the patient with regard to death and the complications of the state of health. The results obtained can be found in Fig. 4.

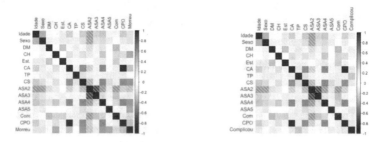

Fig. 4. Results obtained for correlation with death (left) and with complications (right) in patients with ulcers.

5   Discussion

With the analysis of the results obtained is easy to understand that, in the vast majority, the values of accuracy are greater than 90%, with respect to the prediction of death using the PULP score. This fact can be explained by the number of deaths in the whole dataset. There are only 4 cases of death in 69 patients. This fact can influence, negatively, the results obtained, making them less relevant due to the universe of study. However, the results can be improved, by the reformulation of the dataset, especially with regard to case numbers.

Focusing the analysis on the results obtained, more specifically the accuracy values, is easy to conclude that the best value achieved is 97.40%, using a normalized dataset, and with the application of clustering techniques. Concerning the method used, the one that leads to a better result is OneR. The number of clusters was one of the parameters in the study. Despite the fact that the determination of this number by the squared errors leads to a different number, the rational analysis makes us conclude that the ideal cluster number is 2 because the classification according to the target attribute is "Yes" or "No". Therefore, makes sense try to fit the results in only two groups. Another parameter in the study is the influence of oversampling in the results. How can be checked by the analysis of the results, the oversampling produces an increasing on the accuracy result, which can be explained by the simple fact that this approach leads to the replication, in this case, triplication of data, that can lead to the addiction of the model. Crossing the remaining results is possible to infer that the PULP score allows better predictions of death, as it is mentioned in the literature, and as expected because takes in account a much larger number of factors. However, with regard to complications, the results are precisely the opposite. In this situation, Boey can achieve better results, leading to better predictions of complications in the health status of the patients. The last one is a totally new approach, which does not yet have a comparison term with the results obtained.

At last, with regard to the established correlations, as we can see in Fig. 4, it was found that the four main parameters, with a strong positive correlation with the target attribute, that should be incorporated in a new system to predict death effectively, were ASA 4, ASA 5, serum creatinine and liver cirrhosis. In the other hand, the three parameters that most influence the occurrence, or not, of complications are perforation time greater than 24 h, the shock in the patient's admission, and creatinine.

6   Conclusion and Future Work

With this work is possible to conclude that Data Mining can be a very important step in trying to solve some problems that can be found in health systems nowadays and, in trying to create new approaches and protect patients from some situations like death and health complications. In this specific case, is possible to conclude that PULP is a better scoring system to predict that than Boey. However, Boey shows up to be superior when it comes to complications. Regarding the correlations, there are 4 major attributes, in the prediction of death, ASA 4, ASA 5, serum creatinine and liver cirrhosis, and 3 in the prediction of complications, perforation time greater than 24 h, the shock in the patient's admission, and creatinine. The results obtained, however, are not conclusive due to the dimension of the dataset. Once it has few arguments, the conclusions are not totally reliable. This fact, and overwhelming the possibility of future work, can be bypassed by the increase of the dataset, increasing the number of patients in it, and balancing the number of patients in each possible value of the target.

Acknowledgments. This work has been supported by Compete: POCI-01-0145-FEDER-007043 and FCT within the Project Scope UID/CEC/00319/2013.

References

1. Sandler, R.S., Everhart, J.E., Donowitz, M., et al.: The burden of selected digestive diseases in the United States. Gastroenterology **122**(5), 1500–1511 (2002)
2. Wang, Y.R., Richter, J.E., Dempsey, D.T.: Trends and outcomes of hospitalizations for peptic ulcer disease in the United States, 1993 to 2006. Ann. Surg. **251**(1), 51 (2010)
3. Irabor, D.O.: An audit of peptic ulcer surgery in Ibadan, Nigeria. West Afr. J. Med. **24**(3), 242 (2005)
4. Sánchez-Delgado, J., Gené, E., Suárez, D., García-Iglesias, P., et al.: Has H. pylori prevalence in bleeding peptic ulcer been underestimated? A meta-regression. Am. J. Gastroenterol. **106**(3), 398 (2011)
5. Lohsiriwat, V., Prapasrivorakul, S., Lohsiriwat, D.: Perforated peptic ulcer: clinical presentation, surgical outcomes, and the accuracy of the Boey scoring system in predicting postoperative morbidity and mortality. World J. Surg. **33**, 80–85 (2009)
6. Møller, M.H., Engebjerg, M.C., Adamsen, S., Bendix, J., Thomsen, R.W.: The Peptic Ulcer Perforation (PULP) score: a predictor of mortality following peptic ulcer perforation. A cohort study. Acta Anaesthesiol. Scand. **56**(5), 655–662 (2012)
7. Neto, C., Peixoto, H., Abelha, V., Abelha, A., Machado, J.: Knowledge discovery from surgical waiting lists. Procedia Comput. Sci. **121**, 1104–1111 (2017)
8. Zhao, Y.: R and Data Mining: Examples and Case Studies. Academic Press, San Diego (2012)
9. Pereira, J., Peixoto, H., Machado, J., Abelha, A.: A data mining approach for cardiovascular diagnosis. Open Comput. Sci. **7**(1), 36–40 (2017)
10. Cohen, W.W.: Fast effective rule induction. In: Machine Learning Proceedings 1995, pp. 115–123 (1995)
11. Kohavi, R.: The power of decision tables. In: European Conference on Machine Learning, pp. 174–189. Springer, Heidelberg, April 1995
12. Najm, W.I.: Peptic ulcer disease. Prim. Care Clin. Off. Pract. **38**(3), 383–394 (2011)
13. Parsania, V., Bhalodiya, N., Jani, N.N.: Applying Naïve Bayes, BayesNet, PART, JRip and OneR Algorithms on Hypothyroid Database for Comparative Analysis (2014)
14. Wu, X., Kumar, V., Quinlan, J.R., Ghosh, J., Yang, Q., Motoda, H., McLachlan, G.J., Ng, A., Liu, B., Yu, P.S., Zhou, Z.H., Steinbach, M., Hand, D.J., Steinberg, D.: Top 10 algorithms in data mining. Knowl. Inf. Syst. **14**(1), 1–37 (2008)
15. Quinlan, J.R.: C4.5: Programs for Machine Learning. Elsevier, Amsterdam (2014)
16. Arthur, D., Vassilvitskii, S.: k-means ++: the advantages of careful seeding. In: Proceedings of the Eighteenth Annual ACM-SIAM Symposium on Discrete Algorithms, pp. 1027–1035. Society for Industrial and Applied Mathematics, January 2007
17. Chapman, P., Clinton, J., Kerber, R., Khabaza, T., Reinartz, T., Shearer, C., Wirth, R.: CRISP-DM 1.0 Step-by-step data mining guide (2000)
18. Reis, R., Peixoto, H., Machado, J., Abelha, A.: Machine learning in nutritional follow-up research. Open Comput. Sci. **7**(1), 41–45 (2017)
19. Morais, A., Peixoto, H., Coimbra, C., Abelha, A., Machado, J.: Predicting the need of Neonatal Resuscitation using Data Mining. Procedia Comput. Sci. **113**, 571–576 (2017)
20. Hand, D.J.: Principles of data mining. Drug Saf. **30**(7), 621–622 (2007)

21. Cios, K.J., Moore, G.W.: Uniqueness of medical data mining. Artif. Intell. Med. **26**(1–2), 1–24 (2002)
22. Yoo, I., Alafaireet, P., Marinov, M., Pena-Hernandez, K., Gopidi, R., Chang, J.F., Hua, L.: Data mining in healthcare and biomedicine: a survey of the literature. J. Med. Syst. **36**(4), 2431–2448 (2012)
23. Thorsen, K., Søreide, J.A., Søreide, K.: Scoring systems for outcome prediction in patients with perforated peptic ulcer. Scand. J. Trauma Resusc. Emerg. Med. **21**(1), 25 (2013)

Predicting the Length of Hospital Stay After Surgery for Perforated Peptic Ulcer

José Machado[1(✉)], Ana Catarina Cardoso[2], Inês Gomes[2], Inês Silva[2],
Vítor Lopes[3], Hugo Peixoto[1], and António Abelha[1]

[1] Algoritmi Research Center, University of Minho, 4710 Braga, Portugal
{jmac,hpeixoto,abelha}@di.uminho.pt
[2] University of Minho, Campus de Gualtar, 4710 Braga, Portugal
{a74464,a75115,a76457}@alunos.uminho.pt
[3] Centro Hospitalar do Tâmega e Sousa, 4650 Penafiel, Portugal
72777@chts.min-saude.pt

Abstract. The management of peptic ulcer disease usually implies an urgent surgical procedure with the need of a patient's hospital admission. By predicting the length of hospital stay of patients, improvements can be made regarding the quality of services provided to patients. This paper focuses on using real data to identify patterns in patients' profiles and surgical events, in order to predict if patients will need hospital care for a shorter or longer period of time. This goal is pursued using a Data Mining process which follows the CRISP-DM methodology. In particular, classification models are built by combining different scenarios, algorithms and sampling methods. The data mining model which performed best achieved an accuracy of 87.30%, a specificity of 89.40%, and a sensitivity of 81.30%, using JRip, a rule-based algorithm and Cross Validation as a sampling method.

Keywords: Perforated peptic ulcer · Length of hospital stay · Data mining ·
CRISP-DM · Classification · Decision Support Systems

1 Introduction

The annual incidence of uncomplicated peptic ulcer disease (PUD) is approximately one case per 1000 persons-years in the general population, with the incidence of complicated disease being 0.7 cases per 1000 persons-years [1]. Possible complications of peptic ulcer disease include haemorrhage, perforation and obstruction. Its relative prevalence varies geographically – in developed countries, haemorrhage is the most common cause (up to 73%), followed by perforation (9%) and obstruction (3%). Although not being the most common complication, perforations stands out as being the complication with the highest mortality rate [2]. A different pattern on the incidence of complications occur in developing countries, with obstruction being the most common cause of complication (56%), followed by perforation (30%) and bleeding (10%) [3]. Although medical treatment is possible, treatment of a peptic ulcer perforation usually implies a surgical procedure. The time between admission and the surgical procedure, the duration of the surgery and the period of the patient's recovery are

© Springer Nature Switzerland AG 2019
Á. Rocha et al. (Eds.): ICITS 2019, AISC 918, pp. 569–579, 2019.
https://doi.org/10.1007/978-3-030-11890-7_55

all aspects which contribute to the patient remaining in the hospital for a few days. The duration of hospitalisation can be influenced by a wide range of different factors, like comorbidities, the surgical approach (open versus laparoscopic) or postoperative complications [4].

An approximated prediction of the length of hospital stay can be an important information for a health institution. If a hospital knows for how long a patient will need to be admitted, the easier it becomes to improve the management of its resources, thereby reducing costs [5]. To enable this prediction, the patients' data collected in the hospital can be used in Decision Support Systems (DSS), and Data Mining (DM). DSS are computer-based solutions used to assist processes of decision-making and problem-solving [6]. To meet the aimed goals, these systems need to acquire knowledge through raw data, which can be attained using DM techniques. Despite being an old method, DM has been increasingly used in recent years and it leans on the discovery of new facts through the analysis of large amounts of data [7]. Although it is widely used in different areas, the application of DM in healthcare sector has shown extreme relevance [8]. Hospital information systems (HIS) store huge amounts of patients' data and by studying it from different perspectives helps on discovering unknown patterns and trends on patients' profiles. This new knowledge can be very useful for health professionals to diagnose and find treatments for patients or to improve the management of healthcare institutions [9, 10].

The present paper focuses on the application of DM techniques to predict how long patients, who underwent surgery for perforated peptic ulcer, will have to be hospitalised. Since the goal is to classify this period as short or long (categorical response), predictive models with be used with a classification purpose.

2 Background and Related Work

Gastric and duodenal ulcers, also known as peptic ulcers, develop due mainly to Helicobacter pylori infections and the use of non-steroidal anti-inflammatory drugs (NSAIDs) that contribute to the destruction of the protection mechanisms of the gastrointestinal mucosa. Without this protection, the mucosa becomes more susceptible to the effect of acids and enzymes presents in gastrointestinal tract, allowing the appearance of mucosa breaks. The resulting ulcers are commonly found by endoscopy, in the stomach (gastric ulcers) or in the upper area of the intestine (duodenal ulcers) [11].

Whereas several non-operative treatments can be successful in treating haemorrhage from a peptic ulcer, perforation is usually treated by a surgical procedure [12]. In addition to conventional open approach, this surgery can be done by laparoscopy. Instead of large incisions, this last method performs small cuts where a video camera and surgical tools are inserted [13]. Despite the low number of complications associated with laparoscopic surgery, when this approach is used, unpredictable conditions may lead to conversion to open surgery [14].

The success of a surgery can also be influenced by the health status of the patient before surgery. To evaluate this factor, the most commonly used classification is the American Society Anaesthesiologists (ASA) score. The first level (I) of this scale represents healthy patients and the highest (VI) refers to brain death patients [15].

Although several studies have been carried out in order to predict the mortality or morbidity of patients undergoing surgery, only a few focus on the prediction of the length of hospital stay.

In 2014, Caetano et al. studied the possibility of predicting the time of hospitalisation of patients submitted to surgery, based on indicators of the hospitalisation process and DM techniques. A regression approach in data of patients from a Portuguese hospital allowed the implementation of several learning methods: Average Prediction, Multiple Regression, Decision Tree, Artificial Neural Network, Support Vector Machine and Random Forest. However, the best learning model was obtained by the Random Forest method, revealing a higher influence of inpatient clinical process attributes than patient's characteristics [5].

With a similar purpose but a different methodology, Li et al. (2009) used Statistical Package for the Social Sciences (SPSS) software to perform a statistical analysis of collected data on patients with PUD who survived surgery. Through a univariate analysis, the attributes most related to hospitalisation time were found. Combining these attributes (preoperative and postoperative factors), a multiple regression model was built, which allowed to predict the number of days of hospitalisation [4].

3 Materials, Methods, and Methodology

The dataset being studied is composed by medical records of hospitalised patients who had surgical intervention for perforation of either a gastric or duodenal ulcer. These records date from 2010 to 2013, and all surgical events were performed in a hospital placed on the north of Portugal.

As for the methods used, six different data mining techniques were explored: NaïveBayes (NB), Logistic Regression (LR), Support Vector Machines (SVM), Rule-Based (RB), Decision Trees (DT), and Decision Trees Ensemble (DTE). All experiments were conducted using the data mining software Weka, and an algorithm was chosen to represent each technique, accordingly to Weka's available collection. The NaïveBayes algorithm exemplifies the NB technique and it is based on the Bayes rule of conditional probability, analysing each attribute individually and independently when classifying a new instance [16, 17]. SimpleLogistic is introduced as a LR linear classifier, used to estimate the binary class probabilities, based on independent variables identified as relevant. In other words, this algorithm can measure the correlation between the class labels and the values of selected attributes [18]. Sequential Minimal Optimization (SMO) is an improved extension of SMV, which solves the issue of large quadratic programming (QP) by breaking it down to several smaller QP problems. These are solved analytically instead of numerically, which reduces significantly the computing time [16, 19]. The RB algorithm implemented was JRip, which is based on an iterative process, generating a set of rules covering a subset of the training set. For each iteration, instances covered by a rule are removed, until there are instances left to cover [20]. J48 algorithm represents the DT technique and it is an implementation of the C4.5 decision tree learner [21]. While training the data, a decision tree is built based on values learned for each attribute. For new instances, the algorithm routes the data through the nodes (attributes), depending on their attribute values, until a leaf is reached (class). Lastly, Random Forest (RF) exemplifies the DTE technique. As its name suggests, RF is an ensemble of decision trees, which uses random feature selection during the process. This algorithm usually performs better than a single tree classifier [22].

The present data mining process was guided by the Cross-Industry Standard Process for Data Mining (CRISP-DM) methodology.

4   Data Mining Process

The CRISP-DM methodology comprehends six phases: Business Understanding, Data Understanding, Data Preparation, Modeling, Evaluation and Deployment [17, 23]. Yet, only the first five were considered, as described in the present section.

4.1   Business Understanding

The business goal is to study which factors influence on length of hospital stay of patients who underwent surgery as a treatment for peptic ulcer perforation. Therefore, the data mining goal is to build models capable of extracting relevant information from real clinic cases, and identifying patterns on the available data, to ultimately help on the medical decision on finding the best path to successful surgery and a faster patient's recovery.

4.2   Data Understanding

Having already set the goal, the next phase involved looking at the data available for mining and on its quality [23]. The data on a total of sixty-nine patients had been previously collected from databases of a hospital in the north of Portugal, and efforts were made to identify and keep solely relevant information for the proposed goal. The dataset gives insight on the profile and health condition of each of all sixty-nine patients before undergoing surgery, along with information on the type of surgery and approach used, and also on the outcome.

For a simpler and better overview on the available data, an analysis was performed on the statistical distribution of all numeric and nominal attributes. The percentages for each value of a given attribute were shown in Table 1 (if numeric) or Table 2 (if nominal).

Table 1. Statistical distribution of numeric attributes.

Attribute	Minimum	Maximum	Mean	Standard deviation
Admission age (years)	21	90	55.14	17.65
Length of hospital stay (days)	3	37	10.03	6.71

Table 2. Statistical distribution of nominal attributes.

Attribute	Value	%	Attribute	Value	%
Gender	Female	31.9	Ulcer Type	Gastric	26.1
	Male	68.1		Duodenal	73.9
Waiting Time Until Surgery	Less than 6 h	11.6	Abdominal Defilement	Localised peritonitis	20.3
	6–12 h	30.4		Generalised peritonitis	79.7
	12–24 h	30.4		No peritonitis	0
	More than 24 h	27.5			
Complication1/Complication2	Sepsis	4.3/0	Other Complication	Ischemic stroke	1.4
	Septic shock	1.4/1.4		Hypovolemic shock	1.4
	Abdominal wall infection	2.9/1.4		Anastomotic stenosis	1.4
	Evisceration	0/0		Duodenal stenosis	1.4
	Respiratory infection	8.7/2.9		Gastric fistula	1.4
	Suture rupture of the lesion	0/0		Splenic laceration	1.4
	Bleeding	4.3/0		Hepatorenal syndrome	1.4
	Recurrence of injury	0/0		Seroma	1.4
	Intra-abdominal abscess	1.4/0		Blank values	88.4
	Urinary infection	0/0			
	Hydroelectric unbalance	2.9/1.4			
	No complication	65.2/89.9			
	Other	8.7/2.9			
Surgery Type	Suture	18.8	Conversion Motive	No lesion identification	2.9
	Omental Patch	0		Lesion size	1.4
	Suture + Omental Patch	69.6		Margins of friable lesion	0
	Subtotal Gastrectomy (SG)	4.3		Peritonitis severity	4.3
	SG + Vagatomy	0		Anesthetic complications	1.4
	Pyloroplasty	7.2		Not applicable	89.9

(*continued*)

Table 2. (*continued*)

Attribute	Value	%	Attribute	Value	%
Approach	Open	58.0	Drain	No	39.1
	Laparoscopic	42.0		Yes	60.9
ASA	I	1.4	Patient Died	No	94.2
	II	56.5		Yes	5.8
	III	24.6			
	IV	15.9			
	V	1.4			
	VI	0			

4.3 Data Preparation

This phase was the most time consuming amongst all five phases implemented of the CRISP-DM process, since a proper preparation of the data can greatly improve the performance of a data mining model.

Firstly, a cleaning process was performed to preserve the consistency of the data and to remove any incomplete information: rows with inconsistent data and columns with blank values were removed; redundant columns were also excluded, along with irrelevant attributes. Some of the remaining columns had information which could be transformed into new and more useful data for the data mining goal. For example, by combining all three columns giving information on types of complications from the surgical intervention, an attribute derivation was built: rather studying the type of complication which had higher incidence and influence on the outcome, the number of complications per patient was considered instead; by narrowing the former information from three to a single column, its weight would be equally comparable to the weight of any other attribute. Other columns were also transformed: the attribute regarding the motive of conversion during surgery was simplified by giving only information on whether conversion occurred during surgery or not, and the column for admission ages, initially numeric, had its values grouped into intervals, for an easier analysis on patients' age range.

The initial aim was to predict whether a patient would die or not after the surgical intervention. However, the number of deaths were very little, and therefore inconclusive. The target was later defined as predicting if the length of hospital stay of a patient would be either short or long. Information on patients who died had to be dismissed from this study, not to influence the results. Having this done, the mean value of days was re-calculated considering the sixty-three remaining instances, obtaining an average length of hospital stay of 9.73 days. Thus, 10 days were established as a comparison value. For values lower than 10 days, the length of hospital stay of a patient would be classified as "short", while for values equal and greater than 10 days, the class value would be "long". For the new target, the class distribution obtained was 74.6% for "short" and 25.4% for "long".

All attributes, except the class attribute, were then converted into values between 0 and 1 using the Min-Max Normalisation, in order to get better results.

4.4   Modeling

After transforming the data, several data mining models (DMM) were created based on six parameters, as seen in the formula below:

$$DMM_n = \{A_f, S_i, DMA_y, SM_c, DA_b, TG_j\} \qquad (1)$$

All models were built regarding a classification approach (A), and no oversampling or undersampling was performed (DA). Six data mining algorithms (DMA) were tested – NaïveBayes, SimpleLogistic, SMO, JRip, J48, and RF – and the two sampling methods (SM) implemented were Cross Validation with 10 folds and Percentage Split with 1/3 of data used for testing. A single target (TG) was considered, and used to determine whether the length of hospital stay would be short or long. As for the scenarios (S), five were defined to identify the factors which have a higher influence on the length of hospital stay. The first scenario (S1) includes all attributes, while the second scenario (S2) relies on all factors not determined by the patient's demographic information. On the other hand, the third scenario (S3) is focused solely on the patient's profile and health condition before undergoing surgery. The forth scenario (S4) includes attributes related to the surgery and complications, while the last scenario (S5) is based on factors highly expected to have a huge effect on the surgery outcome.

S1 = {All attributes}

S2 = {Abdominal Defilement, ASA, Waiting Time Until Surgery, Approach, Surgery Type, Conversion, Drain, Number of Complications}

S3 = {Gender, Admission Age, Ulcer Type, ASA, Abdominal Defilement, Waiting Time Until Surgery}

S4 = {Ulcer Type, Approach, Surgery Type, Conversion, Number of Complications}

S5 = {Admission Age, ASA, Waiting Time Until Surgery, Number of Complications}

In short, 60 models were included in this study, as represented below:

DMM = {1 Approach, 5 Scenarios, 6 DM Algorithms, 2 Sampling Methods, 1 Data Approach, 1 Target}.

4.5   Evaluation

Being this a binary classification problem, in order to evaluate the predictions of each of the 60 DMMs, the chosen criteria was the confusion matrix. This predictive classification table shows the number of True Negatives (TN), False Negatives (FN), False Positives (FP) and True Positives (TP) and these can be combined to measure how well the DMM performed, accordingly to the proposed goal. Three different metrics were calculated to do so: accuracy, sensitivity (also known as true positive rate), and specificity (also known as true negative rate), as represented in Eqs. 2, 3 and 4, respectively [24].

$$Accuracy\,(ACC) = (TP + TN)/(TP + TN + FP + FN) \qquad (2)$$

$$Sensitivity\,(TPR) = (TP)/(TP + FN) \qquad (3)$$

$$\text{Specificity (SPC)} = (TN)/(TN + FP) \tag{4}$$

To summarise the results and pursue the best data mining model obtained amongst all of the 60 proposed, a threshold was set to 80% on accuracy, specificity, and sensitivity metrics. All models which met the threshold value and above, were included in Table 3.

Table 3. Best accuracy, specificity, and sensitivity results achieved for threshold of at least 80% on each metric.

DMM	Scenario	DM algorithm	Sampling method	Accuracy (%)	Specificity (%)	Sensitivity (%)
1	S1	NaïveBayes	Percentage Split	85.71	87.50	80.00
2	**S1**	**JRip**	**Cross Validation**	**87.30**	**89.40**	**81.30**
3	S1	JRip	Percentage Split	85.71	87.50	80.00
4	S2	JRip	Percentage Split	85.71	87.50	80.00
5	S4	SimpleLogistic	Cross Validation	85.74	87.20	81.30
6	S4	SimpleLogistic	Percentage Split	85.71	87.50	80.00
7	S4	JRip	Cross Validation	85.74	87.20	81.30
8	S5	JRip	Cross Validation	85.74	87.20	81.30

5 Discussion

As observed in Table 3, the best achieved accuracy and specificity had percentages of 87.30% and 89.40% respectively, and both were performed by the first scenario, being JRip the data mining algorithm and Cross Validation the implemented sampling method. As for the sensitivity, four data mining models (2, 5, 7, and 8) accomplished the percentage of 81.30%, representing the scenarios S1, S4, and S5. When comparing these metrics, it is shown that sensitivity had slightly lower percentages. However, this can be explained by the low number of cases in the study representing a longer length of hospital stay. Despite this behaviour being expected, in terms of the patient's hospital experience, and on a hospital administration level, sensitivity is a very important metric: it is more serious to wrongly classify a longer length of stay than the opposite, so higher percentages of sensitivity are desired.

Scenarios S1 and S4 had a huge influence on the results, implying that the more information the better (S1) and the surgical event has a determining weight on a patient's length of stay in a hospital. Scenarios S2 and S5 also had good results, as both consider factors such as the waiting time and the number of complications during surgery. On the other hand, the third scenario (S3) did not perform as well as others, ending up not being displayed on the best results obtained. This suggests that the information on the patient's profile and health condition before surgery is not enough to make an accurate prediction on whether the patient will need a shorter or longer hospital care after surgery. As for the data mining algorithms, only half of the six tried made it to the best results: NaïveBayes, JRip, and SimpleLogistic. On the contrary, SMO algorithm did not make it to the top 3, perhaps because it fits problems with larger amount of data better. Along with SMO, J48 and RF algorithms did not achieve great results either, suggesting that techniques regarding Decision Trees were not appropriated for data being studied and the goal.

In terms of sampling methods used, both presented good results in general, as seen in Table 3. However, by comparing DMM 2 and 3, in which the same scenario (S1) and algorithm (JRip) were used, but different sampling methods, it is observed that Cross Validation helps on producing a higher percentage on all metrics. This might be explained by the sampling of data each one uses for training and testing. While Cross Validation iterates through all data to train and test, Percentage Split only uses a portion of data for training and the remaining for testing, therefore learning less.

In short, the data mining model defined by the first scenario (S1), the JRip algorithm, and Cross Validation as sampling method – DMM 2, marked in bold – reunited the best metrics combination above all. It was expected to have a decent performance on this DMM, since the represented scenario considers all the attributes which were selected as relevant in the first place to make an accurate prediction.

Considering the proposed business goal, this study presented satisfactory results. In a clinical environment it could give health professionals and hospital administrators a solid confidence of approximately 81% on taking decisions, regarding a patient's length of stay in a hospital unity. Of course, all metrics could be improved, however for more accurate results, a larger number of cases should have been studied.

6 Conclusions and Future Work

In this study, the data mining process was applied to real data regarding patients with peptic ulcer perforation who had surgical intervention. The aim was to build accurate data mining models to predict the length of hospital stay of those patients and ultimately help on decision making for better quality services in hospital units.

Several scenarios were tested, along with different DM algorithms and sampling methods, and all data mining models achieved an accuracy greater than 80%.

The DMM with better performance featured a classification approach, a scenario containing all attributes (S1), a rule-based technique using the JRip algorithm, Cross Validation as sampling method. The best metrics combination achieved was an accuracy of 87.30%, a specificity of 89.40% and a sensitivity of 81.30%.

In a Future Work, the goal would be integrating new data (a larger number of instances and exploring new attributes and how they are correlated), in order to refine the data mining model, and thus achieving better results.

Acknowledgments. This work has been supported by Compete: POCI-01-0145-FEDER-007043 and FCT within the Project Scope UID/CEC/00319/2013.

References

1. Lin, K.J., García Rodríguez, L.A., Díaz, S.H.: Systematic review of peptic ulcer disease incidence rates: do studies without validation provide reliable estimates? Pharmacoepidemiol. Drug Saf. **20**, 718–728 (2011)
2. Wang, Y.R., Richter, J.E., Dempsey, D.T.: Trends and outcomes of hospitalizations for peptic ulcer disease in the United States, 1993 to 2006. Ann. Surg. **251**, 51–58 (2010)
3. Irabor, D.O.: An audit of peptic ulcer surgery in Ibadan. Nigeria. West Afr. J. Med. **24**, 242–245 (2005)
4. Li, C.H., Bair, M.J., Chang, W.H., Shih, S.C., Lin, S.C., Yeh, C.Y.: Predictive model for length of hospital stay of patients surviving surgery for perforated peptic ulcer. J. Formos. Med. Assoc. **108**, 644–652 (2009)
5. Caetano, N., Cortez, P., Laureano, R.M.S.: Using data mining for prediction of hospital length of stay: an application of the CRISP-DM methodology. In: Cordeiro, J., Hammoudi, S., Maciaszek, L., Camp, O., Filipe, J. (eds.) International Conference on Enterprise Information Systems, pp. 149–166. Springer, Cham (2015)
6. Pereira, J.J.R.: Modelos de data mining para multi-previsão: Aplicação à medicina intensiva (2005)
7. Morais, A., Peixoto, H., Coimbra, C., Abelha, A., Machado, J.: Predicting the need of Neonatal Resuscitation using Data Mining. Procedia Comput. Sci. **113**, 571–576 (2017)
8. Pereira, J., Peixoto, H., Machado, J., Abelha, A.: A data mining approach for cardiovascular diagnosis. Open Comput. Sci. **7**, 36–40 (2017)
9. Reis, R., Peixoto, H., Machado, J., Abelha, A.: Machine learning in nutritional follow-up research. Open Comput. Sci. **7**, 41–45 (2017)
10. Milovic, B., Milovic, M.: Prediction and decision making in health care using data mining. Int. J. Public Health Sci. **1**, 69–78 (2012)
11. Malfertheiner, P., Chan, F.K.L., McColl, K.E.L.: Peptic ulcer disease. Lancet **374**, 1449–1461 (2009)
12. Sanabria, A., Villegas, M.I., Uribe, C.H.M.: Laparoscopic repair for perforated peptic ulcer disease. Cochrane Database Syst. Rev. (2013)
13. Bhogal, R.H., Athwal, R., Durkin, D., Deakin, M., Cheruvu, C.N.V.: Comparison between open and laparoscopic repair of perforated peptic ulcer disease. World J. Surg. **32**, 2371–2374 (2008)
14. Bertleff, M.J.O.E., Lange, J.F.: Laparoscopic correction of perforated peptic ulcer: first choice? A review of literature. Surg. Endosc. **24**, 1231–1239 (2010)
15. American Society of Anesthesiologists: ASA physical status classification system (2014)
16. Deshmukh, B., Patil, A.S., Pawar, B.V.: Comparison of classification algorithms using WEKA on various datasets. Int. J. Comput. Sci. Inf. Technol. **4**, 85–90 (2011)
17. Rodrigues, M., Peixoto, H., Esteves, M., Machado, J., Abelha, A.: Understanding stroke in dialysis and chronic kidney disease. Procedia Comput. Sci. **113**, 591–596 (2017)
18. Landwehr, N., Hall, M., Frank, E.: Logistic model trees. Mach. Learn. **59**, 161–205 (2005)
19. Platt, J.: Sequential minimal optimization: a fast algorithm for training support vector machines (1998)

20. Bhargava, N., Jain, A., Kumar, A., Dac-Nhuong, L.: Detection of malicious executables using rule based classification algorithms. In: Jaiswal, A., Solanki, V.K., Lu, Z. (Joan), Rajput, N. (eds.) Proceedings of the First International Conference on Information Technology and Knowledge Management, pp. 35–38. PTI (2018)
21. Solanki, A.V.: Data mining techniques using WEKA classification for sickle cell disease. Int. J. Comput. Sci. Inf. Technol. **5**, 5857–5860 (2014)
22. Stojanova, D., Panov, P., Kobler, A., Džeroski, S., Taškova, K.: Learning to predict forest fires with different data mining techniques. In: Conference on Data Mining and Data Warehouses, pp. 255–258 (2006)
23. IBM: IBM SPSS Modeler CRISP-DM Guide (2011)
24. Doreswamy, H.K.S.: Performance evaluation of predictive classifiers for knowledge discovery from engineering materials data sets. CIIT Int. J. Artif. Intell. Syst. Mach. Learn. **3**, 162–168 (2011)

Big Data Analytics and Applications

Benefits of Applying Big-Data Tools for Log-Centralisation in SMEs

Vitor da Silva[1(✉)], Francesc Giné[1], Magda Valls[1], David Tapia[2],
and Marta Sarret[2]

[1] Polytechnic School, University of Lleida, 25001 Lleida, Spain
{vdasilva,sisco}@diei.udl.cat, magda@matematica.udl.cat
[2] LleidaNetworks Serveis Telemàtics, 25003 Lleida, Spain
{dtapia,mgatnau}@lleida.net

Abstract. The benefits of big-data have been proven to ensure more
control over the data, adding improvements in security and complex
query capabilities across many datasets. However, a problem faced
by many companies, especially by small and medium-sized companies
(SMEs), is to define when it is necessary to apply big-data tools. Log
management becomes a relevant challenge when the volume starts to
grow. This paper aims to define the benefits of applying big-data tools
to dealing with log-management. In addition, it provides implementation of log-centralisation based on a cluster made up of commodity
nodes for medium-volume data environments using big-data technologies. The proposed system is tested on a real study case, in particular on a medium-sized telecommunication company. The results show
that the implemented system brings efficiency in storing and analysing
medium-volume datasets. Furthermore, the proposed solution scales the
performance based on the data size and number of nodes, providing
improvements in data security, data analysis and data storage.

Keywords: Big-data · SME · Log-centralisation · Log-management

1 Introduction

Every process running on an information system generates digital traces,
often stored in files called log-files or just "log". Developing an adequate log-
management mechanism may be critical for a company to detect errors or mis-
behaviours efficiently. A common approach is to set up a log-centralisation system. However, this problem is generally not seen as a crucial point by small
and medium-sized companies (SMEs) [1]. Controlling and managing logs could
become a challenge due to many factors: complexity of different log formats,
data volume, complex infrastructure or even ignorance of the available tools.

This work is supported by projects MTM2017-83271-R, TIN2017-84553-C2-2-R and
2016DI090.

Big-data technology appears as a good candidate for processing this growing amount of log data [2,3]. Unfortunately, this technology is perceived as unaffordable for SMEs due to the misconception that it is only aimed at processing the large amounts of data commonly associated with big companies [4]. In addition, big-data technologies can be considered expensive and hence a SME must understand what their needs are before investing in this [5]. Helping these companies to identify the benefits of applying big-data for log-management is the main motivation of this paper, and consequently the right momentum to start.

This paper aims to compare commonly tools used for searching log in files with a generic on-premises solution for log-centralisation using common big-data tools to answer the question raised above. In order to evaluate performance and suitability, the system is developed in the framework of a real SME. The implementation is built according to three strict requirements: open-source tools, on-premises hosting and commodity hardware. These requirements fit a wide range of SMEs, hence the solution can be extrapolated straightforwardly.

The solution implemented lies in a cluster made up of six nodes running a search engine and data storage tool. The experimental results reveal that the proposed centralised system outperforms the commonly used tools and it scales linearly in relation to the volume of data and the number of nodes in the system. Likewise, the centralised log processing ensured the security and efficiency of the log procedure in the target company.

The remainder of this paper is structured as follows. Section 2 contextualises the research into log processing and the big-data challenge focused on SMEs. Next, the developed generic solution is described in detail in Sect. 3. Section 4 explains our study case using an SME as the experimentation platform. Section 5 evaluates the performance of the proposed system against non-big-data tools in relation to the data volume and number of nodes, analysing the operational benefits. Finally, Sect. 6 concludes the paper and discusses future directions.

2   State of the Art

In the literature, many papers describe the advantages of implementing new methodologies for building log-management infrastructures [6,7], understanding and automating log analysis [8] and establishing decision models from process event logs [9]. Of special interest for us is the work by Calvanese et al. [10]. This emphasises the relevance of the legacy data management in SMEs. In contrast with our approach, they present an ad-hoc framework for extracting event logs from legacy information systems, using an SME case.

A Gartner consulting report [11] estimated that by 2017, approximately 15% of companies would be actively using IT Operations Analytic (ITOA) technologies to provide insights into both business execution and IT operations, showing clearly the need to manage and analyse the large amount of data they generated. This necessity has sparked the interest of the research community and, as a consequence, some works in the literature address the challenge of using big-data in companies [4,12]. Regardless of government statistics and financial reports demonstrating that SMEs represented over 99% of all companies in the main economies in 2016–2017 [15,16], few works on the big-data topic have related to this type of company. Coleman et al. [17] discuss the benefits of big-data being used by SMEs, proposing a model of evolution towards the adoption of big-data. Kalan and Ünalir [5] argue about the high demand for technical skills in order to apply big-data technologies, defining opportunities based on business intelligence concepts. Sena et al. [18] present the need for SMEs to define their requirements in order to apply the correct big-data approach, and to have a clear vision of what they want and which problems should be solved.

According to [6,7], the efficiency of log processing can be improved by a log-centralisation system for the following reasons: Efficient Daily Operations, Insights and Data Monitoring, Security and Data Storage. However, few of these works are focused on processing log events with big-data tools. We can find only generic discussions about application of models for big-data using logs [1,13] or describing the kind of data inside them [5] or analysing the aggregated benefits of using big-data [14].

To the best of the authors' knowledge, this is the first work on analysing the benefits of a big-data implementation oriented towards dealing with log datasets in SMEs.

3   General System Overview

The most widespread definition of big-data is the 3 V approach, firstly introduced by Laney in 2001 [19] describing the three dimensions: Volume, Velocity and Variety. Later on, such definition evolved to 5 Vs, adding Veracity and Value [20]. In this paper, we use the 3 V definition, doing a step-back in the big-data 5 Vs approach for the sake of the SMEs. The idea is to use a simplified and well-defined base with the 3 V approach in order to provide a clear comparison between different tools.

To compare commonly used programs for query log files we implemented a generic solution using big-data tools. The non-big-data programs chosen to make the comparisons were LESS and EGREP, both very popular tools, to compare with the proposed solution. EGREP is intended to search for a pattern using extended regular expressions inside a text file. LESS is a filter for file paging and text searching, commonly used when a text file cannot be read by editors such as nano, vi or notepad due its big size.

In order to embrace the major number of SMEs, the following requisites were taken into account throughout the development of the proposed solution: efficiency, low cost (preferable using open-source tools) and On-premises.

Keeping in mind these prior requisites, a model stack depicted in Fig. 1, made up of three independent layers was proposed. The stack bottom layer is acting as the long-term global storage and centralising all files in one place. The middle layer provides the big-data tools used to search for data in the data-storage layer, and this is the connector between the bottom and top layers. The top layer allows users to interact with the data, providing tools to analyse and understand these data, making it possible to create plots and dashboards.

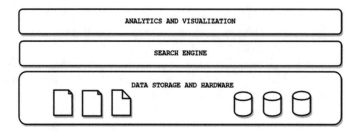

Fig. 1. Architecture stack

4   Study Case

In order to evaluate its performance and benefits, the system presented in this paper was implemented in the framework of a particular company. This study case company was chosen taking into account the requirements of an SME (fewer than 250 employees and a turnover lower than 40 million dollars) [16]. Based on these requirements, the company selected was a small international telecommunications business in Spain, with a staff of around fifty workers and offices spread over the world.

Currently, the data volume stored by the company is approximately 35 TB. Of this data, 22% are log files from different servers. A preliminary analysis of the company in relation to log-data processing demonstrates that 80% of the data stored was poorly used and without any structured methodology to process it into the files. Figure 2 shows the annual log evolution from the last 3 years. According to this figure, we can see a linear increase in the annual log data.

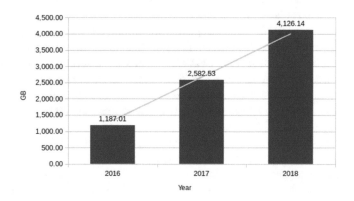

Fig. 2. Evolution of log generation

4.1 Big-Data Tools

The Proposed solution combines two different open-source systems: Hadoop and Elasticsearch. Elasticsearch integrated with Hadoop ensures easy implementation, low-cost maintenance and immediate results.

Hadoop is a well-known framework applied in many big-data scenarios, while Elasticsearch can perform fast queries in the log data, making it possible for the SME to evolve from small data to big-data.

Elasticsearch is designed for log processing using standard protocols, whose flexibility turns it into an appropriate choice for implementing the stack middle layer. Kibana was chosen as the visualisation software due to its maximum compatibility with Elasticsearch, convenience and usability. Additionally, the Logstash tool was defined as the environment data ingest component.

The decision to integrate Hadoop and Elasticsearch, storing data in two different systems, was intended to add robustness and flexibility to the environment. Thus, if a complex query is needed and unavailable in the Kibana query suite, it is possible to use the Hadoop query tools, although it may provoke a penalty in the query time. Hence, Hadoop is used to do batch queries over long-term storage files.

4.2 Implementation

Given that both Hadoop and Elasticsearch follow a master-slave model, a cluster with this paradigm was implemented. Specifically, the cluster was made up of 6 nodes: one master, one master backup and four slaves.

Figure 3 shows the final configuration of the cluster. Following the dataflow of this figure, Logstash inserts log data into both systems, one devoted to search operations with Elasticsearch and the other intended for long-term log storage,

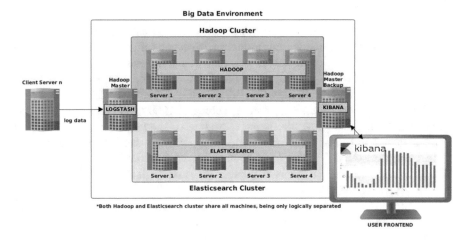

Fig. 3. Cluster implementation.

keeping up to the six last months' accumulated. The choice of the latest six months' log data is due a characteristic of the study case company, having logs information older than six months hardly demanded.

The hardware specification is defined according to the annual volume of data ingestion shown in Fig. 2. From this real data, we estimate the minimal disk space needed to work properly for six months, which would be 2TB in Hadoop (last 6 month's storage) and 500 GB in Elasticsearch system (the last month's storage). Thus, six commodity servers were used, obtaining the cluster configuration described in Table 1.

Table 1. Final cluster setup

Server	QTY	CPU	RAM memory	Hard disk
Master and data node	1	8 × Intel(R) Xeon(R) CPU E5335 @ 2.00 GHz	32 GB	1 TB
Master backup and data node	1	8 × Intel(R) Xeon(R) CPU E5335 @ 2.00 GHz	32 GB	1 TB
Data node	4	8 × Intel(R) Xeon(R) CPU E5335 @ 2.00 GHz	16 GB	1 TB

5   Experimental Results

This section analyses performance results comparing the commonly used LESS and EGREP tools with Elasticsearch and Hadoop. The experiments cover the three axes of big-data: volume, velocity and variety.

5.1   Performance Results

Figure 4 shows, in log-scale, the response time of each tool against the volume axis, comparing the same query parameters with seven different sized files. These results show that Elasticsearch was the fastest engine with a maximum time response of 9 s for 30 GB. On the other hand, EGREP and Hadoop GREP were mostly located int the range between 100 and 600 s, while the LESS command offered the worst performance exceeding values of 1000 s, and reaching a maximum of 4272 s in the biggest file. Elasticsearch gave the fastest performance due to the inverted indexing mechanism which allows very fast full-text searches, unlike LESS or EGREP, which are full-scan search programs. The performance of EGREP and Hadoop GREP was quite similar, although Hadoop GREP, which uses the Map Reduce mechanism, offered better results as the volume grew.

The query response time of the proposed solution, described in Sect. 3, was tested over the complete cluster environment using Elasticsearch and Hadoop. Experiments were run over different log data sizes, ranging from 100 GB to 800 GB, and composed of 739 sample files in the largest test (average file size

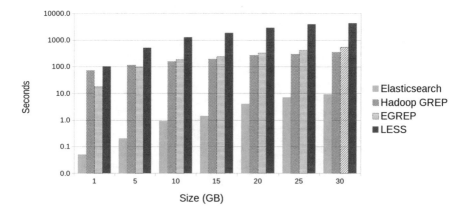

Fig. 4. Comparison of query performance

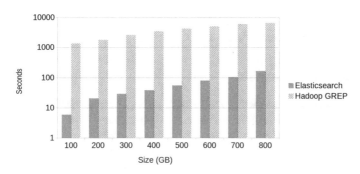

Fig. 5. Query response time by size

of 1.08 GB). Figure 5 shows average response time which were below 160 s over 800 GB using Elasticsearch, while Hadoop return responses between 1372 and 6574 s. The disparity in scale of the results was due to the Elasticsearch schema-on-write designed to query strings. Elasticsearch maps the data before ingesting, and leads to improved performance, whilst Hadoop ingests raw data, and hence the response time is penalised. Nevertheless, these results can be considered good response times in a large cluster, enabling fast queries for SMEs in all company log data using Elasticsearch or even more sophisticated queries using Hadoop suite tools.

In order to check the velocity parameter, the scalability of the system was tested in relation to the number of nodes, which was scaled from three up to six. For each trial using Elasticsearch, four different sample log datasets (between 100 GB and 400 GB) were used in order to obtain the effect of data size. For each dataset, 10 different queries were launched and the average response times are shown in Fig. 6. This figure shows that the response time grew with the data size and decreased with the number of nodes. Comparing the 3 nodes against the 6 nodes execution, Speedup obtained was 2 (200-GB case), while the best Speedup

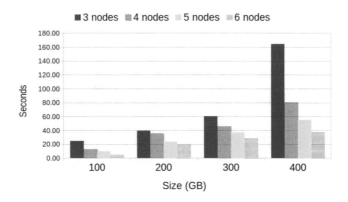

Fig. 6. Query response time by node number

was 4.2 (400-GB case). The experimental results reveal the good scalability of the proposed solution, meaning a control over the velocity axis.

5.2 Operational Benefits

The usage of the two most commonly-used tools (LESS and EGREP) may generate some security and usability issues inside a company. For instance, in a digital service company, technicians in the client support service may receive concerns or complaints from the customers. To detect and verify a reported incident, they may need to perform a query in the log file, which may be highly time-consuming. Furthermore, in terms of security, these customer service agents may need permission to access directly, which increases the risk of unintentional changes in the file or even deletion in the worst case.

The implementation of the centralised cluster reinforces security, since the agents will not need direct access to data in order to detect an incident. Furthermore, usability is significantly improved, offering a user interface which is properly designed to search inside log files. Hence, the search experience evolves into an easier procedure and may take just a few steps instead of the many steps previously required when using non-big-data tools. With the new operation, few steps are necessary to find a specific log line, and this is more user-friendly.

Moreover, one of big-data advantages lies in its capacity to process any kind of data. Thereby, it is impossible to compare non-big-data tools in the variety aspect. The proposed solution implements the Hadoop framework, capable of ingesting any kind of raw data. Thus, when it is necessary to process non-text data or unconventional information, Hadoop becomes the available option enabling the SME cross different data types. Table 2 summarises the main benefits of using big-data tools for log-centralisation compared with the conventional ones.

Table 2. Operational benefits summary

Feature	Non-Big-Data	Big-Data
Response time for files ≤ 1 GB	≥ 100 s	≤ 0.1 s
Response time for files ≤ 10 GB	≥ 100 s	≤ 1 s
Response time for files ≤ 30 GB	≥ 600 s	≤ 10 s
Scalable	✘	✔
Built-in alerts	✘	✔
Fault tolerance	✘	✔
Visual interface	✘	✔
Cross information with external datasets	✘	✔
Process multimedia files	✘	✔
Security certification compliance	✘	✔

6   Conclusion

This paper presents the benefits of log-centralisation using big-data technologies in small and medium-sized companies (SMEs) using on-premises solutions and open-source software.

The proposed solution demonstrates that low-cost machines can be used to create a unified cluster combined with such big-data tools as Elasticsearch and Hadoop, centralising all the logs of a SME. It allows two different levels of search engine, a first one characterised by efficient searches through the short-term log storage files and a secondary searching mechanism, which allows more complex but less efficient queries, through long-term log files.

The experimental results indicate a substantial improvement in log query performance, compared with the tools commonly used in the case-study company. Likewise, our results reveal that the system implemented scales perfectly with the log-data size and the number of nodes. Thus, it makes it easy to expand the cluster according to the increase in the log-data needs, as adding more low-cost machines is sufficient to maintain proper functionality. In addition, at the operational level, the technicians can avoid direct access to the log file, leading them to use specific analytics tools in compliance with security certifications. With this substantial change, the risk of any illegal file changes or even accidental deletion is mitigated with the corresponding security benefits. Another benefit pointed out by the experimental results is the capability to cross information between different datasets, powering the benefits of big-data inside an SME.

In conclusion, big-data is a mainstream for all companies eager for a competitive place in the market, since nowadays data control is the key to improving company operations. This project extends these benefits to SMEs, bringing the possibility of a new information era for them, all this on a low budget.

As future work, we aim to import relational database data and store in the same model described in this paper to compare the benefits. Thus, a complete solution for data analysis focused in SMEs will be generated. We also aim to compare on-premises solutions with cloud solutions and improve the data-ingestion mechanism.

References

1. Miranskyy, A., Hamou-Lhadj, A., Cialini, E., Larsson, A.: Operational-log analysis for big-data systems: challenges and solutions. IEEE Softw. **33**(2), 52–59 (2015)
2. Chen, M., Mao, S., Liu, Y.: Big-data: a survey. Mob. Netw. Appl. **19**(2), 171–209 (2014)
3. Xia, X.G.: Small data, mid data, and big-data versus algebra, analysis, and topology. IEEE Signal Process. Mag. **34**(1), 48–51 (2017)
4. Ardagna, C.A., Ceravolo, P., Damiani, E.: Big-data analytics as-a-service: issues and challenges. In: IEEE International Conference on Big-Data (Big-Data), pp. 3638–3644 (2016)
5. Kalan, R.S., Ünalir, M.O.: Leveraging big-data technology for small and medium-sized enterprises (SMES). In: 6th International Conference on Computer and Knowledge Engineering (ICCKE), pp. 1–6 (2016)
6. Chuvakin, A., Peterson, G.: How to do application logging right. IEEE Secur. Priv. **8**(4), 82–85 (2010)
7. Anastopoulos, V., Katsikas, S.K.: A methodology for building a log-management infrastructure. In IEEE International Symposium on Signal Processing and Information Technology (ISSPIT), pp. 301–306 (2014)
8. Nagappan, M., Vouk, M.A.: Abstracting log lines to log event types for mining software system logs. In: 7th IEEE Working Conference on Mining Software Repositories (MSR 2010), pp. 114–117 (2010)
9. Bazhenova, E., Buelow, S., Weske, M.: Discovering decision models from event logs. In: Business Information Systems (BIS 2016), Lecture Notes in Business Information Processing, vol. 255 (2016)
10. Calvanese, D., Kalayci, T.E., Montali, M., Tinella, S.: Ontology-based data access for extracting event logs from legacy data: the onprom tool and methodology. In: Business Information Systems (BIS 2017), Lecture Notes in Business Information Processing, vol. 288 (2017)
11. Gartner Inc.: Apply IT Operations Analytics to Broader Datasets for Greater Business Insight, June (2014)
12. Shokri, R., Osman, M.: Leveraging big-data technology for small and medium-sized enterprises (SMEs). In: 6th International Conference on Computer and Knowledge Engineering (ICCKE 2016) (2016)
13. Amar, M., Lemoudden, M., El Ouahidi, B.: Log file's centralisation to improve cloud security. In: 2016 2nd International Conference on Cloud Computing Technologies and Applications (CloudTech), pp. 178–183 (2016)
14. Sharma, S., Mangat, V.: Technology and trends to handle big-data: survey. In: Fifth International Conference on Advanced Computing & Communication Technologies, pp. 266–271 (2015)
15. United States Small Business Profile: Office of Advocacy, United States Small (2016). Business Administration
16. Muller, P., Julius, J., Herr, D., Koch, L., Peycheva, V., McKiernan, S.: Annual Report On European SMEs 2016/2017. Entrepreneurship and SMEs. European Commission, Internal Market, Industry (2017)

17. Coleman, S., Göb, R., Manco, G., Pievatolo, A., Tort-Martorelle, X., Reisf, M.S.: How can SMEs benefit from big-data? Challenges and a path forward. Qual. Reliab. Eng. Int. **32**(6), 2151–2164 (2016)

18. Sena, D., Ozturkb, M., Vayvayc, O.: An overview of big-data for Growth in SMEs. In: 12th International Strategic Management Conference, ISMC 2016, 28–30 October 2016, Antalya, Turkey. Procedia - Social and Behavioral Sciences, vol. 235, pp. 159–167 (2016)

19. Laney, D.: 3D Data Management: Controlling Data Volume, Velocity and Variety. Technical report, META Group (2001)

20. Demchenko, Y., Membrey, P., Grosso, P., de Laat, C.: Addressing big-data issues in scientific data infrastructure. In: First International Symposium on Big-Data and Data Analytics in Collaboration (BDDAC 2013). Part of The 2013 International Conference on Collaboration Technologies and Systems (CTS 2013), 20–24 May, San Diego, California, USA (2013)

Big Data and Advanced Analytics

Arshiya Begum[1(✉)], Farheen Fatima[2(✉)], and Rabia Haneef[3(✉)]

[1] Computer Science Department,
King Khaled University, Abha, Kingdom of Saudi Arabia
arshiyabegum@ymail.com
[2] Analyst, Accenture Private Limited, Hitech City, Hyderabad, India
fatima.farheen711@gmail.com
[3] Deloitte100 Private Limited, Hitech City, Hyderabad, India
rabia.haneef@gmail.com

Abstract. Today a huge amount of data is collected and added in modern information system each day which become difficult to manage as it keeps on growing. To manage such type of data, Big Data and its emerging technology have been used. Big Data with their potential have attracted substantial interest both in academics and practitioners. Many organizations have adopted big data analytics which has become the trending practice to construct valuable information from data. The Analytic process uses big data Tools, and an organization uses such tools to improve operational efficiency to derive new revenue streams and to get competitive advantage over business rivals.

Also, Business Intelligence has captured more market in academic, e-commerce, and business over the last decades. Business Intelligence (BI) is facing challenges due to the latest development of big data which uses big data analytics to enhance BI.

Given the significant nature of both big data and big data Analytics, The main objective of this paper is to discuss the advanced analytics techniques and challenges.

Keywords: Big Data · Business Intelligence · Advanced analytics

1 Introduction

Gartner (2012) defines big data as *"Big data is high-volume, high velocity and high variety information assets that demand cost-effective, innovative forms of information processing for enhanced insight and decision making"*.

The data includes textual content (such as structured, semi-structured as well as unstructured), to multimedia content (such as videos, audio, images) on multiple platforms [10].

Big data has the capacity to store, manage and process it efficiently. There are three main characteristics of big data such as volume, variety, and velocity. The first characteristics is a great volume of data second is data cannot be structured into table and third is velocity which means the speed at which data is generated, Thus need to be processed & analyzed fast. Further to it, two new characteristics have been added to it named as veracity and value as shown in Fig. 1.

© Springer Nature Switzerland AG 2019
Á. Rocha et al. (Eds.): ICITS 2019, AISC 918, pp. 594–601, 2019.
https://doi.org/10.1007/978-3-030-11890-7_57

Fig. 1. 5v's of Big Data

However, apart from structured and unstructured, there are many other new data types that can be processed to yield into insight of business and condition. For example data collected from twitter feeds, network data, and call detail report is not stored directly into a warehouse until it is preprocessed, the main idea of this is to reduce the data and put it in structured form, later it can be compared with the rest of data and analyzed with traditional business Intelligence Tools [12].

1.1   Characteristics of Big Data

Volume: It is a large amount of data which is generated continuously [1] from multiple sources such as text, audio, video, and images. This data can be in size of Zettabyte, Yottabytes or Brontobytes. The Table 1 below shows the datasets volume size information [2].

Table 1.   Dataset volume size

Value	Name	Abbreviation
1000_1	Kilobytes	KB
1000_2	Megabytes	MB
1000_3	Gigabytes	GB
1000_4	Terabytes	TB
1000_5	Petabytes	PB
1000_6	Exabytes	EB
1000_7	Zettabytes	ZB
1000_8	Yottabytes	YB
1000_9	Brontobytes	BB
1000_{10}	Geopbytes	GEB

Velocity: It refers to the rate at which data are generated and the speed at which it is analyzed. Data sets are produced fast and moved around such as image, audio, and video etc.

Variety: This refers to the numerous types of datasets that contain structured, unstructured and semi-structured data. Structured data is considered to be most organized data can be stored in a table with rows and columns. Unstructured data can be in the form of text, images, video, etc., these types of data are both human and machine generated and difficult to store into a database. Semi structured data is not stored in table besides they possess some properties which make them convertible into structured data [12].

Veracity: IBM coined veracity as fourth V of big data, which represents the unreliable inherent of data for example customer tweets on social media are uncertain yet they contain valuable information. Thus to deal with uncertain of data due to data inconsistency and incompleteness is another facet of big data [3]. Big data analytics allow us to work on such type of data.

Value: Oracle introduced value as the attribute of big data. Based on such definition Big data sets have a relatively "low-value density". That is, the date in original form has a low value relative to its volume. A high value can be obtained by analyzing the large volume of data.

1.2   Big Data Analytics

Data analytics are of two type passive and active which can further be classified as Descriptive, Diagnostic, predictive and prescriptive.

Descriptive Analytics: It is a set of techniques to understand data and analyze performance. It identifies the attribute, evaluate and estimate the magnitude of each attribute to the final solution.

Diagnostic Analytics: It is a set of techniques to determine what has happened and why. It begins with the descriptive analytics where it extracts the pattern from large data qualities, correlates data types and estimates linear and non-linear behavior.

Predictive Analytics: It is a set of techniques to analyze current and historical data to determine what has happened. It begins with Descriptive and Diagnostic Analytics where it chooses right data based on knowledge domain and relationship among variables.

Prescriptive Analytics: It predicts what is likely to happened, when it will happen along with why it will happen.

Decisive Analytics: It is a set of techniques for visualizing information and course of action to facilitate human decision making when presented with a set of alternatives.

2 Advanced Analytics

Advanced Analytics is an application of multiple analytic methods that addresses the diversity of big data; the outcome of such data can be structured or unstructured. It also provides descriptive results to yield actionable predictive and prescriptive results that facilitate decision making. These are a set of techniques that require new soft architecture and application framework to solve complex problems.

Hence, big data analytics is where advanced analytic techniques are applied on big datasets [4]. There are thousands of big data tools for data analysis where it inspects the data, cleanses the data, transform and model the data to discoverable useful information. Advanced analytics include data such as data mining, machine learning, pattern matching, forecasting, visualization, and semantic analysis etc. [5].

2.1 Edge Device Analytics

Analytics at the edge is different It means carrying out analysis of data from a non-central point in a system such as a network switch, peripheral node or connected device or sensor. But today edge capabilities are still relatively unsophisticated, lacking anything that a cloud computing power can provide. Therefore new approaches need to be found [6].

Edge analytic performs analysis at the point where data is generated or closer to the "thing" that are the source of the new data. For example, sensors in a train or stop light which provides intelligent monitoring of data based on their analysis. Another such example is a car, a washing machine or a fitness device, or could be some agricultural equipment or any other industrial device [6].

Edge analytic has emerged as time and resource saving. With this Analytic technique filtering and compression, processes are tied to the downstream analytical requirement which indicates filtering and compression algorithm must be dynamic.

2.1.1 Edge Analytic in Action: Use Case

Critical Failure

Failure is a natural phenomenon which can have a catastrophic result when occur. By monitoring, analyzing data at conceptual level edge analytics can identify cause before it effect and enable earlier problem detection.

LSA Edge Model is a sophisticated simulation that provides an accurate prediction of impact and failure. Upon early prediction of failure, user can trigger automatically initiate action.

Environmental Constraints

Edge Model is effective in environmental constraints such as processing capability, connectivity and bandwidth. Example of such constraint is In-flight aircraft analysis, ranges from early warning of minor maintenance to critical system failure.

2.2 Streaming Analytics

Streaming data analytics must occur in real time as the data passes through the sensing or collecting device. It identifies and examines the pattern of interest as the data is being created which yield to instant insight and immediate action.

2.2.1 Streaming Analytics in Action: Use Case

Call Center Monitoring
Real-time customer interaction is a good example of streaming platform. Call center proved the quantifiable business benefits from streaming analytics such as productivity gain, customer complaint result speed and customer satisfaction.

A successful analytics allow the call center to process a large number of calls on a distributed network, also it provides monitoring platform which allows a unified view and analysis of even in real time.

Manufacturing Sector
Manufacturer in the area such as automotive, electronic, durable goods and chemicals are employing data analytics and streaming analytics to improve their production.

Smart Manufacturing is the way to success in the manufacturing industry where it installs sensors in various division if companies to collect data, check for alarming factors and take appropriate decision.

2.3 Location Analytics

Location Analytics focuses on thematic mapping and special analysis for the business analytics. This solution involves simple mapping and special analysis capabilities that work directly with BI and enterprise data.

Esri has built a new simple with a powerful solution for the location analytics known as Esri maps. This map supports easy to use mapping and spatial analytics [8].

2.3.1 Location Analytics in Action: Use Case

Baking
Banking is a location- based business. A bank can better understand and manage its network and merchants by understanding the competition with others. With the increase in competitors, a bank realizes its need to optimize its own network, grow customer and deliver improved products and services.

Based on the analysis of its customer information, the bank can quickly identify, visualize and analyze key factors influencing the performance of branches.

Mapping Visualization
Representation of a data in area of visualization has many practical applications. Mapping technology is innovative tool for the visual representation that we use every day. Google maps help the user to locate places while Google Earth is a tool to explore the home planet. Visual Technique in Google Earth allows user to visualize data onto 3D mapped views.

2.4   Web Analytics

It is a study of behavior of web user and how society makes decisions. Commercially it is a collection and analysis of data from a website to determine which aspects of website achieve the business objectives.

2.4.1   Web Analytics in Action: Use Case

Website
Every operation on the web is a process and it is a complex decision making especially when it comes to forming customer-centric web strategy [7]. Web Analytic tools were used to improve sales of the website of IT services

Some of the challenges of web analytics in designing a website are as follows:
How to design a web page to influence the opinion of people
How to measure the influence of web page content on user opinion
How to measure difficulty or complexity of the material presented on the web page
How to design a good web page to increase page Rank

Research-Oriented Online Community
In Web-based online community users can go inactive which may lead to the difficulties in justifying researcher's continuous support to create and maintain it. Dwyer [14] discovered that the users are active in a passive online community. Web analytics is applied on research-oriented virtual community to expose browsing pattern from web server logs. This approach gives us a clear picture of member activity at web- based online communities.

2.5   Visual Analytics

According to Thomas [15] Visual Analytics is defined as "science of analytical reasoning facilitated by interactive visual interface". It is an evolving discipline which is driving a new way to present data and information to the user.

2.5.1   Visual Analytics in Action: Use Case

Business Analyst at Management Services
Business Analyst provides a variety of services related to business management such as financial fraud and forensic investigation. A company may receive a large amount of both structured and unstructured data from his client.

Business analyst task is to examine unstructured data from the database of client and identify linkage between people or companies to investigate financial fraud such as suspicious supplier's invoices, or the fictitious customer to boost revenue [13].

Software Product Assessment
Software Projects encounter bottlenecks due to many factors: architecture, code size, bad coding style, team structure or wrong estimation of the requirements. Using visualization tools for supporting decision making in the framework of the software project can be highly cost and time effective in helping decision making in the management of the software projects.

3   Challenges of Advanced Analytics

According to Kailser, Armour, Espinose, Money [9] summarize challenges as

- Can analysis improve in a better system and environment model?
- How to measure value of analytic
- As the data grows large what is the limit for value of big data
- Can a good algorithm, overcome data quality problem.
- As the more and more data is added and analyzed, can big data improve decision making and how do we measure it

Some of the other challenges of Advanced Analytics are

- Data Analysis
- Ethical problems
- Data Annotation
- End to End system
- Data Sources
- Metrics

4   Conclusion

The objective of this paper is to describe big data concepts. The paper first defined what big data means and attributes of Big data, highlighting the fact that size is only one dimension of big data. Other dimensions are velocity, variety is equally important. The paper primary focuses on analytics to gain valid and valuable insights from big data. Also, it discusses various advanced analytics and challenges.

The Major innovation in big data analytical technique have not yet taken place, one anticipates the emergence of such analytics in the near future. For example real-time analytics will likely become a popular field of research.

References

1. Mukherjee, S., Shaw, R.: Big data–concepts, applications, challenges and future scope. Int. J. Adv. Res. Comput. Commun. Eng. 5(2), 66–74 (2016)
2. Chawda, R.K., Thakur, G.: Big data and advanced analytics tools. In: Symposium on Colossal Data Analysis and Networking (CDAN) (2016)
3. Gandomi, A., Haider, M.: Beyond the hype: big data concepts, methods, and analytics. Int. J. Inf. Manag. 35(2), 137–144 (2015)
4. Elgendy, N., Elragal, A.: Big data analytics: a literature review paper. In: ICDM 2014: Advances in Data Mining. Applications and Theoretical Aspects, pp. 214–227 (2014)
5. https://www.gartner.com/it-glossary/advanced-analytics/
6. https://www.accenture.com/.../Accenture-Insight-Mobility-Edge-Analytics.pdf
7. Singal, H., Kohli, S., Sharma, A.K.: Web analytics: state-of-art & literature assessment. In: 2014 5th International Conference- Confluence the Next Generation Information Technology Summit (Confluence) (2014)

8. http://www.esri.com/news/arcnews/fall12articles/location-analytics-the-next-big-step-in-business-analysis.html
9. Kaisler, S., Armour, F., Espinosa, J.A., Money, W.: Big data: issues and challenges moving forward. In: HICSS 2013, Proceedings of the 2013 46th Hawaii International Conference on System Sciences, pp. 995–1004 (2013)
10. Sivarajah, U., et al.: Critical analysis of Big Data challenges and analytical methods. J. Bus. Res. **70**, 263–286 (2017)
11. Big Data Analytics. An Oracle White Paper, March 2013
12. Arora, Y., Goyal, D.: Big data: a review of analytics methods & techniques. In: 2nd International Conference on Contemporary Computing and Informatics (IC3I) (2016)
14. Kang, Y., Stasko, J.: Examining the use of a visual analytics system for sensemaking tasks: case studies with domain experts. IEEE Trans. Vis. Comput. Graph. **18**, 2869–2878 (2012)
14. Dwyer, C., Zhang, Y., Hiltz, S.R.: Using web analytics to measure the activity in a research-oriented online community. In: Tenth Americas Conference on Information Systems, New York, August 2004
15. Thomas, J., Cook, K.: Illuminating the Path: Research and Development Agenda for Visual Analytics. IEEE Press, Los Alamitos (2005)

Performance Data Analysis for Parallel Processing Using Bigdata Distribution

Iván Ortiz-Garcés[✉], Nicolás Yánez, and W. Villegas-Ch

Facultad de Ingenierías y Ciencias Aplicadas, Universidad de Las Américas,
Quito, Ecuador
{ivan.ortiz,nicolas.yanez,
william.villegas}@udla.edu.ec

Abstract. The following document presents metrics and pointers for datacenter performance evaluation, whose production workflow will be improved by a parallel computing software, each cluster instance was virtualized providing for scalability and availability for every person who access to the system at different locations. Apache spark will be used as parallel processing distribution through different scenarios, each one will handle workload on physical and virtual nodes, after the collection of time response a comparations will be realized for determinate if the parallel distribution is an ideal solution for guarantee processing requirements.

Keywords: Performance analysis · Scalability · Parallel computing · Quality of services

1 Introduction

Real time process and high concurrency data requires high capacity processing at the minimum time response, a paradigm who compromise communications with a fault tolerance services. Since virtual machines are available at production lines is necessary to know if the computing capacities solve the computing requirements a problem of performance who points to resolve the acquisition of new devices with better prestaciones at the front office or back office stations. In terms of data processing the parallelism works as a workload controller whose traffic levels impact on performance and prevents downtime a solution whose Phil Colella propose at 2004 with the seven dwarfs mathematical models, these are basically models who aim for data analysis over different situations. The software who will be tested for our performance analysis is called Spache spark, this one will provide of statistical analysis of data inputs, in other words this software can initiate sessions and track the computing processing in every work node showing a schematic graphic of time responses and which parallel task was completed or who was skipped by redundancy.

In more simple terms the objective of this paper will be the implementation of parallel processing load framework based on a cluster of devices, physical and virtualized for optimizing the performance and resources of a LAN network. As result of

© Springer Nature Switzerland AG 2019
Á. Rocha et al. (Eds.): ICITS 2019, AISC 918, pp. 602–611, 2019.
https://doi.org/10.1007/978-3-030-11890-7_58

data analysis the IT administrator could provide pointers to request more computing data or redistribute computer resources to critical areas providing of quality of services and efficiency at resources administration.

2   Previous Studies

The parallelism was presented as a performance solution, in the year 2005 the company SUN was released to the market IBM processors of the Niagara denomination whose premise was the work of multiple threads under an architecture composed of several processors, in detail the functionality of the tasks resides in the thread sequentially and stop according to their criticality.

At this point, processes tend to be linear because of their sequential execution, but the problem arises of wanting to reuse already existing processes or results or with too much workload for only one processor. In processes where the response time is not dispensable, the critical points of the system are only subject to operations, but to processes in real time, performance is a fundamental part of business continuity. This is where the paradigm of parallelism formulates the integration of models for the simultaneous processing of data for new applications in real time.

Since 2004, Intel together with IBM and SUN have given way to high-performance processors based on three approaches of parallelism, level-of-instruction parallelism, thread-level parallelism, data-level parallelism, of which the latter are oriented to programmers.

In 2009 at UC Berkeley, the computer network the term cluster was presented as a framework with capabilities to manage resources more globally and whose control operates up to 50,000 operative nodes [2], regardless of the operating system the administrator becomes a potential solution for several components for the problem posed.

In 2016 Apache Spark is released as an open source software with the premises of distributed storage, information compression techniques through redundant mapping and filtering, in-memory data storage, eliminating read and write operations in rigid discotheques.

2.1   Spark Framework

The framework selected for performance analysis would be Apache Spark, this provides a high-level API development, the advantage of a HLAPI are:

- Real time data processing, UDP protocol for transmission, TCP control data.
- Session layer will synchronize communication and reserve resources for data flow.

One of spark premises for memory allocations is the possibility of using the swap partition, all memory data will be allocated on memory so if database provide of a rack of 24 gigabytes all that memory capacity will be used for spark memory without operative system limitations, this solution could be useful is the node requires more data processing over critical areas, consequently IO operations over hard-drive won't be needed.

A collection of data is threatened as RDD, this collection of data also known as a dataset, will be an array of data that spark will be parallelize over the operative cluster.

3 Data Collection

In every stage the data was collected with different loads of information but only the result will be measure by the time of processing the data allocation in memory or hard-drive devices.

For evaluate the performance over spark distribution each instance will initiate sessions with 1[Gb] of memory and one core this will be compared with three different stages like a benchmark with virtual and psychical devices.

All workload at first was created by a script whose create iterations will be parallelized over the cluster in a data set.

3.1 Program Code

The data who will be tested over spark framework was generated by Scala language, this will create an arrangement of arrays for spark context parallelization, at first eighty-thousand rows were created, error was injected for processing filters then the array is inserted over parallelization context.

```
nRows=80000
nStart=0
codeList=["101T","109T","111T","161T","144T","106T"]
listT=[]

#generate data
while nStart < nRows:
codeRow="AB000"
codeRow+=str(nStart)
selectedRow=codeList[random.randint(0,len(codeList)-1)]
listT.append([selectedRow,codeRow,round(random.random()*100,2)]
)
nStart+=1

#insert mistake
nStart=0
mistakeE=math.ceil(nRows*0.02)
while nStart < mistakeE:
nAzar=random.randint(0,nRows)
codeRow="AC000"
codeRow+=str(nStart)
if nStart%2==0:
listT[nAzar][2]=-round(random.random()*100,2)
else:
listT[nAzar][1]=codeRow
nStart+=1

#Array and parallelization context
numPartitions = 80
parallelizationRDD=spark.sparkContext.parallelize(listT,
numPartitions)

#filter
rowOverCero=parallelizationRDD.filter(_.split(",")(1).toDouble
>0))

#Data collection
parallelizationRDD.collect()
rowOverCero.collect()
```

The Table 1 present a workload of 80000 rows for one core processor, those throughput times are significantly small that is because of volatile memory storage, when the partitions were increased the one core device didn't show any response, all the partition of hard drive were filled keeping some of the data, the device didn't shut down the node keep working with its task but no IO operations were allowed on the hard drive.

For the second stage the same workload was tested but this time every process where running with three nodes over vSphere, each node with one core and one gigabyte of volatile memory, at Table 2 every process where executed, the parallelization process also delays time operation for 16000 partitions.

Table 1. Parallelization context, one instance production with one core, one gigabyte of memory.

Array (transactions)	#Partitions	Time response [seconds]
80000	80	4,4 [s]
80000	800	22,7 [s]
80000	8000	No data
80000	16000	No data

Table 2. Parallelization context, virtual repository working with three instances and each one with one core and one gigabyte of memory.

Array (transactions)	#Partitions	Time response [seconds]
80000	80	10,9 [s]
80000	800	31 [s]
80000	8000	211 [s]
80000	16000	527 [s]

Third stage collet data from three physical devices connected at the same LAN network, at Table 3 the time throughput where delayed for latency and process double the times with the half of workload.

Table 3. Parallelization context, physical repository working with three instances and each one with one core and one gigabyte of memory.

Array (transactions)	#Partitions	Time response [seconds]
40000	40	2,6 [s]
40000	400	9,6 [s]
40000	4000	228 [s]
40000	8000	975 [s]

The last stage was planned for start processing data at the same time, all nodes were start sessions at master node, Table 4 shows how slave nodes has priority over computing resources, therefore master node take delay over all operations.

Table 4. Parallelization context, virtual repository working with three instances and each one with one core and one gigabyte of memory.

Array (transactions)	#Partitions	Time response [seconds]		
		Slave 1	Slave2	Master
80000	80	3 [s]	2,6 [s]	6,2 [s]
80000	800	13,5 [s]	13,4 [s]	42 [s]
80000	8000	252 [s]	264 [s]	493 [s]

More test was done to help more for reach more behaviors with data throughput, therefore pointers were calculated at the next section (Table 5).

Table 5. Average time collected for four stages, transactions variable defining workload.

Array (transactions)	Time response [seconds]			
	Stage 1	Stage 2	Stage 3	Stage 4
50	8,7 [s]	3 [s]	2,8 [s]	3,1 [s]
100	13,2 [s]	13,5 [s]	3,5 [s]	4,4 [s]
10000	17,05 [s]	252 [s]	16,02 [s]	15,5 [s]
20000	78,5 [s]	40 [s]	138 [s]	45,5 [s]
40000	579 [s]	158 [s]	405 [s]	100 [s]
80000	No data	256 [s]	No data	121 [s]

4 Metrics

From this point basic pointer will be presented for analyze parallel performance, such as the gain in Eq. 1 whose determinate how the process become more faster if the time for n processors increase, eventually this comes to zero if the number of processors are bigger than the problem itself, although it's a waste of resources.

$$G(n) = t(1)/t(n). \tag{1}$$

In our first stage the one core processor response time will determinate t(1) and the parallel implementation would be represented as t(n), at the next section this pointer will increase eventually but present a limit cause data with tiny information can't be distributed for its complexity.

$$E(n) = G(1)/G(n). \tag{2}$$

As second factor the efficiency or E(n) will provide if the computing resources are enough for the processing task, this factor should be comprehended as lower values for more needed resources in the other hand values above 1 as resources could handle more data computing.

$$R(n) = O(n)/O(1). \tag{3}$$

As third factor the redundancy R(n) which value grows if the parallelization is and the amount of data is high. The O(n) mean the number of operations used in the n processors against the number of operations used in one operation, this value can be calculated counting the number of iterations of code and the level of partitions assigned for the parallelization context.

$$U(n) = R(n)/G(n). \tag{4}$$

System utility U(n) which shows the percent of system in use, this will provide to IT administrators pointers for purchase, update or change hard-drive devices or increase datacenter resources, if the utility value shows a lower percent it means that the resources are not in use and new nodes could be implemented at the solution, otherwise if the percent is high it means all resources are in use.

$$Q(n) = [G(n)/E(n)]/R(n). \tag{5}$$

As last pointer the quality of system will be measured with the gain, efficiency, redundancy values. This value returns high value if the solution is ideal for the problem itself. In the other hand if the value is close to zero the solution will be a waste of resources.

5   Analysis of Results

Since response time at Table 1 present high delay at operations, the operative system takes all memory location in swap partition, stop running memory allocation for processing cause all data was stored on it, as we could see at Table 6, there are three stages the second shows all nodes working over the datacenter instance, the stage 3 shows the gain at physical nodes where the latency on network was present, the stage 4 all nodes where working at the same time, there's an increase at gain at stage 4 where all nodes work at the same time in the other hand there's a decrease of gain where all nodes works on different physical devices as stage 4. In summary the computing resources at stage 2 and 4 are enough for their task and don't need to be upgraded, there's a decrease at stage 3 but isn't below the value of 1 so there's no need for more data capacities.

Table 6. Data gain collected from previous stages.

Array (transactions)	Gain [times]		
	Stage 2 (Virtual)	Stage 3 (Physical)	Stage 4 (Virtual)
50	3,08	3,06	2,78
100	2,1	3,77	3
10000	0,9	1,06	1,08
20000	1,9	0,5	1,73
40000	3,6	1,4	5,75

The Table 7 shows the efficiency in spark solution, this value will always decrease cause the increase of workload, another factor who decrease this pointer is the bandwidth of the network who works with delay at the communications, redundancy at operation also decrease this value, if the efficiency measure is close to zero the resources aren't enough for the data workload.

Table 7. Data efficiency of spark solution.

Array (transactions)	Efficiency		
	Stage 2 (Virtual)	Stage 3 (Physical)	Stage 4 (Virtual)
50	1,08	1,06	0,92
100	0,7	1,25	1
10000	0,3	0,35	0,3
20000	0,65	0,18	0,57
40000	0,21	0,47	1,91

Redundancy at spark solution always be the cause of decrease of efficiency cause for execute a parallel processing there's need to compute a process more than once, this is for a loss of data at communications, or use of libraries such as mapReduce who remap datasets for find a key and match this key for get a result, is one weakest like but a useful tool is the workload isn't so big.

As we can appreciate over Table 8 the redundancy at stage 4 shows a high value because every node starts a session at the same time, it means every node duplicate it computing process at the master node, but, almost stage 2 and stage 4 operates at the same master node but stage one just processes one session with one slave node.

Table 8. Data redundancy, main factor for decrease parallel operations.

Array (transactions)	Redundancy [times]		
	Stage 2 (Virtual)	Stage 3 (Physical)	Stage 4 (Virtual)
50	1,15	1,15	3,45
100	3,4	3,45	10
10000	3,4	3,45	10
20000	4,5	4,56	13,7
40000	5,7	5,7	13,7
80000	4,6	4.6	10,3

The utility equation at Table 9 shows the percentage of computing processing used for task resolution, at more workload more percent of computing resource will be required, at stage 4 all node request a workload with 80000 rows so it can handle the request, our production nodes reach it operation threshold.

As last the quality of solution is based on how much redundancy was injected to obtain a precise solution, this pointer define if the design and operational environment is ideal for a constant workload.

Table 9. Data utility as percent unit.

Array (transactions)	Utility [percent]		
	Stage 2 (Virtual)	Stage 3 (Physical)	Stage 4 (Virtual)
50	3,55%	3,5%	9,64%
100	7,2%	13,06%	31,16%
10000	3,2%	3,68%	11%
20000	8,9%	2,58%	23,7%
40000	20,8%	8,16%	78,8%
80000	No data	No data	No data

As Table 10 shows values above 1 are an ideal solution for the problem itself but just for low workload, for stage 3 where the latency, redundancy and delay were always present so the latency over network affect time responses.

Table 10. Data quality for chosen paradigm.

Array (transactions)	Utility [percent]		
	Stage 2 (Virtual)	Stage 3 (Physical)	Stage 4 (Virtual)
50	2,74	2,71	0,74
100	0,42	1,3	0,28
10000	0,08	0,1	0,03
20000	0,27	0,02	0,07
40000	0,77	0,11	0,8
80000	No data	No data	No data

6   Conclusions

Performance of the system will show a final impact over the quality of services this is because the redundancy generated using filters over array generates a remapping process over operations, the throughput time collected over all stages provide spark as a good developing software which decrees process operating due memory storage.

The framework tested also prove a good solution for scalability this is because all performance could be adjusted just changing the memory to one with more capabilities, there's no need to buy expensive architectures.

The gain reached in the collected data prove that a process could be done 5 time faster by processing the workload in parallel, just three nodes could balance high workload data.

The system met the performance requirements to balance the workload in the operating nodes for large volumes of information, achieving that the system utilization reached only 78% for the tasks assigned.

The performance of the system in parallel was only influenced by network latency generated affected the efficiency of the system in half in a LAN, this is because the parallelism collects results and if there are delays in the collection the quality of parallelism decays.

In conclusion, the framework used is a potential tool for data processing that does not use licenses to provide a quality service, optimizing the resolution of problems based on operating memory and providing an interface layers to establish sessions in any operating node, this means that if more nodes with limited features are implemented, they can initialize sessions and through the hadoop component obtain a replica of the results provided.

References

1. Hennessy, J., Patterson, D.: Computer architecture: a quantitative approximation, San Francisco, pp. 3–5 (2007)
2. Ghodsi, A., Joseph, A., Randy, K., Scott, S., Ion, S.: A platform for fine-grained resource sharing in the data center. In: IEEE Access, California, pp. 1–12 (2009)
3. Asanovic, K., Bodik, R., Catanzaro, B., Gebis, J., Husbands, P., Keutzer, K., Patterson, D., Plishker, W., Shalf, J., Webb, S., Yelick, K.W.: The Landscape of Parallel Computing Research: A View from Berkeley. Universidad de Berkeley, California (2006)
4. Oliker, L., LiGerd, X., Biswas, H.: Ordering Unstructured Meshes for sparce matrix computations on leading parallel system. Berkeley, California (2000)
5. Langer, U., Paule, P.: Numerical Methods and Symbols of Scientific Computing: Progress and Prospects. Mathematical Computing Institute, Australia (2011)
6. Intel IT Center: Planning Guide: Getting Started with Hadoop. Steps IT Managers Can Take to Move Forward with Big Data Analytics (2012). http://www.intel.com/content/dam/www/public/us/en/documents/guides/getting-started-with-hadoop-planning-guide.Pdf
7. Singh, S., Singh, N.: Big Data analytics. In: International Conference on Communication, Information & Computing Technology Mumbai India. IEEE (2011)
8. Kossmann, D., Kraska, T., Loesing, S.: An evaluation of alternative architectures for transaction processing in the cloud. In: Proceedings of the 2010 International Conference on Management of Data, pp. 579–590. ACM (2010)
9. Dean, J., Ghemawat, S.: Mapreduce: simplified data processing on large clusters. Commun. ACM **51**(1), 107–113 (2008)
10. Xu, Y., Kostamaa, P., Gao, L.: Integrating hadoop and parallel DBMs. In: Proceedings of the 2010 International Conference on Management of Data, pp. 969–974. ACM (2010)
11. Jiang, D., Tung, A., Chen, G.: Map-Join-Reduce: toward scalable and efficient data analysis on large clusters. IEEE Trans. Knowl. Data Eng. **23**(9), 1299–1311 (2011)
12. Villegas-Ch, W., Luján-Mora, S., Buenaño-Fernandez, D., Palacios-Pacheco, X.: Big Data, the next step in the evolution of educational data analysis. In: International Conference on Information Theoretic Security, pp. 138–147. Springer, Cham, January 2018
13. Villegas-Ch, W., Luján-Mora, S.: Analysis of data mining techniques applied to LMS for personalized education. In: IEEE World Engineering Education Conference (EDUNINE), pp. 85–89. IEEE, March 2017
14. Villegas-Ch, W., Luján-Mora, S., Buenaño-Fernandez, D.: Towards the integration of business intelligence tools applied to educational data mining. In: 2018 IEEE World Engineering Education Conference (EDUNINE), pp. 1–5. IEEE, March 2018
15. Villegas-Ch, W., Luján-Mora, S., Buenaño-Fernandez, D.: Data mining toolkit for extraction of knowledge from LMS. In: Proceedings of the 2017 9th International Conference on Education Technology and Computers, pp. 31–35. ACM, December 2017

Cognitive Security for Incident Management Process

Roberto Andrade[1(✉)], Jenny Torres[1], and Susana Cadena[2]

[1] Escuela Politécnica Nacional, Quito, Ecuador
roberto.andrade@epn.edu.ec
[2] Universidad Central del Ecuador, Quito, Ecuador

Abstract. This work presents the literature review about the process of handling security incidents to identify standards or guidelines published by international organizations. Based on this research we identified the phases of the incident management processes with the goal of analyze automation proposals for improve efficiency and response times. Finally, we analyzed the contribution of cognitive security to enhanced the cognitive skills of security specialists in the execution of tasks that are associated with the detection phase in the incident management process.

Keywords: Cognitive security · Incident response · Detection · Big-data

1 Introduction

The lack of an adequate incident management process has a negative effect on the organizations to provide an efficient action to resolve the security incident [1]. The process of incident management has been proposed by international organizations, but there are still organizations that cannot adequately handle a computer attack. The success of a response to incidents requires not only of technological solutions, also needs of personnel able to execute correctly the tasks defined in the process in the shortest time possible. Some limitations that may affect this objective are:

- Limited personal.
- Large amounts of information.
- Administrative tasks that reduce the time for post-incident analysis.
- Lack of flexibility and scalability of the current security solutions in front of the dynamic environments that are generated with the appearance of the cloud, IOT, 5G networks.

According to NIST, new technologies offers an opportunity to process relevant data security in structured and unstructured formats, allowing the analysis of situational awareness in infrastructure conformed by large and complex networks [7]. In this work we analyze alternatives that have been proposed to

© Springer Nature Switzerland AG 2019
Á. Rocha et al. (Eds.): ICITS 2019, AISC 918, pp. 612–621, 2019.
https://doi.org/10.1007/978-3-030-11890-7_59

improve response times and effectiveness in the incident response processes. Concepts such as cognitive security that include the contributions of artificial intelligence, data analytics, and bigdata can contribute to this objective of improving the response processes to resolve incidents. Our motivation is analyze the proposals related to the status of the incident management processes and the future strategies for enhanced the incident response process.

The document is organized as follows. Section 2 presents background on the challenges of cybersecurity. Section 3 focuses on a literature review of security incident response processes, analyzing standards and guidelines used in this context and alternatives to improve the incident management processes. Section 4 analyzes the cognitive security proposal for the incident security responses. Finally, Sect. 5 makes the conclusions and proposals for future work.

2   Background

2.1   Challenges of Incident Response Process

In the analysis presented in the work "Rethinking Security Incident Response: The Integration of Agile Principles" [8], presents some problems about the response of security incidents process, we can highlight some of them:

- Linear incident response models that does not adapt efficiently to the capabilities needed to handle attacks.
- Processes focused on containment, eradication and recovery, omitting important phases such as preparation, detection and lessons learned.
- Current uses do not allow a true understanding or knowledge of the causes of the incident.
- Reduced use of digital forensic skills.

Complementary, IBM indicates that only 8% of security knowledge is used to establish security defenses, organizations focus on security alerts, network activity, configuration logs, without consider research documents, analysis reports, forensic reports, web pages, and presentation of conferences [9]. According to IBM through cognitive security it is possible to achieve the following aspects [9]:

- Maintain the currency of security knowledge.
- Remove human error and dependency on research skills.
- Reduce time required to investigate and respond to security incidents.

3   Literature Review

The goal of incident response is reduce the impact of a security attack and return to an acceptable security posture [10]. Some organizations to handle security incidents deploy incident response teams called CSIRT, CERT or SIRT [11]. International organizations such as National Institute of Standards and Technology, European Network and Information Security Agency, SysAdmin Audit,

Networking and Security Institute, and British Standards Institution, establish some standards or guidelines to handle security incidents. In order to improve the efficiency and effectiveness of incident response processes the use of automation has been chosen as a strategy.

3.1   Research Methodology

The proposed research methodology consists of the following steps [12–14]:

1. Define the research questions.
2. Establish the selection of scientific databases.
3. Establish the inclusion and exclusion criteria.
4. Analyze and synthesize.
5. Report and use the results.

The following research questions were defined for this work: What standards and guidelines proposed by international organizations have been used in scientific works related to security incident management processes? What proposals have been established to improve the effectiveness and performance of security incident response processes?

In the present work for literature review we selected the following academic databases: IEEXplore, ACM, Scopus, Science Direct, and Web of Science for obtain specific information about standards used in the security incident management process. The research has been limited to publication dates from 2014 to 2018, taking into consider that the Morris virus was in 1988 and started the development of preventive and reactive measures for the response of incidents and the creation of the first computer incident response team CERT/CC [16], so we consider that approximated ten years since this date, the standards or guidelines had a greater development. According to the inclusion and exclusion criteria established we defined the following search strings:

- "Security" AND "Incident" AND "Management".
- "Security" AND "Incident" AND "Response".
- "Security" AND "Incident" AND "Handling".
- "Security" AND "Incident" AND "Standards" AND "Guidelines".

To determine the contributions in automated incident management process we defined the following search strings:

- "Security" AND "Incident" AND "Automated".
- "Security" AND "Incident" AND "Response" AND "Automated".
- "Incident" AND "Response" AND "Automated".
- "Security" AND "Incident" AND "Detection".
- "Security" AND "Incident" AND "Dynamic".

According with the research criterion of this work, 19 relevant papers were found, 8 related to incident response process standards or guidelines and 11 about strategics for security incident response. The analysis of document RFC 2196 - Site Security Handbook of the Internet Engineering Task Force [15] has been excluded. The following results were obtained based on the literature review: Five international organizations have been identified, which proposes handling security incidents standards or guidelines.

1. Computer Emergency Response Team Coordination Center - CERT/CC. Six phases: preparation, identification, containment, eradication, recovery, and lessons learned.
   - The Handbook for Computer Security Incident Response Teams (CSIRT).
   - The State of the Practice for CSIRTs [11].
   - The Organizational models for CSIRT [18].
   - Defining Incident Management Processes for CSIRTs: A Work in Progress [17].
2. National Institute of Standards and Technology- NIST. Four phases preparation, detection and analysis, containment, eradication and recovery, and post-incident activity
   - NIST Special Publication (NIST SP 800-61) [21].
3. European Network and Information Security Agency -ENISA. Six phases: incident report, report registration, triage, incident resolution, incident closure, and post-analysis
   - Good Practice Guide for Incident Management [23].
4. SysAdmin Audit, Networking and Security Institute - SANS. Six phases: preparation, identification, containment, eradication, recovery, and lessons learned.
   - SANS' Incident Handler's Handbook [19].
   - An Incident Handling Process for Small and Medium Businesses [22].
5. International Organization for Standardization and International Electrotechnical Commission. Five phases: plan and prepare, detection and reporting, assessment and decision, responses, and lessons learned.
   - ISO/IEC 27035:2011 - Information technology - Security techniques [24].

In Table 1, we present a consolidated list of topics that cover the proposed standards or guidelines on security incident management processes, indicating their year of publication.

The following relevant aspects have been found in the context of the strategics of security incident response processes. According to [25], you can have three levels of automation:

- Manual response system
- Automatic response system
- Notification system

Dynamic Models. In [2], proposes a methodology to automatically determine incident response based on mathematical models to generate cognitive maps. The cognitive map is created by a consensus of the security team to decide how resolve an incident in basis of attack scenarios using different paths that can be taken by the attackers or the security analyst.

Table 1. Topics about incident response process

Organization	N.Doc.	Topics	Year
CERT	4	Organizational Model	2003
		Incident handling process	
		CSIRT Services	
ENISA	1	Incidents handling phase	2010
		Roles	
		Workflows	
		Polices	
SANS	1	Incident handling process	2011
NIST	1	Incidents handling phase	2012
ISO	1	Incident handling process	2012

Game Theory Model. The use of game theory is based on establishing two attacking and defending roles to predict possible sequences of attacks that can be executed. Game theory proposals are:

– Nash equilibrium strategies [26]
– stochastic Models [5]
– Bayesian learning [27]

Decision-Making Models. Based on the literature review the following proposals for decision-making models are:

– Fuzzy decision making and risk assessment [3]
– Ontologies-Based [6]
– Hierarchical task network planning [28]
– Risk impact tolerance [29]
– Attack damage cost [30]
– Markov decision [31]

Multi-agents Models. In [4], a multi-agent system (MAS) use an autonomous and decentralized architecture in order to execute tasks and solve problems related to the handling of incidents.

The proposal defines different roles of agents:

– User agent: is the system user who can view the details of the incident.
– Administrator agent: responsible for the administration of the system.
– Supervisor agent: in charge of supervising IT services to detect anomalies or issues.
– Incident agent: is responsible for handling the incident.
– Diagnostic agent: responsible for evaluating the impact of the incident.
– Support agents: works in coordination with the incident agent.

In the Fig. 1, we present a summary of the strategies of security incidents responses.

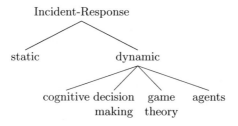

Fig. 1. Incident response strategics map

4 Cognitive Security for Incident Response

4.1 Cognitive Security Tasks in Incident Response

NIST, proposed the Cybersecurity Framework [33], based on risk management that considers the standards analyzed previously for determine, test and implement controls to reduce risk. NIST considers that a workforce should be adapt, design, develop, implement, maintain and continually improve cybersecurity practices within critical infrastructure environments. Cognitive security provides the ability to process large amounts of data from external sources, internal, structured and unstructured, enhancing the ability of the security analyst to detect and respond to threats [32]. Cybersecurity Framework [33], includes the following categories for detection:

- Detect anomalies and events, and their potential impact.
- Implement continuous monitoring capabilities and verify proactive measures.
- Maintain detection processes to provide awareness of anomalous events.

The following five subcategories are defined for "Detect anomalies and events" category.

- DE.AE-1: Baseline for network operations and data flow expected by users and systems.
- DE.AE-2: Detected events are analyzed to understand the methods and objectives of the attack.
- DE.AE-3: Data events are aggregated and correlated by multiple sources and sensors.
- DE.AE-4: Impact of events is determined.
- DE.AE-5: Thresholds of incident alerts are established.

At the RSA conference in 2017, IBM [32] presented the cognitive tasks that must be performed by a security analyst in the investigation of an security incident.

1. Understanding the local context of the incident.
   - TC-1. Review incident data.
   - TC-2. Review the events by aspects of interest.

- TC-3. Pivot in the data to find atypical values or outliers.
- TC-4. Expand the search to find more data.
2. TC-5. Investigate the threat to develop experience.
   - TC-6. Discover new threats.
   - TC-7. Determine indicators of commitment in other sources.
3. TC-8. Apply intelligence to investigate the incident.
   - TC-9. Discover IPs potentially infected.
   - TC-10. Qualify the incident based on the knowledge generated based on the investigation of the threat.

In the Table 2, we present the correlation between the detection subcategory of the CSF and the cognitive tasks required by the analyst.

Table 2. Cognitive tasks in CSF

Subcategory	Cognitive tasks
DE.AE-1	TC-3, TC-4, TC-6
DE.AE-2	TC-1, TC-2
DE.AE-3	TC-7, TC-9
DE.AE-4	TC-7, TC-8, TC-10
DE.AE-5	TC-5, TC-10

4.2 Cognitive Security Tools for Incident Response

For implement the tasks of the subcategories of the "detect anomalies and events" category of CSF, is required the contribution of the security specialists. The tasks for identify and classify events is based on knowledge and experience from security specialist and external sources (logs, security news, blogs or indicator of compromise-IOC). As mentioned in this work, full-time security specialists in the practice is limited and additionally faces large amount of information that exceeds their human capacity. In this context, the cognitive security allows increase the cognitive skills of the security specialists for support the execution of detection tasks. The authors of this work consider that cognitive security is a set of technological solutions, procedures and data analysis techniques, which allow accomplish the objective of enhanced the cognitive skills of the security specialist and provides the organization of the situational-awareness in real time. In the Fig. 2 we present a proposal of cognitive security model that includes the technological solutions that could be used to enhanced cognitive skills of security analyst. The model presents a relationship of technological solutions with states of awareness of cybersecurity (situation and self-awareness).

Using cognitive security could generate the knowledge within each system (self-awareness) and of the environment (situational awareness). For example, data analysis allows strengthening the cognitive skills of security experts, to execute the tasks that allow accomplish the goals of detection subcategories proposed by the CSF in the NIST. Through the use of bigdata, the cognitive tasks TC-3, TC-4, TC-7 and TC-9 could be executed, and accomplish the subcategories DE-AE-1 and DE-AE-3 of the CSF.

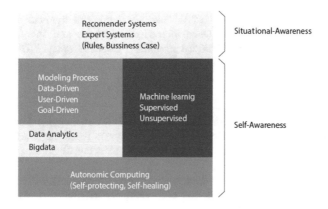

Fig. 2. Cognitive Security Model

5   Conclusions and Future Works

From the literature review conducted in this work, different standards and guidelines for the incident management process were identified, proposed by five organizations: CERT, NIST, ENISA, ISO, and SANS, which includes topics as the establishment of a incident response team (CSIRT), the description of roles and functions in the CSIRT, the definition of communication protocols between CSIRT and the process flows for the management security incidents.

Triage is one the process more important in the management of security incident response process and this requires the greatest attention of security analysts for prioritize an incident based on the identification, classification and impact of an incident in the shortest time possible using information of logs, security bulletins and consult with other specialists.

Large amount of information is generated by complex and large networks, even more with the use of solutions like IoT, cloud and mobile, that could exceed the attention and cognitive processing of security analyst. The cognitive security could improve the cognitive skills of the security analysts making use of technological solutions such as: bigdata, machine learning and data mining to generate states of cybersecurity awareness.

The state of cybersecurity awareness is generated from the cognitive processes performed by the security analyst associated with the detection processes of anomalies and incidents. Our proposal of cognitive security model is based on relating the cognitive processes that are executed trough of cognitive security tasks that contribute to the fulfillment of the activities defined in the subcategory of security detection of incident and anomalies proposed in the CSF by NIST.

A future work is to propose a methodology of evaluation of the cognitive processes executed by the security analyst to measure the improvement in performance when using cognitive security solution.

Acknowledgment. The authors would like to thank the financial support of the Ecuadorian Corporation for the Development of Research and the Academy (RED CEDIA) for the development of this work, under Project Grant GT-II-2017.

References

1. Nugraha, A., Legowo, N.: Implementation of incident management for data services using ITIL V3 in telecommunication operator company. In: 2017 International Conference on Applied Computer and Communication Technologies (ComCom), pp. 1–6 (2017)
2. Krichene, J., Boudriga N.: Incident response probabilistic cognitive maps. In: 2008 IEEE International Symposium on Parallel and Distributed Processing with Applications, pp. 689–694 (2008)
3. Berenjian, S., Shajari, M., Farshid, N., Hatamian, M.: Intelligent automated intrusion response system based on fuzzy decision making and risk assessment. In: 2016 IEEE 8th International Conference on Intelligent Systems (IS), pp. 709–714 (2016)
4. Latrache, A., Nfaoui, H., Boumhidi, J.: Multi agent based incident management system according to ITIL. In: 2015 Intelligent Systems and Computer Vision (ISCV), pp. 1–7 (2015)
5. Kundu, A., Ghosh, S. K.: Game theoretic attack response framework for enterprise networks. In: Distributed Computing and Internet Technology, pp. 263–274 (2014)
6. Lanchas, V.M., González, V.A.V., Bueno, F.R.: Ontologies-based automated intrusion response system. In: Herrero, Á., Corchado, E., Redondo, C., Alonso, Á (eds.) Computational Intelligence in Security for Information Systems 2010. AISC, vol. 85. Springer, Heidelberg (2010)
7. NIST: Roadmap for Improving Critical Infrastructure Cybersecurity (2014). https://www.nist.gov/
8. Grispos, G., Glisson, W., Storer, T.: Rethinking security incident response: the integration of agile principles. In: 20th Americas Conference on Information Systems, Journal, Dagstuhl Reports (2014)
9. IBM: Applied Cognitive Security: Complementing the Security Analyst (2017). https://www.rsaconference.com/writable/presentations
10. Johannes, W.: Limits to Effectiveness in Computer Security Incident Response Teams (2005). https://resources.sei.cmu.edu/library/
11. Killcrece, G., Kossakowski, K.-P., Ruefle, R., Zajicek, M.: State of the Practice of Computer Security Incident Response Teams. Software Engineering Institute, Carnegie Mellon University (2003)
12. Denyer, D., Tranfield, D.: Producing a systematic review. In: The Sage Handbook of Organizational Research Methods, pp. 671–689 (2009)
13. Costa, E., Soares, A.L., de Sousa, J.P.: Information, knowledge and collaboration management in the internationalisation of SMEs: a systematic literature review. Int. J. Inf. Manag. **36**(4), 557–569 (2016)
14. Guilera, G., Barrios, M., Gómez-Benito, J.: Meta-analysis in psychology: a bibliometric study. Scientometrics **94**(3), 943–954 (2013)
15. IETF: RFC2196 Site Security Handbook (1997). http://www.rfc-editor.org/rfc/pdfrfc/rfc2196.txt.pdf
16. IETF: RFC2235 Hobbes' Internet Timeline (1997). https://tools.ietf.org/html/rfc2235. Accessed 9 Sept 2018

17. Alberts, C., Dorofee, A., Killcrece, G., Ruefle, R., Zajicek, M.: Defining Incident Management Processes for CSIRTs: A Work in Progress. Carnegie Mellon University, Software Engineering Institute (2004)
18. Killcrece, G., Kossakowski, K.-P., Ruefle, R., Zajicek, M.: Organizational Models for Computer Security Incident Response Teams. Software Engineering Institute, Carnegie Mellon University (2003)
19. SANS: Incident Handler's Handbook (2005). https://www.sans.org/
20. ITIL: Information Technology Infrastructure Library (2011). https://www.axelos.com/best-practice-solutions/itil
21. NIST: Computer Security Incident Handling Guide (2012). https://csrc.nist.gov/publications
22. SANS: Incident Handler's Handbook (2005). https://www.sans.org/reading-room/whitepapers
23. ENISA: Good Practice Guide for Incident Management (2010). https://www.enisa.europa.eu/publications/
24. ISO: ISO/IEC 27035:2016 Information technology - security techniques – information security incident management (2016). https://www.iso.org/
25. Shameli-Sendi, A., Ezzati-Jivan, N., Jabbarifar, M., Dagenais, M.: Intrusion response systems: survey and taxonomy. Int. J. Comput. Sci. Netw. Secur. (IJCSNS) **12**, 1–14 (2012)
26. Zonouz, S., Khurana, H., Sanders, W., Yardley, T.: RRE: a game-theoretic intrusion response and recovery engine. J. IEEE Trans. Parallel Distrib. Syst. **25**, 395–406 (2014)
27. Luo, Y., Szidarovszky, F., Al-Nashif, Y., Hariri, S.: A fictitious play-based response strategy for multistage intrusion defense systems. J. Secur. Commun. Netw. **7**, 473–491 (2014)
28. Chengpo, M., Yingjiu, L.: An intrusion response decision-making model based on hierarchical task network planning. J. Expert Syst. Appl. **37**, 2465–2472 (2010)
29. Shameli-Sendi, A., Dagenais, M.: ARITO: cyber-attack response system using accurate risk impact tolerance. Int. J. Inf. Secur. **13**, 367–390 (2014)
30. Shameli-Sendi, A., Louafi, H., Cheriet, M.: Dynamic optimal countermeasure selection for intrusion response system. IEEE Trans. Dependable Secur. Comput. **15**, 755–770 (2018)
31. Iannucci, S., Abdelwahed, S.: Model-based response planning strategies for autonomic intrusion protection. ACM Trans. Auton. Adapt. Syst. **13**, 4 (2018)
32. IBM: Applied cognitive security complementing the security analyst (2017). https://www.rsaconference.com
33. NIST: Cybersecurity Framework (2018). https://www.nist.gov/cyberframework/

Human-Computer Interaction

Gesture Elicitation and Usability Testing for an Armband Interacting with Netflix and Spotify

Robin Guérit[1], Alessandro Cierro[1], Jean Vanderdonckt[1(✉)] (iD),
and Jorge Luis Pérez-Medina[2] (iD)

[1] Université catholique de Louvain (UCL),
LouRIM - Place des Doyens, 1, 1348 Louvain-la-Neuve, Belgium
{robin.guerit,alessandro.cierro}@student.uclouvain.be,
jean.vanderdonckt@uclouvain.be
[2] Intelligent and Interactive Systems Lab (SI² Lab),
Universidad de Las Américas (UDLA), Quito, Ecuador
jorge.perez.medina@udla.edu.ec

Abstract. Controlling home entertainment devices, like music and video, via an armband could free the user from using remote controls, but assessing their overall usability with mid-air and micro-gestures still represents an open research question today. For this purpose, this paper reports on results gained by jointly conducting and comparing two studies involving participants using a Thalmic Myo armband to control a NetFlix SmartTV and Spotify: (1) a gesture elicitation study to explore a richer set of user-defined gestures, to measure their effectiveness and the user subjective satisfaction of gesture interaction; (2) a System Usability Scale (SUS) to assess the overall usability of this setup and the subjective satisfaction for user-defined gestures.

Keywords: Gesture elicitation study · Mid-air gestures ·
Myo armband · Netflix · SmartTV · TV interaction ·
Wearable computing

1 Introduction

With modern technologies, new devices came up and are used in many daily life contexts of use, such as TV, recorder, communicating objects. There has been much research on how to control a TV using different modalities of interaction, ranging from the traditional physical remote control, to full-body gesture interaction [20]. With traditional remote controls the user is forced to always manipulate a physical device that may not be understandable, which is often inconsistent with another one, and which can be easily lost. When a new function is added to the TV, the remote control cannot be updated accordingly. *User-defined gestures* can be proposed by the end users themselves in order to

© Springer Nature Switzerland AG 2019
Á. Rocha et al. (Eds.): ICITS 2019, AISC 918, pp. 625–637, 2019.
https://doi.org/10.1007/978-3-030-11890-7_60

Double Top	Wave Right	Wave Left	Spread Fingers	Rotate	Make a Fist	
unlock	Forward	Rewind	Play and Pause	Turn Volume Up And Down	Control Volume	Myo Information interface (visual feedback)

Fig. 1. Thalmic Myo Armband system-defined gestures.

come up with a consensus set of gestures [21]. It is not because these gestures emerge based on some consensus that they are simultaneously usable.

In this paper, we would like to investigate whether another interaction modality, i.e. an Armband based on electromyographic sensors (EMG) [7], could become a realistic alternative for interacting with common tasks based on a TV by combining a usability testing and a gesture elicitation study. The Myo Armband [7], is an off-the-shelf motion control device. It includes 8 EMGs able to measure electrical activity in the forearm muscles, thanks to which it detects and identifies the user's arm movements and/or gestures.

Two plugins were developed for this study. We also incorporate the control of Netflix thanks to the Netflix Connector: a free additional app built into Myo Connect (standalone Myo software) provided by Thalmic Labs and available in the official Myo Market. The Netflix Connector enables the control of the Netflix interface into the web browser. The control is set up through 7 standardized and non-changeable *system-defined gestures* (Fig. 1), which will be used for the navigation through both Netflix and Spotify. Our practical results consist in (1) an analysis of the agreement rates among the gesture proposals of the participants, (2) gestures comparison between TV interaction and music control with different positions, (3) usability testing scores about the predefined gestures of the Myo armband, (4) a comparison between the agreement rate pre- and post-experiment with the Myo armband, and (5) limits and discussion.

2   Related Work

This section discusses prior work on the Myo armband technology and reviews the principles of the gesture elicitation methodology in order to collect and understand the users preferences for mid-air gestures. Previous studies show several technological aspects of the Myo Armband in several domains of applications, such as: healthcare [16], virtual reality [17], music [5], and sports. For example, MyoGym [10] consists of an interactive application capturing up to 30 different physical exercises, monitor the movements of the end user, and reports on the progression over time. Another system [13] also relies on EMG to train the patient's muscles when they received a prosthetics. Squeeze gestures can be captured by the Armband in [14] by children suffering from cerebral palsy when they manipulate objects in virtual or mixed reality. Another research compared the use of the Nova bio-medical base sensor (a non-invasive wearable system which allows the measure of complex physiological phenomena) [13] and the

Myo Armband for muscle fatigue detection. The Myo armband could even be used in some limited medical scenarios where high accuracy levels of EMG are not required [13]. To expand the set of armband gestures, another gesture recognizer has been tested [8], which outperforms the original Myo algorithm with an overall accuracy of 95% compared to 68% for this gesture set. Instead of point-based recognizers, vector-based gesture recognizers also appear [18].

Understand, collect, and analyze users preferences with an interactive technology is the purpose of gesture elicitation. End users are not always consulted by designers when they create gesture interfaces [4]. Gesture elicitation, also known as "participatory design" [2] or "guessability studies" [21], is suitable for a large variety of sensing devices and domains [19], which could then be mapped into user interface design artefacts [12]. The outcome of a gesture elicitation study consists in a characterization of users' gesture input behavior with valuable information for designers, practitioners, and end users regarding the consensus levels between participants (computed as agreement or co-agreement scores [19–21]). The most recent formalization of the elicitation methodology was proposed in [20]. Gesture elicitation study also worked in a smart home as a larger environment [19]. For controlling television interaction showed, gestures are carried out in mid-air [4].

3   Research Questions

We divided our experiment into four phases in order to understand and test our research questions and hypothesis defined below: **RQ1**. What are the user's preferred gestures with an interactive armband for the TV interaction in their comfortable viewing position? Which gestures created by the participants are the most adapted to the requested action (referent)? Analysis and evaluation of gestures proposed under objective indicators (agreement rates) and a subjective questionnaire. **RQ2**. What are the user's preferred gestures for the use of an interactive armband to control music in a different active position? How many changes? Hypothesis 1: Users use the same gestures to control music as for the TV interaction. **RQ3**. What is the Myo usability? Learning and testing. What is the usability of the Myo for the Netflix interaction in a comfortable position? **RQ4**. What are the actions ultimately taken after the experiments? Compromise and justification of the use of the Myo bracelet.

4   Experiment and Methodology

We conducted a gesture elicitation experiment by following the methodology from the review of literature and a usability test was also conducted [9,15,20].

4.1   Participants

We interviewed 21 participants (10 female), aged between 15 and 71 years old ($M = 35.86$, $SD = 18.3$), who volunteered for our experiment. Two pilots and 19 for the main study. We had to exclude 2 outliers who did not succeed the pretest. The majority of the participants (16) were right-handed.

Table 1. Set of referents used for the elicitation experiment.

No.	Referent	Description	No.	Referent	Description
1	Unlock	Turn on the TV set	5	Rewind	Turn off the volume
2	Play	Turn off the TV set	6	Volume Up	Go to the next channel
3	Pause	Increase the volume	7	Volume Down	Go to the previous channel
4	Forward	Decrease the volume			

4.2 Apparatus

The experiments took place in two rooms set up to make the participant feel almost at home. In each setup there were a two-seat sofa and some blankets and cushions to enable a usual comfortable viewing position. We interviewed each participant at a time and the experiment lasted about 1 h. We used *Okja*, the South Korean-American movie, for all the Netflix phases and Michael Jackson's *Don't Stop 'Til You Get Enough* for the Spotify phases in order to stay neutral and avoid both choking scene and explicit lyrics. The participant wore armband on the dominant arm throughout the experiment. We informed the participant about the sensors included in the armband without showing any gesture and about the possibility to perform mid-air and finger gestures.

4.3 Procedure

Phase 1. After introducing the Myo Armband, we asked the participant to take a comfortable viewing position on the sofa with all the cushions needed (see Fig. 2). We presented the 7 referents they were going to elicit (see, Table 1). We allowed the participant to view all referents at once (after presenting the "unlock" referent alone) and they could revisit their gestures at any time. In doing so, we encouraged the participant to try out several possibilities for each referent. The participant had 5 min to think about his gestures. Then, we asked him to perform each referent by using his own suggested gestures and in the same time we controlled the computer to allow him to foresee the results. At the end, the participant was asked to complete a gesture questionnaire for each of his 7 suggested gestures and several subjective questions were asked.

Fig. 2. Positions.

Phase 2. The participant had to quit their comfortable position and stand up. They were asked to use the Myo Armband (using the same 7 referents) to navigate in Spotify instead of Netflix. They had 5 min to find out if they were going to use the same gestures as in the phase 1, change all the gestures or change some of them. Once again, we simulated the Spotify system to allow the participant to imagine the interactions related to their gestures. The aim of this phase is to observe the transferability of the suggested gestures during the first phase from one context of use to another.

Phase 3. We introduced the standardised Myo gestures for each referent and the armband manager gestures interfaces (see Fig. 1, right section). To allow the best recognition with the armband, we created a new profile for each participant by using a calibration step to teach the standardized gestures. Then, the participant was asked to lay back in their previous comfortable viewing position and to perform all the gestures for training. During the practice time we asked the participant to perform each action (randomly chosen) when asked, this phase lasted 5 min. The participant was asked to perform the standardized gestures following their own desires for 5 min. Then the participant completed a gesture questionnaire for each referent and a SUS test. During this phase we used the Armband instead of only simulate the results as previously.

Phase 4. The participant was asked to stand up as we asked them to use Spotify one last time. We asked the participant to figure out (after suggesting their own gestures, during the phase 1 and 2, and learning the standardized Myo gestures) what gestures they wanted to keep to perform the referent actions for Spotify. They also were aware they could suggest completely new gestures. Having taken 5 min to consider the matter, we asked the participant to perform each referent while we simulated the results (as it was the case during the phase 2). An IBM Computer Satisfaction Usability (IBM CSUQ) Questionnaire [11] was completed by the participant, and final open questions were asked.

4.4 Qualitative and Quantitative Measures

We employed measures to evaluate and understand the users preferred gestures and the users' performances in using the Myo Armband: (**1**) We asked the participants to assume a VIEWING POSITION that they usually employed when watching TV at home. According to the results of the [5], we included the influence of a natural, comfortable position during the interactive test with Netflix. To assure a great ecological validity in our research, we will describe all different participants' positions and include their remarks if they had difficulties related to their positions. (**2**) A CREATIVITY TEST was asked before the beginning of the experiment. It is made of 40 questions and gives a free assessment of one's level of creativity through 8 different metrics (abstraction, connection, perspective, curiosity, boldness, paradox, complexity and persistence)[1]. The participants needed a score higher than 55 to be part of the experiment. We rejected 2 participants because of a too low score. (**3**) We also conducted a MOTOR SKILL test in order to select people for the elicitation step, we measured motor ability levels in fingers. The test was inspired by NEPSY test [9] and consisted of touching their

---

[1] http://www.testmycreativity.com.

thumb with the other fingers tip 8 times. (4) We calculated for each referent the central measure or the agreement rate (AR) as defined in [20] to understand the extent to which the participants agreed on each gesture. Proposed gestures for specified tasks are analyzed and used in comparison in order to predict the most suitable gestures. (5) A questionnaire with 7 subjective questions, on a 7-point Likert scale, was developed to allow the participants to rate the quality of a gesture. They could indicate the ease of execution, the memorability, the good correspondence or the enjoyability in performing [6]. (6) We used the SUS questionnaire [1,3] to measure the system usability. (7) Finally, the IBM CSUQ questionnaire [11] was used to relate the user satisfaction.

5   Results

5.1   Participants Viewing Position

We distinguished 16 different positions on 5 dimensions (Table 2). Some were in a semi upright position, some used different items like cushions and blanket and some put their feet on the couch. Participants were free to choose their own position as categorizing relaxed viewing positions is challenging.

Table 2. Overview of the participants viewing position.

Upper body	Feet	Back	Arm	Items
Sitting upright, sitting semi upright, lying on the couch, semi lying on the couch, sitting lying on the couch	On the couch with knees to right, on the couch with knees to left, on the couch with legs straight, on the couch with legs crossed, on the ground	On the back couch	Released on a cushion, released on the couch, released on the legs	On cushion and a blanket, sitting on a cushion, sitting with a blanket

5.2   Phase 1

We computed the agreement scores using the AGreement Analysis Toolkit tool (AGATe) [19,20]. The resulting gesture set – taking the gesture with the highest agreement rate for all participants – is presented in the left graph of Fig. 3. There are three referents for which there are very low agreement scores; for these referents, we have provided two or three elicited gestures in the Table 3. Except for the volume referents, we do not have a consensus on gestures proposals for other referents. These results could be explained by our observations. The participants were very creative during the elicitation test. Overall, we elicited 57 gestures for the 7 referents by the 19 people. Most of the elicited gestures are performed with the hand and the arm, only 4 out of 19 participants performed

Table 3. Set of referents used for the elicitation study.

Referent	Description of the elicited gesture (Phase 1)	Description of the elicited gesture (Phase 4)
Volume Up	Raise the arm vertically with the palm up	(1) Raise the arm vertically with the palm up; (2) Make a fist and rotate right
Volume Down	Put the arm down vertically with the palm down	(1) Put arm down vertically with the palm down; (2) Make a fist and rotate left
Rewind	(1) Move the arm and the hand to the left; (2) Wave right	(1) Wave left; (2) Move the arm and the hand to the left
Forward	(1) Move the arm and the hand to the right; (2) Wave right	(3) Wave right; (4) Move the arm and the hand to the right
Pause	(1) Extend your arm and point your hand face to the TV; (2) Get your hands down by tapping; (3) Make a fist	Spread fingers
Unlock	Snap	Double tap
Play	Extend your arm and point your hand face to the TV	Spread fingers

micro-gestures (with only fingers). Even if most of the gestures are spontaneous (less than 5 min for the conception) and natural, the variety of gestures is too wide and we cannot predict any agreement. Depending on their viewing position, the majority of the proposed gestures have been validated as easy to execute. The participants adapted their gestures to their position without any imposed constraint.

5.3   Phase 2

The vast majority kept all the gestures elicited for this test, 7 people changed their gestures to control music on Spotify. 18 gestures changed on all proposed gestures (133). According to the results provided by the elicitation of the gestures, we can confirm our hypothesis that people keep the same gestures set for the same referents for the navigation in both the video or the music.

5.4   Phase 3

When using the Myo system, the participants followed all the instructions and progressively (practiced) all the tasks we gave them. Calibration requires in

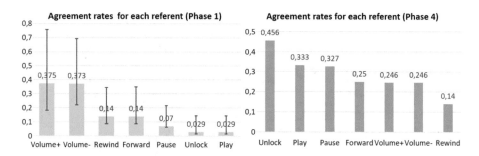

Fig. 3. Agreement scores as computed by AGATe [20]. Error bars show a CI of 95%.

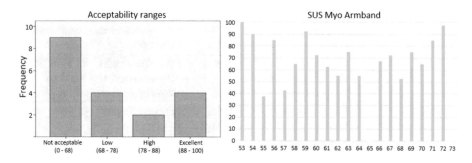

Fig. 4. Acceptability ranges and SUS score for Myo Armband.

general 5 min to define the profile. Thanks to the gesture memory cards, the participants used the predefined gestures easily. In each interview, we followed the whole experiment of the learning process. We completed the whole experiment with all the participants even if some signal recognition troubles appeared. In some cases, the gestures recognition's were not as precise as the natural users' interactions. Furthermore, we discovered a particular inversion in some gestures. For the Rewind and Forward referents as well as the Volume Up and Volume Down, it occurred that the actions were in the opposite direction to the one expected, even if the Myo gesture recognition interfaces (see right section of Fig. 1) were correct. In other words, in some cases the correct Rewind referent (by waving the hand to the right) and the correlated gesture recognition interface activated the Forward action. That inversion effect for the Rewind and Forward referents occurred for 14 participants, and for the Volume Up and Down it occurred 19 times (for all the participants). The total score of the SUS questionnaire is illustrated in the right graph in Fig. 4. The maximum score is 100 when the minimum score achieved is 37.5. The average score is 70.92. We considered score below 68 as not acceptable [1]. Thus, left graph in Fig. 4 demonstrates that 9 out of 19 participants had a not acceptable score. Half of the participants were not convinced by the Myo system in the proposed context. The other half of the participants sometimes very much appreciated the armband as presented. SUS results are to be associated with the series of subjective measures calculated

Table 4. Result of the subjective measures questionnaire.

Referent: Gesture	Good match		Ease of execution		Memorability		Enjoyable to execute	
	M	SD	M	SD	M	SD	M	SD
Unlock: Double tap	5.47	1.83	5.39	1.81	6.16	1.42	5.39	1.88
Play: Spread fingers	5.68	1.49	6.11	1.34	5.84	1.38	5.66	1.55
Pause: Spread fingers	5.47	1.38	6.21	1.02	6	1.45	6.08	1.06
Volume up: Make fist, turn left	3.42	2.48	5.05	1.91	4.11	2.26	5.05	1.51
Volume down: Make fist, turn right	5.47	1.39	6.21	1.02	6	1.45	6.08	1.05
Rewind: Wave right	3.74	2.33	4.92	2.16	5.26	1.93	4.08	1.76
Forward: Wave left	3.84	2.19	4.68	1.85	5.32	1.97	4.08	1.66

on each gesture. In addition, the qualitative data and comments given during the experiment are also included in the analysis of the different criteria.

Good Match. Gestures proposals seem to have a good match for some referents like Unlock, Pause and Play. Right graph in Fig. 3 shows that referents like volume down-up and rewind-forward in a video sequence are to be discussed. Indeed, the vast majority of people were disturbed by the direction that worked. "Oh but it's upside down" was a common expression recorded for 9 participants during the experiment. Going back in the sequence linked to the wave left gesture was not interpreted in a natural way, as much as the gesture given to increase the sound by turning the fist to the left, which disrupted many participants. This comes from confusion about which gestures does what explained previously. Those gestures that actually worked during the experiment were in the opposite direction of what Myo had planned in the setup.

Ease of Execution. The participants measured their effort in the performance of predefined gestures and repeatedly highlighted the difficulty of wrist twist to the left or to the right. This average is lower than the other gestures (see, Table 4). Moreover, the participants repeatedly tried not to go too far in waving their wrist. Therefore it became more difficult to detect the executed gesture and made the users even more frustrated. About the execution of other gestures, they were rather positive. No difficulties were significant in the performance even if the armband sometimes did not have a precise answer or when the system misinterpreted the executed gesture. The participants just tried again until the system figured it out.

Memorability. When asked whether or not the gestures used in Netflix were memorable, participants did not really need extra help to be able to recall them. However, making a fist and turn left (to volume up) is less memorable for some participants (Table 4). Indeed, this gesture requires two particular movements with the hand to perform a single action.

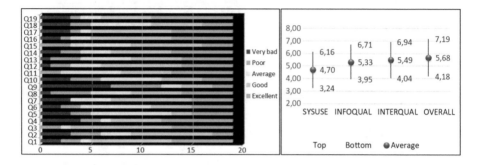

Fig. 5. Results from the IBM CSUQ questionnaire.

Enjoyable to Execute. A similar result was found with the ease of execution. It was the rewind and forward navigation gestures that did not entertain the participants. Stress for the execution of these gestures was constantly present. Figure 5 depicts the IBM CSUQ results. The study suggests that the global subjective satisfaction of the participants involved in the experiment follows a rather positive trend since Q19 ($M = 5.68$, $MD = 6$, $SD = 1.60$) is interpreted positively. Q9 ($M = 4.11$, $MD = 4$, $SD = 1.76$), the most negatively assessed question, raises a particular concern about the information quality to solve the users problems. During the experiment, the visual information guidance to solve problems recognition and to readjust the armband are not clear and helpful for the users. However, regarding the Q10 to Q15 the information quality system is evaluated positively. About the interaction quality (Q16-Q18, $M = 5.49$, $MD = 6$, $SD = 1.45$): all questions have some negative answers, but indicate that there is a positive agreement among the respondents. Finally, Q1 ($M = 5.68$, $MD = 6.0$, $SD = 1.45$) proposes that participants are satisfied with the process supported by the Myo system. Moreover, the most positively assessed question Q3 ($M = 6.05$, $MD = 6.0$, $SD = .91$) demonstrates that participants think that they completed the work correctly using the tool.

5.5 Phase 4

We computed specific agreement scores in which the participants would include all the Myo's system-defined gestures and the user-defined from the study. They exhibit higher agreement scores for each referent than the first one computed (right graph in Fig. 3). Myo's gestures were the most elicited for the Unlock, Play, Pause, Forward and Rewind referents (Table 3). System-defined gestures can feed an effective gestures set to satisfy most people using armband. People insisted on restoring the direction for the gestures: Rewind, Forward, Volume Up and Volume Down. They proposed a simplified gesture for the volume control to avoid making mistakes and created a new version of Wave gesture (left or right) to have less effort and more ease to execute the command. In our sample, 3 users out of 19 kept their own gestures that they defined at the beginning, without taking any system-defined gesture. They liked so much their own gestures that they could not choose any other. Other users reused one or many system-defined gestures. To reduce impact on recognition, participants often decided not to be as precise as the Myo system would recommend.

6 Conclusion

We presented the results of a gesture elicitation study combined with in an entertainment environment with an interactive Myo Armband. We also conducted a usability test in order to align the results between the gesture elicitation study (which returns the most preferred gestures for the tasks envisioned) and the usability study (which returns the most usable gestures for the tasks performed). During the experiment, we shaped a new hypothesis about the age and the ease of execution. The participants under 22 years have a deeper understanding of the proposed gestures and we suppose they adapted themselves to improve the gesture recognition's of the armband while the older users (+58) did not seem to adapt their own gestures. It requires future research to test this new proposal. We confirmed the usage of the same gestures for music and video control. Future studies should find out whether it is still the case when different rooms are used and whether the participant interacts the same way with the presence of other people near them, in a multitask context or in a smart environment. Again, this is the traditional debate of performance versus preference: participants may want to elicit some gestures that are not performant or not usable enough to be really incorporated in the final consensus set. Similarly, participants may return usable gestures (thus maximizing usability), but that are not also returning from the guessabiity study (thus minimizing the preference). Our findings revealed low agreement rates among our participants gesture proposals. Conversely, the participants went to a global higher agreement rate and satisfaction of use after learning and performing the predefined Myo gestures. The final consensus set resulting from this study expands the set of initial system-defined gestures. This new set could therefore become a candidate for an improved gesture recognition.

References

1. Bangor, A., Kortum, P., Miller, J.: Determining what individual SUS scores mean: adding an adjective rating scale. J. Usability Stud. **4**(3), 114–123 (2009)
2. Bergold, J., Thomas, S.: Participatory research methods: a methodological approach in motion. Historical Social Research, pp. 191–222 (2012)
3. Brooke, J., et al.: SUS-A quick and dirty usability scale. In: Usability Evaluation in Industry, vol. 189, no. 194, pp. 4–7 (1996)
4. Chan, E., Seyed, T., Stuerzlinger, W., Yang, X.D., Maurer, F.: User elicitation on single-hand microgestures. In: Proceedings of the Conference on Human Factors in Computing Systems, pp. 3403–3414. ACM (2016)
5. Dalmazzo, D., Ramirez, R.: Air violin: a machine learning approach to fingering gesture recognition. In: 1st ACM SIGCHI International Workshop on Multimodal Interaction for Education, pp. 63–66. ACM (2017)
6. Dong, H., Danesh, A., Figueroa, N., El Saddik, A.: An elicitation study on gesture preferences and memorability toward a practical hand-gesture vocabulary for smart televisions. IEEE Access **3**, 543–555 (2015)
7. Hewitt, J.: The myo gesture-control armband sense your muscle's movements. ExtremeTech (online magazine) (2013)
8. Kerber, F., Puhl, M., Krüger, A.: User-independent real-time hand gesture recognition based on surface electromyography. In: 19th International Conference on HCI with Mobile Devices and Services, p. 36. ACM (2017)
9. Korkman, M.: NEPSY. A developmental neuropsychological assessment. Test materials and manual (1998)
10. Koskimäki, H., Siirtola, P., Röning, J.: Myogym: introducing an open gym data set for activity recognition collected using myo armband. In: International Joint Conference on Pervasive and Ubiquitous Computing, pp. 537–546. ACM (2017)
11. Lewis, J.R.: IBM computer usability satisfaction questionnaires: psychometric evaluation and instructions for use. Int. J. Hum. Comput. Interact. **7**(1), 57–78 (1995)
12. Montero, F., López-Jaquero, V., Vanderdonckt, J., González, P., Lozano, M., Limbourg, Q.: Solving the mapping problem in user interface design by seamless integration in idealxml. In: Gilroy, S.W., Harrison, M.D. (eds.) Interactive Systems. Design, Specification, and Verification, pp. 161–172. Springer, Heidelberg (2006)
13. Montoya, M., Henao, O., Muñoz, J.: Muscle fatigue detection through wearable sensors: a comparative study using the myo armband. In: 18th International Conference on Human Computer Interaction, p. 30. ACM (2017)
14. Munroe, C., Meng, Y., Yanco, H., Begum, M.: Augmented reality eyeglasses for promoting home-based rehabilitation for children with cerebral palsy. In: The Eleventh ACM/IEEE International Conference on Human Robot Interaction, p. 565. IEEE Press (2016)
15. Rajavenkatanarayanan, A., Surathi, Y.V., Babu, A.R., Papakostas, M.: Myodrive: a new way of interacting with mobile devices. In: 9th ACM International Conference on PErvasive Technologies Related to Assistive Environments, p. 18. ACM (2016)
16. Sathiyanarayanan, M., Rajan, S.: Myo armband for physiotherapy healthcare: a case study using gesture recognition application. In: 8th International Conference on Communication Systems and Networks. pp. 1–6. IEEE (2016)
17. Tsai, W.L., Hsu, Y.L., Lin, C.P., Zhu, C.Y., Chen, Y.C., Hu, M.C.: Immersive virtual reality with multimodal interaction and streaming technology. In: 18th ACM International Conference on Multimodal Interaction, p. 416. ACM (2016)

18. Vanderdonckt, J., Roselli, P., Medina, J.L.P.: !FTL, an articulation-invariant stroke gesture recognizer with controllable position, scale, and rotation invariances. In: 20th International Conference on Multimodal Interaction, ICMI 2018, Boulder, CO, USA, October 16–20, 2018
19. Vatavu, R.D.: A comparative study of user-defined handheld vs. freehand gestures for home entertainment environments. J. Ambient Intell. Smart Environ. **5**(2), 187–211 (2013)
20. Vatavu, R.D., Wobbrock, J.O.: Formalizing agreement analysis for elicitation studies: new measures, significance test, and toolkit. In: 33rd ACM Conference on Human Factors in Computing Systems, pp. 1325–1334. ACM (2015)
21. Wobbrock, J.O., Aung, H.H., Rothrock, B., Myers, B.A.: Maximizing the guessability of symbolic input. In: CHI'05 Extended Abstracts on Human Factors in Computing Systems, pp. 1869–1872. ACM (2005)

Accessibility Evaluation of Mobile Applications for Monitoring Air Quality

Patricia Acosta-Vargas[1]([⊠]), Rasa Zalakeviciute[1],
Sergio Luján-Mora[2], and Wilmar Hernandez[1]

[1] Intelligent and Interactive Systems Lab,
Universidad de Las Américas, Quito, Ecuador
{patricia.acosta, rasa.zalakeviciute,
wilmar.hernandez}@udla.edu.ec
[2] Department of Software and Computing Systems,
University of Alicante, Alicante, Spain
sergio.lujan@ua.es

Abstract. This research evaluates the accessibility of mobile applications with the Accessibility Scanner tool. The evaluated applications are related to the quality of the air we breathe. As a matter of fact, nowadays there exists a high number of free applications for mobile devices that provide information about the level of contamination that is affecting cities, as well as the concentration of each significant pollutant. However, not all those mobile applications are accessible. This study uses Accessibility Scanner of Google, applying the accessibility guidelines for mobile applications of Web Content Accessibility Guidelines 2.1. In this study, 10 mobile applications were evaluated, and it can be said that the obtained results can be seen as a useful, practical example of how to develop inclusive mobile applications.

Keywords: Accessibility evaluation · Air quality · Mobile applications ·
Web Content Accessibility Guidelines 2.1 (WCAG 2.1)

1 Introduction

The current human population of over 7.6 billion has doubled since 1970 and is projected to surpass 9.8 billion by 2050, mostly due to the urbanization in the less developed parts of the world [1]. Due to the rapid growth of urban areas, where most of the air pollution is produced and concentrated, we are facing a trend of increasingly more severe health problems [2]. Currently, 92% of the world's urban population live in cities that do not comply with the air quality guidelines of the World Health Organization (WHO) [3]. This information shows air pollution to be the fourth cause of premature mortality and the primary environmental risk [4].

Health studies relate short and long-term exposure to air pollution, increased hospital admissions and emergency room visits, and even to death from heart or lung diseases [5–9]. Therefore, it is essential to avoid exposure to elevated air contamination. While most of the major cities have some level of air quality monitoring, the data reported on the official websites are not easily accessed. A better way to keep the

© Springer Nature Switzerland AG 2019
Á. Rocha et al. (Eds.): ICITS 2019, AISC 918, pp. 638–648, 2019.
https://doi.org/10.1007/978-3-030-11890-7_61

population aware of the risk of being exposed to bad air quality in their city is a mobile application that can be accessed from anywhere.

The accessibility is the degree to which all people can use an object, or access a service, regardless of their technical, cognitive or physical capabilities [10]. Accessibility describes techniques, methods, and theories to make media, in its multiple forms, more accessible to people with disabilities.

This study presents an evaluation of the accessibility of mobile applications that permit real-time monitoring of air quality index in surrounding areas. Air quality is a concern for human and environmental health, climate change and other real problems that we face every day. Therefore, the question is: What can be done? It is important to know what quality of air we breathe, and an answer to the above question can be given by using mobile applications, instantaneously estimating pollution, and alerting the users. Nevertheless, not all applications are accessible, which is why an evaluation will be carried out to identify the main access errors.

For these applications to be accessible, it is essential to apply the Web Content Accessibility Guidelines (WCAG) 2.1 [11]. In this research, the Accessibility Scanner was applied. It is an automatic accessibility validator for Android applications that allows to scan any application installed on a mobile device.

Accessibility Scanner is a free app for Android offered by Google. In this study, version 1.2, updated on March 29, 2018, was used. When applying the evaluation tool, there are some limitations, due to the fact that it does not validate according to a specific standard, and that it only automatically evaluates a small number of aspects related to the accessibility of the application.

The rest of this paper is structured as follows: Details on background and related work are presented in Sect. 2; the topic of accessibility in mobile applications is discussed in Sect. 3; the method used in this research is presented in Sect. 4; Sect. 5 is devoted to the results and discussion of this investigation; finally, Sect. 6 aims to draw the conclusions of this document and to introduce future research works.

2 Background and Related Works

An increasing number of applications and users make the development of mobile applications one of software engineering most promising fields. Therefore, the development of mobile applications faces a serious challenge [12]. In recent years, the use of smart devices and mobile applications has increased considerably [13]. However, accessibility is a relatively neglected property that is seldom tested, because access barriers are not explicitly considered for people with disabilities [14].

The lack of research on accessibility tests is attributed to the lack of precise accessibility definitions in mobile applications. On the one hand, WCAG 2.1 [11] provide guidelines for developing more accessible web applications, from these guidelines can be adapted for mobile applications according to the World Wide Web Consortium [15]. Alternatively, Google provides the Accessibility Scanner tool that applies some accessibility guidelines for Android [16]. Although there are recommendations to improve the development of applications, most developers ignore them frequently. In this paper, we adopt its definition of "Mobile accessibility" [17] to make applications more accessible for people with disabilities when they are using mobile phones and other devices.

A mobile application is a software, according to the technologies involved, and it can be a native application, either a web or hybrid application [18]. The mobile applications have a design to be used in smart mobile devices or tablets; they are downloaded from a distribution platform managed by the manufacturer of the device. The installation process and updates are simple. The applications have a small size, to adapt to the technical limitations of the devices, the mobile applications have better communication function than desktop programs.

3   Accessibility of Mobile Applications

At present, on the subject of accessibility in mobile applications, several previous research works contribute to this study.

In 2018, a study carried out by Krainz et al. [19] indicated that the Android Play Store has more than 3.5 million different applications, but most of them have accessibility problems.

Damaceno et al. [20] explained that the accessibility of mobile devices refers to the ability to interact correctly with the operating system of mobile devices. This work [20] provides a macro view of the accessibility barriers experienced by people with visual disabilities when using mobile devices and proposes a series of accessibility recommendations to guide future studies.

Eler et al. [21] said that it is important to make mobile applications accessible, without excluding people with disabilities, but the lack of support tools for developers to verify if the user interface components can be used by screen readers. Therefore, the development and evaluation of mobile applications are challenging.

Ross et al. [22] analyzed the accessibility of 5,753 free Android mobile apps, based on tags and accessibility barriers. The analysis shows that accessibility barriers are still a widespread problem and suggests continuing to collect and analyze large-scale data to improve accessibility violations in mobile applications.

El-Glaly et al. [23] indicated that mobile applications must be accessible to everyone. However, many of the most popular applications are not. In [23] the authors propose a set of modules that define the accessibility problem and simulates the effects of the accessibility barrier.

Carvalho et al. [24] proposed that the creation of accessible mobile applications involves design decisions. The results of [24] showed that the prototypes developed with the accessibility guidelines of WCAG 2.1 and with the Talkback screen reader were more compatible with the accessibility criteria.

Pichiliani et al. [25] emphasized that mobile applications based on user interface elements require alternative access to visual elements. The results of the evaluation in this study can influence mobile developers to review the access of their applications and provide necessary support for other contexts.

4   Method

In this preliminary research work, 10 mobile applications for the Android system were evaluated. The accessibility evaluation used the Accessibility Scanner tool from the Google Play Store; this tool applies some guidelines from WCAG 2.1 [11].

The data collection was carried out on August 20, 2018. Table 1 summarizes the free tools used to measure air quality and it contains the identifier when assigning the application, the name of the tool, the logo, the version of the tool, the Android version, Android requirements and the name of the company that offers the software.

Table 1. Tools that are used to measure air quality by using Google Play Store

Id	Tool	Logo	Tool version	Android version	Requirements	Company
A	Air Pollution		March 6, 2018	1.0.0	4.1 and later	Technical Library
B	AirVisual		June 1, 2018	4.2.1-19	4.1 and later	AirVisual
C	Plume Air Report		July 9, 2018	It varies according to the device.	It varies according to the device.	Plume Labs
D	Air Quality Index BreezoMeter		July 11, 2018	2.02.00	4.0.3 and later	BreezoMeter
E	Air quality		February 28, 2018	1.3	5.0 and later	FFZ srl
F	Air Quality: Monitor AQI		July 26, 2018	1	4.4 and later	Mind IT Systems
G	Air! World Air Quality		June 18, 2016	It varies according to the device.	It varies according to the device.	inside
H	Clean Air Cast		June 23, 2017	1.0.23	4.1 and later	Aitch3
I	Air Quality Meter - PM10 & AQI		January 9, 2018	2.0.1	4.0 and later	Gaia Consulting
J	Air Pollution Check		June 28, 2017	1.0.0	4.1 and later	evolution technologies

The accessibility test offers suggestions for extending touch elements, increasing contrast or including content descriptions. In this way, users who need greater accessibility can use a mobile application more efficiently.

The method used to evaluate the accessibility of the mobile applications consisted of six sequential phases, as shown in Fig. 1.

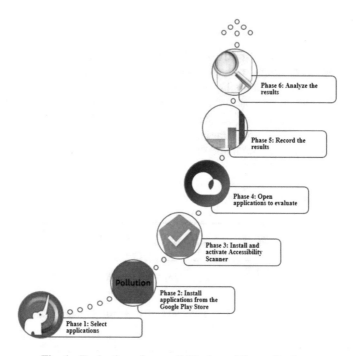

Fig. 1. Evaluation of accessibility in mobile applications

Phase 1: Select Applications. In this phase, mobile applications were selected that help user to know the quality of the air they breathe. These mobile applications were chosen from the Google Play Store. The tools selected to evaluate the accessibility of mobile applications are recorded in Table 1.

Phase 2: Install Applications from the Google Play Store. In this phase, the applications, selected in Phase 1, were installed in a smartphone with version 7.0 of Android. Installed applications are shown in Table 1.

Phase 3: Install and Activate Accessibility Scanner. In this phase, the Accessibility Scanner tool developed by Google[1] was installed from Google Play Store. This tool allows evaluating the accessibility of mobile applications to measure air quality. Accessibility Scanner applies the standard of WCAG 2.1.

The accessibility test of Accessibility Scanner indicates the actions that will be carried out and requests the permission to capture the screens. When activating the tool, a blue button will always be visible on the screen to place it in any area.

Phase 4: Open Applications to Evaluate. In this phase, each application is run to evaluate with the Accessibility Scanner tool. In this case, it is necessary to configure and activate the location of the mobile device so that the application automatically detects the geographical location to measure the air quality.

---

[1] https://support.google.com/accessibility/android/faq/6376582?hl=en.

To start with the accessibility evaluation of mobile applications, do the following: open the application to evaluate, for example, select "Air Pollution" and then press the blue button, which will always be available when activating the tool. Afterwards, the result of the accessibility evaluation will show a report with the elements that present faults in the evaluated mobile application.

Phase 5: Record the Results. In this phase, the evaluation data of the mobile applications with Accessibility Scanner are recorded in a Microsoft Excel sheet.

Phase 6: Analyze the Results. Finally, with the data obtained, located in the dataset, statistical analysis was made using the analysis tools of Microsoft Excel.

5   Results and Discussion

Table 2 shows the relationship that exists between WCAG 2.1 and the criteria applied to the accessibility guidelines in mobile applications. It contains the principle, the success criterion that is most related and the level[2].

Table 2. Principle with success criterion that is most related

WCAG 2.1 principle	Success criterion that is most related	Level
Perceivable		
Small screen size		
Zoom/magnification	1.4.4 Resize text	AA
Contrast	1.4.3 Contrast which requires a contrast of at least 4.5:1 (or 3:1 for large-scale text) and	AA
	1.4.6 Contrast which requires a contrast of at least 7:1 (or 4.5:1 for large-scale text)	AAA
Operable		
Keyboard control for touchscreen devices	2.1.1 Keyboard	A
	2.1.2 No Keyboard Trap	A
	2.4.3 Focus Order	A
	2.4.7 Focus Visible	AA
Touch target size and spacing	2.5.5 Target Size	AAA
Touchscreen gestures		
Device manipulation gestures	2.1.1 Keyboard	A
Placing buttons where they are easy to access		

(*continued*)

[2] The dataset and the analysis of this research are deepened and are available in Microsoft Excel format in the Mendeley located at http://dx.doi.org/10.17632/57w6yj673y.2.

Table 2. (*continued*)

WCAG 2.1 principle	Success criterion that is most related	Level
Understandable		
Changing screen orientation (portrait/landscape)		
Consistent layout	3.2.3 Consistent Navigation (Level AA)	AA
	3.2.4 Consistent Identification (Level AA)	AA
Positioning essential page elements before the page scroll		
Grouping operable elements that perform the same action	2.4.4 Link Purpose (In Context) (Level A)	A
	2.4.9 Link Purpose (Link Only) (Level AA)	AA
Provide clear indication that elements are actionable		
Provide instructions for custom touchscreen and device manipulation gestures	3.2.3 Consistent Navigation (Level AA)	AA
	3.2.4 Consistent Identification (Level AA)	AA
Robust		
Set the virtual keyboard to the type of data entry required		
Provide natural methods for data entry		
Support the characteristic properties of the platform		

Table 3 shows the evaluation of the main screens of the mobile applications, with Accessibility Scanner that scans the screen of mobile applications based on the following: content labels, touch target size, clickable items, and text and image contrast.

Touch target refers to the elements of the screen that must be large enough to have a constant interaction in accordance with the principles of WCAG 2.1 [11]. It is suggested that the target is available through an equivalent link or control on the same page that is at least 44 by 44 CSS pixels.

The color contrast between text and images was evaluated according to the recommendations of WCAG 2.1, that is, the small text must have at least a contrast of 4.5:1 and the large text (14 points in bold or 18 points in the normal) a contrast of at least 3:1. According to WCAG 2.1, the combination of colors selected for the interface of a mobile application affects the ease with which users can read and understand it [11].

Table 4 shows the elements with failures of the main screens of the mobile applications detected by Accessibility Scanner and contains the tool, the number of elements that have failures, and the percent failures. From the data obtained in relation to elements with failures, it is observed that the average value is 38.3, the typical error is 16.2, the median is 19.5, the standard deviation is 39.7, the minimum value is 4.0, and finally, the maximum is 91.0.

Table 3. Evaluation with Accessibility Scanner

Id	# elements	Touch target	Text contrast	Item label	Item descriptions	Clickable items	Image contrast
A	2	2	2	0	0	0	0
B	45	7	22	11	2	3	0
C	13	0	6	0	2	0	5
D	12	0	9	0	1	0	2
E	12	4	1	5	1	0	1
F	13	1	1	1	8	1	1
G	9	0	9	0	0	0	0
H	7	4	2	1	0	0	0
I	113	68	39	1	5	0	0
J	1	1	0	1	0	0	0

Table 4 shows the most significant number of failures corresponds to "Air Quality Meter - PM10 & AQI", with 113 failures corresponding to 49.8%; followed by "AirVisual," which has 45 failures and corresponds to 19.8%, and a third is the "Plume Air Report tool" that corresponds to 5.7%. The tools with the least accessibility problems are "Air Pollution," with two faults corresponding to 0.9%, and "Air Pollution Check," with a failure corresponding to 0.4%.

Table 4. Elements with failures detected by Accessibility Scanner

Tool	# elements	%
Air Quality Meter - PM10 & AQI	113	49.8
AirVisual	45	19.8
Plume Air Report	13	5.7
Air quality: monitor AQI	13	5.7
Air quality index breezometer	12	5.3
Air quality	12	5.3
Air! world air quality	9	4.0
Clean air cast	7	3.1
Air pollution	2	0.9
Air Pollution Check	1	0.4

Figure 2 presents the most frequent failures in the accessibility evaluation of mobile applications. The most frequent failures correspond to "Text contrast" with 91 failures (39.6%), "Touch target" with 87 failures (37.8%), "Item label" with 20 failures (8.7%), "Item descriptions" with 19 failures (8.3%), "Image contrast" with 9 failures (3.9%), and "Clickable items" with 4 failures (1.7%). Therefore, the most repeated failures are related to the text contrast followed by the touch target. When evaluating the accessibility of an application, experts usually use automatic tools with which it is possible to identify some of the accessibility problems that it presents. However, the results of this type of tools usually involve both false negatives and false positives, since they are not capable of detecting all failures. Thus, for example, an automatic evaluation tool can detect if an image has alternative text but cannot interpret if said text is representative concerning the content of the image. This lack of reliability implies that additional manual analysis must be undertaken, in which the expert evaluator must not only carry out a complete examination based on the relevant requirements but also must review all the failures detected by the automatic analysis tools. Therefore, automatic tools will help in the evaluation process, but their use should never be understood as a complete analysis. Also, each automatic analysis tool will have its functions, defects, and strengths, so it will be necessary to use several of them to achieve optimal results.

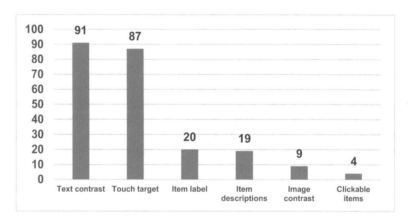

Fig. 2. Most frequent accessibility failures in mobile applications

6   Conclusions and Future Works

The results of this document are available in the Mendeley dataset[3] and also be replicated, which showed that the mobile applications evaluated did not reach an acceptable level of accessibility. Therefore, it is necessary to rectify the failures to comply with the level of accessibility recommended by the World Wide Web Consortium. In addition, it is necessary to include accessibility measures in the development of applications estimating air quality by applying a checklist to correct the identified problems. Furthermore, the results obtained in the evaluation process will serve as a reference point to compare the development of an accessible mobile application to measure air quality. Moreover, these results can provide ideas on how to develop and improve mobile applications to make them more accessible and inclusive.

Future works may propose better methods for using mobile applications by applying the guidelines proposed by WCAG 2.1. In the evaluation process of the 10 mobile applications, it was identified that the most frequent failures correspond to "Text contrast" followed by "Touch target." In the future, the accessibility assessment

---

[3] https://data.mendeley.com/datasets/57w6yj673y/2.

can be carried out in mobile applications considering the type of barriers of accessibility for different types of users, such as to blind people, low-vision users, deaf users, color-blind users, motor impaired users, and cognitive disabilities.

Acknowledgments. The authors thank CEDIA for funding this study through the project "CEPRA-XII-2018-13" and the UDLA project "ERI.WHP.18.01".

References

1. United Nations (Department of Economic and Social Affairs/Population Division 2017): World Population Prospects: The 2017 Revision, Key Findings and Advance Tables. ESA/P/WP/248. https://esa.un.org/unpd/wpp/Publications/
2. World Health Organization (WHO): Air pollution levels rising in many of the world's poorest cities. http://www.who.int/mediacentre/news/releases/2016/air-pollution-rising/en/
3. Limb, M.: Half of wealthy and 98% of poorer cities breach air quality guidelines. Br. Med. J. **353** (2016). https://doi.org/10.1136/bmj.i2730
4. World Health Organization (WHO): 7 million premature deaths annually linked to air pollution. http://www.who.int/mediacentre/news/releases/2014/air-pollution/en/
5. Environmental Protection Agency (EPA): Are you at risk from particles? How can particles affect your health? http://depts.washington.edu/wildfire/resources/particles.pdf
6. Pope III, C., Dockery, D.: Health effects of fine particulate air pollution: lines that connect. Air Waste Manag. Assoc. **56**, 709–742 (2006). https://doi.org/10.1080/10473289.2006.10464545
7. Brook, R.D., Rajagopalan, S., Pope III, C.A., Brook, J.R., Bhatnagar, A., Diez-Roux, A.V., Holguin, F., Hong, Y., Luepker, R.V., Mittleman, M.A., Peters, A., Siscovick, D., Smith, S.C.J., Whitsel, L., Kaufman, J.D.: Particulate matter air pollution and cardiovascular disease: an update to the scientific statement from the American Heart Association. Circulation **121**, 2331–2378 (2010). https://doi.org/10.1161/cir.0b013e3181dbece1
8. Adar, S.D., Sheppard, L., Vedal, S., Polak, J.F., Sampson, P.D., Diez Roux, A.V., Budoff, M., Jacobs, D.R.J., Barr, R.G., Watson, K., Kaufman, J.D.: Fine particulate air pollution and the progression of carotid intima-medial thickness: a prospective cohort study from the multi-ethnic study of atherosclerosis and air pollution. PLoS Med. **10** (2013). https://doi.org/10.1371/journal.pmed.1001430
9. Künzli, N., Kaiser, R., Medina, S., Studnicka, M., Chanel, O., Filliger, P., Herry, M., Horak, F., Puybonnieux-Texier, V., Quénel, P., Schneider, J., Seethaler, R., Vergnaud, J.C., Sommer, H.: Public-health impact of outdoor and traffic-related air pollution: a European assessment. Lancet **356**, 795–801 (2000). https://doi.org/10.1016/S0140-6736(00)02653-2
10. Acosta-Vargas, P., Luján-Mora, S., Acosta, T., Salvador-Ullauri, L.: Toward a combined method for evaluation of web accessibility. In: Advances in Intelligent Systems and Computing, pp. 602–613 (2018)
11. World Wide Web Consortium (W3C): Web Content Accessibility Guidelines (WCAG) 2.1. https://www.w3.org/TR/WCAG21/
12. Vaupel, S., Taentzer, G., Gerlach, R., Guckert, M.: Model-driven development of mobile applications for Android and iOS supporting role-based app variability. Softw. Syst. Model. **17**, 35–63 (2018). https://doi.org/10.1007/s10270-016-0559-4
13. Choudhary, S.R., Gorla, A., Orso, A.: Automated test input generation for Android: are we there yet? In: Proceedings - 2015 30th IEEE/ACM International Conference on Automated Software Engineering, ASE 2015, pp. 429–440 (2016)

14. Zein, S., Salleh, N., Grundy, J.: A systematic mapping study of mobile application testing techniques. J. Syst. Softw. **117**, 334–356 (2016). https://doi.org/10.1016/j.jss.2016.03.065

15. World Wide Web Consortium: Mobile Accessibility: How WCAG 2.0 and Other W3C/WAI Guidelines Apply to Mobile. https://www.w3.org/TR/mobile-accessibility-mapping/

16. Google: Make apps more accessible. https://developer.android.com/guide/topics/ui/accessibility/apps

17. Mobile Accessibility at W3C | Web Accessibility Initiative (WAI) | W3C. https://www.w3.org/WAI/standards-guidelines/mobile/

18. Jabangwe, R., Edison, H., Duc, A.N.: Software engineering process models for mobile app development: a systematic literature review. J. Syst. Softw. **145**, 98–111 (2018). https://doi.org/10.1016/j.jss.2018.08.028

19. Krainz, E., Miesenberger, K., Feiner, J.: Can we improve app accessibility with advanced development methods? In: International Conference on Computers Helping People with Special Needs, pp. 64–70. Springer (2018)

20. Damaceno, R.J.P., Braga, J.C., Mena-Chalco, J.P.: Mobile device accessibility for the visually impaired: problems mapping and recommendations. Univ. Access Inf. Soc. **17**, 421–435 (2018). https://doi.org/10.1007/s10209-017-0540-1

21. Eler, M.M., Rojas, J.M., Ge, Y., Fraser, G.: Automated accessibility testing of mobile apps. In: Proceedings - 2018 IEEE 11th International Conference on Software Testing, Verification and Validation, ICST 2018, pp. 116–126 (2018)

22. Ross, A.S., Zhang, X., Wobbrock, J.O., Fogarty, J.: Examining image-based button labeling for accessibility in Android apps through large-scale analysis. In: ACM SIGACCESS Conference on Computers and Accessibility (ASSETS 2018) (2018)

23. El-Glaly, Y.N., Peruma, A., Krutz, D.E., Hawker, J.S.: Apps for everyone: mobile accessibility learning modules. ACM Inroads **9**, 30–33 (2018). https://doi.org/10.1145/3182184

24. Carvalho, L.P., Freire, A.P.: Native or web-hybrid apps? An analysis of the adequacy for accessibility of Android interface components used with screen readers. In: Proceedings of 16th Brazilian Symposium on Human Factors in Computing Systems, pp. 362–371 (2017). https://doi.org/10.1145/3160504.3160511

25. Pichiliani, M., Hirata, C.: Evaluation of the Android accessibility API recognition rate towards a better user experience. In: International Conference on Universal Access in Human-Computer Interaction, pp. 340–349. Springer, Cham (2015)

Designing Usable Bioinformatics Tools for Specialized Users

Chanaka Mannapperuma[1]([✉]), Nathaniel Street[1],
and John Waterworth[2]

[1] Department of Plant Physiology,
Umeå Plant Science Centre, 901 87 Umeå, Sweden
{Chanaka.Mannapperuma,Nathaniel.Street}@umu.se
[2] Department of Informatics, Umeå University, 901 87, Umeå, Sweden
John.Waterworth@umu.se

Abstract. Visualization - the process of interpreting data into visual forms - is increasingly important in science as data grows rapidly in volume and complexity. A common challenge faced by many biologists is how to benefit from this data deluge without being overwhelmed by it. Here, our main interest is in the visualization of genomes, sequence alignments, phylogenies and systems biology data. Bringing together new technologies, including design theory, and applying them into the above three areas in biology will improve the usability and user interaction. The main goal of this paper is to apply design principles to make bioinformatics resources, evaluate them using different usability methods, and provide recommended steps to design usable tools.

Keywords: User experience designing · Participatory design · Bioinformatics

1 Introduction

Over the past few years, we have been designing and developing bioinformatics tools on the Plant Genome Integrative Explorer [1] (hereafter PlantGenIE) platform. Most of them are online tools. Bioinformatics consists of two subfields; the developments of tools and databases and the application of these tools in generating biological knowledge better understanding living systems [2]. In this article, we concentrate mainly on designing bioinformatics tools for specialized users.

During the last decade, the availability and generation of new data has increased exponentially due to the advancement of new technologies and reducing costs associated with sequencing machines. As a consequence, researchers are getting more and more data. It is important to have efficient and usable sets of tools to identify and understand the biological insights available from this wealth of data. According to the International Organization for Standardization (ISO), usability is the extent to which product can be used by specified users to achieve specified goals with effectiveness, efficiency, and satisfaction in a specified context of use (ISO 9241-110:2006) [3].

PlantGenIE tools have been mainly used by plant biologists and bioinformaticians – not by those with specialized training in how to design tools optimized for usability. Several studies have been conducted to understand the importance of applying user

© Springer Nature Switzerland AG 2019
Á. Rocha et al. (Eds.): ICITS 2019, AISC 918, pp. 649–670, 2019.
https://doi.org/10.1007/978-3-030-11890-7_62

centered design and usability engineering to optimize the usability of bioinformatics online tools that enable researchers to search, interact with, share, synthesize, visualize, and manipulate data more effectively and efficiently [4–11]. Combining KDD-HCI (knowledge discovery in databases and Human-Computer Interaction) approaches can significantly increase the capacity and efficiency of candidate gene discovery while reducing costs and time [12].

2   Applying Design Principles

PlantGenIE is a collection of interoperable web resources for searching, visualizing, and analyzing genomics and transcriptomics data for different forest tree species. Currently it includes dedicated web portals for enabling in-depth exploration of poplar [13] (http://PopGenIE.org), spruce [14] (http://ConGenIE.org), and Arabidopsis [1] (http://AtGenIE.org). Not surprisingly, PlantGenIE has been mainly used by Plant Biologist and Bioinformaticians.

Here is an example of how we have redesigned a one of our PlantGnIE tool based on design principles [15]. The Comparative analysis of co-expression networks tool [16] (hereafter ComPlEX) (Appendix F) provides an interactive comparison of co-expression networks across.

Since there is no example use case, help guide or demonstration, several users could not discover how to utilize the tool. They were also worried about the various type of buttons and their functionalities. The most common questions were 'What does the tool do?' and 'How can I use this tool?'. To answer the above questions, we started redesigning the tool from a low fidelity prototype (Appendix G) and then started applying design principles [15] to make a more usable tool.

Perceivable design elements such as call to action buttons, different cursor movements, relevant links and icons have been introduced in the new version of the ComPlEX tool. These possible actions between an object and an individual are called **Affordances** [18].

Clear and meaningful buttons, labels, notifications and icons are helpful to understand possible actions. These signals are called **Signifiers**. Signifiers are often a reduced number of possible interpretations and make affordances clearer.

Continuous **Feedback** using pre-loaders or corresponding waiting notifications have been introduced to the new tool. Feedback gives information about what action has actually been done and what result has been accomplished. Most of the analyses are calculated on the server side. Therefore, the tool has to inform the users about the current status of the backend analyses ceaselessly. Otherwise users will simply leave the tool or refresh the page.

Mapping is the relationship between two things, the controls and their movements and the results in the world. For example; researchers who are working at the greenhouse might know one of the bug herbivores of the species. We use a similar metaphor for reporting bugs in the tools (Fig. 1).

There are physical, logical, semantic, and cultural **Constraints** for designing tools. As an example, the new version of the tool was designed using bright colors and symbols commonly used by plant biologists. Eliminating design constraints will lead to

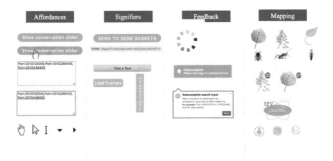

Fig. 1. Affordances, signifiers, feedback, and mapping are some of the basic design principles components.

enhancement of accessibility. For instance, when we design the tools using color blind safe palettes a color-blind user can use the system. Since color-blind users are within the subset of average users, average users can then more easily use the system. This is an example of the concept known as the inclusive design [19].

Design principles have been applied while keeping the consistency of fonts, color, contrast thought the entire tool (Appendix H).

Active user statistics (Fig. 2) showed a significant increase in active users right after redesigning of the tool. This could be due to users starting to understand the tool functions and features. We therefore used several usability methods to evaluate the PlantGenIE platform from the users' perspective. In addition, this provided the possibility to identify drawbacks, issues, and bugs and so further improve PlantGenIE tools.

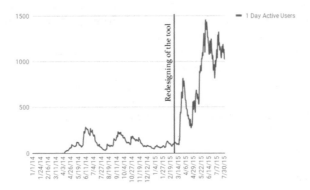

Fig. 2. Some active user statistics before and after redesigning the tool based on Google Analytics. The red line shows the exact time point where we shifted from the old complex tool to the new version.

3 Methodology

Qualitative and quantitative methods have been used to measure the usability of the PlantGenIE tools. The qualitative research was conducted using expert reviews and observational analysis. The quantitative research was conducted using Google Analytics (Google Inc., 2005) and surveys to help understand user requirements and issues that users experienced while using the tools for their analyses.

3.1 Remote Usability Study

The power of the Internet was used to conduct a usability survey, since users can be located almost anywhere in the world and sometimes have quite different sets of requirements. Usability survey questions (Appendix A - Survey questions) were extracted from an earlier published study [20]. All selected questions were divided into five usability quality components [21].

- Learnability: How easy is it for a user to complete a basic task at their first use of a system?
- Efficiency: How quickly can a user familiar with the system perform tasks?
- Memorability: How easy is it for a returned user to reestablish proficiency regarding the system?
- Errors & Feedback: How many errors does a user make using the system? How severe are the mistakes, and how difficult or easy is it to recover from the mistakes?
- Satisfaction: How satisfactory is it to use the product?

Survey questions were tested with a selected set of users at our department to understand their relevance and reduce the complexity of the questions. Then some questions were removed or adjusted according to their suggestions. Finally, the survey was opened to the entire community and results were recorded in an Excel database.

57 users with different profile characteristics participated in the usability survey from 01/11/2015 to 31/12/2015 and evaluated the PlantGenIE tools. The survey was comprised of the five usability components described above, and each component consisted of approximately 10 questions. For each, the user could select one out of five options, ranging from strongly agree to strongly disagree. Appendix K shows the professions of all survey participants. Appendix L shows the survey users by their occupational type and Appendix M shows their overall average usage.

3.2 Analytics

Web analytics is a cornerstone in creating customer satisfaction among website visitors [22]. Web analytics store a variety of data, including user referred sources, web traffic, web server performance, usability studies, demographic data, technical information and user behavior patterns to understand the online user experience. Web analytics data has been collected by most of the companies to conduct market research, find key performance indicators (KPI's) and ultimately increase sales and customer satisfaction.

There are a number of web analytics tools available [23], but Google Analytics (Google Inc., 2005) (hereafter GA) is used by over 85% of the commonly used websites that use traffic analysis tools [24]. In addition, several scientific articles have analyzed GA and evaluated its usefulness as a web analytics tool [25, 26]. GA is free, easy to use, and is remarkably full-featured for conducting quantitative user analytics. Hence, we chose to use GA to compare and further validate the survey results. Dimensions and metrics are two building blocks in GA. Geographic locations, demographics, traffic sources, landing pages, and behavioral flow are examples of dimensions. Metrics are individual elements of a dimension that are measured by a sum or a ratio. For example, conversion rate, number of visits, pages per visit, bounce rate, etc.

3.3   On-Site Usability Study

PlantGenIE practical questions to be answered by those who visited our office seeking bioinformatics help, were prepared based on the PlantGenIE basic tool functions.

Since all participants were plant biologists and were already familiar with some version of PlantGenIE tools, all of them were asked to complete one of the following tasks (Appendix N). Eight participants took part and the test run from December 2015 until March 2016.

Reasons for success or failure were requested at the end of each trial. For example: 'What were the difficulties that you encountered to answer one of the above questions?' and 'How did you know the correct steps to find the answer?'.

3.4   Expert Review

Expert review methods (Appendix O) rely on experts who have deeper knowledge in fields such as Human Computer Interaction and webpage design. Since the experts are non-biologists, they are only concerned about such things as the web design, layout, call to actions, overall quality of the content, color scheme and typography rather than actual functionality of the tool.

Three usability experts evaluated the GeneList tool based on Nielsen's usability principles (Appendix O) [27]. The strength of these checkpoints is that they are based on user research and what has been proven to work on the web, not personal opinions. Prior to starting the review, we introduced the PlantGenIE GeneList tool to the reviewers to ensure that they fully understood its purpose.

4   Results

Here are the results from the above four methods.

4.1   Remote Usability Study

The survey comprised five usability components (Learnability, Efficiency, Memorability, Errors & Feedback and Satisfaction) and each component included approximately 10 questions. Users responded on a five-point scale from strongly agree to strongly disagree. Most of the users (81.54%) chose strongly agree for all five usability components (Appendix P)

The survey included five free-text questions (Appendix B - Questions and Answers to free text survey questions) in addition to the above rating questions. This allowed users to describe user specific problems and suggestions. Answers to the free-text questions were unique and most of them had no relationship with each other.

The Heatmap (Fig. 3) shows five usability components (Learnability, Efficiency, Memorability, Errors & Feedback and Satisfaction) in terms of their answers. Survey participants (81.54%) answered strongly agree option. Therefore, the heat-map shaded toward blue color.

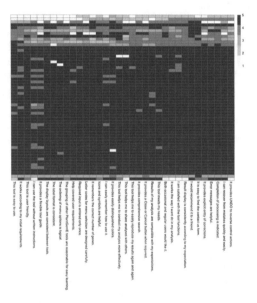

Fig. 3. Heatmap of survey questions and answers. X-axis shows the questions and Y-axis shows the participants. Color represent the strong disagree to strong agree options.

4.2 Analytics

Data was collected from the Google Analytics platform to identify user behaviors and their performances. This included demographic, screen tracking and custom reports methods. The following image (Appendix I) shows the PopGenIE screen track for a twelve-month period.

Different parts of the website received unique user attention. The search input box received only 0.3% user attention. Meanwhile, the project menu link and logo icon showed higher user attention. Some of the tools in the navigation menu received very low user attention while others had significantly higher attention (Appendix I).

Appendix J shows the average user behavioral flow for the same one-year period. Most of the users used the main page as a landing page whereas some users came directly to the gene or transcript pages. 65% of users followed the site path until they found the information about the gene or completed their analysis.

4.3   Usability Study at Our Office

We received different but favorable answers to 10 questions from the people who participated in our on-site usability study. 12 users answered the questions and all the results are attached as Appendix D - Answers to questions asked in the office. Some users suggested the importance of having PlantGenIE Quiz lessons and test, Help center and FAQ pages.

As shown in Table 1, seven out of eight users successfully accomplished the given task. Some users did not use the right tool to do their analysis but they found an alternative procedure. In other words, some users found different ways to accomplished a task.

Table 1. Trial task results at the bioinformatics office.

Tasks	Success/Failure
Typed cellulose synthesis inside main search box	Success
Found the best hit using genelist tool	Success
User did not find the share table button	Failure
Needed some help to find Potri.001G266400 protein sequence	Success
Easily found the answer using exImage tool	Success
Saved using Add to genelist button	Success
User did not find the sequence search tool instead download the sequence from gene information page	Success
Easily generated the heatmap using exHeatmap tool	Success
Typed cellulose synthesis inside main search box	Success
Found the best hit using genelist tool	Success

4.4   Expert Review

Below are the results from the three expert reviewers (Table 2).

Table 2. Checklist questions for expert reviewers.

Usability heuristics example	Reviewer 1	Reviewer 2	Reviewer 3
Loading animation when searching	Good	Agree	Yes
Understandable labels, names and links	Understand	Agree	Fair enough
User has freedom to search whatever they want	Good	Suggested an alternative	Yes
Consistency and web standards	To be improved	Agree	Yes
Continuous Error notifications	Good	Agree	Yes
Results shows all necessary information	Hard to understand	Agree	No
Flexibility and efficiency of use	Agree	Agree	Yes
Aesthetic and minimalist design	Agree	Yes	Yes

(continued)

Table 2. (*continued*)

Usability heuristics example	Reviewer 1	Reviewer 2	Reviewer 3
Help users recognize, diagnose, and recover from errors	Yes	Agree	No
Help and documentation	Yes	Disagree	No

5 Discussion

PlantGenIE comprises a number of different tools. Due to practical difficulties in evaluating each and every tool, we selected a few tools heavily used among plant biologists for our usability evaluations [1].

Preliminary usability studies were conducted using the survey based on initial prototypes. The majority (>80%) of survey participants checked the strongly agree option for scale based questions which comprised five usability components of Learnability, Efficiency, Memorability, Errors & Feedback and Satisfaction. However, the scale based question responses do not help to understand the real problems or provide a starting point to improve the tools. For instance, user feedback on the usability survey suggested the survey was too long, therefore the user may get frustrated and not completed it.

On the other hand, free-text questions were helpful to understand the practical user problems, requirements and their suggestions. Users used this opportunity to describe their views openly according to their own experiences. Most of them mentioned the importance of having more screencasts and detailed help pages to understand the functionalities. They also requested sample scenarios and examples along with the tools and help pages. Some of them experienced intermittent issues with exNet [1], BLAST [28] and GBrowse [29] tools but they did not mention the specific issue nor provide their email addresses to enable follow-up.

The overall results to the scale based questions gave positive feedback about the entire design process and interactivity of the site. However, answers to free text questions showed a demand for some extra functionalities in addition to help pages as listed below:

- Adding version 1.1 access to *Populus trichocarpa* genome.
- Adding SNP information into gene pages.
- Integrating PU numbers with Potri v3 gene models.
- Availability of cDNA and protein expression data(blast) for all Potri.

However, some previous studies have suggested that surveys truly measure user preferences, not product usability [27] and also there are challenges to interpret the survey results. For instance; when a user rates an answer, what does it really mean? Were all the rating scales similar to each other? Was it the users' true judgment based on how they felt about the tool? Are measures really quantifiable? What do the resulting statistics mean?

Despite the possibly low effectiveness of the ratings, surveys are the most widely used usability testing method due to their efficiency in reaching a large audience quickly [30].

Secondly, the analytics reports were used for usability studies. Analytics reports include a large amount of data related to user performance and behaviour. GA results were utilized to improve the usability of our tools. For instance; see Appendix I: the screen track shows comparatively high usage of some parts of the page, while some important components like autocomplete search box received less user attention (0.3%). This information suggests redesigning or highlighting the autocomplete search input as a main starting point of the site.

Some user behavioral flows include loops or dead ends. In other words, the user finds difficulty to navigate through the workflows (Appendix J) or they just leave the site. Therefore, the redesigning of the workflow will be needed for those user flows with loops and more help guides and example scenarios will be added to help avoid users encountering dead end behavioral flows.

Thirdly usability studies were conducted at our office with participants who were were familiar with the functionalities of our in-house tools. They requested help documents, tutorials of advanced tool functionalities. These results help to get a clearer idea of what real user thinks and what they wanted to accomplish using the tool. However, results might be different if we assess an unfamiliar tool instead of GeneList tool with the same participants.

Finally, expert reviews were used as our forth usability method. Molich and Jeffries [31] commented on expert review that: "The only thing you can say about it is that it doesn't require users other than the reviewers". The problems found with usability testing are practical problems encountered by real users. On the other hand, problems found with expert evaluations are potential problems based on design theories and principles. However, our usability experts suggested fundamental yet important usability issues. For instance, moving autocomplete search box into the middle of the header and highlighting it as starting point of the site, include contact us and report a bug feature within the bottom of the webpage, use the same font type, contrast and layout with a similar color scheme.

Tool requirements vary on their functionality, purposes, and given research questions. Thus, it is not possible to satisfy all the users by fulfilling complete set of tool functionalities that suits their set of requirements.

However, the results of applying usability evaluation methods confirm that the design principles are important to make usable bioinformatics web resources to many of the users. We understand that the importance of conducting usability studies during the entire design process while listening to real users and having one or more real users within the design team contributes enormously to making more usable tools.

Based on our findings we recommend combining survey results with GA. This allows validation of the results with each other. There is no best method among the four usability methods that we discussed in this paper. However, combining survey results with GA and expert reviews gave us positive results and allowed us to explore new dimensions to be considered - such as speed of the internet, user's location, look and feel of the tool and other technical details.

6   Conclusion and Future Work

The recommended steps along with usability studies and design principles to be taken during the design process are listed below and illustrated in Fig. 4.

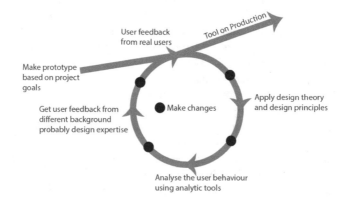

Fig. 4. PopGenIE user behavioral flow for last year (2017/02/31–2018/02/31).

1. Make an initial prototype to suit with the project goals.
2. Start to develop the tool while listening to continuous user feedback.
3. Apply design therories and design principles and improve the usability of tool.
4. Analyse the user behaviour using analytic tool and implement the necessary changes.
5. Get user feedback from different background, probably design expertise would help to improve the tool functionalities from different perspective.
6. Iterate the process from 2nd step to 5th step until the users are satisfied and project goals are accomplished.

These recommended steps can be used by future designers and developers as a checklist for the development of bioinformatics tools.

In future, our work will compare findings with other similar biological websites and report further on how usability studies can contribute making usable bioinformatics tools for different contexts. This will potentially lead to the discovery of more opportunities for the improvement of bioinformatics applications by leveraging design principles and proven practices in usability design.

Acknowledgements. Special thanks to all the participants assisting voluntarily with the survey, observation and interviews. We thank Anna Croon Fors and Hannele Tuominen for extensive support and feedback. We are also very grateful to PlantGenIE and UPSC Bioinformatics teams for helping with the design and development of tools and Niklas Mähler for the design of Figure 3.0.

Appendix A - Survey Questions

A. Learnability

1. This tool is easy to use.
2. It works according to my initial requirements.
3. This tool is user friendly.
4. I can use the tool without written instructions.
5. It provides a flexible tour guide.
6. The display layouts are consistent between tools.
7. The label format is consistent.
8. The ordering of menu options is logical.
9. The grouping of other PlantGenIE tools are reasonable for easy learning.
10. Help covered user requirements.

B. Memorability

1. Required input is entered only once.
2. Letter codes for menu selection are designed carefully.
3. It remembers the correct number of genes.
4. Icons and symbols are helpful.
5. I can easily remember how to use it.
6. It provides easily distinguished colors.

C. Efficiency

1. This tool helps me to conduct my analysis more effectively.
2. This tool helps me to share analyses with others.
3. This tool helps me to easily reproduce my results again and again.
4. It provides comprehensive search.
5. It provides a Close or Cancel button when required.
6. Results of my analysis are compatible with my expectations.

D. Satisfaction

1. This tool meets my needs.
2. Both occasional and regular users would like it.
3. It works the way I want do in my analysis.
4. I am satisfied with the tool functions.
5. Result display is consistently according to my expectation.
6. I would recommend it to a friend.

E. Error and Feedback

1. It is easy to find the contact us form.
2. It provides explicit entry of corrections.
3. Error messages are helpful.
4. Completion of processing is indicated.
5. I can recover from mistakes quickly and easily.
6. It provides UNDO to reverse control actions.

Suggestions (Free text fields)

1. What can we do to improve the functionality of the tool?
2. What can we do to improve the appearance of this tool?
3. Do you have any other suggestions?
4. What, if anything, do you find frustrating or unappealing about this tool?
5. What is your main reason for using this tool today?

Appendix B - Questions and Answers to Free Text Survey Questions

(1) What can we do to improve the functionality of the tool?
exNet doesn't work sometimes.
Improvement in Blast link to gbrowse is expected since sometimes it does not work.
Better to have more help tutorial videos.
Unsure
Identify if any suggestions are arised in surveys by different users.
Better to have more user guidelines
Could you please provide more detailed information?
Write some good help tutorial
It's great if snp information can be added into gene pages.
It's great if you can improve this tool by enabling to add more species
Functionality is excellent. Just I will add some quick instruction for firs step for use the app.
Is it possible to add version 1.1 access to populus trichocarpa genome?
Replacement with Adobe flash tools with JavaScript is great adding to the site
Sometimes Exnet export and sample lists are not functioning. Please note for improvement.
More user instructions are needed
Need to improve the list of example gene list for different aspen assemblies which is missing.
Would be great if you could make it possible to match PU numbers also to Potri v3 gene models without the intermediate step going through v2 models.
If this can be improved to integrate Galaxy into Plantgenie that would be nice.
I would like you to integrate CDNA and protein expression data (blast) for all likes like for Potri.
(2) What can we do to improve the appearance of this tool?
It's OK
Fonts. Another one and more bigger (like this form). Improve general design. Looks very technical (not designed).
NA
Looks great!
(3) Do you have any other suggestions?
Sometimes searching for different aspen genome is not working. Please fix this.
Nope
try with different web browsers

It's nice if there is a latest news and updates page
Please have a introductory course online.
(4) What, if anything, do you find frustrating or unappealing about this tool?
write an email
gene lists sometimes does not work.
My big frustration is that I don't know nothing about genes and plants.
Im satisfied
Nothing so far!
I'm happy with using this tool.
(5) What is your main reason for using this tool today?
User friendly and helpful for efficient research results.
Very supportive for my work
Look at a gene annotation
This tool makes my research analysis work easy and it saves time.
easy and time saving analysis of data
Effective and efficient
Accuracy of analysis of results
Easy analytical support
check genes in my lists
Excellent
Test
Especially conversion between different gene accession numbers. This is the best tool
there is for it! Great work!
genelist
phylogenetic studies not today but generally. and sometimes expression information.
Work!

Appendix C - Usability Questions used in the Office

1. Have you use this tool before?
2. What is the first impression regarding the tool?
3. Could you identify what the tool is all about?
4. Could you find this as an useful tool?
5. How intuitive & helpful is the navigation?
6. Do you find any Issues with functionality and design?
7. Was anything too well hidden in the tool interface?
8. Do you find problem with the color scheme?
9. Is it easy to read (font style & size)?
10. What are the things you like most in this tool?
11. What can be improved?

Appendix D - Answers to Questions asked in the Office

8 people have answered to the usability questions.

Have you use this tool before?
8 out of 8 people have use the people have used the PlantGenIE

What is the first impression regarding the tool?
To do the analysis
Beautiful
Confusing
Need more help pages
Easy to use compare to other tools
No connection between different sites

Could you identify what the tool is all about?
6 out of 8 have identified the functionalities of different tools (exImage, exNet, exPlot and GeneList). Others could not find some of the functionalities.

Could you find this as an useful tool?
6 out of 8 said yes other 2 are not sure about the functionalities or usability.

How intuitive & helpful is the navigation?
Helpful
to some extent
good to have more help guidance

Do you find any Issues with functionality and design?
Some tools are not adopted to different screen sizes

Was anything too well hidden in the tool interface?
Icons are not clear
good to have tooltips

Do you find problem with the colour scheme?
Why green only
Its good for eye
I have no problem

Is it easy to read (font style & size)?
Fonts. Another one and more bigger.
Improve general design.
Looks very technical.
Looks great!

What are the things you like most in this tool?
Very supportive for my work
Look at a gene annotation
Accuracy of analysis of results
This is the best tool there is for it! Great work!

What can be improved?
gene lists sometimes does not work.
Im satisfied
Nothing so far!
I'm happy with using this tool.
conversion between different gene accession numbers.

Appendix E – Initial Expert Review Without a Checklist

January 10, 2016

Website review for http://popgenie.org

"This has to be one of the most interesting sites I have been to in a long time. Honestly, this type of information is beyond my knowledge of this particular subject. Today, January 11, 2016, I introduced this website to a co-worker, who is the Jr. High Science teacher at the school where I work. I asked him to go through it as he had time and tell me his thoughts about the site later in the week.

Upon landing on the home page, I was intrigued with all the colors and how the blank space allowed the information to be broken into clusters that anyone can follow through. The most important thing I noticed, throughout the entire site, the option to "take a tour" of the present page is available- of which I used very frequently. The reason I used this option was to learn more about the way the website was intended for use by the user. There are many options for one to increase their knowledge in this topic with the use of this site.

The PDF poster is magnificent! I studied it for quite some time. This is a piece of information worth printing and using as a reference. It may not be intended for that purpose for a person who is very familiar with this topic, but a someone new- I always print and take notes in order to gain knowledge of the topic and this is a poster I would want to have on hand.

I had fun playing with the different genes and plugging in information. Thank you for the sample workflow page. It provided me an example to go by when plugging in various bits of information. I went from page to page, clicking on a variety of links just to see what I might do next.

I believe this site is ideal for those who are interested in learning about plant genes and the breaking down of the information. This would be something our science teacher may want to incorporate within in class when it comes to the plant life chapter. It will provide an extra hand on opportunity to do research with his students.

During my time on the popgenie site, I found no glitches. The site is user friendly in regards to the tours and easy to read information presented in small clusters. I do not know who did your site, but it moves effortlessly, even with much information to be displayed, it loads quickly.

Thank you so much for allowing me to take part in this. Please keep in touch to update me on your study and research."

Screencasts can be found at YouTube®[https://youtu.be/pXLtBFAYf6k].

Two interface professionals evaluated the PopGenIE tool and recorded screencasts. Screencasts can be found at YouTube®[https://youtu.be/pXLtBFAYf6k].

Appendix F - Initial Version of ComPlEX Tool and Active User Statistics Before and After Redesigning the Tool

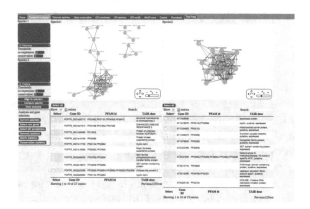

Appendix G - A Low Fidelity Prototype has been used to Understand the Visual Representations and Locations of Essential Tool Components in ComPlEX

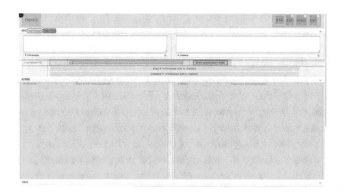

Appendix H - Affordances, Signifiers, Feedback, Mapping, and Constraints are Some of the Basic Design Principles Components

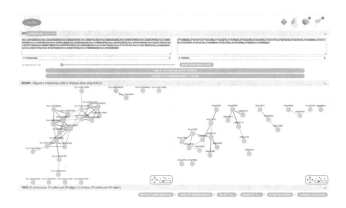

Appendix I - Heatmap of Survey Questions and Answers. X-axis Shows the Questions and Y-axis Shows the Participants. Color Represent the Strong Agree to Strong Disagree Options.

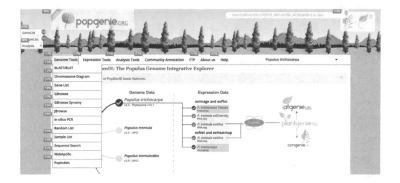

Appendix J - PopGenIE User Behavioral Flow for Last Year (2017/02/31–2018/02/31).

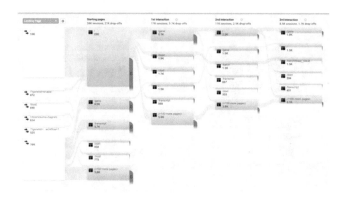

Appendix K - Survey Users by User Profession

Audience type	Number of participants
Plant Biologist/Botanist	22
Biologist - General	16
Bioinformatician	12
Computational Biologist	5
Plant Data Curator	1
Other	1
Total (participants)	57

Appendix L - Survey Users by Occupational Type

Users by profession	Number of participants
Graduate researcher	21
Postdoctoral researcher	19
Research scientist	8
Bioinformatics analyst	5
Undergraduate researcher	3
SW Developer/Engineer	1
Total (participants)	57

Appendix M - Average PlantGenIE usage by survey participants

Average usage	Number of participants
Daily basis	31
Weekly	21
Monthly	2
A few times a year	1
Rarely or never	2
Total (participants)	57

Appendix N - Tasks for the Tests at the Bioinformatics Office

Number	Tasks
1	How many "cellulose synthesis" genes can be found in *Poplulus trichocarpa*?
2	What is the *Populus tremula* best Blast hit for "Potri.001G266400" gene
3	Can you share current gene search results with someone else?
4	Can you BLAST "Potri.001G266400" protein sequence and find the Best Blast hit in *Poplus tremula*?
5	Are roots highly expressed in Potri.013G082200 gene in exImage *Populus trichocarpa* tissue?
6	How to save "cellulose synthesis" genes into PopGenIE GeneList?
7	How to download protein sequence for Potri.013G082200 gene?
8	How to generate heatmap for current genelist?

Appendix O - Usability Heuristics Selected for PopGenIE GeneList Tool Review Based on Jakob Nielsen's Usability Principles

Number	Usability heuristics
1	Visibility of system status
2	Match between system and the real world
3	User control and freedom
4	Consistency and standards
5	Error prevention
6	Recognition rather than recall
7	Flexibility and efficiency of use
8	Aesthetic and minimalist design
9	Help users recognize, diagnose, and recover from errors
10	Help and documentation

Appendix P - Summary Results of Five Usability Components

Learnability	Percentage
Strongly agree	77.9%
Agree	13.1%
Neither agree nor disagree	5.7%
Disagree	1.4%
Strongly disagree	1.9%
Memorability	Percentage
Strongly agree	82.6%
Agree	9.2%
Neither agree nor disagree	4.4%
Disagree	2.9%
Strongly disagree	0.9%
Efficiency	Percentage
Strongly agree	81.8%
Agree	10.2%
Neither agree nor disagree	4.4%
Disagree	3.0%
Strongly disagree	0.6%
Satisfaction	Percentage
Strongly agree	80.9%
Agree	13.6%
Neither agree nor disagree	2.8%
Disagree	2.4%
Strongly disagree	0.3%
Errors and Feedback	Percentage
Strongly agree	84.5%
Agree	8.8%
Neither agree nor disagree	4.4%
Disagree	1.9%
Strongly disagree	0.4%

References

1. Sundell, D., Mannapperuma, C., Netotea, S., Delhomme, N., Lin, Y.-C., Sjödin, A., de Peer, Y., Jansson, S., Hvidsten, R.T., Street, R.N.: The plant genome integrative explorer resource: PlantGenIE.org. New Phytol. **208**(4), 1149–1156 (2015)
2. Xiong, J.: Essential Bioinformatics. Cambridge University Press, Cambridge (2006)
3. ISO 9241-110:2006: Ergonomics of human-system interaction – Part 110: Dialogue principles

4. Al-ageel, N., Al-wabil, A., Badr, G., Alomar, N.: Human factors in the design and evaluation of bioinformatics tools. ScienceDirect (2015)
5. Pavelin, K., Cham, J.A., de Matos, P., Brooksbank, C., Cameron, G., Steinbeck, C.: Bioinformatics meets user-centred design: a perspective. PLoS Comput. Biol. **8**(7), e1002554 (2012)
6. Javahery, H., Seffah, A., Radhakrishnan, T.: Beyond power. Commun. ACM **47**(11), 58 (2004)
7. Mirel, B., Wright, Z.: Heuristic evaluations of bioinformatics tools: a development case. In: Human-Computer Interaction, Pt I (2009)
8. Mirel, B.: Usability and usefulness in bioinformatics: evaluating a tool for querying and analyzing protein interactions based on scientists' actual research questions. In: 2007 IEEE International Professional Communication Conference, pp. 1–8 (2007)
9. Douglas, C., Goulding, R., Farris, L., Atkinson-Grosjean, J.: Socio-Cultural characteristics of usability of bioinformatics databases and tools. Interdiscip. Sci. Rev. **36**(1), 55–71 (2011)
10. Bolchini, D., Finkelstein, A., Perrone, V., Nagl, S.: Better bioinformatics through usability analysis. Bioinformatics **25**(3), 406–412 (2009)
11. Bolchini, D., Finkestein, A., Paolini, P.: Designing usable bio-information architectures. In: Human-Computer Interaction. Interacting in Various Application Domains, pp. 653–662 (2009)
12. Hassani-Pak, K., Rawlings, C.: Knowledge discovery in biological databases for revealing candidate genes linked to complex phenotypes. J. Integr. Bioinform. **4**, 1–2 (2017). https://doi.org/10.1515/jib-2016-0002
13. Sjödin, A., Street, N.R., Sandberg, G., Gustafsson, P., Jansson, S.: The populus genome integrative explorer (PopGenIE): a new resource for exploring the populus genome. New Phytol. **182**, 1013–1025 (2009)
14. Nystedt, B., Street, N.R., Wetterbom, A., Zuccolo, A., Lin, Y.-C., Scofield, D.G., Vezzi, F., Delhomme, N., Giacomello, S., Alexeyenko, A., Vicedomini, R., Sahlin, K., Sherwood, E., Elfstrand, M., Gramzow, L., Holmberg, K., Hällman, J., Keech, O., Klasson, L., Koriabine, M., Kucukoglu, M., Käller, M., Luthman, J., Lysholm, F., Niittylä, T., Olson, Å., Rilakovic, N., Ritland, C., Rosselló, J.A., Sena, J., Svensson, T., Talavera-López, C., Theißen, G., Tuominen, H., Vanneste, K., Wu, Z.-Q., Zhang, B., Zerbe, P., Arvestad, L., Bhalerao, R., Bohlmann, J., Bousquet, J., Garcia Gil, R., Hvidsten, T.R., de Jong, P., MacKay, J., Morgante, M., Ritland, K., Sundberg, B., Lee Thompson, S., Van de Peer, Y., Andersson, B., Nilsson, O., Ingvarsson, P.K., Lundeberg, J., Jansson, S.: The Norway spruce genome sequence and conifer genome evolution. Nature **497**, 579 (2013)
15. Norman, D.: The design of everyday things (1988)
16. Netotea, S., Sundell, D., Street, N.R., Hvidsten, T.R.: ComPlEx: conservation and divergence of co-expression networks in A. thaliana, Populus and O. sativa. BMC Genom. **15**, 106 (2014)
17. Mannapperuma, C.: Try to understand design and design process (2010)
18. Gibson, J.J.: Gibson theory of Affordances.pdf, Chap. Eight. In: The Theory of Affordances (1986)
19. Benyon, D., Crerar, A., Wilkinson, S.: Individual differences and inclusive design. In: User Interfaces for All: Concepts, Methods, and Tools (2001)
20. Lin, X.H., Choong, Y.-Y., Salvendy, G.: A proposed index of usability: a method for comparing the relative usability of different software systems. Behav. Inf. Technol. **16**(4–5), 267–277 (1997)
21. Nielsen, J., Pernice, K.: Eyetracking web usability (2010)
22. Croll, A., Power, S.: Complete web monitoring: watching your visitors, performance, communities, and competitors (2009)

23. Bekavac, I., Praničević, D.G.: Web analytics tools and web metrics tools: an overview and comparative analysis. Croat. Oper. Res. Rev. **6**, 373–386 (2015)
24. W3Techs - extensive and reliable web technology surveys
25. Bhatnagar, A.: Web analytics for business intelligence beyond hits and sessions. Online **33**, 32–35 (2009)
26. Probets, S., Hasan, L., Morris, A.: Using google analytics to evaluate the usability of e-commerce sites
27. Neilsan, J.: Designing web usability: the practice of simplicity. Interact. Mark. **1**, 8–22 (2001). https://doi.org/10.1057/palgrave.im.4340116
28. Altschul, S.F., Gish, W., Miller, W., Myers, E.W., Lipman, D.J.: Basic local alignment search tool. J. Mol. Biol. **215**, 403–410 (1990)
29. Stein, L.D.: The generic genome browser: a building block for a model organism system database. Genome Res. **12**(10), 1599–1610 (2002)
30. Rosenbaum, S., Rohn, J.A., Humburg, J.: A toolkit for strategic usability. In: Proceedings of the SIGCHI Conference on Human Factors in Computing Systems, CHI 2000 (2000)
31. Molich, R., Jeffries, R.: Comparative expert reviews (2003)

A Serious Game to Learn Basic English for People with Hearing Impairments

María José Fernández[1], Angel Jaramillo-Alcázar[1(✉)], Marco Galarza-Castillo[1], and Sergio Luján-Mora[2]

[1] Facultad de Ingenierías y Ciencias Aplicadas, Universidad de Las Américas, Quito, Ecuador
{maria.fernandez,angel.jaramillo,marco.galarza}@udla.edu.ec
[2] Departamento de Lenguajes y Sistemas Informáticos, Universidad de Alicante, Alicante, Spain
sergio.lujan@ua.es

Abstract. In recent years, higher education has faced the need to apply educational changes due to the increase of students with special needs in the classroom. Day after day, we see changes in technology that can be exploited and mobile devices are an example of this trend. Mobile devices can be a tool used to improve the education of students with special needs, and people with hearing disabilities are not a exception. Thus, this research focuses on the development of accessible mobile serious games for education. Serious games allow people with hearing disabilities to learn by combining a tool such as a video game with a learning methodology. In this paper, we present My First English Game (MFEG), a video game whose main objective is to teach basic topics of the English language in an interactive way. Additionally, the video game includes accessibility features that make it inclusive for people with hearing disabilities. The final goal of this work is to contribute to the improvement of English learning for people with disabilities.

Keywords: Video games · Accessibility · English learning · Hearing impairments · Serious game

1 Introduction

Nowadays, technology is an active part of people's daily lives, including teaching-learning processes. Many educational institutions use technology as a tool to reinforce the teaching process, and teaching the English language is no exception [22].

It is important to consider that students with special needs are looking for options to advance in their learning. Especially hearing impaired individuals who use these spaces as a basic tool to accomplish their objectives. This kind of students use sign language as a mean of communication, some of them also use Spanish, English and others languages. With the influence of migration, it is

© Springer Nature Switzerland AG 2019
Á. Rocha et al. (Eds.): ICITS 2019, AISC 918, pp. 671–679, 2019.
https://doi.org/10.1007/978-3-030-11890-7_63

important for them to learn another language to be able to communicate with the rest of the world.

Students with disabilities have special needs inside and outside the classroom due to the difficulties that learning a language may generate. These needs have to be satisfied individually. According to each case, this would imply a change of the teacher's strategic focus at the time of his lecture.

In addition, thanks to technology a variety of study techniques have been developed to ease learning. Some examples are different serious games whose objective includes teaching, practice and reinforce of what was learned in the classroom. Serious games is a category of video games designed with the purpose of support the educational process [15]. One of the main advantages of serious games is that they enable learning through entertainment [11]. It is estimated that the development of this kind of games will reach $5,400 million in 2020, at a compound annual growth rate of 16.38% between 2015 and 2020 [14]. In many cases, serious games allow teachers to apply new teaching methods [19]. This kind of video games are designed to understand the necessities of students in the acquisition of skills, knowledge [4] and the achieving of learning outcomes [5]. Thus, in this article a serious game to support the learning of basic English for people with hearing impairments is proposed.

There are many applications to develop video games. One of them that has had great growth is Roblox [20]. Roblox is a platform which offers on-line multi player video games where users can create their own games or can play the games available and designed by other users. This platform allows the use of three-dimensional worlds. The player can customize attires, compile and exchange objects which belong to a collection or limited edition [18].

On the other hand, Roblox Studio is a tool that allows video game development and construction. It is a free program that allows the developer to build any object using a variety of materials, colors and worlds. This platform offers varied templates which presents the different environments that could be developed and changed by each user. It also offers access from any of these devices: Amazon, Mac, iOS, Android, Xbox One and PC.

The rest of this article is organized as follows. In Sect. 2, we present a review of hearing impairments and previous works about accessibility in video games for people with hearing impairments. In Sect. 3, we explain about the serious game development process and the methodology used. Next, in Sect. 4, we test the serious game and discuss about the results. Finally, in Sect. 5, we conclude the research and we outline our future works.

2   Hearing Impairment Accessibility

More than one billion people live around the world with some form of disability, about 15% of the world population [23]. In addition, in 2010 around 360 million people worldwide had disabling hearing loss, and 32 million of these were children [25]. These people usually live with social, educational and entertainment limitations because their impairments.

Furthermore, we can define accessibility as the ability of an object or service to be used in spite of the condition or disability of a person [1]. Accessibility in video games is a factor that is beginning to be considered by software developers. However, there exist several players with hearing disabilities. It is the case of Chris Robinson [12], known as Phoenix, who is deaf and he is a fighting game player. Another case is Adam "Loop" Bahriz who is legally deaf and blind Counter-Strike gamer [17]. These cases exist because some video games are accessible to people with different disabilities.

According to [24], there are four grades of hearing impairment based on a decibel scale representing hearing loss:

- Slight/mild hearing loss (26–40 dB).
- Moderate hearing loss (41–60 dB).
- Severe hearing loss (61–80 dB).
- Profound hearing loss (over 81 dB).

People with mild, moderate and severe hearing loss are grouped under the term *hard of hearing*; whereas people with profound hearing loss are *deaf*. Both, hard of hearing and deaf together represent the total number of cases of hearing impairments [25].

Each disability has its own particularities and therefore its own accessibility parameters. There are accessible video games for people with hearing impairments. This is the case of MusicPuzzle, a video game for encouraging active listening among deaf and hard of hearing people [13]. Also Memosign [2] and Robostar [16] are video games that promotes learning sign language. There are mobile video games that have accessibility features for people with hearing impairments. GameOhm [10], for example, is a video game that includes subtitles in all of the video game's conversations. This allows the video game to be used by deaf or hard of hearing people.

Before designing a video game for people with disabilities, it is necessary to understand their characteristics and the accessibility parameters. There are some authors that describe these accessibility parameters and the paths that video games should have, considering the perspective of impairment and/or the type of device the individual uses.

According to [8], it is necessary to consider three levels of accessibility, from a minimum to an ideal scenario:

- *Low Level - Good:* It refers to simple implementation complexity and good accessibility features.
- *Medium Level - Better:* It refers to a medium implementation complexity and better accessibility features.
- *High Level - Best:* It refers to a high implementation complexity and the best accessibility features.

These levels include basic access elements to advanced ones that must be considered when developing a video game.

When a video game is developed, it is important to consider the benefits that a game needs to have, that is why the International Game Developers Association (IGDA) and its Special Need Group for Video Games Access states: "Customer satisfaction needs to reach the highest target to offer new opportunities to learn skills" [7].

In [9], the authors propose a compilation of accessibility guidelines for people with hearing impairments for mobile devices. Some of them were considered as a requirement of the video game presented in this article and they are listed in Table 1, classified by their respective level of accessibility.

Table 1. Accessibility guidelines

Level	Video games guidelines
Low level - good	Provide subtitles
	Use simple language
	No essential information in audio alone
	Adequate interface for the player age
	Appropriate words-per-minute
	Easy installation
Medium level - better	Use explicit visual feedback
	Background noise to minimum during speech
	Possibility for repetition
	Visual indication of who is currently speaking
	Pause while text is being read
High level - best	Allow several different input and output devices

3   Video Game Development

For the development of our video game, we require a methodology that helps us to do it in the shortest possible time, for that reason we select Scrum. Scrum is a methodology whose main goal is to reduce the complexity in product development. "Scrum teams work is based on a variety of requirements and technologies to generate functional products using empiric" [3].

Scrum executes a process using short and fixed term cycles that are developed in two weeks, maximum 4 weeks, including feedback on the product development. Scrum calls "sprint" to each developed interaction which is recommended to carry on for only 4 weeks. Sprint is, then, an interval of time with a maximum duration of a month for the presentation of a product. For this reason, sprint is the core of a Scrum process since it creates the growth of a product.

Scrum has some artifacts involved in a project:

1. Meetings
   a. Sprint schemata: Before beginning each sprint, it is necessary to plan the project to determine the objectives and all the interaction stages.
   b. Daily meetings: It is necessary to review each step including an achievement percentage of what has been developed, making sure that deadlines are met. If there is something pending, it should be included in the next phase to be able to achieve the goals.
   c. Sprint review: Analysis and review of software increment.
2. Elements
   a. Product list: Requirements stated by the user to perform the first product review and to acknowledge the type of final product.
   b. Sprint list: Types of work that need to be developed during the sprint to generate the so awaited growth.
   c. Increase: Result of each sprint.

The definition of the vision of the game was made from interviews with English teachers and students with hearing disabilities. Some of the teachers had previous experience working with people with special educational needs. Both parties shared their knowledge and experience to detect the needs that the video game will satisfy.

The student with hearing impairment commonly has a low level of English due to the difficulties for communication, the type of personality (shy), and their previous experiences [6]. Currently, the tool to support this situation is to offer one-to-one tutoring programs. However, the video game can replace the tutoring with the teacher as a tool to practice the topics taught in a class.

By investigating and analyzing the different tech tools, having meetings with users (students from first level with hearing impairments) and English teachers experts, the findings are:

- Teachers' description (interviews).
- English students survey.
- Video game presentations (meetings with leader teacher).
- User testing processes (students from first level with hearing impairments).
- The video game is published for Roblox users so that it can be tested.

The level of English chosen for the video game is starter or beginner, focusing on the following topics:

- Colors
- Subject pronouns
- Verb to be
- Yes, no questions, and short answers
- Functional language (Introduce yourself)
- Use of expressions (Thanks, you're welcome, well done)
- Spelling rules (Plural nouns).

3.1 Video Game Scenes

The video game scenarios are described as follows:

1. *Beginning:* Once the video game has been selected and downloaded, the character that represents the student is placed in a reception where a helper shows him what he is supposed to do. After this, the character goes into a classroom where he will observe some objects. Once the student is out of this environment, the character walks across a rainbow bridge that ends at a helper position who will give the character information about each color which will be used throughout the game. To finish the first level the character needs to go down a slide to find a teacher asking a question about the colors and giving examples of them.
2. *Subject pronouns and Verb to be:* In this section, there is a teacher that shares information about subject pronouns and verb to be which are needed to start all the challenges in the game. This section combines skills with abilities.
3. *Yes/No questions, and short answers:* After passing through the previous challenges, the character reaches a different section where there is a teacher that offers a test to make sure that the player remembers what a subject pronoun is. The player practices answering short affirmative and negative questions.
4. *Plural nouns:* Finally, the character finds a maze where he will have to find the exit. There are a lot of riddles that need to be solved to find the teacher that finally evaluates all knowledge. If the answer is correct, the door will be opened and the character will receive medals as rewards.

4   Results and Discussion

Once the video game was implemented, the accessibility features were validated. In the Low Level - Good, the video game has subtitles in all conversations. Likewise, each dialog uses a simple language and no important information is presented in audio alone. In Fig. 1, we present some of these characteristics.

In the Medium Level - Better, the video game presents indicators of who is talking. Likewise, the video game reduces ambient noise to a minimum and uses specific feedback. In addition, the video game gives the option of repetition when there is an erroneous answer in the control questions.

Finally, in High Level - Best, Roblox allows the player to connect a variety of devices to its platform. This is the case of Oculus Rift, which transforms the video game player's experience into a three-dimensional environment.

The video game was tested by students of first semesters, some of them with hearing impairments as we can see in Fig. 2. After the tests, all the students indicated that the video game had features and challenges that helps them to learn the basic principles of English. In the case of students with hearing impairments, they wanted to use the video game as a tool to practice in home and reinforce what they learnt in the classroom. Students also indicated that it is necessary to address more topics in the game through more levels and greater difficulty.

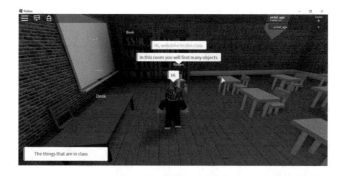

Fig. 1. Low level - good accessibility

Fig. 2. Tests with students

After the application of this tests, we discovered which topics are the most important and how to improve them. Now, we recognize how challenge the game is to improve the design and the content of the project. All the benefits were analyzed as well as the opportunities to develop other projects similar to this for other subjects. These results allow us to understand that video games need to improve accessibility features for people with special needs.

5   Conclusions and Future Work

This research aims to improve the entertainment and learning conditions of people with hearing impairments that cannot access video games because of their condition, especially serious games. Due to the growth of this type of games, and their contribution to the teaching process, accessibility parameters should be considered in their design and in their implementation. The use of non-accessible serious games goes against the Article 24 - Education of the United Nations Convention on the Rights of Persons with Disabilities [21], considering that avoids people with impairments having access to education on equal terms with a person without disabilities.

This video game tries to satisfy the needs generated by English learning at a beginners' level. The game offers information about each topic and asks questions to confirm that the student is understanding each part of the language class. This research is important to involve students with disabilities in an effective process of English learning, by using video games as a tool that also offers fun and knowledge.

By applying Scrum, the project met the planned deadlines. All the steps in creation are presented on time and feedback is also valuable to continue. The use of a software development methodology allows to maintain an adequate follow-up to complete the project satisfactorily.

This video game is only the beginning and basis to continue researching and developing more levels along with the active participation of students, becoming a great tool to be added up to the work in class. The continuous incorporation of new accessibility guidelines will make the video game more accessible.

References

1. Accessible University: Defining Accessibility (2016). https://goo.gl/Hwr2vC
2. Bouzid, Y., Khenissi, M.A., Jemni, M.: Designing a game generator as an educational technology for the deaf learners. In: International Conference on Information Communication Technology and Accessibility, pp. 1–6 (2015)
3. Francia, J.: What is Scrum?—Scrum.org (2017). https://goo.gl/VMjJEx
4. Ghannem, A.: Characterization of serious games guided by the educational objectives. In: International Conference on Technological Ecosystems for Enhancing Multiculturality, pp. 227–233 (2014)
5. Guillén-Nieto, V., Aleson-Carbonell, M.: Serious games and learning effectiveness: the case of It's a Deal!. Comput. Educ. **58**(1), 435–448 (2012)
6. Harmer, L.: Health care delivery and deaf people: practice, problems, and recommendations for change. J. Deaf Stud. Deaf Educ. **4**(2), 73–110 (1999)
7. International Game Developers Association (IGDA)—Game Accessibility SIG: Accessibility in Games: Motivations and Approaches (2004). https://goo.gl/3gUaV2
8. Jaramillo-Alcázar, A., Luján-Mora, S.: Mobile serious games: an accessibility assessment for people with visual impairments. In: International Conference on Technological Ecosystems for Enhancing Multiculturality, pp. 1–6 (2017)
9. Jaramillo-Alcázar, A., Luján-Mora, S.: An approach to mobile serious games accessibility assessment for people with hearing impairments. In: International Conference on Information Technology and Systems, pp. 552–562 (2018)
10. Jaramillo-Alcazar, A., Guaita, C., Rosero, J.L., Lujan-Mora, S.: Towards an accessible mobile serious game for electronic engineering students with hearing impairments. In: IEEE World Engineering Education Conference, pp. 1–5 (2018)
11. Koster, R.: Theory of Fun for Game Design, 2nd edn. O'Reilly Media, Inc. (2013)
12. Kotaku: Deaf Gamer Founds A Fighting Games Team For Players Like Him (2017). https://goo.gl/y5cWMQ
13. Li, Z., Wang, H.: A mobile game for encouraging active listening among deaf and hard of hearing people: comparing the usage between mobile and desktop game (2015). https://goo.gl/tkUC3r

14. Markets and Markets: Serious Game Market worth $5,448.82 Million by 2020 (n.d.). https://goo.gl/WbdR7t
15. Michael, D.R., Chen, S.L.: Serious Games: Games That Educate, Train, and Inform. Muska & Lipman/Premier-Trade (2005)
16. Özkul, A., Köse, H., Yorganci, R., Ince, G.: Robostar: an interaction game with humanoid robots for learning sign language. In: IEEE International Conference on Robotics and Biomimetics, pp. 522–527 (2014)
17. PCGamesN: Legally deaf-blind CS:GO player Loop offered pro streamer contract after community support (2017). https://goo.gl/MwKzeJ
18. Roblox: Whats Roblox? (2018). https://corp.roblox.com/
19. Sauvé, L., Sénécal, S., Kaufman, D., Renaud, L., Leclerc, J.: The design of generic serious game shell. In: International Conference on Information Technology Based Higher Education and Training, pp. 1–5 (2011)
20. Takahashi, D.: At 10, Roblox surpasses 30 million monthly users and 300 million hours of engagement—VentureBeat (2016). https://goo.gl/S9SgiS
21. United Nations General Assembly: Convention on the rights of persons with disabilities (2008). https://goo.gl/ZuFucZ
22. Wang, K.: Application of information technology in cultivating college English learning motivation. In: International Conference on Logistics, Informatics and Service Sciences, pp. 1–5 (2016)
23. World Health Organization: World report on disability (2011). https://goo.gl/C1YBTw
24. World Health Organization: Grades of hearing impairment (2016). https://goo.gl/1pWAZd
25. World Health Organization: Deafness and hearing loss (2017). https://goo.gl/35F73p

Ethics, Computers and Security

The Art of Phishing

Teresa Guarda[1,2,3]($\boxtimes$), Maria Fernanda Augusto[1,4],
and Isabel Lopes[3,5]

[1] Universidad Estatal Península de Santa Elena – UPSE, La Libertad, Ecuador
tguarda@gmail.com, mfg.augusto@gmail.com
[2] Universidad de las Fuerzas Armadas-ESPE, Sangolqui, Quito, Ecuador
[3] Algoritmi Centre, Minho University, Guimarães, Portugal
[4] BITrum-Research Group, C/San Lorenzo 2, 24007 León, Spain
[5] UNIAG (Applied Management Research Unit),
Polytechnic Institute of Bragança, Bragança, Portugal
isalopes@ipb.pt

Abstract. Nowadays there are many threats that a company needs to protect itself. Everyone knows someone who has fallen for a coup by using an email, message or phone. People who pass by someone they trust, to extract data and money from the victim. These three ways are used to try convince someone to deliver accounts, credit card and document data in companies and at a particular level. According to Symantec, more than 6 hundreds of companies per day are targeted for Phishing, specifically Business E-Mail Compromise (BEC). In it, criminals pass through a central figure in the company, usually the CEO, and try to extract information or get employees to transfer money. This type of attack has generated in the last years losses of billions of dollars for the businesses affected. It is urgent that all company employees and individuals know as soon as possible what Phishing is and what steps to take.

Keywords: Phishing · Spear phishing · SMiShing · QRishing · Vishing · Threats · Security · IoT

1 Introduction

The Internet has become an integral part of our lives on a personal and professional level. With the rapid technological advancement many types of fraud have emerged seeking obtain confidential data and use them for personal profit. The main ones being Phishing, vishing, SMiShing, and QRishing.

Symantec 2018 Internet Security Threat Report (ISTR) shows that the targeted attack industry continues its expansion, including a 600% increase in IoT attacks. According to Symantec, innovation, organization and sophistication are the tools of cyber-attackers, because they work even more efficiently to discover new vulnerabilities [1].

The increase in the use of Internet services, produces that we provide personal information to different platforms, applications, sites and web portals. This is not done in a controlled way, on the contrary, every time we pay less attention, so it is expected that there are people inspired by various motivations that want to get our data.

Á. Rocha et al. (Eds.): ICITS 2019, AISC 918, pp. 683–690, 2019.
https://doi.org/10.1007/978-3-030-11890-7_64

Phishing arise from this thought of subtraction of information, and currently is considered one of the most common crimes, and unfortunately the crime in which we fall more easily [2].

Phishing is a process of obtaining confidential information illegally from users, such as usernames and passwords, for fraudulent purposes [3]; which is operationalized through the sending of emails as a bait, with the objective of capturing a potential victim, creating the necessary scenarios to make credible the whole scheme.

Lately phishing is one of the most common attacks on the Internet [4], because it is a relatively easy attack to apply and reaches multiple users at the same time. And it is enough that the user clicks on the malicious link to have their personal data stolen.

Phishing attacks are executed by performing the following steps (Fig. 1):

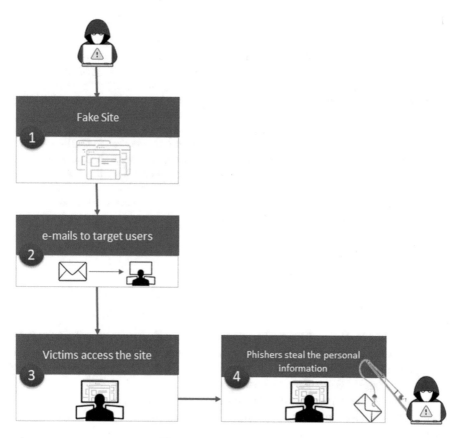

Fig. 1. Phishing attacks steps.

- In a first step the phisher creates and makes available a false site equal to the legitimate site;
- After in the second step, false emails are sent with the link from the legitimate site to the potential victim on behalf of legitimate companies and organizations, trying to convince potential victims to visit the sites;
- In the third step, victims visit the fake site by clicking on the link and entering their useful information;
- Then in the fourth step, phishers steal personal information and thus begin to perform fraud.

2 Phishing

The term Phishing comes from the English word fishing which means fishery. This threat consists of false messages sent by criminals who pass through reputable companies or people [1]. They cast the bait and often the victims are hooked. Among the most common baits we have: the most urgent email from the bank warning that your account would be blocked; and the offer of a well-known company in which it only gives you a few hours to decide [2].

Spear Phishing is a highly localized Phishing attack, preceded by a thorough study of the target by the attacker [5–7]. Spear Phishing has a ten steps cycle:

(1) The phisher defines its target within an organization, using social networks or others Internet information sources to find the company employee's with access to the data\systems;

(2) The attacker's work reveals the patience of a fisherman, he will follow the digital footprint of the potential victim (employee), identifying other people he may know;

(3) A false but recognizable email is created to impress a colleague or boss;

(4) An email is sent to the false e-mail employee, with a link or file (s) attached;

(5) The fake email skips the spam filter and arrives at the employee's inbox;

(6) The employee opens the email because he knows (he thinks he knows) the sender;

(7) The employee trusts the sender, and clicks on the link, or opens the attachment;

(8) The open website causes the credentials to be stolen, allowing malware to be installed. When opening attachment the malware is installed infecting the system \network;

(9) The attacker use the backdoor to stolen information, which if used intelligently by the attacker can guarantee him enough knowledge to assimilate the identity of someone from the organization with more power [3].

In 2017 the phishing rate in South Africa was the highest, where 1 in 785 emails was a phishing attack [1] (Fig. 2).

Rank	Country	1 in
1	South Africa	785
2	Netherlands	1,298
3	Malaysia	1,359
4	Hungary	1,569
5	Portugal	1,671
6	Austria	1,675
7	Taiwan	1,906
8	Brazil	2,117
9	Indonesia	2,380
10	Singapore	2,422

Fig. 2. Phishing rate by country (source Symantec [1]).

For Symantec "Spear phishing is the number one infection vector employed by 71% of organized groups in 2017. The use of zero days continues to fall out of favor. In fact, only 27% of the 140 targeted attack groups that Symantec tracks have been known to use zero-day vulnerabilities at any point in the past" [1]. In 2017 71% of targeted attack infection vectors were caused by spear phishing emails [1] (Fig. 3), spear phishing is the most popular.

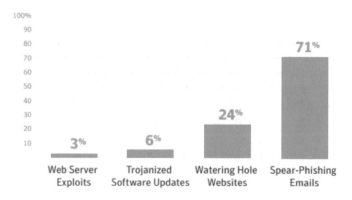

Fig. 3. Targeted attack infection vectors (source Symantec [1]).

We can consider another's forms of Phishing, like SMiShing, QRishing and Vishing; which we will describe in the following subsections (Fig. 4).

Fig. 4. Phishing forms. ❶ Phishing, ❷ SMiShing, ❸ QRishing,❹ Vishing.

2.1 SMiShing

SMiShing means Phishing by SMS. SMiShing is similar to Phishing, a text message is sent to the user of the phone instead of an email. In the text message, the user is asked to call a certain phone number or to go immediately to the website to take immediate action, and a personal information system (passwords, credit card data) is requested through an automatic response system. These messages ask to use it and click on a link and fill out a form or reply to the message. It could be for example, about a need to update registration or the opportunity to redeem a prize not to be missed. SMiShing is characterized by having two attack vectors [8]:

(1) When the attacker sends an SMS with information about changes, a purchase, changes, refunds or cancellations. The user is alarmed and is deceived, the message includes a phone number of the attacker, and thus could communicate with the attacker requesting personal information of the user or codes necessary for a micro-payment online. This form of attack is summarized by the theft of information through a telephone conversation between the victim and the attacker;

(2) Through the sending of an SMS that includes a URL that when visited by the user, a malware is installed on his phone. Then it replicates all the user information stored on the phone to the attacker's server.

2.2 QRishing

QR codes are becoming more and more present in our daily lives (shoppings, train stations, marketing promotions, touristic spots, apps that can be used to read and decipher codes and in many cases can lead directly to a website, among others). Once QR codes are used for legitimate purposes, they can also be developed or manipulated for illicit purposes and have the same effect as phishing e-mail. QRishing uses socially engineered bait to make potential victims scan the code [9].

2.3 Vishing

The Vishing attacks, abbreviated to Voice phishing, is an electronic fraud technique. Vishing is applied over phone calls rather than through messages or email. These calls are primarily intended to obtain bank details or other important personal information from victims, usually by automatic calls or equipment that modify the voice of the fraudster [8].

When a visher creates an automatic voice system to make voice calls to telephone users requesting private information, it is called Vishing or Voice Phishing. The intent is the same as phishing by email or phishing by SMS, The voice call creates a sense of urgency for the user to take action and provide additional information [10].

3   Attack and Prevention

Company's usual security policy includes anti-virus software, firewall, e-mail protection, among others. While prevention is always the best option, a company should not rely only on best practices on the part of its users, it should make sure that all its employees are fully aware of what Phishing is and the risks to which they can expose themselves if they are not attentive.

The most common form of phishing attack is through email. Sometimes phishers use domains similar to real emails or even common domains like hotmail. If someone you know has become a victim of a phishing attack, attackers can gain access to that person's account and fire e-mails through it or him. What the attacks usually have in common is the alarming tone. Phrases like "your bank account was blocked" or "your email account will be deleted soon" are used to attract attention and make the victim act without thinking [11].

It is not enough to know what Phishing is to be free of it. You need to always be alert and instruct all your team members to be aware of this type of threat. The main tips are:

- Always check the sender's email address;
- Observe the contact information and signature of the email;
- Never write your access data and passwords on forms or pages sent by email
- Do not click on links unless you're sure the sender is trusted;
- Do not download attachments without first checking with the antivirus;
- If you receive an unusual request to transfer files or money from an acquaintance, manager or co-worker, confirm with the person if the request is true.

In the case of SMiShing when clicking or replying to the message, you can be directed to malicious sites, where, once you enter your data, they will end up in the hands of criminals. The targets are usually personal information such as address or CPF, credit card data and bank access passwords, social networks and emails. To avoid SMiShing, never click sent links and never provide your data on unknown forms or sites. Whenever you make purchases or transactions that require typing your data, make sure the URL starts with https instead of http.

With QRishing, just as it is possible to attack users, it is possible to attack the systems that will make use of the content of this QR Code. Depending on the content inserted in the QR Code and the security of the application that makes use of the scanned content it is possible to cause of the most diverse frauds. It is also possible to dominate the victim's device if the QR Code content exploits some vulnerability like Buffer Overflow in the application that interprets the QR Code. For attackers, one of the advantages of using QR Code to trigger access to a URL is the fact that the user does not have to type the URL and often end up only holding onto the displayed content, thus becoming a victim in potential for Phishing attacks that lead to cloned-looking Web sites. Users are often the weakest link in a system, so instructing them is critical. When we're talking about targeted attacks on businesses, where curiosity drives users to risk exposures. To avoid QRishing users should only scan secure sources with a reliable scanner, and disable any kind of automatic action on the part of the reader. In the case of information systems there should be a duly updated whitelist and verify that the content size is as expected [12].

In Vishing attacks, calls can be made directly by a person or using recordings or automated voices. It is common for the fraudster to pretend that he has to confirm personal data to update the register at the bank, authorize a purchase on the credit card or give away some credit. Do not provide personal information over the phone under any circumstances. If you do have a pending purchase, you can call the central bank whose number is usually written on the card. Do not call the numbers passed by the person who called you, as they are also part of the coup.

4   Conclusions

On the Internet, attacker's try to get users to access malicious links through various tactics, including email persuasion, or by requesting confirmation of cadastral data, by false letters sent via mail taking advantage of the name of entities, by SMS and any other means that can lead the innocent user to bite the bait in this attack called Phishing. While all these means involve some kind of persuasion, none is so simpler and probably effective as the Quick Response Code (QR Code).

Today's diversity of computing devices and mobile devices allow users to download and install applications without realizing that they may not be a copy of legitimate official applications,

Attackers are using phishing as a highly profitable activity. In recent years there has been an increase in the technology, diversity and sophistication of these attacks Phishing differs from traditional scams at the level of fraud that can be achieved, being a higher level.

Online users should be informed and undertake a periodic vulnerability analysis to identify and mitigate weaknesses that can lead to a successful phishing attack.

References

1. Symantec Internet Security Threat Report. Symantec, vol. 23 (2018)
2. Alsharnouby, M., Alaca, F., Chiasson, S.: Why phishing still works: user strategies for combating phishing attacks. Int. J. Hum.-Comput. Stud. **82**, 69–82 (2015)
3. Brill, J.A., McGeehan, R., Muriello, D.G.: U.S. Patent No. 9,578,499, Washington, DC (2017)
4. Jing, Q., Vasilakos, A.V., Wan, J., Lu, J., Qiu, D.: Wirel. Netw. **20**(8), 2481–2501 (2014)

5. Mohammad, R.M., Thabtah, F., McCluskey, L.: Tutorial and critical analysis of phishing websites methods. Comput. Sci. Rev. **17**, 1–24 (2015)
6. Ariu, D., Frumento, E., Fumera, G.: Social engineering 2.0: a foundational work. In: Proceedings of the Computing Frontiers Conference (2017)
7. Gascon, H., Ullrich, S., Stritter, B., Rieck, K.: Reading between the lines: content-agnostic detection of spear-phishing emails. In: International Symposium on Research in Attacks, Intrusions, and Defenses (2018)
8. Chiew, K.L., Yong, K.S.C., Tan, C.L.: A survey of phishing attacks: their types, vectors and technical approaches. Expert Syst. Appl. **106**, 1–20 (2018)
9. Rzeszut, E., Bachrach, D.: 10 don'ts on your digital devices: the non-techie's survival guide to cyber security and privacy. Apress (2014)
10. Dhiman, P., Wajid, S.A., Quraishi, F.F.: A comprehensive study of social engineering - the art of mind hacking. Int. J. Sci. Res. Comput. Sci. Eng. Inf. Technol. **2**(6), 543–548 (2017)
11. Oliveira, D., Rocha, H., Yang, H., Ellis, D., Dommaraju, S., Muradoglu, M., Weir, D., Soliman, A., Lin, T., Ebner, N.: Dissecting spear phishing emails for older vs young adults: on the interplay of weapons of influence and life domains in predicting susceptibility to phishing. In: Proceedings of the 2017 CHI Conference on Human Factors in Computing Systems (2017)
12. Falkner, S., Kieseberg, P., Simos, D.E., Traxler, C., Weippl, E.: Usable cryptographic QR codes. In: IEEE International Conference on Industrial Technology (ICIT) (2018)

Health Informatics

Comparison of Atrial Fibrillation Detection Performance Using Decision Trees, SVM and Artificial Neural Network

Szymon Sieciński[✉], Paweł S. Kostka, and Ewaryst J. Tkacz

Department of Biosensors and Biomedical Signal Processing,
Silesian University of Technology, Zabrze, Poland
szymon.siecinski@polsl.pl

Abstract. Atrial fibrillation (AFib) is a supraventricular tachyarrhythmia characterized by uncoordinated atrial activation and ineffective atrial contraction. AFib affects 1–2% of the general population, its prevalence increases with age and may remain long undiagnosed. Due to costs of hospitalization and treatment related to AFib and increasing prevalence, effective methods of detecting atrial fibrillation are needed.

In this study we compared AFib classification using support vector machine (SVM), artificial neural network (ANN) and binary decision trees on 10 ECG signals. We considered 8 parameters associated with RR intervals: mean RR, SDNN, RMSSD, PLF, PHF, LF/HF, SD1 and SD2. In this comparison the best performing AFib classifier was binary decision tree with maximum number of splits equal to 100 and the worst case was SVM classifier with medium Gaussian kernel and using only one feature. Achieved result should encourage further studies using decision trees.

Keywords: Atrial fibrillation · Classification · SVM ·
Classification trees · ANN

1 Introduction

Atrial fibrillation (AFib) is a supraventricular tachyarrhythmia characterized by uncoordinated atrial activation and ineffective atrial contraction. On the electrocardiogram (ECG), AFib is characterized by irregular RR intervals, absence of P waves, replaced by fibrillatory waves and irregular atrial activity [2,6,10]. Atrial fibrillation affects 1–2% of the general population and is associated with increased rates of death, stroke, heart failure, thromboembolism and degraded quality of life. AFib may remain long undiagnosed [2]. The prevalence of atrial fibrillation increases with age [2,6,10]. Diagnosis is based on patient history and clinical examination, confirmed by 12-lead electrocardiography, sometimes using Holter monitoring [2,21].

© Springer Nature Switzerland AG 2019
Á. Rocha et al. (Eds.): ICITS 2019, AISC 918, pp. 693–701, 2019.
https://doi.org/10.1007/978-3-030-11890-7_65

Due to the costs of hospitalization and treatment related to AFib and increasing prevalence in aging societies, development of accurate and effective methods of atrial fibrillation detection is crucial [2, 6, 10, 15]. There are two main groups of algorithms based on traces of atrial fibrillation in electrocardiograms: algorithms based on RR interval irregularities and algorithm based on detecting the P-wave absence [4, 8, 14, 16]. Due to the noise level found in ambulatory electrocardiographic (ECG) recordings and the small amplitude of P wave, methods based on RR irregularities are preferred [8, 14]. Our approach is based on features derived from RR intervals.

The purpose of this study was to compare the performance of atrial fibrillation classifiers based on decision trees, Support Vector Machine (SVM) and artificial neural network (ANN) on 10 signals from MIT-BIH Arrhythmia Database (MITDB).

2   Materials and Methods

2.1   Signal Database

In this study the authors used 10 fully annotated signals from MIT-BIH Arrhythmia Database (MITDB) [17] available on PhysioNet.org [20] containing AFib episodes: 201, 202, 205, 208, 210, 215, 217, 219, 220, 221. Each signal lasted 30 min 5 s, was acquired with sampling frequency of 360 Hz and was annotated independently by two physicians.

2.2   Signal Pre-processing

Signal pre-processing and classification was performed off-line using MATLAB (The MathWorks, Inc., Natick, MA, USA). The first step was reading signals and reference annotations. The next step was QRS complex detection using Pan-Tompkins algorithm [19] implemented by D. Wedekind [26]. According to Larburu et al. [14], the R wave is the most prominent feature of ECG, making it relatively simple to detect. The next step of proposed method was calculating RR intervals. Then, the signal was interpolated and divided into non-overlapping segments of 128 samples. For each segment we calculated the following parameters:

1. Mean RR interval ($\overline{RR}$),
2. Standard deviation of RR interval (SDNN),
3. Root mean square of successful differences (RMSSD),
4. Power of low frequency spectrum of HRV (PLF),
5. Power of high frequency spectrum of HRV (PHF),
6. PLF over PHF ratio (LF/HF),
7. SD1 (short term heart rate variability parameter),
8. SD2 (long term heart rate variability parameter).

Mean RR interval, SDNN, RMSSD, PLF, PHF and LF/HF are parameters describing heart rate variability (HRV). Heart rate variability is the phenomenon of occurring RR intervals (inter-beat intervals) over time associated with autonomous nervous system activity [7]. The power of the low frequency band (PLF) was computed in the band 0.04–0.15 Hz, and the power of the high frequency band (PHF) was computed in the band 0.15–0.4 Hz. The LF/HF ratio was computed as the PLF/PHF ratio.

SD1 and SD2 are parameters associated with RR interval scatter plot. Scatter plot is a diagram in which each RR interval is plotted as a function of the previous RR interval [11]. It is a geometrical and non-linear method usually used in HRV analysis. The scatter plot of normal ECG seems as comet-shaped, in which nearly all points are centralized along the diagonal line. However, the scatter plot of AFib seems as unfolded fan-shaped, in which all the points are dispersed around the whole plot [21].

SD2 measures the dispersion of points of scatter plot along the diagonal line and SD1 measures the dispersion of points perpendicular to the diagonal line [24]. SD1 is calculated as the standard deviation of the distances of points from $y = x$ axis (see Eq. 1), measuring the width of the ellipse and indicating the short-term variability. SD2 is calculated as the standard deviation of the distances of points from $y = -x + 2\overline{RR}$ axis (see Eq. 2), measuring the length of the ellipse and indicating the long-term variability [1]. They can be defined as:

$$SD1 = stddev\left(|RR_{n+1} - RR_n|/\sqrt{2}\right) \qquad (1)$$

$$SD2 = stddev\left(\left|\frac{RR_{n+1} - RR_n}{\sqrt{2}} - 2\overline{RR}\right|\right) \qquad (2)$$

where RR_n is a RR interval with $n = 1, 2, \ldots N - 1$, RR_{n+1} is the RR interval next to RR_n and $stddev()$ is the standard deviation of equation within the parenthesis [21].

2.3   Classification

We performed AFib episodes detection using three classifiers: artificial neural network (ANN), support vector machine (SVM) and binary decision trees. For all classifiers, we tested their performance by comparing classified segments from testing set with reference annotations divided into 128-sample segments.

Support vector machine (SVM) is a classifier proposed by Vapnik et al. at AT&T Bell Laboratories [25] which separates the classes by constructing a hyperplane [3]. SVM has been applied in areas such as face recognition, text categorization, time series prediction [18] and handwritten characters recognition [22].

In this study we used feed-forward ANN consisting of input layer, hidden layer and output layer. Hidden layer used sigmoid function of activation and consisted of 8 neurons. Output layer used softmax function of propagation and consisted of one neuron. Training of the proposed artificial neural network was based on scaled conjugate gradient backpropagation.

We tested the performance of classification of AFib episodes by SVM and decision trees using 10-fold cross-validation and Classification Learner MATLAB application. For binary decision trees Gini's Diversity Index was used as the split criterion and tested three categories of number of splits: fine (100 splits), medium (30 splits) and coarse (5 splits). We considered linear, quadratic, cubic and Gaussian SVM kernels with three different scale categories (coarse, medium and fine). For ANN, we considered 5, 10 and 20 neurons in hidden layer and we divided randomly 70% of data as the training set, 15% of data was testing set and 15% of data was validation set.

3 Results

We present classification performance as accuracy (Acc), area under curve (AUC), true positives (TP), true negatives (TN), false negatives (FN), false positives (FP), true positive rate (TPR), false negative rate (FNR), positive predictive value (PPV), false discovery rate (FDR) and false positive rate (FPR) for binary decision trees classifier and SVM. For ANN we consider the number of neurons in hidden layer, TP, TN, FN, FP, accuracy and false omission rate. Total number of segments considered in this study is 50790.

Table 1. AFib Classification performance of binary decision trees.

Number of features after PCA	Acc	AUC	TP	TN	FN	FP	TPR	FNR	PPV	FDR	FPR
Fine trees (maximum number of splits: 100)											
No PCA	95.8%	0.99	21265	27391	676	1458	97%	3%	94%	6%	5%
7	94.5%	0.99	20787	27206	1154	1643	95%	5%	93%	7%	6%
6	94.8%	0.99	20792	27332	1517	1149	95%	5%	93%	7%	5%
5	94.8%	0.99	20970	27195	1654	971	96%	4%	93%	7%	5%
4	92.1%	0.97	20232	26542	1709	2307	92%	8%	90%	10%	8%
3	86.4%	0.93	17971	25925	2924	3970	82%	18%	86%	14%	10%
2	83.9%	0.92	16140	26483	5801	2366	74%	18%	86%	14%	10%
1	83.9%	0.79	13583	22896	8358	5953	62%	38%	70%	30%	21%
Medium trees (maximum number of splits: 30)											
No PCA	93.6%	0.98	20664	26874	1277	1975	94%	6%	91%	9%	7%
7	90.7%	0.96	19854	26221	2087	2628	90%	10%	88%	12%	9%
6	91.6%	0.97	20392	26138	1549	2711	93%	7%	88%	12%	9%
5	91.3%	0.96	20740	25645	1201	3204	95%	5%	87%	13%	11%
4	87.8%	0.93	19189	25397	2752	3452	87%	13%	85%	15%	12%
3	82.8%	0.88	17829	24209	4112	4640	81%	19%	79%	21%	16%
2	80.6%	0.87	16775	24140	5166	4709	76%	24%	78%	22%	16%
1	71.7%	0.73	11956	24437	4412	9985	54%	46%	73%	27%	15%
Coarse trees (maximum number of splits: 5)											
No PCA	81.3%	0.73	17122	24177	4819	4672	78%	22%	79%	11%	6%
7	80.3%	0.84	17027	23732	4914	5117	78%	22%	77%	23%	18%
6	80.1%	0.84	16912	23756	5029	5093	77%	23%	77%	23%	18%
5	79.5%	0.82	17520	22875	4421	5974	80%	20%	75%	25%	21%
4	80.7%	0.84	18133	22843	3808	6006	83%	17%	75%	25%	21%
3	74.6%	0.80	10831	27082	11110	1767	49%	51%	86%	14%	6%
2	74.8%	0.79	18110	19869	3831	8980	83%	17%	67%	33%	31%
1	67.8%	0.73	8631	25789	13310	3060	39%	61%	74%	26%	6%

Table 2. Classification performance of support vector machine.

Number of features after PCA	Acc	AUC	TP	TN	FN	FP	TPR	FNR	PPV	FDR	FPR
Linear kernel											
No PCA	81.5%	0.89	17376	24024	4565	4825	79%	21%	78%	22%	17%
7	67.2%	0.70	13646	20500	8295	8349	62%	38%	62%	38%	17%
6	67.6%	0.70	13296	21034	8645	7815	61%	39%	63%	37%	27%
5	66.0%	0.66	8140	25366	13801	3483	37%	63%	70%	30%	12%
4	55.0%	0.55	3425	24493	18516	4356	16%	84%	44%	56%	15%
3	48.7%	0.48	9028	15686	12913	13163	41%	59%	41%	59%	46%
2	52.9%	0.92	6282	20564	15659	8285	29%	71%	43%	57%	10%
1	49.9%	0.5	10875	14458	11066	14391	50%	50%	43%	57%	50%
Quadratic kernel											
No PCA	83.7%	0.91	17366	25120	4575	3729	79%	21%	82%	12%	13%
7	87.1%	0.94	18652	25561	3289	3288	85%	15%	85%	15%	11%
6	66.2%	0.74	10163	23475	11778	5374	46%	54%	65%	35%	19%
5	55.2%	0.74	8128	19918	13813	8931	37%	63%	48%	52%	31%
4	54.0%	0.51	4823	22602	17118	6247	22%	78%	44%	56%	8%
3	54.0%	0.51	4823	22602	17118	6247	22%	78%	44%	56%	22%
2	51.9%	0.49	7866	18506	14075	10343	36%	64%	43%	57%	36%
1	71.7%	0.51	10	28821	21931	28	0%	100%	26%	74%	0%
Cubic kernel											
No PCA	55.5%	0.58	13731	14437	8210	14412	63%	37%	49%	51%	50%
7	62.8%	0.65	10840	21036	11101	7813	49%	51%	58%	42%	27%
6	55.3%	0.57	10115	17958	11826	10891	46%	54%	48%	52%	38%
5	55.8%	0.56	10594	17722	11347	11127	48%	52%	49%	51%	39%
4	55.9%	0.54	4510	23880	17431	4969	21%	79%	48%	52%	17%
3	55.3%	0.50	2877	25224	19064	3625	13%	87%	44%	56%	13%
2	47.4%	0.79	16140	26483	5801	2366	56%	44%	42%	58%	59%
1	48.8%	0.50	13065	11744	8876	17105	60%	40%	43%	57%	59%
Fine Gaussian kernel (scale < 1)											
No PCA	93.6%	0.98	21033	26525	908	2324	92%	8%	90%	10%	4%
7	94.5%	0.98	20788	27194	1153	1655	95%	5%	93%	7%	6%
6	93.8%	0.98	20865	26787	1076	2062	95%	5%	91%	9%	7%
5	94.8%	0.98	20907	27220	1034	1629	95%	5%	93%	7%	6%
4	93.5%	0.98	20523	26943	1418	1906	94%	6%	92%	8%	7%
3	65.3%	0.93	17971	25925	2924	3970	82%	18%	86%	14%	10%
2	67.4%	0.72	9617	24593	12324	4256	44%	56%	69%	31%	15%
1	57.9%	0.51	1007	28420	20934	429	5%	95%	70%	30%	1%
Medium Gaussian kernel (scale = 1 ÷ 3)											
No PCA	92.6%	0.97	20872	26169	1069	2680	95%	5%	89%	11%	5%
7	92.6%	0.98	20863	26162	1078	2687	95%	5%	89%	11%	9%
6	92.2%	0.99	20870	25965	1071	2884	95%	5%	88%	12%	5%
5	92.2%	0.97	20874	25962	1067	2887	95%	5%	88%	12%	10%
4	85.9%	0.93	19267	24369	2674	4480	88%	12%	81%	19%	16%
3	76.6%	0.93	11700	27181	10241	1668	53%	47%	88%	12%	6%
2	74.8%	0.79	16140	26483	5801	2366	74%	18%	86%	14%	10%
1	57.1%	0.48	513	28476	21428	373	2%	98%	58%	42%	1%
Coarse Gaussian kernel (scale = 4 ÷ 11)											
No PCA	80.8%	0.92	18181	22849	3760	6000	83%	17%	75%	25%	21%
7	84.3%	0.93	19450	23358	2491	5491	89%	22%	79%	11%	19%
6	82.8%	0.88	20363	21683	1578	7166	93%	7%	74%	26%	25%
5	85.3%	0.93	20891	22422	1050	6427	95%	5%	76%	24%	22%
4	70.6%	0.83	13200	22668	8741	6181	60%	40%	68%	32%	21%
3	65.3%	0.61	5427	27737	16514	1112	25%	75%	83%	17%	4%
2	57.7%	0.49	2256	27064	19685	1785	10%	90%	56%	44%	6%
1	56.9%	0.46	284	28626	21657	223	1%	99%	56%	44%	6%

Area under curve (AUC) is the area under receiver operating characteristic (ROC) curve, which is equal to the probability the classifier will rank a randomly chosen positive instance higher than negative instance [5].

Table 1 presents classification performance of decision trees and the influence of dimensionality reduction using principal component analysis (PCA). Table 2 shows the influence of dimensionality reduction using PCA and different kernels on classification performance of SVM. In Table 3 we present parameters of classification performance derived from all samples confusion matrix after training the ANN.

For all analyzed types of trees (coarse, medium and fine) the best results of classification were achieved without reducing the number of features using PCA. Increasing number of features used in classification and number of splits in decision trees improves the overall results, except for FDR when using coarse tree and 1–3 features after PCA and medium tree and one feature.

In case of linear and medium Gaussian the highest accuracy was achieved using all considered features, without using PCA. For SVM with cubic kernel higher TPR and FNR was achieved using PCA and 7 features. The lowest FDR for all considered types of kernel was achieved for observations with highest accuracy, except the case of cubic kernel and 7 features. Increasing the number of features increases the accuracy, TPR, PPV and AUC with the exception of coarse Gaussian Kernel, fine Gaussian Kernel, despite small differences between the case "No PCA" and the best achieved result, and for cubic kernel. The best case for SVM was fine Gaussian kernel and using 7-8 features for classification.

Table 3. Classification performance of pattern recognition neural network.

After first training						
Neurons in hidden layer	TP	TN	FN	FP	Acc	False omission rate
5	19500	23980	4869	2441	85.6%	14.4%
10	19817	25795	3054	2124	89.8%	10.2%
20	20360	26092	2757	1581	91.5%	8.5%
After retraining						
5	18834	24788	4061	3107	85.9%	14.1%
10	20137	26073	2776	1804	91%	9%
20	20360	26092	2630	1173	92.8%	7.2%

For pattern recognition neural network classifier, higher number of neurons in hidden layer has better performance (increase of TP, TN, accuracy and decrease of FN, FP, false omission rate). Retraining slightly improves overall classification performance.

4   Discussion

Calculated results of classification show small differences between the best cases for each considered classifier: fine decision tree and no PCA (Acc = 95.8%, AUC = 0.99, TPR = 97%, FNR = 3%, PPV = 94%, FDR = 6%, FPR = 5%), fine Gaussian SVM and 7 features after PCA (Acc = 94.8%, AUC = 0.98, TPR = 95%, FNR = 5%, PPV = 93%, FDR = 7%, FPR = 6%) and ANN using 20 neurons in hidden layer, after retraining (Acc = 92.8%, False omission rate = 7.2% for all segments).

The best performance withing decision trees was achieved for fine trees, within SVM classifiers the best performance has SVM with fine Gaussian kernel (scale <1). Its performance was higher than reported in papers [12,13] in terms of true positive rate (TPR) and specificity (Sp) defined as $Sp = \frac{TN}{TN+FP}$: in this paper we achieved TPR = 95% and Sp = 94% on MITDB database (for comparision in paper [12] TPR = 87%, Sp = 84% and in paper [13] TPR = 87%, Sp = 86%).

The best performing classifier was the binary decision tree with maximum number of splits equal to 100 using all eight considered features (mean RR, SDNN, RMSSD, PLF, PHF, LF/HF, SD1, SD2). However, using SVM classifier with Gaussian kernel with scale <1, decision trees with maximum number of splits 30, feed-forward artificial neural network with 20 neurons in hidden layer, SVM with medium Gaussian kernel (scale between 1 and 3) and SVM with quadratic kernel have slightly worse performance. The worst performance was achieved for SVM classifier with cubic kernel in comparison with other kernel types.

5   Conclusion

We compared the performance of atrial fibrillation detection of artificial neural network, SVM and binary decision trees on ECG signals from MITDB database. Binary decision tree with a maximum number of splits of 100 (Acc = 95.8%, AUC = 0.99, TPR = 97%, FNR = 3%, PPV = 94%, FDR = 6%, FPR = 5%) was the best of compared classifiers. Achieved results should encourage further tests on different data sets.

References

1. Brennan, M., Palaniswami, M., Kamen, P.: Do existing measure of poincaré plot geometry reflect nonlinear features of heart rate variability? IEEE Trans. Biomed. Eng. **48**, 1342–1347 (2001)
2. Kirchhof, P., et al.: ESC Guidelines for the management of atrial fibrillation developed in collaboration with EACTS. Eur. Heart J. **37**, 2893–2962 (2016). https://doi.org/10.1093/eurheartj/ehw210
3. Cortes, C., Vapnik, V.: Support-vector networks. Mach. Learn **20**, 273 (1995). https://doi.org/10.1007/BF00994018

4. Dash, S., Raeder, E., Merchant, S., Chon, K.: A statistical approach for accurate detection of atrial fibrillation and flutter. Comput. Cardiol. **36**, 137–140 (2009)
5. Fawcett, T.: An introduction to ROC analysis. Pattern Recogn. Lett. **27**, 861–874 (2006)
6. Fuster, V., et al.: ACC/AHA/ESC 2006 guidelines for the management of patients with atrial fibrillation: a report of the American College of Cardiology/American Heart Association Task Force on Practice Guidelines and the European Society of Cardiology Committee for Practice Guidelines (Writing Committee to Revise the 2001 Guidelines for the Management of Patients With Atrial Fibrillation). Circulation **114**, e257–e354 (2006)
7. Camm, A.J., Malik, M., Bigger, J.T., Breithardt, G., Cerutti, S., Cohen, R., Coumel, P., Fallen, E., Kennedy, H., Kleiger, R.E., Lombardi, F.: Heart rate variability standards of measurement, physiological interpretation, and clinical use. Task force of the European Society of Cardiology the North American Society of Pacing Electrophysiology. Circulation **93**, 1043–1065 (1996). https://doi.org/10.1161/01.CIR.93.5.1043
8. Huang, C., Ye, S., Chen, H., Li, D., He, F., Tu, Y.: A novel method for detection of the transition between atrial fibrillation and sinus rhythm. IEEE Trans. Biomed. Eng. **58**(4), 1113–1119 (2011)
9. Hulley, S.B., et al.: Designing Clinical Research, 3rd edn, pp. 189–190. Lippincott Williams & Wilkins, Philadelphia (2007)
10. January, C.T., Wann, L.S., Alpert, J.S., Calkins, H., Cigarroa, J.E., Cleveland Jr, J.C., Conti, J.B., Ellinor, P.T., Ezekowitz, M.D., Field, M.E., Murray, K.T., Sacco, R.L., Stevenson, W.G., Tchou, P.J., Tracy, C.M., Yancy, C.W.: 2014 AHA/ACC/HRS guideline for the management of patients with atrial fibrillation: a report of the American college of cardiology/American heart association task force on practice guidelines and the heart rhythm society. Circulation **130**, e199–e267 (2014)
11. Karmakar, C.K., Gubbi, J., Khandoker, A.H., Palaniswami, M.: Analyzing temporal variability of standard descriptors of poincaré plots. J. Electrocardiol. **43**, 719–724 (2010)
12. Kostka, P.S., Tkacz, E.J.: Feature based, extraction, on time-frequency and independent component analysis for improvement of separation ability in atrial fibrillation detector. In: 30th Annual International Conference of the IEEE Engineering in Medicine and Biology Society, Vancouver, BC, Canada, pp. 2960–2963 (2008)
13. Kostka, P.S., Tkacz, E.J.: Feature extraction in time-frequency signal analysis by means of matched wavelets as a feature generator. In: 33rd Annual International Conference of the IEEE Engineering in Medicine and Biology Society, Boston, MA, USA, 30–3 August–September 2011, pp. 4996–4999 (2011)
14. Larburu, N., Lopetegi, T., Romero, I.: Comparative study of algorithms for atrial fibrillation detection. In: 2011 Computing in Cardiology, Hangzhou, pp. 265–268. IEEE (2011)
15. Markides, V., Schilling, R.J.: Atrial fibrillation: classification, pathophysiology, mechanisms and drug treatment. Heart **89**, 939–943 (2003)
16. Mohebbi, M., Ghassemian, H.: Detection of atrial fibrillation episodes using SVM. In: 30th Annual International Conference of the IEEE Engineering in Medicine and Biology Society, Vancouver, BC, Canada, pp. 177–180 (2008)
17. Moody, G.B., Mark, R.G.: The impact of the MIT-BIH arrhythmia database. IEEE Eng. Med. Biol **20**(3), 45–50 (2001). (PMID: 11446209)
18. Osuna, E., Freund, R., Girosi, F.: Training support vector machines: an application to face detection. In: Proceedings of CVPR 1997, Puerto Rico (1997)

19. Pan, J., Tompkins, W.J.: A real-time QRS detection algorithm. IEEE Trans. Biomed. Eng. (BME) **32**(3), 230–236 (1985). https://doi.org/10.1109/TBME.1985.325532

20. Goldberger, A.L., Goldberger, A.L., Amaral, L.A.N., Glass, L., Hausdorff, J.M., Ivanov, P.C.H., Mark, R.G., Mietus, J.E., Moody, G.B., Peng, C.-K., Stanley, H.E.: PhysioBank, PhysioToolkit, and PhysioNet: components of a new research resource for complex physiologic signals. Circulation **101**(23), e215–e220 (2000). http://circ.ahajournals.org/content/101/23/e215.full

21. Ruan, X., Liu, C., Wang, X., Li, P.: Automatic detection of atrial fibrillation using R-R interval signal. In: 4th International Conference on Biomedical Engineering and Informatics, vol. 2, pp. 644–647 (2011)

22. Schölkopf, B., Burges, C., Vapnik, V.: Extracting support data for a given task. In: Proceedings of First International Conference on Knowledge Discovery & Data Mining, Menlo Park, pp. 252–257 (1995)

23. Tateno, K., Glass, L.: A method for detection of atrial fibrillation using RR intervals. Comput. Cardiol. **27**, 391–394 (2000)

24. Tulppo, M.P., Makikallio, T.H., Takala, T.E., Seppanen, T., Huikuri, H.V.: Quantitative beat-to-beat analysis of heart rate dynamics during exercise. Am. J. Physiol. **271**, H244–H252 (1996)

25. Vapnik, V.: The Nature of Statistical Learning Theory. Springer, New York (1995)

26. Wedekind, D.: qrsdetector. TU Dresden, Institute for Biomedical Engineering, Biosignal Processing Group (2014). https://github.com/danielwedekind/qrsdetector

27. Wiesel, J., Fitzig, L., Herschman, Y., Messineo, F.C.: Detection of atrial fibrillation using a modified microlife blood pressure monitor. Am. J. Hypertens. **22**, 848–52 (2009)

Visualizing the Daily Physical Activities and Nutrition Information of High School Athletes

Rahul Patel and Chris Scaffidi$^{(\boxtimes)}$

Oregon State University, Corvallis, OR 97330, USA
scaffidc@engr.oregonstate.edu

Abstract. Tracking food intake, sleep, and physical activity, then generating consolidated peer visual reports for participants in large-scale health studies, can be challenging. This paper presents an integrated health study dashboard that enables scientists to conduct large-scale field studies involving hundreds of subjects organized into dozens of treatment groups, including the collection and analysis of data from devices and surveys. The system also provides a feature for the scientist to perform bulk processing and editing of subject demographic information, as well as to generate health reports. The system was used in six rounds of reports for 750 high school students. A preliminary qualitative evaluation was conducted with the user responsible for generating those reports for all the students, as well as five other graduate students who could use the system for other field studies in health science. Their answers revealed that the system was considered both usable and effective.

Keywords: Health informatics · Information technology · Field studies

1 Introduction

Ours is a 5-year research project, funded by the United States government, which focuses on building healthy nutrition, physical activity and life skills for healthy weight maintenance. The intervention includes face-to-face sport nutrition, physical activity and life-skills lessons, assessments of diet, body composition and physical activity on/off the soccer field, and on-line immersive learning to reinforce the lessons [1]. Subjects are 750 high school soccer players (aged 14–19) from over two dozen schools located in the United States, coordinated through their coaches.

One of the components of the project is a website that we created, enabling our team scientists to log in, to create studies, to enroll subjects into those studies, to assign Fitbit trackers [2], to send periodic Qualtrics surveys to subjects [3], to retrieve the results of these surveys as well as information obtained from subjects' devices (steps taken and calories burned per minute), and finally to export or analyze these results [1]. Thus, within this browser-based system, scientists could run a study and collect data for a field study. Historically, our team's researchers have collected information from participants for a short time, generally a seven-day period that our student subjects have

© Springer Nature Switzerland AG 2019
Á. Rocha et al. (Eds.): ICITS 2019, AISC 918, pp. 702–711, 2019.
https://doi.org/10.1007/978-3-030-11890-7_66

known as the "7-Day Challenge," then analyze the data to create reports. Such data-collection and reporting periods occur several times per year.

These analysis and reporting steps have in the past required significant expenditures of effort and time. Although our existing web-based system—which we refer to as the "baseline" system in this paper—could automate the *gathering* of survey and physical activity data into a consolidated dataset, it still was necessary to manually download these data, to write scripts for cleaning and analyzing the data, and to manually construct reports visualizing the results.

We sought a better approach that would achieve three goals. First, we required informatics features that could automate the analysis and reporting steps on our project website to reduce the amount of manual work for scientists. Second, we saw an opportunity to generate better reports that presented *personalized* data in a form that could help to show individual participants their data from the 7-Day Challenge. Third, we wanted these reports to compare each participant's results versus those of his or her peer group (i.e., the other students at the same school) as well as versus the recommendations of nutrition and exercise experts, to provide a context for students to interpret their own reports.

We recognized that these enhancements to our baseline system could point the way toward construction of similar features in other systems utilized for conducting large-scale health field studies. In Sect. 2, we present related work on other existing systems for generating visual reports in health studies, then turn to our own system in Sect. 3. We present the preliminary evaluation of our system in Sect. 4 and summarize conclusions in Sect. 5.

2   Related Work on Visual Reports for Health Field Studies

As systems for tracking physical activity and nutrition have become widespread, visual reporting has become a common method for depicting personal changes over time. It is widely recognized that visualizations can be useful elements of health informatics systems because they can enable users to understand their own exercise and nutrition habits. Particularly in the context of physical activity tracking, many consumer-oriented devices (e.g., Fitbit [2] and Dailyburn [4]) use visualizations such as bar, line and pie charts to represent physical activity as a form of feedback to users. Such systems facilitate giving users accurate feedback about calorie expenditures as well as personalized, automated coaching about exercise habits [5, 6]. Beyond the use of such visualizations in consumer-oriented systems, research studies have presented data to study participants and others as a means of providing feedback. For instance, one research study used visualizations to help diabetic patients explore the relationship between their blood glucose level, their diets, and their daily routines [7]. Providing feedback to users about their fitness-tracking data can give them a sense of control [8].

Visualizations in such systems are often designed with the intent of *influencing* users to engage in healthy behaviors. For instance, a project with the goal to develop a "system to promote an active lifestyle for individuals and to recommend to them valuable interventions by making comparisons to their past habits" developed a prototype system "to promote an active lifestyle and a visual design capable of engaging users in the goal of increasing self-motivation" [9]. The main idea was to use visual-izations to study past habits to improve a person's health and well-being in a proactive manner. They used a platform called ATHENA (activity-awareness for human-engaged wellness applications) to visualize a subject's daily, monthly and weekly physical routines. These visualizations can be shared with social media so the subject can earn appreciation and be motivated [9].

The primary limitation of informatics systems potentially suitable for health field studies, to date, is that *they have provided limited information to users about how they compare to others in their peer group*. For example, FoodWatch is a mobile app for recording diet in field studies. It analyzes and depicts the user's intake of carbohy-drates, fats and proteins but does not compare these to other users, nor does it track sleep or physical activity [10]. Similarly, Shut Eye is an app that is focused on sleep-tracking, but it does not compare with other users; nor does it track diet or activity [11]. Likewise, Fitbit [2] and Google Fit [12] track activity, MyFitnessPal [13] tracks diet and exercise, Apple Health [14] tracks all three—but none of these provide a built-in means of comparing to peers.

The lack of peer-group comparisons in systems for health field studies is limiting because such comparisons could help motivate healthy choices among research sub-jects in terms of getting adequate nutrition, sleep, and physical activity. The potential value of peer-group comparisons has long been recognized in an *academic* setting, where visual reports are a well-established method for enabling *students* to compare themselves to one another. For example, one literature review examined the usage of data visualizations related to Course Management Systems (CMS), which instructors use to manage or distribute course materials and conduct online learning activities [15]. The literature review found that CMS logs were being used to track the learning activity of the students, as well as to enable students to understand how their own progress compared to others in their classes [15]. Another study, in the context of training in writing skills, showed that enabling students to more easily perceive the contributions of peers to group writing assignments was effective at promoting higher levels of student engagement [16]. Thus, although the short-term focus of the current paper is on reducing the manual effort required of our scientists to generate reports, the long-term goal is ultimately to use these personalized visualizations to drive behavior, analogous to how teachers use peer comparisons to motivate students.

3   System for Personalized Reporting

Reducing the effort of generating personalized reports with peer-group comparisons required adding multiple features to our baseline system. In the subsections below, we divide these enhancements into two main groups: those related to data acquisition, and those related to data presentation.

3.1   Data Acquisition Features

At the outset, we recognized the need to generate printed reports on pieces of paper for our students, because not all of them own a personal computer or smartphone. This would require printing contact information of students and their parents on the individual reports, so that the right report could be sent to each student. We also saw the potential value of showing body mass index (BMI) on the report so that students could observe changes resulting from their health choices.

The baseline version of our system, unfortunately, lacked fields for storing all of the information above. Thus, although scientists could create lists of students in our website and link those students to their Qualtrics surveys and Fitbit data, our team would then have to turn to a separate collection of messy spreadsheets managed by graduate students for managing students' contact and BMI information. Therefore, the first set of features that we added to our baseline system were fields for team scientists to record within our system each subject's contact information, their parents' contact information, their date of birth, height and weight.

Furthermore, motivated by the desire to further reduce the work of managing data, we also added new functionality for bulk manipulation of all contact and demographic data. With these new features, our team can log into the website, download a spreadsheet (in comma-separated value format, with one row per student in our study), edit as many fields for as many subjects as desired, and upload the revised data all at once. This greatly reduced the number of tedious mouse-clicks involved in editing subjects' data and enabled us to discard spreadsheets after upload.

Each subject in our study has a unique 8-digit number that we assign when a student enrolls into one of the two dozen groups. Each of these peer groups has slightly different demographics, such as being located at a different school, or is receiving a slightly different intervention; for example, some of them receive educational lessons on healthy eating and exercise. Each row of the spreadsheet contains a column indicating the student's identification number (in addition to another column indicating the peer group), making it possible to associate the right data with the right subject.

When our team scientist uploads a new spreadsheet, the system checks the values in the identification column for validation purposes. If it cannot automatically and uniquely identify a student associated with a given row, it issues a warning to the user with a note about which rows will not be automatically imported into the system. The system provides a means whereby our team member can indicate that new subjects should be added to the database, and/or they can correct and re-upload the spreadsheet if desired. Together, these data acquisition features have greatly simplified the steps required for collecting the demographic information needed in the system's reports.

3.2 Data Reporting and Visualization

A team scientist can generate a visual report by indicating a time period, typically 7 days, over which to generate a peer group's individual personalized reports. All these reports are then displayed on the screen for the scientist's review. The reports, for all the students in the group, can be sent to the printer in a single step, with each report appearing on a separate sheet of paper for distribution to the students in the study. Figure 1 depicts the side of the sheet showing information about the student's actual data, which we discuss in detail below.

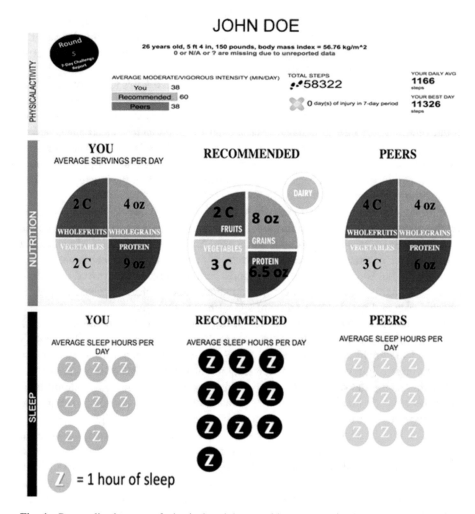

Fig. 1. Personalized report of physical activity, nutrition and sleep, showing individual-level data compared to the government's nutritional recommendations and the aggregate average results of the individual's peer group.

Physical Activity. Every subject is assigned a Fitbit synced periodically to the database, with a granularity indicating the number of steps per minute (i.e., 1440 entries per day). Two key measures of physical activity are tracked: the total number of steps and the total number of minutes of moderate/vigorous intensity. The latter are categorized based on a threshold of 100 steps per minute, following a commonly-applied physical activity promotion heuristic [17–19].

The physical activity portion of the report displays the average number of minutes per day of moderate/vigorous activity for the individual student in a bar chart that also shows the peer group average as well as the level of exercise recommended by health experts. It shows the total number of steps for the week, the daily average, and the total steps on the student's best day. Finally, it shows the number of days of injury, which is computed from the student's Qualtrics survey data.

Nutrition. The next section of the visual report uses a picture of a dinner plate, with segments representing the intake quantities of Whole Fruits, Whole Grains, Proteins, and Vegetables. These are based on daily Qualtrics surveys answered by the students. As with physical activity statistics, the report shows the individual student's data alongside recommended levels of food intake as well as data from the peer group. Dairy is shown for the expert-recommended diagram but happened to be omitted from the survey for this project and thus is not shown in the other two diagrams.

Sleep. The final section of the visual report contains three sets of circles, where each circle represents one hour of sleep. It also contains three columns: one for the subject, one for recommended sleep, and one for the average sleep of subjects in the peer group. All these values are computed and rounded to the nearest whole number of hours of sleep, based on responses to the daily Qualtrics survey.

Missing Values. It is frequently the case that some students forget to complete some surveys. The new analysis and reporting features automatically compute individual daily averages (such as average number of hours of sleep) based on those days for which data exist, and they compute peer averages based on person-days for which data are available. If no data are available at all for a given computation, then that section of the report will display "N/A" or 0. For example, if a student answers no surveys at all during the week, and if he or she forgets to wear a Fitbit, then that student's average food intake values and hours of sleep will show "N/A" in the respective section, and the total steps per week will show a 0.

4   Preliminary Qualitative Evaluation

One of our team members, a graduate student, used the new features of our system for nearly a year. Specifically, this involved using the new bulk-editing features to populate demographic data of the students, running six rounds of the 7-Day Challenge, and then generating the reports. Overall, approximately 750 students participated in this study, spread over approximately 24 peer groups.

We conducted a preliminary qualitative evaluation of our new system with this team member to assess its usability and usefulness. In addition, we also invited five other novice scientists to use the system and to give feedback, thus enabling us to gather additional insights about its strengths and weaknesses for people who were not yet experts at using the system.

4.1   Study Design

Recruitment. The six participants in the preliminary evaluation were graduate students in health sciences, who were familiar with conduct of health field studies at our university. To recruit these participants, an email with the description of the user study was broadcasted to graduate students in the college and, out of all the respondents to the email, six English speaking students above the age of 18 years were selected. None of them had any programming experience. Each participant was emailed a copy of the consent form to read. After each confirmed that he or she wanted to participate, then a time was scheduled for the participant to use the system and participate in the study. Each was offered $25 for participating.

Methodology. The preliminary evaluation had two main parts. The first involved having each participant get experience with the new system features. The second was completion of a brief questionnaire to gather feedback on the system.

In the first part of the evaluation, each participant given a brief introduction about the evaluation, then given an opportunity to ask questions. Afterward, a sheet with instructions was handed to the participants regarding an assigned task. This task simulated the basic elements of collecting and reporting on data in a field study. Of course, it was not possible in the timespan of 30 min to have the participant recruit hundreds of subjects and to run a 7-Day Challenge, so we prepopulated the database with simulated data spanning multiple peer groups. In addition, so that the participant understood the nature of the data involved in the reports, we had the participant complete the same Qualtrics survey that subjects would complete in our real health field study.

Thus, the task had three steps:

1. Complete the daily survey
2. Generate visual reports for the groups of subjects in the simulated study
3. Download the data of each group and compare the results with the visual reports

The second part of the evaluation invited participants to answer a short questionnaire, based on their experiences with the system. The questionnaire covered the topics shown in Table 1. It began with three questions aimed at verifying that participants were graduate students with experience in health science research (who, therefore, were representative of potential users for our system), and then the questionnaire moved on to questions about our system's strengths and weaknesses.

4.2   Results

Each participant completed the assigned task in under 20 min, less than the 30 min that had been anticipated for the preliminary evaluation. All of them successfully completed the task, and none appeared to encounter any significant difficulties along the way. All of them confirmed that they were graduate students with experience in health science research. In response to the questionnaire's second half, about the system's strengths and weaknesses, their feedback was generally positive, including answers from the one participant who had used the system during the project study, whose responses were very similar to those of the others in the group. Therefore, we report data for all the participants in summary below, rather than calling out results for individuals.

In response to the questionnaire, 4 of the 6 participants confirmed that the Qualtrics survey enabled them to record the information efficiently. The remaining 2 said that it "Mostly" sufficed, and they pointed out the need for collecting additional data on physical activity in addition to the information gathered by the survey on injuries. We

Table 1. Summary of the feedback questionnaire

Background and demographic questions to verify participants' background
• What is your job or year of study?
• What is your field or major?
• What is your experience with collecting data, analyzing data, or doing other research related to the following? (Check all that apply: Research on nutrition, exercise, sleep, other health)

Questions about our new system features to evaluate strengths and weaknesses
• Did the survey enable you to effectively record information?
• (Options for this question and the next: Yes, Mostly, Somewhat, Not at all.)
• One key purpose of the system is to speed up the creation of a report that summarizes diet, physical activity and sleep. Do you think the goal was achieved?

• How easy was it to understand the following parts of the report? (Options for each of these three sub-questions: Very easy, Somewhat easy, Somewhat difficult, Very difficult)
    – Nutrition
    – Exercise
    – Sleep

• What aspects of the software or the report do you like the most?
• What suggestions do you have about improving the software or the report, if any?

view their comments as a positive for our system design, in that this feedback highlights the crucial value of our system's integration with physical activity tracking devices (Fitbits) to complement self-report data.

The next question asked about how well the system sped up creation of a report, to which 5 of 6 participants agreed that the goal was achieved. One person "Mostly" agreed and noted that it would be somewhat time-consuming for somebody using the report to compute other information, such as number of calories burned, and that automatically calculating this information could also be beneficial.

In response to the third set of questions, concerning understandability of the report sections, 5 out of 6 agreed that the physical activity was easy to follow. One participant (the same as above) again repeated that the survey data alone was insufficient and therefore maybe not easy to understand.

Likewise, 5 out of 6 agreed that the nutrition data was very easy to understand. One person said that the nutrition section was difficult to understand because nutrients such as fiber and fats were not specifically identified on the report. We conclude from this that the section of the report should be labeled "Food Intake" rather than "Nutrition," due to the distinction between food consumption and the specific nutrients within the foods.

Of the 6 participants, 5 said that the sleep section was very or somewhat easy to understand. One indicated that the section was somewhat difficult to understand due to the large number of circles that required counting. Based on this feedback, we could try switching to a bar chart, as we used for the physical activity section.

Thus, overall, every question above generated the highest possible rating from 4 or 5 of the 6 participants. In addition, every respondent was able to identify at least one aspect of the software or report that he or she liked. Most of their answers focused on the visual representation of data and included the following statements. For example, one of them said he liked the "Graph + Charts" the most, another liked the "Summary in a visual format," while a third said the "Graphics make things clear." One said, "Color-coding was helpful, and slide-scale for 'number of servings' in food categories worked well." The others also said, "Easy to follow nutritional and sleep habits" and "I like that it summarizes the data in a friendly-visual manner through the colors and graphs for each category." One concluded, "It makes it easy to read and understand."

Of the 6 participants, 1 had no suggestions for improvement, and of the others, 4 gave suggestions for improving the exercise section of the report. One requested "additional exercise/physical activity information," another noted "there are other ways to record exercise, the third said "relating steps to Moderate to Vigorous Physical Activity might help make it easier to understand the recommendation," and the last said that it was "difficult to see separate questions." In addition to these comments, one suggested increasing the text size on BMI and decreasing the text size on charts and graphs.

In short, on every dimension above, most participants gave the highest rating, yet there remained a few areas for improvement—chiefly in regard to expanding the exercise section and making it more detailed. We view these results as highly positive and useful for guiding the future development path of the system.

5   Conclusions and Future Research Opportunities

This paper has presented an integrated health study system that enables scientists to conduct large-scale field studies involving hundreds of subjects organized into dozens of treatment groups, including the collection and analysis of data from devices and surveys, as well as to generate health reports. Thus, it addresses the time-consuming challenge of tracking food intake, sleep, and physical activity, then generating consolidated peer visual reports. The system also provides a feature for the scientist to perform bulk processing and editing of subject demographic information. The preliminary study gave promising results, and further research, such a random-assignment experiment, could evaluate if the system is both usable and effective.

Our system is a useful starting point for brainstorming similar features for other systems, and it provides a basis for additional enhancements. For example, in terms of supporting field studies, we can further enhance our own system to support statistical analyses comparing results of one treatment group to another, thereby further supporting health science. This would facilitate evaluating, in future studies, the extent to which providing individual students with personalized reports motivates them to engage in healthy behaviors.

References

1. Wong, S.S., Manore, M., Patton-Lopez, M., Schuna, J., Dorbolo, J., Skoog, I., Scaffidi, C., Chiang, P., Johnson, T., Curiel, C.: The wave ripples for change: obesity prevention in high-school soccer players (year 3 of 5). In: Society of Nutrition Education and Behavior Annual Conference (2016)
2. Diaz, K., Krupka, D., Chang, M., Peacock, J., Ma, Y., Goldsmith, J., Schwartz, J., Davidson, K.: Fitbit: an accurate and reliable device for wireless physical activity tracking. Int. J. Cardiol. **185**, 138–140 (2015)
3. Qualtrics survey system. https://www.qualtrics.com/
4. Dailyburn system. http://www.dailyburn.com/
5. Stephens, J., Allen, J., Himmelfarb, C.: Smart coaching to promote physical activity, diet change, and cardiovascular health. J. Cardiovasc. Nurs. **26**(4), 282–284 (2011)
6. Zhan, A., Chang, M., Chen, Y., Terzis, A.: Accurate caloric expenditure of bicyclists using cellphones. In: Conference on Embedded Network Sensor Systems, pp. 71–84 (2012)
7. Frost, J., Smith, B.: Visualizing health: imagery in diabetes education. In: Conference on Designing for User Experiences, 1–14 (2003)
8. Gabbiadini, A., Greitemeyer, T.: Fitness mobile apps positively affect attitudes, perceived behavioral control and physical activities. J. Sports Med. Phys. Fitness (2018)
9. Fahim, M., Idris, M., Ali, R., Nugent, C., Kang, B., Huh, E., Lee, S.: ATHENA: A personalized platform to promote an active lifestyle and wellbeing based on physical, mental and social health primitives. Sensors **14**(5), 9313–9329 (2014)
10. Yang, J.: Master's thesis, School of Design, Rochester Institute of Technology (2017)
11. Bauer, J., Consolvo, S., Greenstein, B., Schooler, J., Wu, E., Watson, N., Kientz, J.: ShutEye: encouraging awareness of healthy sleep recommendations with a mobile, peripheral display. In: Conference on Human Factors in Computing Systems, pp. 1401–1410 (2012)
12. Menaspa, P.: Effortless activity tracking with Google Fit. Br. J. Sports Med. **49**(24), 1598 (2015)
13. Evans, D.: MyFitnessPal. Br. J. Sports Med. **51**(14), 1101–1102 (2017)
14. Welch, C.: Apple HealthKit announced: a hub for all your iOS fitness tracking needs. The Verge (2014)
15. Zhang, H., Almeroth, K., Knight, A., Bulger, M., Mayer, R.: Moodog: tracking students' online learning activities. In: EdMedia: World Conference on Educational Media and Technology, pp. 4415–4422 (2007)
16. Liu, M., Liu, L., Liu, L.: Group awareness increases student engagement in online collaborative writing. Internet High. Educ. **38**, 1–8 (2018)
17. Davis, M., Fox, K., Hillsdon, M., Coulson, J., Sharp, D., Stathi, A., Thompson, J.: Getting out and about in older adults: the nature of daily trips and their association with objectively assessed physical activity. Int. J. Behav. Nutr. Phys. Act. **8**(1), 116 (2011)
18. Hayes, L., Van Camp, C.: Increasing physical activity of children during school recess. J. Appl. Behav. Anal. **48**(3), 690–695 (2015)
19. Marshall, S., Levy, S., Tudor-Locke, C., Kolkhorst, F., Wooten, K., Ji, M., Macera, C., Ainsworth, B.: Translating physical activity recommendations into a pedometer-based step goal: 3000 steps in 30 minutes. Am. J. Prev. Med. **36**(5), 410–415 (2009)

Unwanted RBAC Functions over Health Information System (HIS)

Marcelo Antonio de Carvalho Junior$^{(\boxtimes)}$ and Paulo Bandiera-Paiva

Health Informatics Department, Federal University of Sao Paulo,
São Paulo, Brazil
{carvalho.junior,paiva}@unifesp.br

Abstract. Objective: This article describes unwanted existing role based access-control (RBAC) standard functions over Health Information Systems (HIS) for overall accountability purposes and highlights potential information security policy violation. Methods: RBAC standard study and functions mapping to use-case scenarios is used. Results: Administrative RBAC Core commands are redesign to cope with the need of continuous accountability from HIS users'. Actual function issues, proposed adaptation and inner RBAC reflexes are discussed.

Keywords: Information systems (L01.700.508.300) · Information security · Access-control (N04.452.758.849.350) · Standards (E05.978.808) · RBAC · Privacy (SP9.130.010.010)

1 Introduction

When it comes to protecting Health Information Systems (HIS) information through access-control, there always a difficult balance between restrictions and accountability that needs to be taken in consideration. Differently from other industries, HIS access limitation to a minimum (least privilege concept) based on user profile it's not so simple. That's due to specific health-care professional interaction, medical procedure and supported decision making that uses sheer HIS information as base. That is, a medical decision is only fully supported if enough HIS information is made available to his specific user on the system while performing a given task. On the other hand, full access goes against every corporate security policy. If not all, what is enough then for responsible medical decision? Access-control policies [1] that limit or restrict health-professional users based on type or health-care specialty are difficult to achieve and cumbersome when it comes to balancing this access limit. A well-known health-care professional role-based approach used to dictate what slice of HIS patient contained information a user should access is based on ISO 21298 [2] and ISO 21091 [3] standards. That, in theory, can be used by an HIS security administrator to restrict user's responsibility to that amount of information. However, a common allegation from health-care professional users that are held responsible for misconducted procedure or even aggravated illicities via HIS is that only the whole view of patient data on system would allow for correct conduct to be taken. This goes against most types of information security policies that enforce constraints based on roles or compartmentalizes information to level

Á. Rocha et al. (Eds.): ICITS 2019, AISC 918, pp. 712–719, 2019.
https://doi.org/10.1007/978-3-030-11890-7_67

access according with user interaction type. As RBAC [1] is predominant on HIS and operates based on this very principle for role limitation, this accountability concern is an important aspect of security policy construction and access-control decisions for the health-care scenario.

Accountability also means the ability to correlate actions performed on HIS to the authenticated user. The system should be able to bind a user ID to the related person at all times, either by access-control authorization imposition before actions to take place and afterwards via audit trail (logs). The main objective is to provide for means to investigate performed actions that breaks information security policy, ethics or health-care professional obligation burden to a person. There are some few known scenarios that make this accountability difficult to achieve on health-care service provision. One is the lack of administration control over user creation and the other is the HIS limited scope per health-care institution. The user creation has two important aspects for that matter related to accountability that is: being capable of univocally distinguish users and to keep track of related person's actions during all system-use life cycle.

The first problem with this regard is that not all systems require unique binding to a person. Notice that this requirement cannot be achieved by simply controlling unique ID associated to user. This requires a strong bound to the user outside system scope, like a government issued ID or some real-life permanent relationship [4]. In a digital format, this could be arguably Public Key Infrastructure (PKI) related ID or nowadays even blockchain issued. Considering that a health-care professional can operate different HIS in different health-care institutions, that means this control must be sufficient to map that professional system-independently. Although meaningful use from Health Information Technology for Economic and Clinical Health Act (HITECH) encourages sharing patient records among health-care providers, this complicates users tracking on this federated architecture.

The second problem arises when systems allow a user deletion, breaking subject's history and binding capabilities. In this scenario, a user that broke a security rule could disappear and even reappear as another user in a future moment. That implicates not only that this user is no longer existing but that other person could now be tied to the same username.

This unwanted scenario is possible not only because HIS implementers fail foreseeing this events and deeds, but because strictly following RBAC instructions for user administration capability does not suffice for this level of accountability control.

In this article, we describe AddUser(RC-04) and DeleteUser(RC-26) RBAC Core administrative functions and their limitation to address the accountability problems early mentioned. We then discuss both orientative and function description amends to the RBAC standard to prevent the early described scenario.

The remainder sections of this article are divided as following: (a) Related Work, where we cite existing work addressing access-control accountability issues and main differentiation from this article contribution; (b) Objective and Methods, where we declare document analysis methodology and state the identification of RBAC limitations to control users life-cycle as main objective; (c) RBAC Unwanted Core Functions describes actual administrative functions and points accountability limitations; (d) New Functions proposes function changes; (e) Conclusion.

1.1   Related Work

Most of the user accountability studies are abridged to the authentication portion of access-control. For the authorization portion like in focus here, studies were aimed at obligations related to user tasks, better audit over system-task' authorization/negation or strong bond to user identification as recent examples below. Focused on accountability from an administrative perspective, Rajkumar and Sandhu [5] proposed Administrative Role-Based Access Control (ARBAC). The improvement was focused on user management tasks (assign and revoke functions), including the reporting and requesting approval for co-administrators.

Addressing accountability indirectly, Baracaldo and Joshi [6] work was focused on ranking users based on obligation fulfillment related to their role assignments. The developed trust model is helpful predicting illicitudes, insider attackers or simply incompetent users based on action history. Advantages from predecessors cited in their work are related to the weighting scheme and better prediction on user intent.

Wainer and Kumar [7] presented an updated version of user delegation privileges that affect accountability control. Delegation and delegation acceptance are discussed for that matter.

Also addressing delegation issues and related accountability weaknesses, Rabin and Gudes [8] discusses the misuse of PKI in ABAC functions that validate user's certificates.

Exploring audit capabilities for a posteriori accountability checking, Azkia et al. propose Policy-Oriented Log Processor tied to the authorization decision module. The main idea is to use a new defined quadruple (subject, action, object and time) that provide for policy compliance verification over log information for past events.

On our contribution, we focus on accountability bottom-line in RBAC from a simplistic perspective, the ability to track users permanently via the access-control.

2   Objectives and Methods

This study objective is to highlight RBAC implementation problems of AddUser(RC-04) and DeleteUser(RC-26) functions according with 2012' standard publication. More specifically, the potential information security policy violation and lack of accountability impacts related to those administrative commands considering actual semantic and command interpretation. This paper describes the functions and the interpretation problems through INCITS 359-2012 ANSI' standard textual analysis. Then proposes a newly designed set of functions and instructions to HIS developers to overcome pointed limitations. Substitutive Z notations for functions are stated.

3   RBAC Unwanted Core Functions

3.1   User Deletion Issue

Users at (USERS) construct, in terms of RBAC standard definition, is defined as a person (human being). Nonetheless, other systems and electronic devices can also interact as such with RBAC aware HIS. There are basically two inner concepts that use USERS directly on RBAC authorization vetting decisions: (a) user assignment (UA) and (b) user' sessions. The first associates a specific user to one or more roles in the HIS. The second links USERS to a temporary permission set via active roles during HIS interaction.

Depending on the RBAC model component and set of RBAC functions on HIS that are put in action, the USERS instantiation may vary. "Administrative" and "Supporting" functions set is used to deploy and define allowed behavior for a specific user while "Review" and "Advanced Review" set are used to support access-control decisions based on user' authorization using RBAC Core component. For instance, using the Administrative set of functions the USERS can be manipulated directly during commissioning and deletion (RC-04 and RC-26). Also, indirectly from the same functions set that deal with roles and permission assignment (RC-06, RC-22, RC-10, RC-18, and RC-32) via "assigned_users" parameter. Review and Advanced Review uses both methods during a user's session.

USERS\{user} instruction from RC-26 function represents major concern described in this article. The main problem is that per standard instructions, that implementation results in user complete removal from USERS data set. The remaining control instructed by the standard is insufficient as the only lock after deletion that prevents the same user to be recreated and allocated to a different person is based on erasable NAME abstract data-type on RC-04. That is, the command is valid only if the new user is not present at USERS (user:NAME $\notin$ USERS). Therefore, as the only suggested bond from user to a person from this person is this nickname contained in NAME, there's no RBAC inner control that prevents breaking person's history interaction on HIS. Hence, affecting directly the accountability capabilities on the system.

3.2   Other Function Reflexes

Although HIS RBAC compliant means that at least Core set of functions are met, several systems implement the full stack or even mixed access-control models today [10]. More complete RBAC aware HIS can implement Core, Hierarchical, Static Separation of Duties (SSD) and Dynamic Separation of Duties (DSD). These additional components can be added independently, onto Core. When using this more complete component scenario, some of the Core functions are redefined to accommodate the extra complexity. To better demonstrate RC-26 limitation function earlier described on RBAC as a whole, Fig. 1 below illustrates the relationship with USERS, user_sessions, assigned_users, and authorized_users among components. Red color on functions indicates dependency to the two focused functions depicted in blue. A total of 17 of underlying function are impacted.

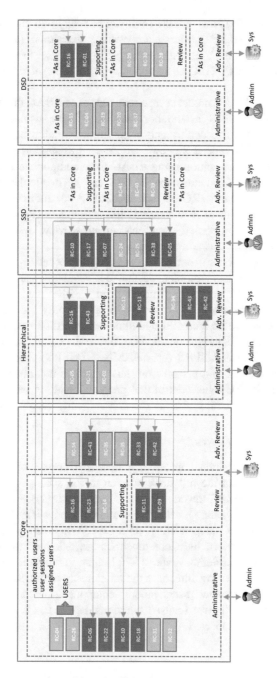

Fig. 1. Representation of function RC-26 and reflexes on RBAC components

4   New Functions

Administrative errors can occur. A mistaken RC-04 command can generate a wrong username or any other feature associated with user's record on HIS. Therefore, is naturally useful to allow RC-26 execution for correction purposes. In this case, we advocate that the legal command operation must consider that the created user never logged-in and performed no action on the system or at least was not commissioned for use by UA – PA assignment. That is to ensure a wrongly configured user did not influence system content by any means and consequently no accountability link will be miss by user deletion. A verification statement on this function for that matter could be achieved by; [$\forall$s: SESSIONS • s $\notin$ SESSIONS_HISTORY] and/or; (user $\mapsto$ role) $\notin$ UA. Please notice that the first verification is feasible only if another undeclared RBAC capability is present on HIS, that is user session history. User history is currently absent from RBAC standard [1]. This session history was previously advocated on past work while discussing improvement on separation of duties via DSD control [11]. Therefore, our suggestion considers a simpler approach based on UA association only. By actual function re-significance, new abstract-type user_status is turned off instead of user removal. The multi-functional new function deletes users only if not previously set for use. The xample uses a time-based constraint using HIS server reference to indicate active user_status if valid time-stamps are to be found. This can be populated by RC-04 as proposed here, but also by AssignUser(RC-10) during administrative UA. Discrete values (ie. "active" - "inactive") can also be implemented accordingly. Implementers who wish to add the new user status control over ongoing systems can undergo seamlessly updating USER database as per current standard interpretation all existing users are actually active. RC-26 proposition:

DeleteUser (user: NAME) $\lhd$ (1)
user $\in$ USERS
[$\forall$s $\in$ SESSIONS • s $\in$ user_sessions (user) $\Rightarrow$ DeleteSession (s)]
θS = [(user $\mapsto$ role) $\in$ UA $\wedge$ (user: user_status $\leq$ sysdate())]
UA' = UA \ {r : ROLES • user $\mapsto$ r}
assigned_users' = {r: ROLES • r $\mapsto$ (assigned_users(r) \ {user})}
USERS' = {(S • true $\Rightarrow$ u: USERS • u: user_status $\mapsto$ $\emptyset$)$\wedge$($\neg$S • true$\Rightarrow$USERS \ {user}) }$\rhd$

RC-04 also gains similar additional control to check for user existence and current status. Thus, the same function can be used to create or activate a user. RC-04 proposition:

AddUser (user: NAME) $\lhd$ (2)
θS = [(user $\mapsto$ role) $\in$ UA $\wedge$ (user: user_status $\leq$ sysdate())]
USERS' = { ($\neg$S • true $\wedge$ user $\notin$ USERS $\Rightarrow$ USERS $\cup$ {user: user_status $\mapsto$ $\emptyset$})$\wedge$(S • true $\Rightarrow$ user: user_status $\rightarrow$ sysdate())}
user_sessions' = user_sessions $\cup$ { user $\mapsto$ $\emptyset$} $\rhd$

4.1 Ambiguity Clarification

Although we believe that newly designed functions better describe USERS administration for user creation and deletion, they are not an absolutely necessary amend needed in the standard. That's because more precise instruction and description to aid implementers on a correct interpretation of the standard can serve as good.

Sessions, users and other objects may be in fact abstractions and not a real part of the access-control structure. Like we understand that sessions are in reality server controlled by diverse means (unique ID, randomization, replay-control, etc.) and the access-control borrows it to embed user and role information, the same understanding could be done for USERS data set. That is, as long as an implementer manages the users in a separate table applying means for accountability and strong bind to person's attached, there is no real issue accepting that inner access-control table will have user instance erased keeping original functions as it is. In short, the new functions could be replaced by clearer textual guidance to implementers. Some clarifications are already present in the standard. Functional specification informative annex gives a broad view and informs expected behavior at some point. However, basic clarification is clearly missing. For instance, NAME abstract on function shall be univocal, Support functions shall consider only active and unique sessions identifier, and so on.

5   Conclusions

Native RBAC functions that create and deletes users on the system does not provide for user's accountability and strong bond for actions history. RC-26, as is on standard, is an unwanted function on HIS. New abstract data-type are suggested to better bind the person behind username declaration (subject) and to establish a revoked status on the user not allowing permanent deletion. Real use-case scenarios discussion illustrates potential damage to accountability link loss. Redesigned functions suggested circumvent that and can atone for better accountability on RBAC aware HIS.

Acknowledgments. We thank CAPES and its partnership with Sao Paulo Federal University (Unifesp) sponsorship for this project.

References

1. INCITS 359-2012 Information Technology - Role Based Access Control. ANSI 2012. http://webstore.ansi.org/
2. ISO 21298:2017 - Health informatics – Functional and structural roles. International Organization for Standardization (2017)
3. ISO 21091:2013 - Health informatics – Directory services for healthcare providers, subjects of care and other entities. International Organization for Standardization (2017)
4. Carvalho, M.: Bonds to the subject. In: Proceedings - International Carnahan Conference on Security Technology, Medelin, pp. 1–10 (2014). https://ieeexplore.ieee.org/document/6922035/

5. Rajkumar, P.V., Sandhu, R.: POSTER: security enhanced administrative role based access control models. Comput. Netw. **112**, 1802–1804 (2016). https://doi.org/10.1016/j.comnet.2016.11.007

6. Baracaldo, N., Joshi, J.: Beyond accountability: using obligations to reduce risk exposure and deter insider attacks. In: Proceedings of the 18th ACM Symposium Access Control Models and Technologies - SACMAT 2013, p. 213 (2013). https://dl.acm.org/citation.cfm?id=2462411\. http://www.scopus.com/inward/record.url?eid=2-s2.0-84883108231&partnerID=tZOtx3y1

7. Wainer, J., Kumar, A.: A fine-grained, controllable, user-to-user delegation method in RBAC. In: Proceedings of the Tenth ACM Symposium on Access Control Models and Technologies, p. 66 (2005). http://portal.acm.org/citation.cfm?id=1063991

8. Rabin, A., Gudes, E.: Secure protocol of ABAC certificates revocation and delegation. In: Foundations and Practice of Security, FPS 2017. Lecture Notes in Computer Science, vol. 10723 (2017)

9. Azkia, H., Cuppens-Boulahia, N., Cuppens, F., Coatrieux, G., Oulmakhzoune, S.: Deployment of a posteriori access control using IHE ATNA. Int. J. Inf. Secur. **14**(5), 471–483 (2015)

10. de Carvalho Junior, M.A., Bandiera-Paiva, P.: Acces-control authorization model for health information system (HIS) in Brazil. J. Health Inform. **10**(3), 79–82 (2018)

11. de Carvalho Junior, M.A., Bandiera-Paiva, P.: Evaluating ISO 14441 privacy requirements on role based access control (RBAC) restrict mode via colored petri nets (CPN) modeling. In: Proceedings - International Carnahan Conference on Security Technology (2017)

Applications to Help Local Authorities to Support Community-Dwelling Older Adults

Ana Isabel Martins, Hilma Caravau, Ana Filipa Rosa,
Alexandra Queirós, and Nelson Pacheco Rocha[✉]

Institute of Electronics and Informatics Engineering of Aveiro,
University of Aveiro, Campus Universitário de Santiago,
3810-193 Aveiro, Portugal
{anaisabelmartins,hilmacaravau,
filiparosa,alexandra,npr}@ua.pt

Abstract. Local authorities have an increasing importance both in safeguarding the interest of the older adults and in providing direct assistance when needed. This paper reports an experimental study aiming to implement information applications to help local authorities to support community-dwelling older adults. For that, it was applied: (i) good practices pertaining to new technological developments; and (ii) healthcare standards to guarantee the interoperability of the resulting social care information. The results point that Fast Healthcare Interoperability Resources (FHIR) presents several resources that might be used to model the required social care information.

Keywords: Community care · Local authorities · Older adults · FHIR

1 Introduction

The continuity of care [1, 2] requires the surpassing of problems such as the fragmentation of the information of the care receivers, often dispersed and isolated in different silos, such as hospitals, primary health centers, clinics or laboratories. Each of these silos might contain valuable information, which usually presents low interoperability levels and, therefore, is inadequate in terms of patient-centered care. Therefore, it assumes a paramount importance the use of healthcare interoperability standards, such as the Fast Healthcare Interoperability Resources (FHIR), developed by HL7, to provide longitudinal records of patient's health and healthcare [3].

Moreover, it is also necessary to consider the need to optimize the use of the available information related to other care and assistance services, such as social care services, in addition to health care services, to manage the societal challenges arising from the ageing populations.

Within the community care networks to support older adults (i.e. the care networks composed by formal caregivers, informal caregivers and assistance providers, namely to support domestic activities, personal care, nursing care, administrative tasks or transporting [4]), mechanisms to register, access and share information are non-existent or substantially inferior to those available for health care networks. In fact, on the one hand, the procedures associated with community care networks are substantially less

© Springer Nature Switzerland AG 2019
Á. Rocha et al. (Eds.): ICITS 2019, AISC 918, pp. 720–729, 2019.
https://doi.org/10.1007/978-3-030-11890-7_68

structured than similar mechanisms of the health care networks and, on the other, there are significant technological differences between health care and social care institutions, which makes it very difficult for the social care providers to communicate across institutional and professional boundaries, and consequently to access essential information.

The experimental study reported in this article aimed to contribute to the optimization of information technology to support assistance and social care services related to community care networks. In particular, one of its aims was to investigate the role of health care standards for the communication of social care information. In this paper, as a case study, the authors focused on social services provided by local authorities [5].

2 Background

Considering the population ageing, and the consequent increase in the consumption of resources in the age group over 65, a major concern is the continuity of care in space and time, which requires the involvement and coordination of a great diversity of institutions and care providers [1].

Although in some countries it is difficult to identify coordinated policies and strategies, as well as coherent design of the implemented structures, it is possible to conclude that once it has been accepted the existence of a social problem that cannot be of exclusive responsibility of the families, most countries have opted for institutionalization models [1, 2]. However, later on, institutionalization was complemented by community-centered services [1]. In this respect, it is paradigmatic the example of the home care services, that are seen as a potentially effective way of maintaining assisted persons autonomy and independence [6, 7].

The increasing number of older adults requiring care provision, as well as the decrease of the availability of informal caregivers, pressure the expansion of formal care services [6], which is not compatible with scarce economic and financial resources. In order to adequately meet the demanding needs and to ensure equal access and treatment with sufficient resources, there is the need to promote, in a sustained manner, the effectiveness and efficiency of health and social care services, as well as their articulation within the community care networks, in a continuum that should go from the prevention and management of lifestyles to the institutionalization.

The use of technological solutions to provide information applications to mediate different actors involved in the continuity of care has been proposed by several authors [8–10] using different technologies [11–13], namely for the management of chronic diseases [12–15]. Nowadays, the need for retrieving, managing and delivering large amounts of health care information is addressed by Electronic Health Records (EHR) [10, 16]. For the interconnection of different EHR systems it is necessary to use health care standards to guarantee the interoperability of patient's information. Particularly, the FHIR has become increasingly important because it combines the best features of other known specifications, such as HL7 v2, HL7 v3, and Clinical Document Architecture.

However, in terms of continuity of care, it is important to consider other types of information in addition to clinical parameters, such as family support, habitability conditions, habits, activities or routines, which are information types that are not covered by EHR systems. Therefore, in 2003, in the United Kingdom, it was introduced an additional concept, the Electronic Social Care Record (i.e. a comprehensive individual record within social services) [3]. Moreover, the scientific literature reports the implementation of shared care plans in accordance with European standard EN 13940-1 for continuity of care [17], and a considerable effort to develop appropriate services [14, 18–20] to manage psychosocial information [21], information provided by the assisted persons and their informal caregivers (e.g. Personal Health Records) and information related to automatic data collection, namely using innovative monitoring devices [22–24].

However, most technological solutions supporting the continuity of care [9, 17, 25] do not present comprehensive approaches for the integration of information related to assisted persons outside the boundaries of health care facilities, nor guarantee interoperability with the information retrieved by health information systems. Often, dedicated databases are used for the persistence of information, and the available standards to support the technical and semantic interoperability of health care information are not considered. These promote the existence of information silos, which is incompatible with the continuity of care vision.

This situation is due to various factors such as the inadequacy of policies to support older adults, the separation between health care and social care (in many countries, dependent of different ministries and financed in different ways), the different nature of health care and of social care, the still widespread idea of considering social care services as the lower hierarchical level of the whole line of formal care [1, 2], which is sometimes referred to in the scientific literature as social and health care divide [1, 26], and the conflicting views on ethical principles, privacy, confidentiality, data access, as well as public and professional concerns over access to personal records or government surveillance.

In Portugal, the social sector includes more than four thousand institutions exclusively dedicated to the provision of social care. To these must be added other institutions, such as groups of volunteers, or formal institutions which are normally not associated with the care provision, but which play an important role in the proximity to the citizen, namely the local authorities. Contrary to what happened to health care institutions after the turn of the century, social care institutions have not yet been able to make a real technological leap. Indeed, although there has been an increase in investment on information systems in recent years, most of these investments have been directed to administrative management applications and less to applications aiming to promote personalized care, namely due to the non-existence of solutions or to the existence of clearly inadequate solutions.

3   Study Design

The project Social Cooperation for Integrated Assisted Living (SOCIAL) [27], whose fundamental element is a platform of services, aims to contribute to the ambitious goal of continuity of care, by providing innovative applications to support communication and information sharing within community care networks to support older adults.

The assistance and social care services provided by local authorities assume a paramount importance, since the local authorities are one of the resources for the first contact with the population and are involved in organizing and planning activities or providing financial support and direct assistance when needed [28]. Therefore, the first scenario being considered within the SOCIAL project was related to the social support provided by local authorities to older adults. The two city councils that have been cooperating in the study reported in this paper have in common social support offices, which support older adults living in their domiciles (most of them living alone or living with another older adult) and programs offices, which intends to offer several programs and activities aiming to promote the quality of life of older adults living independently in the community.

Important concerns of the experimental study were to identify what information models are required and to evaluate the possibility to use health care standards to guarantee the interoperability of the resulting social care information. Moreover, it was also considered important that the applications should have user interfaces adapted to the different users, not only care providers but also assisted persons. In fact, user interfaces are a vehicle to transport the information to the different users, namely in terms of data collection or data visualization (e.g. dashboards or information maps) and they should provide an enjoyable user experience, irrespective of the user's proficiency or the information being made available.

Therefore, the study design combined several methods to conform the gold standard for research pertaining to new technological developments, specifically in terms of the active and ongoing participation of stakeholders throughout the different development phases [29, 30].

First, a literature review was performed. This literature review focused on existing evidence about successful ageing, active aging, well-being, care approaches, and stakeholder's needs. Moreover, it was also examined relevant policy documentation and research studies upon the advancement of technologies that might allow older adults to live safely, independently, autonomously, and comfortably, without being required to leave their own residences, but with the necessary support services to their changing needs [15, 31].

Afterwards, face-to-face interviews were conducted with relevant stakeholders from formal care networks that are experts in the social care services. The participants were selected according to their relevance in city councils, borough councils, and social care institutions.

Following the interviews, data analysis was performed. This data analysis, focused on understanding care approaches and specific stakeholder requirements, and was the basis for the creation of several personas (i.e. archetypes of hypothetical users, including their needs, goals and tasks) and scenarios (i.e. stories with settings, personas who have needs and goals and a sequence of events and tasks) [32, 33].

The personas and scenarios were used to support problem solving around the different requirements and to design the applications, namely their user interfaces.

The first phase of the evaluation occurred after the implementation of the initial prototypes. The evaluation involved representatives of the different stakeholders and was focused on eliciting feedback regarding the high-level product concept and a subset of interfaces. Participants were invited to review prototypes to explicitly comment on usability issues (e.g. interaction style, nomenclature, iconography or tasks workflow).

All procedures performed in this study involving human participants were in accordance with the ethical standards of the institutional and/or national research committee and with the 1964 Helsinki declaration and its later amendments [34]. All the necessary authorizations were obtained, the study had the approval of an Ethical Committee, and all participants signed an informed consent prior to data collection.

4 Applications

According to the objectives of the experimental study, the first applications were developed targeting the requirements of the city council's social support offices and programs offices.

A social support office intends to provide various types of assistance to requests related to older adults that live in their domiciles. In this respect, a request for assistance must be understood as the possibility for the caregiver or the assisted person to create attendance or occurrence requests for specific needs or to report an anomalous situation. Moreover, a programs office intends to deliver several programs (i.e. initiatives composed by different types of activities, such as physical activity - swimming or trekking -, or sociocultural activities - theatre or dance) aiming to promote the quality of life of older adults living independently.

In a simple way, the following actors and respective uses cases were considered for the social support office: (i) assisted persons and their informal caregivers (i.e. request for assistance, provision of information, and access to plans and notifications); and (ii) responsible for the social support office and case managers (i.e. assisted person's management, information retrieval, and access to plans and notifications).

In turn, for the programs office the following actors and respective uses cases were considered: (i) assisted persons and their informal caregivers (i.e. information access, applying, registration, withdrawal, or provision of feedback about programs and activities); (ii) operational assistants (i.e. provision of information of certain activities to potential participants, enrolment and registration of participants, provision of feedback about the progress of programs and activities, and access to plans and notifications); (iii) technician of the city council (i.e. creation, allocation of human, material and economic resources, calculations of the costs, monitoring and submission of programs and activities, and access to notifications); and (iv) member of the city council (i.e. approval of programs and activities, evaluation of ongoing programs and activities, and access to notifications).

Therefore, three different applications were considered: (i) MySocial, a mobile and web application targeting assisted persons and their informal caregivers, aiming to provide a comprehensive but configurable set of functions; (ii) SocialOperationManagement, a web application targeting care providers; and (iii) SocialPlanning, a web application focused on the secondary use of information.

The MySocial application includes: (i) the management of demographic data (e.g. name, date of birth, gender, marital status, profession, schooling, household, nationality, identification document, address, telephone or email), personal factors (e.g. physical conditions, lifestyles, habits, facing problems skills, social background, or past and present life events), and management of data that characterize specific situations or events; (ii) assistance requests; (iii) self-association to establish a direct relationship to assist another person; (iv) application, registration and withdrawal of programs and activities; (v) evaluation of programs and activities (e.g. enumeration of positive and negative aspects); (vi) plans and tasks (e.g. management of individual plans or tasks or monitoring of daily activities, such as, for example, eating habits or planned physical activity); (vii) notifications; and (viii) to do list.

In turn, the SocialOperationManagement application includes: (i) management of assisted person's personal data; (ii) management of assistance requests (e.g. assignment of a case manager to a particular assisted person, profiling the assisted person, identification of needs or significant aspects of the assisted person's health conditions); (iii) referral of an assisted person to a specific provider, service or institution; iv) sharing of information (e.g. information exchange between providers with distinct roles, without replication, but with differentiated access); (v) creation and approval of programs and activities; (vi) management of programs and activities (e.g. management of human, physical and economic resources, association of assisted persons to specific programs or activities, provision of information to potential participants, enrolment and registration of participants, or provision of feedback about the progress of ongoing programs and activities); (vii) planning, approval and scheduling of programs and activities; (viii) notifications; and (ix) to do list.

Finally, SocialPlanning is an analysis and decision support application based on the secondary use of anonymized data related with assisted persons.

In terms of the information model to support the various applications, the fundamental entity is the Person. The individuals interacting with the SOCIAL platform (e.g. an assisted person, a caregiver, or even an employee of an institution) have identities and, therefore, are considered Persons that might have several roles. Regardless of the role they play on the platform, their demographic data should be registered. Moreover, in the case of formal care providers, they might belong to a given institution, including city councils, borough councils, and social care institutions.

The FHIR resource Person allows the association of several instantiation resources belonging to the same individual, allowing the different roles that each individual can assume within the platform. In the case of the individuals assuming the role of care or assistance provider, they can be characterized by the following FHIR resources: Practitioner (i.e. a person who is directly or indirectly involved in the provisioning of care and assistance); Practitioner Role resource (i.e. a specific set of roles, locations, specialties or services that a care or assistance provider might perform at an institution during a period of time; and Organization resource (i.e. a formally or informally recognized grouping of people or institutions organized for the purpose of achieving some form of collective action).

Additionally, in the case of assisted persons it is also necessary to consider personal factors, and other data characterizing specific situations (e.g. assistance request) including health conditions and care plans. Information related to personal factors or to the characterization of specific situations that might result from the application of assessment instruments (e.g. functioning assessment or the assessment of the individual economic situation for the evaluation of the need for social support). On the other hand, tasks such as home visits are considered to be an integral part of a care plan and are themselves relevant to generate notifications or to fulfil the to do list.

The different types of assessments can be covered by the composition of two FHIR resources: the Questionnaire Response resource (i.e. a structured set of questions and respective answers) and the Observation resource (i.e. measurements and simple assertions made about an assisted person, device or other subject).

Additionally, each assisted person has an individual plan that consists of personal tasks and tasks that involve care and assistance providers of one or more institutions. These entities can be mapped to FHIR resources, namely Care Plan (i.e. description of how one or more care or assistance provider intend to deliver care for a particular assisted person, group or community for a period of time, possibly limited to care for a specific condition or a set of conditions) and Task (i.e. a task to be performed).

Finally, programs must also be considered. Each program is made up of a set of activities (which can be mapped to the FHIR Task resource) that are usually planned to run throughout a predefined period. Each program is identified with a type (e.g. sport, travel, among others), which allows assisted persons and informal caregivers to easily find programs of their interest. On the other hand, it is also necessary to add that programs form calendars (i.e. activities and the respective schedules).

Thus, several FHIR resources are required to support programs, namely: (i) the Value Set resource, which is used to indicate the program type; (ii) Care Plan resource to store all attributes of a schedule of a specific assisted person; and (iii) Appointment resources for the activities (i.e. each activity will be associated to Appointments, which corresponds to its scheduling for the respective subscribers).

Other requirement of the programs, such as advertisement indications or registration limits, are easily mapped via FHIR extensions.

5   Discussion and Conclusion

The persistence strategies are based on the concept of FHIR resource, a basic unit of interoperability defined to represent clinical concepts such as patients, observations or adverse reactions.

Although the FHIR is a specification for communication and interoperability, the use of FHIR resources presents several advantages, namely compliance with established international standards, and information modelling using entities that are already mature and prepared to guarantee interoperability between applications. In addition, the implementation using noSQL databases and the JSON format, provides flexibility in the data structure, allowing the definition of volatile models and the optimization of data searches, since the information is no longer dispersed in several relational tables. Like any other approach, this one also has its drawbacks, in particular because it is necessary an extra attention to aspects such as references between resources (i.e. each FHIR resource includes references to other resources, so it is necessary to ensure the consistency of these references to prevent duplication of information, unsupported content and transactions, and invalidated data structures and their data types).

FHIR resources are adequate to persist the information of the applications being developed for the provision of social care by local authorities. Therefore, the results indicate that FHIR, a health care interoperability standard, can be used for the combination of heterogeneous information, including social care information related to older adults living in the community. A prototype system was successfully implemented and showed the applicability of the approach. Nevertheless, to provide semantic interoperability, further work should be performed to achieve the normalization of key concepts.

However, some specific situations are not covered. This can be easily solved using extensions, which the FHIR provides. Nevertheless, it should be noted that extensions are not as generalizable as the FHIR features, so that further research is needed.

Further application domains (e.g. home care) have also been considered and similar methods are being used for the development of other prototypes. Moreover, consolidated versions of the applications are being prepared to be evaluated in a real conditions pilot study.

Finally, it is also clear the need for information consolidation mechanisms (e.g. big data) to support the planning and monitoring of social care services being provided by local authorities. In this sense, it is essential to aggregate, synthesize and filter anonymized data related to assisted persons, in order to provide relevant statistical information. This raises a set of difficulties because, on the one hand, there is the need to optimize the information queries, in accordance with FHIR features, and, on the other, there is the need to prevent non-authorized analyses.

Acknowledgments. This work was partially supported by COMPETE - Programa Operacional Competitividade e Internacionalização (COMPETE 2020), Sistema de Incentivos à Investigação e Desenvolvimento Tecnológico (SI I&DT), under the project Social Cooperation for Integrated Assisted Living (SOCIAL).

References

1. Leichsenring, K.: Developing integrated health and social care services for older persons in Europe. Int. J. Integr. Care **4**(3) (2004)
2. Leichsenring, K., Billings, J., Nies, N.: Long-Term Care in Europe: Improving Policy and Practice. Palgrave Macmillan, Basingstoke (2013)
3. Baines, S., Hill, P., Garrety, K.: What happens when digital information systems are brought into health and social care? Comparing approaches to social policy in England and Australia. Soc. Policy Soc. **13**(4), 569–578 (2014)
4. Jacobs, M., Groenou, M., Aartsen, M., Deeg, D.: Diversity in older adults' care networks: the added value of individual beliefs and social network proximity. J. Gerontol. – Ser. B Psychol. Sci. Soc. Sci. **73**(2), 326–336 (2018)
5. Waverijn, G., Groenewegen, P., de Klerk, M.: Social capital, collective efficacy and the provision of social support services and amenities by municipalities in the Netherlands. Health Soc. Care Community **25**(2), 414–423 (2017)
6. Genet, N., Boerma, W., Kringos, D., Bouman, A., Francke, A., Fagerström, C., Melchiorre, M., Greco, C., Devillé, W.: Home care in Europe: a systematic literature review. BMC Health Serv. Res. **11**, 207 (2011)
7. Santana, S., Dias, A., Souza, E., Rocha, N.: The domiciliary support service in Portugal and the change of paradigm in care provision. Int. J. Integr. Care **7**(1) (2007)

8. Hägglund, M., Chen, R., Koch, S.: Modeling shared care plans using CONTsys and openEHR to support shared homecare of the elderly. J. Am. Med. Inform. Assoc. 18(1), 66–69 (2011)

9. Warren, I., Weerasinghe, T., Maddison, R., Wang, Y.: OdinTelehealth: a mobile service platform for telehealth. Procedia Comput. Sci. 5, 681–688 (2011)

10. Häyrinen, K., Saranto, K., Nykänen, P.: Definition, structure, content, use and impacts of electronic health records: a review of the research literature. Int. J. Med. Inform. 77(5), 291–304 (2008)

11. Ekonomou, E., Fan, L., Buchanan, W., Thuemmler, C.: An integrated cloud-based healthcare infrastructure. In: Proceedings of the 2011 IEEE Third International Conference on Cloud Computing Technology and Science (CloudCom). IEEE (2011)

12. Meier, C.A., Fitzgerald, M.C., Smith, J.M.: eHealth: extending, enhancing, and evolving health care. Annu. Rev. Biomed. Eng. 15, 359–382 (2013)

13. Capurro, D., Ganzinger, M., Perez-Lu, J., Knaup, P.: Effectiveness of ehealth interventions and information needs in palliative care: a systematic literature review. J. Med. Internet Res. 16(3), e72 (2014)

14. Spitzer, W.J., Davidson, K.W.: Future trends in health and health care: implications for social work practice in an aging society. Soc. Work Health Care 52(10), 959–986 (2013)

15. Queirós, A., Pereira, L., Dias, A., Rocha, N.P.: Technologies for ageing in place to support home monitoring of patients with chronic diseases. In: Proceedings of the 10th International Joint Conference on Biomedical Engineering Systems and Technologies. SciTePress (2017)

16. Eichelberg, M., Aden, T., Riesmeier, J., Dogac, A., Laleci, G.B.: A survey and analysis of electronic healthcare record standards. ACM Comput. Surv. 37(4), 277–315 (2005)

17. Li, S.H., Wang, C.Y., Lu, W.H., Lin, Y.Y., Yen, D.C.: Design and implementation of a telecare information platform. J. Med. Syst. 36(3), 1629–1650 (2010)

18. Atkins, D., Cullen, T.: The future of health information technology: implications for research. Med. Care 51, S1–S3 (2013)

19. Bossen, C., Christensen, L.R., Grönvall, E., Vestergaard, L.S.: CareCoor: augmenting the coordination of cooperative home care work. Int. J. Med. Inform. 82(5), e189–e199 (2013)

20. Memon, M., Wagner, S.R., Pedersen, C.F., Beevi, F.H., Hansen, F.O.: Ambient assisted living healthcare frameworks, platforms, standards, and quality attributes. Sensors (Basel) 14(3), 4312–4341 (2014)

21. Alvarelhão, J., Silva, A., Martins, A., Queirós, A., Amaro, A., Rocha, N., Lains, J.: Comparing the content of instruments assessing environmental factors using the international classification of functioning, disability and health. J. Rehabil. Med. 44(1), 1–6 (2012)

22. Rantz, M.J., Skubic, M., Alexander, G., Popescu, M., Aud, M.A., Wakefield, B.J., et al.: Developing a comprehensive electronic health record to enhance nursing care coordination, use of technology, and research. J. Gerontol. Nurs. 36(1), 13–17 (2010)

23. Reeder, B., Meyer, E., Lazar, A., Chaudhuri, S., Thompson, H.J., Demiris, G.: Framing the evidence for health smart homes and home-based consumer health technologies as a public health intervention for independent aging: a systematic review. Int. J. Med. Inform. 82(7), 565–579 (2013)

24. Knaup, P., Schöpe, L.: Using data from ambient assisted living and smart homes in electronic health records. Methods Inf. Med. 53(3), 149–151 (2014)

25. Haritou, M., Glickman, Y., Androulidakis, A., Xefteris, S., Anastasiou, A., Baboshin, A., Cuno, S., Koutsouris, D.: A technology platform for a novel home care delivery service to patients with dementia. J. Med. Imaging Health Inform. 2(1), 49–55 (2012)

26. Glasby, J., Dickinson, H.: Partnership Working in Health and Social Care: What is Integrated Care and How Can We Deliver It? Policy Press, Bristol (2014)

27. Sousa, M., Arieira, L., Queirós, A., Martins, A.I., Rocha, N.P., Augusto, F., Duarte, F., Neves, T., Damasceno, A.: SOCIAL platform. In: Advances in Intelligent Systems and Computing, vol. 746, pp. 1162–1168 (2018)
28. Segurança Social: Redes Locais de Intervenção Social (RLIS). Segurança Social, Lisbon (2016)
29. Cahill, J., McLoughlin, S., Wetherall, S.: The design of new technology supporting wellbeing, independence and social participation, for older adults domiciled in residential homes and/or assisted living communities. Technologies **6**(1), 18 (2018)
30. Venable, J., Pries-Heje, J., Baskerville, R.: FEDS: a framework for evaluation in design science research. Eur. J. Inf. Syst. **25**(1), 77–89 (2016)
31. Connelly, K., Mokhtari, M., Falk, T.H.: Approaches to understanding the impact of technologies for aging in place: a mini-review. Gerontology **60**(3), 282–288 (2014)
32. Cooper, A.: The Inmates are Running the Asylum: Why High Tech Products Drive us Crazy and How to Restore the Sanity. Pearson Higher Education, Upper Saddle River (2004)
33. Queirós, A., Cerqueira, M., Martins, A.I., Silva, A.G., Alvarelhão, J., Teixeira, A., Rocha, N.P.: ICF inspired personas to improve development for usability and accessibility in ambient assisted living. Procedia Comput. Sci. **27**, 409–418 (2014)
34. World Medical Association: World Medical Association Declaration of Helsinki. Ethical principles for medical research involving human subjects. Bull. World Health Organ. **79**, 373 (2001)

A Critical Analysis of Requirements and Recommendations for Multi-modal Access Control in Hospitals

Mapula Elisa Maeko and Dustin van der Haar$^{(\boxtimes)}$

University of Johannesburg, PO Box 524, Auckland Park,
Johannesburg 2006, South Africa
elisa.maeko@gmail.com, dvanderhaar@uj.ac.za

Abstract. Lack of user awareness and user acceptance of the importance of practising a good access-control approach, particularly in the healthcare sector of South Africa, is a challenging issue in the area of the security of information. The challenge results in breaches to information stored in medical information systems, fraud, and medical identity theft. The paper proposes the hospital user awareness and user acceptance framework for multimodal access control in the medical information systems. The survey results from participants in Charlotte Maxeke Academic hospital, based in Johannesburg showed a response rate of 86%. Focusing on the objectives of the study the results showed that 76.7% of users' lack knowledge in the security of information and that there is a lack of security awareness education and training in the hospital. The results of the study are presented and analysed through IBM SPSS statistical software that is used for data analysis.

Keywords: Access control · User awareness · User acceptance ·
Fingerprint biometrics · Smart cards · Medical information systems

1 Introduction

Technological advances such as the implementation of electronic medical information systems have been extensively used in the health care sector of South Africa. The security of individual information is a big concern where a significant amount of patient information is stored, processed, retrieved and transferred electronically daily [19]. Unauthorised access to medical information poses a great deal of security risk and threatens the privacy of confidential information [3]. The solution is thereof required within the healthcare environment to protect patient medical information better and to control access in the medical information systems [17]. The health care industry uses the information to achieve immediate and future goals, the information further used for research purposes such as epidemiology studies which study the causes of sickness, the effects and various patens in sickness, for coming out with new potentials to prevent sicknesses. The security of medical information means securing data from unauthorised access; security may be in any form including securing the equipment's physically,

© Springer Nature Switzerland AG 2019
Á. Rocha et al. (Eds.): ICITS 2019, AISC 918, pp. 730–743, 2019.
https://doi.org/10.1007/978-3-030-11890-7_69

securing our networks, providing a password for gaining access, logically protecting our systems and applications.

The focus of the study is the security of patients' medical information in medical information systems. By focusing on good and efficient awareness programs through many types of training and education, better security can be achieved [6]. Running awareness and training is an important aspect of securing Information in any environment and if implemented effectively it has the benefits of saving the organisation from errors and even save money. The awareness sessions have the benefit of informing the users the areas required to be aware off and if facing a threat what procedure they need to follow to prevent an attack. Working in an environment with a significant amount of data need the healthcare sector to have solutions such as a policy, standards and procedures to deal with any issue arising and a good access control infrastructure is the starting point to secure information from malicious acts. Making employees of an organisation know and understand the company policy on information security is very crucial and the disciplinary action taken if the security of information compromised in any manner [6]. To solve this organisation can come with a solution such as creating posters to put in the corridors, lifts, passages and wall to remind and inform the users on the importance of securing data primarily in a big data environment. They can also create information on their website and even e-mail the information to the staff [22].

The use of training sessions is very crucial, as it will demonstrate how to solve a problem confronted with and addressed. "The Tennessee University in the United States has implemented a good and impressive information security website for raising awareness with complete videos, helpful tips and examples". The website is a source of providing tips and those hints to enhance training sessions conducted. For example, using posters keep users reminded always to make the security of information a priority.

2 Problem Background and Computational Details

Theft of patients' identity is rapidly becoming a problem in the healthcare industry. The problem is how the confidentiality and privacy of patient information are protected and maintained [11]. Critical data often suffer from the misuse and abuse by unauthorised individuals. Medical records are often vulnerable and at risk, since they contain personal information, such as the patients' identity, the treatment they are currently on the history of employment, and income. Medical records are often vulnerable and at risk, since they contain personal information [7]. Looking at the information security threat that faces the healthcare industry, a problem does exist where user awareness education and training are considered important security measures to protect sensitive information.

The problem is that there are insufficient technology resources, and poor access control measures in the hospital, due to users sharing usernames and passwords to access the medical information system. It is an information security risk to share credentials to access sensitive information. The challenges put the healthcare industry at risk of attacks [3]. The patient information should be secured at the highest level, lest they fall into the hands of an unauthorised individual who has malicious intentions. Table 1 indicate that 11.6% of users perceive very harmless the issue of login credentials exposed to an unauthorised individual without their knowledge. The table shows that the users need information security training to educate them about the risk of information security breaches. Perception on exposed credentials without knowledge revealed user is considering that should their login credentials be exposed without their knowledge is very harmful, hence the need for secure access control in the medical information system like multimodal access control.

Table 1. Perception on exposed credentials without knowledge

		Frequency	Percent	Valid percent	Cumulative percent
Valid	Very harmful	29	67.4	67.4	67.4
	Somewhat harmful	4	9.3	9.3	76.7
	Neither	4	9.3	9.3	86.0
	Slightly harmless	1	2.3	2.3	88.4
	Very harmless	5	11.6	11.6	100.0
	Total	43	100.0	100.0	

The unauthorised individuals can gain access to the system then steal patient information to commit fraud. The health care industry suffers the loss of finance and even reputation. It has costly financial implications for both the individual affected and the health care provider [8]. Furthermore, medical records safeguarded so that they are accurate, complete and reliable, while free from errors. Medical records need to be readily available anytime when required. Preventing unauthorised access to records will ensure the hospital's render quality services to citizens. Lack of user awareness and acceptance to control access to medical information systems leads to various problems that include loss of medical records stolen to commit fraud [11]. Thereby showing that by controlling access to environments such as the hospital's, where an outsized amount of confidential medical information kept, is essential to protect the privacy of electronic medical records.

2.1   The Research Questions

2.1.1   What are the Problems in Access Control Management in Medical Information Systems?

The research question explores the existing limitations to access control in medical information systems in the hospital. We investigate and identify the gaps focusing on security of information, meaning the CIA of information, which are the Confidentiality, Integrity and Availability [21].

2.1.2   How do Users Think and Trust the use of Smart Cards and Fingerprint Technology to Access Medical Information System

The research question analyses the user's attitude and perception on the use of smart card and fingerprint-based access control technology to securely access medical information systems. The aim is to understand and meet the requirements for a successful implementation of a multimodal access control technology (Smartcard and fingerprint).

2.1.3   How will the Problems of User Awareness and Acceptance Solved to Improve the Security of Patient Information Stored in Medical Information Systems

The results of the question provide the solution for the improvement of access control through user awareness and acceptance of information security achieved in the hospital. The literature review followed by a survey, discussions and observations address the research question.

The research questions lead to the Objectives of the study, they follow next.

1. To investigate the information security problems that exist in controlling access to the medical information system in the hospital environment.
2. To explore the current awareness and acceptance of the security of patient confidentiality and privacy of information while users' access medical records.
3. Explore and analyse the information security vulnerabilities, threat and risk that exist in the hospital, and then based on the results of the study the recommendation provided.

Considering the various problems identified above, the analysis of a more understanding of the contributing factors affecting effective access to the system being adequate education and training. An analysis of the study follows in the Research methodology section.

3   Research Methodology

The research methodology used followed a Qualitative, interpretive with Literature Review, Survey and Framework [5]. The target population was drawn from the healthcare industry in South Africa, more specifically from Charlotte Maxeke Academic Hospital in Johannesburg. It included a survey distributed to staff in the Medical Records, Information management, and ICT departments. Fifty participants were selected randomly from the entire hospital staff population. Survey questions consisted of varyingly structured open-ended and closed-ended questions.

The 43 completed and returned surveys were analysed. They were selected as being relevant because they use patient health records in their daily activities, and are believed to have the relevant information on the importance of securing medical. The survey results presented and analysed using IBM SPSS program, a software program for statistical data analysis. The results tabulated on frequency tables and graphs. The process followed is outlined using the requirement discovery in conducting the research on multi-modal (smart card and fingerprint-based) access control. The requirement discovery defines and analyses the user awareness and acceptance of access control requirements by using a smart card and fingerprint-based access control at Charlotte Maxeke Academic Hospital for safeguarding the information stored in the medical information systems.

4   Requirement Discovery

The requirement discovery identified the requirements for user awareness and acceptance in the organisation that influence the successful implementation of an access control technology. Role-based access control is referenced where a user can carry out all activities which the user is authorised to. The role of access enables administrators to assign roles and privileges to user activities [13]. They allow a user to access only authorised tasks, and they help minimise the risk of damage to information due to an intruder pretending to be a legitimate user [9]. Additionally, the role of duties, such as no user is allowed authority over all activities, and that the user who executes a task should not be the same user who authorises the task. Appropriate management of incidents is required in the organisation to ensure that the organisation has a plan of action in place to prevent unauthorised access.

Successful evaluation is performed through survey feedback from employees, requiring the participating employee to share their comments or fill in a questionnaire to help the organisation improve awareness training. The requirement discovery process can help give the awareness programme direction and address concerns with the aim of maintaining the integrity of Awareness of information security is promoted through education and supporting the implementation of information security policies which ensure acceptable use of and access to information [6]. Furthermore, awareness and acceptance policies will ensure a proper response to security incidents and industry compliance. The non-functional requirements are the medical record information systems properties and constraints, such as the systems' reliability, response times, and ease of use among users. They include equipment such as computers (and other devices required for secure access control), connectivity of such equipment over the network, safety regarding terms of stability of power supply, and protection of databases with anti-virus software [10].

In order to achieve this goal, there is a focus on the following elements: how to conduct user awareness and system acceptance in health care, how multimodal access control is deployed, and how to conduct education and training, so that a more relevant framework is created. The requirement discovery unpacks the components for the creation of a framework. Playing a central role in the organisation allowing the monitoring and assessment of security controls, through ongoing monitoring and evaluation, to provide sufficient mitigation of security risks. First, the background on the healthcare industry and information system is outlined, focusing on the scope, method, functional, and non-functional requirements, the role of the survey in the study, question formulation, objectives, and the impact of requirement discovery. Discovery of requirements impacts awareness before implementing security access control policies in the organisation in providing guidance for employees to know what are acceptable actions, and what may result in fines and disciplinary action when non-compliance is reported [10]. The survey played a role in collecting the required information by outlining the objectives of the study to answer the research question. The results of the study, showing the actual outcomes and findings.

5 Results and Discussion

The statistical results from requirements discovery are drawn and analysed, which are the contribution of the study. The results show that the appropriate user awareness and the correct attitude of users towards the system, together with the necessary skills and excellent management involvement to communicate and engage the users may positively influence the successful implementation of a multimodal access control technology (Smartcard and fingerprint). Electronic health records maintained in the hospital they are collected from paper records that hold the entire history of patients' consultations, diagnosis and treatment received. Since the use of electronic medical records is progressing quickly the need to train and educate user's increases as well to ensure the user is aware of the risks and vulnerabilities to information, and that they accept the responsibility to securely access information and protect the confidentiality and privacy of information [18].

The hospital is currently using usernames and passwords to secure access to medical records. The literature review the method of authentication has been found to be ineffective for the healthcare sector that holds a significant amount of information that is confidential [4]. The usernames and passwords are often lost or forgotten impacting on the user being unable to render services on time (Table 2).

Table 2. The breakdown for the attendance of security awareness training

		Frequency	Percent	Valid percent	Cumulative percent
Valid	Yes	10	23.3	23.3	23.3
	No	33	76.7	76.7	100.0
	Total	43	100.0	100.0	

Concerning the attendance of security awareness training, the results show 76.7% of respondents have not attended any security education and training and only 23.3% of respondents have attended training. A 23% show a low level of information security awareness among participants and the solution is there required to mitigate the information vulnerabilities due to most staff show a lack of knowledge (Fig. 1).

The results show a lack of attendance of security awareness training is a risk to the organisation due to users being the target of security attacks. Furthermore, a mean value of 1.77, Std. Dev. = .427 and N = 43, showing a high frequency of users' non-attendance of awareness training. The hospital needs to take the results of the study seriously for the security of medical information. The hospital needs to invest in security awareness programs and training for all users to increase the culture of security consciousness among everyone (Table 3).

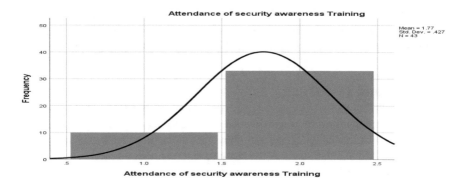

Fig. 1. Attendance of security awareness training

Table 3. The breakdown for the frequency of training offered

		Frequency	Percent	Valid percent
Valid	0	1	2.3	2.3
	Weekly	3	7.0	7.0
	Monthly	2	4.7	4.7
	Quarterly	4	9.3	9.3
	Annually	32	74.4	74.4
	5	1	2.3	2.3
	Total	43	100.0	100.0

The results show that times training for users often offered annually, the problem is that the users not reminded and knowledge enhances on the security of information, a case where new staff members will not know of information security concerns, measures and compliance within the hospital. It's a security concern not to educate and train users on security, the privacy and confidentiality of information in the medical information systems. The results show the frequency of training offered is currently low; we see 74.4% of respondents stated training offered annually and 9.3% indicates attendance of training quarterly (Fig. 2).

The frequency of security awareness within the organisation needs to be improved with frequent user awareness conducted. Furthermore, a mean value of 3.53, Std. Dev. = 1.054 and N = 43 are observed; the results show the hospital needs to increase security awareness on access control to achieve acceptance of multi-modal access control technology effectively. User awareness and acquired knowledge make a positive impact on user behaviour, ensuring the hospital ability to mitigate and prevent the security risks timely.

Figure 3 shows the results on a bar chart, it illustrates among the users who have not attended any security awareness training 0.25% of respondents to indicate the importance of raising awareness before implementation of access control technology. Through raising user awareness in the hospital, wrong user security behaviour is addressed at the initial stage, helping users have a positive attitude towards the

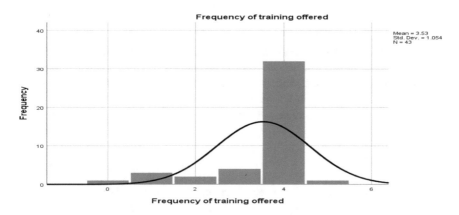

Fig. 2. The frequency of training offered

technology before its implementation. Improved access control to the hospital medical information systems and security culture within the hospital increased. Noted on the results the users regarding the importance of raising awareness before implementation of technology a crucial factor, a mean value of 1.47, the standard deviation of 0.909 and N-value of 43 observed in the results of the study.

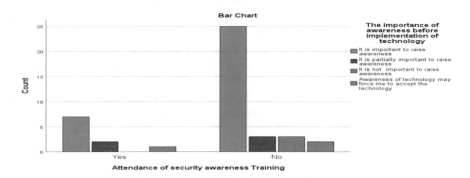

Fig. 3. Attendance of security awareness training * The importance of awareness before implementation of technology

The employees often share login information to access medical record; they are not fully aware of the potential risks of this practice. The user's reasons for non-compliance are that often they need to share computers and credentials to be able to complete their daily task; it may also be to train learners. Users' lack of awareness of information security is often a high-risk level to security risks and a threat to information [18]. Loss of patients records and duplicate medical records affected the patients, medical professionals and right diagnosis and treatment of patient negatively impacts on overall service delivery. Awareness of access control procedures required to educate users on the importance of safeguarding medical records and protecting the privacy of information. Training programmes and awareness sessions are a requirement for employees to access patient medical records safely.

Furthermore, the results show there is a lack of in-depth theoretical framework guiding users on how information security awareness training enhances the user's knowledge, attitude and, behavioural perception towards security. Information security awareness and acceptance is a fundamental part that could positively influence a change in the user perception and acceptance of technology for implementation.

6 Framework

The process followed for requirements discovery included the formal techniques and processes used to identify and collect system problems and the identify solution requirements using research, surveys, sampling, interviews, and other techniques for data collection [16]. The survey used in the study is used to assess and determine user awareness and acceptance of smart card and fingerprint-based access control technology within the healthcare sector in South Africa. For administering the survey to the health care participants, the participants were informed via e-mail of the availability of the survey online with a description of the purpose of the questionnaire. Furthermore, the survey was distributed in person to participants. A bipolar scaling method was used on a five-point Likert scale, covering the background information, and knowledge of smart card and fingerprint technology. The current use of technology in the organisation, efforts of the organisation to investigate the technology, and user perception concerning the technology in question were also covered [20]. The information was required to ensure the participants are the relevant target group for the study.

The question is used to explore the current security risks to information, and the users' attitude toward access control technology. It is essential in this phase that missing access control requirements, conflicting requirements, and requirements that overlap is identified, and solutions b provided to solve these problems so that a more relevant framework is created. The research framework discusses the roadmap, structural components and processes of user awareness and acceptance of smart card and fingerprint-based access control. Hospital User Awareness and User Acceptance Framework (HUA) framework is a supporting guide and presented in layered structures and discussion of processes. The HUAUA framework is intended to address the access control challenges the healthcare organisation to prevent security access control risks posed by unauthorised access to medical information systems [14]. It is evident that users are the weak source of information security, vulnerable to system misuse, intentional or unintentional errors and exposing access credentials [3]. Educating users with regards to secured access plays a role in preventing security risk regarding access to medical information systems. The framework on user awareness and user acceptance support, previous research literature in information security [6], affirming that the awareness of information security through education and training is an integral part in an organisation to ensure the security and privacy of information is maintained.

Management uses various techniques to ensure the protection information from vulnerabilities. The hospital management often uses the following measure to investigate the security risk and threat to sensitive medical information. They include access control, authentication, the encryption of information, risk management, disaster recovery and continuity plan. Different models have been developed to examine the factors affecting user awareness and acceptance of the technology. We explored different access control model such as the UTAUT model for adoption of technology before implementation in the hospital environment; the models are explored to determine the useful model suitable for use in the healthcare industry for protecting sensitive information. User acceptance of the technology needs evaluation during the early stages of the system design.

An initial performance and acceptance test of the system is required; the assessment can predict the future user acceptance of the system, and significantly reduce the possibility of rejection in the process. From the results of the study, it was illustrated that 74% of the respondents state it is important to raise awareness of access control technology before its implementation. Only a few respondents are aware of the use of a multimodal access control technology. Furthermore, an awareness of the technology and the perceived benefits of the systems may influence acceptance of access control technology. It is identified that the lack of system ease of use and user-friendliness is a barrier to user acceptance of the system [7]. The significance of HUA framework for secure access control in the medical information systems is that users are concerned about the protection of private information stored in the authentication systems. Smart cards and fingerprint-based methods of authentication seem to offer an added level of security to sensitive information [12]. Based on the results, the framework may contribute to the improvement and enhancement of access control in the medical information systems in the hospital. In cases where a user is considered such as raising user awareness to positively impact the belief system of the user as well as the knowledge in using the technology. However, it is essential for the users in the hospital to be equipped with the skills and knowledge to use the technology. Though there are various mechanisms for the high level of access control to systems, the requirement for access control policies is an obligation for an organisation to implement. Hence, security policies are required for user awareness and acceptance as the policies provide responsible behaviour. All parties involved need to possess knowledge of the healthcare security policies, understand the level of protection as well as their responsibilities outlined in the policies [1].

There is a challenge that users may demand to access any system resources anytime, so such access needs control to enable secure and authorised access. For example, a clinician at the hospital can read and write to patient medical records assigned to in the hospital. It is therefore required for user awareness and acceptance of policies set in the hospital system that restrict or permit a clinician to read/write on patient medical records whether based on daily services provided, the level of services such as the case of an emergency, the ward assigned to and the location. Models of access control define the relationships between objects, subjects, operations and permissions [15]. Organisations object include the files, printers, software or hardware settings, the subject may be the end user, the process functions. Operations are the end user's processes of modifying, adding or deleting an object. To implement security system administrators, need to set up access control measures to protect organisations information through the Role-based access control (RBAC).

6.1   The Significance of Role-Based Access Control

A focus is on the Role-based Access Control model (RBAC) as the need for user awareness and acceptance continues to be central to access medical information systems in the healthcare sector securely. In this case, the user needs to be aware of the resources and level of permissions they should have to access the system, according to their roles in the hospital. The RBAC allows additional restrictions in the systems to allow and restrict users based on the roles they have been assigned [13]. Role-based access control is nondiscretionary; it uses the tasks or roles assigned to the subject to grant or deny access to objects [9].

The Role-based access control model is significant for the hospital environment due to its relevance and ability to mitigate the risks of insider threat, where user access resources restricted from unauthorised activities, in this case, a user is only authorised to the information he can access. Hence, the HUA framework created for user awareness and user acceptance of multimodal-access control for information security. The User awareness and user acceptance of multimodal-access control for information security.

6.2   User Awareness and User Acceptance of Multimodal-Access Control for Information Security

HUA framework provides a plan and procedure to implement access control security through user awareness and acceptance. A tool for realising the information security goal through appropriate access control and measure the success in managing human risk through awareness and acceptance [2]. A way to benchmark against how awareness education and training in another organisation are. Figure 4 below illustrates the user awareness and user acceptance of multimodal-access control for information security.

At the initial stage, Phase I, management needs to be involved, to ensure communication of the desired goal to all role-players. The second phase, Phase II ensures the Identification of the problem area, the Information security gaps and the access control gaps in the hospital environment. Phase III, the critical factor is to develop

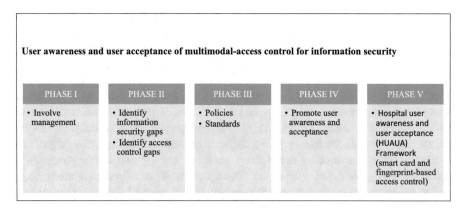

Fig. 4. User awareness and user acceptance of multimodal-access control for information security

Security policies on protecting the confidentiality, integrity and availability of information stored in medical information systems to ensure employees protected from being personally liable. We manage information security through information security standards such as ISO/IEC 27000, to help an organisation manage patient sensitive information and keep it secured. In Phase IV, now users need to take responsibility through compliance with security measures in place, user awareness through educational programmes and training mandatory in the hospital [6].

The programs ensure users are knowledgeable and understand the security risk to information and are equipped act appropriately when issues arise. The awareness programmes enhance the users' compliance levels and increase acceptance level before implementation of access control technology [20]. Phase V on the hospital user awareness and user acceptance (HUAUA) framework developed to mitigate the information security risk and threat in the hospital through multimodal access control technology (Smartcard and Fingerprint) A basis for sustainability through continuous learning. The HUA framework may improve deployment, incorporating intentions of the user to accept and use the technology voluntarily. Hospital employees need to understand the use of multimodal access control technology to access the medical information systems, and the importance of protecting sensitive information. By improving the security culture in the hospital through user awareness of technology, awareness sessions may be developed in a way to lower the perceived challenges and improve acceptance [18].

7 Recommendations

The current access control in the medical information systems security in the hospital addresses some issues in protecting patient information. When focusing on the problem of the study, it noted that the security of medical information systems needs to be improved through user awareness and acceptance so that the authorised employees have access to the information system as required. The improvement required in the hospital due to the confidential information it holds. The hospital needs to provide high-security access control to protect the privacy of electronic medical records against the information security risks and threat. A secure access control method required in healthcare so the doctors, nurses, records management employees, Information technology unit can render quality health care services to the citizens [22].

The implementation of a more secure access control method required so that the doctors and nurses can retrieve the correct medical record of a patient timely and give the patient the right diagnosis and treatment based on the medical record history. The method of authentication should unauthorised access that results in medical errors, where the need arises to prevent duplicate medical files and an incorrect retrieval of a patient medical record; it will enable the prevention of incorrect diagnosis and treatment, thus preventing medical risks [8]. Hence, we initiate the hospital user awareness and user acceptance framework with user identification through smartcard and fingerprint-based access control. The recommendation for future work includes a focus on the following areas: Issues of user awareness in the public health care, issues of user acceptance, secure access control issues, implementation and cost, hygiene, government policies, the standard and legislative framework. The work initiates the need for access control policies defined in the hospital to guide employees to practice secure access to medical information systems.

The study recommends that future research conducted to determine the required processes in developing user awareness and acceptance of technology before its implementation. An analysis of the requirements of each stakeholder required, here the stakeholders are Politicians (Union), top management, healthcare employees, service providers and patients. Noted from the limitations that the hospital's stakeholders play a huge role, in this case, the politicians (unions) and top management are always involved before any decision made. Raising awareness to this group of stakeholders (unions and top management) to understand the technology, the impact the technology will offer in the overall level of service delivery will ensure successful implementation of smart cards and fingerprint.

- Further research needed to analyse the best practices for adopting awareness sessions and training on a regular basis in the hospital.
- There is a need to develop the guidelines and policies that are legally binding by law within health care that govern access control to the medical information system. The requirement on implementing information the privacy law to protect the privacy of sensitive medical information [17].

Furthermore, the hospital needs to conduct a broader study on the technical issues, the infrastructure before adopting smart card and fingerprint-based access control method in the medical information systems security in the South African hospitals.

8 Conclusion

The use of Hospital User Awareness and User Acceptance Framework is applicable for user awareness and user acceptance in the hospital environment. The framework argued to be used for proper access control of the user in the medical information systems. The results show that the awareness level pertaining the appropriate access control in the medical information systems is low, the South African healthcare sector needs to appropriate awareness across all stakeholders to ensure knowledge of issues concerning confidentiality, the privacy and security of medical information. Is required of management to focus on the security of patient information while ensuring availability of such information to the authorised user when medical records are required.

The proposed model on user awareness and acceptance model for access control ensure that the issues regarding access control addressed timely and that all users targeted per the levels exposed to the use of medical information system. For access in the medical information systems, users' identification and verification done through smartcard and fingerprint-based access control to ensure the availability of patients' medical records securely from unauthorised access. The study proposes that the use of smartcard and fingerprint-based access control may enhance the security of medical records stored in medical information systems. A positive contribution will be to enforce the implementation of the plans outlined in the framework to ensure only authorised individuals are granted access to medical records. In the study a better way to restrict access to the medical information system such as the use of smartcard and fingerprint biometric systems to gain access to confidential patient information.

References

1. Abubaka, F.M., Ahmad, H.B.: Mediating role of technology awareness on social influence–behavioural intention relationship. Res. J. **2**(1), 119 (2014)
2. Becker, J., Knackstedt, R., Pöppelbuß, J.: Developing maturity models for IT management. Bus. Inf. Syst. Eng. **1**(3), 213–222 (2009)
3. Besnard, D., Arief, B.: Computer security impaired by legitimate users. Comput. Secur. **23**(3), 253–264 (2004)
4. Boyinbode, O., Toriola, G.: CloudeMR: a cloud-based electronic medical record system. Int. J. Hybrid Inf. Technol. **8**(4), 201–212 (2015)
5. Bryman, A.: Social Research Methods. Oxford University Press, Oxford (2015)
6. Caballero, A.: Security education, training, and awareness. In: Computer and Information Security Handbook, 3rd edn, pp. 497–505 (2017)
7. D'Arcy, J., Hovav, A., Galletta, D.: User awareness of security countermeasures and its impact on information systems misuse: a deterrence approach. Inf. Syst. Res. **20**(1), 79–98 (2009)
8. Drake, T., Kanu, A., Silverman, N.: Health care fraud. Am. Crim. Law Rev. **50**, 1131 (2013)
9. Ferraiolo, D., Kuhn, D.R., Chandramouli, R.: Role-Based Access Control. Artech House, Norwood (2003)
10. Glinz, M.: On non-functional requirements. In: 15th IEEE International Requirements Engineering Conference, RE 2007, pp. 21–26. IEEE, October 2007
11. Lafferty, L.: Medical identity theft: the future threat of health care fraud is now. J. Health Care Compliance **9**(1), 11–20 (2007)
12. Omotosho, A., Adegbola, O., Adelakin, B., Adelakun, A., Emuoyibofarhe, J.: Exploiting multimodal biometrics in e-privacy scheme for electronic health records. arXiv preprint arXiv:1502.01233 (2015)
13. Osborn, S.: Mandatory access control and role-based access control revisited. In: Proceedings of the Second ACM Workshop on Role-Based Access Control, pp. 31–40. ACM, November 1997
14. Peltier, T.R.: Information Security Risk Analysis. Auerbach Publications (2010)
15. Sandhu, R.S., Coyne, E.J., Feinstein, H.L., Youman, C.E.: Role-based access control models. Computer **29**(2), 38–47 (1996)
16. Smyth, R.: Exploring the usefulness of a conceptual framework as a research tool: a researcher's reflections. Issues Educ. Res. **14**(2), 167 (2004)
17. Solove, D.J., Schwartz, P.: Information Privacy Law. Wolters Kluwer Law & Business (2014)
18. Spears, J.L., Barki, H.: User participation in information systems security risk management. MIS Q. **34**(3), 503–522 (2010)
19. Terry, N.P., Francis, L.P.: Ensuring the privacy and confidentiality of electronic health records. Univ. Ill. Law Rev. 681 (2007)
20. Venkatesh, V., Morris, M.G., Davis, G.B., Davis, F.D.: User acceptance of information technology: toward a unified view. MIS Q. **27**(3), 425–478 (2003)
21. Von Solms, R., Van Niekerk, J.: From information security to cyber security. Comput. Secur. **38**, 97–102 (2013)
22. Ward, R., Stevens, C., Brentnall, P., Briddon, J.: The attitudes of health care staff to information technology: a comprehensive review of the research literature. Health Inf. Libr. J. **25**(2), 81–97 (2008)

Analysis of Medical Documents with Text Mining and Association Rule Mining

Ruth Reátegui[1,2(✉)] and Sylvie Ratté[1]

[1] École de technologie supérieure, Montreal, Canada
Sylvie.Ratte@etsmtl.ca
[2] Universidad Técnica Particular de Loja, Loja, Ecuador
rmreategui@utpl.edu.ec

Abstract. Text mining techniques extracts meaningful information from large amounts of semi-structured and unstructured texts. In this work, the MetaMap tool was used to extract medical entities like diseases and syndromes from discharge summaries. Also, association rule mining algorithms such as Apriori and FP-Growth were applied to the extracted entities in order to find associations between them. The dataset used consists of 1237 discharge summaries obtained from the 2008 i2b2 Obesity Challenge. The rules that have a principal diagnosis as antecedent showed that the cardiac disease frequently occurred with other diseases like hypertension and diabetes. Most of the rules describe associations between diabetes and other diseases like hypertension, dyslipidemia, nephropathy, heart disease, lung diseases, and arthritis. These rules have a confidence parameter of above 0.5.

Keywords: Clinical text · Text mining · Association rule mining

1 Introduction

Electronic Health Records (EHR) provide a vast amount of information related to a patient's health. Clinical texts contained within EHRs have either a narrative or an unstructured format with some characteristics different from the general text and use a variety of natural language [1]. These texts contain information including symptoms, disease, treatments [2] that could be useful to researches, health professionals, patients [3] and others.

The extraction and analysis of information in clinical texts allows researchers to predict an association between diseases, to observe a chronological progression of medical problems, to identify the treatments for a disease, to view lab results, and more.

Text mining extracts meaningful information from large amounts of semi-structured and unstructured text data [4]. This process uses many techniques to analyze, process and transform text into numerical data in order to convert it into structured data [5, 6]. Some of these techniques are statistical, machine learning, artificial intelligence and natural language processing (NLP) [4]. In the clinical domain text mining is used to extract clinical entities, identify relations between entities, segment the text in sentences or sections, or classify clinical concepts.

© Springer Nature Switzerland AG 2019
Á. Rocha et al. (Eds.): ICITS 2019, AISC 918, pp. 744–753, 2019.
https://doi.org/10.1007/978-3-030-11890-7_70

Moreover, association rule mining (ARM) is a data mining technique that aims to extract correlations, frequent patterns or casual structures inside of data repositories. ARM has been used to market basket analysis [7] and provides insight about which products tend to be purchased together and which are most amenable to promotion [8]. Therefore, these rules applied to clinical texts could show correlations among diseases, symptoms, and drugs. Also, these rules facilitate an analysis of any associations that might exist between two diseases or among three or more comorbidities [9].

In the clinical field, both text mining and ARM, have been used by some researchers. For example, [9] determined associations among diseases that accompany type-2 diabetes mellitus. In this work, the researchers applied ARM to clinical data. Furthermore, [8] with a aim to identify association between diagnoses, they applied association rules to ICD-9 diagnoses' codes obtained from 10,000 patients' records. Moreover, [10] used natural language technique to extract diseases, symptoms and drugs from 309 medical transcription files. Then, the researchers utilized ARM to discover a correlation among diseases, among diseases and symptoms, and among diseases and drugs. Also, [7] used frequencies of co-occurrence, Bayesian networks and ARM to found connection between comorbidities in hypertensive patients. They extract the information from text description in EHR.

Following the researches above mentioned, our work attempts to automatically extract diseases from discharge summaries and uses ARM algorithms to find associations between the diseases present in two summary sections: the past medical history and the principal diagnosis. Discharge summaries are clinical notes written by a health professional at the conclusion of a hospital stay or series of treatments [11]. Then, these notes provide relevant information about the patient's health. Some sections in these notes include: past medical history, social history, history of present illness, physical examination, principal diagnosis, and so on.

The data used was the i2b2 Obesity Challenge dataset. This dataset consists of 1237 discharge summaries from the Partners HealthCare Research Patient Data Repository whose patients are overweight or have diabetes.

2   Methodology

2.1   Dataset

In this work, the Informatics for Integrating Biology and the Bedside (i2b2) Obesity Challenge dataset was used. This consists of 1237 discharge summaries from the Partners HealthCare Research Patient Data Repository whose patients are overweight or diabetic. The summaries provide important information, such as principal diagnosis, history of present illness, pre-admission medications, past medical history, family history, social history, allergies, hospital course by problems, complications, physical examination on discharge, discharge medications, and follow-up appointments.

Furthermore, the dataset has two types of expert annotations. One of them is a textual annotation that experts made considering the explicitly-documented information in the discharge summaries: the other is an intuitive annotation that experts made according to their intuition and judgment [12].

The annotations consist of evaluating obesity and its comorbidities according to the following reasoning:

- Present: the patient has/had the disease.
- Absent: the patient does/did not have the disease.
- Questionable: the patient may have the disease.
- Unmentioned: the disease is not mentioned in the discharge summary.

The obesity comorbidities are: asthma, atherosclerotic cardiovascular disease (CAD), congestive heart failure (CHF), depression, diabetes mellitus (DM), gallstones/cholecystectomy, gastroesophageal reflux disease (GERD), gout, hypercholesterolemia, hypertension (HTN), hypertriglyceridemia, obstructive sleep apnea (OSA), osteoarthritis (OA), peripheral vascular disease (PVD), and venous insufficiency.

2.2 Tools

For this project, the following tools were considered:

- WEKA[1] is a collection of machine learning algorithms for data mining tasks. This software includes tools for data pre-processing, classification, clustering, association rules, regression, visualization. This was used in the data mining task.
- MetaMap is a program used to map biomedical concepts to the unified medical language system (UMLS) Metathesaurus [13]. Accordingly, UMLS is a large thesaurus that provides a representation of biomedical concepts [14]. UMLS represent the concepts with a unique and permanent concept identifier (CUI). MetaMap was used to extract and codify the disease and syndrome entities mentioned in the discharge summaries.
- The UMLS Terminology Service (UTS[2]) is a web service of the National Library of medicine (NLM). This browser was used to identify families of UMLS codes that correspond to disease and syndrome entities.

2.3 Association Rule Mining (ARM)

ARM is a technique of data mining that aims to extract correlations, frequent patterns or casual structures from data repositories [15]. The association rule algorithms were developed primarily to analyze transactional databases and market baskets analysis. The purpose of the analysis is to derive association rules that identify the items and co-occurrences of different items that appear frequently [5].

A formal way to define the association rules is:

Let $I = I1, I2, ..., Im$ be a set of m distinct attributes. An association rule is an implication in the form of $X \Rightarrow Y$, where $X, Y \subset I$ are sets of items called itemsets, and $X \cap Y = \varnothing$ [15]. X is the antecedent while Y is the consequent. A rule does not necessarily mean cause and effect, but it identifies simultaneous occurrences between

---

[1] http://www.cs.waikato.ac.nz/ml/weka/.

[2] https://uts.nlm.nih.gov/home.html.

items in antecedent X and consequent Y [9]. A rule X $\Rightarrow$ Y means X implies Y [10, 15] or if item X exists, item Y coexists [9].

To describe the strength in the relation between two or more terms, the following statistics are used [5]:

- Support: denotes the relative frequency with which an item or term occurs across the documents. This reflects how often a term or combination of terms co-occurs.
- Confidence: is a conditional probability. Given a rule such as: if A, then B, given all cases with A, the confidence is the proportion of those cases that also have B.

The thresholds of support and confidence are predefined by users. The two thresholds are called minimal support and minimal confidence respectively [15].

Two algorithms that can be used to determinate association rules include Apriori and FP-Growth. Apriori is more efficient in the candidate-generation process. This uses pruning techniques to avoid measuring certain terms and guaranteeing completeness [15]. The FP-Growth mines the complete set of frequent patterns by pattern-fragment growth [16].

In text mining, that is based on a document-term matrix, it is useful derive association rules by taking into account the presence of specific terms or phrases. The co-occurrences of specific terms may provide insights into the meaning of the text by identifying related themes [5].

2.4   Pretreatment

In this step, MetaMap was used to obtain the CUI of the disease or syndrome entities present in the discharge summaries. 211 summaries from the training dataset were used. These summaries have CUIs in the past-medical-history section and the principal-diagnosis section.

Next, the R[3] tool was used to create a document-term matrix with 211 rows and 495 columns. The CUIs are the features (columns) in the matrix. The columns have 342 CUIs in the past-medical-history entities and 153 CUIs in the principal-diagnosis entities. The cell of the matrix has only two values: 0 if a CUI does not exist in the document, and 1 otherwise.

Given the large number of CUIs, UTS was used to identify families of CUIs and allow for a reduction of the matrix. Using UMLS to aggregated CUIs has demonstrated to improve the result of the entity extraction process [2]. Therefore, in this work UTS was used to the identify families of CUIs. In this way, the new matrix has 293 columns (203 for past medical history and 90 for principal diagnosis). To reduce the matrix even further, we eliminate those CUIs that appeared less than five summaries. We considerate five summaries because it represents less than 3% of the total of analyzed summaries. The final matrix has 211 rows and 54 columns (38 for past medical history and 16 for principal diagnosis). The CUIs and the names of the diseases and syndromes are shown in Table 1.

---

[3] https://www.r-project.org/. R is free software for statistical computing and graphics. Also, this tool has a list of packages to work with text mining and natural language processing.

Table 1. CUIs and names of diseases and syndromes extracted with MetaMap.

Num	CUI	Disease or syndrome	Num	CUI	Disease or syndrome
1	dc0002871	Anemia	28	c0030920	Peptic ulcer
2	dc0010068	Coronary heart disease	29	c0149871	Deep vein thrombosis
3	dc0009450	Infection, NOS (Communicable diseases)	30	c0232197	Fibrillation
4	dc0018801	Heart failure	31	c0242339	Dyslipidaemia (Dyslipidemias)
5	dc0020538	Hypertensive disease	32	c0442874	Neuropathy
6	dc0232197	Fibrillation	33	c0038454	Cerebrovascular accident
7	dc0243026	Sepsis	34	c0917805	TIA (Transient Cerebral Ischemia)
8	dc0878544	Cardiomyopathies	35	c1260873	Aortic valve disorder
9	dc0003864	Arthritis	36	c3714636	Pneumonitis
10	dc0024115	Lung diseases	37	c0019163	Hepatitis B
11	dc0011847	Diabetes	38	c0020676	Hypothyroidism
12	dc3714636	Pneumonitis	39	c0039483	Polymyalgia rheumatica (giant cell arteritis)
13	dc0039082	Syndrome	40	c0085096	Peripheral vascular diseases
14	dc0022658	Nephropathy (Kidney diseases)	41	c0040961	Regurgitation, tricuspid (tricuspid valve insufficiency)
15	dc0038454	Cerebrovascular accident	42	c0149721	Left ventricular hypertrophy
16	dc0085096	Peripheral vascular diseases	43	c0152107	Postmyocardial infarction syndrome
17	c0002871	Anemia	44	c0235480	Paroxysmal atrial fibrillation
18	c0240066	Iron deficiency	45	c1847014	Pulmonary disease, chronic obstructive, severe early-onset
19	c0003864	Arthritis	46	c3714496	COPD (Chronic obstructive pulmonary disease of horses)
20	c0878544	Cardiomyopathies	47	c0017168	Gastroesophageal reflux disease
21	c0010068	Coronary heart disease	48	c0012634	Disease
22	c0011847	Diabetes	49	c0008679	Chronic disease
23	c0017152	Gastritis	50	c0028754	Obesity
24	c0015397	Disorder of eye	51	c0009450	Infection, NOS (Communicable Diseases)
25	c0018801	Heart failure	52	c0039082	Syndrome
26	c0020538	Hypertensive disease	53	c0022658	Nephropathy (Kidney Diseases)
27	c0024115	Lung diseases	54	c0026265	Diseases of mitral valve

Letter d at the beginning of the CUIs indicates it was extracted from principal diagnosis section

2.5   Modeling

Two main experiments were performed in Weka to find association with the entities extracted from the sections: past medical history and the principal diagnosis. The Apriori and FP-Growth algorithms were used. Both algorithms were configured to obtain 100 rules with a confidence parameter of 0.5. The value of this parameter was selected in order to explore a greater number of rules and associations. As we can see in the Table 2, the rules obtained are sorting from greatest to least according the confidence value.

3   Results and Discussion

The maximum number of rules obtained was 38 for the Apriori and 37 for the FP-Growth algorithms. The results are similar using both algorithms. Table 2 shows results from the Apriori algorithm, which include one more rule than the FP-Growth. As we mentioned above, the dataset used has information about patients that could have obesity, diabetes, or other comorbidities related to obesity.

The results obtained showed associations between diseases that are clearly explained in the medical literature. Diabetes is the disease present in most of the rules. This disease is a metabolic disorder affecting people worldwide. The type-2 DM is the most common form of diabetes, and it is characterized by insulin resistance [17].

The rule "Heart failure, hypertensive disease → Diabetes" had the highest confidence value (0.9). Also, the rules show that heart disease frequently occurred with other diseases like hypertension and diabetes. Coronary heart disease or coronary artery disease (CAD) is a major determinant of the long-term prognosis among patients with diabetes mellitus (DM) [18]. Diabetes is related with hypertension, coronary heart disease, heart failure, and left ventricular hypertrophy, which influences to atrial fibrillation [19]. Patients with diabetes have a higher risk of cardiovascular diseases [20], and DM increases mortality after an MI and especially with a CAD prognostic [18].

Some of our rules describe the association between diabetes and other diseases like hypertension, dyslipideamia, nephropathy and cardiac disease. Adult patients with type-2 DM present a long list of comorbidities such as hypertension, obesity, hyperlipidemia, CAD, heart failure, chronic kidney diseases (CKD), chronic obstructive pulmonary diseases, arthritis, and neuropathy [21]. These diseases are clearly represented in the rules found in our work.

Furthermore, hypertension and diabetes present overlapping complications. Also, hypertension and type-2 DM share other risk factors including obesity, physical inactivity, insulin resistance [22]. Hypertension and DM have a significant impact on macrovascular (e.g. CAD, MI, stroke, peripheral vascular diseases) and microvascular disorders (e.g. retinopathy, nephropathy, neuropathy) [17, 20]. Also, familiar predisposition to diabetes and hypertension seems to be polygenic in origin [20]. In the work of [22] blood pressure and blood glucose values showed correlations across a wide spectrum from normal to elevated levels. The authors worked in a cross-sectional data and biracial cohort.

Table 2. Apriori results.

Num	Apriori rules	Conf	Lift
1	Heart Failure, Hypertensive disease ==> Diabetes	0.9	1.41
2	Heart Failure ==> Diabetes	0.89	1.41
3	Dyslipidaemia ==> Hypertensive disease	0.86	1.44
4	Diabetes, Dyslipidaemia ==> Hypertensive disease	0.86	1.42
5	Dyslipidaemia ==> Diabetes	0.83	1.31
6	Coronary heart disease, Hypertensive disease ==> Diabetes	0.83	1.3
7	Hypertensive disease, Dyslipidaemia ==> Diabetes	0.82	1.3
8	Hypertensive disease, Nephropathy ==> Diabetes	0.81	1.27
9	Nephropathy ==> Diabetes	0.8	1.27
10	Diabetes, Syndrome ==> Hypertensive disease	0.79	1.31
11	Diabetes, Coronary heart disease* ==> Hypertensive disease	0.78	1.3
12	Coronary heart disease ==> Diabetes	0.77	1.22
13	Diabetes, Heart Failure ==> Hypertensive disease	0.76	1.27
14	Heart Failure ==> Hypertensive disease	0.76	1.27
15	Hypertensive disease, Syndrome ==> Diabetes	0.76	1.19
16	Hypertensive disease ==> Diabetes	0.75	1.18
17	Hypertensive disease, Coronary heart disease* ==> Diabetes	0.74	1.16
18	Syndrome ==> Hypertensive disease	0.73	1.2
19	Dyslipidaemia ==> Diabetes, Hypertensive disease	0.71	1.58
20	Diabetes ==> Hypertensive disease	0.71	1.18
21	Coronary heart disease* ==> Hypertensive disease	0.71	1.18
22	Diabetes, Nephropathy ==> Hypertensive disease	0.71	1.18
23	Nephropathy ==> Hypertensive disease	0.71	1.17
24	Syndrome ==> Diabetes	0.7	1.1
25	Heart Failure ==> Diabetes, Hypertensive disease	0.68	1.52
26	Obesity ==> Hypertensive disease	0.68	1.13
27	Cardiomyopathies* ==> Diabetes	0.68	1.06
28	Coronary heart disease* ==> Diabetes	0.67	1.05
29	Coronary heart disease, Diabetes ==> Hypertensive disease	0.66	1.1
30	Lung disease ==> Diabetes	0.64	1.01
31	Arthritis ==> Diabetes	0.64	1.01
32	Coronary heart disease ==> Hypertensive disease	0.61	1.02
33	Nephropathy ==> Diabetes, Hypertensive disease	0.57	1.26
34	Lung disease ==> Hypertensive disease	0.56	0.92
35	Syndrome ==> Diabetes, Hypertensive disease	0.55	1.22
36	Arthritis ==> Hypertensive disease	0.53	0.88
37	Coronary heart disease* ==> Diabetes, Hypertensive disease	0.52	1.16
38	Coronary heart disease ==> Diabetes, Hypertensive disease	0.51	1.13

Diseases with * were found in the principal diagnosis section

Diabetic nephropathy (DN) is a common complication in diabetic patients, and complications such as elevated glucose levels, obesity, high blood pressure, dyslipidemia, etc., increases the development and progression of DN [20, 23]. DN often leads to end-stage renal diseases, and DN patients have very high cardiovascular risk compared with patients with coronary heart diseases [24].

Dyslipidaemia is a preferable term to hyperlipidaemia, this abnormality, together with elevated triglycerides, are features of the metabolic syndrome [25]. Insulin resistance and hyperglycemia are associated with diabetes dyslipidemia pathogenesis [17]. A positive-associated factor of diabetes includes hyperlipidemia and hypertension [26].

4 Conclusion

Clinical notes contain significant amounts of information that can be explored by physicians and researchers. MetaMap was used to extract diseases and syndromes mentioned in the past-medical-history and principal-diagnosis sections of the discharge summaries.

In textual data, association rules are used to identify correlations among terms [10]. Association rules used in clinical texts reveal correlations between medical entities like diseases and syndromes.

As we can see, some of the rules extracted in this work show comorbidities related to diabetes and obesity disease. The association rules that have a diagnosis as antecedent show that heart disease frequently occurred with other diseases like hypertension and diabetes. Most of the rules describe associations between the diabetes and other diseases like hypertension, dyslipideamia, associations between hypertension diseases and dyslipideamia, heart diseases, and diabetes. These confirm that hypertension and diabetes present overlapping complications [20].

As was mentioned in [8], the extracted association rules, not only confirm what is already known from medical literature, but also generate some interesting associations that can be explored and explained by medical researchers. Similarly to [9], our work found that hypertension has an important role in the association between diabetes and some of its comorbidities such as dyslipidaemia, coronary heart disease and nephropathy.

We used the UMLS codes (CUI) as features that represent the disease and syndrome mentioned in the past-medical-history and the principal-diagnosis sections. For future works, information such as lifestyle, physical activity, dietary habits as well as symptoms and treatments could be considered. Furthermore, other data mining techniques could be used in order to identify groups of patients with similar characteristics, tailored treatment for a disease, or a new classification for a disease.

References

1. Chiaramello, E., Paglialonga, A., Pinciroli, F., Tognola, G.: Attempting to use MetaMap in clinical practice: a feasibility study on the identification of medical concepts from italian clinical notes. Stud. Health Technol. Inform. **228**, 28–32 (2016)
2. Reategui, R., Ratte, S.: Comparison of MetaMap and cTAKES for entity extraction in clinical notes. BMC Med. Inform. Decis. Mak. **18**, 74 (2018)

3. Pradhan, S., Elhadad, N., South, B.R., Martinez, D., Christensen, L., Vogel, A., Suominen, H., Chapman, W.W., Savova, G.: Evaluating the state of the art in disorder recognition and normalization of the clinical narrative. J. Am. Med. Inf. Assoc.: JAMIA **22**, 143–154 (2015)
4. Sun, W., Cai, Z., Li, Y., Liu, F., Fang, S., Wang, G.: Data processing and text mining technologies on electronic medical records: a review. J. Healthc. Eng. **2018** (2018). 4302425
5. Miner, G., Delen, D., Elder, J., Fast, A., Hill, T., Nisbet, R.A.: Practical Text Mining and Statistical Analysis for Non-structured Text Data Applications. Elsevier Inc., New York (2012)
6. Weiss, S., Indurkhya, N., Zhang, T., Damerau, F.: Text Mining Predictive Methods (2005)
7. Bukhanov, N., Balakhontceva, M., Krikunov, A., Sabirov, A., Semakova, A., Zvartau, N., Konradi, A.: Clustering of comorbidities based on conditional probabilities of diseases in hypertensive patients. Procedia Comput. Sci. **108**, 2478–2487 (2017)
8. Kang'ethe, S., Wagacha, P.: Extracting Diagnosis Patterns in Electronic Medical Records using Association Rule Mining (2014)
9. Kim, H.S., Shin, A.M., Kim, M.K., Kim, Y.N.: Comorbidity study on type 2 diabetes mellitus using data mining. Korean J. Internal Med. **27**, 197–202 (2012)
10. Lakshmi, K.S., Vadivu, G.: Extracting association rules from medical health records using multi-criteria decision analysis. Procedia Comput. Sci. **115**, 290–295 (2017)
11. Raghavan, P.: Medical Event Timeline Generation from Clinical Narratives. Doctor of Philosophy, The Ohio State University (2014)
12. Uzuner, Ö.: Recognizing obesity and comorbidities in sparse data. JAMIA **16**, 561–570 (2009)
13. Aronson, A.R., Lang, F.-M.: An overview of MetaMap: historical perspective and recent advances. JAMIA **17**, 229–236 (2010)
14. Aronso, A.: Effective mapping of biomedical text to the UMLS metathesaurus: the MetaMap program. In: AMIA Annual Symposium Proceedings 2001, pp. 17–21 (2001)
15. Kotsiantis, S., Kanellopoulos, D.: Association rules mining: a recent overview. GESTS Int. Trans. Comput. Sci. Eng. **32**, 71–82 (2006)
16. Han, J.W., Pei, J., Yin, Y.W.: Mining frequent patterns without candidate generation. SIGMOD Rec. **29**, 1–12 (2000)
17. Kavakiotis, I., Tsave, O., Salifoglou, A., Maglaveras, N., Vlahavas, I., Chouvarda, I.: Machine learning and data mining methods in diabetes research. Comput. Struct. Biotechnol. J. **15**, 104–116 (2017)
18. Aronson, D., Edelman, E.R.: Coronary artery disease and diabetes mellitus. Cardiol. Clin. **32**, 439–455 (2014)
19. Aune, D., Feng, T., Schlesinger, S., Janszky, I., Norat, T., Riboli, E.: Diabetes mellitus, blood glucose and the risk of atrial fibrillation: a systematic review and meta-analysis of cohort studies. J. Diabetes Complications **32**, 501–511 (2018)
20. Long, A.N., Dagogo-Jack, S.: Comorbidities of diabetes and hypertension: mechanisms and approach to target organ protection. J. Clin. Hypertens. (Greenwich) **13**, 244–251 (2011)
21. Lin, P.J., Kent, D.M., Winn, A., Cohen, J.T., Neumann, P.J.: Multiple chronic conditions in type 2 diabetes mellitus: prevalence and consequences. Am. J. Manag. Care **21**, e23–e34 (2015)
22. Edeoga, C., Owei, I., Siwakoti, K., Umekwe, N., Ceesay, F., Wan, J., Dagogo-Jack, S.: Relationships between blood pressure and blood glucose among offspring of parents with type 2 diabetes: prediction of incident dysglycemia in a biracial cohort. J. Diabetes Complications **31**, 1580–1586 (2017)
23. Wang, Y.-Z., Xu, W.-W., Zhu, D.-Y., Zhang, N., Wang, Y.-L., Ding, M., Xie, X.-M., Sun, L.-L., Wang, X.-X.: Specific expression network analysis of diabetic nephropathy kidney tissue revealed key methylated sites. J. Cell. Physiol. **233**, 7139–7147 (2018)

24. Tziomalos, K., Athyros, V.G.: Diabetic nephropathy: new risk factors and improvements in diagnosis. Rev. Diabet. Stud. **12**, 110–118 (2015)

25. Thompson, G.R.: Management of dyslipidaemia. Heart **90**, 949–955 (2004)

26. Anderson, A.E., Kerr, W.T., Thames, A., Li, T., Xiao, J.Y., Cohen, M.S.: Electronic health record phenotyping improves detection and screening of type 2 diabetes in the general United States population: a cross-sectional, unselected, retrospective study. J. Biomed. Inform. **60**, 162–168 (2016)

Customized Walk Paths for the Elderly

João Amaral[1,2], Mário Rodrigues[1,3(✉)], Luis Jorge Gonçalves[1],
and Cláudio Teixeira[2,3]

[1] ESTGA, University of Aveiro, Aveiro, Portugal
mjfr@ua.pt
[2] DETI, University of Aveiro, Aveiro, Portugal
[3] IEETA - Institute of Electronics and Informatics Engineering of Aveiro,
Aveiro, Portugal

Abstract. Nowadays a large number of applications rely on appropriate path planning stressing the importance of such algorithms. Usually the path planning can be adjusted by the transportation type (walking, cycling, driving) and special restrictions as roads blockages. However these adjustments are not enough for other tasks such as planning a walk path of a given size for the elderly, a tourist office application for planning a journey based on desired points of interest to visit, among other.

A project named Smartwalk dedicated to helping the elderly in taking an appropriate amount of steps each day, for improving their autonomy, independence and functional capacity, needs to take into account several factors in the path planning. Some are related to avoiding problematic areas such as stairs or slippery streets, also, to increase the chance of long term adoption, it is important that people exercise in familiar environments. So, the objective is to allow plan routes through places where people live, including points of interest such as gardens and touristic places, and avoiding areas more difficult or less desirable for people (e.g. streets with stairs).

In this paper we describe how the project is being deployed in Águeda - Portugal, a city with less than 50 k inhabitants and featuring a data communication infrastructure in its urban area. A demonstration of the algorithm working and some performance tests are presented and discussed.

Keywords: Customized walk paths · Elderly activity · Smart city

1 Introduction

Taking into account that the elderly population in Portugal, and worldwide, is increasing, new tools are necessary in order to provide conditions for that same population to maintain an adequate level of physical activity. Practice physical

Research funded by CENTRO-01-0145-FEDER-024293, project SmartWalk: Smart Cities for Active Seniors.

Á. Rocha et al. (Eds.): ICITS 2019, AISC 918, pp. 754–763, 2019.
https://doi.org/10.1007/978-3-030-11890-7_71

exercise regularly have several benefits for the elderly and their health, especially if the physical exercise is appropriate for their personal physical condition. This work is being developed as a complementary part of the Smartwalk project, in which the physical condition of the elderly are monitored and where a health professional suggests walk paths in the same area people are already familiar with, for increasing the chances of its adoption, and that are based on the elderly condition, encouraging in this way a more active life for the elderly people.

Stock path planning algorithms are not adequate for this purpose as they do not take into account features, whether positive or negative, that are relevant for the elderly. Some are related to avoiding problematic areas such as stairs or slippery streets, other are about places where people usually stroll, including points of interest such as gardens and touristic places, and avoiding areas more difficult or less desirable for people (e.g. streets with stairs). After conducting a research on path planning a solution has been devised and developed, that aims to create the best possible routes taking into account a customized cost function based of several parameters identified by health professionals that work with the elderly. For the moment, the suggested routes are inspected and selected by someone responsible for the path planning.

This article is divided into 5 sections. The next section presents the information about previous work on path planning in various contexts containing their respective analysis and solutions/implementations. Section 3 is where the main problem is discussed including the algorithms studied. The last two sections contain the results of the tests carried out and the conclusions, respectively.

2   Background and Related Work

The importance of having the best path plan led to the proposal of several algorithms, with impacts on efficiency and on travel cost reduction whether in terms of time or resources saved. Here will be mentioned the ones more similar to the present proposal.

The Tourist Trip Design Problem (TTDP) [7] is a recurring problem for when visiting a city with the intend to visit as many tourist points or points with special interest as possible. The problem discussed in [7] is based on the question of how to plan the same route taking into account the preferences of the tourists in terms of the order of points of interest (POI) and other limits imposed by them. The article classified the different variants of TTDP as single tour and multiple tour. The difference between these two is that the single tour variant maximizes the profit of the trip while respecting the conditions of the tourists and attributes of the POIS and can be modeled using the Traveling Salesman Problem with Profits (TSPP) and The Orienteering Problem [18]. While the multiple tour in his turn looks for the best combination of tours based on the restriction of the duration of the visit in terms of days, and can thus use the TSPP extension called Vehicle Routing Problem with Profits (VRPP) [3].

In the article [9] it is indicated that when using the Time Dependent Team Orienteering Problem with Time Windows (TDTOPTW) it can solve the path planning problem of the tourists when using the public transportation of the cities. TDTOPTW extends of the Team Orienteering Problem With Time Windows (TOPTW) [19] which by itself is a extension of the Orienteering Problem (OP) [18]. While there is another extension to the OP that considers multiple routes called Team orienteering problem (TOP) [10] TOPTW considers that it is necessary to visit the maximum possible places within a certain time. Returning to TDTOPTW in which we need to take into account the temporal restriction between the nodes when using the public transports. It is necessary a directed graph in which it has to have a set of POIS to visit, the number of days of travel and the time that the person wants to spend at every location. In the article was indicated the importance of the time that the person leaves the current place to proceed to the next, taking into consideration the type of transport (taxi, bus or even walking). The strategy presented in [9] used to solve this problem is the use of algorithms based on the CSCRoutes presented in [8].

In article [17], it was studied a planning problem in a weighted directed graph in which they looked for the least expensive path from point A to point B passing through a group of non-repeating predefined nodes. What makes it different from the other general problems of path planning is that the generated path has to pass through a group of intermediate nodes defined and end up in the last node whose node is not fixed. This makes path finding more complicated and complex since each intermediate node would be a temporary target in the search process.

As far as the heuristic research of the algorithm A* and the link-stat algorithm for routing computation served as inspiration to solve this problem. The article explains how the OSPF routing computation protocol works. Also explains that each router uses the Dijkstra algorithm, with the objective to calculate the shortest path in the tree between the router in question as a root to all other routers in the area inside of the link-state database (LSDB), using also the algorithm A*.

In the article [4] the path planning problem was resolved with a improved Ant Colony Optimization algorithm with custom parameters that were obtain the by studying the paths that were most used by the use of a road performance personalized quantization algorithm evaluation and also studying the habits and characteristics of the user by using a log fuzzy C-means clustering algorithm (LFCM) witch uses the log data of the trajectories taken.

However, currently on the market there are both web and mobile applications that solve the problem finding path for several contexts such the following applications: mtrip [15], JiTT Coimbra [11], Voyager [16], and GPS Map Route Planner [2]. During this research, we are faced with the fact that there is a lack of tools (web or mobile) that use some of the parameters that are relevant for the elderly when it comes to road planning based on their physical condition and preferences.

3 Custom Walk Paths

In this work an application programming interface (API) was planned and developed to be used in a wide range of path planning applications such as path planning with the focus on the elderly population, a touristic/leisure walk path with point of interest (POI) to be used in a wine route planner, for instance, or even a peddy-paper. It can also be used by a tourism agency which can plan bicycle rides, or walking paths. The API allows to manipulate the cost function of the planning algorithm using multi-parameters which will influence the decision about the optimal path.

The objective of this work is to find the path with the minimum cost which intersects all selected POI, avoiding obstacles such as blocked roads due to maintenance and also avoiding areas marked as undesirable. In order to develop the route customization component in accordance with the user's limitations and preferences, it is necessary to create a function that will customize the default cost along the route so that the most appropriate routes can be selected. This function works by appending specific and customized cost to the adequate streets according to selected parameters. For instance, is possible to increase the cost of a street with stairs to prevent including that street in a path for an elderly person who has difficulty climbing stairs. Even if the deviation takes longer than the original route, the easier route for the person in question is the most important aspect. Table 1 shows the parameter set thought to date for the developed cost function. The current cost (CC) is the base value which will be manipulated by the values presented in the third column of the table, and represent the elderly preferences. In the category named security level there are multiple variables to calculate the cost since there are multi types of road since a given road can allow cars but is a pedestrian route while another is a cicleway and a pedestrian route and so on. Every road have a combination of users that must take into account on road type basis.

The locations of the POIS will also determinate the path generated, even considering the elderly users, if they choose pass by that point. However, selecting POIs will not affect the parameters of the cost function as they are only determined by the choices that are relative to the profile of the people who will use a given path.

The first challenge addressed was the need to create a method that, based on a certain number of geographic points, returns an optimal path (or as close as possible) that passes once and only once for all POIs and that is closed, meaning that the end point coincides with the start point. The problem of Traveling Salesman [13] is one of the classic problems in which the objective is to find the shortest route with the condition that the seller has to pass through all the cities without repetition at least once when the last city is the where did it start from.

Two well known algorithms for the Traveling Salesmen Problem (TSP) were studied: the Ant Colony Optimization and the Simulated Annealing Other algorithms that also can be used for solving TSP but need some adaptation or supporting heuristics were not considered as our goal if to study how to combine parameters of distinct nature (from people, from city, and other) in the

Table 1. Table of parameters for the cost function.

Category	Parameter	Custom Cost Values
Pavement type	– Tarmac pavement, – Dirt pavement, – Cicleway, – Limestone/Concrete pavement, – Wood/Metal pavement	– CC * 1, – CC * 1.1 – CC * 1.2 – CC * 1.3 – CC * 1.4
Sidewalk width		– Max width: 2 m, – Min width: 0.5 m, – if $0.5 < MW < 2$: CC * 1.2
City wifi network access	– Connection with the hot-spot network	– CC / 1.01
Security level	– Allow cars, – Allow bicycles	– Multi variable – Multi variable
Path inclination	– Descent, – Ascent, – Steps, (Number of steps, height of the steps with or without handrail)	– CC / 1.02 – CC * 1.25 – CC / 1.03
Route status	– Inundation, – Wet floor, – Dry floor, – Lack of maintenance, – Leveling the path, – Roads in maintenance	– CC * 10, – CC * 2 – CC * 1 – CC * 1.5 – CC * 1.2 – CC * 2
Cover	– Provides shade, – Protects from rain	– CC / 1.02 – CC / 1.02
Route location	– Incorporated in a park, – Existence of POI for the individual	– CC / 1.03 – CC / 1.20

cost function, and not so much in having the most optimized algorithm for now. In this section will be first explained the two algorithms considered, each in a dedicated subsection, followed by our decision and respective justification.

3.1 Ant Colony Optimization

Ant Colony Optimization (ACO) is a probabilistic technique to solve several static or dynamic computational problems that includes finding acceptable routes through a graph [12]. The algorithm was developed based on the behavior of the ant during its search for food [6]. The ant is a blind insect which has led it to develop natural methods of navigating. Basically when the ants start looking for food, they drop a pheromone that modifies the environment allowing the communication between them and the colony, also serving as a memory of the return path to the starting point causing the ants to choose the path with a greater pheromone concentration.

The behavior of the ants in the choice of path is explained in 3 moments. The moment when the ants leave the colony, the detection of a obstacle witch leads the formulation of the best solution to circumvent the obstacle in the more efficient way possible and return to the colony.

The images and mathematical formulas that explain the probabilistic methods that involve the choice of paths by the ants, as well as the calculations that determine the evaporation of the pheromone levels are explained and represented in [12].

3.2 Simulating Annealing

Simulating Annealing [12] is an optimization technique based on annealing. Annealing by himself, it is a heat treatment that certain materials which upon being subjected to an increase and decrease of temperatures from a gradual way until their reach their thermal equilibrium, they become more solid and consistent.

Annealing is an algorithmic simulation of this physical phenomenon. By using a probabilistic meta-heuristic, we can find a good solution to global optimization problems such as Traveling salesman, Task Assignment and Scheduling, Circuit Layout and others. At high temperatures the search process verifies the whole panorama but as the temperature decreases the space in search decreases to the area of the current result.

Like as in Ant colony Algorithm, in [12] we can verify the flowchart of the algorithm as well as an explicit explanation and representation of the mathematical formulas and a great analysis for the presented algorithm.

3.3 Our Approach

These two algorithms were compared in [5] with 3 distinct setups, and each setup was ran 5 times. They conclude that the algorithm Ant Colony is consistently faster than the Annealing. However the annealing algorithm is the default algorithm in pgRouting, the software used by the first project were this path planning API will be used. Thus some tests were conducted in order to assess if the performance of the Simulated Annealing algorithms witch results were classified as adequate to that project requirements (reported in Sect. 4).

To generate the path needed, the user should click in the area of the map were the path should begin and end. After the click the geographical coordinates are gathered so the system can calculate and select the nearest internal graph vertex from that position. Figure 1 depicts the actions triggered by the users' actions. The map is renderer to support user interactions and the path planning is done using the graph system that is generated using OpenStreetMap's data and stored in the database. The database have the information about the points of interest (table pois), custom cost for each user using dynamic temporary tables and personal preferences (table utilizador) which will influence the cost between each vertex in the graph.

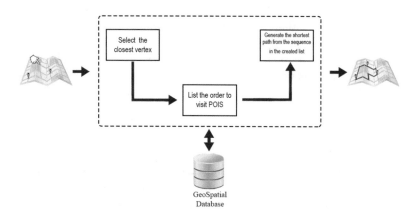

Fig. 1. Actions triggered by users' clicks.

After knowing the starting location, it is necessary to determine the sequence of POIS to visit and create a list where the chosen location must be the first and last position. This sequence is the answer of the Traveling Salesman Problem where the last position is the same as the first and where every POIS must be visited at least once. The heuristic to determinate the POIS sequence is the euclidean distance between the nodes. The solution may be not be the best and optimal but is guaranteed after a number of iterations.

Once with the list completed with the right order of visit, we calculate the shortest path between each position using the Dijkstra algorithm. However there are cases that the shortest path between POIS can use a road that were used previously, this situation must be tackle with care to minimize but not avoid it completely. There are situations that a POI is in a dead-end street, in this case it is necessary to use a already taken road till it reaches a new unused road. This condition can be achieved by increasing the cost of every road used dynamically. After this calculation the result is returned as a geoJSON format suitable to be rendered as a path in the map.

As the map uses the data acquired through the service osm2po that gave us real information about the streets of the city of Águeda. Osm2po is a free project that allows to generate SQL scripts from the OpenStreetMap database repository. It is possible to get various informations like the roads that are exclusive for cars, bicycle or only the pedestrian paths of the pretentious city area or for whole country if necessary [14]. The data gathered using this tool includes the maximum speed in each edge of the graph, the length from point A or B, the cost from the direction A to B and the reverse cost that is from B to A because there are roads with one way (in this case the reverse cost value must have a bigger value than the cost). This informations is compatible with databases that have the PostGis and pgRouting extension installed and can be used with Open Street Maps that is a editable map of the world which has been built by volunteers from scratch [1].

4   Tests and Results

The path is personalized to accommodate the needs of the user, the POIs that should be visited, and the areas that are to avoid in the path generated. Thus tests were specifically designed to assess if the algorithm behavior was adequate in the presence of blocked streets or areas to avoid. Figures 2 and 3 depict these cases, respectively, by presenting at the left-hand side the path planned with POIs and without restrictions, and at the right-hand side the path planned when those restrictions are introduced. To insert or remove an existing restriction the user needs to right click on the map to open a dialog box (see left-hand side of Fig. 3), select the desired option, and then click on the map for choosing the point where the restriction should be added or removed. Blocked roads and areas to avoid are implemented by multiplying the cost of the respective segment by 5.

In Fig. 2 is visible that when two road blocks are inserted in segments that previously were part of the optimal path, the algorithm selects a new path without using those roads. If a point is marked as unwanted POI, as the one marked by the black hexagon at the lower right side, then the selected path will avoid that point as depicted in Fig. 3.

In addition of the tests cases presented in the figures there were created multiple paths with different sources and POIS to validate the algorithm in terms of execution time.

4.1   Performance Tests

Along of the previous results presented, the performance was measured since the path planning algorithm is been prepared to be integrated in concrete projects such as SmartWalk.

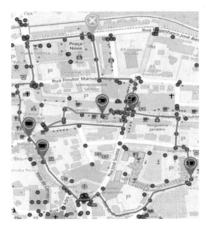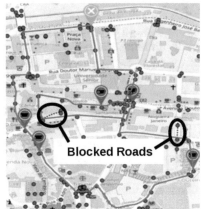

Fig. 2. Test #1: 5 POIs selected and 2 road blocks.

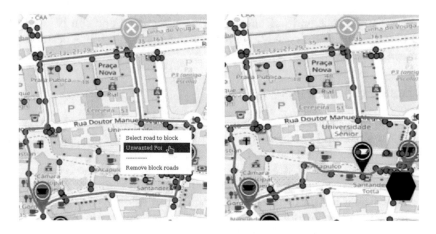

Fig. 3. Test #2: 2 POIS selected and 1 unwanted POI.

The worst case scenario was considered on the most up-to-date information on city of Águeda - Portugal where were considered that the path traverses the city graph, that is composed by approximately 140 nodes.

We used pgRouting default values for all parameters of the algorithm such as the initial temperature and the cooling factor. Considering 10 distinct starting points, running for each the algorithm 5 times on the SmartWalk server, a computer featuring an Intel® Xeon® Processor E5-2440 running at 2.40 GHz, 2.0 GB of RAM, and SCSI HDD running at 10 k RPM, the worst case scenario takes approximately 7 s to find the sub-optimal sequence of POIS to visit for each case. This time interval is considered adequate since there will only used a small fraction of those 140 points.

5   Conclusion

The path planning problem and software solution takes into account the preferences and profile of the people who will use the paths so the path generated reflex the personification for each individual. It was created a special case for the elderly population as the elderly person have specific problems and requirements that will influence the path, for example the use of familiar locations as landmarks for their paths. Nevertheless, the solution presented here can be applied in more contexts that exemplified previously.

With the tests conducted it was clear that is relatively easy to dynamically generate the path having into account the requirements that imply the planning of walk paths while avoiding obstacles which will increase the difficulty or even blocking problematic areas. The preferred path which is generated gives an overall less costly alternative taking into account the user specific profile.

The utilization of tools such as Postgis and pgRouting allowed the use of the algorithm Simulated Annealing along with the data gather from the OpenStreetMap project. With all this tools it was possible to create near real scenarios with real data that gave us the possibility to test the API and the algorithm developed altogether in a situation were the user could not use certain roads, exclude some areas, and avoiding unwanted point of interest, giving alternatives to all of the above situations.

References

1. OpenStreetMap. https://www.openstreetmap.org. Accessed 04 July 2018
2. Appsomania: GPS Map Route Planner. https://play.google.com/store/apps/details?id=com.gps.route.planner.map. Accessed 11 June 2018
3. Archetti, C., Hertz, A., Speranza, M.G.: Metaheuristics for the team orienteering problem. J. Heuristics **13**(1), 49–76 (2007)
4. Chen, P., Zhang, X., Chen, X., Liu, M.: Path planning strategy for vehicle navigation based on user habits. Appl. Sci. **8**(3) (2018). https://doi.org/10.3390/app8030407, http://www.mdpi.com/2076-3417/8/3/407
5. Chmait, N., Challita, K.: Using simulated annealing and ant-colony optimization algorithms to solve the scheduling problem. Comput. Sci. Inf. Technol. **1**(3), 208–224 (2013)
6. Dorigo, M., Gambardella, L.M.: Ant colonies for the travelling salesman problem. Biosystems **43**(2), 73–81 (1997)
7. Gavalas, D., Konstantopoulos, C., Mastakas, K., Pantziou, G.: A survey on algorithmic approaches for solving tourist trip design problems. J. Heuristics **20**(3), 291–328 (2014)
8. Gavalas, D., Konstantopoulos, C., Mastakas, K., Pantziou, G., Tasoulas, Y.: Cluster-based heuristics for the team orienteering problem with time windows. In: International Symposium on Experimental Algorithms, pp. 390–401. Springer (2013)
9. Gavalas, D., Konstantopoulos, C., Mastakas, K., Pantziou, G., Vathis, N.: Heuristics for the time dependent team orienteering problem: application to tourist route planning. Comput. Oper. Res. **62**, 36–50 (2015)
10. I-Ming, C., Golden, B.L., Wasil, E.A.: The team orienteering problem. Eur. J. Oper. Res. **88**(3), 464–474 (1996)
11. JiTT.travel: JiTT Coimbra. http://www.turismodecoimbra.pt/aplicacoes/. Accessed 11 June 2018
12. Jones, M.T.: Al Application Programming (2003)
13. Lin, S., Kernighan, B.W.: An effective heuristic algorithm for the traveling-salesman problem. Oper. Res. **21**(2), 498–516 (1973)
14. Moeller, C.: osm2po. http://osm2po.de. Accessed 04 July 2018
15. de Pardieu, F.: Redefining Mobile in Travel. https://www.mtrip.com. Accessed 11 June 2018
16. Sensis: Voyager. https://play.google.com/store/apps/details?id=com.sensis.voyager.app. Accessed 11 June 2018
17. Sun, Q., Wan, W., Chen, G., Feng, X.: Path planning algorithm under specific constraints in weighted directed graph. In: 2016 International Conference on Audio, Language and Image Processing (ICALIP), pp. 635–640. IEEE (2016)
18. Tsiligirides, T.: Heuristic methods applied to orienteering. J. Oper. Res. Soc. **35**(9), 797–809 (1984)
19. Vansteenwegen, P.: Planning in tourism and public transportation. Ph.D. thesis. Springer (2009)

Standardizing a Shoe Insole Based on ISO/IEEE 11073 Personal Health Device (X73-PHD) Standards

Hawazin Badawi[✉], Fedwa Laamarti, Faisal Arafsha,
and Abdulmotaleb El Saddik

Multimedia Communications Research Laboratory (MCRLab),
University of Ottawa, Ottawa, ON K1N 6N5, Canada
{hbada049,flaam077,farafsha,elsaddik}@uottawa.ca

Abstract. Personal healthcare systems play fundamental role in shaping the future of healthcare. With the explosion of Digital Twin [1] including wearable technology, many healthcare systems depend on personal health devices (PHD) to perform their functions. It is thus important to standardize the utilized PHDs. In this paper, we propose the X73-PHD standard for a smart shoe insole as this promising device is not yet standardized. We provide a background on shoe insoles and their importance in gait analysis. The main contribution of this paper is providing the detailed design of X73-PHD standard for the shoe insole (SI) through the three main parts of X73-PHD standards, which are: domain information model (DIM), service model (SM), and communication model (CM). Besides,we explain the hardware architecture and the functional model of the SI used in this work. Finally, the standard implementation for the SI is presented.

Keywords: Shoe insole · Standardization · Personal health device · X73-PHD

1 Introduction

Personal healthcare systems represent a convenient solution complementing medical healthcare systems. They aim to enhance people awareness about their health statuses. They target both healthy individuals and patients. Personal health systems depend on PHDs and wearable technologies in general to perform their functions efficiently. Consequently, standardizing the PHDs is critical to provide a unified language that can be used to interact with such systems globally. It facilitates the development and expanding of personal health services given the plug-and-play real time interoperability feature of the 11073 standards. Moreover, using standardized PHDs can help provide smart health services in different environments such as smart cities.

In response to this need, the International Organization for Standardization (ISO), IEEE and a group of manufacturers have collaborated and proposed the X73-PHD standards in 2008. This family of standards aims to facilitate the communication between PHDs and managers such as computer systems and cell phones in a direct manner and away from complex point-of-care systems used in clinics [2]. Therefore,

© Springer Nature Switzerland AG 2019
Á. Rocha et al. (Eds.): ICITS 2019, AISC 918, pp. 764–778, 2019.
https://doi.org/10.1007/978-3-030-11890-7_72

implementing the X73-PHD standards provide plug-and-play real time interoperability that facilitate data exchange between the agent and the manager.

Currently, the IEEE standard development working group (IEEE-WG) for PHD [3] contains twenty-two active standards. Many more PHDs, which can be utilized in a wide range of such systems are not yet standardized, the smart insole is one of them.

Smart shoe insole is defined as a personal health device that is equipped with different sensors to monitor vital signs such as the plantar pressure through foot pressure points and collect such valuable data in real time. As a PHD, SI is responsible for collecting and transmitting data to an associated manager to analyze it in order to infer information such as foot or physical activity disorders. Therefore, SI represents a source of valuable data for personal healthcare applications. This importance is shown clearly through several SIs that are available commercially such as [4–6]. Consequently, standardizing the SI according to the X73-PHD standards can be very useful, and we are building a workgroup application for this standard in order to submit it to the IEEE-WG [3]. In this paper, we propose a detailed standard for the SI.

In the proposed standard, we aim to meet the specific requirements stated in the communication profile defined in IEEE 11073-20601 for the personal health agents and managers, which are typically used outside a clinical setting, such as mobile or in a person's home systems [2]. Table 1 shows the list of the requirements and the extent to which the current implementation of the standard for the SI has met these requirements. Specifically, the second column called "Status in the SI", which illustrates whether the requirement in the current implementation of the standard has been considered or not. The value "Yes" means this requirement has been fully met whereas "No" means the requirement has not been met yet but the work is under progress to accommodate it.

2 Background

Different devices and approaches have been used in the past to quantify gait, including force plates, accelerometers, and camera systems [7]. Wired insoles with embedded sensors have been used for gait analysis successfully over the most recent decades [8, 9]. In recent years, however, studies have been moving more towards a non-restrictive solution such as gait measurement mats. However, these mats limit the walking distance to a few meters [10, 11]. The advancement of embedded components in these mats, such as pressure sensors and microcontrollers, have advanced rapidly in recent years towards smaller and more reliable solutions. These components can fit inside shoe insoles to provide a wireless and non-restrictive device that can monitor gait parameters. Recent studies [12–18] have been testing and validating these smart insoles for different health applications to explore the possibility of using them as replacement for mats. The main advantages of using wireless gait measurement devices is that they can be used in natural and non-restrictive way, and they can monitor change over long periods away from hospital or laboratory settings. Moreover, these devices can be utilized in personal health systems developed for monitoring physical activity such as systems in [19, 20]. Consequently, the usage of wireless gait measurement devices can enhance the quality of applications that depend on tracking physical activity to perform their tasks such as systems in [21, 22].

Table 1. Requirements of personal health agents and managers according to IEEE 11073-20601 [2] and current implementation status.

Requirement	Status in the SI		Note
	Yes	No	
"Personal health agents typically have very limited computing capabilities"	✓		
"Personal health agents typically have a fixed configuration, and they are used with a single manager device"	✓		This requirement has been met in the standard implantation in Sect. 5.3
"Personal health agents are frequently battery powered, mobile devices, using a wireless communication link. Therefore, energy efficiency of the protocol is an important aspect"	✓		
"Personal health agents are often not permanently active. For example, a weighing scale may provide data only once or twice a day. An efficient connection procedure is needed for minimum overhead for such devices"	✓		This requirement has been met by having the SI activated once an individual steps on it which triggers the pressure sensors
"Personal health managers tend to have more processing power, memory, and storage space so the protocol intentionally places more load on the managers"			This requirement has been met by implementing the manager on a cloud and being accessible through the computer as well as a mobile device
"Personal health agents and managers convey information that could be useful to clinical professionals. As such, the quality of the data may be considered to have clinical merit even if acquired in a personal health or remote monitoring environment"		✓	This requirement has not been met yet. However, we will consider it in the future applications

3   Design of the X73-PHD Standard for SI

We designed the proposed standard based on the communication profile defined in IEEE 11073-20601 [2] to fulfill the specific requirements for the personal health agents and managers. Then, we referred to the existing standards of three personal health devices: Basic ECG (Part 10406), Pulse Oximeter (Part 10404), and Weighing Scale (Part 10415) found in [3]. We reviewed these standards starting with the communication model (CM). We listed the attributes and values of the main communication states: association, configuration, operation (GET MDS, data reporting) and disassociation. We identified the similarities and differences in the attributes and values for the three standardized devices. We found that all devices use the same attributes names in the CM, and the change occurs in the values. The majority of the attributes values are similar as well, and the main change occurs in the device-specific attributes. For example, in data reporting procedure for the ECG, the attribute "ScanReportInfoFixed. obs-scan-fixed.count = 3" because the standardized basic ECG is 1- to 3-lead ECG. Therefore, in our proposed standard, the value is 12 because the SI has 12 pressure sensors. Applying the same strategy on DIM and SM, we proposed the DIM shown in Fig. 1, described in the following subsection.

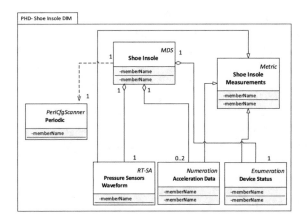

Fig. 1. Proposed Domain Information Model (DIM) for the SI

3.1 Domain Information Model (DIM)

According to [2], the DIM illustrates the agent's information as a set of classes. The attributes of each class represent the data that is delivered to or retrieved by the manager through objects instantiated from each class. The manager control agent's behavior and report its status by utilizing the methods of each class. In this standard, each class of the personal health device DIM is defined by the following information:

- **Nomenclature code** that is used to identify the class. This code plays critical role in the configuration event to report the class of each object and enable the manager to specify the class type.
- **Attributes** that are defined in the class.
- **Methods** that are defined in the class.
- **Events** that are possibly generated by objects instantiated from the class.
- **Services** that are available for the class such as setting or getting.

Following the personal health object and class definitions in [2], the SI is defined as a set of classes as depicted in Fig. 1, showing the structure and capabilities of this device's components. The objects instantiated from the proposed DIM, and the classes' types in the diagram are described as follow:

- **Shoe Insole (MDS object):** This object is instantiated from the Medical Device System (MDS) class. The definition of this object is shown in (Sect. 5.1.2)
- **Shoe Insole Measurements (Metric Class):** According to the definition of metric class in [2], Shoe Insole Measurements is the base class for all objects representing measurements and status in the SI agent, which are Pressure Sensors Waveform, Acceleration Data and Device Status. The definition of this class is shown in (Sect. 5.1.3)
- **Pressure Sensors Waveform (RT-SA object):** This object is instantiated from the real-time sample array (RT-SA) class. The definition of this object is shown in (Sect. 5.1.4)
- **Acceleration Data (Numeration object):** This object is instantiated from the numeric class. The definition of this object is shown in (Sect. 5.1.5)
- **Device Status (Enumeration object):** This object is instantiated from the enumeration class. The definition of this object is shown in (Sect. 5.1.6)
- **Periodic (PeriCfgScanner):** This object is instantiated from the scanner class. It is used to facilitate the reporting of agent-initiated data transfers. It imports the information about agent (SI) status from the Shoe Insole object. A periodic scanner is chosen because enabling it will cause sending the pressure sensors waveform continuously. The definition of this object is shown in (Sect. 5.1.7).

3.2 Service Model (SM)

According to [2], SM defines "the conceptual mechanisms for data exchange services. These services are mapped to messages that are exchanged between the agent and manager". Thus, the service model implementation interprets in form of exchanged messages between the agent and the manager as explained in Sect. 5.2.

3.3 Communication Model (CM)

According to [2], CM is used mainly to describe the communication between one or more agents and a single manager as point-to-point connection. A connection state machine is used for this purpose where the communication between a single agent and the manger initiated by the connection state, followed by association, configuration, operation and disassociation states, which are the four states of CM. Also, CM is used to describe how to handle error conditions and convert DIM abstract data into binary messages using medical device encoding rules (MDER). Section 5.3 present the implementation of CM.

4 Hardware Architecture and Functional Model

4.1 Hardware Architecture

In order to collect detailed gait information, 12 Force-Sensitive Resistors (FSR), also referred to as pressure sensors, were placed on the insole in areas where pressure force can be measured. FSR sensors have been reliably used in several studies to measure gait pressure [7]. Multiple studies suggested that the force pressure on the medial midfoot (arch) is minimal/negligible [8], and thus are not measured in this design. The twelve FSR sensors are distributed such that they cover the remaining areas of the foot. High quality sensors were used (Interlink FSR-402 [9]), which can measure force between 0–100 N. ADC outputs were normalized to values between 0–100 representing applied force pressure.

An Inertial measurement unit (IMU) is also placed in the insole's medial midfoot area where minimal pressure is observed. The IMU is used to calculate foot kinetics and orientation through capturing real time 3D angular acceleration, 3D rotation (via gyroscope data), and 3D orientation (via magnetometer data). For this purpose, we used MPU-9250 which is a 9-axis IMU with high precision, and capable of sampling at a frequency of 50 Hz.

All FSR sensors and the IMU are embedded in the insole, which will be placed inside the user's shoe to read gait measurements. Figure 2 shows the designed Printed Circuit Boards (PCB) layout of the insole and the actual implementation of the device. This single device provides information that can be used for calculating the different modes of gait, temporal characteristics, center of mass, orientation, and force distribution.

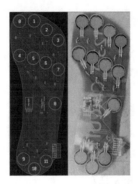

Fig. 2. Shoe insole used in developing the system in this paper.

4.2   Functional Model

A smart insole as a PHD transmits sensory information by going through four main stages shown in Fig. 3:

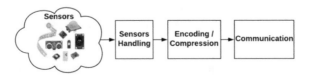

Fig. 3. Smart shoe insole functional model.

- **The first stage** is the physical interaction stage where the actual sensors have direct contact with the sensed environment. In the context of the shoe insoles, the direct contact to the sensed environment is done through FSR. FSR sensors receive voltage and respond with a percentage of this applied voltage depending on the applied pressure/force.
- **The second stage** is sensors handling. This is responsible for direct communication with the connected sensors, and translating the received digital or analog voltages into information. For example, applying a pressure of 30 N on a 5 V powered FSR sensor that can measure up to 100 N should make the sensor respond by producing voltage equal to 30% of the 5 V, which is 1.5 V. The sensors handling stage in this example should translate the received voltages into a unit of force, i.e. Newton.
- **The third stage** is encoding/compression. After converting digital and analog voltage signals into data, the encoding/compression stage performs algorithms to pack the received sensory data and perform compression techniques to minimize the overall stream that is about to happen in the communication stage.
- **The fourth stage** is the communication. This stage can then transmit the compressed/encoded data using the X73-PHD standards communication protocol described in Sect. 5.3.

5   X73-PHD Implementation

This section describes the implementation of the proposed X73-PHD standard for the SI in light of optimized exchange protocol [2].

5.1   Domain Information Model (DIM)

All the classes in the DIM follow the same nomenclature code structure discussed in following:

5.1.1   Nomenclature Usage in DIM

According to [2], the definition of nomenclature is "A set of context-dependent partitions. The nomenclature code in each context-dependent partition is defined by a 16-bit code that supports up to 65 536 independent terms per partition. The partitions are referenced by a 16-bit partition code." Thus, according to the standard, a reference identifier (ID) will be used to refer to nomenclature term codes within a specific partition and nomenclature partition codes for different partitions. The general form of the reference ID is MDC_XXX_YYY where MDC stands for Medical Device Communication. According to Table 1 in [2], the SI assigned to partition number 129, where the nomenclature category is personal health devices health and fitness. This assignment implies the usage of a 16-bit term code to propose the SI standard.

5.1.2   Shoe Insole (MDS Object)

It is the top-level object that acts as a root object for the SI. It represents agent's identification and status through its attributes, which justifies the fact that each agent has only one MDS object. The nomenclature code to identify the MDS class is MDC_MOC_VMS_MDS_SIMP. The list of mandatory attributes is shown in the following table (Table 2). In the table, the first column shows the attribute name according to the standard while the second column shows the attribute nomenclature reference ID, which is used in class definition in the code. The last column shows the attribute qualifier, which identifies when the attribute is to be implemented in the object. There are three qualifiers, which are M for Mandatory, C for Conditional and O for Optional. According to the standard [2], an attribute can be further qualified as S for static, D for Dynamic, Ob for Observational. These qualifiers consider the change of attribute's value during the life of an association. All the tables in this paper follow the same structure.

Table 2. Shoe insole mandatory attributes.

Attribute Name	Attribute ID	Qualifiers
Handle	MDC_ATTR_ID_HANDLE	M/S
System-Model	{"Manufacturer", "Model"}	M/S
System-Id	MDC_ATTR_SYS_ID	M/S
Dev-Configuration-Id	MDC_ATTR_DEV_CONFIG_ID	M/S

5.1.3   Shoe Insole Measurements (Metric Class)

According to the definition of metric class in [2], this class contains common and shared attributes, which are inherited by Pressure Sensors Waveform, Acceleration Data and Device Status objects in addition to methods, events and services. Since the metric class never instantiated, it is never part of the agent (SI) configuration. Due to this fact, we will consider a basic/simple DIM for the SI in this proposed standard, which is shown in Fig. 4. This DIM illustrates the metric object (Shoe Insole Measurements) instances. Regarding the methods, events and services, referring to 20601 [2] shows that there are currently no metric object methods defined in this standard and no uses of the metric SET or GET services. Regarding the events, the standard states that objects drive from metric class do not report their observations directly; rather, they reported observations through another object, such as the MDS object and a scanner object.

In our proposed standard, Pressure Sensors Waveform, Acceleration Data, and Device Status objects report their observations through Periodic object that is instantiated from a scanner class as discussed in Sect. 5.1.7. The nomenclature code to identify this Metric class is MDC_MOC_VMO_METRIC. However, this nomenclature code is not used in an agent or a manager implementation as the metric class is just a base class for other classes.

5.1.4   Pressure Sensors Waveform (RT-SA Object)

This object represents the waveform measurement generated from different pressure sensors and associates with Shoe Insole object. It is specified as a RT-SA object because this object type represents continuous samples or waveforms, which is the case

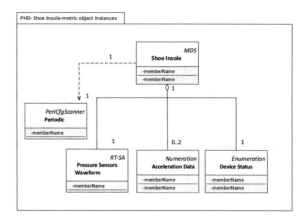

Fig. 4. Basic/Simple Shoe Insole (SI) Domain Information Model (DIM).

in the generated waves. The values of the RT-SA object are sent from the agent to the manager using the EVENT REPORT service through Periodic object. The nomenclature code to identify this RT-SA class is MDC_MOC_VMO_ METRIC_SA_RT. This object inherits the attribute from the metric class (Shoe Insole Measurements), in addition to other specific attributes defined for this object. Regarding the methods, events and services, this class inherits the same current situation from the super class, which is the metric class explained in the previous section. The list of inherited mandatory attributes and this object's specific attributes is shown in the following table (Table 3).

Table 3. Pressure sensors waveform attributes.

Attribute Name	Attribute ID	Qualifiers
Handle	MDC_ATTR_ID_HANDLE	M/S
Type	MDC_ATTR_ID_TYPE	M/S
Metric-Spec-Small	MDC_ATTR_METRIC_SPEC_SMALL	M/D
Enum-Observed-Value-Basic-Bit-Str	MDC_ATTR_ENUM_OBS_VAL_BASIC_BIT_STR	C/Ob

5.1.5   Acceleration Data (Numeric Object)

This object represents the acceleration values and associates with Shoe Insole object. It is specified as a numeration object because numeric objects represent episodic measurements, which is the case in the acceleration data that change occasionally. The nomenclature code to identify this numeric class is MDC_MOC_VMO_METRIC_NU. This object inherits the attribute from the metric class (Shoe Insole Measurements), in addition to other specific attributes defined for this object. Regarding the methods, events and services, this class inherits the same current situation from the super class, which is the metric class explained in Sect. 5.1.3. The list of inherited mandatory attributes and this object's specific attributes is shown in the following table (Table 4).

Table 4. Acceleration data attributes.

Attribute Name	Attribute ID	Qualifiers
Handle	MDC_ATTR_ID_HANDLE	M/S
Type	MDC_ATTR_ID_TYPE	M/S
Metric-Spec-Small	MDC_ATTR_METRIC_SPEC_SMALL	M/D
Basic-Nu-Observed-Value	MDC_ATTR_NU_VAL_O	C/Ob
Accuracy	MDC_ATTR_NU_ACCUR_MSMT	O/S

5.1.6 Device Status (Enumeration Object)

This object represents events occurred while operating the agent and associates with the Shoe Insole object. It is specified as an enumeration object because enumeration object type represents status information and/or annotation information. The values of the enumeration object are coded in one of two forms: normative codes or free text. Similar to the Pressure Sensors Waveform (RT-SA object), the EVENT REPORT service is used to send the values of the enumeration object from the agent to the manager using the EVENT REPORT service through Periodic object. Losing the connection with the SI is an example of the events handled by this object. The nomenclature code to identify this enumeration class is MDC_MOC_VMO_METRIC_ENUM. This object inherits the attribute from the metric class (Shoe Insole Measurements), in addition to other specific attributes defined for this object. Regarding the methods, events and services, this class inherits the same current situation from the super class, which is the metric class explained in Sect. 5.1.3. The list of inherited mandatory attributes and this object's specific attributes is shown in the following table (Table 5).

Table 5. Device status attributes.

Attribute Name	Attribute ID	Qualifiers
Handle	MDC_ATTR_ID_HANDLE	M/S
Type	MDC_ATTR_ID_TYPE	M/S
Metric-Spec-Small	MDC_ATTR_METRIC_SPEC_SMALL	M/D
Sample-Period	MDC_ATTR_TIME_PD_SAMP	M/S
Simple-Sa-Observed-Value	MDC_ATTR_SIMP_SA_OBS_VAL	M/Ob
Scale-and-RangeSpecification	MDC_ATTR_SCALE_SPECN_I16	M/D
Sa-Specification	MDC_ATTR_SA_SPECN	M/S

5.1.7 Periodic (PeriCfgScanner Class)

This class is inherited from CfgScanner class, which in turns inherited from Scanner Class. The CfgScanner is an abstract class that defines the communication behavior of a configurable scanner object of both types: EpiCfgScanner and PeriCfgScanner. Therefore, it cannot be instantiated.

Starting from the root, and according to IEEE 20601 in [2], scanner class is used for two purposes: allowing the manager to control the flow of data and facilitating the reporting mechanism as an optimized packaging. The latter purpose is achieved by enabling the collected sets of changed attribute values from one or more metric objects to be packaged together in a single event report and send it to the manager in an efficient manner better than doing this using MDS class event. In details, a scanner class collects AttributeChangeSets and maps them to ObservationScans in scan event report. Then, it dispatches the scan event report according to its specific type. The scanner object can be Episodic or Periodic. If it is Episodic, it dispatches scan event reports when an episode completes according to the application definition. If it is Periodic, it dispatches scan event reports when the period reported in the Reporting-Interval attribute expires, and in this case, more than one AttributeChangeSet may come from the same object, which is not allowed in the Episodic scanner.

In the proposed standard for SI, we consider the periodic scanner since we have multiple pressure sensors that may experience several value changes in a specific interval period. Thus, Periodic object imports the information about agent (SI) status from the Shoe Insole MDS class. The nomenclature code to identify the periodic configurable scanner class is MDC_MOC_SCAN_CFG_PERI.

The following table (Table 6) shows the mandatory attributes of the periodic configurable scanner class. Regarding the scanner methods and services, there are currently no scanner object methods, GET or SET service defined in this standard. There are six events for this class [2].

Table 6. Periodic mandatory attributes.

Attribute Name	Attribute ID	Qualifiers
Handle	MDC_ATTR_ID_HANDLE	M/S
Operational-State	MDC_ATTR_OP_STAT	M/D
Confirm-Mode	MDC_ATTR_CONFIRM_MODE	M/D
Reporting-Interval	MDC_ATTR_SCAN_REP_PD	M/D

It is worth mentioning that due to the limited storage resources in the currently implemented SI, there is no persistent metric store (PM-store) object along with PM-segments that are used to provide a persistent storage mechanism for metrics to be accessed by the manager at a later time. However, extending the storage resources in the SI is part of our future plan.

5.2   Service Model (SM)

The 20601 protocol structure [2] differentiates between two types of messages: connection-management-related messages (association services) and object-related messages (object access services). Thus, the main association messages are Association Request from the agent (A) to the manager (M): (A→M), Association Response

(M→A), Release/Disassociation Request (A→M), Release/Disassociation Response (M→A), and Abort message that terminates the association immediately without response. It sent as a result of a failure either on agent or manager side. Figure 5, which depicts the current implementation of the proposed X73 standard for the SI, illustrates these messages in red solid and dotted arrows.

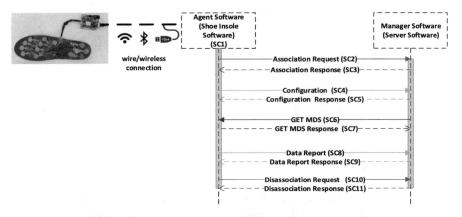

Fig. 5. The proposed service and communication models for the shoe insole.

The data access services used to exchange DIM data between the agent and the manger. GET, SET, Action and Event Report are the fundamental services in each SM. As stated in [2]:

- **GET service** is "used by the manager to retrieve the values of the agent MDS object and PM-store attributes". In our implementation shown in Fig. 5, the GET service and its response are illustrated in blue solid and dotted arrows.
- **SET service** is "used by the manager to set values of attributes of the agent's object. Currently, only the scanner objects support the SET service to control the agent."
- **Event Report service** is "used by the agent to send configuration updates and measurement data to the manager". As explained in the previous Sect. 5.1.7, the periodic class is responsible about this service in the proposed standard for the SI. In our implementation shown in Fig. 5, the Data Report service and its response are illustrated in green solid and dotted arrows.
- **ACTION service** is "used by the manager to invoke actions (or methods) supported by the agent." MDS-Data-Request action is an example of actions that is used to request measurement data from the agent.

5.3 Communication Model (CM)

Based on the specifications in the X73-PHD standards-Part 20601 [2], we implemented the proposed standard for the SI as a set of classes that represents the CM four states shown in Fig. 5. We decided to call them X73-standard classes (SCs), in addition to other supportive classes.

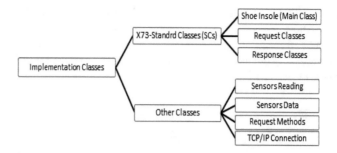

Fig. 6. Classification of classes in the standard implementation for the SI.

Figure 6 shows the classification of classes in the standard implementation. The X73-Standard classes in the current implementation are:

- **Shoe insole:** it is the main class SC1 that contains all classes attributes, in addition to the main methods such as methods that handle the received instructions from users.
- **Request classes:** they are the classes used to send requests from the agent (SI) or the manger. SC2 (association request), SC4 (configuration request), SC8 (data report), and SC10 (disassociation request) are the classes used by the agent to send requests to the manager. SC6 is the class used by the manger to request the SI information. In the current implementation, these classes contain the attributes with their values and act as data storage because they do not contain methods.
- **Response classes:** they are the classes used to respond to the requests sent by the agent (SI) or the manger. SC3 (association response), SC5 (configuration response), SC9 (data report response), and SC11 (disassociation response) are the classes used by the manager to respond to the requests sent by the agent. SC7 is the class used by the agent to respond to the requests sent by the manger. These response classes cause assigning values in the agent.

Other classes are used to facilitate the implementation of the SI standard. In the current implementation, they are four classes:

- **Sensors reading:** this class is used to get data from sensors to be ready for manager requests.
- **Sensors data:** this class is used to save the read data from sensors.
- **Request:** this class contains all requests methods to be called/accessed by request classes.
- **TCP/IP connection:** this class is used to send data from the shoe insole to the manager. We decided to use TCP transmission protocol to guarantee the reliability in the data transmission.

6 Conclusion and Future Work

In this paper, we proposed the X73-PHD standard for the SI. A thorough design of the standard for SI as a PHD has been provided and discussed. The implementation of the three sections of the standard: DIM, SM and CM is explained. For the future work, we plan to extend the storage resources in the currently used SI and implement it as a persistent metric store (PM-store) object along with PM-segments according to the standard. Moreover, we plan to utilize the collected data in personal healthcare applications.

References

1. El Saddik, A.: Digital twins: the convergence of multimedia technologies. IEEE Multimedia **25**(2), 87–92 (2018)
2. Committee of the IEEE Engineering in Medicine, S. and Society, B. 1107. IEEE Std 11073-20601TM-2014. Standard for Health informatics—Personal health device communication—Application profile—Optimized Exchange Protocol (1107)
3. IEEE Standards Association. https://standards.ieee.org/develop/wg/PHD.html
4. GPS SmartSole. http://gpssmartsole.com/gpssmartsole/. Accessed 8 Nov 2018
5. DIGITSOLE. https://www.digitsole.com/. Accessed 8 Nov 2018
6. FEETME. http://www.feetme.fr/en/index.php. Accessed 8 Nov 2018
7. Muro-de-la-Herran, A., García-Zapirain, B., Méndez-Zorrilla, A.: Gait analysis methods: an overview of wearable and non-wearable systems, highlighting clinical applications. Sensors **14**(2), 3362–3394 (2014)
8. Hausdorff, J.M., Ladin, Z., Wei, J.Y.: Footswitch system for measurement of the temporal parameters of gait. J. Biomech. **28**(3), 347–351 (1995)
9. Fraser, S.A., Li, K.Z., DeMont, R.G., Penhune, V.B.: Effects of balance status and age on muscle activation while walking under divided attention. J. Gerontol. B, Psychol. Sci. Soc. Sci. **62**(3), P171–P178 (2007)
10. Cutlip, R.G., Mancinelli, C., Huber, F., DiPasquale, J.: Evaluation of an instrumented walkway for measurement of the kinematic parameters of gait. Gait Posture **12**(2), 134–138 (2000)
11. van Uden, C.J., Besser, M.P.: Test-retest reliability of temporal and spatial gait characteristics measured with an instrumented walkway system (GAITRite®). BMC Musculoskelet. Disord. **5**(1), 13 (2004)
12. Arafsha, F., Laamarti, F., El Saddik, A.: Development of a wireless CPS for gait parameters measurement and analysis. In: International Instrumentation and Measurement Technology Conference (I2MTC), May 2018
13. Howell, A.M., Kobayashi, T., Hayes, H.A., Foreman, K.B., Bamberg, S.J.M.: Kinetic gait analysis using a low-cost insole. IEEE Trans. Biomed. Eng. **60**(12), 3284–3290 (2013)
14. Hafidh, B., Al Osman, H., El Saddik, A.: SmartInsole: a foot-based activity and gait measurement device. In: 2013 IEEE International Conference on Multimedia and Expo Workshops (ICMEW), pp. 1–4 (2013)
15. Crea, S., Donati, M., De Rossi, S., Oddo, C., Vitiello, N.: A wireless flexible sensorized insole for gait analysis. Sensors **14**(1), 1073–1093 (2014)
16. Tan, A.M., Fuss, F.K., Weizman, Y., Woudstra, Y., Troynikov, O.: Design of low cost smart insole for real time measurement of plantar pressure. Procedia Technol. **20**, 117–122 (2015)

17. Jagos, H., Pils, K., Haller, M., Wassermann, C., Chhatwal, C., Rafolt, D., Rattay, F.: Mobile gait analysis via eSHOEs instrumented shoe insoles: a pilot study for validation against the gold standard GAITRite. J. Med. Eng. Technol. **41**(5), 375–386 (2017)
18. Arafsha, F., Hanna, C., Aboualmagd, A., Fraser, S., El Saddik, A.: Instrumented Wireless SmartInsole system for mobile gait analysis: a validation pilot study with Tekscan Strideway. J. Sens. Actuator Netw. **7**(3), 36 (2018)
19. Badawi, H.F., Dong, H., El Saddik, A.: Mobile cloud-based physical activity advisory system using biofeedback sensors. Future Gener. Comput. Syst. **66**, 59–70 (2017)
20. Park, K., Lim, S.: A multipurpose smart activity monitoring system for personalized health services. Inf. Sci. **314**, 240–254 (2015)
21. DesClouds, P., Laamarti, F., Durand-Bush, N., El Saddik, A.: Developing and testing an application to assess the impact of smartphone usage on well-being and performance outcomes of student-athletes. In: International Conference on Information Theoretic Security, 2018, pp. 883–896. Springer, Cham (2018)
22. Badawi, H., El Saddik, A.: Towards a context-aware biofeedback activity recommendation mobile application for healthy lifestyle. Procedia Comput. Sci. **21**, 382–389 (2013)

Information Technologies in Education

Analysis of Relevant Factors to Measure the Impact of Investment in e-learning Ecosystems in Public Universities

Doris Meza-Bolaños[1]([⊠]), Patricia Compañ Rosique[2]([⊠]),
and Rosana Satorre Cuerda[2]([⊠])

[1] Central University of Ecuador, Quito, Ecuador
dmeza@uce.edu.ec
[2] Alicante University, Alicante, Spain
{patricia.company, rosana.satorre}@ua.es

Abstract. This paper analyzes different perspectives used in different models to assess the impact of online learning ecosystems in order to support the decision to make an investment of resources in any context. By means of an analysis of the scenario, it is possible to detect factors of influence that allow not only to evaluate the ecosystems in the financial area, but also can be carried out in a social environment and even determining affective aspects. The application of heat maps has helped to determine that the analyzed models focus their interest in evaluating the student but not in determining their interest, well-being and compliance with their expectations, which are important factors to later measure the return on investment (ROI). From the representation of the heat map, it is possible to determine which areas are being analyzed and which are not. This will allow us to focus on a new model that considers how to improve those deficiencies.

Keywords: e-learning ecosystems · SROI · SLR · Evaluation models · Heat maps

1 Introduction

The decision to make an investment of resources in any context requires a thorough study of the current situation. Through an analysis of the scenario, it is possible to detect the strengths, threats, opportunities as well as any other important factors influencing that decision. When Public Universities investing in technology arises the need to know what the improvement entails and if it is possible to assume that investment will improve learning [6]. Obviously, it does not seem reasonable to assume that such an increase in technology resources will lead to an improvement in learning, if a prior study of how to measure the impact of online teaching and learning ecosystems has not been done before [8]. To measure efficiency in the use of these ecosystems; as a support to the teaching - learning strategy in the face - to - face classes of higher education, seeks to measure the total impact of this tool in its environment, this means to measure a space of value broader than economic, including economic dimensions, also social dimensions that contribute to show the true value of these tools,

© Springer Nature Switzerland AG 2019
Á. Rocha et al. (Eds.): ICITS 2019, AISC 918, pp. 781–790, 2019.
https://doi.org/10.1007/978-3-030-11890-7_73

showing the benefits of its activity in the information society in which it operates and, at the same time, giving relevance to the needs, aspirations and expectations of its actors.

It should be emphasized that the term ecosystems in the learning environment refers to technological ecosystems: that is, ecosystems defined as the evolution of traditional information systems to support the management of knowledge in heterogeneous environments and when talking about measuring [7], as a result of the fulfillment of the objectives of a program aimed at improving the teaching-learning process by applying these tools.

In this paper we study different models to try to find out which are the relevant aspects to study to measure the impact of the investment of learning ecosystems. Thanks to the boom in the use of TIC'S in classrooms over the last few years, there is not only a growing participation of higher education institutions (HEIs) in techno-logical issues, but organizations are more aware of their effect on the knowledge society and the impact it produces [16]. In this sense, HEIs realize that the value generated by the use of technological platforms is not only economic, financial, but also social.

The use of learning platforms generates changes in the organization that use them [9]. By building knowledge, autonomous learning, virtual group work, virtual contact with the tutor, transfer of information, broadening access to the resources and services of the platform, tools impact on the lives of users, Their capabilities, their future opportunities and their knowledge-building styles. Thus, by assessing their socio-economic impact, HEIs can reduce costs and risks and create new opportunities for action that minimize negative impacts and maximize positive ones. Equally, impact assessment can help HEIs show their interest groups the generation of (socio-economic) benefits for the environment in which they operate.

In the evaluation process, and in particular in the measurement phase, HEIs can understand the needs, aspirations, resources and incentives of the actors, allowing them to develop new knowledge that in turn will generate new products and services or improve existing ones [13].

In summary, efficient use of online learning ecosystems through a value assessment process can produce important benefits such as: improving investment efficiency, demonstrating results beyond purpose and involving all stakeholders. To evaluate the impact on the use of online learning platforms, a series of criteria to be measured should be established in order to verify the efficiency of these tools.

1.1   Analyzed Aspects

The study seeks to determine the impact of two aspects:

- Economic aspect focused on:
  - Growth (access to global knowledge, innovation generation and knowledge transfer).
  - Return on investment (operational efficiency, level of development of stakeholders).
  - Risk management (operational, regulatory and operational risk).

– Social aspect focused on the changes produced in the lives of users of this tool, measuring the following aspects [11]:
- Increase of users with access to this service.
- Increases of new types of devices to access e-learning spaces.
- Reduction of time spent learning a topic.
- Level of perceived knowledge transfer.
- Level of achievement of objectives of each of the actors.
- Number of trained, sensitized and trained beneficiaries on the use of infrastructures.

The economic and financial value is reflected in income and expenses, balance sheets and income statements, but human value is not reflected in any state of the organization. If it is necessary to describe the value generated by learning ecosystems, it is necessary to have a method to measure and reflect human value, considering all the actors that are part of the platform implementation and use environment [10].

Two main problems arise in order for these platforms to be evaluated and to determine their true value: the existence of a large number of measurement methodologies, each of which are based on dissimilar assumptions and functionalities, focusing on different types of impact; adapted to different purposes [12].

After reviewing the two models that will serve as the starting point of this work, we present the possibility of evaluating the common and specific situations of each of these models by means of heat maps, representing the information in a simple and intuitive form, without reducing the characteristics of the original information. By means of heat maps it is sought to graphically represent each of the indicators of the models, in order to facilitate the understanding of the elements and their relationships. The addition of a color code to represent the level of influence of each element allows the construction of the map so that, in addition to the structure given by these maps, it is also possible to visualize the state of each model [1].

Heat maps are powerful tools of representation that have the following main objectives:

- Provide a systematic way of structuring the components of the models studied.
- Evaluate the characteristics of the selected models.
- Identify relevant and common aspects of these models.

2   Methodology of Research

There are numerous and very diverse works published in the different databases. In this situation, it is difficult to obtain an objective view of a topic since many works are published with different points of view and different results. A Systematic Literature Review (SLR) [4] helps to obtain objective information on a research topic by identifying, critically assessing and integrating the results of all relevant studies in that subject. An SRL has several objectives. Among them we can highlight: existing research advances, identification of relationships as well as formulations of general concepts. Therefore, an SLR [14] is in itself a research work and serves much more than to obtain a mere compilation of published works on a topic. In the initial phase of application of the systematic literature review the following questions have been raised:

RQ1: What methodologies have been proposed to evaluate the impact of e-learning?
RQ2: What methodologies quantify impact considering return on investment, expectations and welfare?

The answers to these questions mark the starting point of the research and, in a certain way, conditions the following stages in our research. The queries made in the databases have been:

Q1: e-learning OR elearning OR blearning OR b-learning OR technology - enhanced AND ((Effectiveness, impact, evaluation) OR (metrics, "return on investment", "return of expectation", "return of welfare"))

In the next phase the criteria for determining the inclusion or exclusion of the work resulting from the consultations have been established. These criteria are based on the type of database, date of publication and area of knowledge. The application of these restrictions allows us to select the most interesting works for our study.

The databases handled in the review of systematic literature have been: Scopus, Web of Science (WoS) and Google Scholar. Due to the characteristics of the research, it is also interesting to consult other sources, considered of gray literature. The gray literature [3], also called unconventional literature, semi-published literature or invisible literature, it is any type of document that is not disseminated through ordinary commercial publication channels, and therefore poses problems of access.

The period of time that has been considered relevant covers from 1996 to 2016. This restriction is based on the fact that the first concepts of e-learning appeared in early 1996. As for the areas considered to restrict the search have been: education, engineering and social sciences. Although the concept of e-learning is used in many and varied areas, in which no metrics or indicators are mentioned, it is only discussed and described its use, for that reason, it is of no interest for our study. Therefore, the work will focus on the determination of relevant variables that make feasible the faithful representation of the factors included in a model that seeks to measure ROI, ROE and welfare, in order to construct an instrument to measure variables considered appropriate and define standards for testing and measurement.

Once all the documents of interest have been collected and in this phase of the investigation, we have been able to study in detail some of the relevant variables of the process.

As mentioned, the available models are diverse, so it is necessary to identify which are the most suitable. For this study several of them have been reviewed and the work is focused on two that are estimated to be oriented to assess the human and financial part (Duart and Kirkpatrick's models) [2, 12] from which relevant indicators will be extracted. From these two models four perspectives are observed: Improvement of affective aspects, quality and efficiency of the teaching-learning process, transfer of knowledge and infrastructure [5].

Within these perspectives, influence factors have been defined for each model. In future work, the indicators will be defined in a specific way and how their measures will be carried out.

In order to determine the influencing factors of online learning ecosystems, the SROI (Social Return on Investment) impact chain concepts are applied. This concept has great importance since the value of what is created day by day goes far beyond what can be measured in monetary terms, this is, in most cases, the only type of value that is measured and quantifies. Therefore, many decisions that are made may not be as adequate, because they are based on incomplete information, regarding their true impact.

Around SROI there is a framework for measuring and quantifying a broader concept of value, an account of how change is generated by measuring social, environmental and economic outcomes using monetary terms to represent results. SROI deals with value rather than money; as a common unit [15].

A SROI analysis can take different forms. You can group the social value generated by the whole organization, or just focus on a specific aspect of your work.

As can be seen in Fig. 1, the chain of impact creation involves a set of concepts that start from the definition of the resources that contribute the ecosystems in the process of teaching - learning, definition of new activities that are carried out with contribution of learning ecosystems, to determine the measurable results and then to determine the quantifiable changes produced, this will allow to define results attributable directly to the ecosystems.

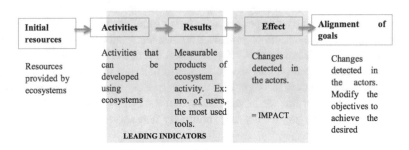

Fig. 1. Chain of impact creation.

Within the idea of impact is taken into account four concepts:

- Displacement: what percentage of the change has shifted other changes.
- Dead Weight: reflects if the changes could have been achieved without the use of the platform.
- Attribution: is the percentage of changes that is not attributable to the management of the organization.
- Decrements: is the deterioration of a change over time.
- All this will be applied to carry out the calculation of the SROI. At this stage all changes will be added and any negative impacts will be subtracted.

In addition to ROE, ROI and SROI, heat maps [1] have been used to visually define common and important indicators of the two models. The objective is to continue incorporating other models of interest to the heat map so that it reflects in an intuitive way information from different perspectives. The map thus generated will provide a global and immediate vision.

The reason for using these two-dimensional graphical representations of data is the ease of symbolizing the values of a variable with colors, its intuitive nature of the color scale in relation to temperature facilitates its comprehension. From the experience, by A. Bojko and according to the same nature the intense color is perceived as hotter than the medium intensity, and the medium intensity as hotter than the less intensity. So it is not difficult to understand that the amount of heat is proportional to the level of the represented variable. In addition the heat maps show the data directly on the variable to which they are representing. According to Bojko: "Because the data could not be closer to the elements to which they belong, it takes little mental effort to read a map of heat".

The reason for using these maps is to evaluate the influence factors of the two models chosen, from each model have been considered common and relevant characteristics. The criterion applied in the development of the preliminary phase to the heat maps is shown in Table 1. As can be seen, each perspective is assigned a color and each influence factor an X, if the influence factor has been considered in the model and the cell is left blank if this factor has not been considered in the model. From the values assigned, the above table is obtained, with the analysis of the presence or not of the influence factors in the two models.

A third map is generated with the crossing of the two previous ones where it is possible to visualize more clearly, what are the influence factors that the two models have not approached. The areas painted with lower tonalities such as expectations and the affective perspective of both organization and teachers and innovation in the quality perspective are examples of what is being affirmed. From the third heat map of Figs. 2 and 3, important information has been obtained to measure the impact of ecosystems applied to higher education, through the collection of existing information and analysis of the importance and influence of each factor On the agents involved. The heat map shows this information graphically and intuitively and serves as a basis for the following analyzes.

As can be seen in Fig. 2, Kirkpatrick's model works in the four perspectives, specifically focused on students, not so with the other actors (organization and teachers). Neither model is oriented to measure affective aspects so factors such as motivation and interest, critical behavior, attitudes and satisfaction, and fundamentally academic prestige will be of particular interest in the study.

Table 1. Heat map of influence factors of the models studied. Affective and quality perspectives of the teaching - learning process.

Model	Actors	Prospects	Factors of influence	Improvement of affective aspects				Quality and efficiency teaching-learning process				Knowledge transfer				Infrastructure		
				MI	CB	AS	AP	FX	AC	IT	IN	C	DB	CR	DS	AV	S	EQ
Kirkpatrick	Org		Expectations - ROE						X	X	X	X	X	X		X	X	X
			Wellness SROI		X			X	X	X	X	X				X	X	X
			Costs - ROI									X	X	X		X	X	X
	Student		Expectations - ROE	X	X	X	X	X	X	X	X	X	X	X	X			
			Wellness SROI	X	X	X	X	X	X	X	X	X	X	X	X	X	X	X
			Costs - ROI	X	X	X	X	X	X		X	X	X	X	X			
	Teacher		Expectations - ROE	X	X	X		X	X	X		X	X	X		X	X	X
			Wellness SROI		X			X	X	X			X	X		X	X	X
			Costs - ROI	X	X	X	X	X	X	X	X	X	X	X				
Duart	Org		Expectations - ROE	X				X	X	X			X	X	X			
			Wellness SROI	X	X	X		X	X			X			X	X	X	X
			Costs - ROI					X	X	X	X	X	X	X	X	X	X	
	Student		Expectations - ROE	X	X	X	X	X	X		X	X	X	X	X	X		X
			Wellness SROI	X	X	X	X	X	X	X	X	X	X	X	X	X	X	X
			Costs - ROI	X	X	X		X	X	X	X	X	X	X	X		X	X
	Teacher		Expectations - ROE	X	X	X		X	X	X	X	X	X			X	X	X
			Wellness SROI					X	X	X	X					X	X	
			Costs - ROI	X	X	X		X	X	X	X	X	X	X		X		

Note: MI = Motivation and interest; CB = Critical behavior; AS = Attitudes and satisfaction; AP = Academic prestige; FX = Flexibility (space and time); AC = Accessibility; IT = Interactivity; IN = Innovation; C = Collaboration; DB = Creation of databases and new courses; R = Creation of open repository of knowledge area; AV = Availability; S = SUPPORT; EQ = Equipment (h, s, network) and services; DS = Development of skills;

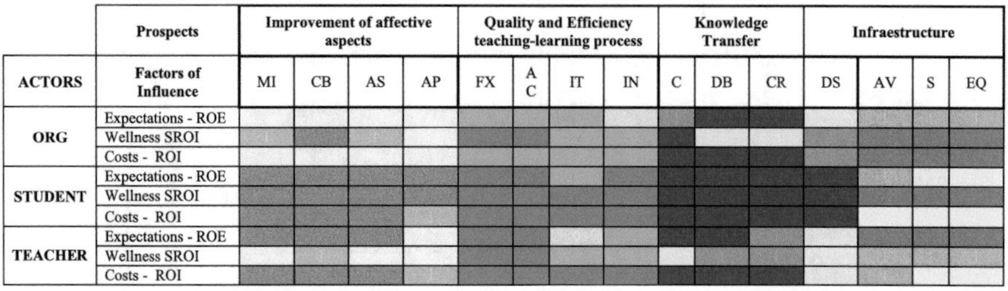

	Prospects	Improvement of affective aspects				Quality and Efficiency teaching-learning process				Knowledge Transfer			Infraestructure			
ACTORS	**Factors of Influence**	MI	CB	AS	AP	FX	AC	IT	IN	C	DB	CR	DS	AV	S	EQ
ORG	Expectations - ROE															
	Wellness SROI															
	Costs - ROI															
STUDENT	Expectations - ROE															
	Wellness SROI															
	Costs - ROI															
TEACHER	Expectations - ROE															
	Wellness SROI															
	Costs - ROI															

Note: MI= Motivation and interest; CB=Critical behavior; AS=Attitudes and satisfaction; AP=Academic prestige; FX=Flexibility (space and time); AC=Accessibility; IT=Interactivity; IN=Innovation; C=Colaboration; DB=Creation of databases and new courses; R=Creation of open repository of knowledge area; DS=Develop-ment of skills;AV=Availability; S= SUPPORT; EQ = Equipment (h, s, network) and services

Fig. 2. Heat map of influence factors of the models studied. Perspectives transfer of knowledge and infrastructure

3   Results and Future Works

Regarding the transfer of knowledge, as can be seen in Fig. 3, the two models approach the subject in a superficial way, with respect to this perspective is a set of aspects that can be of influence and that has not been taken in counts as the interactivity between the user and the way in which the innovation of each one of the actors can be valued. On the infrastructure, in the two models there are no aspects that allow to evaluate the services that the actors receive and the availability, through the heat maps it is possible to establish the influence factors that are going to be considered as the basis for the proposal of a new model.

Having analyzed perspectives and factors of influence between the two models, such as: improvement of aspects of affectivity, quality and efficiency in the teaching-learning processes, knowledge transfer and infrastructure, it is concluded that there are some influence factors that can be approached in greater depth and from these to define indicators that allow to estimate the real impact of online learning ecosystems within the higher education environment.

Heat maps are a technique that allows determining the factors that must be worked more strongly and that have not yet been taken into account by currently recognized models, this will allow to consolidate a robust and valid model to measure the impact of ecosystems Technological advances in higher education and the perspectives and factors with which they should be considered.

For now, two models have been chosen to perform this test, but other models will be analyzed with the same technique and the next phase will be defined where the indicators and the methodology that each will apply for their measurement will be defined.

Therefore, the model that will be implemented later, the accepted and accepted perspectives, the students and the organization as welfare and the return to expectations, and these results will be like the values for the return of the investment. In addition, this map shows that universities should improve the tools that students provide to facilitate the knowledge transfer process with simple infrastructure, testing the support and quality services that are available to all users.

MODELS	ACTORS	Factors of Influence	Prospects	Knowledge Transfer			Infrastructure		
			Collaboration	Creation of databases and new courses	Creation of open repository of knowledge area	Development of skills	Availability	Support	Equipment (h, s, network) and services
KIRKPATRICK	ORG	Expectations - ROE	1	1	1	0	1	1	1
		Wellness SROI	1	0	0	0	1	1	1
		Costs - ROI	1	1	1	0	1	1	1
	STUDENT	Expectations - ROE	1	1	1	1	0	0	0
		Wellness SROI	1	1	1	1	1	1	1
		Costs - ROI	1	1	1	1	0	0	0
	TEACHER	Expectations - ROE	1	1	1	0	1	1	1
		Wellness SROI	0	1	1	0	1	1	1
		Costs - ROI	1	1	1	0	0	0	0
DUART	ORG	Expectations - ROE	0	1	1	0	0	0	0
		Wellness SROI	1	0	0	1	1	1	1
		Costs - ROI	1	1	1	1	1	1	1
	STUDENT	Expectations - ROE	1	1	1	1	1	0	0
		Wellness SROI	1	1	1	1	1	1	1
		Costs - ROI	1	1	1	1	0	0	0
	TEACHER	Expectations - ROE	1	1	0	0	1	1	1
		Wellness SROI	0	0	0	0	0	0	0
		Costs - ROI	1	1	1	0	1	0	0
	ORG	Expectations - ROE	1	2	2	0	1	1	1
		Wellness SROI	2	0	0	1	2	2	2
		Costs - ROI	2	2	2	1	2	2	2
	STUDENT	Expectations - ROE	2	2	2	2	1	0	0
		Wellness SROI	2	2	2	2	2	2	2
		Costs - ROI	2	2	2	2	0	0	0
	TEACHER	Expectations - ROE	2	2	1	0	2	2	2
		Wellness SROI	0	1	1	0	1	1	1
		Costs - ROI	2	2	2	0	1	0	0

Fig. 3. Heat map of influence factors of the models studied. Perspectives transfer of knowledge and infrastructure

References

1. Bojko, A.: Informative or misleading? Heatmaps deconstructed. In: Jacko, J.A. (ed.) Human-Computer Interaction. New Trends, pp. 30–39. Springer, Heidelberg (2009)
2. Duart, J.M.: ROI y e-learning: más allá de beneficios y costes. Documento en línea (2002). http://www.uoc.edu/web/esp/art/uoc/duart0902/duart0902.Html
3. Ferreras-Fernández, T., et al.: Visibilidad de la literatura gris científica a través de repositorios. El caso de las tesis doctorales en GREDOS. In: XV Workshop de REBIUN sobre proyectos digitales y VI Jornadas de OS-Repositorios 11–13 de marzo de 2015, Córdoba, Spain (2015)
4. Ferreras-Fernández, T., Martín-Rodero, H., García-Peñalvo, F.J., Merlo-Vega, J.A.: The systematic review of literature in LIS: an approach. Paper presented at the Proceedings of the Fourth International Conference on Technological Ecosystems for Enhancing Multiculturality (2016)
5. Garcia Mestanza, J., Gómez, J.L.: MODELO DE MEDICIÓN DE LA CALIDAD DOCENTE DE LA UNIVERSIDAD DE MÁLAGA EN ENTORNOS VIRTUALES DE APRENDIZAJE AJUSTADO A LOS NUEVOS PLANES DE ESTUDIO
6. García-Peñalvo, F.J.: La universidad de la próxima década: la universidad digital (2011)
7. García-Peñalvo, F.J.: Ecosistemas de aprendizaje adaptativos. Grupo de investigación en Interacción y eLearning (GRIAL) Instituto de Ciencias de la Educación, Departamento de Informática y Automática, Universidad de Salamanca (2016)

8. Llorens, F.: La tecnología como motor de la innovación educativa. Estrategia y política institucional de la Universidad de Alicante. Arbor, 185(Extra) (2009)
9. Llorens, F., Molina, R., Compañ, P., Satorre, R.: Technological ecosystem for open education. In: Neves-Silva, R., Tsihrintzis, G.A., Uskov, V., Howlett, R.J., Jain, L.C. (eds.) Smart Digital Futures, vol. 262, pp. 706–715. IOS Press, Amsterdam (2014)
10. Meza-Bolaños, D.V., Compañ-Rosique, P., Satorre-Cuerda, R.: Designing a model to estimate the impact and cost effectiveness of online learning platforms. In: Proceedings of the Fourth International Conference on Technological Ecosystems for Enhancing Multiculturality, pp. 1109–1114 (2016)
11. Planella, J., et al.: Del e-learning y sus otras miradas: una perspectiva social. Revista de Universidad y Sociedad del Conocimiento (RUSC), vol. 1 (2004)
12. Rubio, M.J.: Enfoques y modelos de evaluación del e-learning (2003)
13. Peña Zepeda, H.H.: El Rol de la Evaluación en la Tecnología Instruccional. In: Congreso Virtual sobre Tecnología, Educación y Sociedad (2015)
14. Siddaway, A.: What is a systematic literature review and how do I do one? https://www.stir.ac.uk/media/schools/management/documents/centregradresearch/How%20to%20do%20a%20systematic%20literature%20review%20and%20meta-analysis.pdf
15. Roux, H.N.: El SROI (social return on investment): Un método para medir el impacto social de las inversiones. Análisis financiero 113, 34–43 (2010)
16. Wiepcke, C.: 360-degree evaluation of e-learning measures: taking into account phases, levels, stakeholders and methods. Ubiquitous Learn. 3, 57–68 (2011)

On-Ramps to Learning: The Progression of Learners Through Topics in the Online LabVIEW Forum

Christopher Scaffidi[✉]

Oregon State University, Corvallis, OR 97330, USA
scaffidc@engr.oregonstate.edu

Abstract. Online forums can facilitate collaborative learning in situations where instructors impose structure promoting constructive interaction among students. This paper presents an investigation of how well an online forum, such as the LabVIEW programmer forum, supports learning in the absence of instructor-imposed structure. This study focuses first on whether specific topics served to draw users into the community and second on whether users displayed evidence of learning over time. Unsupervised machine learning on 475,094 posts in the LabVIEW forum identified 974 topical clusters among these posts, and statistical analysis confirmed that a minority (30%) of clusters accounted for over 70% of users' initial posts. Linear regression revealed that subsequent posts by each user were indeed more likely to be flagged by the community as valuable, offering potential evidence of learning. However, this trend was not strong or uniform, suggesting the need for additional innovations in information technologies to support independent learning.

Keywords: Information technologies in education · Online forums

1 Introduction

Online forums play a crucial role in supporting independent- and distance-learning. For example, with programing skills becoming integral to many occupations such as engineering and science [1, 2], online forums can help novice programmers to overcome barriers to skill development in independent-learning settings [3, 4]. In addition, distance education courses can leverage online forums as a means of facilitating discussion among students to reinforce learning from live remote lectures, recorded videos and other technical resources (e.g., virtual simulation environments) [5]. One study found that the highest levels of learning can occur when users take advantage of forums' support for "bringing different ideas together" [6]. With the rising importance of forums in distance education, tools now exist for automatically grading students on how well they contribute to online discussions [7].

Analyses of forum data have provided insights into the broad range of ways in which users take advantage of these resources. For example, students in distance education courses most commonly use them for getting help with practical applications and with understanding abstract concepts [6]. Getting help with specific problems is

© Springer Nature Switzerland AG 2019
Á. Rocha et al. (Eds.): ICITS 2019, AISC 918, pp. 791–801, 2019.
https://doi.org/10.1007/978-3-030-11890-7_74

also the primary use of forums even outside of structured courses [8, 9]. At their best, forums enable participants to access fast, direct, immediately actionable help from experts [10], and new tools are being developed to help users quickly identify experts within the community who can aid in solving problems [11].

While the studies and other research cited above confirms the *interest* of teachers and students in using forums, but to what extent do online forums aid in *learning*?

In distance education, research has demonstrated that online forums can facilitate collaborative learning, especially to the extent that they aid discussions among small groups of students [12]. In such settings, the amount of learning has correlated with the extent to which people act as "repliers" who answer other students [13].

Unfortunately, concerning independent learning *outside* of a structured context such as distance education courses, the evidence is somewhat discouraging. For examples, studies of one online forum for an animation programming language have found that the code posted by most users grew in sophistication only slowly [14, 15]. Likewise, an online forum exists for LabVIEW, a visual programming language enabling engineers and scientists to draw structured diagrams that compile into executable programs [16]; a study of this forum, too, found low rates of growth in the sophistication of code posted by users over time [17].

Online forums frequently manifest high rates of dropout, where users stop participating quickly after joining, in the absence of an instructor forcing students to keep using the forum [9, 14, 18]. In one study of two forums' users, those who did not receive replies to their questions (or who received insulting replies) typically never returned to the forum [9]. Even those that did return rarely transitioned into a role of helping other users [9].

Given these concerns about the high rate of dropout and potential impacts on learning, the current paper examines user data from the LabVIEW forum, with a goal of understanding the extent to which users exhibited evidence of learning as they engaged with different topics within the forum. This will help to reveal whether users are simply coming in, getting help with one topic and then leaving—or, alternatively, entering the forum and accumulating some expertise before dropping out.

Specifically, the current study addresses two questions. RQ1: To what extent do *specific topics serve as predominant entry points* to the online community? RQ2: As users gain experience in the community, *do subsequent contributions manifest increasing value?*

2   Materials and Methods

2.1   Data Source

A data set was obtained as described in prior work [17], wherein a program was written that automatically retrieved all LabVIEW online support resources from the websites of National Instruments (makers of LabVIEW). These included all threads in the Lab-VIEW forum; each of these consisted of one or more post, which the program parsed out of the threads. The words contained in each resource were then identified. In addition, any LabVIEW code attached to each resource was parsed, in order to determine what programming primitives they contained (e.g., loop, open socket, read from socket, write to file, etc.) LabVIEW is a visual programming language: the code is pictures. Therefore, parsing LabVIEW code including parsing screenshots of code embedded in resources' textual content as images (i.e., the program doing this parsing, as explained in prior work, could automatically examine screenshots of code and thereby determine what programming primitives the code contained [17]).

Each resource was then represented as a vector in a multi-dimensional space, with one dimension per distinct word or programming primitive. Following a common construction based on the TF-IDF model [19], each vector entry equaled the product of one factor indicating how frequently the word or programming primitive appeared in the resource, times a second factor indicating how rarely the word or programming primitive appeared across all resources [4, 17]. The TF-IDF model is an appropriate model, as past work has shown it effectively and accurately summarizes the content of online forum threads [3, 4, 17, 20].

The program then used k-means unsupervised machine learning to assign each resource to one of 1000 different clusters (see [4] for details and specifics of quality-assurance during clustering). Prior studies have confirmed that clustering online Lab-VIEW threads in this fashion does tend to gather threads together into a cluster that addresses an identifiable topic [3, 4, 17]. For example, one cluster might relate to implementation of a hardware control loop, while another cluster might relate to configuration of an XY scatter plot.

Finally, the program iterated through the list of users and, for each, selected the first 100 posts in sorted order (or fewer for users who lacked 100). In the process, it also took note of the topic cluster where each post's thread had been assigned. In the following sections, **cluster[u, i]** will refer to the topic cluster where post i of user u was placed, where i = 0 refers to the initial post of a user.

2.2   Data Analysis

To answer the first research question—whether clear topical entry points exist in the forum—the distribution of initial posts across distinct clusters was computed, thereby making it possible to identify which clusters contained the most initial posts. The proportion of all posts contained in these clusters was then calculated, to ascertain whether a small set of clusters accounted for most of the initial posts. Clusters containing exceptionally large numbers of initial posts were then examined, in order to understand what topics they involved. Furthermore, the distribution of non-initial posts was also calculated and compared to the distribution of initial posts using Pearson's chi-squared test, thereby providing a statistical test of whether clusters accounted for a disproportionate amount of initial posts relative to what would be expected based on the distribution non-initial posts.

To answer the second research question, the analysis examined the relationship between user experience and the likelihood that the user's next post would be valuable. This analysis took advantage of the fact that the LabVIEW forums offer multiple ways for other users in the community to explicitly flag posts as valuable. Two measures of experience were considered: the user's total number of posts up to the point of the next post, and the total number of distinct clusters into which the user had previously posted. These data were analyzed with linear regression.

3 Results

3.1 Specific Topics as Predominant Entry Points

A total of 180,881 pages including 171,274 forum threads were retrieved from the National Instruments website, which included 963,940 posts. Unsupervised learning was performed on these threads to produce 1000 clusters. Of the forum posts, 475,094 were among each user's first 100 posts and thus retained for further analysis in the current study. Of the 1000 clusters, 974 of contained at least one thread that included at least one post that appeared among the first 100 posts of a forum user. Thus, these 974 clusters thus contained the entirety of all users' early histories in the LabVIEW forum since its inception. Users made a median number of 40 posts each (i.e., for half of the users, i ranges only from 0 to 39).

Almost all of the initial posts (i = 0) appeared within relatively few clusters. For example, 30% of the clusters (280 of 974) together contained 70% of users' initial posts. The top 1% of the clusters (10 of 974), in fact, accounted for over 10% of users' initial posts.

These top ten clusters generally related to topics that users might plausibly experience when just starting out with LabVIEW. Examples included getting drivers installed, reading data from hardware, and calling existing code written in another language (Table 1). Topics appearing in clusters with relatively fewer initial posts—and a higher proportion of non-initial posts—generally involved somewhat more complex topics. For example, the middle five clusters shown in Table 1 related to invoking command-line scripts, performing computations with complex numbers or structured data, and performing more sophisticated functions involving hardware than simple data acquisition. Finally, clusters that contained the most non-initial posts but zero initial posts discussed other advanced topics. These included, for example, integrating programmatically with simulation systems, building LabVIEW code into standalone executables (which generally makes sense only to do once a program is already working in LabVIEW), and designing as well as implementing relatively advanced algorithms (including signal processing).

The descriptive data above suggest some clusters play the role as *"entry points"* where initial posts have tended to appear, whereas other clusters have played a subsequent role where later posts have appeared with relatively higher prevalence. Are these apparent differences in roles significant? Or put in statistical terms, has the distribution of initial posts among clusters differed from the distribution of non-initial posts among clusters?

To answer this question, a chi-squared test was performed on the second and third columns of Table 1, thus testing whether the distribution of initial posts differed from the distribution of non-initial posts. The result was significant (P < 0.0001, χ^2 = 7748.3, df = 973, N = 974 distinct clusters), confirming initial posts were

Table 1. Representative clusters, including the 4 that contained the most number of users' initial posts (clusters 1–4), the 4 that contained zero initial posts (i.e., i = 0) but had the highest number of non-initial posts (clusters 971–974), and 4 from the middle of the distribution (clusters 278–281, near the point at which 30% of clusters together contained 70% of all initial points).

Cluster	Number of initial posts	Number of non-initial posts	Topic discussed by posts whose threads appeared in this cluster
1	170	4491	Reading from VISA device via serial port
2	135	1330	Installing and configuring Agilent drivers
3	124	1797	Reading from VISA device via USB
4	100	2015	Invoking existing code via ActiveX interfaces
.	.	.	.
.	.	.	.
.	.	.	.
278	11	353	Running DOS shell commands from LabVIEW
279	11	374	Setting up robotics hardware kits
280	11	410	Computations with complex numbers
281	11	479	Detecting hardware & operating system events
.	.	.	.
.	.	.	.
.	.	.	.
971	0	489	Integration with Simulation Interface Toolkit
972	0	567	Errors while building executables
973	0	628	Diverse algorithm problems involving loops
974	0	1108	Signal processing on waveforms

distributed in a significantly different way than non-initial posts. Thus, certain clusters and associated topics have indeed served as entry points to the online community.

It is worth noting that even though certain clusters served as *entry* points into the community, users did not stay in those entryways very long. On average, each user's next post (in cluster[u, 1]) had only a 33% chance of being in the same cluster as that user's initial post (cluster[u, 0]). By the next post, users already had only a 15% chance of still posting into the same cluster. In other words, entry clusters were useful for drawing users into the online forum, where they then began interacting within threads associated with topics in other clusters.

Recall, however, that not all users posted 100 messages; they contributed to a median of only 40 posts each. In fact, 10% of users never contributed beyond their initial post, and 30% posted a dozen times or less. As a result, 17% of users posted to only 1 cluster ever, 22% to 1 or 2 clusters, 29% to 6 or fewer clusters, and 55% to 20 or fewer. To summarize, certain clusters served as entryway doors to the community, and a typical user rapidly moved out of those clusters into several additional clusters, but typically not more than 20.

3.2   Making Increasingly Valuable Posts with Experience

The second research question was whether users contributed increasingly valuable posts with increasing experience after the initial post of each. A post was considered valuable if it was explicitly marked as a solution (i.e., the person who started the thread considered the post to be a solution to the problem that motivated the thread) or if it received any kudos (i.e., indications of approval from other members of the community).

For a given post, two measures were used to represent the corresponding user's experience.

The first measure of experience was post number—i.e., the number of prior posts for that user. The proportion of posts marked as valuable rose almost monotonically and approximately linearly with the post number (Fig. 1). Specifically, it rose from approximately 4% at i = 1 to approximately 6% at i = 40 (i.e., the median number of posts per person, as noted in Sect. 3.1), and then as high as 10% at i = 100.

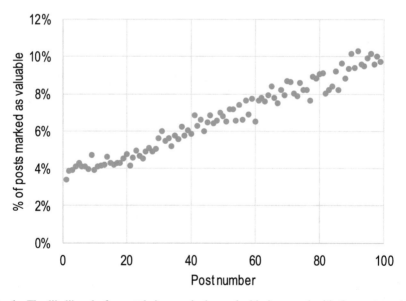

Fig. 1. The likelihood of a post being marked as valuable increased with the post number.

The second measure, indicating in some sense the breadth of experience, was the number of distinct clusters into which the user had previously posted. The average value of posts, for a given number of distinct prior clusters, showed little change until the number of clusters rose to 30—and, thereafter, displayed wide scatter (Fig. 2).

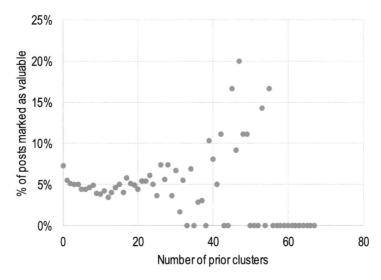

Fig. 2. The likelihood of a post being marked as valuable did not follow a consistent trend with respect to the number of prior clusters into which the corresponding user had previously posted.

This increased scatter is attributable to the fact that fewer than half of users ever reached this high number of 30 clusters (Sect. 3.1). Fewer than a third of users ever contributed this much. Consequently, most posts occurred in the portion of Fig. 2 that evidences a flat relationship between average estimated value and the number of prior clusters (i.e., $0 < i < 30$).

The analysis of the post's value, versus number of prior posts and number of prior clusters, was significant ($P < 0.0001$, $F(2, 465120) = 1649$, $t = 54.1$ for the coefficient on post number, $t = -27.0$ for the coefficient on number of prior clusters, $N = 465123$ posts after the initial post). Although this confirms that users did make increasingly valuable contributions with number of prior posts, from the standpoint of statistical significance, the exceptionally low $R^2 = 0.007$ indicates that other factors explain the vast majority of variance in posts' value. In addition, the fact that the second coefficient was *negative* (i.e., post value versus prior number of clusters) is troubling and suggests that prior experience with other topics offers little information for predicting the value of future posts.

4   Discussion

RQ1: To what extent do specific topics serve as predominant entry points to the online community? The analysis revealed that the distribution of initial posts differed significantly from the distribution of non-initial posts. Approximately 1% of clusters accounted for 10% of initial posts and 30% of clusters accounted for 70% of initial posts. The clusters with the highest proportions of initial posts did seem more topically relevant to the anticipated experiences of new users, while clusters with zero initial posts and high numbers of non-initial posts did seem relevant to more advanced topics. It is thus reasonable to believe that specific clusters—and their corresponding topics—do serve as entry points to the online community. Users quickly spread out from their initial topics within a matter of a few posts. On average, users stopped posting around the point when they had posted into approximately 20 clusters.

RQ2: As users gain experience in the community, do subsequent contributions manifest increasing value? The value of posts rose moderately with the number of prior posts, improving by half (from 4% to 6%) by the point where approximately half of users stopped contributing posts. On the other hand, the value of posts did not show any consistent relationship relative to the number of topic clusters into which the corresponding user had previously posted—and, in fact, if any relationship did exist, it was that of a negative correlation.

5   Conclusions and Future Work

Overall, this study indicates that LabVIEW users generally enter through specific topics, from which they rapidly branch out, gaining some skill as they gain experience. This increase, however, was moderate: by the time a median user dropped out (around post 40 and topic 20 on average), posts were still only half again as likely to be marked as valuable as when the user started. The conclusion is users contribute to a wide range of topics in the forums, but the evidence of their learning is limited.

This conclusion is consistent with that of prior work [17] examining the complexity of users' posted code as a function of time spent in the forums. That study found only a slight long-term upward trend in the size of code attached to forum posts and in the diversity of different LabVIEW primitives used. Moreover, it found that posted code showed a decline in complexity during the period right after users began contributing to the forums. The work is also consistent with prior studies noting the difficulty of learning LabVIEW solely from available online resources, both due to the lack of organization among these resources [3] and the limited effectiveness of the forum search engine for finding resources [4].

The current study may hint at another possible factor that could limit users' ability to build expertise. Specifically, the negative correlation between post value and number of prior clusters (Sect. 3.2) suggests that the rapid broadening of users' focus might be somewhat counter-productive in terms of enhancing their ability to make valuable posts. Perhaps there would instead be some benefit to users if they were to maintain focus in a relatively narrow range of topics long enough to build expertise. Further research will need to investigate this possible explanation for the results. In addition, studies to address other potential avenues of investigation.

First, a key limitation of the current research is its focus on user activity and explicit indications of forum posts' value, which provide only an indirect and limited view into the process and manifestation of learning. Although the conclusion is largely consistent with that of other work (above), future research could draw upon a rich array of other methods for obtaining other views into learning of LabVIEW. For example, a longitudinal study of users (potentially using periodic questionnaires, tests and interviews) could further establish what common sequences of topics are often encountered, how users' understanding of LabVIEW concepts develops over time, and how proficient they become at performing activities not reflected in the forums (e.g., designing systems and creating code).

Second, future work could aim to help users develop expertise with LabVIEW faster and more successfully. This could involve providing users with better scaffolding [21] to help them identify, focus on, and synthesize learning from appropriate resources. For example, the LabVIEW programming tool (the IDE) could offer a user interface that helps users to *systematically* work through examples that incrementally help the user accumulate expertise in a chosen area. These resources could include examples that already ship with the IDE as well as resources posted online. This user interface could guide novices to, and walk them through, resources relevant to the topics commonly encountered by new users. They could then allow each user to define personalized learning goals and proceed through additional resources that build mastery in the selected direction.

Finally, research could also investigate the appropriateness and effectiveness of alternate technology that goes beyond the thread-based forum model so dominant right now. Possibilities include multi-cast tutoring systems [22] and interactive program visualizations [23]. Unfortunately, all of these approaches can cost more than forums to populate with content, to the extent that they typically depend on paid staff rather than users' peers. Developing a more crowd-sourced approach that combines low cost and high effectiveness could pay substantial dividends to users of LabVIEW and other programming languages.

Acknowledgements. National Instruments funded this research and gave permission to download the contents of the online forums. Any opinions, findings, and conclusions or recommendations expressed in this material are those of the authors and do not necessarily reflect the views of National Instruments.

References

1. Guo, P.: Why scientists and engineers must learn programming. Commun. ACM (2013). https://tinyurl.com/me7fu4s
2. Nguyen-Hoan, L., Flint, S., Sankaranarayana, R.: A survey of scientific software development. In: International Symposium on Empirical Software Engineering and Measurement (2010)
3. Scaffidi, C.: What training is needed by practicing engineers who create cyberphysical systems? In: IEEE Euromicro Conference series on Software Engineering and Advanced Applications (SEAA), pp. 298–305 (2015)

4. Scaffidi, C., Chambers, C., Surisetty, S.: A code-centric cluster-based approach for searching online support forums for programmers. In: IEEE International Conference on Machine Learning Applications, pp. 1032–1037 (2015)
5. Fonseca, P., Juan, A., Pla, L., Rodriguez, S., Faulin, J.: Simulation education in the internet age: some experiences on the use of pure online and blended learning models. In: Winter Simulation Conference, pp. 299–309 (2009)
6. Çakıroğlu, U.: Analyzing the effect of learning styles and study habits of distance learners on learning performances: a case of an introductory programming course. Int. Rev. Res. Open Distrib. Learn. 15(4), 161–185 (2014)
7. Nunes, B., Tyler-Jones, M., de Campos, G., Siqueira, S., Casanova, M.: FAT: a real-time (F) orum (A)ssessment (T)ool to assist tutors with discussion forums assessment. In: ACM Symposium on Applied Computing (2015)
8. Rekha, V., Venkatapathy, S.: Understanding the usage of online forums as learning platforms. Procedia Comput. Sci. 46, 499–506 (2015)
9. Singh, V.: Newcomer integration and learning in technical support communities for open source software. In: ACM International Conference on Supporting Group Work, pp. 65–74 (2012)
10. Teo, H., Johri, A.: Fast, functional, and fitting: expert response dynamics and response quality in an online newcomer help forum. In: ACM Conference on Computer Supported Cooperative Work & Social Computing, pp. 332–341 (2014)
11. Kardan, A., Omidvar, A., Behzadi, M.: Context based expert finding in online communities using social network analysis. Int. J. Comput. Sci. Res. Appl. 2(1), 79–88 (2012)
12. Shaw, R.: The relationships among group size, participation, and performance of programming language learning supported with online forums. Comput. Educ. 62(1), 196–207 (2013)
13. Shaw, R.: A study of the relationships among learning styles, participation types, and performance in programming language learning supported by online forums. Comput. Educ. 58(1), 111–120 (2012)
14. Matias, J., Dasgupta, S., Hill, B.: Skill progression in Scratch revisited. In: ACM CHI Conference on Human Factors in Computing Systems, pp. 1486–1490 (2016)
15. Yang, S., Domeniconi, C., Revelle, M., Sweeney, M., Gelman, B., Beckley, C., Johri, A.: Uncovering trajectories of informal learning in large online communities of creators. In: ACM Conference on Learning@ Scale, pp. 131–140 (2015)
16. Kodosky, J., MacCrisken, J., Rymar, G.: Visual programming using structured data flow. In: IEEE Workshop on Visual Languages, pp. 34–39 (1991)
17. Scaffidi, C.: Changes in LabVIEW programs posted to an online forum as users gain experience. Int. J. Softw. Eng. Appl. 9(1), 1–10 (2018)
18. Velasquez, A., Wash, R., Lampe, C., Bjornrud, T.: Latent users in an online user-generated content community. Comput. Support. Coop. Work 23(1), 21–50 (2014)
19. Chakrabarti, S.: Mining the Web: Discovering Knowledge from Hypertext Data. Elsevier, San Francisco (2002)
20. Awasthi, P., Hsiao, I.: INSIGHT: a semantic visual analytics for programming discussion forums. In: International Learning Analytics and Knowledge Conference, pp. 24–31 (2015)
21. Hogan, K., Pressley, M.: Scaffolding Student Learning: Instructional Approaches and Issues. Brookline Books, Cambridge (1997)

22. Guo, P.: CodeOpticon: real-time, one-to-many human tutoring for computer programming. In: ACM Symposium on User Interface Software & Technology, pp. 599–608 (2015)

23. Guo, P., White, J., Zanelatto, R.: Codechella: multi-user program visualizations for real-time tutoring and collaborative learning. In: IEEE Symposium on Visual Languages and Human-Centric Computing, pp. 79–87 (2015)

Learning Objects Evaluation from User's Perspective

María de los Ángeles Alonso$^{(\boxtimes)}$, Verónica Martínez, Iliana Castillo, and Yira Muñoz

Autonomous University of Hidalgo State, Computing and Electronics Academic Area, Carretera Pachuca-Tulancingo Km 4.5, 42184 Hidalgo, Mexico {marial, vlazcano, ilianac, yira}@uaeh.edu.mx

Abstract. Inside the learning process, the Learning Objects (LO) are the most innovative didactic materials that can be used in distance education, face-to-face teaching, organizational knowledge management, training processes and continuing education or curriculum innovation, because its capacity for reusability and application in diverse contexts and circumstances. LO's evaluation is a very important task that allows to select the object based on their quality and relevance. Currently, there are several instruments for this purpose which perform the evaluation focused on different aspects such as, for example, the content of the LO, its interface and some of the technological aspects but without deepening in the pedagogical aspect of learning. In the present work, an evaluation is proposed from the user's perspective, where the criterion or indicators are based on the pedagogical dimensions of Reeves and Bloom's taxonomy focused on the digital age.

Keywords: Learning objects evaluation · Pedagogical indicators of quality · Quality in didactic resources

1 Introduction

One of the main challenges faced during the formation process is to facilitate the reutilization of Learning Digital Resources (LDR) which refers to all of multimedia documents such as text, image, sound or video used to meet an educational objective.

On this regard, there is a strong tendency to use Learning Objects (LO) as basic exchange units through the different virtual platforms. These resources ensure the reusability and interoperability of the LDR without losing its functionalities and characteristics [1].

A Learning Object is described as: *A unit with a learning objective, characterized by being digital, independent, with a single or few related ideas, accessible through the metadata and aimed to be reutilized in different contexts and platforms* [2, p. 2].

High quality educational resources shall have measurable characteristics. This is especially the case of the LO since, as a didactic resource, needs to comply with pedagogic objectives and ensure learning.

© Springer Nature Switzerland AG 2019
Á. Rocha et al. (Eds.): ICITS 2019, AISC 918, pp. 802–813, 2019.
https://doi.org/10.1007/978-3-030-11890-7_75

The assessment of the LO is based on different aspects such as the following:

- *Technological*: to evaluate if a software product has a high-quality design and concordance.
- *Interface*: to qualify everything related to the metadata selection by identifying what is appropriate and the elements to be integrated (explanatory texts, graphs, photographs, music, spoken text, video, animation, evaluations, experimentation modules, hyperlinks) in order to effectively use them in the learning process of different contexts where they are applied.
- *Content*: considers the level of compliance with the standards and recommendations defined by the LO.
- *User*: evaluates whether the learning objectives are met and to what extent acquisition of knowledge in the student takes place.

1.1   Related Works

The evaluation of didactic materials and especially of the learning objects is not a new subject. Several proposals have been raised to establish different dimensions and criteria to be considered when evaluating their quality.

A highly addressed dimension has been the pedagogical one with a view to measure whether the content and activities of the objects are suitable to the student's developmental stage and to ensure their complexity does not hinder understanding of the theme presented [3], if the object facilitates learning [4, 5] and if it allows for knowledge acquisition [6, 7] or competencies acquisition [8] in some cases, include the pedagogical and didactic-curricular categories, where the former measures the generated student's attention and motivation as well as the interactivity and creativity that promotes, and the latter considers if the created object is suitable to the educational level in which is used, the learning timing, the activities, and if the object provides feedback [8, 9].

Similar to the pedagogical dimension is the educational one which is proposed to evaluate if the object is interesting for the student [5]. Some further criteria considered are the learning objectives and goals, specifically to observe if objects are aliened to them [6, 7, 10]. It also evaluates the object structure [11], its relevance and veracity [8].

Another not less important dimension is the technological one, where technical criteria of the object are assessed. It refers to the integration of means and materials such as visuals, audios, etc. [3], however, in some cases they are referred in a general way and do not specify the kind of criteria that is being considered [6, 12], specifically refer to the design and presentation of the object [6, 8–11, 13], to usability [6, 8, 14], reusability [8, 10, 15] and mobility, which basically consists of the interoperability among different platforms.

More recently, the Evaluareed project was created [14] with similar objectives to the previously described but considering the most utilized criteria according to the research done by the authors. Finally, based on the Information Systems approach, it is proposed the quality evaluation of LO's through the services theory [14] which, adapted to the context of didactic resources, retakes the pedagogical and technological dimension described above.

A further aspect considered in the quality evaluation of didactic resources has is the evaluator himself, who can be related to the development process or the usage of the resource. Therefore, developers, theme experts, pedagogues, teachers, students and general users may also be evaluators.

Most proposals consider the user or student [3–10, 13, 14] which promotes student's engagement in his own learning process, feels motivated and recommends the already developed objects for other students to be used [7], other proposals consider only the teacher [16], and in some cases both [7, 9, 13], in order to identify whether the object allows for the student's acquisition of knowledge and if it focuses on the objective planned by the teacher [7].

The work Analysis of Evaluation Criteria of Didactic Digital Materials states that additionally to students and teachers, researchers and IT/communication specialists should also evaluate objects [13]. In [8] propose external evaluators, in [11] object creators and reviewers are also evaluators and in [5] the administrator of the repository where objects are stored, and experts are also considered in the evaluation process.

2   Methodology

The aim of this research is to evaluate the quality of didactic digital resources from the user's perspective in compliance with the learning objective and the knowledge acquisition. For that, the outcome resulted from the application of two evaluation instruments is analyzed from the student's and teacher's perspective respectively. These instruments involve aspects of different dimensions such as epistemological, pedagogical philosophy, previous and new knowledge, motivational, structural and the ones related to the Bloom's taxonomy. This section describes the steps followed in the research methodology during the evaluation of didactic resources from the definition of the criteria applied to the analysis of the outcome resulted from the evaluation.

The following are the methodology steps used to evaluate the digital didactic resources:

(1) Define the quality assessment criteria considering students and teachers as users.
(2) Generate questionnaires addressing the established criteria.
(3) Identify a digital didactic resource being used in a given subject.
(4) Identify the random sample of users interacting with the LO.
(5) Apply the instruments to the identified sample.
(6) Analyze the questionnaires outcome according to the established criteria and in concordance with the objective of this research.

2.1   Criteria for Quality Assessment

Considering the approaches utilized so far and without losing sight of the technological aspect of the resource evaluation, this work deepens further in the pedagogical aspects, especially the ones regarding the impact of the resource on the student's learning. Therefore, the definition of the user's aspects was performed based on Reeves' pedagogical dimensions [17] and addresses the opinion of its main users: student and teacher. Additionally, and as distinctive feature of this kind of evaluations, includes the

Bloom's taxonomy oriented to the digital age [18], in order to classify the LO according to the competency level to be developed in the student with this material.

Reeves' pedagogical dimensions refer to the capacities of computer education to generate formative interactions, control student's progress, empower effective teachers, adapt to individual differences or promote cooperative learning. The pedagogical dimensions are based on a given aspect of a didactic theory or criterion to evaluate different computer education forms in which the technological aspect is intrinsically evaluated. On the other hand, Bloom's taxonomy does not focuses on tools or ICT themselves since they are the means, but rather on the use of all of them to remember, understand, apply, evaluate and create which he called *Thinking Skills*.

Table 1 depicts the defined indicators that address the pedagogical dimensions proposed by Reeves, understood from the student's and teacher's role.

Table 1. Indicators to measure the quality of the LO

Dimension	Indicator	User
Epistemological	The LO allows student to develop his own conclusions through criteria and reasoning	Student and Teacher
	The LO promotes learning acquisition through interactive activities that stimulate the student's senses	Teacher
Pedagogical Philosophy	–The LO follows an instructive approach which means the teacher explains why and how to learn the subject –The LO follows a constructivist approach where the student is autonomous and decides the way in which knowledge is acquired	Teacher
	The LO helped to build student's knowledge	Student
Underlying Philosophy	The introduction stimulates students learning	Student and Teacher
	The student is asked while using the LO	Student and Teacher
	The LO provides feedback on the given answers	Student and Teacher
	The LO presents suitable strategies for the knowledge and abilities to be acquired	Teacher
Goal Oriented	The LO was a useful didactic support to the student	Teacher
	The LO is appropriate to the educational level of the student (bachelor)	Teacher
Experience Value	The LO presents interesting/engaging cases and problems	Student and Teacher
	The LO allowed you to have a relevant knowledge experience in your life	Student
Teacher's Role	The LO allows you to perform as facilitator	Teacher
Flexibility	It is possible to choose sections of the LO according to the lesson strategies or activities	Teacher

(continued)

Table 1. (*continued*)

Dimension	Indicator	User
Error Value	You learnt from your errors while performing the evaluations of the LO	Student
Motivation Source	The LO activities provide some kind of stimuli or reward (grade) to motivate the student to answer them correctly	Student
	The LO evaluations provide some kind of stimuli or reward (grade) to motivate you to answer them correctly	Student
	The LO elements (interfaces, multimedia, games, etc.) motivate you in your learning	Student and Teacher
Suitability to individual differences	The LO establishes diverse learning strategies for students who learn in different ways	Student and Teacher

The Likert scale is used to answer the instrument with values ranging from 0 to 4 where the value 0 means It Can't be rated, the values ranging from 1 to 4 represent compliance with the asseveration from smaller to greater, being the values: Nothing, Little, Quite and Totally, respectively.

In addition to the evaluation of the quality of the Learning Object as didactic resource and teaching support, the teacher survey is completed with a question aimed at classifying the resource according to its educational objective, based on the thinking skills; remember, comprehend, apply, analyze, evaluate and create, defined by Bloom's taxonomy. This question is presented in Table 2.

Table 2. Indicators to classify the knowledge type of the LO according to Bloom's taxonomy.

Level	Indicator
Remember	Utilizes textual, graphic and/or multimedia elements that allow student to remember the content of this material
Comprehend	Contains elements such as bullets, steps, procedures or some strategies that allow student to understand content easily
Apply	Includes activities where the student can apply the acquired knowledge
Analyze	The theme leads to the analysis on the part of the student
Evaluate	The content leads student to evaluate something from his perspective
Create	Promotes the creation of new ideas, interest for knowledge and leads student to generate his own methods and techniques to solve problems

In order to assign a final grade to the LO, the following procedure is applied:

1. Following the grading equivalence formula showed below a value was assigned to each criterion from 0 to 4.

$$Equivalence = \frac{(Test\ Item\ grade * 10)}{4} \quad (1)$$

2. To grade the item with value 0 its equivalence is 0, for 1 is 2.5, for 2 is 5, for 3 is 7.5 and for 4 is 10.
3. The average of the above values is calculated.
4. Depending on the value obtained in the previous step, a qualitative grade is assigned to the learning object (excellent, good, regular or bad) using for that the Table 3.

Table 3. Final score of a LO's quality

Qualitative grade	Average
Excellent	9–10
Good	8–8.9
Regular	6–7.9
Bad	0–5.9

2.2   Selection of the Digital Teaching Resource

Regarding the digital didactic resource, it was chosen a Learning Object from Constructors in C++ theme from the Object-Oriented Programming subject for the IT Bachelor's degree at the Autonomous University of Hidalgo State. It is important to highlight that this is a presence-based modality program and the educational materials as well as the usage of research are educational support. This LO was developed following the Development Methodology for Learning Objects (MEDOA for its acronym in Spanish) [19]. Therefore, it has a structure, indicating its own methodology, being the same: cover, learning objective, introduction, theme development, examples, activities evaluations, glossary, references and credits.

2.3   Sample Selection

The object student population corresponds to a random sample of 24 students from a total of 85 forming the third semester students group registered in the above mentioned subject representing a 28.24% sample, considered enough to perform the research. Students are divided into four groups from which two attend the morning schedule and the other two the evening schedule.

On the other hand, for the selection of expert evaluators the description in [6] was followed where is stated that the greater the experience of the evaluators, the greater the number of problems they are able to detect. Other authors establish that the size of the effective sample ranges between 3 to 5 individuals, since the execution of the test itself with additional individuals is not likely to reveal new information, since most relevant problems are detected by the first ones [20]. Taking into account the above mentioned, four evaluator teachers were chosen from the educational program mentioned.

2.4   Application of Didactic Resource

Once the LO was identified and the teacher and student samples were defined, the following was performed:

(1) The students utilized the LO in order to reinforce learning of the Constructors in C ++ and were asked to perform the activities and integrated evaluations aimed at generating meaningful learning for them. After using the LO, they were contextualized in the importance of the process and quality evaluation of the material. Then, they were asked to answer the instrument which was created for them and aimed at this purpose.

(2) On the other hand, teachers were contextualized in the importance and intention of the educational material evaluation. Then, they were given access to the educational resource to be explored. Once they finished their work, they were given the instrument to answer the questions addressed to the assessment of the LO.

2.5   Application of the Evaluation Instrument

The quality evaluation of the Learning Object *Constructors in C++* was performed through the application of the instruments proposed in this work and implemented in the Google Forms tool for the advantages it provides. The general results of the LO evaluation done by students and teachers regarding its quality are shown in Fig. 1.

A detailed analysis of the answers provided to the instrument by the students in relation to Reeves' pedagogical dimensions is shown in Table 4, while in Table 5, the same information is presented but provided by the teachers.

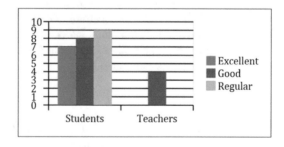

Fig. 1. Quality evaluation of the LO according to students and teachers.

From the results shown in Table 4, it can be observed that students evaluated the LO in general terms as Excellent and Good quality, which is a satisfactory outcome. In order to deepen in the results, the average value was calculated, obtaining Excellent in the dimensions *Pedagogical Philosophy* and *Error Value*, while in the rest it was Good. In order to assess the relevance of the average value, the standard deviation was calculated in each of the dimensions, whose values were in a range of 1.27 to 1.78, which suggests that the evaluations have been very close to the average value.

However, attention needs to be paid to the *Experience Value* dimension, where students rated *presents cases and problems which are interesting/stimulating for the student* 75%, which is not low but should be taking into account for further LO maintenance.

Table 4. Quality of the LO by pedagogical dimension according to students.

Dimension	Aspect to be evaluated	Indicator	Excellent	Good
Epistemological	How do individuals learn?	The LO allows student to develop his own conclusions through criterion and reasoning	43.5%	**47.8%**
Pedagogical Philosophy	Is knowledge acquired?	The LO helped the student to build knowledge?	**58.3%**	37.5%
Underlying Philosophy	Does it provide any stimulus?	Does the introduction stimulate student learning?	41.7%	**45.8%**
	Does it require an answer?	Student is enquired while learning with the LO	33.3%	**58.3%**
	Is feedback provided?	The LO provides feedback to the answers given by the student	**52.2%**	39.1%
Experience Value	Does it provide a focal case or problem to retrieve and increase knowledge?	The LO provides interesting/stimulating cases or problems to the student	29.2%	**45.8%**
	Is the student aware of what he is capable of performing and is able to acquire useful knowledge for his personal life?	The LO allowed you to have a knowledge experience relevant to your life	**45.8%**	37.5%
Error Value	Does it provide opportunities to learn from errors?	You learnt from your errors while performing the evaluation of the LO	**54.2%**	41.7%
Motivation Source	Were rewards satisfactory?	The LO activities provide some kind of stimuli or reward (grade) that motivated student to answer them correctly	**45.8%**	50%
		The LO evaluations provide some kind of stimuli or reward (grade) that motivated student to answer them correctly	**45.8%**	**45.8%**
	Does it provide educational materials that include interactive multimedia?	The elements of the LO (interfaces, multimedia, games, etc.) motivate you to learn	37.5%	**45.8%**
Adjustment to individual differences	The no homogeneity of the students to learn is considered	The LO establishes diverse learning strategies for students with different learning styles	**45.8%**	41.7%

Table 5. Quality of the LO by pedagogical dimension according to teachers

Dimension	Aspect to be evaluated	Indicator	Excellent	Good
Epistemological	How do individuals learn?	The LO allows student to develop his own conclusions through criterion and reasoning	25%	**75%**
	Does it include mechanisms or tools that help build your own knowledge?	The LO promotes learning acquisition through interactive activities that stimulate the student's senses	**75%**	25%
Pedagogical Philosophy	Approach used for the knowledge acquisition	−The LO follows an instructive approach which means the teacher explains why and how to learn the subject. −The LO follows a constructivist approach where the student is autonomous and decides the way in which knowledge is acquired	**75%**	25%
Underlying Philosophy	Does it provide any stimulus?	The introduction stimulates students learning	**100%**	0%
	Does it require an answer?	Student is enquired while learning with the LO	25%	**75%**
	Is feedback provided?	The LO provides feedback to the answers given by the student	**100%**	0%
	Does it have a variety of learning strategies?	The LO presents the appropriate strategies for the knowledge or skill that must be acquired	0%	**100%**
Goal orientation	Does it present knowledge with efficacy?	The LO was useful as a didactic support for the student	**75%**	25%
		The LO is appropriate for the educational level of the student (Bachelor's degree)	**100%**	0%
Experience Value	Does it provide opportunities for learning from mistakes?	The LO presents interesting/stimulating cases or problems for the student	**75%**	25%
Teacher Role	Does it focus on giving the role of facilitator?	The LO is a tool that act as a facilitator	**75%**	25%
Flexibility	Is there freedom to modify the activities of the program?	The student can choose the sections of the LO according to the class strategies or activity	**100%**	0%
Motivation Source	Does it provide educational materials that include interactive multimedia?	The elements of the LO (interfaces, multimedia, games, etc.) motivate you to learn	**75%**	25%
Adjustment to individual differences	The no homogeneity of the students to learn is considered	The LO establishes diverse learning strategies for students with different learning styles	**50%**	**50%**

According to the Table 5, it can be seen the rigor in the evaluation by the teachers, who stated the student should be asked to a greater extent and the strategies to acquire knowledge should be improved in order to achieve the excellence of the Learning Object. In this case, the average value of all the dimensions was Excellent, there being a tie between Excellent and Good in the *Epistemological, Underlying Philosophy* and *Adjustment to individual differences* dimensions. Likewise, the standard deviation for each of the evaluated dimensions was calculated, and the results obtained were in a range of 0 to 1, which confirms that the values are very similar to the average value.

Finally, based on Bloom's taxonomy, it was identified that knowledge type that the LO promotes in students is *Analysis* which means that the Learning Object theme leads to an analysis process on part of them.

2.6 Results Analysis

The outcome analysis of the quality evaluation of the Learning Object based on the criteria from different dimensions is performed in three parts considering the coincident criteria by students and teachers and the no-coincident ones independently; the ones evaluated by students and the ones evaluated by teachers.

For the 8 coincident criteria the result was Excellent (5 for 62.5%) for the Pedagogical Philosophy, Underlying Philosophy, Error Value, Motivation Source and Suitability to individual differences. Good (2 for 25%) in the Epistemological dimension and no coincident evaluations (1 for 12.5%), where teachers rated Excellent and the student Good.

In the 4 criteria exclusively for students, the result was (1 for 25%) for Experience Value. Good (2 for 50%) and for Motivation Source and similar evaluations Excellent and Good (1 for 25%).

For the 6 criteria exclusively for teachers the result was Excellent (5 for 83.3%) for the Epistemological dimension and Goal Oriented and Good (1 for 16.7%) for the Underlying Philosophy dimension.

These results show that from the Teacher's and Student's perspective, there is a coincidence regarding rating the Learning Object Evaluation as Excellent by 72.22% and Good by 38.88%, which shows a very good level of acceptance. However, and considering from a case study perspective, it highlights the need to pay attention to the aspects referred in the *Motivation Source* dimension, which evaluates if educational materials including interactive multimedia are provided; rated as good by students but leading to the objective of reinforcing this aspect in order to achieve excellence. On the other hand, 100% of the teachers validated a question from the *Underlying Philosophy* dimension, especially the one regarding the use of a variety of strategies rated as good, which also highlights the need to revise this aspect.

3 Conclusions

The developed works show the need to evaluate educational materials with the intention of favoring teaching and learning processes, as well as having techniques and tools that help to determine their value and usefulness. In this sense, by the end of this research we conclude that the evaluation of didactic materials is an essential task within the formation process as it validates their level effectiveness for the purpose they are intended; in other words, if their use aids to achieve the expected knowledge. In this work, the Reeves' pedagogical dimensions were especially valuable, allowing assessing the quality of the resource in depth from the pedagogical perspective, but intrinsically considering technological aspects so as to achieve a more effective use of the resource.

It should be noted that the most of the authors have made proposals focused only on certain dimensions, in some cases giving greater importance to pedagogical, in others to technological and some more in to usability, even, some of them consider specific indicators of only one dimension. However, in this work all the pedagogical dimensions proposed by Reeves were considered and questions are presented in each indicator that allow to evaluate them in greater detail, causing the results of the evaluation to be so much punctual and objective to be considered in the LO's improvement process.

Even though there are several revised proposals where the evaluation process of educational materials falls on students, a smaller amount considers together students and teachers, and only one considers just teachers. In this project, both actors were considered, because of the importance and transcendence of the role that they play in an educational process since the teachers are the ones who select the most appropriate materials to teach their classes and the students must be able to choose the sources of information that best support them in their learning process.

Regarding the LO classification based on Bloom's taxonomy, even though it does not value the object from the quality perspective, this outcome will have very relevant role when choosing a resource based on the type of thinking skill aimed to be developed, which would come to facilitate and favor teaching.

A contribution of the present investigation is that, through an evaluation process, focused mainly on pedagogical aspects, it is evident that an LO satisfies the learning needs of the students and the teaching of the teachers, but not in the same degree.

References

1. Menéndez, V.H., Castellanos, M.E., Prieto, M.E.: e-Learning y Objetos de Aprendizaje. Facultad de Matemáticas-UADY, Mérida, Yucatán. México (2012)
2. Morales, E., García, F.J., Barrón, Á., Berlaga, A.J., López, C.: Propuesta de Evaluación de Objetos de Aprendizaje. In: IV Simposio Pluridisciplinar sobre Diseño, Evaluación y Desarrollo de Contenidos Educativos Reutilizables, SPDECE. Bilbao, España (2005)
3. García, A.: Evaluación de recursos tecnológicos didácticos mediante e-rúbricas. RED-Revista de Educación a Distancia, 49, art. 13 (2016). http://dx.doi.org/10.6018/red/49/13
4. Massa, S., De Giusti, A., Pesado, P.: Objetos de Aprendizaje: Metodología de Desarrollo y Evaluación de la Calidad. 315. La Plata, Buenos Aires, Argentina: Universidad Nacional de la Plata: Facultad de Informática (2012)
5. Tabares, M.V., Duque, M.N.D., Ovalle, C.D.A.: Modelo por capas para evaluación de la calidad de Objetos de Aprendizaje en repositorios. Revista Electrónica de Investigación Educativa 19(3), 33–48 (2017)

6. Pinto, M., Gómez, C., Fernández, A., Vinciane, A.: Evaluareed: desarrollo de una herramienta para la evaluación de la calidad de los recursos educativos electrónicos. INVESTIGACIÓN BIBLIOTECOLÓGICA 31(72), 227–248. México (2015)

7. Baldassarri, S., Álvarez, P.: M-eRoDes: Involucrando a los estudiantes en la creación y evaluación colaborativa de objetos de aprendizaje. In: Actas de las XXII Jornadas sobre la Enseñanza Universitaria de la Informática, pp. 195–202, Universidad de Almería (2016)

8. Insuasty, E., García, M., Víctor, A., Insuasti, J.: Comparación de tres metodologías de evaluación de objetos de aprendizaje virtuales. Teoría de la Educación. Educación y Cultura en la Sociedad de la Información, 15(2), 67–85. Universidad de Salamanca, España (2014)

9. Massa, S., Rodríguez, B.D.: Objetos de Aprendizaje: propuesta de evaluación de calidad pedagógica y tecnológica. In: Actas del Congreso Iberoamericano de Ciencia, Tecnología, Innovación y Educación. Bs. As. Argentina. Organización Estados Iberoamericanos (OEI) (2014)

10. Gordillo, M., Barra, A., Quemada, V.: Estimación de la Calidad de Objetos de Aprendizaje en Repositorios de Recursos Educativos Abiertos Basada en las Interacciones de los Estudiantes. Educación XXI 21(1), 285–302 (2018)

11. Cepeda, O., Gallardo, I., Rodríguez, J.: Aspectos e indicadores para evaluar la calidad de los OA Revista de Universidad y Sociedad del Conocimiento (2017)

12. Barroso, J., Cabero, J.: Evaluación de objetos de aprendizaje en Realidad Aumentada: estudio piloto en el Grado de Medicina. Enseñanza & Teaching 34(2), 149–167 (2016)

13. Aguilar, I., Ayala, J., Lugo, O., Zarco, A.: Análisis de criterios de evaluación para la calidad de los materiales didácticos digitales. Revista Iberoamericana de Ciencia Tecnología y Sociedad 9(25), 73–89 (2014)

14. Velázquez, C., Álvarez, F., Muñoz, J., Cardona, P., Silva, A., Hernández, Y., Cechinel, C.: Un Estudio de la Satisfacción Obtenida con el Uso de Objetos de Aprendizaje. IX Conferencia Latinoamericana de Objetos de Aprendizaje y Tecnologías para el Aprendizaje (LACLO2014), Manizales, Colombia, pp. 248–256 (2014)

15. Collaguazo, P., Padilla, A., Chamba-Eras, L.: Propuesta de un Modelo Genérico para el Diseño y Valoración de Objetos de Aprendizaje basado en Estándares E-Learning. LACLO (2015)

16. Triana, M., Ceballos, J., Villa, J.: Una dimensión didáctica y conceptual de un instrumento para la Valoración de Objetos Virtuales de Aprendizaje. El caso de las fracciones Entramado 12(2), 166–186. Universidad Libre Cali, Colombia (2016)

17. Coll, C., Reeves, T., Hirumi, A., Petters, O.: Manual: Del docente presencial al docente virtual. Procesos formativos de enseñanza- aprendizaje on-line. Athabasca University-Canada's Open University (2003)

18. Churches, A.: Taxonomía de Bloom para la Era Digital (2007)

19. Alonso, M., Castillo, I., Martínez, V., Muñoz, Y.: MEDOA: Metodología para el Desarrollo de Objetos de Aprendizaje. In: 12a Conferencia Iberoamericana en Sistemas, Cibernética e informática: CISCI (2013)

20. Nielsen, J., Landauer, T.: A mathematical model of finding of usability problems. In: Proceedings of CHI 1993, Proceedings of the INTERACT 1993 and CHI'93 Conference on Human Factors in Computing Systems (1993)

ICT Integration in the Teaching/Learning Process of Natural Sciences for Seventh Grade Elementary Students'

Fanny Román[1]([✉]), Ramiro Delgado[1]([✉]), Christian Ubilluz[1]([✉]), and Cesar Bedón[2]([✉])

[1] Computer Science Department, Universidad de las Fuerzas Armadas ESPE, Av. General Rumiñahui S/N y Ambato, Sangolquí, Ecuador
fannyromans@gmail.com,
{rndelgado, cmubilluz}@espe.edu.ec
[2] Science School, Universidad Central del Ecuador, Av. America, Quito, Ecuador
cbedon@uce.edu.ec

Abstract. The teaching-learning process and all the technologies have evolved alongside, which has allowed new educative methods based on Information and Communications Technologies (ICT), and also has allow the fulfillment of the established educational institution's goals in their curricular blocks by using computers, internet, interactive classrooms, social networks, audio, video, images, among others. In brief, this research shows ICT integration at the learning process of a Natural Science Class for Seventh Grade Elementary Students' on "La Inmaculada" Franciscan Elementary School; in addition, it evaluates ICT contribution to validate the changes that have being made over students and teachers during years 2015–2016 and 2016–2017.

Keywords: TIC's · Technology · Communication · Teaching · Learning · Basic education

1 Introduction

At present, people live inside a changing life, also education and technology have changed, and they have innovated [1]. For instance, January 8 2008, UNESCO had published ICT's competence standards for teachers, in order to guarantee an educational process with transcendence and social impact [2].

Currently, people need a customized, didactic, critical and ubiquitous learning process, in pursuance of subjects' concepts, and it can be made through search and exchange of data, using technological resources with net connection. This gives the possibility of arguing and define concepts integrated with knowledge [3]. Moreover, it develops technological research competences, and it allows the thinking skills to transcend classroom boundaries [4].

Ecuador's Basic General Education (EGB) is composed of ten levels in which students must learn and develop skills of language, interpretation, solve problems, and to understand natural and social life [5].

© Springer Nature Switzerland AG 2019
Á. Rocha et al. (Eds.): ICITS 2019, AISC 918, pp. 814–822, 2019.
https://doi.org/10.1007/978-3-030-11890-7_76

According to the needs of the country, as mandatory laws; the principles of Ecuadorian educational are unity, continuity, sequence, flexibility and permanence; in the perspective of a democratic, humanistic, investigative, scientific and technical orientation. Moreover, it has a moral, historical and social sense, inspired by nationality, peace, social justice and the defense of human rights [6].

Activities and evaluations used in this experimental research were designed and executed over two working groups at the field of Natural Sciences. With the first group, it was used the traditional learning method and with the second group was used the learning integrating ICT's. The data has shown that ICTs education is effective, efficient, and it has helped students to achieve academic goals and improve their understanding.

2 Information and Communication Technologies ICTs

"ICT Information and Communication Technologies are conceived as the universe of two sets. The one set is represented by the traditional Communication Technologies (TC), consisted mainly of radio, television and conventional telephony. The second set is represented by Information Technologies (IT). They are in charge of the design, development, promotion, the maintenance and administration of information through computer systems [7]." "These technologies are tools that allow the acquisition, production, storage, processing, communication, recording and presentation of information, in the form of voice, images and data that are contained in signals of acoustic, optical or electromagnetic nature [8]."

It is important to mention that twenty years ago radio, television and telephony were a good support for traditional education that used technological infrastructure based on transistors, analogical tv and traditional mail. On the other hand, the 21st century technologies based on smartphones, internet, high speed networks, social networks an cloud services are effective instruments for learning process in all education levels.

2.1 MOODLE – ICT Virtual Environment for Teaching and Learning

Modular Object-Oriented Dynamic Learning Environment (MOODLE) is a free software tool which feedback from multiple institutions and participants that collaborate in network. Nowadays, there are 330,000 registered courses around 196 countries and 70 languages [9].

MOODLE is a Learning Management Systems (LMS), a platform specialized that manages activities, resources and learning of an organization, a course as a group or a student individually, proposing collaborative models, synchronous or asynchronous communication and considering interactivity as a decisive criterion [10].

MOODLE works on Linux, Mac and Windows. It is not necessary to know how to program to use it and it also has a user-friendly environment as shown in Fig. 1. All files are encrypted and there are continuous automatic backups in order to prevent information courses loss, documents nor files. MOODLE has an excellent documentation, online support and user communities that can solve any doubts.

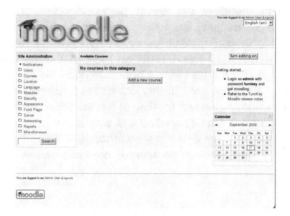

Fig. 1. MOODLE work environment

Used in teaching, Moodle's strengths are:

- Total management of the Virtual Learning Environment.
- Assignment and control of roles and access.
- Increase students' motivation.
- Ease the production and distribution of subjects.
- Subjects' sharing between courses or institutions.
- Application of evaluation types.
- Technological support for subjects' availability adapted to the pedagogical model and instructional design of the organization.

Moodle is a courses generator that facilitates communication (synchronous or asynchronous), also it allows the organization and transmission of subject and materials to support classes, which contribute in a positive way to face-to-face, distance and mixed learning. In brief, Moodle is presented as a form of effective communication between the student and the teacher.

By promoting a space for online collaboration, the collective construction of knowledge is allowed, which is evidenced in the communication alternatives, interaction. Thus, it promotes autonomy and makes students responsible for their learning process [11].

3 Elementary School Description

"La Inmaculada" Franciscana Elementary School is located on General Enriquez avenue N°3366, Montufar street, Sangolquí, Rumiñahui, Ecuador. It includes all the educational cycles, from first to tenth grade. Educational staff includes: 42 teachers; 6 in the administrative area; and also, it has 543 students.

This research has taken the sample from Seventh Grade Elementary Students universe, the average age of this group was eleven years; school cycle 2015–2016 and 2016–2017 with 2 courses each year. Each group was composed of 2 courses with 48 students divided on 24 students each. They were tested before and after this research. In each case one course (24 students) was the experimental group and the other one was the control group. The experiment was execute in the second part of the academic year which corresponds to five months since February to July in each period of classes.

4   Curriculum Description

To develop the experiment, it was chosen the Natural Sciences subject. The block 1, named: The living Beings, according to the curriculum from the Ecuadorian educational system, detailed below:

> **Block 1**
>> **Thematic Unit 1 - Living beings**
>>> **Content**
>>> - Living beings
>>>   - o   The Animal Kingdom Classification
>>>   - ▪ Vertebrates
>>>   - • Amphibians

4.1   Conducting the Experiment

After defining the experimental and control groups, it was applied a pre-test and post-test. The analyzed results were based on the following parameters:

- School performance of the two groups of students;
- Acceptance of the inclusion of ICT by the students of the experimental group;
- Sociability and communication between the teacher and the students of the 2 groups.
- Increase of inventive and creative Skills by using ICT.

The instruments used for data collection, were different types of test, which were applied during the school year 2015–2016 and 2016–2017. Those pre-test and post-test were used to compare ICT contribution on the two research groups, whose results allowed to define the level of achievement found in the development of competences in Natural Sciences.

5   Experimental Evaluation Results

In order to evaluate the application of the students' concepts, Pre-Tests and Post-Tests, of multiple choice were taken. The subjects' unit covered concepts of Living Beings - Classification of the animal kingdom - Vertebrates - Amphibians, both the experimental group and the control group, in both periods, obtaining the following results:

The pre-test has measured the level of knowledge of the students using multiple-choice questions about previous concepts related to living beings. Table 1, Fig. 2 have shown these results, emphasizing that the experimental group used ICTs.

Table 1. Previous concepts - percentage of handling and application

Group	Experimental	Control
Control at required learning (10–9)	20,00%	23,33%
Achieve the required learning (8.99–7)	40,00%	40,00%
It is close to achieving the required learning (6.99–4)	36,67%	30,00%
Students does not achieve the required learning (<4)	3,33%	3,67%

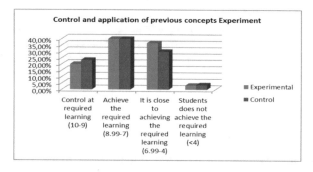

Fig. 2. Control and application of previous concepts experiment

The post-test has measured the knowledge reached in each group. In the control group was followed the traditional model and, in the experimental group was used ICT. The results are presented in Table 2, Fig. 3. It is evidenced that the use of ICTs helped notably the students of the experimental group.

Table 2. Control and application of post concepts experiment.

Group	Experimental	Control
Control at required learning (10–9)	43,33%	30,00%
Achieve the required learning (8.99–7)	46,67%	40,00%
It is close to achieving the required learning (6.99–4)	10,00%	26,67%
Students does not achieve the required learning (<4)	0,00%	3,33%

The results have shown that, there is an increase in the percentage of comprehension of the subject in the experimental group with 13% in grades between 9–10, 6% in grades between 7–8.99, a decrease of 16.6% in grades between 4–6.99 and the absence of grades <4, which indicates that the use of ICTs is an important factor that stimulates and improves students learning.

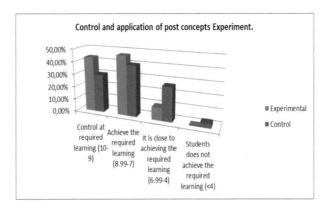

Fig. 3. Control and application of post concepts experiment

Also, the results allow to conclude that the ICT can be used in the learning process of others subjects such as: mathematics, physics, chemistry, literature, etc. The main part of this change is the design methodology which includes new types of resources, activities and an effective Learning Management System like moodle.

On the other hand, the acceptance of the inclusion of ICTs in the teaching-learning environment is reflected in Table 3, Fig. 4. It is important to note that even in the control group there is a high interest in the inclusion of ICTs, since 80% considers the inclusion of ICTs are important or very important.

Table 3. Acceptance of the inclusion of TIC's

Group	Experimental	Control
Very important	53,33%	40,00%
Important	26,67%	23,33%
Relatively important	6,67%	20,00%
Slightly important	6,67%	10,00%
Not important	6,67%	6,67%

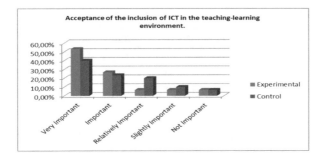

Fig. 4. Acceptance of the inclusion of ICT in the teaching-learning environment.

Table 4 and Fig. 5 present the results on sociability and communication between the teacher and the students.

Table 4. Sociability and communication between the teacher and students.

Parameter	Experimental		Control	
	Pre-test	Post-test	Pre-test	Post-test
Exist sociability and communication	60.00%	96.67%	53.33%	53.33%
No exist sociability and communication	40.00%	3.33%	46.67%	46.67%

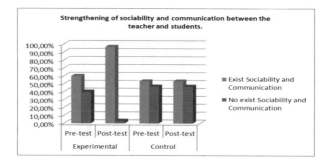

Fig. 5. Strengthening of sociability and communication between the teacher and students.

It was asked the student if they perceived the existence of Communication and Socialization between him/her and the teacher at the beginning and also at the end of the unit. The results show that in the control group the percentages remained the same, whereas in the experimental group the perception increased by 36% and the negative perception decreased in the same range. This result shows that the ICT can be used in administrative issues because it is a good communication and socialization channel.

Table 5 and Fig. 6 show the results on the evaluation of inventive and creative skills with and without the inclusion of ICTs in the groups.

Table 5. Inventive and creative skills for technological use.

Group	Experimental		Control	
	Pre-test	Post-test	Pre-test	Post-test
Inventive	53.33%	90.00%	56.67%	60.00%
Creative	46.67%	45.00%	43.33%	46.67%

The results show an increase of approximately 45% in the experimental group; whereas in the control group only 4% is increased. So we can conclude that the ICT provides an increase of inventive and creative skills.

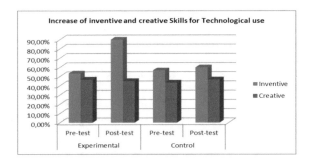

Fig. 6. Increase of inventive and creative skills for technological use.

6 Conclusions and Future Work

After the post-test, it is concluded that the school skills performance of the experimental group improved with respect to the control group in 13.33% thanks to the use of ICT.

There is evidence of acceptance in the process of teaching natural sciences by using ICT, what makes the teacher a facilitator and mediator.

Communications and sociability between the teacher and students improve thanks to ICT, in fact, it built new bonds of trust with the teacher, because it allows students to experience and acquire knowledge in an experiential way.

There was an increase in inventive and creative skills thanks to the use of ICT.

To offer adequate training for teachers in the area of Information Technology with emphasis on the use of ICT, so that their theoretical training arrives to virtual classrooms successfully. Furthermore, this adequate training, enables a quality teaching-learning process for the different areas.

References

1. Carlos, M.: Las tecnologías para la innovación y la práctica docente. Revista Brasileira de Educação **18**(52), 25–26 (2013)
2. UNESCO: Estándares de competencia en Tic para Docentes. Organización de las Naciones Unidas para la Educación la Ciencia y la Cultura. http://www.eduteka.org/EstandaresDocentesUnesco.php
3. Gómez, J.: Sistema de Aprendizaje Basado en Internet de las Cosas como apoyo a los procesos de Enseñanza. In: Aprendizaje en Estudiantes de Ingeniería. Encuentro Internacional de Educación en Ingeniería ACOH 2014, pp. 1–7 (2014)
4. Molina, A.: Mejores Prácticas de Aprendizaje Móvil para el Desarrollo de Competencias en la Educación Superior. IEEE-RITA **5**(4), 175–183 (2010)
5. Ministerio de Educación del Ecuador. http://educacion.gob.ec/educacion-general-basica/
6. OEI - Sistemas Educativos Nacionales: Estructura del Sistema Educativo. http://www.oei.es/quipu/ecuador/ecu04.pdf
7. Samillán, L.: Tic's en la Informática. http://www.scribd.com/doc/3285023/TICS-EN-LA-INFORMATICA

8. González, M.: Parte II: Tecnologías de la Información y las Comunicaciones (TIC). http://www.eumed.net/libros/2006a/mga-01/2b.htm
9. Ros, I.: Moodle, la plataforma para la enseñanza y organización escolar. http://www.ehu.es/ikastorratza/2_alea/moodle.pdf
10. Baumgartner, P.: Cómo elegir una herramienta de gestión de contenido en función de un modelo de aprendizaje. http://www.elearningeuropa.info
11. Santana, E., Gomes, A., Leitao, A, Sales, G.: LMS em Contexto Escolar: estudo sobre o uso da Moodle pelos docentes de duas escolas do Norte de Portugal. http://www.eft.educom.pt/index.php/eft/article/view/74/52

An Investigation and Presentation of a Model for Factors Influencing the Agility of Human Resources a Case Study of Yazd Electricity Distribution Company to Springer Proceedings

Hamed Alvansaz[1], Maryam Hakakian[1],
Mohammadfarid Alvansazyazdi[2,5],
Nelson Esteban Salgado Reyes[3(✉)],
and Alejandro Miguel Camino Solórzano[4]

[1] Yazd Electricity Distribution Company, Yazd, Iran
[2] Facultad Ingeniera Ciencias Físicas y Matemática, Carrera Ingeniera Civil,
Universidad Central del Ecuador, Quito, Ecuador
alvansaz@gmail.com
[3] Facultad de Ingeniería, Escuela de Sistemas,
Pontificia Universidad Católica del Ecuador, Quito, Ecuador
nesalgado@puce.edu.ec
[4] Facultad de Arquitectura,
Rector en la Universidad Laica Eloy Alfaro de Manabí, Manta, Ecuador
camino2h@yahoo.es
[5] Facultad Ingeniera, Carrera Ingeniera Civil,
Universidad Laica Eloy Alfaro de Manabí, Manta, Ecuador

Abstract. Agility resembles the idea of speed and change through business environment. Agility focus on responding to dynamic and turbulent markets and customer's demands. It is while the implications of agile competition dependent upon competitive contexts which are particularly active within an organization. The present study is an attempt to find out which factors are more influential on attaining agility with regard to human resources. The sample for this study includes the staff working in Yazd Electricity Distribution Company. Random sampling method was used in this study and the data were gathered through the use of a questionnaire (after examining its reliability and validity as an instrument of data collection). Furthermore, in order to investigate this issue, the fuzzy approach was used as one of the beneficial instruments for measuring those problems with vague indices. Structural Equation Modeling was also applied to experiment research hypotheses. The results of this study show that in order to face such changes, agility based on human resources' knowledge, human resources' skills, the existence of cooperative culture in the organization, and access to information is necessary. Besides, regarding the importance of agility of human resources for organizations, it is suggested to investigate other influential factors on agility in future research to contribute to the further development of the model presented within this research.

Keywords: Human resources agility · Fuzzy approach ·
Human resources knowledge · Cooperation culture ·
Staff's access to information

© Springer Nature Switzerland AG 2019
Á. Rocha et al. (Eds.): ICITS 2019, AISC 918, pp. 823–834, 2019.
https://doi.org/10.1007/978-3-030-11890-7_77

1 Introduction

Agility resembles the concept of change and speed within the working environment (Shahaee 2006). According to Van Hoek et al. (2001), agility is such a management concept which is focused on turbulent and dynamic markets and customer's demands (Sukatia et al. 2012). Goldman et al. (1995a, b), have defined agility as a reaction toward the changes in the working environment. According to Kid (1996), an agile organization is a speedy, compatible and conscious business which is capable of quickly adjusting itself to unpredictable changes and events, market's opportunities and customers' demands. According to Vipul et al. (2008) definition, agility is a quick reaction toward changes in customers' demands in both dimensions of volume and variety.

On the other hand, the implications of agile competition are highly dependent on competitive areas which are particularly active within an organization. Agile companies will face changes aggressively. For agile competitors, change and lack of confidence of the source of reviving opportunities are considered as permanent success methods.

Therefore to confront changes that are previously rare, agility is dependent upon innovation, skills, human resources knowledge and individuals' access to information (KhoshSima 2007; Shahaee 2006).

2 Statement of the Problem

Today's world is a place of constant change and an era of instability. Nowadays, the ways of organization manage have become far more complex and multi-dimensional, and integration of their external structure is possible only when both suppliers and key customers are simultaneously merged. In the process of growth for Global Competitive Economics, most of the company tries hard to focus on presentation of a better customer value compare with their counterparts. Agility is the successful exploration of competitive bases (speed, flexibility, innovation pro-activeness, quality and profitability) through the integration of reconfigurable resources and best practices in a knowledge-rich environment to provide customer-driven products and services in a fast changing market environment (Sukatia et al. 2012). This brand new concept called "Organizational Agility". For an organization, it is only agile when it is capable of coordinating by using both knowledge and cooperation (both from within the organization and from outside) and it will as efficiently and quickly as possible will create its required resources, produce and support them. Accordingly, the development and improvement of methods for evaluation of agility have become more valuable. All the modern measurement methods for agility are accompanied with some shortages. In fact, the intensity of environmental changes and the volatility of customers' requirement are among determining factors for the degree of agility required for that firm. And based on its nature and different amount of such changes, the desired degree of agility will be different for organizations too. It is like when production systems are handled and used by people. Therefore, it seems necessary that measuring their degree of agility is also possible by using human's knowledge and perceptions (KhoshSima et al. 2007). Thus, the human resource has become one of the significant elements in determining the degree of agility as required by the organization and providing a pre-action response toward the changes based on the existing capabilities.

3 Background of the Study

Considering that the concept of agility and accordingly, the concept of human resources' agility are regarded as up to date concepts in the management area, but those studies investigating the agility of human resources, in particular, are scant so far. On the other hand, in most of the studies regarding the investigation of agility within organizations (as one of the most vital and influential factors on the survival of the company), the human dimension of agility was considered and evaluated.

1. Sukatia et al. (2012) were mainly focused on investigating the impact of organizational activities on the agility of supply chain (Case study: Malaysian industries).
2. Olfat and Ranjirchi (2009) study, entitled as "A Model for Organizational Agility in Iran's Electronic Industry".
3. Hamidi et al. (2010) study, entitled as "The role of human resources management in organizational agility".
4. Fathian et al. (2010) study which took a new perspective toward evaluation of agility through supply chain using fuzzy approach.
5. Madelline and Youssef (2003) study, entitled as "The human dimension of organizational agility". In this study, the leader of the organization, its members, customers, and suppliers are considered as the four dimensions existing within an organization. The results of the study illustrate that the interaction among these four human dimensions of organization will result in enhancing the quality and speed and lowering costs.

4 Agility of Human Resources

Agility mostly defined as "organizational agility is created through a flow from outward to inward". This pushed to created agility through the application of some special ingredients of organizational features and properties. While these features are variously defined, most of them are quite similar. Gold man et al. has mostly focused on proper structure, information technology and human resources (Goldman et al. 1995a, b). Cotter emphasized on the structure, business continuum and human resources (Cotter 1995). However, in Overhelt's model, the emphasis was put on common values, structure or information technology, work processes and behavior (related to human resources) (Hamidi et al. 2010). According to Fathian et al. (2010), Organizational instruments are at the service of realizing their agility including their flexible structure, human resources, information technology and innovations well as creativity (Fathian et al. 2010).

As it can be seen here, human resources are considered as the most valuable asset of every organization among its other elements and instruments (Ziaee et al. 2013). Agility of human resources is defined as the ability of its staff to present a strategical reaction toward lack of confidence (Khosravi et al. 2012). Agility can be defined as a result of quick reaction of an organization toward the changes in the market as well as customers' demands. For an organization to achieve desired agility, the contributors of agility, should be present in the organization (Molahosseini and Mostafavi 2008). We can argue that the agility of human resources is the ability of the staff to provide a strategic reaction toward lack of confidence (Khosravi et al. 2012). Considering what has been said so far, four main factors, and 16 sub-factors are determined.

4.1   Individuals' Access to Information

Displayed Huge changes are observed in scientific areas such as electronics and information technology. The growth in these areas are so fast, and therefore most of the organizations even don't have required time to adjust themselves to some changes (Khosravi et al. 2012). The continuous changes in technology are seen as the pre-requisite for becoming competitive and to survive in today's global competition (Olfat and Ranjirchi 2010). Technology is defined by researchers as the knowledge, processes, instruments, methods and systems which are exploited for creation and production of goods and services (Khalil 2000). Every organization makes use of particular kinds of technologies in order to fulfill its tasks, one of them is information and communication technology (Khosravi et al. 2012). From the perspective of scholars in this area, one of the important factors in the effective interaction of organization members with suppliers is to possess a kind of information technology which is compatible with organization's circumstances. Also, such technology will be capable of providing up to date and detailed information for its customers (Hamidi et al. 2010).

Breu et al. (2002) have emphasized the importance of information technology in agility of human resources (Breu et al. 2002). Communication and exchange of information among the staff will facilitate learning, change, and adjustment. Agile organizations require both kinds of vertical and horizontal flow of information. Therefore, the staff should be aware of their informational needs and they should also share proper information (Hamidi et al. 2010).

Regarding what has been said so far, the main dimension of individuals' access to information is accompanied by some sub-dimensions such as information technology, staff's use of the internet, superiority in using updated technology, and the availability of communication among various levels of the organization.

4.2   Cooperation Culture

Robins (2003) believes that organizational culture consists of a system of common perceptions which members of an organization held it toward that organization. In his idea, organizational culture is a method through which members of an organization consider its features (not to mention whether they like it or not), i.e. it's a descriptive term (Robins 2003).

Culture consists of a set of common values within organizational environments which includes beliefs regarding the current methods or the best methods for doing something, and it also includes the current or the best value method that everyone should have (including social values). Such values are like the glue that holds different components and pieces of an organization stick together (Hamidi et al. 2010). Considering the spread of organizational culture, also to avoid any interference with the current study, cooperation culture is studied as one of the main features of agile organizations and investigating other aspects of organizational culture, and its influence on agility was avoided.

Cooperation culture can be discussed in three main areas of study:

1. Individual's relationship within an organization
2. Group and team relationships
3. Organization's leadership

From the perspective of Goldman et al. (1995a, b) and Overholt (1996), agile culture requires trust, innovation, flexibility, risk-taking, teamwork, cooperation, empowerment and mutual responsibility (Hamidi et al. 2010).

In this regards, the identified sub-dimensions of cooperation culture include a sense of unity, teamwork, increasing collaborative relationships for improvement and attempting toward achieving organization's objectives (leadership).

4.3 Human Resources Knowledge

Most often knowledge is defined as the "justified belief of oneself" (King 2008). Knowledge has always been considered as the major factor in the success of every organization. However, nowadays that creativity is introduced continuously as the key factor in the liberation of world economic, goods and services are mostly dependent upon knowledge level, organizations have become more flexible, and more advanced technologies incorporated toward the integration of organizations. Therefore, knowledge is considered as one of the most important resources for an organization (Graham 2003).

Knowledge doesn't exclusively refer to the transmission of information, but it refers to creativity, and mobility as well. Goldman et al. suggest that agile companies usually possess a kind of permanent culture of learning (1995). Therefore, we can conclude that agile companies can be considered as knowledge-based organizations. Quat and Zender, develop this idea and suggest that organizations can integrate a combination of capabilities simultaneously. They can also merge current and acquired knowledge together, and they can use it to achieve organization's objectives (Graham 2003).

The main dimension of human resources knowledge has been identified with sub-dimensions such as aware adjustment with changes, the ability to predict changes and exploiting changes to the benefit of that organization.

4.4 Human Resources Skills

Agile resources do their best to achieve a situation where their staff in all levels blames themselves as well as their colleagues for the results (not only blaming the one in a particular position for the final results). To quote Bridger (1994) "in a highly dynamic organization, its activities vary with time and simultaneously, individuals' responsibilities may also vary accordingly. Such changes and variations include a multiplicity of leaders in monitoring the activities, applying various programs for working, doing activities in various places that require various activities as well and only those staffs are able to do various activities that possess relevant skills in that area. Therefore, agile organizations specify a high amount of time and energy to the development of teaching programs. Such organizations perceive permanent learning and the acquisition of skills by their staff as one of their fundamental values (Hamidi et al. 2010).

Sumukades and Sawhney have developed a hierarchical and theoretical model that shows how various activities taken by the management of human resources (such as involving their staff in the task, development and training their staff, multi-skilled staff, etc.) will result in the agility of human resources (Sumukades and Sawhney 2004). Iravani and Krishnamurthy (2007) also suggests that multi-skilled staffs are able to do a wide range of activities. Therefore, they possess a high degree of flexibility in counteracting the lack of confidence (Ziaee et al. 2013).

In the current model, the main dimension of human resources skill is identified to have some other sub-dimensions such as the ability to adjust oneself to unexpected changes, possessing multi-skilled and flexible staff, responding to changes by using proper solutions, responding to changes in the right time and the ability to quickly adjust and transform. In the following section, the current model is presented. In the following section, the current model is presented (Fig. 1).

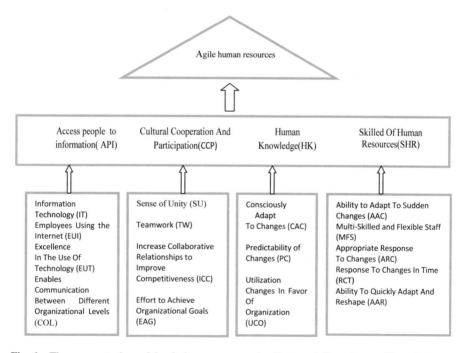

Fig. 1. The conceptual model of the present study (factors influencing agility of human resources).

Description about questionnaire: The questionnaire consists of two major parts:

A. General questions: In general questions, demographic information (age, education, etc.) has been gathered in relation to respondents.

B. Specialist Questions: In designing this section, the Five Likert Range has been used, which is one of the most commonly used measure comparisons. This questionnaire has been adapted from Lin Ching-Torng's article (Ching-Torng et al. 2006)

5 Methodology and Research Hypotheses

Regarding its objectives, the present study is applied research, while in terms of methodology, it is descriptive and is considered as a field study. Also, regarding the issue that in this study we're seeking to find a relationship among various variables of the study (not a causality), thus the present study is of correlation type.

The statistical sample of this study includes the whole group of staff who are working in Yazd Electricity Distribution Company. Since the main objective of this study is to investigate the factors contributing to the agility of the human resources of an organization, all the staff were selected for the purpose of the study. Furthermore, regarding the fact that the number of participants in the sample is known, random sampling method was used so that the chance of each staff is the same to be included in the investigation.

Considering the related literature on this subject four hypotheses were generated for this study. They are:

H1: The accessibility of the information to the staff is influential on the agility of human resources in Yazd Electricity Distribution Company.

H2: The degree of existing cooperation culture is influential on determining Yazd Electricity Distribution Company's human resources level of agility.

H3: The knowledge possessed by human resources is influential on the level of agility for Yazd Electricity Distribution Company staff.

H4: The skill level of human resources is influential on the level of agility required by Yazd Electricity Distribution Company.

5.1 Reliability and Validity of the Questionnaire

A questionnaire was used in this study to investigate the hypotheses of the study. In order to test its validity, face validity was used. Also, in order to test reliability, Cronbach's alpha was used. Using the data obtained from the questionnaires (127 questionnaires) and SPSS software, v.19, confidence interval was computed through the use of Cronbach's alpha. Its value was equal to 0.909 for all the items and for every variable was more than 0.7 (Table 1).

Table 1. Cronbach's Alpha test for computing reliability of each variable as well as the whole questionnaire.

Variable	Number of items	Cronbach's Alpha value
Staff's access to information	11	0.785
Cooperation culture	9	0.767
Human resources knowledge	7	0.765
Human resources skills	6	0.768
Total	33	0.909

5.2   Testing Hypotheses

Structural equation modeling (SEM) is a form of causal modeling that includes a diverse set of mathematical models, computer algorithms, and statistical methods that fit networks of constructs to data. SEM includes confirmatory factor analysis, path analysis, partial least squares path modeling, and latent growth modeling. Therefore, investigation of research's hypotheses was done based on SEM and help of Lisrel software. At first, the compatibility and fit of the sub-criteria of the model has been studied:

The results are presented in Figs. 2, 3, 4 and 5.

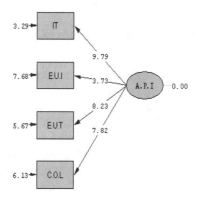

Chi-Square=2.64, df=2, P-value=0.26651, RMSEA=0.051

Fig. 2. t values for staff's access to information dimension (API)

5.3   Fitness of Study's Model

Regarding the approval of all sub-dimensions of the model, the final model was evaluated regarding its fitness. The results of testing by Lisrel Software confirms the fitness of this model based on fitness indices. When the value of Goodness of Fit Index (GFI) will be more than 0.9, we can conclude that the model is a good fit.

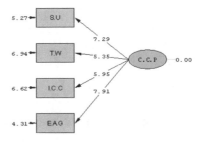

Chi-Square=3.28, df=2, P-value=0.19435, RMSEA=0.071

Fig. 3. t values for cooperation culture dimension (CCP)

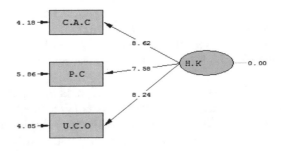

Chi-Square=0.2, df=1, P-value=0.005, RMSEA=0.005

Fig. 4. t values for human resources knowledge

On the other hand, RMSE, which is Root Mean Square of Root Mean Square Error of Approximation (RMSEA) Estimations and is defined as the value per each degree of freedom, should be less than 0.1. Further, the values of RFI, CFI, and NFI are higher

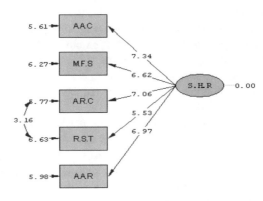

Chi-Square=3.91, df=4, P-value=0.41822, RMSEA=0.000

Fig. 5. t values for human resources skills dimension (SHR)

than 0.9 which illustrate the fitness of the applied model. The proportion of $x2$ degrees of freedom as the most important index in determining the fitting is lower 2 in this model which is located in a proper range.

6 Discussion and Conclusion

Hypothesis 1: The results confirm that staff's access to information is influential on the level of agility possessed by staff in Yazd Electricity Distribution Company.

In Hamidi et al. (2010) the mutuality of human resources and information technology was approved, and it was argued that proper application of information technology would lead to autonomy in occupational decisions, the creation of justice and trust atmosphere at the workplace, group work and simple and flexible flow of information within the organization. The results of the present study also emphasize issue this in order to increase the agility of human resources.

Hypothesis 2: The results show that the degree of possessing cooperation culture is influential in determining the level of agility of human resources in Yazd Electricity Distribution Company.

This illustrates the fact that specifying the objectives of the organization as a guide and road map for staff will help them to decide better in the event of crises based on organizational objectives. Also, the existence of justice behavior among staff will improve the sense of unity among them, so that they will consider themselves as a member of a larger organization and when they make decisions, they give priority to organizational objectives. Joseph Cricketo (2033)'s model also emphasizes the role of flexibility and responsive culture on enhancing agility in an organization.

Hypothesis 3: The results show that human resources knowledge is influential on the agility of human resources in Yazd Electricity Distribution Company.

Some terminology such as knowledge management, organizational learning, learning organization, etc. emphasizes the need for continuous creativity as a key factor for the salvation of world economic in the contemporary era. That's because most of the goods and services are based on the level of knowledge and information and organizations are more flexible and more advanced technologies are used for their integration. Therefore, nowadays knowledge is regarded as the most important resource in organizations. The present study shows that employing knowledgeable staff is really important because of its impact on the agility of human resources.

Hypothesis 4: The results illustrate that the skill level of human resources is influential in specifying the level of agility required by human resources in Yazd Electricity Distribution Company.

Iravani and Krishnamurthy (2007) found out that multi-skilled staffs are able to do various tasks. Therefore, to diminish lack of confidence, they possess the required flexibility. In his study, Bridger (1994) suggest that "in an organization with a high degree of dynamism that activities may vary with time, staff's responsibilities and tasks will also vary. Such changes require the staff to work under the supervision of various team leaders, using multiple work programs, doing activities in various locations and doing various activities and tasks which require their skills and specifications.

Investigating the hypotheses of the study reveals that every organization require agile staff in order to respond quickly to changes in the environment or conditions of the crises require attending to both aspects of workforce, i.e. their knowledge and skills (experience) and organization, i.e. improving the cooperation culture among organization's members and creation of proper platforms for exchange of information both with staff and with organizational environment. Ignoring one of these aspects will result in lack of competency of the staff to respond effectively and instantaneously at the time of crises.

Acknowledgment. This research was supported by Facultad de Arquitectura, Universidad Laica Eloy Alfaro de Manabí, Ecuador, and Pontificia Universidad Católica del Ecuador.

References

Sukatia, I., Hamida, A.B., Baharuna, R., Yusoff, R.M., Anuar, M.A.: The effect of organizational practices on supply chain agility: an empirical investigation on Malaysia manufacturing industry. Soc. Behav. Sci. **40**, 274–281 (2012)

Jain, V., Benyoucef, L., Deshmukh, S.G.: A new approach for evaluating agility in supply chains using fuzzy association rules mining. Eng. Appl. Artif. Intell. **21**, 367–385 (2008)

Madelline, C., Youssef, M.: The human side of organizational agility. Ind. Manag. Data Syst. **103**(6), 387–397 (2003)

Goldman, S., Nagel, R., Preiss, K.: Agile Competitors and Virtual Organizations. Van Nosteand Reinhold, New York (1995a)

Cotter, J.: The 20% Solution: Using Rapid Redesign to Create Tomorrow's Organization Today. Wiley, New York (1995)

Khalil, T.: Management of Technology: The Key to Prosperity & Wealth Creation. McGraw-Hill, New York (2000)

Breu, K., Hemingway, C.J., Strathern, M., Bridger, D.: Workforce agility: the new employee strategy for the knowledge economy. J. Inf. Technol. **17**(1), 21–31 (2002)

King, W.R.: Knowledge management and organization learning. Omega **36**(2), 167–172 (2008). http://www.sciencedirect.com

Graham, K.: The Impact of Knowledge Management Technologies on Learning Within Organizations: An Empirical Analysis. The Florida State University College of Business (2003)

Sumukadas, N., Sawhney, R.: Workforce agility through employee involvement. IIE Trans. **36**(10), 919–940 (2004)

Iravani, S.M.R., Krishnamurthy, V.: Workforce agility in repair and maintenance environments. Manuf. Serv. Oper. Manag. **9**(2), 168–184 (2007)

Shahaee, B.: The human dimension of organizational agility. Tadbir J. **17**(175), 21–24 (2006)

KhoshSima, G., Jafar-Nejad, A., Mohaghar, A.: Agile manufacturing system, frameworks and nablers. In: Second National Conference on Industrial Engineering, pp. 163–169 (2007)

Olfat, L., Ranjirchi, M.: A model for organizational agility in Iran's electronic industry. J. Iran. Manag. Sci. **13**, 47–74 (2009)

Olfat, L., Ranjirchi, M.: Data envelopment analysis: a new approach to measure the agility of organizations. Manag. Res. Iran (Modares Hum. Sci.) 14(2), (66) 21–44 (2010)

Hamidi, N., Hasanpoor, A., Kiaee, M., Mosavi, S.M.: The role of human resources management in organizational agility. J. Ind. Manag. (8), 111–127 (2010)

Ziaee, M., Hasngholipur, T., Abaspur, A., Yarahmadzehi, M.: In the Tehran University Accounting and Auditing review (2013)

Molahosseini, A., Mostafavi, S.: Evaluation of organizational agility using fuzzy technique. Tadbir. Monthly Mag. **186**(8), 3–5 (2008)

Khosravi, A., Abtahi, S.H., Ahmadi, R., Salimi, H.: Factors enabling agility manpower Delphi electronics industries. Q. Improve Manag. **6**(4, 18), 129–153 (2012)

VanHoek, R.I., Harrison, A., Christopher, M.: Measuring agile capabilities in the supply chain. Int. J. Oper. Prod. Manag. **21**(1/2), 126–147 (2001)

Goldman, S., Nagel, R., Preiss, K.: Agile Competitors and Virtual Organizations. Van Nosteand Reinhold, New York (1995b)

Overholt, M.H.: Building Felixible Organizations. Kendall/Hunt Publishing Company, Dubuque (1996)

Bridges, W.: Job Shift. Addison-Wesley, Reading (1994)

Ching-Torng, L., Hero, C., Yi-Hong, T.: Agility evaluation using fuzzy Logic. Int. J. Prod. Econ. **101**(2006), 353–368 (2006)

Kid, P.T.: Un paradigma del siglo XXI en la fabricación ágil: forjar nuevas fronteras. Addison Wesley, Wokingham (1996)

Fathian. M., Seyedhosseini, S.M., Ali Ahmadi, A., Fekri, R.: Agile new product development model using path analysis method for iranian auto industries. Int. J. Ind. Eng. Prod. **4**, 77–89 (2010). http://IJIEPM.iust.ac.ir/

Robins basic pathology, 7th edn. (2003)

Higher Education Challenge Characterization to Implement Automated Essay Scoring Model for Universities with a Current Traditional Learning Evaluation System

José Carlos Machicao[1,2(✉)]

[1] GestioDinámica, Lima, Peru
jcmachicao@gestiodinamica.com
[2] Universidad Continental, Lima, Peru
jmachicao@continental.edu.pe

Abstract. Higher education is currently challenged to respond to a massive interest in learning with a current model that shows increasing evidence of too much cost, effort and decreasing efficacy of the operational learning process. Artificial intelligence has gained presence as a solution, but the integration process is already reporting problems and will not be implemented easily, in particular for universities with low degree of automation integrated to their systems. Universities need to quickly adapt and develop organizational and individual competencies, and clarity about the elements for new learning evaluation systems. This work contributes to propose a model to help universities to define these new systems making the most of artificial intelligence tools for academic essays scoring.

Keywords: Higher education · Automated essay scoring ·
Artificial intelligence · Knowledge management

1 Introduction

Higher education is changing rapidly, and despite the complexity of this process, it can be understood as a transition from a traditional vertical model towards a mainly horizontal concept [1, 3].

The current concept of good university is defined by a system generating competencies guaranteed by the closeness of the academic relationship between teachers and students, making quality easily verifiable through a complex educational and ethic tradition helping to find a future benefit for the student and the transmission of experience from the teachers, using a background of knowledge from the global academic system [2].

The future configuration of higher education is probably going to be much more student-centered, decentralized, massive in terms of number of students for the same course, for very specific topics, based on atomized content certifications, probably not based in complete careers but in competencies [1, 4].

© Springer Nature Switzerland AG 2019
Á. Rocha et al. (Eds.): ICITS 2019, AISC 918, pp. 835–844, 2019.
https://doi.org/10.1007/978-3-030-11890-7_78

This means that universities should be faster and more accurate facilitating knowledge without reducing quality. It is not difficult to make some estimates about how many teachers, tutors and personnel volume in general would be necessary to complete all tasks ensuring quality and accuracy building competencies for students.

In Peru, for instance, it is easy to see that the number of students is growing, the cost for the university to bring the educational service per student is increasing, and the percentage of graduated students in decreasing [5] (Fig. 1).

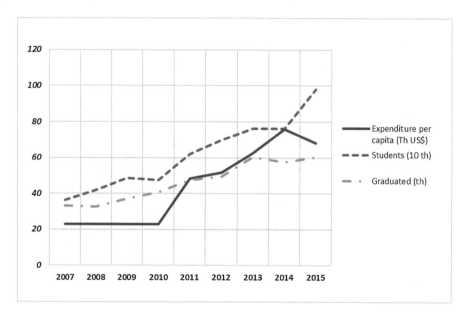

Fig. 1. University expenditure per student in thousands of US$, number of students in 10 thousand, and number of graduated students from higher education in Peru between 2007 and 2015. The graph shows the first two increasing and the third showing a growing gap between graduated and registered students, in Peru [5].

Many studies already identified the need to revise university budgets in order to avoid expensive fees for students who progressively don't find worthy their investment as before [3, 6, 7].

Despite the difficulties of this transition, universities were and are a critical element in society, and its value is clearly demonstrated and should be preserved [3]. But in order to maintain value delivery for society, universities need to develop a complete new operational model, adapted for the new conditions, which among others, are represented by massive demand for higher education, demand atomization of capacities, lack of interest in current careers, high production costs for universities, or teaching staff overload [1, 4].

A careful feasibility analysis should be performed about this shift because the challenges that emerge for the feasibility of higher education and more specifically the operational learning process.

2   Definition of the Challenge

Universities with a traditional learning evaluation system (LES) share some common characteristics: are mainly manual or semi-automatized, its operation is expensive, and are beginning to report frequent failures. In the near future likely universities would have a fully automated LES. For some approaches, the change would appear as an easy and straight forward task, simply leave the challenge to the artificial intelligence (AI) solutions. For other approaches AI could be seen as a hazard. For many universities the change will involve the need to adapt very diverse and large cultural contributions which would not be included in standardized AI systems and algorithms.

First, this work proposes the definition of the challenge for universities and then for its alternative solutions.

2.1   Description of Basic Concepts for Modeling

In order to generate a model, a system and its basic components need to be defined.

Learning Evaluation System (LES). Given certain learning outcomes that are established by a university for the student, the LES defines how the connections between the direct interaction with the student through educational activities and those outcomes work [8]. There are many kinds of activities within this system, some of them are evaluation activities. In a traditional approach, one typical activity is the written exam, asking the student to write down descriptive paragraphs on certain topic and the subsequent reading from the teacher. The mark the teacher assigns represents the evaluation of learning outcomes and is part of the evaluation system.

Academic Essay. In this work, it will be understood as an around-two-thousand-words text, developed by a member of the academic community in certain time, describing or explaining ideas related to certain topic in an articulate way, which contributes to learning outcomes [10]. A university describes "writing an academic essay means fashioning a coherent set of ideas into an argument" [15].

Essay Scoring. Within a LES, and regarding specifically students, scoring is a process that measures how aligned is a learning product with the learning outcomes, usually expressed through a quantitative or ordinal scale. The concept defines scoring applied to an essay [10]. In order to raise the reliability in a scoring scale, there are some criteria used to have certain level of complexity in order to cover different aspects of academic creation. In higher education, usually essay scoring is performed directly by a teacher or tutor in a university course.

Automated Essay Scoring (AES). If essay scoring is not directly performed by a member of the teaching staff, it can be performed by an algorithm. AES is a topic present in academic literature for some decades now [11, 12]. Even when it is so simple, the method of multiple-choice questions is already an initiative to "automatize" the recount of certain choices. But the current concept is much more related to artificial intelligence solutions. In fact, an ongoing debate is what the role of teachers will be in the future if AES is accepted universally [16].

Natural Language Processing (NLP). Is the use of symbolic methods where knowledge about language is explicitly encoded in rules or other representations towards the solution of problems to automatize human languages [19].

Artificial Intelligence (AI). An integrated set of computational complex algorithms and systems able to perform tasks normally requiring human intelligence, such as visual perception, speech recognition, decision-making, and translation between languages [9].

2.2   Benefit/Cost Analysis for the Current Operational Learning System

Given a learning process, learning quality and opportunity are based on how universities can supply learning results with realistic organizational effort, cost and time. The resource estimation about the real process is already meaning deficit for universities, therefore increasing teaching staff is discarded. The only way to make the learning process feasible for universities and accessible for the demand is by increasing the ability to (i) provide a learning service at a technically possible cost in time, and (ii) be able to have enough features of this service available to all students who demand it [17].

As evidence shows in many other fields, artificial intelligence is demonstrating the ability to allow permanently time, cost and quality stuck processes to find a way to be consistently feasible again [20].

Understanding the learning process dilemma, and considering all current efforts to already automatize a large percentage of the learning process using digitalization, artificial intelligence is by far a suitable and available tool to replicate in time the ability of teachers to reach deadlines to ensure learning quality for a large number of students, complete the remaining percentage of learning features.

Automation is not new to writing in education systems. The very well-known multiple-choice option for questions was already a successful evidence present for decades in education. There are reports about AES since Ellis Page already applied a method to automate essay scoring for the first time in 1966 [11].

This early effort already helped the operational learning process to be less costly, timely and more accurate, but was successful mainly for structured information, which is contained in learning products generated by students under standardized formats (tables, numbers, multiple-choice answers).

But, apparently, the learning process cannot be only standardized and needs to evaluate the student ability to generate educational non-standardized products, which are mainly non-structured. Automated essay scoring was gradually finding a critical position in education [12].

Therefore, even when the usage of AI is almost naturally used for LES and AES, this fact does not mean that there are not problems associated. Early reports about limitations of the experiments already point out potential reduction in scoring quality, misalignment of the human side of a document, hyper-automation that disconnects the conceptual supervision, among others [13, 14, 18].

3   Proposed Model for AES Using AI

In order to model an AES process that solves the current limitations, universities need to contextualize it within a realistic scenario. First, a conceptual background of AES in higher education will be explained. Second, a revision of currently available AI tools and finally a model will be proposed.

3.1   Conceptual Background to Model an AES Process in Higher Education

Based on the approach described above, and in order to reduce the complexity of the challenge, a single problem is taken to be faced in this study. Given that there are many solutions for structured information (for instance, multiple choice questions are highly automated), this study will focus only in unstructured contents to score, and mainly in essay scoring.

The educational objective of writing an essay is to detect complex capabilities from the student, similar to those needed in a real challenge [21]. In the past, this objective was ensured through the personal relationship between the teacher and the student. The student received instructions from the teacher and would write paragraphs trying to demonstrate that the concepts lectured by the teacher have been understood. The teacher reviewed the essay and found potential improvements and mistakes, which would be written as corrections. In earlier periods, the teacher did not generate a set of criteria in order to know what would be corrected or not.

Later models began to include previously designed and communicated criteria, formats and rubrics [25]. It seems obvious that the effort to give more structure to essays originated in the need to score them faster. As recently became clearer in literature, the skills to write in a complex way cannot be taken out from the student curriculum [26].

Undoubtedly, there are many proposals about the essay structure and contents to ensure its quality. As a summary, modern scoring is more related to: (i) Focus, (ii) grammatical correctness, (iii) logic of ideas, (iv) insight level, (v) reference correctness, (vi) benchmark versus expectancies [12].

3.2   Natural Language Processing (NLP) Available Methods Review

The objective of analyzing tools is to link it to AES success. From the many approaches to NLP during latest decade, those using neural networks already reported the best results because of many features, among others: computational speed, conceptual accuracy, flexibility. In order to model a the complete LES for real universities it has to be considered how these features contribute to successful learning results.

From early discussion, even when there are no perfect reports from any approach, the objective is to replicate accurately the way teachers score an essay in a non-highly technologized university.

During the recent five years, a lot of reports arouse with new and more successful approaches, some of them are summarized in Table 1 [22–24].

This information will be used below to propose an operational learning system and subsequently recommend a structure for an AES.

Table 1. Potential contribution from available NLP tools and approaches to learning results through AES [22–24].

Approach or tool	Description	Potential contribution to learning results
Word and sentence statistics	Tools able to extract statistical numeric features from words and sentences, such as length, frequency of words, or frequency of grammatical forms. It helps to obtain word presence diagrams from a document, or the frequency of relationships verb-subject Example of published tool: NLTK	• Support tool for teachers • Testing tool for students • Not able to score automatically
Grammatical tagging	Tools able to analyze the grammatic structure of sentences and identify main or secondary roles of partial meanings. Allows to automatize the structure building and the statistical analysis of more complex phrases or sentences	• Support tool for teachers • Testing tool for students • Not able to score automatically • More concentrated in grammar sense
Complex networks	Tools able to combine in a complex way certain number of variables or features to show them graphicly. It helps to find some level of "understanding" of concepts but still needs human interpretation combined with graph algorithms	• Support tool for teachers • Learning tool for students • Able to score some features such as structure • Manages concepts to some extent
Concept and Word Vectorization	Tools able to generate meaning representations using vectors, usually based on neural networks. Developed using approaches such as CBOW, or skip-grams, to generate models (some of them are word2vec, or doc2vec) which are able to contain meanings testable against human enquires. Beginning to "understand" human concepts. A variance of this method, or a simple application are methods such as topic extraction or topic clustering	• Partial replacement of teachers in scoring or interaction with students • Able to manage concepts • Able to score partially

3.3   Operational Learning System Modeling

In order to achieve something for real, any challenge needs to be technically possible, somebody need to have the skills to get it and the will to do it. Achievements need feasibility, capability and will.

Essays are usually a demonstration of how something (for instance an achievement) becomes real. This demonstration needs to show some signs of truth or certainty, which are expressed through some essay success factors. In academic terms, this is equivalent to a successful evaluation, which in time demonstrates that some capabilities were built.

In time, these concept of success of the essay contains certain concrete components, in which the evidence of learning can be found. A diagram organizing these concepts is shown below in Fig. 2.

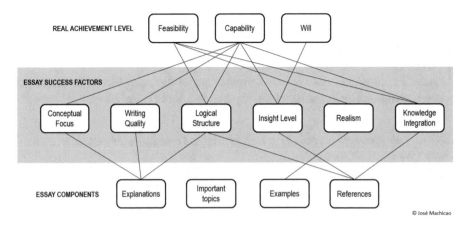

Fig. 2. Different levels of success help to give context to a successful essay. Achievement level needs essay success factor to be evident, and these need essay components to be verified.

Within the current typical operational configuration for the learning process, usually students, teachers and university authorities interact to achieve success. In order to implement an AES system, four main actors are identified as responsible for its operation: (i) students, (ii) teachers, (iii) algorithm engineers and (iv) university authorities or representatives. The roles are somehow different in these two scenarios (Table 2).

Table 2. Actors involved in the essay scoring process and its roles within the learning system under current and future scenario.

Actor	Current role description	AES role description
Student	Generator of essay, consequently generator of essay success factors. Eventually competency receiver. Raise questions to the teacher	Generator of essay, consequently generator of essay success factors. Eventually competency receiver. Should be able to interact with algorithms using cycles
Teacher	Generator of the aim of the competence and capability to be built. Consistency supervisor	Generator of the aim of the competence and capability to be built. Consistency supervisor. Should be able to configure parameters for the algorithms alone or through the algorithm engineer
Algorithm Engineer	No role	Translates the input from the teacher to the algorithm, adjusting parameters. Guarantees the reading of the features generated by the student
University	Entity in charge of learning process, and the management of resources and actions in order to optimize contribution from experts to reinforce student competencies	Entity in charge of learning process, and the management of resources and actions in order to optimize contribution from experts to reinforce student competencies. Should establish policies for algorithm design and supervision

It is clear that there is a noticeable shift between both scenarios. Even when the main aim of the original roles is still present, roles should be complemented with the ability to integrate automation, and therefore interaction with artificial intelligence algorithms.

Also, the concrete basic verifiable components of the essay should be understood within a system including artificial intelligence as a tool (Table 3).

Table 3. Description of concrete elements of the LES and examples within a high-level scoring criteria model.

Element	Description	Verification strategy
Explanations	Paragraphs developing concepts similar to those prioritized by the teacher	Algorithmic network verification Method: Document Vectorization
Important topics	Words or phrases coincident with those prioritized by the teacher	Coincidence con previously prepared list Method: NLP, word and sentence statistics and complex network graphs
Examples	Using important topics and referring to real cases	Coincidence con previously prepared list Method: Document Vectorization and Topics Extraction
References	Authors or entities coincident with those prioritized by the teacher	Coincidence con previously prepared list

Also, the concrete basic verifiable components of the essay should be understood within a system including artificial intelligence as a tool.

4   Conclusions

Universities are currently struggling with high operational costs and the pending challenge to respond to increasing demand, maintaining learning quality in time. This challenge is especially tough in countries with low level of automatization of the learning process.

Higher education in these cases is under the risk for universities to fall in either the extreme of neglecting AI, or expecting AI to solve all challenges and problems.

Particularly for the challenge for universities teaching staff, given the recent growth of AI as a tool for all sectors, a clear opportunity is growing to implement AES using AI tools, which demands to model processes to include them given an existing and often traditional way of scoring.

This challenge demands to build up a better organizational capability. The model should fit teacher accessibility, demanded speed of scoring, student capacity building and university benefit/cost demanded ratio. This work allowed to demonstrate the feasibility of implementing a realistic AES context and process under different scenarios and concludes that there are criteria that brings a high probability of success for traditional universities with low level of automation within the learning process.

The available tools have been identified and classified, and finally characterized to demonstrate its usability for each element of academic essays.

Further research can be done going deep in the human capability of universities to embrace AI as a tool and experiment the implementation of more models to increase the trust from all actors involved.

References

1. QS Quacquarelli Symonds: Understanding the Future of Higher Education. QS Website (2018). http://www.qs.com/understanding-future-higher-education/
2. Moran, H., Powell, J.: Running a tight ship: can universities plot a course through rough seas? Special Report for the Guardian, HSBC, UUK (2018). https://uploads.guim.co.uk/
3. Kerr, C.: The Uses of the University, 5th edn. Harvard University Press, Cambridge (2003)
4. Ferrer-Balas, D., et al.: Going beyond the rhetoric: system-wide changes in universities for sustainable societies. J. Cleaner Prod. **18**(7), 607–610 (2010)
5. INEI: Estadísticas de Educación Superior. Instituto Nacional de Estadísticas Website (2018). https://www.inei.gob.pe/estadisticas/indice-tematico/nivel-de-educacion-alcanzado-8034/
6. Vedder, R.: Seven Challenges Facing Higher Education. Forbes Magazine. The Center for College Affordability and Productivity (2017). https://www.forbes.com/sites/ccap/
7. Bok, D.: Universities in the Marketplace. The Commercialization of Higher Education. Princeton University Press, Princeton (2003)
8. McDonald, M.: Systematic Assessment of Learning Outcomes: Developing Multiple-Choice Exams. Jones & Barlett Learning, Boston (2002)
9. Warwick, K.: Artificial Intelligence: The Basics. Routledge, London (2013)
10. Connelly, J., Forsyth, P.: Essay Writing Skills: Essential Techniques to Gain Top Marks. Kogan Page Publishers, London (2012)
11. Page, E.B.: The imminence of grading essays by computer. Phi Delta Kappa **47**, 238–243 (1966)
12. Shermis, M.D., Burstein, J.C.: Automated Essay Scoring: A Cross-Disciplinary Perspective. Routledge, New York (2003)
13. Ramalingam, V., Pandian, A., Chetry, P., Nigam, H.: Automated essay grading using machine learning. In: National Conference on Mathematical Techniques and its Applications (NCMTA 2018) (2018). https://doi.org/10.1088/1742-6596/1000/1/012030. http://iopscience.iop.org/article
14. Shermis, M.D.: The challenges of emulating human behavior in writing assessment. Assessing Writ. J. **22**, 91–99 (2014)
15. Harvard University: Essay Structure. Harvard College Writing Centre (2017). https://writingcenter.fas.harvard.edu
16. Hoppe, U., Verdejo, M.F.: Artificial Intelligence in Education: Shaping the Future of Learning Through Intelligent Technologies. IOS Press, Amsterdam (2003)
17. Archibald, R., Feldman, D.: Explaining increases in higher education costs. J. High. Educ. **79**(3), 268–295 (2008)
18. Muller, V., Bostrom, N.: Future progress in artificial intelligence: a survey of expert opinion. In: Fundamental Issues of Artificial Intelligence. Springer (2016)

19. Dale, R., Moisl, H., Somers, H.: Handbook of Natural Language Processing. Business and Economics. CRC Press, Boca Raton (2000)
20. Wissner-Gross, A.D., Freer, C.E.: Causal entropic forces. Phys. Rev. Lett. **110**, 168702 (2013)
21. Glaser, R.: Knowing, Learning, and Instruction. Lawrence Erlbaum Publishers, London (1989)
22. Bird, S., et al.: Natural Language Processing with Python: Analyzing Text with the Natural Language Toolkit. O'Reilly Media Inc., Sebastopol (2009)
23. Le, Q., Mikolov, T.: Distributed Representations of Sentences and Documents. Google Inc. (2014)
24. Scikit Learn: Scikit-Learn Website (2018). http://scikit-learn.org
25. Norton, L.S.: Essay-writing: what really counts? High. Educ. J. **20**(4), 411–442 (1990)
26. Mumford, M.D., et al.: Leadership skills for a changing world: Solving complex social problems. Leadersh. Q. J. **11**(1), 11–35 (2000)

Cybersecurity and Cyber-Defense

Cybersecurity Baseline, An Exploration, Which Permits to Delineate National Cybersecurity Strategy in Ecuador

Mario Ron[1,2(✉)], Oswaldo Rivera[1], Walter Fuertes[1],
Theofilos Toulkeridis[1], and Javier Díaz[2]

[1] Universidad de las Fuerzas Armadas ESPE,
Av. General Ruminahui, Sangolquí, Ecuador
{mbron,loriveral,wmfuertes,ttoulkeridis}@espe.edu.ec
[2] Universidad Nacional de la Plata,
Avenida 7 877, 1900 La Plata, Buenos Aires, Argentina
jdiaz@unlp.edu.ar

Abstract. Information and Communication Technologies entails risks inherent to their use, which must be treated by organizations to protect their users from possible damage to their integrity. To determine the way to protect users, a real knowledge of the current situation is indispensable. This study allows to know the current state of cybersecurity in Ecuador through a systematic and organized process, based on the main criterion that should be considered in the establishment of a National Cybersecurity Strategy (NCS), considering its level of maturity. The results obtained as a baseline are a significant contribution to the definition of the NCS in Ecuador and the procedure could be applied by other countries in the region both to prepare their policies and to monitor the impact they have had.

Keywords: Cybersecurity · Baseline · Policy · Organizations · Ecuador

1 Introduction

Cyberspace is completely an unprecedented space in human knowledge [1], where many activities of social, political and economic interest interact in this dimension when interconnecting clients, enterprises and governments. This new space has created new civil rights or modified the way to exercise them both for natural and legal persons. Among them, the right to privacy, to the good name or reputation, to the protection of their property, to secure communication and reliable, to the use of true information, among others. The Ecuadorian State in its Political Constitution, as a fundamental legal norm, has the responsibility to develop policies that protect the rights of the people, in this case to provide protection to the use of cyberspace.

Within this scenario, this study conducts a field research accomplished in a technical and systematic way, using reliable and consistent procedures based in public and private sources and organizations related to cybersecurity in the country, as well as from international or regional sources that has, in some way, performed studies related to cybersecurity in Ecuador.

© Springer Nature Switzerland AG 2019
Á. Rocha et al. (Eds.): ICITS 2019, AISC 918, pp. 847–857, 2019.
https://doi.org/10.1007/978-3-030-11890-7_79

The Organization of American States (OAS) has developed systematic field research processes and support to determine the current state of cybersecurity in many countries and maintains an important knowledge base, which is used in this research, as well as those from regional organizations and information security companies.

The questions that addressing this research are: What are the main criterion that intervene in a NCS? What are the focal organizations related to cybersecurity in the country? What is the current status of the criterion and organizations related to cybersecurity? The assumption of this research is that the definition of a baseline developed according to a systematic and organized process with the most relevant criterion of a NCS, allow to establish an initial agreement for its development.

The rest of the paper is organized as follows: Sect. 1 presents the systematic methodology and the processes developed. Section 2 refers to the information obtained in the field research, organized according to the established aspects. Section 3 presents the analysis of the information obtained and its projection in the NCS. Section 4 explains related works. Finally Sect. 5 contains the conclusions and future work lines.

2   Field Research, Methods, and Processes

2.1   Field Research Process

The process begins with the formation of the stakeholder group in the study. This was conformed of government representatives, academic entities, and social or business groups related to cybersecurity in the country, who determines the important criterion related to the NCS. Furthermore, we have considering the development of similar policies in the region and the world based on their relevance, scope, timeliness, and economic and social impact, as well as those that have served to establish levels of maturity, such as the one presented by the OAS [1]. Its selection is based on an affinity diagram. Then, using a context diagram, the main stakeholders are defined in the preparation of the policy. Afterward, they were selected according to their knowledge of the subject, impact on their organization, relevance, capacity of contribution and availability, which will be provide the relevant information. With these elements, the Field Research Plan is elaborated in which the criterion become research objectives, which are broken down into important topics and basic questions. The basic questions are used to develop the field research instruments, which in this case the major ones were: surveys, questionnaires, interviews, documentary review, and direct observation, which will serve for the analysis, obtaining conclusions and contributions to the NCS.

2.2   Criterion Selection

There are currently several countries that have declared their cybersecurity policies in South America. They have defined aspects of interest for the initial diagnosis and specific strategies in their NCS. As stated before, the OAS also defines criterion to establish the model of cybersecurity maturity level in Latin America [1]. On the other hand, the International Telecommunications Union (ITU) [2], also establishes criterion that guide the search for information of the initial diagnosis. Using an affinity diagram, these proposed elements are grouped to determine the criterion that will be taken into account in the field research, considering the percentage of inclusion of these criterion in the documents analyzed. The result is presented in Fig. 1.

COUNTRY \ ASPECTS	Culture and awareness	Legal and regulatory	Training and education	Critical infrastructure	Response to incidents	Institutionality and organization	Cooperation and assistance	Risk management	Research and Innovation	Coordination and collaboration	Standards and technical criteria	Governance	Fight against cybercrime	Digital Dev. and Telecommunications	Industry and technology	Infrastructure of the information	Development of capabilities	National Strategy	Monitoring and evaluation	Operator Practices	Promotion of rights	Economic impact	Planning and framework	Operational continuity
Chile	·	·	·	·	·	·	·	·	·			·	·	·	·	·								
Colombia	·	·	·	·		·	·	·		·			·				·						·	·
Costa Rica	·	·	·	·	·	·	·	·	·	·			·		·		·		·	·	·	·	·	
Guatemala	·	·	·	·	·		·			·	·			·	·		·							
Jamaica	·	·	·	·	·		·	·			·					·								
México	·	·	·	·			·	·	·	·					·									
Panamá	·	·	·	·	·				·	·					·									·
Paraguay	·		·	·	·				·	·			·	·			·							
República Dominicana	·	·	·	·	·		·				·													
Trinidad -Tobago	·	·			·			·				·												
Maturity levels OAS	·	·	·	·	·		·		·			·	·	·	·	·		·		·				
ITU NCSG	·	·	·		·	·	·			·		·												
% Inclusión	100	92	83	83	75	67	58	58	50	42	42	42	33	33	33	33	25	25	25	25	25	17	17	8

Fig. 1. Criterion selection criteria.

2.3 Field Research Plan

The criterions to be investigated became in objectives, from which the important topics and the basic questions were derived. From the national government structure and with the participation of the private sector [3], the entities to provide the information were selected. With those and other elements the Field Research Plan was elaborated. It design is presented in Fig. 2. As an example, one of the objectives is developed.

Objectives	Important themes	Basic questions	Research instrument	Source	Execution date	Responsable	Coordination	Outcomes
Stakeholder collaboration	1*	2*	3*	4*	5*	6*	7*	8*
Culture and awareness	-	-	-	-	-	-	-	-
....	-	-	-	-	-	-	-	-

Fig. 2. Research plan.

From where: 1 * Public entities participation/Private company participation/Events of participation/General coordination/Results of participation; 2 * Participating entities?/ Events made?/Liability and plans of the Coordination; 3 * Survey Nro.01/Interview

No. 01/Documentary research guide; 4 * Ministry of Telecommunications (MINTEL); 5 * 08/10/2018; 6 * Oswaldo Rivera; 7 * Secretariat MINTEL; 8 * Evidence ENC-01/Evidence ENT-01/Evidence GID-01;

The questions allow to elaborate the research instruments that basically contain the identifying data, execution instructions, body of the document and results. These instruments are applied according to the plan and evidence is collected.

3 Research Findings - Cybersecurity Baseline

3.1 Context Information

Digital Development and Telecommunications. The telecommunications sector has developed consistently. Ecuador is ranked 84 out of 193 countries with the EGDI index [4]. The index and its components that are presented in Fig. 3.

E-Government Development Index (EGDI) 2018						E-Participation Index (EPI) and its utilisation					
Rank	EGDI 2018	Online Service	Telecomm. Infrastructure	Human Capital	Level of Income	Rank	EPI 2018	Total %	Stage 1%	Stage 2%	Stage 3%
84	0.6129	0,792	0.3699	0.7395	Upper Middle	81	0.6742	68.48%	70.00%	78.26%	54.55%
Americas average	0.5898	0.6095	0.4441	0.7157		Americas average	0.6043	61,72%	68,76%	61,74%	54,03%

Fig. 3. Development of the digital and telecommunication sector.

As for the E-Participation index (EPI), the increase compared to the one measured in 2016 is also significant, as presented in Fig. 3 [4].

Industry and Technology. Currently 90% of households have at least one cell phone, 52.0% of the population use a computer, and 58.3% use the Internet and 58.5%. In relation to 2012, in 2017 the equipment of laptops in the home increased by 12.1 points and Internet access increased by 14.7 points, equivalent to 50%. It is worth mentioning that 10.5% of people between 15 and 49 years old are digital illiterates and the use of social networks from a smartphone reaches 31.9% [5].

Concerning the business sector, 98% have at least one computer. 96.6% have access to the Internet and of these 96.8% use fixed broadband. 13.9% make purchases online and 9.2% sales through this channel [6]. Regarding open source, of the companies that have Internet connection, 98.1% used browsers of this type and of the total of the companies 56.5% used office applications, 39% operating systems, and 29.4% other type of open source software. There is an increase in the use of social networks in 12 points since 2012. The Global Innovation Index consigns Ecuador in 96th place among 126 countries, in terms of Information and Communication Technologies (ICT), it is ranked 77th with an index of 52.3 [7].

Information Infrastructure. There is not an integrated National Information System, however the National Public Data Registration System (SINARDAP) has been created that integrates the information of several organisms that process public information, the National Information System for the planning (SNI), the Information System for Decentralized Autonomous Governments (SIGAD), the Quipux Document Management System, SIGEF and other national and sectoral systems [8].

The telecommunications infrastructure has grown every year, as can it be seen in the measurement made by the United Nations for 2018 [9], which qualifies the country with an index of 0.37. As of June 2017, 92.46% of SMA population coverage was achieved in 3G and 4G technologies. [10]. The connectivity is privileged by three submarine cables: PANAM, SAM1 and PCSS, with an average navigation speed of 6.2 Mbps per user [11], above the regional average of 5.6 Mbps. This capacity has allowed the construction of data mega centers with global certifications. There is also a Traffic Exchange Point (NAP.ec) that hosts the Internet infrastructure and allows the exchange of 97% of local traffic. Despite this having an advanced level of deployment of backbone infrastructure, there are 6% cantons to connect with this technology they represent [10].

MINTEL has implemented 854 Info-centers, to support the most vulnerable population that, due to its socioeconomic status, has been excluded from access to ICT. Community Info-centers are also used by public institutions to provide online services.

The government has control of the technological infrastructure, through the National Telecommunications Corporation (NTC). However, networks and systems are concessioned to third parties. There is dependency of other countries on cybersecurity technology [8]. Most companies have their own IT infrastructure (47.77%) and only a minority (8.89%) has been subcontracted, 36.3% use Service Desk [3].

Economic Impact. The expenses in telecommunications services are higher than 7% of the monthly income of the household. Likewise, the 14% of households with lower income have a computer and 11% have Internet access. One of the causes is the application of tariffs that reaches 15% for mobile phones and 10% for computers or tablets. Infrastructure operators have made investments and implemented measures to build trust in online services, due to operational risk regulations of the Superintendence of Banks and Insurance, which has allowed the development financial transactions.

National Strategy. There is no NCS, therefore cybersecurity is handled independently in all sectors of the country. A project has been initiated to define it under the initiative of several public agencies, including the MINTEL, ARCOTEL, Cyber Defense Command of the Joint Command of the Armed Forces (COCIBER), Computer Incident Response Center of Ecuador (EcuCERT), and associations related to private company, academia and international organizations. The intention to participate in the creation of national cybersecurity policies by private companies has been very important [3].

Planning and Framework. The National Development Plan 2017–2021, establishes in its Objective 7, "To improve the electronic government index to 2021". In 2016, MINTEL issued the National Plan for Telecommunications and Information Technologies [12]. This establishes as one of the programs "ensure the use of ICT for the economic and social development of the country". The National Electronic Government Plan 2018–2021 was published, which includes its implementation, governance, monitoring and evaluation. In 2018, the White Book of the Information and Knowledge Society was published, which is a five-axis plan; the fourth one of them is related to Information Security and Protection of Personal Data [10]. 73.10% of companies (73.10%) have been concerned about the preparation of their strategic planning. However, its dissemination and updating have been less than 10% [3].

Institutionally and Organization. There are organizations interested in cybersecurity. However, they have not yet coordinated their actions. There is no formally established national cybersecurity command and control center. The following institutions have been determined with specific responsibilities: MINTEL to govern Telecommunications and the Information Society, COCIBER to respond to incidents on strategic digital critical infrastructure of the Armed Forces and the State. EcuCERT and CEDIA-CERT provide response to computer security incidents teams.

Coordination and Collaboration. The MINTEL has organized working groups in the main cities of Ecuador to involve organizations, companies and members of the public. Furthermore, it establish a commitment to work together in the development of the Plan of the Information and Knowledge Society, the Program of Information Security and responsible use of ICT and support in cybersecurity in the country [8].

Cooperation and Assistance. Ecuador is part of international organizations such as ITU, OAS, ONU and others, which have working groups related to cybersecurity, which should be activated. The assistance of the OAS is foreseen to prepare the NCS.

3.2   Culture and Training

Culture and Awareness. The curricula of primary and secondary education do not contain subjects in science and computation, therefore the development of digital skills, through the educational system, is limited.

MINTEL has carried out several campaigns through social networks, to prevent crimes such as grooming, theft and information, phishing, spam, scam, sexting, vamping and cyberbulling [13]. In the community Infocenters, the Digital Enrollment Plan is fulfilled, with which citizens are trained on the safe and healthy use of the Internet and social networks [13]. The public and private sectors and society in general, are still not aware of the threats and are not yet taking proactive measures to improve their cybersecurity. The financial sector, on the other hand, has taken a lot of interest in security, because the law holds them liable for fraud losses.

Operators' Practices. The operators have developed some security measures in accordance with the ISO/IEC 27000 and other international standards. The national technology infrastructure is in charge of the NTC, which is currently undergoing a process of recertification in ISO 27001. The operators of critical infrastructure, in charge of public companies, still do not advance in consistent processes of securing information systems.

Training and Education. There is a market for education and training in information security. However, there is no specific offer in undergraduate careers. In the case of postgraduate programs related to information security have increased. In terms of professional certification, there is room for subsidiaries of the companies and international institutions of the area such as ISACA, EC-Council and others, however the certification costs are limiting for their access. The number of professional instructors in cyber security is scarce. There are no known train-the-trainer programs. The transfer of knowledge from trained personnel to new staff is limited.

Capacity Development. The size of information security equipment in most companies is small (62.65%), there is 10.84% that has no personnel [3]. The Global Cybersecurity Index (GCI) points out that Ecuador is positioned at a mature level (66 of the global ranking) and sixth in Latin America countries [14]. However, it has serious limitations in relation to the most advanced countries in Cybersecurity, with some progress in the Capacity Building and Cooperation components [8, 10, 13, 15].

Research and Innovation. The Global Innovation Index places Ecuador with the Research & development Index (RDI) of 6.2 in position 71 among 126 countries [7]. It has consistently had a high rate of early entrepreneur activity (TEA), in 2017 it was 29.62% (lower than in 2016). Both the proportion of nascent and new entrepreneurs has been reduced in 2017 [16]. The research and development of cybersecurity tools and procedures is in an initial state, there is not yet adequate infrastructure, neither production of related technology and the number of researchers is really reduced.

3.3   Legislation

Legal and Regulatory Framework. The Organic and Integral Penal Code (OIPC) [15] includes crimes such as harassment, offering of sexual services to minors, fraud, fraudulent appropriation, disclosure and illegal interception of data. Also includes illegal electronic transfer of assets, attacks on the integrity of Information Systems (IS), access not consented by electronic means to IS and crimes against public information. There is progress in criminal legislation, but it is important to strengthen the legal framework, after an analysis of the impact of the law on crime prevention [17].

There is legislation regarding privacy, data protection and freedom of expression, the latter is being modified by the National Assembly, the Electronic Commerce Law, the Telecommunications Law, the General Law of Postal Services, have also been issued. Organic Law of Identity Management and Civil Data. Criminal procedural law applies ad-hoc to cybercrime and the use of electronic evidence in other crimes. Digital evidence is relatively new in this medium.

Standards and Technical Criteria. There are 20 Ecuadorian technical standards adopted from the ISO 27000 series regarding information security. It has been implemented the Government's Information Security Scheme (GISS) in Public Administration entities. However, its application has not been completed. There is no evidence of the use of cybersecurity standards in procurement processes and in the development of the software still they are not applied. In the research conducted, companies are not aware of the importance of security standards and their use in information security, the implementation of cybersecurity policies reaches 20.48% [3].

Fight against Cybercrime. Since the approval of OIPC (2014) 6211 complaints about cybercrimes have been recorded. The most common crimes are: "fraudulent appropriation by electronic means", "Non-consensual access to a computer, telematics or telecommunications system" and "Contact with sexual purposes with minors under 18 years of age by electronic means". Nonetheless, the growth rate of financial crimes has decreased, which shows greater care in the management electronic means. This has been corroborated by the Cybersecurity Observatory in Latin America and the Caribbean [16, 17]. However the training is not adequate to prosecute computer-related crimes. The trained justice operators is limited.

Promotion of Civil Rights. Laws and policies that promote access to personal data stored by public institutions have been enacted. The transparency of information for citizens is greater than 97% in public institutions. Regarding the release of open data for the use of citizens, the Open Data Inventory (ODIN) 2016, places Ecuador in 2016 in position 31 of 173 countries, with a score of 56 [8].

3.4   Technical Processes

Governance. The companies' boards have a limited knowledge of cybersecurity. Therefore they do not exercise technology adequately and leave IT management units that responsibility. The application of the ISMS in public entities has made it possible to start with the knowledge of these aspects by senior management. Then again, there is still no awareness of the specific actions that must be undertaken [3].

Risk Management. The National Secretariat of Risk has focused on environmental risks and natural disasters, not so much on computers. The GISS contemplates a risk analysis for its implementation, however in the evaluation handled by the MINTEL in 2018, only 16.36% obtained a good result [8]. The training of experts in cyber risk is still at an early stage and there are few certified professionals. Crisis management in the area of cybersecurity has not yet been designed and does not have national officials.

Operational Continuity. Despite the government's initiative to implement the ISMS in public institutions, operational continuity plans are not common. The measures of digital redundancy are optional. The financial and commercial entities have started a culture in this regard to maintain the services to their clients and not to cause losses due to lost profits. There is little training of certified professionals in that area.

Critical Infrastructures. The COCIBER has determined the critical infrastructures of the country according to the National Catalog of Critical Infrastructures, of a confidential nature, but the vulnerabilities have not been identified nor is there a formal collaboration mechanism for the response to an attack.

Response to Incidents. The Computer Incidents Response Center of Ecuador (Ecu-CERT) and CERT-CEDIA, act in the prevention and resolution of computer security incidents. However its impact is not significant, due to the limitations in its competences, which the Law establishes [8]. Cyber events or threats have been registered and classified at the national level as incidents. In spite of this, the response capacity is limited and reactive, it is not coordinated, and it is done ad-hoc. Responsibility for responding to incidents at the national level is not established. The emergency response assets have not been identified and lack integration.

Monitoring and Evaluation. The National Secretariat of Planning (SENPLADES) maintains the Integrated System of Public Planning and the Government by Results (GBR) with the monitoring and evaluation in general is conducted. The evaluation of the GISS performed by MINTEL [10] determines that only 16.36% of the entities have a good result. New graduate programs have allowed incorporating professional computer auditors to companies and public and private institutions to help evaluation.

4 Analysis and Discussion of Results

4.1 Definition of Maturity Level Model

In order to assess the current state of the criterion, six levels of maturity are defined, taking into account those presented by the Maturity Model published by the OAS [1]. Those defined by ISACA in its process capacity model and those contained in ISO/IEC 15504-2 [18], the first has 5 levels and the next two have 6 levels, the levels defined for this case study are: Level 0: Incomplete, process not fully implemented, does not fulfil its purpose; Level 1: Implemented, process fulfils its purpose; Level 2: Managed, process planned, supervised and maintained, as well as its products; Level 3: Formalized, the process uses a set of documented and standard activities; Level 4: Consistent, process measured and is carried out in a predictable manner; Level 5: Innovated, the process has the capacity to innovate, be resilient and react quickly to changes in the environment.

4.2 Maturity Level Model Application

The information obtained in the field research is analyzed according to the characteristics established in each level of the defined maturity model and each criterion is graded and assigned a level. This procedure allows to establish the level of the current state and later the level that the country wants to reach after applying a policy. As an example, an agreement could be established to provide formal training by the universities to the justice operators with a specific program documented and planned, in this way the criterion: Fight against cybercrime would go from level 0 to level 1.

Table 1 shows the result of the analysis, includes the criterion, the level and the summary observation of the condition of the criterion, in such a way that it can be compatible with other established models such as the one presented by the OAS [1].

Table 1. Factor's maturity levels.

Factor	Level	Observations
Context information		
Industry and Technology	1	96.6% companies have technology. GII = 96/126
Information infrastructure	1	Privileged connectivity, don't have a national integrated system
Economic impact	1	Expenses in high IT services, start banking
National Strategy	0	Does not exist
Planning	2	There are National and Business Plans
Institutionally and organization	1	Institutions without functional interrelation
Coordination and collaboration	0	Work tables are started
Cooperation and assistance	0	International cooperation available
Culture and training		
Culture and awareness	1	Only the financial sector has taken interest
Operator Practices	0	They do not have consistent processes
Training and education	1	There are no degree courses, the postgraduate ones have increased
Development of capabilities	1	GCI = 66/. There is no technology production
Research and Innovation	0	There is no infrastructure or human capital
Legislation		
Legal and regulatory framework	3	Crime legislation and rights protection framework
Standards and technical criteria	1	ISO 27000 standards, but its use is not consistent
Fight against cybercrime	0	Limited practice by justice operators
Promotion of rights	1	Laws and policies enacted. ODIN = 31/73
Technical processes		
Governance	0	Limited knowledge of senior management
Risk management	0	Only to natural disasters and environmental risks
Operational continuity	0	Just in companies with high financial risk
Critical infrastructure	0	Definition and unstructured protection
Response to incidents	0	Unstructured or integrated incident response
Monitoring and evaluation	1	GBR. Masters in Computer Auditing

5   Conclusions and Future Work

A description of the current state of cybersecurity in Ecuador has been presented, through a methodological process that has enabled the selection of the most widely used criterion at the global level. To carry out, an extensive field research with a diversity of sources that have provided reliable information that allows establishing a baseline with levels of maturity. From these would depart to determine the desired state of cybersecurity in the country, through the definition of a NCS, containing effective actions in the protection of critical information and consequently of business operations and citizens' rights. As future work, we planed the development of the methodology for the construction of NCS in developing countries.

Acknowledgments. The authors would like to express special thanks to all institutions, companies and professionals who participated in the field research as they did before in similar processes to develop the NCS.

References

1. OAS, 2016 Cybersecurity Report (2016)
2. ITU, ITU National Cybersecurity Strategy Guide (2011)
3. Ron, M., Bonilla, M., Fuertes, W., Díaz, J., Toulkeridis, T.: Applicability of cybersecurity standards in Ecuador - a field exploration. In: MICRADS 2018, vol. 94, pp. 27–40 (2011)
4. UN Department of Economic and Social Affairs, E-government Survey 2016 (2016)
5. INEC, "ENEMDU-TIC 2017" (2017)
6. INEC, "Empresas y TIC," Inec, p. 54 (2015)
7. GII, C. University, INSEAD, and WIPO, "Global innovation Index" (2018)
8. MINTEL, Plan Nacional de Gobierno Electrónico, Quito - Ecuador (2017)
9. UNDESA, 2018 E-Government Survey (2018)
10. MINTEL, Libro Blanco de la Sociedad de Ia Información. Ecuador (2018)
11. Akamai, "state of the internet Q1 2017 report" (2017)
12. MINTEL, Plan Nacional De Telecomunicaciones Y Tecnologías de Información del Ecuador 2016–2021. Sect. Telecomunicaciones y Tecnol. la Inf., vol. 1, p. 66 (2016)
13. MINTEL. Rendición de Cuentas 2017, Quito - Ecuador (2018)
14. ITU. Global Cybersecurity Index (GCI) 2017 (2017)
15. Rom, M., Bustamante, F., Fuertes, W., Tulkeredis, T.: Situational status of global cybersecurity and cyber defense according to global indicators. Adaptation of a model for Ecuador. In: MICRADS 2018, vol. 94 (2018)
16. OEA-BID. Observatorio de la Ciberseguridad en América Latina y el Caribe. http://observatoriociberseguridad.com/. Accessed 12 Feb 2018
17. Ron, M., Fuertes, W., Bonilla, M., Toulkeridis, T., Diaz, J.: Cybercrime in Ecuador, an exploration, which allows to define national cybersecurity policies. In: CISTI 2018, June 2018, pp. 1–7 (2018)
18. ISACA. A Business Framework for the Governance and Management of Enterprise IT (2013)

Advancement in Cybercrime Investigation – The New European Legal Instruments for Collecting Cross-border E-evidence

Borka Jerman Blažič[✉] and Tomaž Klobučar

Jožef Stefan Institute, Jamova 39, 1000 Ljubljana, Slovenia
{borka,tomaz}@e5.ijs.si

Abstract. This paper presents the current issues and proposes legal instruments for removing the barriers to gathering cross-border electronic evidence for a more efficient fight against crime and cybercrime investigation. It presents the legal scene in the EU, the efforts related to the implementation of the Directive 2014/41/EU regarding the European Investigation Order in criminal matters, and the provision of the procedures for e-evidence collection proposed within the EU proposal for the Directive for orders to preserve and produce evidence related to criminal acts in a territory different from the location of the crime act. The legal instruments proposed in the new EU regulation for collecting e-evidence are analyzed and the expected removal of jurisdictive barriers for a more efficient justice processing of crime acts in an interconnected world is assessed.

Keywords: Electronic evidence · Crime · Investigation order · Production order · Preservation order · LIVE_FOR

1 Introduction

The interconnectivity of the global economy enables criminals to operate trans-jurisdictionally, with elements connected to their crimes spread widely across the globe in both time and space [1]. More and more crimes are committed online today, facilitated by computing devices such as mobile phones, tablets or PCs. The nature of these devices allows that traces associated with crime are left. That is why electronic forms of evidence for judicial proceedings are increasingly in demand today. Digital devices store, transmit or process a wealth of user data that is integrated with e-mail accounts, cloud-based services, social networking platforms, and synchronized desktop applications. The electronic data can contain detailed sequences of events indicative of criminal intent, interrelationships between organized networks of offenders, and the whereabouts of suspects of criminal activity. A large quantity of electronic evidence (e-evidence) is gathered by various service providers, such as telecommunication services, information society services, the Internet, and cloud services. The statistics from Europe show that 100.000 requests for e-evidence were issued in 2017 by the Member States to the main service providers. More than 60% were sent to Google [2].

© Springer Nature Switzerland AG 2019
Á. Rocha et al. (Eds.): ICITS 2019, AISC 918, pp. 858–867, 2019.
https://doi.org/10.1007/978-3-030-11890-7_80

Despite mutual agreements signed between many countries in the world and extra-territorial legal provisions for criminal acts committed in foreign jurisdictions the practical implementation of the laws enabling provision of e-evidence was assessed as ineffective [3]. Whilst an offender may be apprehended in one jurisdiction, the e-evidence required in an investigation may be located in another country [4] and that requires a special legal form to order its handing-over. For most forms of crimes, in particular in cases of cybercrime whose increasing occurrence we witness in the last decade, e-evidence – such as account subscriber information, traffic or metadata, or content data – has become of ultimate importance as it provides significant leads for investigators, often being the only evidence connected to the crime acts. E-evidence connected to crime acts perpetrated in the interconnected society is often cross-jurisdictional, because the data is stored outside the sphere of influence of the inves-tigating authority in the country where investigation has been launched, or by providers of electronic communication services and platforms whose main seat is located outside the investigating country, resulting in the fact that investigating authorities are not able to use their domestic investigative tools.

Major problem with collecting e-evidence is that the investigators must ensure that they abide by applicable laws or otherwise risk that the seized exhibits will be declared inadmissible at trial. In some jurisdictions there are exceptions that may justify war-rantless search and seizure activities (e.g. consent, 'emergency' terrorist situations, plain view doctrine, search relating to arrest, etc.). The actual search of the data stored on a device usually requires a warrant in common law countries. In circumstances where there is a substantial risk of losing evidence, such as where data sanitizing and other anti-forensics tools are active, some jurisdictions permit law enforcement to perform a limited search of devices without a warrant due to the perceived vulnerability of the data [5]. In addition, many investigators encounter administrative delays in obtaining legal authority to conduct police investigations due to judicial uncertainty. Another problem is connected to the preservation of stored data as different practices are implemented in different countries. The legal treatment of data may differ signifi-cantly from one country to another, posing a minefield of technical and legal com-plexities. As such, it becomes critical that investigators and prosecutors be well informed of geo-specific data-mapping issues including embargoes or prohibitions of data exchange.

This paper presents the new instruments proposed in Europe for removing barriers in collecting cross-border e-evidence to ensure a more efficient fight against crime and cybercrime. In this context the paper briefly presents the current legal scene in the EU, the efforts related to the adoption and implementation of the [6] and provision of the procedures for e-evidence collection proposed within the EU proposal for the Directive for orders to preserve and produce evidence related to criminal acts in a territory different from the location of the crime act. This paper is organised as follows: the introduction section presents the problem, the next section provides a brief description of what kind of tools are available, and the last two sections of the paper present the instruments proposed in the new EU regulation for collecting e-evidence. The con-cluding marks provide brief assessments about the expected removals of jurisdictive barriers for a more efficient justice processing of criminal acts in an interconnected world.

2   E-evidence and the Problem of Cross-border Search

The main legal instrument in the fight against cybercrime within an international dimension is the Council of Europe's Convention on Cybercrime Act [7]. This document is a result of the efforts dedicated to solving the problem of cyber-based crime by the members of the Council of Europe (CoE), Interpol, Europol, the Organization for Economic Cooperation and Development, the G8 Group of States, Commonwealth and the United Nations (UN). The document of the Council of Europe is known as Budapest Cybercrime Convention (hereinafter referred to as the CCC) that entered into force in July 2004. The document was ratified by 43 out of 47 Members (until August 2018) of the Council of Europe (San Marino, Ireland, Russia, and Sweden have not yet ratified it) and by Argentina, Australia, Cabo Verde, Canada, Chile, Costa Rica, the Dominican Republic, Israel, Japan, Mauritius, Morocco, Panama, Paraguay, the Philippines, Senegal, Sri Lanka, Tonga, and the United States of America. The CCC is not limited to matters of cybercrime as it embraces investigatory measures concerning "the collection of evidence in electronic form" for any form of offence where such electronic evidence may be relevant. The CCC with its 61 contracting states constitutes the first and most significant multilateral binding instrument for defense against cybercrime with legal instruments, sometimes considered as "the most comprehensive international standard to date" [8].

The implementation of the CCC which is based on the existence of the MLA (Mutual Legal Assistance) agreement between the signatories has shown that the implementation depends on more detailed specification of the instruments provided within the CCC known as the production order. To solve this problem the Cybercrime Convention Committee (hereinafter T-CY) of CoE, published in March 2017 the document entitled the Guidance Note focused on the interpretation of Article 18 of CCC and the interpretation of the "Production order instrument for getting subscriber information" (hereinafter Guidance Note) [8]. Soon after the adoption of the Guidance Note criticism about the interpretation of the article (18) (1) (b) CCC in the Guidance Note appeared with a detailed analysis of the specified instrument. The major problem was found to be in the definitions within the article "Article 1 (3)" of the CCC where the service provider is defined as "(i) any public or private entity that provides to users of its service the ability to communicate using a computer system, and (ii) any other entity that processes or stores computer data on behalf of such communication service or users of such service."

Article 18 (1) of the CCC defines "production order" for collecting e-evidence as the issuance of production orders by law enforcement bodies to foreign established service providers or persons, without specifying whether the required subscriber information is stored within or outside the ordering state's territory or not. In addition, the exact definition of the types of data to be searched and their potential location is not provided either. The Guidance Note with the provided interpretation is misleading about the way the data should be sought as the mutual agreement of the cooperating states about the respective document does not allow to be classified as consent to the extraterritorial enforcement of domestic production orders for e-evidence collection. This statement can be true only in cases when a production order addresses foreign-based service providers storing data on foreign territory. In addition, this action is not considered to be an exercise of state power in the territory of another state. Otherwise, such a production order would require a permissive rule derived from international custom or convention in order to not conflict with international law. The extension of the scope of Article 18 of the CCC in the proposed way by the Guidance Note represents a challenge to the principle of territoriality and sovereignty, which are essentially interconnected with the protection of fundamental rights and the rule of law. It is also crucial that the CCC and the Guidance Note do not pay any attention to the protection of fundamental human rights and the rule of law as any responsibility in regards to the effective protection of fundamental rights is specified as follows: "the parties to the convention are expected to form a community of trust that respects Article 15 CCC" [7]. The proposed instrument does not clarify what subscriber information actually stands for, although this missing information is of tremendous importance to the evaluation of the impact regarding data privacy has on production orders issued according to Article 18 (1) (b) CCC. Another gap in the Guidance Note is the neglected clarification of a dynamic IP address and whether this data can be considered as subscriber information, or does it fall into the category of traffic data and is consequently outside the scope of Article 18 of the CCC. The lack of consistent definitions of different types of data in the Guidance Note may also result in conflicts of law as regards the scope of envisaged measures. This may lead to misunderstandings between the requesting authority and executing authority or the service provider addressed. Forces dealing with investigation of cybercrime expected the Guidance Note's interpretation of Article 18 CCC to circumvent the established instruments of judicial cooperation, such as the MLA agreements, which allow the cooperation of law enforcement agencies from different countries, while ensuring respect for the interest of sovereignty of each country, but the expectations were not met.

3   European Approach to Investigation of Cybercrime and Collection of Data

Another legal document that addresses the collection of e-evidence is the Directive regarding the European Investigation Order in criminal matters (Directive 2014/41/EU, hereinafter the EIOD), adopted by the Parliament and the Council of the European Union on 3 April 2014. From 22 May 2017, the Directive replaced the corresponding provisions of the following conventions applicable between the Member States bound by this Directive (Art. 34 of the EIOD):

(a) the European Convention on Mutual Assistance in Criminal Matters of the Council of Europe of April 1959, as well as two additional protocols, and bilateral agreements concluded pursuant to article 26 thereof;
(b) the Convention implementing the Schengen Agreement; and
(c) Convention on Mutual Assistance in Criminal Matters between the Member States of the European Union and its protocol.

In addition, the EIOD replaces the Council Framework Decision 2008/978/JHA on the European Evidence Warrant to obtain objects, documents and data to be used in proceedings in criminal matters and it also replaces those provisions of the Framework decision 2003/577/JHA regarding the freezing of evidence (the recitals 25 and 26 of the EIOD).

The European Investigation Order (EIO) is "a judicial decision which has been issued or validated by a judicial authority of a Member State ("the issuing state") to have one or several specific investigative measure(s) carried out in another member state ("the executing state") to obtain evidence in accordance with this Directive" [6]. Regarding the authorities capable of issuing and validating an EIO (Article 2(2)), the EIOD provides details on the meaning and scope of the term "judicial authority" and lists the following entities: judges, courts, investigation judges, and public prosecutors competent in the crime case concerned. While based on the principle of mutual recognition, the EIO also incorporates the flexibility of the traditional system of MLA (Recital 6 of the EIOD) and thus the executing states enjoy the margin of manoeuvre when receiving a request via an EIO. The EIOD applies to any investigative measures directed to the gathering of any type of evidence, including e-evidence, in criminal proceedings, except in the case of joint investigation teams (Article 3 of the EIOD).

The Directive provides specific rules for e-evidence in two cases:

- Requests for the identification of persons holding a subscription of a specified phone number or IP addresses (Article 10.(2).e of the EIOD).
- Requests for the interception of telecommunications (Article 30 of the EIOD).

The EIOD defines the deadlines for the executing authority to decide on the recognition or execution of an EIO and an additional time limit to carry out the investigation measure. By criminal proceedings, the EIOD refers to those that are initiated by judicial and administrative authorities. In case of the latter, this will only apply insofar they give rise to proceedings before courts with criminal jurisdiction (Article 4 of the EIOD).

4 The New European Legal Instruments for Collecting Cross-border E-evidence

Due to the growth of the cybercrime act and the requests for e-evidence issued to the service providers the European Commission was asked to "explore the possibilities for a common EU approach on enforcement jurisdiction in cyberspace in situations where the current frameworks are not sufficient, for example in situations where relevant e-evidence moves between jurisdictions in short fractions of time". The Commission was specially requested to determine "which connecting factors can provide grounds for enforcement jurisdiction in cyberspace" and "whether, and if so, which investigative measures can be used regardless of physical borders". Similar concerns were expressed especially in relation to the MLA process and its reform by the US Department of Justice set up in the document "Cross-Border Law Enforcement Demands: The analysis of the US Department of Justice's Proposed Bill" was issued on August 2016. The answer to these requests came on 17 April 2018 in the form of two documents: The Impact Assessment about the Proposal for a regulation of the European Parliament and of the Council on the Regulation European Production and Preservation Orders for electronic evidence in criminal matters [9].

The main objective of the Regulation is setting an EU legal instrument for investigative measures addressed to a service provider enabling authorities to request ("production request") or order ("production order") a service provider in another Member State to disclose information about a user. The proposed Regulation targets the specific problem created by the volatile nature of electronic evidence and its international dimension. It seeks to adapt cooperation mechanisms to the digital age, providing judiciary and law enforcement tools to address the way criminals communicate today, and to counter modern forms of criminality. The use of such tools depends on their being subject to strong protection mechanisms for fundamental rights. The suggested instrument is designed to co-exist with the current judicial cooperation instruments, e.g. the CCC and EIO, as they are considered relevant and can be used as appropriate by the competent authorities. This applies to bilateral agreements between the Union and non-EU countries, such as the Agreement on Mutual Legal Assistance ('MLA') between the EU and the US and the Agreement on MLA between the EU and Japan.

The regulation sees the preservation and production orders as an action issued to seek the preservation or production of data stored by a service provider located in another jurisdiction that is required as evidence in a criminal investigation or criminal proceedings. Both orders can be served on providers of electronic communication services, social networks, online marketplaces, other hosting service providers and providers of internet infrastructure such as IP address and domain name registries, or on their legal representatives where they exist. The orders refer to the specific known or unknown perpetrators of a criminal offence that has already taken place. The European preservation order allows only preserving data that is already stored at the time of receipt of the order to be asked. The procedure is depicted on Fig. 1.

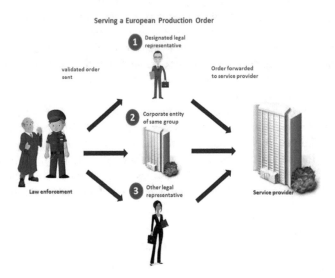

Fig. 1. Serving a European production order [2]

The proposed Regulation and proposed instruments will help in overcoming most of the identified problems in the area of cross-border e-evidence collection in cases of criminal acts as follows:

(a) Provision of better applicability of collected cross-border e-evidence at trial due to the applicability of the regulation to Member States and envisaged agreements with third countries (e.g. the US) with different jurisdictions.

Given the diversity of legal approaches, the number of policy areas concerned (security, fundamental rights including procedural rights and protection of personal data, economic issues), and the large range of stakeholders, the proposed Regulation at Union-level legislation is providing the most appropriate mechanisms to address the identified problems.

Regarding third countries, for example the US, specific policy measures are envisaged. International agreements coherent with EU-internal solutions could be set also to provide a basis for closer international cooperation with safeguards comparable to those of the EU-internal solution with regard to individuals' rights, including judicial redress. These agreements could cover judicial cooperation, direct cooperation and/or direct access.

(b) The exact definition of the entities that can issue a production order for collecting e-evidence is presented in Article 4, while the condition for issuing orders are specified in Article 5. The European Production Order is implemented by issuing a European Production Order Certificate (defined in Article 8), which is translated and sent to the service provider. The order may only be issued if a similar measure is available for the same criminal offence in a comparable domestic situation in the issuing state. The misleading approach in the Guidance Note about the method to look for the data as it claims that agreement by the party states to the respective document will not be classified as consent to extraterritorial enforcement of domestic production orders, is now clearly resolved in the Regulation and explained in the Impact assessment document. For orders to produce transactional and content data, a judge or court is required. For subscriber or access data, this can also be done by a prosecutor.

(c) The category of data that could be searched for in a production order for collecting crime related e-evidence is now clearly specified in the proposed Regulation. These definitions rely on the electronic communication services specified in the proposal for a Directive establishing the European Electronic Communications Code. They include subscriber data, access data, transactional data (the three categories commonly referred to jointly as 'non-content data') and stored content data. They are defined as follows:

- Non-content data:
    - Subscriber data, which allows the identification of a subscriber to a service. Examples: subscriber's name, address, telephone number.

- Metadata, which relates to the provision of services and includes "electronic communication metadata". Examples of these data are: data relative to the connection, traffic or location of the communication.
- Access logs, which record the time and date an individual has accessed a service, and the IP address from which the service was accessed.
- Transaction logs, which identify products or services an individual has obtained from a provider or a third party (e.g. purchase of cloud storage space).
- Content data: For example, text, voice, videos, images, and sound stored in a digital format, other than subscriber or metadata.

The type of data defined above may imply a different treatment by existing rules and different procedures to access it. Each of the above categories may contain personal data and are thus covered by the safeguards under the EU data protection acquis, but the intensity of the impact on fundamental rights varies between them, in particular between subscriber data on the one hand and metadata and content data on the other. The Regulation specifies how they should be applied. All the data specified above refers to electronically stored data that already exists. Intercepted data (i.e. data from the real-time interception of telecommunications) is out of the scope of the proposed Regulation as there are specific and significantly different rules that determine access to that data.

In addition to the specification of the category of data to be sought, the proposed Regulation addresses providers of the following services:

- Both traditional telecommunication services (example: voice telephony, SMS, internet access service) as well as new Internet-based services enabling interpersonal communications, such as voice over IP, instant messaging and web-based email services (Over-the-Top communications services, 'OTTs') are considered. These OTTs are in general not subject to the current EU electronic communications framework (i.e. Directive 2002/21/EC), including the current ePrivacy Directive, which in general applies only to traditional telecommunication services. In line with the expansion of the EU electronic communications and ePrivacy framework, OTTs will be covered by this initiative as well (e.g. Gmail, WhatsApp);
- Information society services as defined in the Directive 98/34/EC that store data at the individual request of a recipient of a service; this includes a variety of known services providers such as social networks (e.g. Facebook and Twitter), cloud services (e.g. Microsoft, Dropbox or Amazon Web Services), online marketplaces (e.g. eBay or Amazon marketplace) or other hosting service providers (e.g. Bluehost);
- Internet infrastructure services such as IP address providers and domain name registries and registrars and associated privacy and proxy services (e.g. GoDaddy) [2].

(d) The issue of human rights protection is addressed in Article 1(2) by a statement that clearly stresses that the Regulation does not have the effect of modifying the obligation to respect the fundamental rights and legal principles as enshrined in Article 6 of the Treaty on the European Union. This addresses the categories of data that contain personal data and are covered by the safeguards under the EU data protection acquis. However, the impact of the Regulation on the fundamental rights varies, in particular between subscriber data on the one hand and transactional and content data on the other hand. These data categories will stay protected according to the EU data protection acquis.

(e) The cooperation of national criminal courts is regulated with the system applied in a particular country and the service providers. Additional policy measures are proposed to cover this problem. Single points of contact are envisaged, both on the public authorities' side and on the service providers' side. This instrument is envisaged for ensuring the quality of outgoing requests and building relationships of confidence with providers, by provision of the direct cooperation between the involved authorities and service providers by helping to clarify the provider's policies. Another proposed measure is the standardisation and reduction of forms used by law enforcement.

Regarding the cooperation with the US practical measures for enhancing the judicial cooperation between public authorities in the EU and the US the following procedures based on the existing mutual legal assistance are envisaged: regular technical dialogues with the US Department of Justice to continue to improve the process, speed and success rate of MLA requests, and maintaining regular contacts between the EU Delegation to the US, the Commission and liaison magistrates of Member States in the US will discuss further the MLA process issues. In addition, the US Department of Justice's Proposed Bill and the Cross-Border Law Enforcement will be analysed and followed according to the proposer's request.

(f) The problem of education and training. The provision of opportunities to exchange best practices and for training of public authorities in the EU on the cooperation with US-based providers is envisaged, as well as additional trainings for law enforcement and judicial authorities to support the functioning of the direct cooperation between judicial authorities and service providers. Platform is being developed in a form of a static repository for service provider policies, but also it can be used as an interactive tool guiding law enforcement authorities in the identification of the relevant service provider.

5   Discussion and Conclusion

The new proposed European legal instruments for digital investigation of cybercrime will certainly contribute to the following:

- Removing of the fragmentation of legal frameworks in Member States, which was identified as a major challenge by service providers seeking to comply with requests based on different national laws;
- A better expediency of judicial cooperation based on the existing Union legislation, notably via the EIO.

Considering the above issues and the diversity of legal approaches, the number of policy areas concerned (security, fundamental rights including procedural rights and the protection of personal data, economic issues), and the large range of stakeholders, EU-level legislation such as the proposed Regulation is the most appropriate method to address the identified problems. The practical instruments, mechanism and policy measures proposed are elaborated in sufficient details and are convincing in their mission to improve the investigation of cybercrime in highly connected society. However, time will be needed for the measures and the instruments to be adopted and applied.

Acknowledgments. This work was co-funded by the European Union's Justice Program (2014–2020) through the LIVE_FOR (Criminal Justice Access to Digital Evidences in the Cloud – LIVE_FORensics) project.

References

1. Herrera-Flanigan, J.R., Ghosh, S.: Criminal regulations. In: Ghosh, S., Turrini, E. (eds.) Cybercrimes: A Multidisciplinary Analysis, pp. 265–308. Springer, Berlin (2010)
2. European Commission: Commission Staff Working Document Impact Assessment Accompanying the document Proposal for a Regulation of the European Parliament and of the Council on European Production and Preservation Orders for electronic evidence in criminal matters and Proposal for a Directive of the European Parliament and of the Council laying down harmonised rules on the appointment of legal representatives for the purpose of gathering evidence in criminal proceedings, SWD (2018) 118 final (2018)
3. Geist, M.: Cyber law 2.0. Boston College Law Rev. **44**(2), 359–396 (2003)
4. Brown, C.S.D.: Investigating and prosecuting cyber crime: forensic dependencies and barriers to justice. Int. J. Cyber Criminol. **9**(1), 55–119 (2015)
5. Dee, M.: Getting back to the fourth amendment: warrantless cell phone searches. New York Law School Law Rev. **56**, 1129–1163 (2012). http://www.nylslawreview.com/wp-content/uploads/sites/16/2012/02/Dee-note.pdf
6. European Parliament and Council of the European Union, Directive 2014/41/EU of the European Parliament and of the Council of 3 April 2014 regarding the European Investigation Order in criminal matters (2014)
7. Council of Europe: Convention on Cybercrime Act, European Treaty Series - No. 185 (2001)
8. Cybercrime Convention Committee (T-CY): T-CY Guidance Note #10 Production orders for subscriber information (Article 18 Budapest Convention), revised version as adopted by the T-CY following the 16th Plenary by written procedure, 28 February 2017. https://rm.coe.int/16806f943e
9. European Parliament and Council of the European Union, Proposal for a Regulation of the European Parliament and of the Council on European Production and Preservation Orders for electronic evidence in criminal matters (COM (2018) 225 final) (2018)

Electromagnetics, Sensors and Antennas for Security

An Overview of RFID Benefits and Limitations: Hardware Solution for Multipath Reduction

Francesca Venneri and Sandra Costanzo[(⊠)]

DIMES – University of Calabria, 87036 Rende, CS, Italy
costanzo@dimes.unical.it

Abstract. RFID benefits and limitations are outlined in this work, showing some future challenges enabling RFID technology to play a key role in the Internet of Things. Some open issues and possible solutions are discussed to improve the reliability of RFID systems. In particular, the multipath phenomena is addressed, proposing metamaterial absorbers as a possible hardware solution for multipath reduction in indoor scenario. The potentialities of a fractal based metamaterial absorber are demonstrated. A very high versatile MA-cell is designed, offering both miniaturized sizes as well as multi-band and/or broadband behavior, within the RFID UHF-band.

Keywords: RFID · IoT · Multipath · Metamaterials

1 Introduction

Nowadays, the use of RFID (Radio Frequency IDentification) systems [1–3] has become widespread in many areas ranging from industry to daily life.

RFID is a form of wireless telecommunication that uses electromagnetic waves to identify and track objects. In the last decade, RFID technology has been applied to several applications ranging from retail, logistics and supply chain management, industrial automation, electronic transaction, baggage tracking, vehicle tolling/identification, healthcare services, scientific research, security, etc.

RFID takes the barcoding concept and, thanks to the use of radio waves, allows to simultaneously identify many items, without direct line-of-sight, within a distance of a few centimeters up to several meters.

Furthermore, RFID represents a key technology of the Internet of Things (IoT). At this purpose, several research works suggest the integration of RFID technology with Wireless Sensor Networks (WSNs) for implementing smart environments [4]. Actually, RFID and WSNs complement each other: RFID systems provide identification and location of items, but usually they are not able of obtaining physical state of objects under observation; conversely, the sensor devices adopted in WSNs allow to monitor the physical state of objects under observation or to sense environments parameters, such as temperature, pressure, humidity. Hence, the integration of these two technologies will expand their overall functionality and capacity. Anyway, more recent researches show how RFID technology is now ripe to provide part of the IoT physical

© Springer Nature Switzerland AG 2019
Á. Rocha et al. (Eds.): ICITS 2019, AISC 918, pp. 871–877, 2019.
https://doi.org/10.1007/978-3-030-11890-7_81

layer [5], through the use of low-cost RFID-tags, able to monitoring some critical environmental parameters such as temperature [6], the presence/level of humidity and some other gases [7, 8], or through the use of wearable and implantable RFID-tags able to produce data about person's health state [5].

Although the enormous benefits deriving from the adoption of RFID technology, there are some open issues to be solved for improving the reliability of RFID systems. The most severe RFID issues are privacy and authentication security. Currently, many researchers work to implement low cost security and privacy protocol, based to traditional cryptography [9]. Furthermore, RFID tag performances are affected by many electromagnetic (EM) factors, including the EM-properties of objects in contact with the tag antenna [10–12], and the multipath propagation [13]. The latter factor is particularly critical for RFID systems operating in indoor locations. Although several multipath propagation models have been developed for mobile communications and for some indoor wireless systems, such as WLAN [14], only few studies have been devoted to multipath propagation in RFID systems [13]. So, practical techniques are necessary to prevent incorrect tags reading due to multipath. In this paper, the use of metamaterial (MA) absorbers is considered as an effective hardware solution to reduce multiple reflection interferences within restricted indoor locations. They consist of a periodic metallic pattern printed on a thin grounded substrate, designed to realize perfect absorption around a given frequency. Since the first demonstration [15], MAs are adopted in several applications, ranging from microwave to optical frequencies [16]. The full potentialities of a fractal based MA configuration, originally introduced by the authors in [17], are illustrated in this paper. Several designs are discussed and proposed as effective solution for multipath interferences reduction in UHF-RFID systems.

2   RFID Systems Specifications

A basic RFID system consists of one or more tags and a radio transmitter/receiver called reader (Fig. 1). The tags are small electronics devices, attached to the objects to be identified, comprising an antenna and a chip with unique identification code.

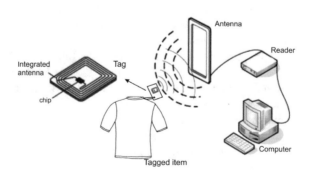

Fig. 1. RFID system.

The RFID reader is a network connected device with an antenna, that sends power as well as data and commands to the tags [18]. The signal emitted by the reader is received and backscattered by the tag. Instantly, the reader decodes the data transmitted by the tagged item and provides the connection with the software management system for data processing and storage (Fig. 1).

RFID technologies can be classified into two main categories: passive and active RFID. The former one has no power sources inside, but it relies on the reader to supply power for wireless communication, and the latter has power sources (i.e. a battery) installed inside. According to the adopted radio frequency, the passive RFID technologies are usually classified into low frequency (LF) RFID, high frequency (HF) RFID, ultra high frequency (UHF) RFID, and microwave RFID. The assigned operating frequency ranges are specified in Table 1, while Fig. 2 shows the worldwide RFID frequency allocation by country. Each frequency band offers several advantages and limitations (see Table 1). For example, an RFID system operating at lower frequencies has shorter read range (i.e. distance from which a reader can communicate with a tag) and slower data rate, but lower sensitivity to radio waves interferences caused by metallic objects or liquid substances. Conversely, a system operating at higher frequencies has fast data rates and longer read ranges, but it is more sensitive to radio waves interferences caused by surrounding objects.

Thanks to their faster data rates and longer read ranges (up to 10 m), UHF-system are gained increased interest in various commercial and industrial applications. Furthermore, passive UHF-RFID tags are cheaper and easier than microwave tags. However, UHF-RFID systems are quite sensitive to signal interferences [19]. In particular, when operating in restricted indoor locations, the multipath phenomena reduces RFID system throughput and performances.

Table 1. RFID frequency ranges and their main applications [18].

Frequency	Approximate read range (tag type)	Data rate	Applications
LF (125 kHz)	<5 cm (Passive)	Low	Animal identification Access control
HF (13.56 MHz)	10 cm–1 m (Passive)	Low to moderate	Smart cards Payment
UHF (860–960 MHz)	3 m–10 m (Passive)	Moderate to high	Logistics and supply chain Pallet and case tracking Retail and apparel Authentication
Microwave (2.45 & 5.8 GHz)	10 m–15 m (Passive) 20 m–40 m (Active)	High	Electronic toll collection Container tracking

For instance, multipath can cause the detection of RFID tags beyond the coverage area of the reader antenna (in-channel interference [20]), whilst under normal operation conditions (i.e. free space), only those items within the coverage area of the reader antenna are detected. The above issue could be mitigated by adopting directive

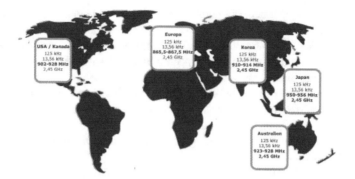

Fig. 2. Worldwide RFID frequency allocation by country.

antennas, whose radiation pattern fit the area of interest, thus minimizing multi-reflections contributions [20]. Furthermore, software-based approaches have been proposed in literature to mitigate multipath interferences [21].

As an alternative, the use of microwave absorbers can represent an effective hardware solution to reduce multipath reflection interferences. Metamaterial absorbers (MAs) are recently proposed as small volume microwave absorbing structures, useful to improve the reliability of UHF-RFID systems: an optically transparent absorber is proposed in [22]; a low-cost MA absorber is proposed in [23, 24] for the European $865 \div 868$ MHz band, a miniaturized MA unit cell with four lumped resistors is discussed in [25].

MA consists of a periodic resonant metallic structure printed on a low-loss and thin grounded dielectric slab, that is designed to achieve a perfect absorption around a given frequency. Thus, by properly positioning a MA panel into the operating environment, undesired multi-reflections from walls and/or other objects can be effectively reduced (Fig. 3).

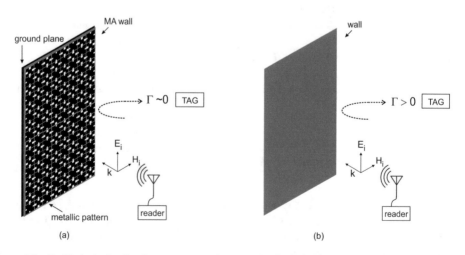

Fig. 3. Undesired reflections representation: (a) with MA panel; (b) without MA panel.

3   Fractal MA Potentialities

The Minkowski fractal geometry, already adopted by the authors for the design of reflectarray antennas [26, 27], is presented in this paper as a versatile solution for multipath mitigation within the RFID UHF-band. As a matter of the fact, the proposed configuration allows to achieve a good compromise in the design of ultra-thin, low-cost and miniaturized MA unit cells. Furthermore, by exploiting the high versatility offered by the adopted fractal shape, which consists of a fixed length L × L Minkowski patch with a scaling factor S (Fig. 4(a)), it is possible to achieve very interesting solutions. Both degrees of freedom inherent to the adopted Minkowski shape, i.e. the patch length L and the inset size SL (Fig. 4(a)), can be properly exploited to satisfy the absorption condition (i.e. $\Gamma = 0$) at the desired frequencies [17]. ·

So, following the design rule given in [17], the results listed below are achieved, considering a common FR4 substrate ($\varepsilon_r = 4.4$, tan$\delta = 0.02$, h = 3.2 mm $\cong$ $\lambda_0/100$):

(a) very small unit cells are designed in [17], where D is down to $0.15\lambda_0$ (about 40–65% smaller than those designed in [22–24]), without using any lumped resistors [25] and adopting a thinner substrate (i.e. $\lambda_0/100$ at the operating frequency of 868 MHz);

(b) a dual band behavior is achieved and discussed in [28], enclosing two pairs of miniaturized Minkowski patches inside the same small unit cell (D = 0.4λ). A proper tuning of fractal shape dimensions (i.e. L_1–S_1, L_2–S_2, Fig. 4) allows to achieve two distinct absorbing peaks.

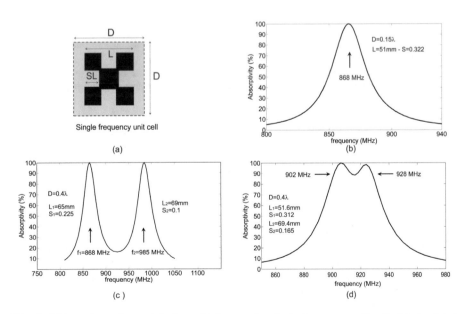

Fig. 4. MA configuration and designs: (a) layout of a single frequency unit cell; (b) simulated absorptivity of a single frequency UHF-cell; (c) simulated absorptivity of a dual-band UHF-cell; (d) simulated absorptivity of a broad-band UHF-cell.

The details of the above designs are respectively depicted in Fig. 4(b) and (c).

The versatility of the proposed fractal structure is further applied in this work to achieve a broadband behavior, by exploiting the dual resonant mode obtained in [28].

A MA-cell covering the whole UHF RFID United States-assigned frequency band (US 902–928 MHz) is designed by properly re-sizing the dual resonant cell in Fig. 4(c). In particular, the separation distances between the MA absorption peaks are properly reduced to achieve an absorptivity greater than 88% within the desired frequency range (see Fig. 4(d)). The following fractal dimensions allow to successfully satisfy the given design requirements: $L_1 = 51.6$ mm – $S_1 = 0.312$; $L_2 = 69.4$ mm – $S_2 = 0.165$.

4 Conclusion

An overview of RFID benefits and limitations has been outlined in this work, giving a brief survey on the state-of the-art of RFID applications and future challenges.

In particular, the multipath phenomena has been addressed, and a useful hardware solution, based on the adoption of metamaterial absorbers, has been discussed and proposed for multipath adverse effects reduction. The potentialities of a fractal-based metamaterial absorber have been illustrated, for operations within RFID UHF-band. A high versatility has been demonstrated, showing the ability of the proposed MA-cell to offer both miniaturized sizes as well as multi-band and/or broad-band behavior.

References

1. Finkenzeller, K.: RFID Handbook: Fundamentals and Applications in Contactless Smart Cards and Identification. Wiley, Hoboken (2003)
2. Vita, G.D., Iannaccone, G.: Design criteria for the RF section of UHF and microwave passive RFID transponders. IEEE Trans. Microwave Theor. Technol. 53(9), 2978–2990 (2005)
3. Chawla, V., Ha, D.-S.: An overview of passive RFID. IEEE Appl. Pract. 11–17 (2007)
4. Liu, H., Bolic, M., Nayak, A., Stojmenovi, I.: Integration of RFID and wireless sensor networks. In: Encyclopedia on Ad Hoc and Ubiquitous Computing, pp. 319–347. World Scientific (2009)
5. Amendola, S., Lodato, R., Manzari, S., Occhiuzzi, C., Marrocco, G.: RFID technology for IoT-based personal healthcare in smart spaces. IEEE Internet of Things J. 1(2), 144–152 (2014)
6. Bhattacharyya, R., Floerkemeier, C., Sarma, S.: RFID tag antenna based temperature sensing. In: Proceedings of IEEE International Conference on RFID, pp. 8–15 (2010)
7. Manzari, S., Occhiuzzi, C., Newell, S., Catini, A., Di Natale, C., Marrocco, G.: Humidity sensing by polymer-loaded UHF RFID antennas. IEEE Sens. J. 12(9), 2851–2858 (2012)
8. Occhiuzzi, C., Rida, A., Marrocco, G., Tentzeris, M.: RFID passive gas sensor integrating carbon nanotubes. IEEE Trans. Microw. Theor. Technol. 59(10), 2674–2684 (2011)
9. Maple, C.: Security and privacy in the Internet of Things. J. Cyber Policy 2(2), 155–184 (2017)

10. Nikitin, P.V., Rao, K.V.S., Lam, S.F., Pillai, V., Martinez, R., Heinrich, H.: Power reflection coefficient analysis for complex impedances in RFID tag design. IEEE Trans. Microw. Theor. Technol. **53**(9), 2721–2725 (2005)
11. Banerjee, S.R., Jesme, R., Sainati, R.A.: Performance analysis of short range UHF propagation as applicable to passive RFID. In: Proceedings of IEEE International Conference on RFID, Grapevine, TX, pp. 30–36 (2007)
12. Griffin, J.D., Durgin, G.D., Haldi, A., Kippelen, B.: RF tag antenna performance on various materials using radio link budgets. IEEE Antennas Wirel. Propag. Lett. **5**, 247–250 (2006)
13. Lazaro, A., Girbau, D., Salias, D.: Radio link budgets for UHF RFID on multipath environments. IEEE Trans. Antennas Propag. **57**(4), 1241–1251 (2009)
14. Saleh, A.A., Valenzuela, R.A.: A statistical model for indoor multipath propagation. IEEE J. Sel. Areas Commun. SAC **5**(2), 128–137 (1987)
15. Landy, N.I., Sajuyigbe, S., Mock, J.J., Smith, D.R., Padilla, W.J.: Perfect metamaterial absorber. Phys. Rev. Lett. **100**(20), 207402 (2008)
16. Watts, C.M., Liu, X., Padilla, W.J.: Metamaterial electromagnetic wave absorbers. Adv. Mater. **24**, 98–120 (2012)
17. Venneri, F., Costanzo, S., Di Massa, G.: Fractal-shaped metamaterial absorbers for multireflections mitigation in the UHF band. IEEE Antennas Wirel. Propag. Lett. **17**(2), 255–258 (2018)
18. https://www.impinj.com/
19. Lazaro, A., Girbau, D., Villarino, R.: Effects of interferences in UHF RFID systems. Prog. Electromagnet. Res. PIER **98**, 425–443 (2009)
20. Álvarez López, Y., Franssen, J., Álvarez Narciandi, G., Pagnozzi, J., González-Pinto Arrillaga, I., Las-Heras Andrés, F.: RFID Technology for management and tracking: e-health applications. Sensors **18**(8), 2663 (2018)
21. Wiseman, Y.: Compression Scheme for RFID Equipment. IEEE International Conference on Electro Information Technology (EIT 2016), North Dakota, USA, 387–392, (2016)
22. Okano, Y., Ogino, S., Ishikawa, K.: Development of optically transparent ultrathin microwave absorber for ultrahigh-frequency RF identification system. IEEE Trans. Microw. Theor. Technol. **60**(8), 2456–2464 (2012)
23. Costa, F., Genovesi, S., Monorchio, A., Manara, G.: Low-cost metamaterial absorbers for sub-GHz wireless systems. IEEE Antennas Wirel. Propag. Lett. **13**, 27–30 (2014)
24. Costa, F., Genovesi, S., Monorchio A., Manara, G.: Perfect metamaterial absorbers in the ultra-high frequency range. In: International Symposium on Electromagnetic Theory, Hiroshima (2013)
25. Zuo, W., Yang, Y., He, X., Zhan, D., Zhang, Q.: A miniaturized metamaterial absorber for ultrahigh-frequency RFID system. IEEE Antennas Wirel. Propag. Lett. https://doi.org/10.1109/lawp.2016.2574885
26. Costanzo, S., Venneri, F.: Miniaturized fractal reflectarray element using fixed-size patch. IEEE Antennas Wirel. Propag. Lett. **13**, 1437–1440 (2014)
27. Costanzo, S., Venneri, F., Di Massa, G., Borgia, A., Costanzo, A., Raffo A.: Fractal reflectarray antennas: state of art and new opportunities. Int. J. Antennas Propag. (2016). https://doi.org/10.1155/2016/7165143. Article ID 7165143
28. Venneri, F., Costanzo, S., Di Massa, G., Raffo, A.: Multi-band designs of fractal microwave absorbers. In: WorldCIST 2018, Advances in Intelligent Systems and Computing, vol. 746, Springer, Cham (2018)

Media, Applied Technology and Communication

Minors and Artificial Intelligence – Implications to Media Literacy

Jussi Okkonen[✉] and Sirkku Kotilainen

Tampere University, 33014 Tampere, Finland
{jussi.okkonen,sirkku.kotilainen}@uta.fi

Abstract. The artificial intelligent systems are reshaping human reality. Especially children and youth are altering their mobile communication and media usage fundamentally in their everyday lives today. During children's use the mobile artificial intelligent applications, the "robot" inside the application is learning from the feedback provided by children and youth through their uses. Based on review on current research and preliminary interviews, this paper addresses the topic from two perspectives. The first perspective is about acceptability of such technologies in general, i.e. to whom or what people are willing to grant right to control content, access and actions in digital sphere. The second perspective is about how users, and their parents too, are affected by the use of pervasive and immersive technology that shapes their media practices. The objective of the pilot study is to gain insightful knowledge of understandings of the artificial intelligence based adaptive media uses among children and youth together with practices of media and information literacies.

Keywords: Artificial intelligence · Minors · Media literacy

1 Introduction

The artificial intelligent systems are reshaping human reality. They engulf our physical and mental bodies and they affect our everyday life and wellbeing via applications and services in general, but especially the wellbeing of minors. During children's use the mobile artificial intelligent (later AI) applications, the "robot" built in the application is learning from the feedback provided by children and youth through their uses. These personalized applications are based on the machine learning as the third generation AI (e.g. [1]). These applications learn from children's behaviour in their regular use. How about the user's trust and privacy? What kind of media and information literacies are needed?

User's trust in AI based applications and services is an ethical and moral question in the current information societies with changing digital environments. Therefore it is important to understand the construction of trust in children's practices with these digital media. For example, the European Parliament has pointed out these ethical questions related to the collection and using of personal data in global companies like Google and Microsoft. There always are new media with advertisers, content producers with their changing strategies to attract the consumers [2], as especially children and young more recently in new models of mobile devices having e.g. AI based

© Springer Nature Switzerland AG 2019
Á. Rocha et al. (Eds.): ICITS 2019, AISC 918, pp. 881–890, 2019.
https://doi.org/10.1007/978-3-030-11890-7_82

applications like Bixby by Samsung corporation. This paper is discussing children's media use focusing on trust, privacy and personal information management together with media and information literacy from the perspective of youth agency (e.g. [3]) based on a pilot study with interviews among youth and their parents in Finland and South-Africa. Motivation for the pilot study is in developing a comparative research agenda.

There is some common conceptual core of trust to be enlightened. The definition of trust has studied in following domains: management, marketing, psychology, and sociology. Nevertheless, most authors working on trust provide their own definition, which is tailored for specific domain [4]. In this paper, we study children's trust in artificial intelligent applications. Artificial intelligent research settings the definition of trust needs to be re-formulated. In addition, there is little of empirical research in trust process of AI and children available.

The earlier research in artificial intelligent systems are written mostly in the technical point of view [5]. Research tradition among computer scientists is strong, only recently it has taken people also into focus. For example, in CHI conference, AI and user experience has started to bloom last few years (cf. www.chi.org). In educational sciences topic is also researched, still it mainly focus on how AI is taken as teaching assistant. For example, Serholt et al. [6] report the results of children's uncritical attitudes towards the teaching robot. Most probably, the minors' uncritical attitudes can been noticed in their everyday uses and practices of media as well.

The objective of the pilot study is to gain insightful knowledge of culture-sensitive understandings of the artificial intelligence based adaptive media uses among children and youth together with practices of media and information literacies. The objectives turn into guiding research questions as such: (1) How young people are using adaptive media including AI from the perspectives of trust and privacy? (2) How young people are practising media and information literacies in their uses of adaptive media including AI? (3) How do the parents understand this new phenomenon in youngsters media use? acceptable.

2   Minors, Media Literacy and Trust

Throughout the history of the research on children, youth and media the moral panics framework appear when new technologies and media is introduced to the markets. Drotner [7] argues that each moral panic tends to follow an expectable route starting with pessimistic elitism associated with calls for technocratic and legalistic measures like censorship moving to optimistic pluralism calling for media literacies. Recent technological developments in media spheres have given space to integrative frameworks, which this study applies as well, respecting and increasing young people's agency from the perspective of the UN Convention of Rights of the Child to access, create, explore, express and learn media and information literacies [8]. The proposed study participates in current discussions of IT skills like coding as literacy [9] for recognizing algorithms and AI, as critical media literacies of today [10].

In this study we take a broad view of media literacy where 'media' refers to all the information technologies which provide public spaces and opportunities for interacting with people as audiences, and for participating in and through the media, embedded within social relationships. The term media includes the aspects of production, expression and reception as a process, which are also regarded as core elements of media literacies [11, 12]. Media literacies has been defined in multiple ways and, more recently as MIL, i.e. media and information literacies, mostly promoted by UNESCO [13]. Media literacies as practices that help participation and engagement through and with the media include attitudes, critical judgments, and ethical reflections on the media among other skills [14]. Our hypothesis is, that these previous formulations of media literacies are not enough when we are dealing with artificial intelligence-based applications. Especially the new working logic of AI applications and services challenge the intellect of parents and minors.

Parental mediations strategies have been studied and defined, for example, in EUKIDS Online research (see, eukidsonline.net). Most probably, the previous suggestions of parents getting interested in children's cultures online or setting up family-based rules for using digital media still are valid, but need some updating.

This pilot study is based on the social constructionist perspective [15] according to which the society shapes technology and its use and, the empowerment and wellbeing of children and youth depends on various factors such as access to technological resources, motivations for internet use, policies and education as forming cultural and social capital for children and young.

In cognitive and social science, there is not yet a shared or prevailing, and clear and convincing notation of trust. Mutti [16, 17] states that the number of meanings attributed to the idea of trust in social analysis is disconcerting. Certainly, this deplorable state of thing is the product of a general theoretical negligence [18]. For instance, social capital theory formed by James Samuel Coleman and Robert Putnam has two characteristic features in relation to trust: norms of reciprocity and trustworthiness. [19, 20]

The trust in guided decision-making is more and more depending on AI, whereas trust is built among children in different cultural backgrounds, variety of AI artefacts, media and commercial enterprises. Together they constitute network based platforms i.e. distributed global trust. Distributed trust is a paradigm shift driven by innovation technologies that rewriting the rules of an all-too-human relationship. We are already putting our faith in algorithms over humans in our daily lives, whether it is Amazon recommendations what to read or Netflix suggestions what to watch. In this context, more objective research and critical view in relation between AI and trust is needed.

The most relevant approaches to trust in the technological domain are: the logical approach, the computational approach and sosio-cognitive approach. In this paper, trust is implied as combination of Multi-Agent Systems (MAS) and socio-cognitive approach. MAS prevails in the distribution of the various attitudes of agents who are possibly considering the emergence of the global phenomenon and sosio-cognitive approach, where trust imply the subjective probability of the successful performance given behaviour, and it is on the basis of the subjective perception/evaluation that the agent decides to rely or not, to bet or not on the trustee i.e. artificial intelligent artefact [18].

This paper also take into account that trust is a relational construct. That involves a subject X as cognitive agent, an addressee Y (trustee) as an agent, which is capable of causing some effect as the outcome or its behaviour and causal process itself and its results, that is an act alfa of Y possibly production of desired outcome 0 [21]. The global communication firm Edelman has been conducting an annual "Trust Barometer" asking more than 30 000 people across twenty-eight countries on the level of trust. The media suffered the biggest blow, now distrusted in 82 per cent of all 28 countries surveyed. In UK, the number of people saying they trusted the media fell from 36 per cent in 2016 to 24 per cent in 2017 [22].

Social capital theory proposes that networks of relationships are a valuable resource for both individuals and organizations. According to Arenius [23] those networks are for interaction and for gaining access to assets required. Nahapiet and Ghoshal [24] have observed that the value of the social capital is in its ability to make possible the achievements, which would be impossible without it or would be very costly, thus the value of social capital relies on its usability. Social capital is invisible, ubiquitous, and hard to pindown, so it is usually studied in terms of its manifestations and effect [24]. Social capital can operate at the level of an individual, a team, an organization, an industry, a community, a nation, or an entire economy (e.g. Putnam 1993). According to Arenius [23] the social capital theory suggests that players are able to gain access to various kinds of resources that accrue to them by virtue of their engagement in various kinds of relationships. Resources are available through the contacts or the connections that networks bring. As well as conveying resources, social capital is a means of enforcing norms of behavior among individuals or corporate actors, and thus acts as a constraint.

In inter-personal relationships, or in this case hybrid relationships, social capital facilitates learning because offers access to knowledge and enables knowledge transfer [24]. More social capital, i.e. interaction between different actors, trust and mutual understanding, more efficient is the process of transferring explicit, and especially tacit knowledge. Moreover, it promotes acceptance of information provided.

In the context of using applications and services social capital should be taken in account as sum of more or less institutionalised relationships of mutual acquaintance and recognition or network of social exchanges between operatives or actors engaging in transactions (cf. e.g. [25–27]). Moreover the three dimensions of social capital affecting any actor in that process are structural, i.e. presence or absence of interaction, dimensional, i.e. mutual trust and trustworthiness, and cognitive, i.e. shared understanding of common goals and proper ways to act [28–30]. Structural dimension of social capital is often emphasized when considered relationships of mutual acquaintance that forms a basis for activities. From the perspective of user the gains are twofold as recognition and knowing different participants operative actions that structural social capital offers access. Without relationships of acquaintance the involvement cannot be fulfilled. One might even claim that structural social capital is the source of the first move in any activity, since at least on party of interaction should acknowledge counterparners.

Dimensional dimension of social capital is most important in early trust building. Participant must be sure that any information she or he submits to will be valued. From the perspective of single user dimensional social capital is the issue of trustworthiness.

It is the issue of commitment to the decided goal or avenue of action. On the other hand user is interested in commitment of the others, as the network based on structural social capital cannot hold if there is commitment to other goals than the (explicitly) agreed.

Cognitive dimension of social capital consist of the social exchanges between actors engaging in the process. It is the issue of understanding on "where we are now, and where we are going, what is desirable". Cognitive dimension is shared understanding on what is the primary goal. It is recognition of meaning or rationale for their interaction. From the perspective of user cognitive dimension is the recognition of the goal for activity and knowledge of legitimate ways to achieve the goal. It is more or less issue of what can and should be done. Moreover, it is also knowledge on who to know, i.e. how the network is build. The cognitive dimension can be considered supportive to other two dimensions, as it contains the shared norms. Having knowledge on how one can and should behave in certain situation is important, because without that knowledge networks can be ruined and trust lost.

As trust is built in interaction the social capital theory is most relevant in sense, that the presented dimensions seem to exist in using adaptive platforms or services. The interaction with such technology is of course more straightforward than in more complex social situations, yet analogies are found. Structural dimension is about acknowledging technology as a counterpart, yet doing so user also recognizes mutual relationship to build trust on. Dimensional social capital in this context is issue of trust, i.e. gaining of technology by added value or pure enjoyment. Cognitive dimension is one sided, as by the trust build people tend to act according to technological determinism. Moreover, such assumption on technology side implicit, since it react on previous queries or actions. The same dimension could be found in interaction with technology as in social interaction. This is an important notion when researching how people use and adapt to digital environments optimized by algorithms.

3 Issues of Artificial Intelligence

In order to better understand the phenomenon a series preliminary interviews were conducted with 16 pair of a minor and a parent with 10–17 years aged children in Finland (9 pairs) and South Africa (7 pairs). The pairs were recruited in casual situations in airport based on observation of using of mobile device. The selection of interviewees was based on convenience sampling and voluntary cooperation. In addition, 5 pairs refused to be interviewed. The unstructured interviews shed light on the issue in general and served as initial data on the topic. People were asked if they could share their opinion on using their device. The interviews had two main questions. One aimed for the minor: *"What do you think about artificial intelligence in your device and the services you use?"* The second was aimed for the parent: *"What do you think about the artificial intelligence in the device and the services your child is using?"*. The interviews were conducted in free form discussions about 10 to 15 min. The aim for such brief interviews was to map attitudes on the topic. Moreover, aim was to build understanding on how people see the role of technology in the devices and services. Even the sampling is biased and the results are just scratching the surface on the topic several relevant notions could be found.

The interviews addressed the topic from two perspectives. The first perspective is about acceptability of such technologies in general, i.e. to whom or what minors and parents are willing to grant right to control content, access and actions in digital sphere. The second perspective is about how users are affected by the use of pervasive and immersive technology that shapes their media practices. This is an issue of media literacy, as initial interviews revealed the minors or parent showed to have no understanding or they were indifferent on how AI shapes the digital landscape.

The minors showed no awareness on technology issues or even situations when they are exposed to AI. Not a single interviewed minor had idea of the technology optimizing content for services such as Facebook, Instagram, Snapchat, Youtube. For them it was ok to grant access to their actions in digital environment and they found service providers trustworthy in sense they granted access to their credentials, locations, contents, contacts for example. Moreover, recommendations by applications and services were welcomed and taken as good service rather than side effect of surrendering their privacy.

"I don't mind." [MF16].

"AI helps you." [MSA1].

"I mostly use my phone for discussion and sharing photos and videos. I like when I get notifications what my friends have posted." [MF2].

"It is nice to know what my circles are doing." [MSA2].

AI, i.e. algorithms, increases attractiveness and immersiveness of applications or services according to minors due to better user experience. Most of them told that they appreciate content that suits their taste and habits. Advertisements are easy to skip, so they do not mind to be exposed to those. 7 out of 16 told that they are sometimes annoyed due to social media content that is offered to them. In most services, content offered to them was found interesting and therefore it increased attractiveness of such service. Most of them also were active to browse recommendations as a source of new vistas.

"You can spend long time in Youtube watching really interesting content e.g. on your hobbies or fashion." [MSA3].

"The best thing is when you google something and you have lot of content." [MF15].

Parents did not pay much attention AI features of applications or services the minors use. They were more aware of such elements in services and applications, yet they trusted service and content provider to honor age limits and privacy. Most of the parents openly told that they are not interested or have no understanding on issue. Parents were more concerned actions taken by other online users, and to some extent commercial motives of corporations while exposing minors to well targeted marketing efforts.

"I'm very skeptical on service algorithms, yet I trust google that Iphone knows my child's age and content is suitable." [PF4].

"The phone is somewhat like an extension of life for my child. He seems to be happy with it. I'm concerned on the privacy, not on the commercial content." [PF9].

"Actually I don't care how digital environments work since they are relatively safe for kids." [PSA9].

"I'm not an engineer; easy use devices makes life much easier." [PF11].

Acceptability of AI, i.e. trust of AI, could be summarized on these findings. Children can make distinction between paid and other content, yet they cannot evaluate the difference on validity and relevance of it to them. Manipulation of the news feeds, search results, etc. was found irritating, yet most of them thought it is impossible to escape it. AI is accepted, even appreciated, as it provides content and information related to previous interest.

"Most [search] results seem useful." [MF4].

"I don't mind if there is advertisements." [MSA7].

"Sometimes, or quite often actually, you are recommended to watch videos of totally random people. If you are in hurry it is very annoying." [MF12].

AI is accepted or tolerated since parents see no way to bypass it. They commented that ever since the google and cookies all actions in digital environment are subject to intelligence and manipulation. Avoiding such technology was found too troublesome. Parents thought their children are not "google"-litterate and it concerns them to some extent, since minors are not critical to content they see.

"It [algorithms] is everywhere." [PF10].

"I think people are very happy if there is quick way to find content on their interests." [PSA4].

"School is [nowadays] about using internet to acquire information for assignments. It is convenient for them [children]." [PF2].

Media practices were issue by the parents since AI changes media practices. Several parents brought about their concern because the content is filtered and optimized according to user. Parents also discussed that AI blocks serendipity, because of it is tempting to grasp the "first" option. They thought that critical attitude towards the content requires effort and minors are a bit lazy. One parent stated that it does not require or leverage critical reading skills, as the most "'relevant' is on the top". Parents also claimed that children cannot make clear distinction between optimized content, e.g. advertisements, and other content. In general AI challenges media literacy, since AI awareness of minors is low, even non existing.

"I'm concerned on how children take the first option and follow the path on Youtube or Google. At least mine [child] does not have a clue on critical media skills." [PF2].

4 Discussion and Conclusions

The first key finding was that minors are indifferent on AI, but it concerns parents. The awareness of AI is low and the effects are ambiguous. Moreover, both minors and parents have no comprehension of AI. It is understood through popular culture artefacts such as science fiction novels for example. The second key finding was that minors trust AI as it acknowledged as a smart helping agent. Parents are somewhat concerned on that trust. The third key finding is that parents thought the minors are not AI literate and attention to it should be paid more in the future as algorithms shape more digital sphere.

The AI and minors is topic that not even existed earlier, therefore there is a need for reformulation of media literacies in the digital societies. Based on this pilot study, it seems that further discussion and integration of media literacies and information literacies, even coding as literacies are needed cf. [9, 31]. Media and information literacies belongs to the core civic skills of current digital societies which rely more and more on the discursive power making through communication. As such, for example, the spreading of disinformation aims to guide thinking and actions of the user with hiding some facts, not telling the whole truth like in so called fake news or in hate speech which is about hateful, discriminative expressions of minorities like ethnic groups in societies. Users understanding of the methods of algorithms and artificial intelligence belong to critical media and information literacies (MIL) of today.

How do children build the trust to AI artefacts? During the play AI artefacts behavioural model is becoming gradually familiar and children are capable to recognise the familiar behavioural models. The problematic is when children start to trust familiar behavioural models and advices of AI artefacts, which might not be trustworthy. It is evident, that children can form a trust to AI artefacts very fast. However, how long it takes that they learn to mistrust them, if the artificial intelligent artefacts betray them in any circumstance, it is expected that distrust appear.

This initial study on the topic opened a new research avenue on human technology interaction and media literacy. The findings support the proposition that new technology enables attractive and easy to use applications and services, yet they change how people use information and media. Based on insights of the interviewed people a straightforward conclusion is that they are happy with user-friendly technology. This somewhat rough conclusions on limited data leads to new research challenges. At the end, children needs to understand trust and AI in cognitive level. They need to understand that AI dynamic algorithms are based on their own actions. And the most important, they need to understand that control and power of media and interaction is in their own hands. Need for media literacy 3.0 is essential. The unique real answer for coping with others' autonomy is to establish a real trust relationship. Finally, the ability to understand and model the trust concept to transfer its utility in the technological cooperation framework, will be in fact the bottleneck of the development of the artificial intelligent that is technology of the future.

One of the key questions then is how can the educational system and educational policy answer to the need of conceptual and pedagogical development? This study is calling for further research together with reformulations of national, pan-european and global educational policies. Moreover, there is a need for a research agenda for contemplating the multi-faceted phenomenon of artificial intelligence and minors.

References

1. Schmidhuber, J.: Deep learning in neural networks: an overview. Neural Netw. **61**, 85–117 (2015)
2. van Dijk, J.: Users like you? Theorizing agency in user-generated content. Media Culture Soc. **31**(1), 41–58 (2009). https://doi.org/10.1177/0163443708098245
3. Frau-Meigs, D., Flores, J., Velez, I.: Public Policies in Media and Information Literacy in Europe: Cross-Country Comparisons. Routledge, London (2017)
4. Castelfranchi, C., Falcone, R.: Trust Theory. A Socio-Cognitive and Computational Model. Wiley Series in Agent Technology (2010)
5. Aly, A., Griffits, S., Stramandinoli, F.: Towards intelligent social robots: current advances. Cognitive robotics (2017)
6. Serholt, S., et al.: The case of classroom robots: teacher's deliberations on the ethical tensions. Artif. Intell. Soc. **4**(32), 613–631 (2017)
7. Drotner, K.: Modernity and media panics. In: Skovmand, M., Schroeder, K.C. (eds.) Media Cultures: Reappraising Transnational Media, pp. 42–62. Routledge, London (1992)
8. Livingstone, S.: Children's digital rights: a priority. Intermedia **42**(4/5), 20–24 (2014)
9. Dufva, T., Dufva, M.: Metaphors of code - structuring and broadening the discussion on teaching children to code. Thinking Skills Creativity **22**, 97–110 (2016). https://doi.org/10.1016/j.tsc.2016.09.004
10. Kotilainen, S., Ruokamo, H.: Opettajankoulutus Haloo: TVT ja medialukutaitojen opetus yhteen! [Hello Teacher Education: ICT and Media Literacies belong together!]. In: Salomaa, S., Palsa, L., Malinen, V. (eds.) Opettajaopiskelijat ja mediakasvatus. Kansallisen Audiovisuaalisen Instituutin julkaisuja, January 2017
11. Buckingham, D.: Media Education: Literacy, Learning and Contemporary Culture. Polity (2003)
12. Jenkins, H., Purushotma, R.: Confronting the Challenges of Participatory Culture: Media Education for the 21st Century. The MIT Press, Cambridge (2009)
13. Frau-Meigs, D., Flores, J., Velez, I.: Public Policies in Media and Information Literacy in Europe: Cross-Country Comparisons. Routledge, London (2017)
14. Livingstone, S.: Engaging with media – a matter of literacy? Commun. Cult. Critique **1**(1), 51–62 (2008). https://doi.org/10.1111/j.1753-9137.2007.00006.x
15. Berger, P.L., Luckmann, T.: The social construction of reality: a treatise in the sociology of knowledge (1966)
16. Mutti, A.: La fiducia. Un concetto fragile, una solida realta. In: Rassegna italiana de sociologia, pp. 223–247 (1987)
17. Putman, R.: Bowling alone: the collapse and revival of American community. J. Democracy **6**(1), 65–67 (1995)
18. Castelfranchi, C., Falcone, R.: Trust Theory. A Socio-Cognitive and Computional Model. Wiley Series in Agent Technology (2010)
19. Putnam, R.D.: Tuning in, tuning out: the strange disappearance of social capital in America. P.S.: Polit. Sci. Polit. [Internet]. **28**(4), 1–20 (1995)
20. Putman, R.: Bowling alone: America's declining social capital. In: Culture and Politics, pp. 223–234. Springer, Heidelberg (1995)

21. Castelfranchi, C.: Towards an agent ontology: autonomy, delegation, adaptivity. AI*IA Notizie **11**(3), 45–50 (1998). Special Issue on "Autonomous Intelligent Agents", Italian Association for Artificial Intelligence, Roma
22. Botsman, R.: Who can you trust? How technology Brought Us Together and Why It Might Drive Us Apart. Hachette Book Croup (2017)
23. Arenius, P.: Creation of Firm Level Social Capital, Its Exploitation, and Process of Early Internationalization. HUT Dissertations 2002/3, Espoo (2002)
24. Naphiet, J., Ghosal, S.: Social capital, intellectual capital and the organisational advantage. Acad. Manag. Rev. **22**(2), 242–266 (1998)
25. Seppä, M., Näsi, J.: Playing with the goose – pushing entrepreneurs across the capital gap – who, why, and how? In: The IntEnt 2001 Conference Proceedings (2001)
26. Sorheim, R.: The pre-investment behaviour of business angels – a social capital approach. In: Euram 2003, Milan, Italy, 3–5 April 2003 (2003)
27. Davidsson, P., Honig, B.: The role of social and human capital among nascent entrepreneurs. J. Bus. Ventur. **5210**, 301–331 (2002)
28. Yli-Renko, H.: Dependence, Social Capital, and Learning in Key Customer Relationships: Effects on the Performance of Technology-Based New Firms. Acta Polytechnica Scandinavica, Espoo (1999)
29. Bourdieu, P., Waquant, L.J.D.: Invitation to Reflexive Sociology. The University of Chicago Press, Chicago (1992)
30. Ilmonen, K. (ed.): Sosiaalinen pääoma ja luottamus. SoPhi, Jyväskylä (2000)
31. Pathak-Shelat, M., Kotilainen, S.: Media and information literacies and the well-being of young people: comparative perspectives. In: Kotilainen, S., Kupiainen, R. (eds.) Reflections on Media Education Futures: Contributions to the Conference Media Education Futures in Tampere, Finland 2014, pp. 147–158. Nordicom, Gothenburg (2015)

Media Competence Inequality in Regular and Flexible/Distance Education Systems

Catalina Gonzalez-Cabrera$^{(\boxtimes)}$, Cecilia Ugalde, and Lorena Piedra

University of Azuay, Av. 24 de Mayo. 7-77 and Hernan Malo,
0101981 Cuenca, Ecuador
{cgonzalez, cugalde}@uazuay.edu.ec

Abstract. One of the goals of the Ecuadorian government is to reduce inequality in its society. With this goal in mind, it has established several public policies aimed at reducing the «digital divide» and at improving media literacy among the population. The aim of this study was to compare the level of media competence of students who attended flexible and distance programs with those who studied in the standard education system. We employed a questionnaire based on six dimensions to measure students' media competence. Results obtained from 1,003 students show that competence levels were higher in five of the six dimensions in the standard education system. Male students showed higher media competence. Those over 18 demonstrated lower media competence. The study suggests the need of more research and programs along this line, with the aim of reducing the digital divide.

Keywords: Media literacy · Media competence · ICT · Distance education ·
Digital divide

1 Introduction

In today's educational context a discussion about the conceptualization of media and digital competence is nothing new. Potter [1] reviewed the literature on media education and recommends that to determine the most useful definitions and the best kind of interventions for increasing individuals' levels of media literacy, progress must be made in its conceptualization, research and instruction, as a way to build a definition that considers skills and knowledge for all cultures.

UNESCO [2] defines media and information literacy as a teaching-learning process to apply critical thinking to the reception and creation of media products. Aguaded [3] adds that we need to understand mass media in an integral way, with a critical assessment of their content, at the time that we generate communication in multiple contexts. Thus, Area and Pessoa [4], in their study of the new literacies for citizenship in the Web 2.0 culture, point out that the proposals for devising a new theory in this respect considers that the attainment of literacy with digital technology is a more complex process than mere instruction in how to handle hardware and software.

In the other hand, Dornaleteche, Buitrago and Moreno [5] understand "competence" as the abilities that people must develop to reach a result of "literacy". They

© Springer Nature Switzerland AG 2019
Á. Rocha et al. (Eds.): ICITS 2019, AISC 918, pp. 891–900, 2019.
https://doi.org/10.1007/978-3-030-11890-7_83

view "education" as a process and start from there to add the "digital" label to each of these three terms if they refer to everything affecting the digital context, and "media" to refer to the education-communication field in the broadest sense.

1.1   The Six Dimensions of Ferres and Piscitelli

Ferres and Piscitelli [6] proposed six dimensions to define and determine media competence in a study with input from experts from different countries.

(i) The ideology and values dimension explores the ability to take advantage of the new tools for responsible citizen participation, which in turn determines the skill and ethical attitude needed to search for information, assess the reliability of the sources, detect underlying intentions or interests in corporate and popular productions, their ideology and explicit values; (ii) the language dimension addresses the ability to interpret, appraise, understand and analyze different messages, and in turn establish the relations between them; (iii) the technology dimension addresses people's understanding of technological innovations, the role they play, as well as the skill to interact with them and to cope in multimodal and multimedia contexts. It also refers to the ability to manipulate images, sound, and represent reality; (iv) the reception and interaction dimension serves to determine how students receive and interact with messages, whether they exercise their rights and duties in their use and consumption of media; (v) the aesthetics dimension determines students' sensibility concerning media and media products. In addition, whether they can recognize if a media production fits certain minimal requirements of quality; (vi) and finally, the production and dissemination dimension allows us to determine the knowledge students have of how to produce audiovisual messages. It indicates students' ability to select messages and transform them into new producers of new meaning.

1.2   Media Education

Gutiérrez and Tyner [7] warn that if we reduce media education to just developing digital competence in its technological dimension, then attitudes and values may become lost in the process. If media form part of every citizen's daily life, then training in critical thinking about this new and diverse context must prevail, since in addition to new and improved possibilities it also entails dangers and threats. The European Commission [8] recommended that State members study new ways to make citizens aware of the risks related to the dissemination of personal information, the function of copyright, the use of search engines, the protection of personal data and the right to privacy. As a result, media literacy should be included in the compulsory curriculum.

Digitally literate minors benefit from the new technological environments and know how to deal with online risk [9]. Several studies show that citizens' level of media and digital competence is far from ideal [5, 10, 11].

About the new challenges of educating the "most exposed and least educated" audiences, Sandoval and Aguaded [12] underscore the responsibility of training young people to be producers of responsible contents and be knowledgeable about their consequences. Media literacy cannot remain in the conceptual field; it has to be part of the classroom experience. García, Seglem and Share [13] posit that teachers should be moving towards a more critical and productive pedagogy of media literacy. In this way, students can become citizens who can express their own voice within a participative culture.

Members of this participative culture are "prosumer" citizens, with a series of competences that allow them to carry out actions "both as consumers of media and audiovisual resources, and as producers and creators of messages and contents that are critical, responsible and creative" [14].

1.3 Public Policies and the Digital Divide

Pippa Norris [15] described the term "digital divide" as the gap existing between the technologically rich and poor, rooted in the information poverty of rural areas and poor neighborhoods. This author emphasized that access to digital networks could strengthen democratization processes and improve the endemic problems of poverty.

We can consider digital divide a new social inequality that has emerged in the 21st century owing to the inequality between different social groups when it comes to access, knowledge and competences regarding ICT, as well as to differences in the institutional conditions that capacitate individuals to develop and participate in an Information and Knowledge Society [16].

The Heads of State and Presidents of Latin America and the Caribbean, at the Fourth Summit of the Community meeting of Latin American and Caribbean States [17], considered factors to reduce this technological divide and the differences in access to ICT among its members.

For years, Ecuador occupied one of the lowest positions in reference to ICT access and use, owing to policies focused solely on connectivity or purchase of equipment [18]. Towards the middle of the last decade, the country, favored by high oil prices, external financing and a fiscal stimulus, invested more in education [19]. Ecuador's Constitution, considers education "a guarantor of equality and social inclusion and an essential condition for living a good life" [20].

Within the curriculum for Basic General Education (BGE) and the Unified General Secondary Studies (GSS) there are subjects that contributes to informational, digital and media literacy: critical readings of messages, education for citizenship, computer science applied to education, research in science and technology, language, literature and aesthetics [21]. Now, the country is in the midst of a process that is expected to lead to the consolidation of a "media society".

For these reasons, it is important to identify young people's media competence in order to develop programs and projects that through media will improve their social participation in a responsible and critical way. In addition, they need to develop skills to use the tools entailed in a digital society. This research study is intended to analyze and compare the level of media competence of students in basic and secondary education (BGE and GSS), which in this study are referred to as "Institutes of Regular Education" or (IRE). Students at the same levels but in the system for Distance and Flexible Education are referred to as "Institutes of Distance and Flexible Education" or (IDFE).

Thus, we formulate our hypothesis:

H1: The level of media competence in IRE students is higher in (a) the language dimension, (b) the technology dimension, (c) the reception and interaction dimension, (d) the production and dissemination dimension, (e) the ideology and values dimension, and (f) the aesthetic dimension than the level of media competence of IED-F students.

In addition, we propose the following research questions: Is age an influential factor for improving competence? Is gender a predictor variable of level of competence?

Analysis of this dimension allows us to determine the knowledge students have of how to produce audiovisual messages. It indicates students' ability to select messages, empower themselves, and transform them into new productions with new meaning.

2   Materials and Methods

The correlational study was carried out with a quantitative focus, using data collection to "prove hypotheses based on numerical measurements and statistical analysis" [22]. We collected information through an online questionnaire in the schools with the appropriate technology, using pencil and paper otherwise. A research group in Spain designed the questionnaire (for more information see: goo.gl/oQ5MVX), which was adapted to the Ecuadorian context. The measuring instrument is reliable with a Cronbach's alpha of 0.77 [23].

The questionnaire was based on the six dimensions proposed by Ferres and Piscitelli [6]. The items on the questionnaire intended to determine the students' level of skills, knowledge and use of digital tools, which together comprise the dimensions of media competence.

2.1   Sample

The IRE students were aged between 14 and 18 years, while the IDFE students were aged between 18 and 41 years, many of them were dropped out students returned to school.

We used a probabilistic and random sample. Participants comprised 707 students from IRE public schools, publicly financed fiscomisional schools, and private schools, representing 70.5% of the total sample, and 296 IDFE students, representing the other 29.5%. Women surpassed men in number: 580 (57.8%) vs. 423 (42.2%). The mean age of the participants was 17 (SD = 3.94).

For the analysis, we took into account the differences between the two educational systems. IRE students usually attend classes from Monday through Friday. The school year lasts nine months and students conclude their secondary education in six years. In contrast, IDFE students take an eleven-month curriculum addressed to students over 15 who have dropped out of school for at least two years. This method allows them to conclude their basic education through an accelerated process.

Design and Procedure

Since in addition to being correlational the study has a certain exploratory nature, we wanted to check the predictive power of the variables gender, age, and type of educational system in the variance of each level of the six dimensions, a multiple linear regression method was employed. We run multicollinearity tests to verify whether the assumptions of the model were fulfilled. Results showed that the tolerance values were close to 1 and the variance inflation factors (VIF) were lower than 5, indicating the absence of multicollinearity between the predictor variables and in all the dimensions.

Subsequently, to gain a deeper understanding of the relation between the variables that turned out to be predictive and those that had not, we applied Student's t test to the variables gender and type of educational system, and Pearson's test to the age variable. Then we compared the correlation between the IRE and the IDFE students. We used the SPSS statistical package to process the data and analyze the results.

3   Analysis and Results

The condition for participating in the sample for IER students was to be aged between 14 and 18, and in 10th year of BGE or 1st 2nd or 3rd year of GSS in the schools selected in Cuenca - Ecuador.

In addition, the school's legal representative had to consent to the student's participation in the research. In the IDFE schools selected, the requirement was being a student in 1st 2nd or 3rd year of secondary education. The age of the participants ranged between 18 and 41 years. In many cases they were students who had dropped out and then returned to school.

We used a probabilistic and random sample. Participants comprised 707 students from IRE schools (public, private and co-funded schools), representing 70.5% of the total sample, and 296 IDFE students, representing the other 29.5%. Women surpassed men in number: 580 (57.8%) vs. 423 (42.2%). The mean age of the participants was 17 (SD = 3.94).

Regarding the ideology and values dimension, the model was significant considering the predictor variable not excluded: Type of educational system: $[F(1,1001) = 8.159, p < .004, R^2 = .007]$. We excluded gender and age from the model since they were not significant predictors: age $(\beta = 0.018, p > 0.05)$, gender $(\beta = -0.016, p > 0.05)$. The overall fit of the model considering only the type of educational system as the predictor variable was $R^2 = 0.008$. This means that the type of educational system explained 0.8% of the variance of the variable level of competence in the Ideology and Values dimension. The effect size is small.

The model was also significant $[F(3,999) = 50.811, p < .001, R^2 = 0.130]$ for the language dimension. The variables gender $(\beta = -0.159, p < .001)$ and type of educational system $(\beta = -0.271, p < .001)$ were found to be significant predictors. Age did not turn out to be a significant predictor $(\beta = -0.059, p > .05)$. The predictors included in the model explain 13% of the variance of the variable level of competence in the Language dimension.

In the technology dimension the model was significant $[F(3,999) = 41.868, p < .001, R^2 = 0.109]$. The gender $(\beta = -0.068, p < .05)$, Age $(\beta = -0.103, p < .05)$ and type of educational system $(\beta = -0.241, p < .001)$ variables were significant predictors. The model explains 10.9% of the variance of the variable level of competence in the Technology dimension.

In regard to the reception and interaction dimension, the model was significant $[F(3,999) = 8.022, p < .001, R^2 = 0.021]$. Gender ($\beta = -0.081, p < .05$) and type of educational system ($\beta = -0.118, p < .01$) were significant predictors, whereas age was not ($\beta = -0.010, p > .05$). The model explains 2.1% of the variance of the variable level of competence in the reception and interaction dimension.

Just as with the other dimensions, the model was significant $[F(3,999) = 2,950, p < .05, R^2 = 0.006]$ for the aesthetics dimension. Gender ($\beta = 0.064, p < .05$) and age ($\beta = 0.90, p < .05$) turned out to be a predictor variable but in this case the type of educational system was not ($\beta = -0.069, p > .05$). The model explains only 0.6% of the variance of the variable level of competence in the aesthetic dimension.

The model was also significant in the production and dissemination dimension $[F(3,999) = 49.137, p < .001, R^2 = 126]$. Neither gender ($\beta = 0.037, p > .05$), nor age ($\beta = 0.073, p > .05$) were predictors, whereas the type of education system was indeed a predictor ($\beta = -0.406, p > .001$). The model explains 12.6% of the variance of the variable level of competence in the production and dissemination dimension.

We analyzed the influence of each predictor variable on the level of competence in each dimension, which provided us with the advantage of more precise estimations.

To learn whether there was a significant difference in the level of media competence in terms of gender and type of educational system, we used Student's t test for independent samples.

The test showed that despite not assuming equal variances ($F = 7.571, p < .05$), there were statistically significant differences $[t(595,623) = 2.951, p < .01, r = 0.10]$ between the mean of the level obtained by the IER students ($M = 2.08, SD = 0.409$) and the mean obtained by the IDFE students ($M = 2.00, SD = 0.378$) concerning the Ideology and Values Dimension. The difference in means is very small, and therefore the effect size is slight. However, the mean of the IRE students is greater. Gender was not significant.

For the Language Dimension the test showed that there were statistically significant differences $[t(1001) = 10.806, p < .001, r = .34]$ between the mean of the level obtained by the IRE students ($M = 2.14, SD = 0.461$) and the mean obtained by the IDFE students ($M = 1.79, SD = 0.476$). The effect size in this case is medium. About gender, significant differences were found $[t(1001) = 5.875, p < .001, r = .18]$ between the mean of the level obtained by the male students ($M = 2.14, SD = 0.493$) and the mean of the level obtained by the female students ($M = 1.96, SD = 0.477$). The mean of the males is higher, but the effect size is small.

When the test was run for the Technology Dimension, the results revealed that despite not assuming equal variances ($F = 6.928, p < .05$) there were statistically significant differences $[t(865,023) = 2.965, p < .01]$ between the mean of the level obtained by the male students ($M = 2.06, SD = 0.333$) and the mean obtained by the female students ($M = 2.00, SD = 0.307$). The difference in means is low and therefor the effect size is small: $r = 0.09$, the higher level of the males being very small in this case.

When applied to the type of educational system, equal variances were not assumed ($F = 39.329, p < .05$) but statistically significant differences were indeed found $[t(743,683) = 11.992, p < .001, r = .37]$ between the mean of the level obtained by the IER students ($M = 2.09, SD = 0.326$) and the mean obtained by the IDFE students

($M = 1.86$, $SD = 0.239$). The effect size in this case is medium, and it is important to point out that the mean is once again higher in the IRE students.

Significance was found in the Reception and Interaction Dimension [$t\,(1001) = 4.151$, $p < .001$, $r = .13$] between the mean of the IRE students ($M = 2.33$, $SD = 0.336$) and that of the IDFE students ($M = 2.24$, $SD = 0.328$), the mean of the IRE students being higher with a small effect size, as can be observed.

Significance was also found about gender, and although the effect size is small [$t(1001) = 2.837$, $p < .01$, $r = .08$], the mean of the male students ($M = 2.34$, $SD = 0.344$) was slightly higher than that of the female students ($M = 2.28$, $SD = 0.329$).

Concerning the aesthetic dimension, the test revealed that despite not assuming equal variances ($F = 7.571$, $p < .05$), there were statistically significant differences [$t\,(821,155) = -2.111$, $p < .05$, $r = .08$] between the mean of the level obtained by the male students ($M = 2.17$, $SD = 0.328$), and the mean obtained by the female students ($M = 2.22$, $SD = 0.280$).

The difference of means is very small (lower in the male students), and therefore the effect size is small as well. As to the type of educational system, and not assuming equal variance ($F = 7.571$, $p < .05$), a significant difference was found [$t(595,623) = 2.856$, $p < .01$, $r = .10$], the mean of the IRE students ($M = 2.08$, $SD = 0.409$) being higher than that of the IDFE students ($M = 2.00$, $SD = 0.378$).

In the production and dissemination dimension, again without assuming equal differences, ($F = 74.308$, $p < .05$) there were statistically significant differences [$t(850,877) = 11.915$, $p < .001$, $r = .41$] between the mean of the level obtained by the IRE students ($M = 1.70$, $SD = 0.581$), and the mean of the IDFE students ($M = 1.26$, $SD = 0.366$), the former mean being higher. The effect size is medium, and for now is the highest effect size found among all the dimensions. We found no significant differences between the means in terms of gender.

Neither the ideology and values dimension nor the aesthetic dimension correlate significantly with age, but the Language Dimension does show a significant correlation ($r(1001) = -.265$, $p < .001$), as do the Technology ($r(1001) = -.279$, $p < .001$), the reception and interaction ($r(1001) = -.101$, $p < .001$), and the production and dissemination dimensions ($r(1001) = -0.208$, $p < .001$), the relation being negative and weak. This result has been interpreted as follows: the younger the students, the higher the level of competence in these dimensions, and the older the students (over 18), the lower the level.

4 Discussion and Conclusions

Overall, the results of the IDFE students show a lower level of competence in the Production and Dissemination Dimension ($M = 1.26$, $SD = 0.36$) when compared to the average of the total sample of 1003 students ($M = 1.57$, $SD = .56$). This difference is even greater with the average of the IRE students ($M = 1.70$, $SD = .58$), and therefore the effect size of the difference, $r = .41$. As a result, it we can conclude that there is a clear difference in the level of competence. The effect size is medium in the dimensions of Technology ($r = .37$) and in Language ($r = .34$) when compared with the results obtained in the same dimensions by the IER students. We did not found a significant difference in the results obtained in the Aesthetics dimension. The effect is weak in the Reception and Interaction (r = .13) and the Ideology and Values dimensions (r = .10). Thus, in all the dimensions except Aesthetics, the level of competence of the IRE students is higher.

In terms of gender, we found a significant difference in the level of competence of the male students in the following dimensions: Language (r = .18), Technology (r = .09), Reception and Interaction (r = .08). The female students performed better in the Aesthetic dimension, although the effect size is weak (r = .08).

Regarding the hypothesis posited, we confirmed inequality in five of the six dimensions studied. In response to the research questions, we found that the level of competence can decrease with age, and that men may have a higher level in some dimensions, but the difference found does not evidence an alarming gender gap. It may be that the advantage men have in terms of media knowledge and use makes them more confident in expressing their position on different topics of general interest.

We confirmed the inequality in the levels of media competence according to type of educational system, but it is not so important if we take into consideration the economic and social inequality in both contexts. Torres and Infante [24], in a study carried out in Ecuador to identify digital inequality in universities, concluded that level of income was what most affected the variables defining the possibilities of access, use and use intensity of online tools and resources. Tirado, Mendoza, Aguaded and Marín [25] identified in Ecuadorian secondary school students a sequence between the relations of the different levels of online access and the accumulative effect of technical resources and levels of digital literacy in academic use. Sánchez and Navarro [26], when analyzing the state of ICT-mediated rural secondary schools in Argentina, found both structural and socio-economic inequality.

Considering the results obtained according to the different dimensions employed in the present study, students scored the worst in the Production and Dissemination dimension. This may be a result of the difficulty of the questions in this category, which demanded that the students demonstrate their knowledge, which entailed negative outcomes.

This leads us to infer that, in Ecuador, and above all, in the milieu of low-income citizens, it is important to be a "prosumer" who consolidates an active participation, who disseminates his or her needs and problems in different ways and through different media, in other words, in a multimedia and multimodal way. In a similar study done with children and young people in Spain, it was found that the level of media competence is not optimal. Those researchers recommend, "continue working so that the school curriculum includes media literacy as a fundamental element in the education of prosumer citizens" [14]. This recommendation should also be followed in Ecuador, and future research should carry out an in-depth analysis of the contents of each curriculum present in the Ecuadorian educational system and measure their effects. To do so, the guidelines set out in the research by Ramírez, Renés and Aguaded [27] can be followed. These researchers reviewed the current regulations for Primary Education curricula in all of Spain's Autonomous Communities and quantified the presence of media competence in the assessment criteria of the different areas of the curriculum in fourth year of Primary Education. They then relate the assessment criteria of the curriculum to the descriptors of media competence and emphasize that all the dimensions are reflected in those criteria. Future research studies should delve deeper into the inequalities and gaps in media literacy of parents and teachers in different educational sectors; it is important for parents to participate in the education of their children as much as possible within their possibilities. Shin and Seger [28] demonstrate how parents and children studying English in an urban school in the US, families of Latino minorities, can support learning using ICT; with digital literacy they were able to use blogs to support the academic goals of their children. The results of the study by Foronda, Martínez and Urbina [29] corroborate the need to work on media literacy both in school curricula and in the different areas of non-formal education of citizens in general.

About teacher research and teachers' levels of media literacy, it is important to know how convinced they are about the advantages of using ICT in the classroom. Sancho and Padilla [30] found in their study that only 15 out of the 38 teachers in their sample use ICT in the classroom. ICT can be employed in learning, as reflected in the state of the art of the present study. Incorporating them signifies an element of democratization and equality of opportunities for underprivileged sectors of the country, and if this is accompanied by public educational policies, training of teachers and society in general, then the inequality, exclusion and marginality that make any social, cultural and educational progress difficult can be powerfully challenged.

References

1. Potter, W.J.: Review of literature on media literacy. J. Sociol. Compass. **7**, 417–435 (2013)
2. UNESCO. http://www.goo.gl/0BSABw
3. Aguaded, J.I.: Media Programme (EU), International Support for Media Education. Comunicar **40**, 7–8 (2013)
4. Area, M., Pessoa, T.: From solid to liquid: new literacies to the cultural changes of Web 2.0. Comunicar **38**, 13–20 (2012)
5. Dornaleteche, J., Buitrago, A., Moreno, L.: Categorization, item selection and implementation of an online digital literacy test as media literacy indicator. Comunicar **44**, 177–185 (2015)
6. Ferrés, J., Piscitelli, A.: Media competence. Articulated proposal of dimensions and indicators. Comunicar **19**(38), 75–81 (2012)
7. Gutiérrez, A., Tyner, K.: Media education, media literacy and digital competence. Comunicar **38**, 31–39 (2012)
8. The European Commission. http://goo.gl/HpPgvN
9. Sonck, N., Livingstone, S., Kuiper, E., Haan, J.: Digital literacy and safety skills. EU Kids Online. London School of Economics and Political Sciense (2011)
10. Ferrés, J., Aguaded, I., García, A.: La competencia mediática de la ciudadanía española: dificultades y retos. ICONO 14, vol. 10, pp. 23–42 (2012)
11. Rodríguez, M., Berlanga, I., Sedeño, A.: Análisis crítico de dimensiones de la competencia audiovisual en la etapa de Bachillerato. Historia y Comunicación Soc. **18**, 703–712 (2014)
12. Sandoval, Y., Aguaded, I.: Nuevas audiencias, nuevas responsabilidades. La competencia mediática en la era de la convergencia digital. ICONO 14, vol. 10, pp. 8–22 (2012)

13. García, A., Seglem, R., Share, J.: Transforming teaching and learning through critical media literacy pedagogy. Learn. Landscapes **6**, 109–124 (2013)
14. García, R., Ramírez, A., Rodríguez, M.: Media literacy education for a new prosumer citizenship. Comunicar **43**, 15–23 (2014)
15. Norris, P.: Digital Divide: Civic Engagement, Information Poverty, and the Internet Worldwide. Cambridge University Press, Cambridge (2001)
16. Alva de la Selva, A.R.: Los nuevos rostros de la desigualdad en el siglo XXI: la brecha digital. Revista Mexicana de Ciencias Políticas y Sociales **60**, 265–285 (2015)
17. CELAC. http://goo.gl/7dLfcU
18. Ramírez, J.L.: Las tecnologías de la información y de la comunicación en la educación en cuatro países latinoamericanos. Revista Mexicana de Investigación Educativa **11**, 61–90 (2006)
19. World Bank. http://goo.gl/JQr8Zh
20. Asamblea Constituyente. http://goo.gl/Y7f7Rn
21. Ecuadorian Ministry of Education. http://goo.gl/o8WOhg
22. Hernández, R., Fernández, C., Baptista, M.: Metodología de la Investigación. McGraw-Hill, México D.F (2014)
23. Ferrés, J., García, A., Aguaded, I., Fernández, J., Figueras, M., Blanes, M.: Competencia Mediática. Investigación sobre el grado de competencia de la ciudadanía en España. Gobierno de España, Ministerio de Educación. Instituto de Tecnologías Educativas, Consell de l'Audiovisual de Catalunya (2011)
24. Torres, J., Infante, A.: Digital divide in universities: internet use in ecuadorian universities. Comunicar **37**, 81–88 (2011)
25. Tirado-Morueta, R., Mendoza-Zambrano, D.M., Aguaded-Gómez, J.I., Marín-Gutiérrez, I.: Empirical study of a sequence of access to Internet use in Ecuador. Telematics Inform. **34**, 171–183 (2017)
26. Sánchez, L., Navarro, M.: Secundarias rurales mediadas por tecnologías de la información y la comunicación en el norte de Argentina: democratización, inclusión y problemas éticos. Innovación Educativa, p. 15 (2016)
27. Ramírez, A., Renés, P., Aguaded, I.: La competencia mediática en los criterios de evaluación del currículo de Educación Primaria. Aula Abierta, pp. 1–8 (2015)
28. Shin, D., Seger, W.: Web 2.0 technologies and parent involvement of ELL students: an ecological perspective. Urban Rev. **48**, 311 (2016)
29. Foronda, A., Martínez, J.I., Urbina, A.: Grado de competencia mediática en alumnado adolescente de Esmeraldas (Ecuador). Pixel Bit. Revista de Medios y Educación **52**, 151–165 (2018)
30. Sancho, J., Padilla, P.: Promoting digital competence in secondary education: are schools there? Insights from a case study. J. New Approaches Educ. Res. **5**, 57–63 (2016)

An Information System to Manage Organizational Internal Communications

Lito García-Abad$^{(\boxtimes)}$, Pablo Vázquez-Sande, and Ana Montoya Reyes

CESUGA University College, Obradoiro, 47, 15190 A Coruña, Spain
{jlgarcia, pvazquez, amontoya}@usj.es

Abstract. The article proposes the creation of an Information System based on four major epigraphs that, in turn, are integrated by a set of variables in each of them that will allow the adoption of decisions on internal communication in organizations. The general indicator is sectoral in nature and is shaped according to the characteristics of the organization, the sector in which it operates and the channels used in communication.

Keywords: Internal communication · Organizational communication · Key performance indicators · Management

1 Introduction

One unresolved problem in organizational communication is to measure the efficiency of the actions included in its Communication Plan. The development of an efficient and useful corporate scorecard with Key Performance Indicators or KPI, which enables the company's communications manager to evaluate their tactical proposal and, given his or her position, to conduct corrective actions of the existing plan, is the ideal to which both researchers and professionals aspire.

The Communication Plan must close the communication loop [1], in other words, it must propose the necessary control system to assess the expected results [2], and, consequently, get to know if the goals set in the plan have been or not achieved. A goal which cannot be assessed nor measured is not a goal itself.

Key performance indicators are the basis for the development of an information system and enable the management of communication in an organization beyond its presence in the media and social networks [3], shown in economic terms, based on advertising data and prices or based on the readership and the audience of a given media. All these indicators are, certainly, necessary but not enough in the sector. Likewise, in the digital world communication is assessed with indicators that ignore the evidence of the fact that we are in a field of knowledge where an effective causal relationship between an action and a particular indicator cannot be extracted. Were this the case, it would not make sense to speak about integrated organizational communication at present or to speak about coordinated actions whose aim is to achieve that the whole is greater than the sum of its parts, as the General Systems Theory states.

© Springer Nature Switzerland AG 2019
Á. Rocha et al. (Eds.): ICITS 2019, AISC 918, pp. 901–908, 2019.
https://doi.org/10.1007/978-3-030-11890-7_84

Statements such as the one that says that Communication increases brand aware-ness or the value of an organization are simply hypothetical unless we can confirm it experimentally [4]. And to achieve this, we need to correlate in some way the elements of such assertions [5]: communication increases productivity, motivation, economic results, work environment, etc. However, publications do not reach these conclusions but they present models or generic discourses lacking accuracy. And even more, under obvious demands such as explaining why other variables have not had an influence on a particular case, there are not accurate arguments which consolidate the suggested positions. In Social Sciences, there are so many variables and it is so difficult to change them into mathematical variables when dealing with similar cases that we can only talk in terms of probability, not in terms of indisputable statement [6]. As a consequence of this, basic research methods in social sciences basically rely on statistics [7].

In advertising there are available tools to assess possible or real results from an advertising campaign (advertising tests). Similarly, although at the moment their quality is highly questionable, there are space and economic indicators in advertising that allow us, even with stablished corrections based on media framing, which measure the results of a communications campaign. In other areas of communication, as it is the case of digital communication, these indicators do not either exist or there are their reductionist counterparts: considering aspects such as followers, accesses, interactions, permanence, pages... Indicators that draw virtual clients' profile and report to the communication or marketing manager how they surf the web, what they see, etcetera. Notwithstanding, these are isolated indicators which must be connected to the rest of the organisation: do a high number of followers in Instagram always imply higher brand recognition and higher sales?

There are thousands of variables that index directly the success obtained. Therefore, thinking that an indicator is the reflection of a given action, as an example of cause-effect linkages, is unreal and impossible to sustain rationally.

In recent years, this situation has been clearly proved but it is necessary to underline the fact that only with the use of complex management structures it is possible to design suitable communication strategies for the proposed objectives. There are some attempts in the Personnel Area of organisations which let us see, for example, the environment of an organisation or their employees' motivations, connecting different indicators which are merely testimonies, even residual, and more of a cause for concern as there are tools that help us in the decision making process within the field of communication: for example, connecting absenteeism at the workplace with the results detected in a survey about work environment.

Yet there are not publications which represent a real progress in the existing methodologies up until now. The search for that Information System, which manages communications in an organization (CIS), similar to the Marketing Information System (MKIS), lacks the necessary supports in the business sector to make the necessary progress for its creation by means of an experimental practice.

The existing literature presents proposals for the analysis of items which affects the communicative process ([8–14], among others), but these proposals do not create a system that makes information for decision-making available at any time experimentally.

2   Internal Motivation

Motivation is probably the most fertile area in the analysis of indicators in internal communication. It has been traditionally stated that the higher the communication, the higher the motivation. However, we are standing here at complex indicators where there is not a Communication variable that we can register quantitatively in an objective manner to, hereafter, be compared with the true datum of another complex indicator such as Motivation and see if there is a direct proportionality.

From the discoveries of the Human Relations School, in the 1920's, about the influence of environmental conditions in the employee productivity of a company, at a lower level than the evaluation of job performance carried out by the given manager, new variables or motivating factors have been provided, such as salary, workplace, training and professional development, based values… All of them have led to new advances in the study of motivation, but scarce results in the study of organizational communication which can be considered as a mere reflection in contrast to the other ones.

Motivational factors are subjective and, as a consequence of this, their treatment in some organisations does not mean that they can be extrapolated to other organisations, not even to those in the same sector. Subjective, changeable and, in short, functioning as statistical aggregates that, as the components of the organisation and its circumstances change, are constantly modified. In fact, Lewin defines behaviour as a function of the person and the environment and Goleman as a group of emotional tendencies that guide the objectives.

Be that as it may, motivation is perceived in relation to a goal. Communication and motivation are interrelated [15–17], and both with productivity [18]. Nosnik classifies information as one of those needs included in people's need to belong as it can contribute to create the image of personal achievement and increase interest in one's job. For this author, communication represents the common area of motivational and productive processes, the place where the organisation's and the individual's interests meet.

The present research aims at providing an internal organisational CIS based on the use of multiple variables that, in turn, could be specified considering every single case through their evolution and the development of qualitative and quantitative techniques, parallel to those set out in this paper.

3   Methodological Proposal

The aggregate index of motivation (AIM) is an index that consists of different indicators classified into four subject areas (Motivation, Reputation, Structure and Management) which is presented as an information system to manage internal communications in organisations. The objective in this index is that it can be used as a barometer to test the situation of internal communication in organisations both sectoral and globally.

The conception of Internal Communication underlying in this indicator implies a global character, in other words, understanding communication as a whole consisting of subsystems which interact with one another. In this way, within Internal Communication there are variables which are specific to other types of Communication and have an influence in it, and vice versa, Internal Communication also has an influence on other types of communication.

Thus understood, AIM considers the result of External Communication at all times as well as the Image and Reputation of the organisation, as these are variables acting on employees' perception. Similarly, as it consists of four groups of indicators, organisations can be evaluated considering every group, arising two new rankings within each group, being one of them general and the other one sectoral. In other words, five general rankings can be obtained from the AIM and as many sectoral as sectors considered multiplied by five. Therefore, given J sectors, the following results would be obtained:

	General	Sector 1	Sector 2	Sector J
IDEM general	Value 1	Value 2	Value 3	Value i 1 j
IDEM motivation	Value 4	Value 5	Value 6	Value i 2 j
IDEM reputation	Value 7	Value 8	Value 9	Value i 3 j
IDEM structural	Value 10	Value 11	Value 12	Value i 4 j
IDEM management	Value 13	Value 14	Value 15	Value i 5 j

General indicators (values 1, 4, 7, 10 and 13 on the table) are obtained as a statistical average of the total values in each row, that is to say,

$$\text{Value 1} = \text{Value 2} + \text{Value 3} + \text{Value j}/\text{Number of values}. \tag{1}$$

And more general,

$$AIMij = Evij/Nj. \tag{2}$$

The evolution of the statistical average in the period chosen as a reference shows the evolution of the communication barometer in the country.

Sectoral AIMs (values ij) are, likewise, statistical averages on the values table recorded from the analysed organisations of every sector. The ranking obtained allows not only a classification of the organisations in order of their scores in each group of indicators but also, in relation to the sector average.

Mathematically, measures of central tendency different from the arithmetic mean can be handled, as it is the case of the mean, or in other words, the value which gives the same number of values above and below that given value (central value of the table).

The group of indicators of the AIM system has a clear advantage: the system allows the incorporation of new group indicators that will make that the resulting AIM improves progressively.

In a generic sense, Motivation indicators will be grouped in any of these three groups: quality, behaviour and performance indicators, which complement each other to obtain the global indicator as well as reputation, structure and management indicators. The last indicators mentioned refer to Intellectual and Financial Capital and they can be either individual (work effort, individual absenteeism, number of claims per employee...) or for groups (financial expenses over sales, net profit, profitability over sales, stability index –average number of employees per business unit divided between the number of employees who have been in the business unit in a given period–...).

Structural indicators are those that use as a reference the structure and the communication net of the group and they include communicograms and both the formal and informal nets of information flow in the organization.

Reputation indicators are those that let us monitor the image and visibility of the different types of public the organisations relate to. They must have the following characteristics: (1) they must mean something for the receiver, not for the sender-organization. It is not a valid indicator because we have chosen it but because the receiver provides the indicator with a certain meaning and leads the receiver to a given reaction. This meaning provided by the receiver is integrated in the total meaning of the message. The indicators, of course, do not have to be universal: they can be influenced by socio-cultural factors. And (2) they have a relevant effect as they contribute to the perception of the final message at different rates in contrast to other segments of the total message that the receiver gets. This implies that the indicator has been chosen after different experimental tests carried out in similar business units within an organization as García-Abad proposed in 2005.

In addition to this, some verbal indicators (use of adjectives, message heading, use of "but", structure of the text...), as well as non-verbal ones such as the use of color, image, links to additional web pages and redundancy could also be considered (Fig. 1).

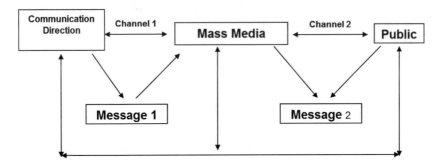

Fig. 1. Each element of the communication process, whether we use mass media or any other intermediary channel, has its own indicators gathering the abundant literature that exists since the middle of the 20th century.

3.1 Determination of Variables

A general framework for the organisation chart has been designed but it is necessary to determine which ones are the singular variables or group variables that we must consider within each group of indicators. For that purpose, we have designed a general proposal for the financial sector designed by García Abad, who from that moment on has lacked an explicit support of an organisation to test it experimentally from within.

The proposal is based on the use of the organisations' intranets which are either medium or big size and also have business units or functional units clearly differentiated.

The indicators considered would be:

1. AIM Motivation. García Abad proposes a total of 18 indicators for the specific case which is presented, which can be extrapolated for a general model: individual absenteeism, training rate, training effort, bureaucratic rate, distributive justice, average wage, average age, average age 80% staff, board's average age, withdrawal rate, success rate, security tax, job insecurity rate, staff turnover, global image of the organisation, staff's opinion, staff's opinion about remuneration.
2. AIM Reputation: corporate image perceived by stakeholder and social reputation.
3. AIM Structural: group structure, multi channels, centrality, leadership, graduate rate, labour seniority, turnover rate.
4. AIM Management: this is divided, in turn, into indicators of traditional communication (aggregate audience, economic value, information turnover, update rate, innovation rate, page views rate, research rate, consultation rates, image, headlines, typeface families...); of digital communication (the typical indicators that can be measured with the control programmes that test the Internet surfer activity); financial (they are, at least, twenty indicators in this section and they present both the financial and commercial output, such as profit, sales, profit over sales, profit per employee, market penetration, customer loyalty, stability index, complaints...)

All these indicators can be implemented in each business unit with the present possibilities that internal networks of organisations offer, in a manner that it is possible to vary individual variables keeping the other ones constant. In this case, it should be necessary to send the same type of information about the organisation to two offices but modifying variables previously determined, such as text format variables included in the Management AIM group.

4 Conclusions

Communication at present implies a greater effort in the development of tools which help managers in the decision making process beyond personal subjectivity or beyond management boards. To do that, it is required to verify proposed plans experimentally, which bring together registered results from different variables as other changes take place in other ones. The development of independent models from business units, which rely on the same general system, would provide the possibility of assessing the impact of each variable in the general system trying to keep constant some of those variables.

The proposed model is born from the consideration of multiple variables classified into four large groups of indicators and a general group which can be contrasted experimentally in organisations with differentiated business units and a digital internal communication network. The four groups of indicators, Motivation, Reputation, Structural and Management, comprise, in turn, multiple variables which will have to be identified in each particular case for every organization. There is the possibility of contrasting each indicator experimentally through the development of communication processes in similar business units that are present in the organization. The action will be taken upon specific elements of the process such as colour, typography, image, active individuals… The use of the intranet opens an enormous experimental field for the development of studies in this area in such a way that those variables which are more relevant at any given moment can be permanently monitored.

References

1. Varona, F.: El círculo de la comunicación. Netbiblo, A Coruña, Spain (2005)
2. Preciado-Hoyos, A., Hincapié-Noreña, C.A., Pabón-Montealegre, M.V.: Los indicadores de medida en la Comunicación Organizacional. Revista de Comunicación 26, 121–131 (2009)
3. Armijos-Maurad, A.P., et al.: Un análisis on-line del turismo comunitario en Ecuador. In: Rúas-Araujo, J., Martínez-Fernández, V.A., Rodríguez-Fernández, M.M., Puentes-Rivera, I., Yaguache-Quichimbo, J., Sánchez-Amboage, E. (eds.) Actas del II Simposio de la Red Internacional de Investigación de Gestión de la Comunicación (XESCOM, Quito – 2016). De los medios y la comunicación de las organizaciones a las redes de valor. Quito, Ecuador, 15–16 September, pp. 623–639 (2016). ISBN: 978-9942-25-054-4
4. Álvarez-Nobell, A.: Medición y evaluación en comunicación. Instituto de Investigación en Relaciones Públicas, Madrid (2011)
5. Álvarez-Nobell, A., Lesta, L.: Medición de los aportes de la gestión estratégica de comunicación interna a los objetivos de la organización. Palabra clave, 14, No. 1 (2011). ISSN 0122-8285
6. Chalmers, A.F.: Qué es esa cosa llamada ciencia? Siglo XXI, Madrid (1982)
7. Popper, K.R.: La lógica de la investigación científica. Tecnos, Madrid (1967)
8. Arias-Montesinos, S., Segarra-Saavedra, J.: Análisis comparativo de Social KPI's en las redes sociales de Pepsi España y Pepsi México. 3C TIC 6(1), 64–76 (2017). http://dx.doi.org/10.17993/3ctic.2017.55.64-76
9. Linares-Herrera, M.P., González-Borges, M.A.: Propuesta de indicadores para la gestión comunicacional del medio impreso. Bibliotecas Anales de investigación 11, 138–149 (2015)
10. González Fernández-Villavicencio, N., Menéndez-Novoa, J.L.; Seoane-García, C., San Millán-Fernández, M.E.: Revisión y propuesta de indicadores (KPI) de la Biblioteca en los medios sociales. Revista Española de Documentación Científica 36(1) (2013). http://dx.doi.org/10.3989/redc.2013.1.919
11. García Abad, L.: La comunicación en las organizaciones. Jornadas sobre los Recursos Humanos en la Administración Pública "Nuevos enfoques de la gestión de RRHH en las Administraciones Públicas". Ayuntamiento de Vitoria-Gasteiz, 24 de junio (2005)
12. García Abad, L.: Comunica: lecturas de comunicación organizacional. Netbiblo, A Coruña (2005)
13. García Abad, L.: La función de comunicación en las infoempresas y sus indicadores de gestión: una aplicación en el sector financiero. Tesis doctoral. Facultad de Ciencias de la Información, Universidad Complutense, Madrid (2005)
14. Ríos Jaruma, J., et al.: Redes sociales como estrategia generadora de valor en las organizaciones. In: Rúas-Araujo, J., Martínez-Fernández, V.A., Rodríguez-Fernández, M.M., Puentes-Rivera, I., Yaguache-Quichimbo, J., Sánchez-Amboage, E. (eds.) Actas del II Simposio de la Red Internacional de Investigación de Gestión de la Comunicación (XESCOM, Quito – 2016). De los medios y la comunicación de las organizaciones a las redes de valor. Quito (Ecuador), 15–16 September (2016). ISBN: 978-9942-25-054-4

15. Schvarstein, L.: Psicología social de las organizaciones. Paidós, Buenos Aires (1992)
16. Flores, F.: Creando organizaciones para el futuro. Dolmen Ediciones, Chile (1997)
17. Morales-Tamaral, A., Pons-Peregort, O.: Influencia de la organización en la motivación laboral. Aplicación al caso concreto de una Administración Pública. Capital Humano, **151**, 26–36 (2002). enero
18. Nosnik, A.: Comunicación, motivación y productividad. In: Martínez de Velasco, A., Nosnik, A. (comp.) Comunicación Organizacional práctica. Manual gerencial. Trillas, México (1988)

The *Ex Ante Test* as a Sign of the Evolution of the European Commission Decision Making in the Field of Public Service Media

Marta Rodríguez-Castro$^{(\boxtimes)}$ and Francisco Campos-Freire

Department of Communication Sciences, Universidade de Santiago
de Compostela, 15782 Santiago de Compostela, Spain
m.rodriguez.castro@usc.es, francisco.campos@usc.com

Abstract. Public Service Broadcasting (PSB) was born and shaped within the national boundaries of each European country. However, as the audiovisual market started to become more transnational and interconnected, the European Union began to play a role in the regulation of such public service. During the past two decades, the main issue concerning European public service broadcasting policy was -and still is- how to regulate the expansion of PSB towards new media. This transition was mostly regulated through the decision-making of the European Commission regarding the State aid granted to PSB by their respective Member States. This paper depicts the evolution of the European Commission's approach to the regulation of PSB, taking the *ex ante* test for the inclusion of new media services within the public service remit as a sign of the turn towards a new paradigm in media policy where the market perspective prevails over the public interest approach.

Keywords: *Ex ante* test · Public Value Test · European Commission ·
2009 Broadcasting Communication · Media policy · New media regulation

1 Introduction

From the latest years of the 1990s until now, European Public Service Broadcasters (PSB) have been struggling with both external pressures and internal needs related to the changing media, political, economic and social environments within which these organizations operate. Economic globalization, political liberalization and technological digitalization [1] led the way to the redefinition of the values and the scope under which PSB should operate. Public service, as a concept that has been in constant change and evolution since the birth of this kind of broadcasters almost one century ago [2], was the focus of an intense debate that involved the public organizations themselves, commercial media lobbing for their own interests and political actors both at the national and at the European level. Although still an ongoing debate, this discussion around the role of public broadcasters reached its highest level of intensity around the latest years of the 2000s, when the European Commission published the Revised Communication on the application of the State aid rules to public service broadcasting [3].

© Springer Nature Switzerland AG 2019
Á. Rocha et al. (Eds.): ICITS 2019, AISC 918, pp. 909–918, 2019.
https://doi.org/10.1007/978-3-030-11890-7_85

This paper depicts the major changes on the European Commission's approach to PSB, evolving from a public interest-oriented perspective towards a market-driven focus. It is considered that the recommendation of the Commission to introduce an *ex ante* test constitutes a critical cornerstone within this evolution, as it has shaped the way that some public service broadcasters link their expansion towards new media services with their public service mission. The results of this research are presented in a chronological order, in an attempt to elucidate the evolution of the European Commission decision-making regarding PSB. The main items on this evolution are graphically represented in Fig. 1.

Fig. 1. Timeline of the evolution of the EU's approach to PSB/PSM regulation Source: processed by the author.

2   Method

The main objective of this paper is to provide an overview on the evolution of the European legal framework for Public Service Broadcasting and its transition towards Public Service Media and the legitimation of the development of new media services. In order to achieve this, the decisions issued by the European Commission regarding the State aid granted to PSB will be analyzed, paying especial attention to the impact of the 2009 Broadcasting Communication and to the reach of the recommendation of an *ex ante* procedure to approve new media services.

The method used for this research was document analysis [4]. The documents that were analyzed were the ones published by different institutions of the European Union regarding public service broadcasting, such as the Council of Europe, the European Court of Justice and the European Parliament. However, the documents issued by the European Commission were the most relevant when approaching the regulation of PSB within the European framework. Besides analyzing all the above mentioned EU-documents, the study was also complemented by a thorough revision of the scientific literature produced regarding the pitfalls presented in the regulation of PSB's new media services and the European Commission decision-making about the funding of PSB.

3   The Evolution of the European Commission's Approach to PSB

The liberalization of the broadcasting market led to a shift towards what Van Cuilenburg and McQuail identified as a "new paradigm in media policy", which is "mainly driven by an economical and technological logic" [5]. The main issues that arise within this paradigm were (and still are) commercial competition, technological innovation, transparency of ownership and choice for consumers (rather than citizens). The European Union (EU) contributed to the establishment of this 'new paradigm' [6] by developing EU regulation aimed at the formulation of competition and State aid guidelines, as well as at the harmonization of broadcasting law. Even if from the 1990s onwards the EU recognized the need to "develop a sociocultural side to complement and make more sustainable its economic side" [7], in the struggle to balance both perspectives, "the Union's substantive output remains centered on economic and competition considerations" [7].

As the power of commercial media started to grow, so did their ability to influence government policy and to introduce their views on the role of PSB into the regulation process [6]. Commercial broadcasters called for a level playing-field, claiming that the mixed funding of the PSB organizations, which received both public funds and advertising income, generates distortion of competition [8]. The main aim of private media consisted mostly in the marginalization of PSB as these public institutions were strong competitors too. Under the influence of neoliberalism, PSB was allowed to exist as long as their activity was limited to those areas where commercial broadcasters could not make profit. The reason behind the existence of PSB would no longer be spectrum scarcity, but market failure. The market failure argument was part of the policy discussion since the very beginning of commercial broadcasting, but the digitalization process, along with the expansion of the new media services offer, provided new grounds for the defenders of this argument [9–11].

The expansion of PSB towards the offering of new media services was in fact one of the most discussed issues in media policy debates since the late 1990s. Despite all the discussion papers, resolutions and guidelines developed during this decade, by the end of the century the EU had not yet established a clear, unambiguous and useful framework for the emerging transition from PSB to Public Service Media [12]. This lack of more specific European policy would be addressed in 2001, when the "Communication from the Commission on the application of State aid rules to public service broadcasting" (henceforth, "the 2001 Broadcasting Communication") was published. This new text was seen as an attempt of the Commission to establish new general guidelines while respecting the variety of PSB systems and the competence of Member States to decide on its organization and funding, as established by the Amsterdam Treaty [13].

The 2001 Broadcasting Communication stated that new media activities can be included within the public service remit, even if they are not considered "programs" in the traditional sense. However, it is also established that if the scope of the remit is extended to include such services, the mandate or entrustment act should be modified accordingly [14].

Nevertheless, the position of the European Commission would change the following year, especially in its decisions concerning the expansion of the PSB services online. Brevini [15] established two different time periods according to the position adopted by the Commission. In the first period, the Commission enabled the activity of traditional PSB broadcasting offer, adopted a technology-neutral approach and explicitly acknowledged the subsidiarity principle and the stipulations of the Amsterdam Protocol establishing the competence of Member States in the field of PSB, limiting its own action to checking for manifest error. By contrast, the second period, starting in 2003 (with the announcement of the Altmark judgement), is characterized by a shift towards a pro-market inclination [15, 16]. The Commission distanced itself from technological neutrality and showed signs of reluctance to support the expansion of PSB to new media activities, as complaints filed by different commercial agents of the industry piled up in the desks of the Directorate General for Competition.

The Commission also started to overstep its competences by suggesting Member States how to define the public service remit of their PSB organizations. One of the most remarkable illustrative examples of this change is found in the case of the BBC Curriculum, a service that the British Broadcasting Corporation was forced to cancel due to the strong scrutiny exerted by the European Commission as a direct consequence of the complaints filed by commercial competitors [17]. BBC Curriculum constitutes a pivotal turning point in media policy regarding the expansion of PSB's activities to new online media, the influence of the commercial sector in decision making and the adoption of a market-oriented perspective by the European Commission. This trend would be developed in many other cases of complaints filed by private operators about PSBs funding.

4   The Idea of an *Ex Ante* Assessment for the PSB-PSM Transition

By 2008, the Directorate General for Competition was overloaded with complaints on State aid granted to PSM organizations. Moreover, the European Commission acknowledged that the fast evolution of new media had led to the obsolescence of the guidelines established by the 2001 Broadcasting Communication. Thus, the Commission decided to open the debate about the renewal of the Broadcasting Communication, launching several consultations with different stakeholders.

The first consultation of the European Commission on the future framework for State funding of public service broadcasting took place between January and March 2008. The Commission received 121 responses, submitted by 17 Member States (in the case of Belgium, Flanders and Wallonia sent their own responses. Moreover, Norway also submitted its comments, as the Broadcasting Communication has EEA relevance), news publishers, different kinds of media enterprises and groups, trade unions, cable and satellite operators, telecom companies, associations of listeners and viewers and some private persons. However, the majority of responses received were submitted by agents of the industry (31%) and public service broadcasters (24%).

The majority of PSB positioned themselves against the review, arguing that the 2001 Broadcasting Communication "has worked fine and already includes the possibility for public service broadcasters to provide new media services" [18]. PSB organizations feared that an update on the Broadcasting Communication would lead to an increased influence of the Commission in broadcasting, as well as to a stricter framework for PSB. On the other hand, private broadcasters, newspaper publishers and other commercial media groups and telecom operators showed their support for such update in order to "set clear boundaries on the possibility for public service broadcasters to offer new media services" [18], as this expansion on their remit implies damages on competition.

The Commission included in this first public consultation some questions regarding *ex ante* evaluations for new PSB services, an approval mechanism that this institution had already required to some Member States (Germany, Ireland and Flanders) in previous decisions [19–21]. By 2008, some countries had already developed such *ex ante* mechanism, namely the United Kingdom, whose *Public Value Test* was the first of its kind and set the main characteristics of the future recommendation of the Commission; and Denmark, which had included this instrument in its most recent contract with its PSB, the DR. Five other countries (Germany, Ireland and Belgium, which had been urged by the Commission to develop this instrument, Norway and Hungary) informed the Commission in their responses that they were planning the adoption of an *ex ante* mechanism.

However, the idea of an *ex ante* evaluation to PSB new services was not received homogeneously by Member States and despite the pro and contra positions adopted by them, there was always room for nuance. The United Kingdom, for instance, suggested the introduction of just minimum requirements of the test. Others, like France, considered that the Broadcasting Communication should include a reference to this *ex ante* test, which then each Member State should define on its own. Finland considered that this procedure was not suitable for the changing media environment and Austria and the Netherlands questioned the inclusion of a market impact assessment in this procedure. Estonia even posed the possibility that this kind of tests would lead to the end of public service provision of some content such as sports.

The European Broadcasting Union (EBU) also submitted a critical perspective against the implementation of an *ex ante* assessment, arguing that the public service remit constitutes a single unit and cannot be assessed according to individual parts, that the State aid rules must be applied considering technology neutrality and that the introduction of an *ex ante* test would lead to delays on the implementation of new media services and to excessive costs [22]. Public service broadcasters, in general, opposed to the introduction of an *ex ante* mechanism in the update of the Broadcasting Communication.

By contrast, commercial broadcasters and print media showed support for the introduction of an *ex ante* evaluation, restricting the activity of public service broadcasters, arguing that "private initiatives should get a chance" [18]. This procedure, they stated, would provide more market certainty and stability. However, the industry also pointed out to the risk of *ex ante* evaluations being used by PSB to approve new services indiscriminately but with a regulatory base. That is the reason why they warned that the introduction of this measure would not stop them from presenting complaints at a national and at a European level, therefore exerting some kind of *ex post* control anyway.

Despite the reluctance of Member States and the manifest opposition of public service broadcasters, the update of the Communication of the Commission on the application of State aid rules to PSB finally included the description of an *ex ante* test. Karen Donders [23] summarized the reasons behind this decision. First of all, the fact that this procedure included a market impact assessment was seen as a way to deal with the complaints from the commercial sector at a national level, therefore reducing the workload undertaken by the Directorate General for Competition. Another reason, deemed flawed by Donders as there is no link between the two regulations, was that the Commission considered that the 2007 Audiovisual Media Services Directive asked for renewed State aid guidelines for PSB. The third reason would be the technological evolutions that the media environment was undergoing, although Donders questions this ground, as the 2001 Broadcasting Communication already considered the introduction of new media services within the mission of PSB. Moreover, if the principle of technological neutrality was actually applied, such evolutions would not require specific guidelines at all.

Thus, the 2009 Broadcasting Communication stipulated that all new media services offered by PSB have to be formally included in its public service remit, as defined in the entrustment Act between the Member State and the undertaking (in the shape of legislation, management contracts or terms of reference). Considering that the modification of such services requires long and exhaustive processes that would delay the launch of new services and hinder PSB innovation, the Commission provides an alternative intended to legitimate new PSB activities in the most agile and effective way: the *ex ante* evaluation. The 2009 Broadcasting Communication states that

> "State aid to public service broadcasters may be used for distributing audiovisual services on all platforms provided that the material requirements of the Amsterdam Protocol are met. To this end, Member States shall consider, by means of a prior evaluation procedure based on an open consultation, whether significant new audiovisual services envisaged by public service broadcasters meet the requirements of the Amsterdam Protocol, i.e. whether they serve the democratic, social and cultural needs of the society, while duly taking into account its potential effects on trading conditions and competition". [3].

The European Commission tried to balance the demands of commercial broadcasters and print media with the demands of Member States and public service broadcasters to respect the subsidiarity principle. The final version of the text was seen as a "compromise", as it did not include that many details on, for instance, the *ex ante* procedure, as the first draft did. Still, most of the enthusiasm shown for the updated Broadcasting Communication came from the sector of private media, as pointed out by Brevini [15].

5   Life After the *Ex Ante* Test

After the entry into force of the 2009 Broadcasting Communication, the European Commission still kept receiving complaints from private media against the funding of public service broadcasters, as the same time that some national authorities also notified the Commission about new services in order to check for State aid incompatibilities. Donders and Moe [1] have highlighted some limitations on the application of the

European Commission's State aid principles to the cases about public service broadcasters. For instance, they point out to the fact that "most decisions were issued only after the General Court (…) had urged the EC to close investigations and decide on the legality of funding to public broadcasters" [1], depicting the reluctance of the Commission to decide on this area. Also argued by these authors, the impact of the decisions of the Commission "did not deal with the scope of public broadcasters' but with the financing method" [1]. In a way, this may be considered to have increased the predominance of the market perspective when developing PSB policies, as more importance was placed on economic aspects that on the social and cultural value of public service broadcasting.

Figure 2 illustrates the evolution of the number of decisions issued by the European Commission regarding the State aid granted to PSB. Austria, the Netherlands, Spain, Denmark, France, Portugal, Belgium and Ireland were the countries whose public media system was under investigation since the approval of the new guidelines for the assessment of State aid to PSB. However, the issue of the introduction of an *ex ante* evaluation only arose in the cases of Austria, Spain, the Netherlands and Belgium. These four countries introduced a prior evaluation procedure within their media laws, although neither Spain nor Belgium have used their *ex ante* test yet.

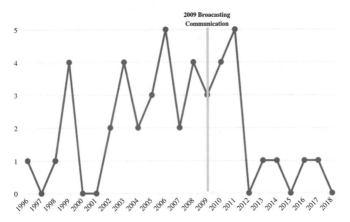

Fig. 2. Evolution of the number of EC State aid decisions concerning PSB Source: processed by the author considering data from the European Commission.

In this regard, besides the decrease in the number of EC decisions, the success of the recommendation of an *ex ante* procedure was quite limited. Up to date, only 10 out of the 28 Member States and two of the three non-Member States of the EEA require a prior assessment procedure to approve new services proposed by their PSB organizations, and the actual use and scope of such procedure is uneven. Moreover, the Commission had explicitly recommended the introduction of an *ex ante* test to four of these countries (including the French-speaking Community and the Flemish Community in Belgium) as part of the commitments required to the Member States in order to clarify the compatibility of the State aid granted to their PSB organizations.

In this regard, it is especially remarkable that this commitment to introduce an *ex ante* test has always been one of the Commission requirements in those cases started by the complaints of commercial competitors. In those processes started by the consultation of the national authorities, such as the many consultations filed by France, this kind of procedure does not seem to be so important. This may lead to the idea that a prior assessment process is more aimed at curbing the protests of the commercial broadcasters, new media outlets and newspapers than to the establishment of improved regulatory and governance guidelines, thus reinforcing the market-oriented approach developed by the European Commission.

6 Conclusions

One decade after its entry into force, the full impact of the 2009 Broadcasting Communication is still to be determined, but some major changes can be observed regarding the decisions of the European Commission, its more market-oriented approach, its attempts to harmonize PSB policy at a European level. Some benefits stemming from the new European State aid framework have to be acknowledged too.

For instance, the number of complaints regarding the State aid granted to public service broadcasting has decreased. The introduction of an *ex ante* assessment at a national level to approve new media services within the scope of PSB was seen as a "panic reaction" [23] of the European Commission in the face of the increasing number of complaints that the Directorate General for Competence had to deal with. By recommending the implementation of a prior evaluation procedure, states Donders (2011:36), the Commission "wanted to silence private sector complaints".

The approach of the European Union to public service broadcasting, particularly to its funding, presents differences both among the institutions and along the years. While the European Parliament and the Council of Europe advocated for the defense of the public interest in public service broadcasting, especially during the decade of the 1990s, the decision-making of the European Commission when dealing with State aid cases was led mostly by a market-oriented perspective, which became more obvious with the publication of the guidelines of the 2009 Broadcasting Communication. Moreover, the recommendation of an *ex ante* procedure to Member States seems like an attempt to transfer the case-by-case approach of the European Commission to each national context, as well as another step on the risky homogenization and integration process of media policy.

As the scope and use of this prior assessment evaluations present major differences between the Member States that have introduced them into their legislation, each national context should be studied independently in order to establish trends and divergences, as well as to establish different categories which could allow us to understand the many ways the *ex ante* test can be used: to legitimate public service media's activities within the online realm, to constraint the scope of its public service remit, or maybe just as a bureaucratic tool reluctantly implemented to satisfy the requirements of the European Commission.

Acknowledgments. The results of this paper correspond to the project "Indicators of governance, funding, accountability, innovation, quality and public service of European PSM applicable to Spain in the digital context" (Reference CSO2015-66543-P), from the State programme "Fomento de la Investigación Científica y Técnica de Excelencia", State Subprogram of "Generación de Conocimiento" from the Ministerio de Economía y Competitividad de España, cofounded by the Fondo Europeo de Desarrollo Regional (FEDER) of the European Union. It is also funded by the activity of Red Internacional de la Investigación de Gestión de la Comunicación (REDES 2016 G-1641 XESCOM), supported by the Consellería de Cultura, Educación e Ordenación Universitaria of the Xunta de Galicia (reference ED341D R2016/019). The author Marta Rodríguez Castro benefits from an FPU Contract (FPU16/05234) from the Ministerio de Educación, Cultura y Deporte.

References

1. Donders, K., Moe, H.: Exporting the Public Value Test. The Regulation of Public Broadcasters' New Media Services across Europe. Nordicom, Gothemburg, Sweden (2011)
2. Raats, T., Pauwels, C.: In search of the holy grail? Comparative analysis in public broadcasting. In: Donders, K., Moe, H. (eds.) Exporting the Public Value Test. The Regulation of Public Broadcasters' New Media Services across Europe, pp. 17–28. Nordicom, Gothemburg, Sweden (2011)
3. European Commission: Communication from the Commission on the application of State aid rules to public service broadcasting (2009/C 257/01) (2009)
4. Karppinen, K., Moe, H.: What we talk about when we talk about document analysis. In: Just, N., Puppis, M. (eds.) Trends in Communication Policy Research, pp. 175–193. Intellect, Bristol (2012)
5. Van Cuilenburg, J., McQuail, D.: Media policy paradigm shifts: towards a new communications policy paradigm. Int. J. Commun. **17**(2), 181–207 (2003)
6. Jakubowicz, K.: A square peg in a round hole: the EU's policy on public service broadcasting. In: Bondebjerg, I., Golding, P. (eds.) European Culture and the Media, pp. 227–301. Intellect, Bristol (2004)
7. Michalis, M.: EU broadcasting governance and PSB: between a rock and a hard place. In: Iosifidis, P. (ed.) Reinventing Public Service Communication: European Broadcasters and Beyond, pp. 36–48. Palgrave Macmillan, Basingstoke (2010)
8. Coppieters, S.: The financing of public service broadcasting. In: Biondi, A., Eeckhout, P., Flynn, J. (eds.) The Law of State Aid in the European Union, pp. 265–279. Oxford University Press, Oxford (2003)
9. Elstein, D.: How to fund public service content in the digital age. In: Gardam, T., Levy, D.A. (eds.) The Price of Plurality: Choice, Diversity and Broadcasting Institutions in the Digital Age, pp. 86–90. Reuters Institute for the Study of Journalism, Oxford (2008)
10. Cave, M.: Competition and the exercise of market power in broadcasting: a review of recent UK experience. Info **7**(5), 20–28 (2005)
11. Armstron, M., Weeds, H.: Public service broadcasting in the digital world. In: Seabright, P., von Hagen, J. (eds.) The Economic Regulation of Broadcasting Markets: Evolving Technology and Challenges for Policy, pp. 81–149. Cambridge University Press, Cambridge (2007)
12. Lowe, G.F., Bardoel, J.: From Public Service Broadcasting to Public Service Media. Nordicom, Gothemburg (2007)

13. Treaty of Amsterdam amending the Treaty on the European Union, the Treaties establishing the European Communities and certain related acts (1997)
14. European Commission: Communication from the Commission on the application of State aid rules to public service broadcasting (2001/C 320/04) (2001)
15. Brevini, B.: European Commission media policy and its pro-market inclination: the revised 2009 Communication on State Aid to PSB and its restraining effect on PSB online. Europ. J. Commun. **28**(2), 183–197 (2013)
16. Michalis, M.: Governing European Communications. Lexington Books, Lanham (2007)
17. Michalis, M.: Balancing public and private interests in online media: the case of BBC digital curriculum. Media Cult. Soc. **34**(8), 944–960 (2012)
18. European Commission: Review of the Broadcasting Communication. Summary of the replies to the public consultation (2008)
19. European Commission: Case E3/2005, Financing of Public Service Broadcasters in Germany. JOCE C/185/2007 (2007)
20. European Commission: Case E4/2005 State financing of RTÉ and TNAG (TG4). JOCE C/121/2008 (2008)
21. European Commission: Case E8/2006 State funding for Flemish public broadcaster VRT. JOCE C/143/2008 (2008)
22. EBU: EBU reply to the Commission's questionnaire (2008)
23. Donders, K.: The public value test. a reasoned response or a panic reaction? In: Donders, K., Moe, H. (eds.) Exporting the Public Value Test: The Regulation of Public Broadcasters' New Media Services across Europe, pp. 29–37. Nordicom, Gothemburg (2011)

TV Archives: Checklist of Indicators of Participation in the Digital Environment

Sara Martínez Cardama[⊠] and Mercedes Caridad Sebastián

Library and Information Science Department, University Carlos III of Madrid,
Calle Madrid 128, 28903 Getafe (Madrid), Spain
{smartil,mercedes}@bib.uc3m.es

Abstract. The audiovisual archives of public television channels constitute an essential element for understanding recent history. This European heritage is managed through different preservation models according to the country. The need to assert the role of the audiovisual archive as a mainstay of public audiovisual content, due not only to its heritage value but also to its exploitation value, is considered. At the methodological level, the foundations of a checklist of assessment indicators with two pillars is proposed: transparency and degree of online access. For the sample, four audiovisual archives of European television channels were used which illustrate different models of conservation and dissemination: British Broadcasting Corporation Archives (BBC) (United Kingdom), Institut National de l'audiovisuel (National Audiovisual Institute) (France), Nederlands Instituut voor Beeld en Geluid (Netherlands Institute for Sound and Vision) (Netherlands), and the Archivo de Radio Televisión Española (RTVE Archive) (Spain). Although European dissemination policies rely on their major role as aggregators of cultural content and increasingly promote their content in the media sphere, the situation of their participation is very irregular. The results reveal very inconsistent dissemination and digitisation policies, low levels of transparency, and policies of limited exploitation of content.

Keywords: TV archives · British Broadcasting Corporation Archives (BBC) ·
Institut National de l'audiovisuel (INA) ·
Netherlands Institute for Sound and Vision · RTVE Archive ·
Digital participation · Transparency · Access

1 Introduction

The value of the audiovisual archives of public television channels in today's society has a dual dynamic: heritage, linked to their undeniable capacity to reflect recent history, and exploitation, for their ability to revalue content and develop new programming. Furthermore, in the digital medium this value is enhanced by new revaluation possibilities for several purposes: to feed own programmes, to serve as a nexus to illustrate the material with past or present-day historic events, and, lastly, to serve as support for expanding the transmedia universes generated, communally, via fiction programmes [1].

Considering that innovation in the area of audiovisual archives has to come from exploitation and showcasing their value through their websites, the need arises to detect

© Springer Nature Switzerland AG 2019
Á. Rocha et al. (Eds.): ICITS 2019, AISC 918, pp. 919–928, 2019.
https://doi.org/10.1007/978-3-030-11890-7_86

how the end user perceives the audiovisual archives, their educational role, services, and participation in terms of in public projects involving digital collections, such as Europeana (in its role as an aggregator), in contributing to European audiovisual heritage.

Research on transparency is closely linked to accountability, governance, and access to information. With regard to this last item, there is no direct correlation between the two. As pointed out by [2], apart from the issue of whether or not this information is accessible, it must be accessible in terms of formats, reusable, and considered relevant and useful to users so as to allow analysis and reflection.

Despite the fact that studies in libraries and archives of their level of transparency have taken on a prominent role, this still has not been pursued comprehensively with respect to audiovisual archives.

It is worth noting that, unlike documentary heritage, the visual archive was not created with the idea of durability. It wasn't until the advent of sound in motion pictures that a need to preserve this cultural expression started to arise. This belated preservation meant that it would not be until the 1980s that international guidelines would start to be established for defining and preserving audiovisual heritage, which includes television heritage. Pioneering in this sense is the UNESCO document entitled Recommendation for the Safeguarding and Preservation of Moving Images of 27 October 1980.

Since this date, international bodies with competences in the area—such as the European Council, the European Parliament, and associations of a professional nature like the ICA/CIA (International Council on Archives), which has a special committee dedicated to audiovisual materials, ICFT (International Council for Film, Television and Audiovisual Communication), and IFTA/FIAT (International Federation of Television Archives)—have started to make recommendations oriented towards conservation and preservation of this heritage.

Nonetheless, these guidelines have come in the form of general recommendations, and a legal vacuum has been the rule for decades. In the case of television heritage, moreover, the preservation model varies from country to country, which complicates its analysis as a whole.

European audiovisual policies originated in the 1980s and are still focussed on a developmental stage of the telecommunications industry and European content. Following this infrastructure development stage, the foundations were laid for the European audiovisual sector, such as the European Audiovisual Observatory (1992) and the successive MEDIA programmes. These programmes were aimed at supporting and building a European content industry. Nonetheless, audiovisual archives as such did not start to be mentioned expressly until the recommendations of the European Parliament and Council of 2002 and 2005, particularly the latter, entitled Recommendation of the European Parliament and of the Council of 16 November 2005 on film heritage and the competitiveness of related industrial activities, which brings together two very important facets of audiovisual heritage: its consideration as cultural heritage and its exploitation.

However, whilst these recommendations urged states to create dissemination policies for their holdings, they did not establish funding for this. Therefore, it is difficult to find policies that combine preservation and conservation. Nonetheless, cooperative projects aimed at managing and unifying systems of description and conservation of European television archives, such as Presto Space, do appear. Following this, the Presto Prime programme aimed to provide access to different archives on diverse web portals, as well as to develop a specific digital preservation portal linked to Europeana [3]. Another relevant project, EUscreen, and the current EUscreenXL, aimed to promote television audiovisual heritage and make it attractive to citizens. This latter portal acts as a pan-European aggregator for Europeana in the audiovisual sphere by bringing together the main European audiovisual institutions and attempting to create a massive audiovisual archive [4]. It should be noted that the Spanish aggregator indicated for EUscreenXL is not TVE but TV3, whilst the RTVE Archive appears only as an associate partner.

As recognised by [3], there have been cooperation projects that directly conflict with the differing intellectual property laws of each country, especially in the area of orphan works. It is precisely the treatment of these works in the audiovisual field that has marked the regulatory debate at the European level [5].

Thus, the regulatory framework of each country in the treatment of its audiovisual heritage configures its policies and, in line with our starting hypothesis, its presence in the digital context. For this reason, four different realities of television heritage conservation models have been selected:

- Two cases, that of Britain's BBC and that of Spain's RTVE, where, as in many countries in our vicinity, corporations create and develop these assets over the years. These institutions also take responsibility for their conservation. In neither country is television heritage subject to legal deposit. In the British case, management of television archives is shared by the BBC itself with the NFA (National Film Archive) [6]. The latter institution is part of the British Film Institute (BFI) and preserves and provides access to both cinematographic films and television programmes. In the Spanish case, the 'audiovisual programmes broadcast by audiovisual communication services' figure among those 'excluded from legal deposit' in the current Legal Deposit Act of 29 July 2011 [7]. Therefore, the responsibility for management and preservation of the heritage falls on each channel.

- The cases of the National Audiovisual Institute (INA) of France and the Netherlands Institute for Sound and Vision are radically different. The French case is paradigmatic, as in France a law exists which guarantees legal deposit for all television heritage. Since its creation, the INA has taken responsibility for guaranteeing the conservation and preservation of radio and television archives. Moreover, it controls all French radio and television documents through legal deposit. The Dutch case, for its part, also delegates television audiovisual production to an institute, the Netherlands Institute for Sound and Vision, for three purposes: heritage, museum use and education.

2 Method

The goal is to test an assessment system based on a checklist of indicators created ad hoc to assess the level of transparency and online access to the audiovisual archives of four models of European television channels. The assessment mechanism, in the form of a checklist, has already been used by [8] for the online presence of the audiovisual archives of European regional television channels. However, indicators were created based on the accessibility of their websites and the search systems provided to the user. This assessment proposal also aims to study the new forms of expression of public channels through social media and, concretely, the participation of the archives from the standpoint of cross-media or multi-platform strategies. Given that specific indicators were not found, the categories established previously in [1] reflecting trends noted in some recent studies on the sector, such as those established by [9] which use multi-platform storytelling to revalue archival audiovisual documents to create a collective participatory memory, were taken into account when establishing indicators to cover this perspective of access. In it, the authors, as mentioned earlier, rely on three dynamisation tasks for audiovisual archives established previously in [10]: expanded access, repackaged content, ancillary content (Table 1).

Table 1. System of indicators

Indicators of transparency and access to audiovisual archives	
Transparency	Access
1.1. Information about the archive (About us, organisational chart, etc.)	2.1. Query system (search capacity, operators, clusters, etc.)
1.2 Content (number of digitised collections, type, coverage, etc.)	2.2. Description of the content (metadata, transcriptions, etc.)
1.3 Contribution to European projects (Europeana content aggregator, EUscreenXL, etc.)	2.3. Transmedia strategies for dynamisation 2.3.1. Expanded access 2.3.2. Repackaged content 2.3.3. Ancillary content
1.4. Information about preservation policies	

3 Results and Discussion

3.1 Transparency

Transparency measures the degree of accessibility of content or collections available to the user. In general, access to collections is more evident in institutions with a heritage purpose than in television archives, where the link to the archive is not found on the main menu but rather requires navigation through various categories or scrolling to gain access. The selected indicators aim to measure transparency in terms of the information about a particular document unit, the visibility of its content (digitised collections, etc.), the degree of collaboration and international collaboration between them. Lastly, it involved perceiving whether or not the digitisation and preservation policies are disseminated to the user in a coherent manner.

3.1.1   Information About the Archive (About Us, Organisational Chart, etc.)

Related to information about the organisation of the archive itself. This is not typically shown explicitly in the archives belonging to television corporations. Nonetheless, in heritage institutions, such as the Netherlands Institute for Sound and Vision and the National Audiovisual Institute (INA) of France, their history, mission and vision are shown in a developed manner. In the case of the INA, a menu is provided ('Galaxie INA') which enables users to understand the entire audiovisual framework maintained by this entity. This menu is present on all of the pages controlled by the institute and, therefore, the user is always situated within a body that contemplates the heritage version as well as the research and educational version. Its public nature supports the presence of information about its organisation, the organisational chart, and employment vacancies at the institute. The same information is found in the Dutch case, with a clear explanation of its three-fold vision: research, museum use and education. This is facilitated by means of a simple, menu consisting of three words—'Visit', 'Collection', 'Knowledge'—which evoke the corresponding functions. Similarly, it offers a list of experts on management of the heritage and audiovisual documentation featuring the individuals' photographs and job titles. This, without a doubt, helps to make the professionals working at the audiovisual archive visible and recognise their experience and areas of specialisation in the eyes of the user.

As for the television corporations, it is surprising that the BBC has stopped updating its traditional archive website, where it formerly included information about the experts who manage its massive collection (considered the largest multimedia archive in the world), and opted for a more decentralised model for its archive, making it even harder for the user to access same (search engines like Google take users to the old version). The new site is presented as merely informative, while the digitised collections and products are divided between their live broadcasting, or streaming, services and the databases developed to search for the audiovisual collections (digital or not). As regards its vision or mission, it dedicates just one paragraph at the collective level and at no time shows its responsible parties or the organisational chart clearly. It does, however, provide information in the form of FAQs. RTVE, for its part, does not provide an 'About us' page.

3.1.2   Content (Number of Digitised Collections, Type, Coverage, etc.)

The systematisation of the digitised collections, as well as their scope and coverage, present uneven results. In the French and Dutch cases, clear coverage of the digitised collections is defined, integrated in the search tool in the latter case. The RTVE archive, for its part, despite supplying a basic categorisation in the archive, does not provide an overview of these. The 'Temáticas en el Archivo' (Subjects) category leads to recent documents selected and available in the 'A la carta' (on-demand) section. It is clear that the on-demand (streaming) section is prioritised, with no real indication to the user of exactly what has been made available or digitised. As for the historical archive, this required a massive job of selective digitisation of its holdings and, on occasion, it has reached out to users through social networks to ask for suggestions regarding the type of programmes they would like to see online.

In the case of the BBC, its previous website offered a selection of subject-oriented collections and materials. Additionally, users were able to search for programmes alphabetically. Currently, access appears fragmented by collection type, separating access to the different products: image databases (both for commercialisation and those available through Flickr), educational content databases, the BBC Motion Gallery (commercialised through Getty Images), and the massive Genome database, which contains records from 1929 to 2009 of both radio and TV. This programme search engine is a powerful tool for researchers and makes it possible to equip the archive with a tool that serves as an inventory for identifying lost or incomplete programmes. This fragmentation is useful for accessing the content in a specific way, however the role of the archive is less visible on the website.

3.1.3   Contribution to European Projects (Europeana Content Aggregator, EUscreenXL, etc.)

In general, the study reveals low visibility and participation in cooperative projects at the European level. Once again, it is the French and Dutch institutions that show a greater presence on their websites. The Netherlands Institute for Sound and Vision and the French National Audiovisual Institute are the two that best reflect participation in cooperative projects. They provide dedicated pages for them in which users can filter current and past projects in which the institutes have participated. The subject matter is quite varied: ranging from metadata or linked data to preservation, digitisation or creation of audiovisual material through different strategies (storytelling, etc.).

As regards the French National Audiovisual Institute, these projects (Presto, Europeana) appear on a specific list, but they are hidden from the user and not visible from the main categories.

Both the BBC and RTVE, on the other hand, only provide news on projects or workshops, but not a query page. An absence of this type of information is perceived, although their professionals are recognised internationally and their presence at both informal meetings and conferences, like those of the FIAT (International Federation of Television Archives), is shown. In the case of the BBC, and although the partners are not other audiovisual archives, its participation in the RES (Research and Education Space)—a network of digital educational collections belonging to such British cultural institutions as the British Library and the British Museum, published as linked open data—is noteworthy.

3.1.4   Information About Preservation Policies

Another interesting aspect was to assess whether these policies are fully transparent to end users. Evidently, the educational capacity offered by the French and Dutch institutions means that these aspects are developed more intensively. In this case, it is worthwhile to highlight the work of the Netherlands Institute for Sound and Vision, the first audiovisual archive in the world to obtain the Data Seal of Approval for the digital management of its collections.

The BBC, on the other hand, does not provide information on its current website. Nonetheless, on the outdated version, one can still consult the 'Meet the Experts' section, where its professionals discuss these processes. Similarly, as mentioned in the introduction, it shares management of the collections with the BFI (British Film Archive), and therefore this institution, as it has a heritage purpose, does inform users about how these audiovisual collections are selected and preserved. As for the RTVE Archive, information about these policies is not available.

3.2   Access

If the indicators grouped under Transparency heading aimed to analyse the type of information the archive provides to its users, those under the Access heading are designed to measure the different options the Internet user has for accessing the contents of the archive. Therefore, this section assesses the search capacity of their systems and the different strategies for showcasing the value of audiovisual archives.

3.2.1   Query System (Search Capacity, Operators, Clusters, etc.)

In general, the functionality of audiovisual archive search systems does not stand out for its user-friendliness due to the impossibility of performing complex searches.

In the case of the INA, the mandatory nature of the legal deposit provision means that the search, as well as the search fields, have to be more exhaustive. It has the Inathèque, a space that brings together all of its catalogues of holdings: television, radio, radio streaming websites, and written assets. All of this is accomplished through advanced search fields that enable the user to combine terms by date, genre, subject, and using boolean operators.

The Dutch search product offers its collection using a search box, and allows filtering by categories or channels. Likewise, it offers different products depending on whether the needs to be met are of media professionals or related to research or education. Thus, for example, products like AVResearcherXL have been developed, a tool considered to fall within digital media archaeology, which enables exploration of multiple resources and visualisation of metadata for both audiovisual material from the institute's collections and the collections of newspaper archives or of the Dutch Royal Library.

In the case of the BBC, it is necessary to assess each one of the products independently. While the system for commercialising images shares the Getty platform and search engine, proprietary products such as Genome have more elaborate search systems with advanced search criteria. In the case of RTVE, the archive itself allows searching by browsing through the subject categories provided, programmes or decades. It is necessary to go to the on-demand service (streaming of full-length episodes) to be able to perform queries.

3.2.2   Description of the Content (Metadata, Transcriptions, etc.)

The degree of description of the audiovisual archives varies. In the heritage institutions, greater detail in the description of the records is perceived. Thus, the catalogue of the legal deposit of the INA includes an extensive description of each item.

The Dutch institute, in its digitised collections, is the centre that provides the most metadata, characterising each record based on information about the programme, people, subject matter, a summary, and about the archive itself (licensing, etc.).

In the case of RTVE, a detailed description of the content of the digitised resources is not provided. The archived website of the BBC collections does provide a synopsis and the main characters or contributors, whilst, through Getty, the BBC Motion Gallery provides a more technical description.

3.2.3   Transmedia Strategies for Dynamization

An analysis of the results was performed from a qualitative perspective and taking into account the three assessment areas established.

3.2.3.1 Expanded Access

As demonstrated in [1], this is the most common strategy for content. As regards the dissemination of full-length, already-broadcast episodes, this depends on the individual centre and its policies. The RTVE Archive is verifiably the only one that shows this complete content.

One of the most common resources within this item was that of utilising the content of the audiovisual archive for a contextual purpose. It has different applications: commemorations, anniversaries or the utilisation of past events to contextualise current news stories to which they have some relationship.

In this case, both the INA and RTVE and BBC archives share these strategies through Twitter. Thus, the use of Twitter hashtags to remember anniversaries (#onthisday) is common in association with current events that provide an opportunity to call up archival images.

3.2.3.2 Repackaged Content

In this point, an attempt was made to analyse whether current productions based on the reuse of archival documents are disseminated. INA does offer thematic dossiers and new materials created with archival documents. However, the Spanish case is the most noteworthy in terms of the reuse of this content. Thus, with programmes like *Cachitos de Hierro y Cromo*, the RTVE Archive utilises its broadcast, during the week and using the appropriate hashtags, to recall key moments of its broadcast.

3.2.3.3 Ancillary Content

With respect to transmedia narration, understood as the generation of ancillary content, the complete lack of strategies which could more effectively help revalue the holdings for a young audience is striking. This question has its origin in the experience of cultural institutions like the National Library of Spain, which aligned its social media posts with the themes of the fictional television series *El Ministerio del Tiempo* (The Ministry of Time), a strategy that served to increase engagement and, above all, publicise its collections.

Cases have been found of transmedia narratives in materials seen on the French INA and the Netherlands Institute for Sound and Vision. In the majority of cases, they are audiovisual products, created using material from their own archives, which are given added value through ancillary content. Likewise, the Dutch institute, through an innovation hub, develops tools and products which at times reuse archival materials.

4   Conclusions and Future Challenges

The study makes it possible to provide a general overview of the situation of audiovisual archives in the digital context. Using the main transparency and access indicators, a measurement system was devised which brings together important elements for measuring the role of these institutions on the Internet as agents of participation in today's communication system.

In this sense, the preliminary results of this study show that the different conservation structures of each country partly determine the policy of dissemination and international cooperation affecting the holdings. Thus, for example, the French and Dutch institutes, as public entities, maintain websites with organisational charts and developed information. They are essential content aggregators in projects like EUScreenXL and participate actively in discussions about preservation and conservation of European television holdings.

Television audiovisual archives have kept up their efforts to develop digitisation strategies for their holdings, although the level of access to same is very limited and subject to search tools that do not enable their discovery. Thus, while the French INA and the Dutch institute develop more detailed search systems, in the case of RTVE, for example, discovery by the user is very limited because of the dilution of the archive in the 'A la carta' section. In the case of the BBC, a portalisation has been detected in the separated access to the audiovisual collections. Moreover, the development of separate tools, increasingly in open access, such as the recent sound effects bank of the BBC, still in its beta version, is noteworthy. In the British case, these products arise within the RES (Research and Education Space) network, which aims to disseminate the content of these archives in the digital environment.

This educational role of audiovisual archives needs to be debated more intensively. Positive cases, both that of the BBC and those of the INA and the Dutch institute, have been detected. On the other hand, in terms of other strategies to revalue holdings in new ways, the use of transmedia narratives is starting to be seen timidly in some archives and associated with a contextual use or one that evokes historic events. Here it is important to highlight the tremendous work done by the RTVE Archive in its Twitter account.

Television audiovisual archives should constitute a partner in the European communications sphere and see themselves not only as a heritage agent but also one of participation and content creation. This value is asserted in the main forums and professional discussions, such as the Netherlands Institute for Sound and Vision's Audiovisual Think Tank, which calls for, on the one hand, greater cooperation between institutions and, on the other, greater involvement by the professionals in today's communication and educational spheres, in addition to discussing the potential roles of these, such as that of constituting a validated fact-checking agent in source-validation projects [11]. The value of this audiovisual heritage must be taken into account to combat disinformation and post-truth culture.

References

1. Caridad Sebastián, M., Morales García, A.M., Martínez Cardama, S.: The television archives: strategies to showcase their value in the transmedia age. Revista Latina de Comunicación Social **73**, 870 (2018)
2. Pacios Lozano, A.R., Rodríguez Bravo, B., Vianello Osti, M., Rey Martín, C., Rodríguez Parada, C.: Transparencia en la gestión de las bibliotecas públicas del Estado a través de sus sedes web. El profesional de la Información **27**(1), 36–48 (2017)
3. Hidalgo, P.: Preservación del patrimonio audiovisual de televisión. El archivo de Televisión Española (TVE): de los orígenes a la digitalización. Universidad Complutense de Madrid, Madrid (2017)
4. del Valle Gastaminza, F.: Patrimonios visuales y audiovisuales: fotografía, sonido, cine, television. In: Ramos, F., Arquero, R. (eds.) Europeana. La plataforma del patrimonio cultural europeo, pp. 109–132. Trea, Gijón (2014)
5. Van Gompel, S., Hugenholtz, P.B.: The orphan works problem: the copyright conundrum of digitizing large-scale audiovisual archives, and how to solve it. Popul. Commun. **8**(1), 61–71 (2010)
6. Pérez Lorenzo, B.: Fuentes para la producción audiovisual. En: Documentación Audiovisual. Madrid: Síntesis. In: Caridad Sebastián, M., Hernández Pérez, T., Rodríguez Mateos, D., Pérez Lorenzo, B. (eds.) Documentación Audiovisual, pp. 167–197. Síntesis, Madrid (2011)
7. Hidalgo, P.: El archivo de RTVV patrimonio audiovisual de la humanidad. MEI **5**(8), 17–30 (2014)
8. Antón, L., Guallar, J.: Análisis de los archivos audiovisuales en internet de las televisiones autonómicas españolas. Revista española de Documentación Científica **37**(1), e033 (2014)
9. Hagedoorn, B.: Towards a participatory memory: multi-platform storytelling in historical television documentary. Continuum **29**(4), 579–592 (2015)
10. Askwith, I.D.: Television 2.0: reconceptualizing TV as an engagement medium, Massachusetts Institute of Technology (2007)
11. Kaufman, P.B.: Towards a new audiovisual think tank for audiovisual archivists and cultural heritage professionals. Netherlands Institute for Sound and Vision, Hilversum (2018)

Concepts and Models of Analysis of Interactive and Transmedia Narratives: A Batman's Universe Case Study

Jorge Ignacio Mora-Fernández[1,2,3]([✉]) [iD]

[1] Research Group IANCED R+D+C+i Interactive Narrative Arts,
Convergences & Emergences in Digital Cultures,
Universidad Nacional del Chimborazo, UNACH, Riobamba, Ecuador
multiculturalvideos@gmail.com,
jorge.mora@unach.edu.ec
[2] Arthur C. Clarke Center for Human Imagination,
University California, San Diego, CA, USA
[3] Laboratorio de Cultura Digital, Universidad Complutense de Madrid,
UCM, Madrid, Spain

Abstract. This paper collects some core concepts of interactive transmedia storytelling to help to develop the discipline of interactive digital narrative studies and to establish functional models for identifying and analyzing interactive narrative elements in transmedia creations. The term "transmedia" was initially used by Marsha Kinder to refer to the intertextuality among films, animation, TV series and toys for children. Later, Henry Jenkins broadened the concept, using "narrative transmedia" first in *Media Convergence* and more recently in his blog http://henryjenkins.org. The present research introduces a model of analysis to study digital storytelling concepts, narrative elements and media characteristics used for generating transmedia using comics, films, webs, fan videos, and videogames. The model is applied to Nolan´s transmedia practices in his Batman's Trilogy to observe and discuss the narrative remixed framework that renew the current DC Universe. The goal was to develop and applied an original transmedia model of analysis to study and draw conclusions about how the narrative elements of the actions, characters, spaces and times converge with coherence and intelligibility, and how immersive, interactive linear, emerging and circular elements serve to engage the fans. The conclusions present how the remix culture applied to interactive narrative design serve as valuable guidelines in the production of transmedia narratives. New but recognizable creations can evolve from the pre-existent modular narrative elements of characters -familiar in their appearance, abilities or personality- as they perform and interact with new actions in original spaces and times.

Keywords: Digital humanities · Interactive linearity ·
Interactive digital storytelling · Transmedia · Convergence ·
Digital culture and edutainment

© Springer Nature Switzerland AG 2019
Á. Rocha et al. (Eds.): ICITS 2019, AISC 918, pp. 929–943, 2019.
https://doi.org/10.1007/978-3-030-11890-7_87

1 Introduction

The purpose of for this research was to study how narrative actions, characters, spaces, and times have been developed at a transmedia level and applied to the design and the creation of interactive narratives. This current research review and collects important theoretical concepts and propose a practical framework for analyzing interactive and transmedia digital narratives. Based on these concepts a model of analysis is created and applied to one successful transmedia case within Batman´s Universe.

1.1   Concepts of Convergence, Transmedia, and Linearity of the Interactivity

While Marsha Kinder initially used the transmedia concept [1] in referring to transmediatic intertextuality, Henry Jenkins is the one who popularized the transmedia narrative term. He describes it in his blog [2] as follows:

> "Transmedia storytelling represents a process where integral elements of a fiction get dispersed systematically across multiple delivery channels for the purpose of creating unified and coordinated entertainment experience. Ideally, each medium makes it own unique contribution to the unfolding of the story."

In his book [3], Jenkins had previously described the concept of convergence as a "paradigm for thinking about media changes". These changes include the stratification, diversification, and interconnection of media as they converge and then influence the decisions of media producers, politicians, and citizens during the production and consumption of the digital culture.

In response to recent communicative and cultural practices, Jenkins broadened the convergence concept to the transmedia term [2]:

> "Transmedia, used by itself, simply means "across media." Transmedia, at this level, is one way of talking about convergence as a set of cultural practices. We might also think about transmedia branding, transmedia performance, transmedia ritual, transmedia play, transmedia activism, and transmedia spectacle, as other logics."

In this regard, the transmedia narrative generates a universe of lineal and/or circular possibilities where the stories start and finish, receiving feedback from one another; each story can be also experienced apart from the others. In this regard, the linearity of the interactivity is understood by Mora-Fernandez [4] as the coherent, communicative, immersive, and multilevel flow: aesthetic, narrative, and emotional.

1.2   Concepts of Immersion, Types of Interaction, and Agency

Transmedia narratives are interactive because they require the participation of spectators' who have been compiling information from different media. In that sense, immersion and agency are generated because the narrative content must be attractive enough for the spectators to become co-authors, as they actively look for disseminated contents in different media. According to Mateas & Murray [5]: "Immersion is the feeling of being present in another place and engaged in the action therein. Immersion is related to Coleridge's *willing suspension of disbelief* when a participant is immersed in an experience."

Mora-Fernández [6] takes this immersion concept to another level, integrating it with different types of interaction:

> "...the identification and responsibility that the user feels about the development of the narrative actions and the process that the character experiments when the user is able to intercede through an interface with the forms and narrative structures, thanks to the different kinds of interaction (selective, transformative, and constructive) that the hypermedia expression offers."

In respect to the concept of agency, Murray [7] remarks:

> "Agency is defined as an aesthetic pleasure characteristic of digital environments, which results from the well-formed exploitation of the procedural and participatory properties. When the behavior of the computer is coherent and the results of participation are clear and well motivated, the person experiments the pleasure of agency" "...agency could be intensified through a dramatic effect."

1.3   Concepts Related to "Narrative Paradox"

In order to integrate the elements of a transmedia story, previous concepts and narrative paradox must be taken into consideration. In doing so, it is possible to analyze the resulting transmedia phenomena. It is possible to understand how transmedia communicative immersion is achieved through by combining narrative spaces, times, actions and characters, as well as the combination of certain types of interactions (selective, transformative, and constructive). Effective combination of these variables facilitates access to the diffused media. The "narrative paradox" is described from the perspective of Bruni [8]:

> "With the advent of new media and its possibilities for interactivity in the generation and reception of narrative structures, the issue of "narrative paradox" arises, in which the relationship between authorship and interactivity is seen as being inversely proportional......The paradox arises in all its implications with the "empowering" possibilities of digital media and presupposes an ideal of "emancipating" the audience from the "tyranny" of the author."

Transmedia Batman's Universe has a narrative paradox, not only because diverse authors created it but because of its encyclopedic nature (comic series, TV, cinema...). It offers the audience not only many ways into a narrative world and its contents, but also the possibility to exchange information and to generate new contents.

Logically, the audience experience is influenced by the contents available through different media, as well as through narrative strategies and digital marketing influences. To provide an interactive dynamic between authors and audiences to the intelligibility and closure, it's also important to give attention to the concept of emergent narratives taken from Jenkins' definitions [3] and described by Hurup Bevensee and Schoenay-Fog [9] as: "... material through a rich environment and intelligent characters, with which the user is able to associate, interpret, and ultimately construct his/her own understanding of the story".

2 Objectives, Research Questions and Hypothesis

The present research analyzes how successful media products use certain transmedia narrative elements in order to use the results later to produce independent edutainment media. This article concentrates on providing with transmedia narrative model of analysis and with a sample of its application to the universe created by the Batman comics, movies, and videogames. The main objective was to study how the narrative elements of the actions, characters, spaces, and times converge coherently and intelligibly to create linear and circular interactive transmedia narratives.

The fundamental question of this research was: What are the combinations of narrative elements (characters, actions, spaces, and times) and structures that generate major interaction and communicative participation in the recipients? Later on, other questions appeared, such as: What narrative elements do creators and transmedia storytellers use, based on the pre-existent narratives in other media, to remix them and generate new original narratives?

The principal hypothesis was that the uses of secondary actions denoting extraordinary abilities of original main and secondary characters, along with new and recognizable scenarios, work together to create more dramatic moments during the resolution of the main action. These storytelling combinations provide a new remix of narrative practices that generates interaction, immersion, convergence and intelligibility in the transmedia narrative.

3 Methodology

This paper provides with the steps to identify a Transmedia product with the following method and below of how to analyze the converge and interactive design between the different aesthetic and narrative elements.

3.1 Transmedia Narrative Characteristics for the Selection and Delimitation of the Object of Study

By all counts, the Batman Universe is considered very rich in its multimedia expressions: comics, TV, animated shows, movies, videogames, toys, etc. In this respect, it qualified as an ideal subject for transmedia studies and meets the characteristics described by Jenkins [10]:

(1) "The integral elements get systematically dispersed through multiple delivery channels." The Batman Universe narrative is realized in the vast circulation of its comics, TV series, animated shows, movies, videogames, toys, etc.

(2) Transmedia storytelling generates a synergy, as it spreads among and within several media and economic conglomerates. Batman has several clear examples, such the release of the campaign http://www.whysoserious.com [11], an alternative reality game, just prior to the opening of the movie *The Dark Knight* [12]. Both cases helped to fuel the Batman Universe, one via the Internet, which centered on the antagonist Joker and on the fans' participations in the alternative

reality game; and the other offering a *preview* of the movie in a comic, the original media where Batman arose. As a result, http://www.whysoserious.com became a viral marketing campaign [11].

(3) Transmedia narratives are encyclopedic; they are not simply based on individual characters or specific plots. Rather, they encompass complex fictional worlds that can sustain multiple interrelated characters and their stories. Such is the case of the Batman Universe where alternative Batmans, Robins, and Catwomans have their own comics, animation series, and/or movies, while their narratives converge and interact.

(4) The different transmedia narrative extensions may serve a variety of different communicative and marketing functions. Their length can provide intuitive information about the characters and their motivations. The narrative extensions in videogames, in comics based on movies, in role games, in toys, etc., encourage audiences to provide feedback, and they complement the release of new adventures in other media. For example, the final confrontation between Batman vs Superman in the animated version of *Batman: The Dark Knight Returns* [13] helped to inspire the videogame Gods Among Us [14] and late on the movie Superman vs. Batman [15], in March 2016. Then the old comics proliferated again, as well as the fans' video clip remixes. Similarly and previously, the video "Why so serious" [11] and its alternative reality game went viral and had the participation of multiple fans within the multitudinous context of the San Diego Comic-Con. This campaign ended up with 10 million participants in 75 cities as a result of its transmedia strategy, which used websites, interactive games, cell-phones, publications, emails, events, videos and collectibles. Along with traditional advertising, this hugely helped in the selling of all the tickets for *The Dark Knight's* [12] premiere, the highest box office of that year and of all time up until then.

(5) Transmedia storytelling practices may expand the potential market for a preexisting media property by creating different points of entry for different audience segments. In the Batman Universe narrative, the access to different audiences is facilitated: the movies' narratives direct audiences to comics' narratives, the comics to the videogames, and the videogames to the generation of *machinima* narratives. These movies created by fans using videogame engines motivate the companies to create new movies and vice versa. With each new media release, the transmedia phenomenon shows up with its expansive qualities. For instance, this happened with the release of the movie *Batman vs. Superman* [15], and more recently with the *Suicide Squad* [16], *The Lego Batman* [17] and the *Justice League* [18] movies. This research accessed the Batman Universe narrative in retrospection through the series of movies: *The Dark Knight* [12], by Nolan; and the two animated movies based on the comic *Batman: The Dark Knight Return* [19].

(6) Jenkins [10] continues saying that "Each individual episode must be accessible on its own terms even as it makes a unique contribution to the narrative system as a whole. Game designer Neil Young coined the term, "additive comprehension," to refer to the ways that each new text adds a new piece of information that forces us to revise our understanding of the fiction as a whole". For example, the back-story of Taila and Bane, shown originally in the movie *The Dark Knight Rises* [20], is

not explicit in the previous media, giving them a new relationship and a past, a new parallel world in the Batman Universe.

(7) "Transmedia storytelling requires a high degree of coordination across the different media sectors, in projects where strong collaboration (or co-creation) is encouraged across the different divisions of the same company, which involves conceiving the property in transmedia terms from the outset [10]." In this regard, important directors, such as Nolan, coming from independent filmmaking, have succeeded in transmedia productions. It seems like the re-edition and remix of narrative elements in his famous movie, *Memento* [21], taught him how to remix pre-existent narratives so that he became able to generate new and original ones in his Batman trilogy. The remix culture creates and provides feedback for complementary stories in other media.

(8) Transmedia storytelling is the ideal aesthetic way for an era of collective intelligence. Consumers become hunters and gatherers moving back across the various narratives trying to stitch together a coherent picture from the dispersed information.

The present research studies how transmedia narratives are generated in the Batman Universe and how it attracts collective intelligence, exploring it. It ignites interest and curiosity in different audiences in different media so that taken together they generate a collective movement of integrating crossmedia knowledge about the Batman Universe. In words of Hayles, highlighted by Kinder [22], "…clarity about the functions of different media is now more crucial than ever", so they can collectively complement each other.

(9) "A transmedia text does not simply disperse information: it provides a set of roles and goals which readers can assume as they enact aspects of the story through their everyday life," Jenkins [10]. The Lego action figures of Batman provided inspiration for the creation of popular videogames developed in several platforms, such as Wii PS2, PS3, etc. Some of these videogames also inspired fans to create *machinima* Lego movies, and later on the Warner Bros Interactive having observed the success of these machinimas created the *Batman Lego* movie [17].

(10) "The encyclopedic ambitions of transmedia texts often result in what might be seen as gaps or excesses incentive to continue to elaborate on these story elements, working them over through their speculations, until they take on a life of their own" Jenkins [10]. Fan videos on YouTube speculated about the plot during the filming of the *Batman vs Superman* [15], *Justice League* [18], and *Wonderwoman* [23] movies. For example, comics have influenced narrative proposals where Superman and Batman interact, as in *Absolute Power* [24], *Hush* [25] and *The Dark Knight Returns* [19] ... The denominated *cosplayers* are fans dressed as their heroes who interact with others at festivals, releases, or related events within the sub-culture of role games. These kind of fans are even included in the sequences of movies such as *The Dark Knight* [12], which included the *cosplayers* as characters who help Batman. The *cosplayers* in this movie had the bad luck of falling into the hands of the Joker, who wouldn't spare their lives until Batman revealed his real identity. Such continuous dialogue

between interact fictional and real practices and the theme of competition between the fans of Batman and the Joker also appeared in an earlier interactive promotional campaign, www.whysoserious.com [11], which generated a Joker-fans recruit. In addition to the premiere, the Batman worlds' accessible narratives complement each other through time with thematic coherence and intelligibility. All these narrative interactions can be denominated as "interactive linearity."

Because of the complexity of the subject of study, it was necessary to do a multimedia content analysis of the transmedia narrative. As a result, the findings and conclusions can be applied to future educative transmedia narratives.

3.2 Transmedia Narrative Model of Analysis

The model proposed is partly based on and synthesized from previous ones by the author Mora-Fernández [4, 6], and it intends to analyze, in a technical way, the original aesthetic and narrative elements, that when integrated, generate immersion in transmedia creations. It is similar to the de-constructivist model of creation of the production logics that the authors and media companies developed through the design of their artworks and productions, and that were used to communicate the transmedia narrative to the public. Finally, this model aims, through a series of comparative close readings and analyses of the elements contained in the transmedia artworks, to describe the relationship among the expressive, narrative, emotional, and value elements that generate a multisensorial,

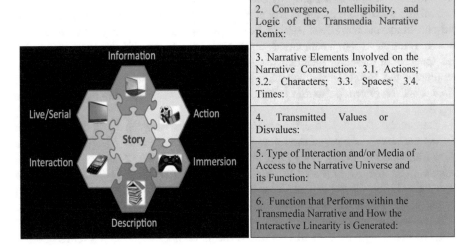

Fig. 1. (Left image) The characteristics of the technological functions serve the central story. (Source: C. Weitrecht). (Right image) Model of analysis for interactive and trans-media narrative elements (Source: Author).

emotional and intellectual immersive narratives, Figure 1 right image. The description of these relationships, between cross-media aesthetics and content creative designs, helps to understand and describe the logical interactive narrative logarithms.

In order to examine more deeply the transmedia combinations of the narrative elements and values communicated through diverse media, a model of transmedia comparative was generated for close readings analysis of the Batman Universe, Figure 1 right image. This model observes and integrates the characteristics indicated by Jenkins [2, 3, 9] that *Christine Weitbrecht* described in her presentation at the TEC Monterrey, in 2013, Fig. 1 left image.

3.3   Application of the Analysis Model to the Subject of Study

Because of the encyclopedic nature of the Batman Universe, it was necessary to select an access point and a delimitation for the observation and analyses of some of the transmedia worlds that have emerged since Batman's first appearance in *The Case of the Chemical Syndicate,* [26] in May 1939. The Detective Comics Mega-Universe has expanded not only within its own universes, with a number of television series including *Arrow, Smallville, Gotham City, The Flash* [27] … but also as the characters generate other joint galaxies such as the one formed by the *Justice League* [18] with *Superman, Batman, Wonderwoman, Flash, Green Lantern, Aquaman,* and *Martian Manhunter.* Anticipation had been created among fans on the Internet for the transmedia of the *Justice League* versus the *Avengers,* a collision between the DC's and the Marvel's Mega-Universes.

Motivated by the premieres of the movies, *Batman vs Superman* [15] on March 2016, *Wonder Woman* [23] on June 2017, and *Justice League* [18] on November 2017, the research focused on one of the most successful transmedia strategies in the DC's Universe by studying the movie trilogy: *Batman Begins* [28], *The Dark Knight* [12], and *The Dark Knight Rises* [20], by director Christopher Nolan.

Nolan's transmedia narrative techniques have served to inspire, provide feedback, and encourage the narratives continuance at the level of comics, toys, TV series, and fans' interactions through campaigns, machinimas, etc. Some of Nolan's trilogy posters and transmedia publicity included the landmarks of the cities where *Batman Begins* premiered: the Eiffel Tower appeared surrounded by bats in Paris' posters and *La Cibeles* and *La Sagrada Familia* in Madrid's and Barcelona's posters respectively, see Fig. 2.

This study looked at the interactive relationships between these films and the diverse narrative elements of the following comics: *Batman Year 1* [29], *Batman: The Dark Knight Returns* [19], *The Killing Joke* [30], *The Man Who Falls* [31], *Birth of the Demon* [32], *The KnightFall* [33], *Contagion* [34], *The Long Halloween* [36], *No Man's Land* [36], *Hush* [25], and *Absolute Power* [24], see Fig. 2 right image. It also looked at the narrative relationships among some of the previously mentioned toys, videogames, publicity campaigns, and shows. The methodology applied the model of analysis, Fig. 1 right side, to all these cross-media in order to develop a series of close readings and comparative analyses of the narrative moments that create transmedia. These cross-media referential moments could be created through a connection, a link, or a convergence, resulting in an intelligible remix that uses one or several narrative or aesthetics elements. From the detailed resulting data, it was possible to create conclusive and coherent descriptions of the transmedia relationships between narrative and aesthetics elements, see the following Sect. 4.

Fig. 2. (Left image) Nolan's Trilogy posters and transmedia publicity including the landmarks of the cities where *Batman Begins* premiered. (Source: Warner Bros Inc.). (Right image) Covers of the analyzed comics that helped to generate transmedia narratives for Nolan's Trilogy in webs, videogames, etc. (Source: D.C Comics).

Given the great number of transmedia narrative moments analyzed and the information collected, the most relevant findings are developed within the discussions and first general conclusions sections.

4   Data Collection, Discussion and Summary of the Findings

Below are summarized the narrative elements of spaces, times, actions, and characters that were most utilized at a cross-media level, from 25 media cases from the Batman's Universe, in order to generate convergence and intelligible interactive transmedia narratives.

In the main action of some sequences, there is general coherence -both in the comic series and the movies. In the comics, *The Man Who Falls* [31], *Batman Year 1* [29], and *Batman: The Dark Knight Returns* [13, 19], a huge bat crashes through similar window panes, making shadows of crosses at the Wayne mansion, and inspiring the birth of the Batman icon. However, there are refinements in the secondary actions: for example, in *Batman Year 1* [29], the bat looms over a bust of Wayne's father, and in *Batman: The Dark Knight Returns* [13, 19], Wayne hears telephonic messages. They always share the following common denominators: the surprise effect of the bat crashing trough glass windows panes, breaking up the human reflections, and triggering Wayne's childhood fears and the hope of being rescued by his father.

Similarly, in the main action of the fall inside the bats' cave, when he's a child, Wayne faces his biggest fear of confronting the bats again and again in all the transmedia narrative sequences: animation, movie, and comics. In *Batman Begins* [28], secondary actions are brought into the narrative: the girl character, Rachel, the persecution, the fight for the arrowhead and the hiding game. *The Dark Knight* [12] movie by Nolan introduces new original characters, props and secondary actions while respecting the primary actions. In *Batman: The Dark Knight Returns* [19] by Miller, Wayne as a child follows a rabbit and falls into the cave, where he sees a lot of bats preceding a gigantic bat; in the animation version, just the big bat's face appears. In each of the cases, the integrity of main actions lends itself to overall coherence, convergence and intelligibility in the transmedia narrative, and the secondary provide innovation.

The comic *Batman: The Dark Knight Returns* [19] makes good use of narrative elements in comics, such as longitudinal framing and typography. While respecting the main actions of the comic and leaving main themes intact, Oliva in the animation version [13] introduces new creative elements such as the rabbit, referring visually to the fall of Alice in Wonderland. The dramatic big animated bat is also introduced as a visual reference to the vampire world. All this contributes to the evolving and expanding qualities of the transmedia Batman Universe. Such dramatizations update and activate new elements for the collective fans' imaginations, while preserving the main actions.

In all the media, the parents' death happens in an alleyway. However, in *Batman Begins* [28] their deaths take place as they leave the opera early through a back door exit. This happened because little Bruce was scared during the opera when he associated the play's characters with his cave fall and the bats' attacks. In both in the comic and the animated versions of *Batman: The Dark Knights Return* [12, 19], the parents' deaths occur after seeing the movie Zorro, as they are walking in an alley.

An anonymous criminal commits the murder in both the comic and the movie *Batman: The Dark Knight Returns* [12, 19]. However, in the movie, the murder occurs during a robbery and is unintentional. In contrast, in Burton's movie, the murderer is a young Joker, who is already enthralled with the theme of death. In the TV show *Gotham City* [27], the main action and the space set (murder and alley) are respected.

Such themes as Catwoman's eroticism and her love-hate relationship with Batman are developed in many different comics. In respect to this transmedia analysis, *The Long Halloween* [35], and *The Knightfall: The Crusade, n503* [33], were the original appearances for the remix done by Nolan in the film *The Dark Knight Rise* [20]. In this movie, Selina Kyle (Catwoman) steals the necklace belonging to Wayne's mother. This prop is part of a brilliant significant romantic and dramatic secondary action, because the necklace reminds Batman of his mother and her assassination. Even more creative is that during the movies the burglary scene and the confrontation between Catwoman and Batman occurs as their public persona (as Selina and Bruce) while in *The Long Halloween* [35] the confrontation is between their secret identities. In addition, the space settings are similar but different; in both, the setting is a private room, but in the movie, it is in Wayne's house, and in the comic, it is in Falcone's house. Another original secondary action that feeds the transmedia narrative is that in *The Long Halloween* [35] comic, both characters dance and kiss each other as Bruce and Selina during the party, before the theft of the jewelry and confrontation, while in the movie the dance takes place after that, during a charity party. Finally, the fighting actions, where Batman and Catwoman protect and seduce each other in a fighting dance, are similarly represented in older comics, *The Long Halloween* [35], *The Knight Fall, & The Knight Quest: The Crusade* [33]. However, in the movie Catwoman saves him.

In *Batman Begins* [28], the remarkable secondary action that boosts the transmedia is the use of an ultrasound artifact that attracts bats. The first time this device is used is in *Batman Year 1* [29], as Batman activates it to escape from the police special forces and to avoid daylight. In the movie, Batman also uses this special bat gadget to save Rachel from the Scarecrow's panic poison, right before taking her to the Bat-cave and injecting her with an antidote, after Gordon enters the building to support him.

In *Batman Begins* [28], Nolan takes Batman's antiheros and secondary actions from the comics, where they appeared separately, and remixes them. He also uses the remix technique in *The Dark Knight* [12] during the interactions between Joker and Two Faces. In *Batman Begins* [28], Nolan integrates the character and actions of Scarecrow, who wants to poison the whole city with his panic gas. This was also inspired by the comic *The Shadow of the Bat-God of Fear* [33], in which Ra's Al Gul wants to destroy Gotham.

There are additional variations and transferences in the space settings, characters and actions between and among different media. In *Batman Begins* [28], Al Gul dies on a metro train, which was the setting for Bane's final battle in the comic, *The KnightFall n666* [33]. In this comic, Bane doesn't die. Another narrative transference is that, in the comic, a new reborn Batman defeats Bane, this is a similar action that takes also places in the movie *The Dark Knight Rises* [20]. During the movie *Batman Begins* [28] Batman lets Al Gul die, a similar action that Batman does with the murdered Arnold Etchison in the comic *The KnightFall: Knight-quest* n508 [33]. In both media scenes, Batman pushes his principle of no killing to the limit. In the movie, when the train is crashing, Batman's action of not saving Al Gul's life for the second time justifies because Wayne previously saved Al Gul from falling off a mountain, which allowed Al Gul's to nearly poison all Gotham later on. This secondary action of a main character being saved on the edge of a mountain was also inspired across media by the comic *The Man Who Falls* [31], which also first introduced Batman's origin. These transmedia dynamics produce curiosity and encourage further exploration of Batman's expanding universe. Fans are invited to broaden their media experience and dive into the different authorial parallel worlds of Batman.

When it comes to the transference combining narrative times, actions and characters' lives, a good example can be observed in *The Dark Knight* [12] movie when Batman saves Gordon's son from Two Faces in a secondary action that increases their friendship. This phenomenon also can be seen in the comic *Batman Year 1*, when a young Wayne jumps from a bridge to catch the Gordon's baby, kidnapped by the Falcones. These narrative transferences and paradoxes maintain the narrative intelligibility, because they are emotionally recognizable by the spectators. The media convergences generate empathy and they invite exploration of the different backgrounds and personalities of the characters. Audiences want to know what happens to whom and the when's and where's among parallel worlds across many media. Convergences make the characters' personalities more complex, as well as more human.

Similarly, transferences and convergences construct lineal narrative interactions by allowing each media to present new creative and meaningful experiences within parallel storylines.

In several comics, Batman's antagonists join forces. When this narrative resource is applied during the transmedia creation, it helps to generate more expectation and the audiences' attention is motivated. Nolan integrates both antagonist characters, Two Faces and the Joker, in his movie. He plays with the narrative structures making the initially co-protagonist, Dent, become co-antagonist, Two Faces, which results in more tension at the end of movie. Similar narrative strategies are used at the level of secondary actions and the resolution changes. In the comic *Batman: The Dark Knight Returns* [18], the Joker dies while in the movie, Two Faces is the one who passes away. Moreover, the movie imports the Joker and Batman's conversation about their alter egos from the comic *The Killing Joke* [30]. In the movie, Joker doesn't die although paradoxically the real actor does as a result of active addiction, getting a posthumous Oscar. With all, the interactive transmedia narratives play off each other as Nolan further remixes in his Trilogy the plots, subplots, characters, and actions from the comic-books, *Hush, The Knightfall, The Killing Joke, No Man's Land, Batman: The Dark Knight Returns* [18, 25, 30, 33, 36], and others, creating something new while honoring the originals.

In remixing the main actions, the communicative immersion of the spectators is increased; at the same time, the original comics get feedback and added value. This immersion intensifies and is more observable when several actions are integrated from epic moments, for example: when Bane in *The KnightFall* [33] tries to generate chaos in Gotham by making the criminals of Arkham Asylum escape through explosions; or when the Scarecrow in the comic *The God of Fear* [33] disperses his hallucinogenic poisonous gas, which boosts the deepest fear of those who inhale it. This last scene also leads to the final transmediatic climax of *Batman Begins* [28], when Ra's Al Gul takes the train to the center of city to explode and evaporate the polluted drain, creating a gas intended to contaminate the city's air, like in the comic *Contagion* [34].

All of these analyzed samples confirm the main hypothesis of this paper: that attractive transmedia narratives mix secondary actions that show extraordinary abilities, main and secondary characters who have complex personalities and creative layers, recognizable original scenarios and new timelines/back-stories. Within the transmedia universe these new remixes add dramatic and new resolutions to the main plot.

Finally, it is necessary to emphasize the acuity of C. Nolan in generating transmedia narratives because:

(1) He brilliantly capitalizes on the words of Gilda, Harvey Dent's wife, on the final page of the comic *The Long Halloween* [35], when she says *"I believe in Harvey Dent!"* by using it in the publicity campaign www.whysoserious.com. During that campaign, Harvey Dent promotes his political career by recruiting followers via a web site called *"I believe in Harvey Dent."* Concurrently Joker tries to attract members to his gang with his own website [11]. Both websites launched before the premiere of *The Dark Knight* [12].

(2) Gilda's quote occurs in the comic right after Dent/Two Faces surrenders to Batman and Gordon on the roof of a building. To maintain the integrity of the

space and the *mise-en-escène,* Nolan uses a similar space setting during the movie *The Dark Knight.* However, during this scene, Dent/Two Faces looses his integrity when he tries to kill Gordon's son. In the comic *The Long Halloween* [35], Dent/Two Faces manages to hold on to some integrity as he keeps fighting the villains and surrenders to the police.

(3) Marony is the prosecutor in the movie, and during Harvey Dent's trial - instead of spraying acid on him as he does in the comic- he tries to shoot him with a broken gun. In these remixes, Nolan creates an intelligible, interactive and linear narrative. The trilogy integrated the best of the developed artistic sub-worlds of the comic series as mentioned before [18, 25, 29–36]. All of these comics have the narrative potentials for further expansion within the media diffusion: cinema, web sites, mobile systems, and videogames.

5 Conclusions

From all these data can be concluded that the model of analysis of interactive transmedia narrative elements (actions, characters, spaces and times) served to reflect and identify in details the construction of effective interactive digital storytelling. It can be also concluded that interactive transmedia narratives provide motivation for the *remix culture* practices between authors and fans using interactive narratives. This generates a *participative culture* around the transmedia characters. The orchestration of the Batman Universe, in the cinema and other media, encouraged fans to participate in the creation of new emergent narratives such as *machinima* movies based on videogames, like *Batman Arkham Archives* [37] or *Injustice Gods Among Us* [38]. This last one had 25 million views on July 24 2018 [38] and its mix of villains inspired the creation of a very crazy Lex Luthor character in the movie *Superman vs. Batman* [15]. His personality was the result of remixing the sharpness of the traditional Lex Luthor with the craziness of The Joker. The universe remixes and the interactive narrative designs continued to evolve within the mentioned last year movies [16–18, 23] and the movies, video games and comics that will follow.

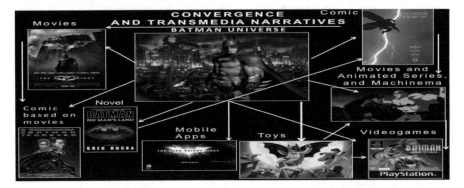

Fig. 3. This scheme summarizes the remixes and dynamics of interactive transmedia narratives across diverse media. More examples at https://vimeo.com/178261749 (Source: Own concept).

Historically, the Batman Universe was inspired initially from the comic-books series, and through the years, it has expanded to include TV-series [27], videogames, and movies. Later on, these cross-media have had other different inspirations thanks to the growing construction of interactive digital storytelling techniques applied to transmedia. Therefore, the pre-existence of serial narratives and wide databases in a media serve as potential narrative worlds in the generation of a new transmedia universe when the *remix culture* and interactive narrative concepts are applied, Fig. 3.

Acknowledgements. This work was supported by the University California, San Diego, Arthur C. Clarke Center for Human Imagination, CA, USA. Universidad Nacional del Chimborazo, UNACH, Research Group IANCED R+D+C+I teractive Narrative Arts, Convergences & Emergences in Digital Cultures, Riobamba, Ecuador. Laboratorio de Cultural Digital, Universidad Complutense de Madrid, UCM, Madrid, & the Prometheus Project, SENESCYT, Secretary of Higher Education, Science, Technology & Innovation of the Republic of Ecuador, by the DIUC, Direction of Research of the University of Cuenca, through the Research Group CICNETART I +D+C, & the School of Philosophy, Letters & Education Sciences, Careers of Social & Digital Communication & Cinema.

References

1. Kinder, M.: Playing with Power in Movies, Television and Video Games, 1st edn. University of California Press, Berkley (1991)
2. Jenkins' Blog. http://henryjenkins.org/2011/08/defining_transmedia_further_re.html. Last Accessed 21 June 2018
3. Jenkins, H.: Convergence Culture. New York University Press, New York (2007)
4. Mora-Fernandez, J.I.: Artecnología en cine interactivo: algunas categorías, interfaces, estructuras narrativas, emociones e investigaciones. In: ArTecnologia. Arte, Tecnologia e linguagens Midiáticas. Buqui, Rio de Janeiro (2013)
5. Mateas, M., Murray, J.: A preliminary poetics for interactive drama & games, & from game-story to cyberdrama. In: Wardrip-Fruin, N., Harrigan, P. (eds.) First Person: New Media as Story, Performance & Game, pp. 2–33. The MIT Press, Cambridge (2004)
6. Mora-Fernández, J.I.: La interfaz hipermedia: el paradigma de la comunicación interactiva. Modelos para implementar la inmersión juvenil en multimedia interactivos culturales. (Videojuegos, cine, realidad aumentada, museos y web). SGAE. Fundación Autor. Ediciones Autor, Colección Datautor, Madrid (2009)
7. Murray, J.: Inventing The Medium. Principles of Interaction, Design as a Cultural Practice. The MIT Press, Cambridge (2012)
8. Bruni, E., Baceviciute, S.: Narrative intelligibility & closure in interactive systems. In: Koenitz, H. et al. (eds.) ICIDS 2013. LNCS, vol. 8230, pp. 13–15, 18, 22. Springer, Heidelberg (2013)
9. Jenkins' Blog. http://henryjenkins.org/2013/02/what-transmedia-producers-need-to-know-about-comics-an-interview-with-tyler-weaver-part-two.html. Last Accessed 21 June 2018
10. Jenkins' Blog. http://henryjenkins.org/2007/03/transmedia_storytelling_101.html. Last Accessed 21 June 2018
11. Warner Bros Inc. Transmedia Campaign. http://www.whysoserious.com and https://www.youtube.com/watch?v=VpuC7HhCPWA. Last Accessed 21 June 2018
12. Nolan, C.: Movie The Dark Knight. Warner Bros Inc., Los Angeles (2009)

13. Oliva, J.: Movie Batman: The Dark Knight Returns I & II. DC Universe Animated Original Movies, Los Angeles (2012)
14. Boon, E., Grinsfelder, D.: Injustice. Gods Among Us. Warner Bros Interactive Entertainment, Los Angeles (2013)
15. Snyder, Z.: Movie Batman vs Superman: Dawn of Justice. Warner Bros Inc., Los Angeles (2016)
16. Ayer, D.: Movie Suicide Squad. Warner Bros Inc., Los Angeles (2016)
17. McKay, C.: The Lego Batman Movie. Warner Bros Inc., Los Angeles (2016)
18. Snyder, Z.: Movie Justice League. Warner Bros Inc., Los Angeles (2017)
19. Miller, F., Constanza, J., Janson, K.: The Dark Knight Returns. DC Comics, New York (1986)
20. Nolan, C.: Movie The Dark Knight Rises. Warner Bros Inc., Los Angeles (2012)
21. Nolan, C.: Movie Memento. Summit Entertainment, Los Angeles (2000)
22. Kinder, M., et al.: Transmedia Frictions: The Digital, the Arts, and the Humanities. University of California Press, Oakland (2014)
23. Jenkins, P.: Movie Wonder Woman. DC Entertainment and Warner Bros Inc., Los Angeles (2017)
24. Loeb, J., Pacheco, C., Merino, J.: Absolute Power. D.C Comics Inc., New York (2005)
25. Loeb, J., Lee, J., Williams, S.: Hush. D.C Comics Inc., New York (2003)
26. Kane, B., Finger, B.: The Case of the Chemical Syndicate, Detective Comics n27. D.C. Comics Inc., New York (1939)
27. D.C. Comics Inc. Television Series Based on DC Superhero. https://en.wikipedia.org/wiki/List_of_television_series_based_on_DC_Comics. Last Accessed 21 June 2018
28. Nolan, C.: Movie Batman Begins. Warner Bros Inc., Los Angeles (2005)
29. Mazzucchelli, D., Miller, F.: Batman Year One. D.C. Comics Inc., New York (1986)
30. Moore, A., Bolland, B.: The Killing Joke. D.C. Comics Inc., New York (1988)
31. O'Neill, D., Giordano, D.: The Man Who Falls in Secrets Origins. D.C. Comics Inc., New York (1989)
32. O'Neill, D., Breyfogle, N.: The Birth of the Demon. D.C. Comics Inc., New York (1992)
33. Dixon, C., Nolan, G., Barreto, E.: The KnightFall series. D.C. Comics Inc., New York (1993)
34. Grant, A., Dixon, C.: Batman Contagion. D.C. Comics Inc., New York (1996)
35. Loeb, J., Sale, T.: The Long Halloween. D.C. Comics Inc., New York (1996)
36. O'Neill, D., Gorfinkel, J.B.: No Man's Land. D.C. Comics Inc., New York (1999)
37. Hill, S., Dini, P.: Videogame Batman Arkham Archives. Warner Bros Interactive Enterntainment, Los Angeles (2009)
38. Boon, E., Grinsfelder, D.: Videogame Injustice Gods Among Us. Warner Bros Interactive Enterntainment, Los Angeles (2013)

Transformation of Andean Cinema in Latin America: Identity and Prostdrama

Miguel Ángel Orosa$^{(\boxtimes)}$, Santiago F. Romero-Espinosa, and Jose A. Fernández-Holgado

Pontifical Catholic University of Ecuador, Ibarra Campus, Ibarra, Ecuador
{maorosal, sromerol}@pucesi.edu.ec, xholgado@gmail.com

Abstract. The object of this research focuses on the cinematography of the Andean countries (Ecuador, Bolivia and Peru, excluding Colombia due to the size of its market) and their architectural transformations. Thus, identity and architecture will be related to the cinema, both by their textual and narrative-audiovisual aspects. To perform the analysis, three models will be exhibited: dramatic, postdramatic and audiovisual postdramatic. The results show that in the three countries there are clear postdramatic manifestations in their cinematography, especially in some of their most outstanding films.

Keywords: Andean cinematography · Andean postdrama · Cinema Ecuador Bolivia Peru · Identity and postdrama in the Andean cinema

1 Introduction and Status of the Issue

1.1 Object of Study

The object of this research focuses on the cinematography of the Andean countries and their architectural transformations. Thus, identity and architecture will be related to the cinema, both in their textual and narrative-audiovisual aspects. Both concepts, identity and architecture, will be the object of study and co-development in this paper. Colombian cinematography will be excluded from this work for two fundamental reasons: firstly, because the Andean identity culture of audiovisual nature is stronger and incisive in Ecuador, Bolivia and Peru than in other countries and geographical areas such as Colombia. Secondly, because the latter's cinematographic market has not the emergent character of the three Andean countries anymore, as it is a segment of enormous proportions and features, that undoubtedly deserves its own study.

We can find works on Andean cinematography from different perspectives: sociological, anthropological, cultural, gender, religious or from the point of view of axiology, philosophical-thematic, national or, of course, commercial and financial; but, so far, we have not observed approaches of this type of studies with respect to the aspects of identity – which are no longer related to the nation-state – and the strictly aesthetic ones [1].

© Springer Nature Switzerland AG 2019
Á. Rocha et al. (Eds.): ICITS 2019, AISC 918, pp. 944–953, 2019.
https://doi.org/10.1007/978-3-030-11890-7_88

1.2   Films Under Study and Their Evolutionary Line

The films that have become the object of analysis are the following (Table 1):

Table 1. Films under study and their evolutionary line

País	Categoría	Año	Nombre película	Autor
Ecuador	90's	1990	La Tigra	Camilo Luzuriaga
		1996	Entre Marx y una mujer desnuda	Camilo Luzuriaga
		2005	1809–1810: mientras llega el día	Camilo Luzuriaga
	Young Ecuadorian Cinema	1999	Ratas, ratones y rateros	Sebastián Cordero
		2006	Crónicas (2004), Rabia (2009) and Pescador (2011)	Sebastián Cordero
		2006	Sara la espantapájaros	Jorge Vivanco
		2011	Qué tan lejos	Tania Hermida
		2011	En el nombre de la hija (2011)	Tania Hermida
		2011	A tus espaldas	Tito Jara
		2013	Con mi corazón en Yambo	Mª Fernanda Restrepo
		2016	La muerte de Jaime Roldós	Manolo Sarmiento and Lisandra I. Rivera
	Short films, Rupai Corporation	1991	En algún meandro de la Estigia, and Árbol de vida	Fernando Mieles
			Kuychi Pucha, Antunio, Malky and Un buen día	Alberto Muenala
	Experimental films	1992	Opus Nigrum	Fernando Mieles
		1995	Metro cúbico	Sebastián Cordero
		2002	L'objetiv	Pancho Viñachi
		2002	Silencio nuclear	Iván Mora
		2003	El correo de las horas	Sandino Burbano
		2015	Quijotes negros	Sandino Burbano
		2010	Prometeo deportado	Fernando Mieles
Perú		1984	Gregorio	Chaski Group
		1988	Juliana	Chaski Group
		2005	Madeinusa	Claudia Llosa
		2009	La teta asustada	Claudia Llosa
		2015	Hija de la Laguna	Ernesto Cabellos Damián
Bolivia		1977	Chuquiago	Antonio Eguino
		1984	Amargo Mar	Antonio Eguino
		1995	Jonás y la ballena rosada	Juan Carlos Valdivia
		2016	Viejo Calavera	Kiro Russo

2   Textual and Audiovisual Methodologies of Drama and the Identity of Postdrama

Regarding the models of analysis that are going to be used, it is worth highlighting that three different paradigms have been created, through which the textual and audiovisual cinematography of the three countries -Ecuador, Peru and Bolivia- will be presented.

The first one is the dramatic paradigm, typical of the classical stage, which will be used to analyze texts or scripts from a unitary point of view, that is, a synthetic point of view. Here, there is a reference to the classical-dramatic composition (Table 2). The second paradigm, the one that corresponds to the textual postdrama, will be used for the purpose of analyzing the scripts of the films under study from the point of view of dispersion or juxtaposition (Table 3). The third will contribute as a model of audiovisual analysis from a perspective that is also postdramatic (Table 4). See in this sense Orosa and López [2].

This model or dramatic paradigm (Model 2) has the purpose of generating order in the field of chaos -postdrama-, giving an organizational sense to its nonsense. Here, the reference is made to the order of the film, a possible internal or inner organization of the script itself. Without order everything is merely a juxtaposition of ideas.

Regarding this model of dramatic canon or Western Dramatic Common Space, which now proceeds to be broken down, it should be noted that it consists of five different segments: the plot, the dramatic organization, the dramatic tension, the narrative dimension in the field of drama and, finally, the technical-formal organization. In this sense, our work can be consulted in Orosa [3, 4].

Therefore, in relation to the plot, the canonical aspects or variables of dramatic use in the field of the Western Common Space are related, first of all, with the selection of the facts of the story [5]. Also the most used tools to build a script are the turning points and the emotional projections. In reference to the dramatic organization, the most significant issues are supported by many authors, see among others García Barrientos [6]. With regard to the dramatic tension, can be consulted in this sense *Poetics*, of Aristotle [7]. This organizational disposition, the dramatic tension, revolves around two fundamental axes: activation and conflict.

The use of narrative elements within the drama takes place from the very origins of the tragedy. There are actions that the author prefers not to show in the scene (external story) but that the viewer must know and, therefore, these events are narrated within the work (not staged), only narrated. This basic use about the narrative elements within drama can be accompanied by entire organizations or intensive narrative arrangements throughout the dramatic play/script as a further element.

Finally, the most abstract of the architectural dispositions within drama is the formal-technical one. It underlies the whole dramatic-narrative organization and would seem like the very foundations of construction. See in this sense the extensive explanations on the formal space in various sections of Orosa [3, 4].

With all these segments, it has been explained what the dramatic composition is, its tools and variables, intended to give a unitary sense, a single emotional and intellectual message to the artistic work, to provide order and meaning to the architecture of the play/script.

Table 2. Dramatic analysis panel

	Category	Component
1	Plot	Plot: turning points. Emotional projections Selection of facts (story): not all the life or legends of hero, but only the events necessary to tell the theme of the work
2	Dramatic organization	External: blocks or acts Middle: themes, movements (exposition, conflict, resolution) Unity of action (a protagonist who seeks an objective)
3	Dramatic tension	Conflict or confrontation (obstacles that separate the protagonist from their objective) Activation: style, spectacular speech, intellectual discourse, characters, music
4	Narrative organization within drama Type of narrative organization	Buildings or narrative organizations involved in the field of drama (arrangements/organization of the "novel" that appear within the drama)
5	Technical-formal organization	Situation, formal drawing, intensity, duration, colors and textures, tempo, story

Source: Orosa y López-López [2]

Precisely, postdrama –as in this study is conceived– would have characteristics that could be considered variables of an open model; its manifestations and parameters would inhabit within the chaotic world of the waters, namely in the field of nonsense, parataxis. Going into the characteristics or patterns of postdrama –see in this sense Orosa [8], and López-Antuñano [9]–, we believe that these variables that belong to chaos are born and have their origin in the chronotope that belongs to this artistic trend. In this trajectory we could mention the following patterns:

Table 3. Postdramatic analysis panel

	Category	Component
1	Time	Absence of chronological or linear time's Temporal fragmentation, of characters, of architectures… Kairos or present time Disregard or controversy over the concept of "reality" Positions in conditional or go to potential ones
2	Theme and text	Textual autarchy (the predominance of the text is denied) Visuality and multidisciplinarity of the show besides the text Thematic fragmentation or dramatic fragmentation of actions on stage Juxtaposition of ideas against organic unity Apparent irrationality, self-referentiality Collage Multiplicity and multiperspectivism Repeating the opposite of what it looked like before

(*continued*)

Table 3. (*continued*)

	Category	Component
3	Characters	Constitution of characters tends to fragmentation or puppets in the hands of the author
4	Scenes	Inconsistency of disciplines and events Non-hierarchical (horizontality) of topics or speeches Sensory communication with the viewer Relations between text and scenes are autonomous and noninterpretive Reiterations of scenes Microscenes Daily routines and micro-stories
5	Dialogues and language	Organizations and non-logical provisions The truth and the exclusivity of the word are questioned No causal logic
6	Participation and result	Sensitive and emotional communication with the viewer Reflection on the world

Source: Orosa y López-López [2]

The third model announced above, related to the audiovisual narrative characteristics of the postdrama, would consist of the following variables:

Table 4. Analysis panel of the postdramatic audiovisual narrative

	Category	Component
1	Point of view	From the use of the image (camera): • subjective shooting • multifragmentation of shooting in the scene • change from the subjective shoot to the objective one in the same shot • rupture of the fourth wall • dream sequences From the use of sound (linked to impossible or innovative chronotopes): • voice off or voice over (to express the character's thoughts) • prolonged silences • sound suspension • change from extradiegetic sound to diegetic sound
2	Dynamism	Camera movements: • tendency to subjective audiovisual and emotional organicism Editing: • multifragmentation • variation in the speed of the sequences (acceleration, slowdown) for reasons linked to the postdrama cronotope Dynamic sound treatment: • music • fluency in the dialogues

(*continued*)

Table 4. (*continued*)

	Category	Component
3	Postproduction	Rupture of the technique of invisible assembly and of the splices. Fragmentation of the chronological line abruptly, associated with a sensation of present time: • ellipsis, flashback, flashforward by cut • intentional crossing the action lines • jump cut • internal rhythm of the shoot • sequence shot and non-academic narratives
4	Image	• Horizontality of the disciplines of image and text • Importance of an anti-classical aesthetic in the composition of the shoots, in the lighting
5	Genders	Hybridization of genres, tendency to self-referentiality, intertextuality and referential citations to other works and disciplines (amateur cinema, avant-garde movies, video clips, TV, literature, etc.), mix of formats, creativity, not copy
6	Reception	The spectator as an autonomous and active subject that gives meaning to the images and interprets the story. Participation of the public in the creation of the work

Source: self made

3   Results and Discussion

3.1   Audiovisual Narrative Analysis

Through the analysis of the Bolivian cinematography produced at the end of the 20th century, certain relevant characteristics were obtained in the audiovisual screening as a result of the application of the model cited in Table 4.

Films like *Chuquiago* (1977), by Antonio Eguino, have a clear influence of the *nouvelle vague*. Like the French authors, the director takes the camera to the street to pay tribute to the city of La Paz through four stories interconnected to the *Rashomon* style (1960). Analyzing the filmic point of view, Eguino uses the technique of hand held camera, following the protagonists and turning this camera into a character. This style of execution gives verisimilitude to the story, it is recorded in real spaces with natural lighting and passers-by are mixed with the actors of the story, increasing with it the sensation of realism. In the editing section, the film uses long-duration shots in which movement and fluid dialogues create an internal rhythm that helps the viewer maintain attention.

Another observed quality is the use of the cutting technique in a perceptible way, the jump cut, popularized by the French filmmaker Jean Luc Godard. The use of archival images of the police, skillfully combined with those of the actors filmed with a hand held camera, manages to generate in the spectator, thanks to the Kulechov effect, the idea of the repression of the police towards the protagonists, even though the security force was never there.

Undoubtedly, the most outstanding film in terms of appropriation of metamodernity and postdrama is *Viejo Calavera*, by Bolivian director Kiro Russo (2016). The author's vision takes us to some abstract formal spaces that, although they come from the techniques of drama, their use is absolutely metamodern and their originality and prodigious talent captures our attention. Fantastic appearances, oniric uses of the editing, spectacular lighting in the technical sense of the term, but of purely postdramatic use. Subjective plans and multifragmentation from different points of view or sequences of prolonged silences, suspension of sound and a new encounter with a very current and authentic production that rarely projects such success in our days.

The Peruvian cinema of the late twentieth century possesses characteristics which are very similar to those of Bolivian cinema of the same period, due to the budgets of film productions. Feature films from the late 20th century, such as *Gregorio* (1982), directed by Fernando Espinoza, Stefan Kaspar and Alejandro Legaspi, members of the Chaski Group, and *Juliana* (1989) by Espinoza and Legaspi, show once again the influence of the *nouvelle vague* on the Latin American cinema. The use of hand held camera, even without hiding the device, and the recording in real locations of the actors mixed among the people of the street, show a clearly documentary style. This hybridisation of genres between fiction and documentary is evident when in the middle of the fiction, interviews to the protagonists who look directly at the camera are introduced, and therefore to the spectator, causing the palpable rupture of the fourth wall.

In the 21st century, a new generation of directors emerges in Peru. They have been trained abroad and have been able to combine the Andean idiosyncrasy with the western vision, in spite of the fact that the main theme continues to be social denunciation [10]. As far as audiovisual narration is concerned, filmmaking moves away from the Hollywood model to become something more personal. The technique of hand held camera is abandoned and, although it is also shot in natural locations, the audiovisual production suffers a stylization since there is a preoccupation for the staging and the use of a tripod, which breaks with the sensation of realism that characterized the Peruvian films of the previous time. The duration of the shots is still long, but with much more harmonic movements, which invite the viewer to reflect; this sensation is reinforced by long silences that repeatedly appear in some sequences. The frames are precious and run away from the classical composition by cutting the figure where it is not habitual and leaving more air than usual. With regard to editing, innovation is sought by abandoning the technique of shot reverse shot in the dialogues. In the documentary genre this stylization is also observed, which benefits the image and which sometimes seems to cross the border of fiction.

As in the case of Bolivia, the work that most accommodates and appropriates the characteristics of postdrama is the documentary *Hija de la Laguna* (2015), by Ernesto Cabellos. All the characteristics of the postdrama announced in Model 3 and 4 appear in this work that exudes talent and authenticity.

The Ecuadorian cinema shares features with the rest of the Andean filmographies, although different properties are observed. It opts for a more Hollywood-style and classical-style, which evolves towards a postmodern independent film style, adapted to its idiosyncrasy. For production reasons, it is usual to use natural locations in the shootings, but without integrating the actors with the passers-by, so the sensation of verisimilitude is lost.

A differentiating feature of Ecuadorian cinema is the greater presence of dream sequences that represent the imagination or the dreams of the characters in a subjective manner; these sequences are illustrated with an audio distortion or suspension of sound that increases the feeling of unreality. In order to achieve the identification of the viewer with the characters, hand held cameras are used, filming a subjective shot from the point of view of the character. In addition, the vicarious experience of the spectator is reinforced through hyperfragmentation; the camera is placed in multiple points of view, thus facilitating the identification of the spectators with different characters. Another characteristic feature of the Ecuadorian films is leaving an open ending to the interpretation of the spectators.

There is a new type of emerging cinema in Ecuador, the Kichwa cinema, made by people who belong to this ethnic group and that seek to represent their worldview. The Andean films treat "indigenismo" in an almost veiled way, the characters are usually secondary and marginal and with them the stereotypes are reinforced. At present, the Ecuadorian non-governmental organization Corporación Rupai, through its production company, breaks with this tendency and empowers the indigenous by granting them a privileged and leading role, representing their traditions. This cultural wealth is not accompanied by the audiovisual production of its works that, at the moment, respond to a simple and academic audiovisual narrative while they continue in the intense search for some feature of differentiating production.

3.2   Textual Analysis of Drama and Postdrama

Viejo Calavera is the most representative work of Bolivian cinema in constructive terms of a postdramatic nature. It plays very well with the viewer's emotional communication, it is a film that exudes a not feigned authenticity, that comes from the heart of the director. The aesthetics, the *mise-en-scène*, the fragmentation of his discourse, his dispersion do not represent life but rather put a piece of life before our eyes, play with the most abstract of art (form, drama) but in absolutely updated terms. It breaks again and again with linearity and textual autarky, dominates the collage effect, with incredible lighting; it does not work in a hierarchical way and masterfully manages the microscenes and contains a lot of daily routines. The use of unreason is splendid and its reflection on the world...

Of great value, in Peru, *Hija de la laguna*, by Ernesto Cabellos, stands out. It is a documentary of 2015. The most significant elements in this documentary are the contemporary variables, described in Tables 3 and 4, which respond to the sensitivity of our time and are used in a masterly way.

The Ecuadorian creation that strongly shows a postdramatic label corresponds to the film *Entre Marx y una mujer desnuda* (1996), directed by Camilo Luzuriaga. Based on the homonymous novel by Jorge Enrique Adoum, this film confronts several temporal levels and different frames of meaning. It is evident that it introduces some notes coming from the dramatic stage into the discourse, that distill order in the composition of the film, such as the strength or intensity of the plot, of a somewhat nostalgic aroma, and the use of various elements of drama tradition such as the conflict (not necessarily used to the way of drama). As the notes of the current postdrama, this movie shows some magical worlds, fantastic spaces, the disintegration of reality and narrative linearity, some symbolic and imaginary landscapes, the intertexts or, of course, metafiction. Its incidence in the nonsense and its constant and ironic question for the reason (logos) is remarkable.

4 Conclusions

Throughout this paper, there have been references to the development of a classical canon of dramatic feature, or Western Common Dramatic Space, which has a tendency to reach an order, a composition (Model 2). The postdrama has its own characteristics (Model 3) that are not present in the previous dramatic model. The fundamental segments belonging to both models have been clearly drawn and serve as a point of reference to analyze (methodology) what happens in the cinematography of the three Andean countries (Ecuador, Boliva and Peru) from the points of view addressed in both models: order and chaos, tradition and metamodernity.

The models of drama and postdrama (audiovisual and textual) applied to the study of the cinematography of the three Andean countries set the basis for future studies to conclude that the audiovisual productions of these three countries enjoy the presence of current postdramatic elements linked to their indigenous and postmodern worldview, especially in the most representative films, from the ones mentioned throughout this paper: *Entre Marx y una mujer desnuda*, *Hija de la laguna* and *Viejo Calavera*.

The models and results used with a methodological nature in this study will also help foresee the future trajectories of the audiovisual industry to enhance the business decision-making processes; also, from an artistic point of view will contribute to the development of future artistic tendencies in concordance with the society of their time and their consequent aesthetic needs.

References

1. Power, K.: Descifrando la glocalización. Editorial Huellas. Búsqueda en Artes y Diseños, Mendoza, Argentina (2003). ISBN 1666-8197
2. Orosa, M.Á., López-López, P.C.: Postdrama culture in Ecuador and Spain: methodological framework and comparative study. Comunicar Revista Científica Iberoamericana de Comunicación y Educación **57**, 39–47 (2018)
3. Orosa, M.Á., López-Golán, M., Márquez-Domínguez, C., Ramos-Gil, Y.T.: El posdrama teleserial norteamerciano: poética y composición (Cómo entender el guion de las mejores series escritas para la televisión en los Estados Unidos). Revista Latina de Comunicación Social **72**, 500–520 (2017). https://doi.org/10.4185/rlcs-2017-1176. http://www.revistalatinacs.org/072paper/1176/26es.html
4. Orosa, M.Á.: El cambio dramático en el modelo teleserial norteamericano. Publicia, Saarbrücken (2012)
5. Alonso, J.L.: La estructura dramática. Las puertas del drama, volumen **10**, 4–9 (2002)
6. García, J.L.: Cómo se comenta una obra de teatro. Síntesis, Madrid (2001)
7. Aristóteles, P. (ed.): trilingüe de Valentín García Yebra. Gredos, Madrid (1974)

8. Orosa, M.Á., Romero-Ortega, A.: Ecuador, the non-communication: postdrama or performance? In: Rocha, Á., Guarda, T. (eds.) Proceedings of the International Conference on Information Technology & Systems, ICITS 2018. Advances in Intelligent Systems and Computing, vol. 721. Springer, Cham (2018)

9. López-Antuñano, J.G.: La escena del siglo XXI. Asociación de Directores de Escena de España, Madrid (2017)

10. Díaz, C.: Cine Peruano, Evolución del cine peruano, relación entre los argumentos y la realidad nacional. (1980–2010). (Final Project). Universidad Nacional de San Agustín. Facultad de Psicología, Relaciones Industriales y Ciencias de la Comunicación. Perú (2016)

Five Ethical Challenges of Immersive Journalism: A Proposal of Good Practices' Indicators

Sara Pérez-Seijo[(⊠)] and Xosé López-García

Faculty of Communication Sciences, Universidade de Santiago de Compostela,
Campus Norte, Av. de Castelao, s/n, 15782 Santiago de Compostela,
A Coruña, Spain
{s.perez.seijo,xose.lopez.garcia}@usc.es

Abstract. Immersive Journalism brings new ethical challenges because some practices and procedures go against conventional journalistic standards. Therefore, the aim of this study was precisely to question decisions and choices taken by professional journalists when they produce and edit immersive content like 360° video reports or virtual reality reconstructions of news. As a result, five main ethical challenges of Immersive Journalism have been identified: image integrity; reconstruction of news; sources and staging; role of the journalist and/or recording team; and sensitive content. Each of these is accompanied by a proposal of good practices to deal with these challenges and to promote reflection among professionals.

Keywords: Immersive Journalism · Ethics · 360-degree video ·
VR journalism · Journalism 360 · Virtual reality

1 Introduction

Journalism cannot be understood without Ethics. Internet opened up new challenges for Journalism [1] and the constant digital and technology innovation forces media to reconsider and to adapt the traditional principles and standards to deal with the new story forms or emerging storytelling technologies: memes, 360-degree video, newsgames, CGI content, GIFs, virtual reality, artificial intelligence and even, among others, publishing on social media. But professional journalists should be honest and act with integrity, regardless of media type, content, technology used or, as the Society of Professional Journalists states in its Code of Ethics [2], format and speed.

Professional conduct and ethics' codes were created to guide journalists in its daily work, to determine ethical limits of professional procedures and decisions and thence to protect audience from malpractices [1]. There are many ethical codes. Some of these include core principles for all media and professional journalists, like the codes of the Society of Professional Journalists [2], The Poynter Institute [3] or the Society for News Design [4]. But there are also other focused on specific fields such as the Radio Television Digital News Association [5] and The National Press Photographers

© Springer Nature Switzerland AG 2019
Á. Rocha et al. (Eds.): ICITS 2019, AISC 918, pp. 954–964, 2019.
https://doi.org/10.1007/978-3-030-11890-7_89

Association [6]. Nevertheless, Filak [7] identifies several common primary tenets: honesty, accuracy, diversity, compassion, independence and accountability.

But new technologies like artificial intelligence, augmented reality or virtual reality (VR) have brought further challenges and demand to adjust the conventional standards to avoid bad practices. Even so, core values are still the same. In this document, 360° videos and VR projects published and produced by media are analyzed from a critical and ethical perspective. The use of these resources and technologies to tell news stories was called Immersive Journalism by De la Peña et al. [8]. They adopt this concept to refer to "the production of news in a form in which people can gain first-person experiences of the events or situation described in news stories".

VR recreations based on actual facts and 360° video reports are examples of immersive contents that are of interest to media. But it should be noted that the most widespread format is the 360° video [9]. User can experience this kind of contents through several means [9–12]: on a desktop by accessing online platforms; or on a smartphone by moving the device around, by touching the screen to explore the scene or with a head mounted display, such as Google Carboard -popular for its low cost-, Samsung Gear VR or Oculus Go.

The aim of Immersive Journalism is to enable viewers experience place illusion. So the head-mounted displays play a fundamental role. Couple with the thought that some immersive works can enhance empathy [13, 14], several ethical challenges arise due to the journalists' practices and procedures to achieve that [12, 15]. Kathleen Bartzen [16] and Tom Kent [17] were among the first to explore the ethics of Immersive Journalism. Issues like the accuracy of animated recreations, digital manipulation and image integrity or even the cost of these productions were put on the table. Some of these ethical challenges, like the image manipulation or video editing, are commonly discussed in Journalism. But others are characteristic of the immersive narrative, on which Lester [18] reflects on taking Kent's factors as the reference.

Marconi and Nakagawa [19], authors of the Associated Press' report *The Age of Dynamic Storytelling. A guide for journalists in a world of immersive 3-D content*, think about the maintaining of journalistic standards beyond the possibilities of VR technology. Nevertheless, the absence of specific standards or ethical guidelines for Immersive Journalism becomes an urgent concern [20].

2 Methodology

Current codes of ethics include standards on image and video production and digital editing. They point out that professional journalists should safeguard graphical integrity of the image and act honestly when they publish the content. Nevertheless, the codes of the main media or professional associations and societies do not yet have specific guidelines for non-fiction 360° videos and VR experiences, although some standards on reconstructions can be applied to both.

This document belongs to an author's larger study about ethical challenges in Immersive Journalism and, taking as a reference a first exploratory study about common malpractices in Immersive Journalism [21], it suggests an initial proposal of good practices indicators by producing and editing non-fiction immersive contents. The objective of this study is to set a series of ethical guidelines to guide professional journalists, or at least to provoke reflection on the challenges and risks of 360° and VR immersive narratives. But some of the indicators proposed clash with the aim of place illusion that Immersive Journalism pursues. It is worth stressing that this document is not intended to be a code of ethics, but a proposal of several recommendations in pursuit of a better and more ethical Immersive Journalism.

The list of indicators proposed is the result of a rhetorical review of several 360° video reports and VR experiences published by media from all over the world. Furthermore, each one of these indicators has been suggested taking as a reference not only the codes of ethics of the above-mentioned professional entities, but also specific media guidelines.

3   Results and Discussion

The aim of this study was to reconsider and to evaluate the ethical limits of media and journalists when they produce and edit non-fiction 360° videos or VR experiences based on true events. As a result, this document presents a proposal of basic indicators of good practices to deal with these challenges and to promote reflection among professionals.

3.1   Image Integrity

Manipulation of images is one of the main ethical debates that concerns visual Journalism; a matter presents since images are used by media. First, pictures; then came videos. The 21st century brought the visual explosion along, but the new technological advances increase the dangers of digital manipulation. By using software like Adobe PhotoShop, journalists can adjust the color or the contrast of the image, but also can add or remove elements from this. Actions like these mean that the reality, as it was captured, changes and this goes against the core value of honesty and the good journalistic practices. For these reasons several ethical codes expressly forbid altering filmed or photographed realities by digitally adding or subtracting elements from the original image.

In 1996, the Society of Professional Journalists (SPJ) included in its code the following standard: "Never distort the content of news photos or video. Image enhancement for technical clarity is always permissible. Label montages and photo illustrations" [22]. It disappears in the review of 2014 and a new one is presented instead: "Never deliberately distort facts or context, including visual information" [2]. The Radio Television Digital News Association (RTDNA) included in its Code of Ethics and Professional Conduct of 2000 the following recommendation: "Professional electronic journalists should not manipulate images or sound in any way that is misleading" [23]. However, RTDNA revised its code in 2015, but the association did not include any specifically point about the image integrity or about the limits while editing photographs and videos [5].

Several ethical codes pay attention to the limits of digital edition, specially to tackle the challenge of fake news. In that connection, The National Press Photographers

Association (NPPA) states that: "Editing should maintain the integrity of the photographic images' content and context. Do not manipulate images or add or alter sound in any way that can mislead viewers or misrepresent subjects" [6]. These standards are also valid for 360° images, both photographs and videos, in terms of image integrity at least.

However, and due to the characteristics of the immersive narrative, more specific standards or ethical guidelines are required to avoid malpractices. Tom Kent (2015), Standards Editor in Associated Press (AP), was one of the first to open the debate about the 360° image integrity [12]. The agency AP published in 2015 the project *The Suite Life* that allowed users to discover virtually some of the world's most luxury suits. Navigation was similar to that Google Maps offers. But the problem was that some rooms had several mirrors, so the recording was difficult. In fact, it is possible to see the camera and so the tripod reflected in one of these. But AP refused to digitally remove the recording equipment because in its statement of News Values and Principles the agency explains that:

"For video, the AP permits the use of subtle, standard methods of improving technical quality [...] and equalization of audio to make the sound clearer-provided the use of these methods does not conceal, obscure, remove or otherwise alter the content, or any portion of the content, of the image" [24].

However, each media shows a different decision, such as digitally removing the tripod or its shadow/reflection or substituting the tripod with an icon, logo or image. But this digital manipulation does not affect the facts, the sources or the story of the immersive piece. So, could it be considered as ethical the digital removal of the tripod and its shadow/reflection from the image if this does not alter the facts and the original scene? Or does this mean lying to public and misrepresent a recording?

But the digital removal of the tripod represents a dichotomy. On the one hand, several codes of ethics forbid manipulate the image content and in the absence of specifically standards for 360° images it could be understood that this includes it. On the other hand, the goal of Immersive Journalism is to allow for experiment place illusion when a user consumes, for example, a 360° video report with a VR headset or a Google Cardboard [15, 25]. To promote this feeling, professionals try to eliminate those elements that could interfere in the user's immersion, as the tripod.

And what about adding elements like people or objects to the original image? There is a particularly controversial case: *Greenland Melting*, produced by Emblematic Group -founded by Nonny de la Peña in 2007- and the programs Frontline and NOVA, both from the Public Broadcasting Service (PBS) American entity. This 360° video, which deals with the fast melting of Greenland's glaciers, includes the hologram of NASA's scientific Eric Rignot; not only that, but he is dressed in a jacket to simulate his presence in the frozen landscape [26]. But is it ethical to add digitally a storyteller, or people in general, to a scene in order to simulate his or her presence there? Is it that ethical if there is a previous notice or warning?

There are other cases in which people have been added digitally to the scene. *Anfiteatro* (2015) is an example. It is a 360° video included in the multimedia project *Ingeniería Romana* from Radiotelevisión Española and that transfers the user to a historical place of Tarraco, a former Roman colony. The difference between this work and the previous one is that in this the storyteller, a man dressed as a Roman, appears with a hexagonal figure under his feet that clearly indicates the digital added. Nevertheless, the risks of this practice are dangerous, specially at a time when fake news are widespread and represent a huge threat.

Having said this, a proposal is made for good practices indicators in Immersive Journalism: (1) Notice viewers if the tripod or its shadow/reflection was removed and, if that, justify why; (2) If the tripod was replaced by a logo, icon or image, explain the reasons; (3) Do not add objects or people to image in order to feign the presence of something or someone in some place.

3.2 Recreation of News

Media sometimes decide to recreate news. There are different ways: through infographics, interactive contents, videos or even VR techniques. But several codes of ethics stress the relevance of labeling and describing the content as a reconstruction of the events. For example, SPJ states the following: "Clearly label illustrations and re-enactments" [2]. A rule also included in the 1996's version: "Avoid misleading re-enactments or staged news events. If re-enactment is necessary to tell a story, label it" [22]. The SPJ stresses the importance of "label" this kind of actions, and similar was the recommendation made by the RTDNA in 2000: "Professional electronic journalists should not present images or sounds that are re-enacted without informing the public" [23].

There are two main formats to produce recreations on Immersive Journalism. On the one hand, simulations of real events recorded using 360° video techniques. On the other hand, experiences generated through computer graphics and based also on true facts. Whatever the format, media should notice that the content is a reconstruction and that the story is based on actual events. Following the above-mentioned standards, journalists should properly label the content. In the absence of specific norms for immersive content, it is understood that media should apply the general ones. Ethical Journalism does not distinguish between formats or even technology. On the basis of that, it proposes three indicators of best practices: (1) The content has been labeled as a recreation; (2) It is noticed that the content is based on true facts; and (3) The resources used to guarantee the accuracy of the reconstruction (interviews, audio recording and so on) are mentioned.

British Broadcasting Corporation has produced reconstructions in both formats: 360° video of real image and VR recreations produced through computer graphics techniques. An example of the first is *Fire Rescue 360°* (2016), an almost six minutes immersive film about a London firefighter and the rescue of six children during a house fire. At the beginning of the piece, a message appears: "what follows is a dramatic reconstruction of a real life event". Although the reconstruction has digital effects, the real firefighter, Paul Rich, appears on some scenes and tells his experience. This production follows the point 3.4.18 about reconstructions included in the BBC's Guidelines: "They should normally be based on a substantial and verifiable body of evidence. They should also be identifiable as reconstructions, for example by using verbal or visual labeling or visual or audio cues, such as slow motion or grading" [27].

An example of a VR reconstruction and produced with CGI is *Inside the horrors of human trafficking in Mexico* (2016). The BBC created this experience to tell the story of Maria, a woman who has been trafficked from Nicaragua to Mexico. In this case, also a message appears in the first seconds to inform the public that the story is based on true facts: "This is the story of Maria; a single mother trafficked from Nicaragua to Mexico. All the events are based on her interviews". It is worth stressing that the BBC includes in its Guidelines a point about production techniques, concretely the 3.4.17, in which states that "any digital creation or manipulation of material, including the use of CGI or other production techniques to create scenes or characters, does not distort the meaning of events, alter the impact of genuine material or otherwise materially mislead our audiences".

3.3 Sources and Staging

Several techniques and resources are applied in 360° videos in order to boost place illusion when users experience content through a VR headset. As noted above, one of these strategies consists in digitally removing the tripod of the image, but there are others that involve the people who take part in the story (storyteller and/or sources). For example, feigning a face to face between viewer and an individual of the story or pretending that the source or even the narrator speaks with the user. In these cases, the professional or the human team influences the behavior of the person and because of that the spontaneity of his/her expression.

Nevertheless, several codes of ethics recommend avoiding the staging in news events. This matter was first discussed by photographers. In this sense, the NPPA maintains in its code of ethics that "while photographing subjects do not intentionally contribute to, alter, or seek to alter or influence events" [6]. The RTDNA is more concrete and in its code says that "staging, dramatization and other alterations -even when labeled as such- can confuse or fool viewers, listeners and readers" [5]. On the other hand, the SPJ refers to "never deliberately distort facts or context, including visual information" [2]; however, SPJ mentioned specifically the staging in its previous review: "Avoid misleading re-enactments or staged news events" [22]. In short, professional journalists should never influence neither scenes they are recording nor condition the behavior of news sources as a means to an end. But the practices found on some 360° videos call the compliance of these ethical standards into question and also thought-provoking: face to face between user and storyteller and/or sources of the information; storyteller and news sources seem to speak to user (direct and spatial allusions); some sources' actions have previously been agreed (begin walking, turn around and leave, etcetera).

The New York Times (TNYT) has applied these strategies in several immersive reports. In *We who remain* (2017), a piece about the war in Nuba Mountains of Sudan, users can experience that the sources talk directly with them because they are staring at them -actually the camera is in front of them, not the user- and on some occasions even they use space references. For example, the journalist Musa John says "when I woke up I found this metal inside my leg" while he shows the object in front of the camera. There are other cases, like *The Displaced* (2015), TNYT's first 360° film and that deals with the displaced people around the world but focused on the story of three children: Oleg (Ukraine), Chuol (South Sudan) and Hanna (Lebanon). In the last sequence, Hanna turns around and starts to walk. This action has been evidently agreed before starting recording.

Although the Immersive Journalism' goal of place illusion is behind these decisions, they go against the ethical standards proposed by several organizations. For this reason, it proposes two indicators of good practices to deal with this challenge: (1) Sources behave naturally, so there seems to be no guidelines about how they should act or even talk in front of the camera; (2) Journalist appears with source/s on the scene instead of hiding and leaving the person/s alone, because this would means that the reporter previously agreed with the source instructions of how and when to start recording.

3.4   Role of the Journalist and/or Recording Team

Every story has a storyteller [28]. A book. An anecdote. A film. News. And of course, a VR news experience or a 360° video report. But in many 360° videos there are an absence on the scene: the journalist or/and the recording team is gone. It does not happen in all pieces, but at least in most of them. When a viewer watches a traditional television documentary, he or she knows that the human team responsible for filming is behind the scene, but there, in the same place. However, and on some occasions, when a user is immersed in a 360° environment thanks to a VR headsets and turns around in order to discover the scene, he or she can not find the journalist or the technicians. How is that possible? In the majority of cases the team is hidden so it is necessary a previous planning: for the crew and also for the sources because they will be alone on scene.

The intentional omission of the journalist and the cameraman, as Kool describes [15], reminds of cinema productions. But whether is the absence of the recording team understood as a strategy for promote user's place illusion? [12]. The user could become a witness because he/she can pass through the window of the device, also known as the fourth wall. In this connection, it could be proposed the following indicators of good practices: (1) the experience includes a notice saying that the professionals responsible for the recording are hidden intentionally; and (2) the recording team appears specified in the credits at the end or even in the text if the video is complemented with a web report.

El campamento de Calais en Francia (2016), a 360° video report produced by the Galician local media Faro de Vigo, is one of the exceptions. In this particular case, the user can see the journalist interviewing a refugee living in the Calais camp. And also, there are credits with information of the team. Nevertheless, the case of the BBC is peculiar. On the publication *Factual Storytelling in 360 Video* posted on November 2016 and updated in March 2017, Zillah Watson, editorial lead on future content and storytelling projects, explains that:

> "You don't have to hide the crew to keep them out of shot as much as we thought. For news films, we left the crew in shot because it felt more transparent to do so. In the *Resistance of Honey* we hid behind trees, or dressed up as beekeepers. But as we progressed we used object removal to erase the crew and tripod from shots [...]. How much digital manipulation of footage is done in post-production may give rise to trust issues in some genres" [29].

3.5   Sensitive Content

There are a lot of ethical discussions about the publication or/and broadcast of graphic or sensitive content. The picture of Alan Kurdi, a three-year-old from Syria drowned in the Turkish coast, taken by photographer Nilüfer Demir in September 2015 was one of them. Was it necessary to publish that photograph to create awareness? Did the public understand better the migrant crisis on the Mediterranean? Is it ethical to show a picture of a dead body -concretely of a minor child- in the media?

Armed conflicts and terrorist attacks have become a regular topic on the news [30]. And a problem is the amount of imagery published of these subjects that includes graphic or sensitive content. Media are getting public used to it. But many codes of ethics warn professionals journalists about the risks and limits of this type of practices. What is the journalistic aim behind the publication of this kind of content? Does the broadcasting of dead body's images clarify the story? Does the use of sensitive content like videos of the exact moment of a bomb attack improve viewers understanding? Maybe the medium could dispense with the most shocking videos or pictures. In this connection, Adornato [31] reflects on the use of warnings to inform people about the content. The author maintains that journalists know what happens in the video, but audience ignores if the images are graphic and sensitive or not. Therefore, should media warn audience about what they are going to see? It could be an ethical and responsible professional procedure.

Immersive Journalism must also deal with these ethical challenges. Social content is one of the main subjects of the 360° video reports and news. So themes like wars, terrorist attacks, displaced people or the migrant crisis on the Mediterranean are common in these pieces [32], in fact in this study were mentioned several. But this would not be different from the conventional audiovisual contents if it were not for the characteristics of immersive narrative. The mediated and simulated transportation and therefore the closeness with the sources and events of the information could favor the feeling of empathy [14], although it depends on the content or even on the topic [13]. In this regard, "the success of VR technology is predicated on using this machine to garner empathy which in turn can make news stories more persuasive and impactful" [15]. Tom Kent, Standards Editor for Associated Press, was one of the first to question the challenges of create empathy: "If the ultimate aim is to create emotion, a journalist could be tempted to omit balancing or inconvenient information that could interfere with the desired emotional effect" [17].

In view of the above, three indicators of good practices are proposed in this section on sensitive content, both for 360° videos and VR works. First, a warning message has been included due to the piece contains graphic or sensitive content. It is the case of *Inside the horrors of human trafficking in Mexico*, a VR recreation based on real interviews and published by the BBC in 2016. The warning reads as follows: "This is a virtual reality experience where you become a woman who has been trafficked into slavery. It contains scenes of threatened and actual violence and is only suitable for people over 18". Also, this norm is included in the BBC's Guidelines. But there are other works that are not properly labeled, like the example mentioned in the next indicator.

Secondly, journalists have avoided recording or showing lurid images. This indicator is based on a SPJ's standard [2]: "Avoid pandering to lurid curiosity, even if others do". *Nobel's nightmare* (2016), a report about the labor of the White Helmets in Syria, does not meet this requirement. The piece, produced by Smart News Agency and published by mediums like the Spanish newspaper El País, presents a controversial scene: the cameraman records a group of people, supposedly White Helmets volunteers, starting to run while the narrator's voice-off explains that they are trying to escape due to a bomb threat.

"The potential for harm is greater with VR than with traditional media. Vulnerable users may watch an intense experience that places them within a situation that triggers a suppressed memory. Imagine if there were visual reporters or citizen journalists with virtual reality cameras filming the aftermath of the terrorist bombing at an arena in Manchester, England in 2017. Watching the carnage through a head-mounted display might be too much to bare for most users" [18].

Furthermore, this scene specifically goes against an ethical advice proposed by RTDNA in 2000: "Professional electronic journalist should refrain from contacting participants in violent situations while the situation is in process" [23].

Finally, the third indicator is as follows: neither music nor sound effects have been added to the images. On the afore-mentioned scene is quite the opposite. In fact, music is present from the beginning to the end of the 360° video report, but it becomes more intense when the bomb threat is impending. What was the journalistic purpose behind this choice? Could this be considered as an attempt to dramatize even more the situation? The NPPA states the following: "Editing should maintain the integrity of the photographic images' content and context. Do not manipulate images or alter the sound in any way that can mislead viewers or misrepresent subjects" [6]. Through music, media and journalists can add editorial tones and intended messages [5]. Music and especial sound effects can be used in films, but Journalism must be objective and present news stories with integrity.

4   Conclusions

Immersive Journalism has opened new possibilities, but some of these conflict with the conventional ethical standards. The aim of this study was precisely to question decisions and choices taken by professional journalists when they produce or edit immersive content like 360° video reports or VR reconstructions of news. Eventually five main ethical challenges of Immersive Journalism have been identified, but a basic and initial proposal of good practices indicators has been set to face them.

Although, there are other risks regarding immersive narrative, both 360° images and VR experiences, which must be also taken into consideration: the use of children and juveniles as sources; the transparency of the news; marketing; and the relation between media and VR technology companies, among others. Matters that will be addressed in later studies.

Acknowledgments. This research has been developed within the project *Uses and informative preferences in the new media map in Spain: journalism models for mobile devices* (Reference: CS02015-64662-C4-4-R) funded by Ministry of Economy and Competitiveness (Government of Spain) and co-funded by the ERDF structural fund. The results of this work correspond also to the project *Indicators related to broadcasters' governance, funding, accountability, innovation, quality and public service applicable to Spain in the digital context* (Reference CS02015-66543-P), belonging to the National Spanish Programme for Encouraging Excellent Scientific and Technical Research of the Spanish Ministry of Economy and Competitiveness, and co-funded by the European Regional Development Fund (ERDF). Furthermore, it is part of the activities promoted through the International Research Network of Communication Management- XES-COM (Reference: ED341D R2016/019), supported by Consellería de Cultura, Educación e Ordenación Universitaria of Xunta de Galicia.

Sara Pérez-Seijo is a beneficiary of the *Formación del Profesorado Universitario* program (FPU16/06156) funded by Spanish Ministry of Education, Culture and Sport.

References

1. Bilbeny, N.: Ética del periodismo: la defensa del interés público por medio de una información libre, veraz y justa. UAB, Barcelona (2012)
2. Society of Professional Journalists Code of Ethics (2014). https://goo.gl/gnL2F4
3. Mcbride, K., Rosenstiel, T.: New guiding principles for a new era of journalism. In: McBride, K., Rosenstiel, T. (eds.) The New Ethics of Journalism: Principles for the 21st Century, pp. 1–6. SAGE Publications, California (2013)
4. Society for News Design: Code of Ethics (2006). https://www.snd.org/about/code-of-ethics/
5. Radio Television Digital News Association: Code of Ethics (2015). https://goo.gl/8ZG1bu
6. National Press Photographers Association: Code of Ethics (2011). https://goo.gl/1LrHWn
7. Filak, V.: Dynamics of News Reporting and Writing: Foundational Skills for a Digital Age. CQ Press, Washington (2019)
8. De la Peña, N., et al.: Immersive journalism: immersive virtual reality for the first-person experience of news. Presence Teleoperators Virtual Environ. **19**(4), 291–301 (2010)
9. Hardee, G.M., McMahan, R.: FIJI: a framework for the immersion-journalism intersection. Front. ICT **4**, 21 (2017)
10. Tse, A., Jennett, C., Moore, J., Watson, Z., Rigby, J., Cox, A.L.: Was i there? Impact of platform and headphones on 360 video immersion. In: CHI Conference Extended Abstracts on Human Factors in Computing Systems (2017)
11. Dooley, K.: Storytelling with virtual reality in 360-degrees: a new screen grammar. Stud. Australas. Cine. **11**(3), 161–171 (2017)
12. Pérez-Seijo, S., López-García, X.: Las dos caras del Periodismo Inmersivo: el desafío de la participación y los problemas éticos. In: Nuevos escenarios en la comunicación: retos y convergencias. PUCE, Quito (2018)
13. Sánchez Laws, A.L.: Can immersive journalism enhance empathy? In: Digital Journalism, pp. 1–26 (2017)
14. Constine, J.: Virtual reality, the empathy machine. TechCrunch (2015). http://goo.gl/VYOK1w
15. Kool, H.: The etichs of immersive journalism: a rethorical analysis of news storytelling with virtual reality technology. Intersect **9**(3), 1–11 (2016)
16. Bartzen, K.: Virtual Journalism: immersive approaches pose new questions. Center for Journalism Ethics (2015). https://goo.gl/9937LN

17. Kent, T.: An ethical reality check for virtual reality journalism. Medium (2015). https://goo.gl/DQUe1Q
18. Lester, P.M.: A Guide for Photographers, Journalists, and Filmmakers. Routledge, New York (2018)
19. Marconi, F., Nakagawa, T.: The age of dynamic storytelling. A guide for journalists in a world of immersive 3-D content. Associated Press (2017)
20. Pérez-Seijo, S., Campos Freire, F.: La ética de la realidad virtual en los medios decomunicación. In: XIII Congreso Latinoamericano de Investigadores de la Comunicación, pp. 167–172 (2016)
21. Pérez-Seijo, S., López-García, X.: La ética del Periodismo Inmersivo a debate. In: X Congreso Internacional de Ciberperiodismo: Profesionales y audiencias en el ecosistema móvil, Bilbao, Spain (2018)
22. Society of Professional Journalists: Code of Ethics (1996). https://goo.gl/nk7aq7
23. Radio Television Digital News Association: Code of ethics and professional conduct (2000)
24. News Values and Principles. Associated Press (2016). https://goo.gl/p9zjFd
25. Pérez-Seijo, S.: Ilusión de presencia en vídeos 360º: Estudio de caso de las estrategias del Lab RTVE.es. Doxa Comunicación, vol. 26, pp. 237–246 (2018)
26. Paura, A.: Virtual reality creates ethical challenges for journalists. International Journalists' Network (2018). https://goo.gl/zyU4YF
27. British Broadcasting Corporation: Editorial Guidelines (2010). https://goo.gl/9DzqXW
28. Genette, G.: Nuevo discurso del relato. Cátedra, Madrid (1998)
29. Watson, Z.: Factual storytelling in 360 video. British Broadcasting Corporation (2016). https://goo.gl/BnSgo6
30. Marthoz, J.-P.: The terrorism story: media learn the hard way from their own mistakes. In: White, A., Elliot, C. (eds.) Trust in Ethical Journalism: The Key to Media Future, pp. 43–45. Ethical Journalism Network, London (2018)
31. Adornato, A.: Mobile and Social Media Journalism: A Practical Guide. CQ Press, California (2018)
32. Sidorenko, P., Cantero de Julián, J.I., Herranz De La Casa, J.M.: Periodismo y realidad virtual: la tecnología al servicio de la información. In: González-Esteban, J.L., García-Avilés. J.A. (eds.) Mediamorfosis: Radiografía de la innovación en el periodismo. Sociedad Española de Periodística, Madrid, pp. 137–150 (2018)

Multidimensional and Multidirectional Journalistic Narrative: From Tumbled Pyramid to Circular Communication

Ana Gabriela Frazão Nogueira[1]([⊠]) and Miguel Túñez-López[2]

[1] Universidade Fernando Pessoa, Porto, Portugal
ana@ufp.edu.pt
[2] Universidade de Santiago de Compostela, Santiago, Spain
miguel.tunez@usc.es

Abstract. The changes in Journalism were the support of how to organize and present its contents. The informative narrative has changed to systemic formulas in which everything is related and is interdependent with everything. The contents passed from being linear to being multidimensional and multidirectional and the structures called the pyramid to move from hierarchical vertical readings to horizontal hierarchical structures. Today's communication is multiple, with narrative models in various formats and with circular flow structures.

Keywords: Circular communication · Multidimensional narrative ·
Immersive journalism · Journalistic information management

1 Introduction

Although its essence is demanded every day, the truth is that the ideals of Journalism have changed: today, Journalism tells society what affects it, what is interesting to it, what happened to itself of it's own protagonism. In this kind of intentional daily life, and in its own context, cyberjournalism represents the current face of Journalism, still in mutability; and represents a textual construction produced, in essence, for the Internet, first and foremost, by *"a professional who accomplishes journalistic tasks within and for an online publication"*[1] and that constitutes a new branch of the profession for writing a text in the characteristics of the medium -that is, interactivity, multimedia and hypertext- and taking the chance of, if it is not so, to block the reader.

And it is undeniable that, in spite of in this relationship between the reader and the text new languages and/or new uses of languages arise, there is an extensive instrumentalization of the medium which, even if still poorly achieved -whether for questions of understanding the potential of the Web or of business investment or both- aims to respond in full to the potentialities of the various structures of digital convergence dynamic organization, exhausting them. Through this requirement, the cyberjournalist assumes that he/she must construct and produce non-linear accounts (and not only

---

[1] http://prisma.cetac.up.pt/artigospdf/ciberjornalismo_e_narrativa_hipermedia.pdf.

© Springer Nature Switzerland AG 2019
Á. Rocha et al. (Eds.): ICITS 2019, AISC 918, pp. 965–974, 2019.
https://doi.org/10.1007/978-3-030-11890-7_90

texts), as Díaz Noci and Tous (2012) [1] assume, characterized by *"multiplatform distribution and multiplicity of authors"*[2]- always of a higher quality in its mixed and multidimensional content and consequent information complexity, eventually adding length, comprehensiveness, communicability and orientation between its options of interactivity, acquired knowledge and current journalistic information, whether in sporadic access or in a direct and personalized control of content information required by the reader.In fact, in this scenario of 'information complexity', which unequivocally states the existence of hypertext, multimedia, hypermedia, intra and intertextuality, and relating these concepts to the fact that, in Journalism, a story is told, and on Online journalism, as Bastos says, it is necessary to create a storyboard to list this non-linear narrative, it is worth presenting the concept of *"transmedia"*[3] as a term and a format for Online journalism because of its *"multimodal and multimodal of many of these reports"*[4], says Ryan (2004) quoted by Díaz Noci and Tous [1].

The information complexity can be integrated between two concepts, Pavlik's *"Contextualized Journalism"*[5], for the cyberjournalist and *"Immersive Reportage"*[6] as a support for total integration and convergence of resources and languages that are faced with the cyber-reader in a three-dimensional environment. In the immersion, the anachronistic elements fit interdimensionally through links that only make sense as the cybernaut chooses his path of reading and multimedia perception. Entering each one of them, the immersive reader defined by Santaella (2004) [2] and Bastos (2005) [3] is able to cope with the speed and intensity of the web.

2 The Inter-dimensional and Multidirectional Narrative

The construction of a journalistic text does not dispense the creative function of the editor, since this is a gift given to the reader and that will cause him to sell his text. Moreover, automation demonstrates that the real value of the journalist's participation is in his/her cognitive contribution against standardized narrative models that allow them to be replaced by robots (Tuñez, Toural & Cacheiro, 2018) [4].

However, if during the twentieth century the news had, by consensus, the base structure of the inverted pyramid - structure by which the qualitative descending reporting of information is initiated by the most important, so that if, for reasons of space, it is to be necessary to 'cut', never eliminate the essential - with the advent of the Internet, the scenario begins to change to a new structure of news. Still, it will be wrong to forget the value and advantages of the inverted pyramid applied to Online journalism, as Salaverria (1999) writes: *"the simplicity of an inverted pyramid can hardly*

---

[2] http://elprofesionaldelainformacion.metapress.com/app/home/contribution.asp?referrer=parent&backto=issue,3,14;journal,6,90;linkingpublicationresults,1:105302,1.

[3] http://henryjenkins.org/2007/03/transmedia_storytelling_101.html.

[4] https://recyt.fecyt.es/index.php/EPI/article/download/epi.2012.sep.03/17883.

[5] http://revistas.ua.pt/index.php/prismacom/article/view/583/536.

[6] http://revistas.ua.pt/index.php/prismacom/article/view/583/536.

be supplanted by multiform structures that compel the journalist to a greater effort of classification of the informative material"[7].

In writing for the printed press, the news-building structure obeys, as we know, to the answer to a set of ordered questions for the Lead (what, who, when, where) and for the Body of Text (how and why) that correspond to that descending order of informative interest. In this inverted format, the writing becomes easier since the journalist assumes as the primary vector of writing the number of characters (and therefore an amount of space) that was granted to report the event, reason why the process is almost exclusively concerned (for of the author styles) in placing as many relevant details in each text unit. So, automating the content of each of these paragraphs in several inverted pyramids that, superimposed in sequence, form, together, only one inverted pyramid, the journalist ensures not only the publication of the most relevant information, as eases, if necessary, as already said, the cutting of the less important text.

Thus, because of the change in the structure of the journalistic information text trough non-linear writing - storyboarding Stovall's laterality[8] -, the possible invalidation of the inverted pyramid in the hypertextual narrative would begin with the fact that, on the Web space is not finite. But, unlike Phillips (in Traquina, 1993:327) [5] who assumes the model as *"an unbalanced device that makes the listing of information unit in descending order of its presumed importance"* we consider that the inverted pyramid tapers as information loses importance and that the writing conforms to this model, leading to reading in a linear editorial structure but also in online journalistic production where detail, content richness and complexity transform and signify the digital narrative in Hypertextuality and to the choosing exercise on immersive reading that, with each click (as redirecting interest), pinpoints to another new world of informative goals, and of knowledge.

So, as Rosental Calmon Alves[9] gives us, *"the inverted pyramid does not mean a linear narrative"*, since a succinct writing style, presents itself as the most important first step for communication in an interactive and rotational environment such as the online. Thus, in an online journalistic production structure, where the text is extended in hypertextual and multi-optional sequences to individual reading/story construction courses, occurs an upgrade of a structure apparently handy and simple, that will totally re-creates the perception of a near or distant reality.

Therefore, the lack of a rule that structures a reading procedure should be taken into account by the cyberjournalist given that this 'unstructured' reading certainly allows the development of various systematizations for hypertextuality. But, in reality, the digital journalist knows that his/her text has two mandatory confluences with the characteristics of the medium: the intra and intertextual relationship and the navigability confer to his/her text in a space-time position with narrative effect through the Media. A process that potentiates, by mutualism, the hypertext and the inverted pyramid. Thus, based on Fig. 1, the pyramid proposed here reintroduces the theoretical

---

[7] http://dadun.unav.edu/bitstream/10171/5186/4/de_la_piramide_invertida_al_hipertexto.pdf.

[8] http://revistas.ua.pt/index.php/prismacom/article/viewFile/583/536.

[9] http://observatoriodaimprensa.com.br/e-noticias/uma-linguagem-em-construcao/.

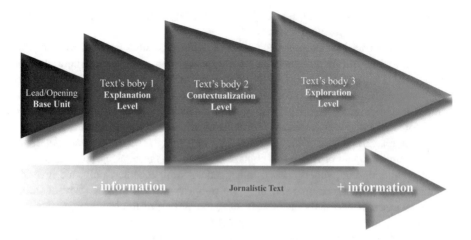

Fig. 1. Reading levels in the journalistic online informative text. (Nogueira and Túñez, 2018, based on Canavilhas, 2010 [6])

characteristics of the inverted pyramid, in the process of writing the online tumbled pyramid.[10]

In this case, the 'Base Unit' will have a connection to the 'lead' or 'opening' scheme, responding to 'what', 'when', 'who', and 'where', where *"this initial text may be a news that depending on developments may or may not evolve into a more elaborate format"*[11]. The 'Level of Explanation' corresponds to the body of the journalistic text with the 'why' and the 'how' revealing factual information necessary to the understanding of the event.

It is at the third level that the multimedia characteristics resulting from the medium begin to take over. In the 'Level of Contextualization', as the name indicates, the link between the different formats and to which we add the Intratextuality and the Hypertext of Complementarity. The 'Level of Exploration', of course, assumes the inter and hypertextual facet of the Web and connects between the text produced and the thematic and informative universe with which each of its sub-themes, characters and facts constitute, through Hypertextuality of Depth.

In this way, by laying the inverted pyramid and extending it to infinity-net, the relation that the informative-news theme comes to have with the reality of both the reader and the journalist is intimate and personal for the first and challenging and enriching to the second, both finding in the Hypertextual, Intertextual and Multimedia Web features, the liberating possibility of not being limited to the textual cell that contradicts the predictability of the form through the consequent amplification of the content. In this sense, Marcos takes a specific position when presenting a third pyramid when evaluating hypertextuality and multitextuality as the Convergent Pyramid[12].

---

[10] http://www.bocc.ubi.pt/pag/canavilhas-joao-webjornalismo-piramide-invertida.pdf.

[11] http://www.bocc.ubi.pt/pag/canavilhas-joao-webjornalismo-piramide-invertida.pdf.

[12] http://www.bdp.org.ar/facultad/catedras/comsoc/cdi/redacmd/2009/03/el_periodismo_ante_la_tecnolog.html.

Knowing the weight of the subjectivity of information given not only by the socio-cultural issues associated with each individual, added to the fact that our senses also influence the organization of information *"giving (also subjective) form to the message, which will fit the intended according to their ability or ability (culture) to do so"* (Peltzer, 1992: 69–70) [7] the cyberjournalist, based on his/her Base Unit, seeks to construct a hypertextualized reading line that intends to organize the chaos in which the reader finds him/herself, in a more comprehensive way possible to fill the entire receptive environment/universe, always remembering Drucker's relation between perception and meaning. And taking into account the Paradigm of Individualization, this process is carried out taking the notion of "objectivity of subjectivity", announced by Peltzer [7], of an abstract reader, first concerned with knowing more and then, with the intertextualised decoding, converging with their knowledge and personal interests.

Therefore, at the 'Base Unit' (the first) Level, the narration of the facts presents the informative-news text to the reader. For the second, the 'Level of Explanation', in the responses to 'How' and 'Why', assumes the description of the facts, contextualizing the reader. The 'Exposition' will be presented at the 'Level of Contextualization' since the interrelation between all the two previous levels will assume the insertion in this space-time context of the content, through a narrative effect and an explanatory action cor-related with this story. The argumentation is added by hypertextuality to the informative-news text in the last level, the one of Exploration. At this point, we fit Salaverria's schematization of the information cell[13] to explain the relationship between this information (the informative-journalistic text) and Hypertextuality.

3   The Inter-dimensional and Multidirectional Narrative

Considering real news, the immediacy in the transmission of facts may end up being a guarantee against information manipulation and misinformation. Thus, if we take into account that the event, at the time of the print, goes through the hands of journalists, from the collection of facts to the disclosure (which presupposes the treatment and choice of information collected from sources), online time, in the vortex of immediacy,

> *"real time would announce the possibility of banning a power to manipulate meaning and freedom of conscience before the facts. This is how it manifests itself as undoubtedly moral and this is also what gives it its absolute and indisputable value of progress* (Curnier, 1999) [8]".

Then, this is the process that concretizes time and dilutes space since this process of cyberjournalistic production ends up compressing both space and, particularly, time, where the horizons of decision making can be shortened. For the reader, borders as obstacles, barriers to knowledge disappear as the world comes to him/her, simultane-ously shared, at any instant, in a space, that is, in fact, virtual, because without real evaluation, basically decoded and interpreted, lived through the individual 'experience' of our minds, a place that is not exhausted but where there is meaning and continuity.

If Interactivity is like a new human capacity that produces, in the user, the action of questioning the computational plot to the point of finding the information it wants and

---

[13] http://www.cienciared.com.ar/ra/usr/3/343/n6_v1pp27_49.pdf.

its way, Hypertext is such "non-sequential writing" (Armanãnzas, Noci e Meso, 1996) [9] shaped in the way of human thought. In this way, the definition passes to a deeper level, that is, the almost tangibility of the written text ceases to exist, creating a new model of writing and association of ideas. A new model of writing because, although the text before us is written, it not only uses new rhetoric as it is carefree, limiting itself within that subject for knowing that does not have to say everything: it is enough to refer or give an idea, and, as in human thought, to bring a new association by changing the order of use of this succession of levels.

It seems to us, then, that the culprit behind this inversion mentioned above is the Hypertext... In fact, society has reached a point of technological advance that strongly implies a change in social classifications. Today, just as the Marconi Constellation becomes Vannevar Bush's[14], the concept of Mass Communication gives way to that of Critical Mass.

In fact, although Manovich considers the mechanics of the hyperlink to be a much more restrictive and compelling structure (favoring mechanically selected information), than the intertextuality that our brain reaches through internal links (in the light of one's memory), or external ones (the search for information on the subject that memory proposes), for this author, the fundamental question was that of the formatting of content since in the New Paradigm of Digital Communication [and individualizing], the Media are more manipulative because they are increasingly more programmable.

However, this author reduces the concept to its existence and does not combine it with its functionality, that is, does not take into account the capacity and the decentralized, global and interactive format of the Web but, above all, does not consider the independence of the user in relation to the media nor as the pivot of the remaining that Castells (2010) defines as 'self-communication':

> "Mass self-communication is "self" because the messages are self-selected, it is that communication that goes from many to many with interactivity with times and variable, controlled spaces. [...] That is, the hypertext is not outside of us, we have the possibility to select images, ideas, messages and build our own text."[15]

Thereby, in this paradigm shift, that is to say, in this alteration of dimensionality, what is really changing is the definition of public space sharing, "where society deliberates, constructs its perceptions and decisions"[16]. We are thus talking about a concept that affects the Journalism that, accustomed to time, hardened with it, and did not pay attention to the theories of aesthetics of the reception (TAR), where a text is not perfected until decoded. Now, a newspaper's Web page is full of hypertext crossings, linking parts of the main text to other texts. By jumping from the required sections, the reader can equally read the newspaper altogether, since texts are also intralinked and, naturally, beyond each topic, with Constructive and Exploratory[17] interlinks - all constructed by the sequential logic of the mind, not by reason.

---

[14] http://www.theatlantic.com/doc/194507/bush.

[15] http://globalizacionydemocracia.udp.cl/wp-content/uploads/2014/03/MANUEL_CASTELLS_2010.pdf.

[16] http://globalizacionydemocracia.udp.cl/wp-content/uploads/2014/03/MANUEL_CASTELLS_2010.pdf.

[17] http://www.educ.fc.ul.pt/hyper/resources/rvaz.htm.

That is why it can be said that the text is no longer what it was: the Online Journal brings the justification that the concept of Mass Communication is outdated and the one that, titled as such, trying to keep it, becomes obsolete until the rapid extinction, by being forgot. The owner of the communication is no longer the sender. Now it is time for the receiver to use the weapons that give him/her the power to demand, set up and make (negotiated, since interative (El-Mir and Valbuena (1995: 29) [10]) decisions on the amount of information read in a strictly personal search sequence. So, TAR is here demonstrated since, if a text is only what each of the recipients wants - individually - to be when he/her decodes it, this is where web page design comes in: if the message can't take the risk of being rejected, must use all possible weapons to seduce, for this reason, its iconic Communicational Image also must counts.

4   Circular Flows for New Journalistic Business Models

The flow of information is circular. For any newspaper, and even if one isn't aware of this gigantic detail, it is always interesting that those who read are interested in informing (source) to be informed (reader), in a faithful and cyclical model to that determined by the medium. In fact, writes Borrat (1989: 67) [11]: *"The conception of the newspaper as a medium of mass communication assumes that this medium is an actor placed in interchange with other actors of the social system"* and the adjustment of the Imposed Image (Media) to the Negotiated Image (born from the interactivity), creates the Communicational Image.

Consequently, the communicational effectiveness of online informative journalistic text should tend towards an expansion of this interaction between Negotiated and Constructed Image, with respect to the affinity that the cyber-reader has with the design and the vertical creative structure, and in depth, of this online journalistic text, that is, it is by the relation that communicational image is, and what it wants to represent, that the purpose is or is not achieved, then making followers: creating seduction. For this reason the online journalistic company can and must comply with the graphic and editorial line designed for the purpose of information, appropriate to the objectives of its online page, in order to create not only a communicative facilitation of its content (because interactive) but also communicative identity in its Institutional Image (which we consider an iconic factor).

For a textual or a sound element to be quickly perceived, linearity and simplicity are the pillars for conception. Even online - or because of it - this must happen. However, the communicational effectiveness of online informative journalistic text is also directly related to the number and complexity of the elements that compose the textual image and the degree of relationship between them. On the other hand, it does not matter either to be simplistic. Concrete must thus reach its peak. We believe that the balance between this negotiation and evaluation of the elements and the objectives that the newspaper wants to achieve, passing them to Online, will create the graphic line for this inter-subjective consensus for Brand Image and the communicative facilitation.

Hence, if the reading is for the sender, the comprehension is for the receiver. As said, in truth, a text is only born when it is decoded and it consists of words that are created by association of images, in two ways: the multiple denotation and the multiple connotation because they are given by each one, individually. This conjugation of elements develops a phenomenon that we call monosemy (emitter) *vs.* polysemy (receiver). In this way, the comprehension occurs from the moment each of the receivers decodes the text as a whole, and each particular word (the *links*). Therefore, is denoting the work of reading and connotative the one of search which, in turn, becomes polysemic: so, only because of the fact that this search exists, gives meanings far beyond what is in the text. We thus close the circle in which, as Drucker (1989: 223) [12] points out, *"for communication to take place there needs to be information and meaning. And meaning requires communion"* however, *"communion requires re-engagement. It requires knowing how to interpret."*

Hypertext and Interactivity have brought a new light that has refreshed the old theories of communication to the point of reformulating Informative Function, that is, unlike Passive Communication Models, this technological advance takes back the theories of the Communication Universals movement to a much more intense mode in the concept of the creation of the informative journalistic text. And it's this abundance and informative ubiquity that leaves no time for a new Journalism Mode, that is, this structural duality for a message ends up creating two images for a Style (printed and online), since the Mode - the Hypertextual Function - is beyond the Channel and that is why the journalist can not write to the Online in the same way as for the Printed.

The Online Journal is, therefore, the result of the elements that, until now, have been described and analyzed, and are the proof that a new era is still being generated. Below, is proposed a systematization of the elements that participate in the construction of the informative online journalistic text, always bearing in mind that, as Araújo et al. (2009: 220) [13], in fact *"the 'newspaper', to be reborn as a news agency of vertical integration and multimedia, will have to give different news"*, will have to bring *"different things from what we already know"* with new *"dynamics of update, contextualization, and all this based on infographics, text, video, etc."*, actions that reposition an 'old school' attitude in the professional of cyberjournalism, because what is at stake is the application of Journalism to a newspaper model that, integrating the characteristics of the medium, allows the totality of the characteristics of the profession that conjugates it, and that results in the production of journalistic content with increased significance, an added value that positions the online newspaper as a central element, rotunda, of the networked social information system because *"the Internet is not only a way to publicize the newspaper, but a media specific"*[18]: writes John Carlin (2009), international correspondent for *El País*, that *"there has never been a better time to do written journalism, and there has never been a worse time to make a living exercising it"*[19].

The truth is that Journalism does not overlap socio-cultural mutations and the economic, dynamic pressures that ultimately ends up breaking paradigms and repositioning the value and relevance of informative content as well as its diffusion method. In the boomerang effect, these changes also influence how society reacts to this new information presentation, in the percentage and functional logic of its use.

---

[18] http://observatoriodaimprensa.com.br/news/view/a_imprensa_escrita_deve_se_reinventar.

[19] http://elpais.com/diario/2009/05/10/domingo/1241927553_850215.html.

Fig. 2. Influences in the structuring of the online informative journalistic text (Nogueira and Túñez, 2018)

Lima Junior assumes that future Journalism should make more use of human senses to improve journalistic informative understanding, a pratice that will result in new content formats with more immersive sensorial impacts.[20] Martini [14] adds that, *"news is more news if you can continue to build information from it for several days. And not only because it allows the deployment on different areas of reality, but also because it facilitates work on an issue already addressed."*[21]

In resume, altering the product will, of course, reshape the journalistic enterprises at the structure and dynamics level to a scenario of that Traquina's (2002) [15] *"journalistic tribe"* being a truthful *"transnational interpretive community"*. But, nowadays, immersed in the advent of New Technologies of Information and Communication (NTIC), as it is possible to verify in Fig. 2 on the Influences to the cyberjournalistic text, the actors of the online media scene interact and affect the informative journalistic product of various forms and directions, because in a circular communicative flow.

---

[20] https://www.academia.edu/1977861/Tecnologias_emergentes_desafiam_o_jornalismo_a_encontrar_novos_formatos_de_conteudo.

[21] http://www.biblioteca-pdf.com/2011/09/periodismo-noticia-y-noticiabilidad-doc.html.

References

1. Noci, J.D., Tous, A.: La audiencia como autor: Narrativas Transmedia y Propriedad Intelectual del Público: Algunas Reflexiones Juridicas (2012)
2. Santaella, L.: Navegar no ciberespaço: o perfil do leitor imersivo. Paullus, São Paulo (2004)
3. Bastos, H.: Ciberjornalismo e narrativa hipermédia. Revista PRISMA.COM, 1 (2005)
4. Túñez-López, J.-M., Toural-Bran, C., Cacheiro-Requeijo, Santiago: Uso de bots y algoritmos para automatizar la redacción de noticias: percepción y actitudes de los periodistas en España. El profesional de la información **27**(4), 750–758 (2018). https://doi.org/10.3145/epi.2018.jul.04
5. Phillps, E.B.: Novidade sem Mudança. In: TRAQUINA, N. Jornalismo: questões, teorias e «estórias». 1ª ed. Lisboa (1993)
6. Canavilhas, J.: Webjornalismo: da pirâmide invertida à pirâmide deitada. BOCC - Biblioteca Online de Ciências da Comunicação (2010)
7. Peltzer, G.: Jornalismo iconográfico. Planeta Editora, Lisboa (1992)
8. Curnier, J.: Tempo real e produção do atraso. Revista de Comunicação e Linguagens «Real vs Virtual», vol. 25–26, pp. 125–135 (1999)
9. Armañanzas, E., Díaz, J., Meso, K.: El periodismo electrónico: información y servicios multimedia en la era del ciberespacio. Ariel Comunicación, Barcelona (1996)
10. El-Mir, A.J., de la Fuente, F.V. (eds.): Manual de periodismo. Prensa Ibérica, Barcelona (1995)
11. Borrat, H.: El periódico, actor político. Ed Gustavo Gili, Barcelona (1989)
12. Drucker, P.F.: As novas realidades: no governo e na política, na economia e nas empresas, na sociedade e na visão do mundo. Livraria Pioneira Editora, São Paulo (1989)
13. Araújo, V., Cardoso, G., Espanha, R.: Da comunicação de massa à comunicação em rede. Porto Editora, Porto (2009)
14. Martini, S.: Periodismo, noticia y noticiabilidad. Editorial Norma, Bogotá (2000)
15. Traquina, N.: Uma comunidade interpretativa transnacional: a tribo jornalística. Media Jornalismo **1**(1), 45–64 (2002)

Correction to: Accessibility and Gamification Applied to Cognitive Training and Memory Improvement

Ana Carol Pontes de Franca, Arcângelo dos Santos Safanelli,
Léia Mayer Eyng, Rodrigo Diego Oliveira, Vânia Ribas Ulbricht,
and Villma Villarouco

Correction to: Chapter "Accessibility and Gamification Applied to Cognitive Training and Memory Improvement" in: Á. Rocha et al. (Eds.): *Information Technology and Systems*, AISC 918, https://doi.org/10.1007/978-3-030-11890-7_43

In the original version of this chapter, the following correction has been incorporated: Rodrigo Diogo Oliveira has been changed to Rodrigo Diego Oliveira.

The updated version of this chapter can be found at
https://doi.org/10.1007/978-3-030-11890-7_43

Author Index

© Springer Nature Switzerland AG 2019
Á. Rocha et al. (Eds.): ICITS 2019, AISC 918, pp. 975–978, 2019.
https://doi.org/10.1007/978-3-030-11890-7

Printed in the United States
By Bookmasters